LLOYD'S AVIATION DEPARTMENT

WORLD SATELLITE SURVEY

ROGER STANYARD BSc MBA

Published by the Corporation of Lloyd's
Printed by Eyre & Spottiswoode, London and Margate

ISBN 1-85044-116-2

FOREWORD

We have attempted in this book to provide a reference source of the Western world's current and proposed satellite systems and an understanding of the overall business of satellite communications. It is explicitly aimed at non-technical readers with a business interest in the field. As a work reference, I have felt it necessary to include summary details of the technical aspects of the field. Therefore a glossary of terms is included at the end. Also included are details of the commercial launch vehicles that are likely to be available to satellite operators over the next decade.

The satellite communications field is changing rapidly. There remain many proposals for new satellite systems in both the developed and developing world; although the industry is maturing and facing competition from new technologies, such as fibre optics, it is already well established and has an interesting future. Although many of the proposed satellite systems are likely to fall by the wayside in the course of time, we have included them for the sake of completeness. Their inclusion also helps to present a complete picture of the communications environment. The reader is reminded that the business of satellite communications is primarily about communications, not satellites. Over the last twenty years, communication has evolved from being one of the world's most boring industries to one which is subject to rapid technical and organisational change and which has a profound impact on the changing economic environment in the Western world.

Section I reviews the current position and likely developments in the satellite communications field, both on a geographic and service by service basis. A qualitative look at the market and its environment is presented. It is concluded from this that satellites are now a well established part of the world's communication infrastructure, but that they need to be seen in the light of rapid developments in other communications technology and the general economic and political environment facing both telecommunications and broadcasting.

Section II surveys each of the current and proposed communications satellite systems, national, regional and international, in the Western world. The systems include those intended for conventional telephony and broadcasting services, Direct Broadcast Satellites, Mobile Communications systems and a new category of Radio Determination Satellite Services. We have attempted to present the main commercial and technical details of each system together with a qualitative review of the project, its market and the business environment facing it.

Section III provides a summary of the commercial launch vehicles currently or expected to be available in the next ten years, including the Soviet Proton vehicle.

Section IV includes a list of acronyms, a glossary of terms, and a summary of the main platforms used for commercial satellites.

The book is designed for easy reference. Satellite systems are presented in alphabetical order according to their commonly used names, rather than the names of the operators. Similarly, launch vehicles are listed according to their names.

ACKNOWLEDGEMENTS:– The author would like to thank all of those individuals and companies who have helped in the preparation of the book. The companies include British Aerospace, Aerospatiale, Eurosatellite, RCA Astro-Electronics, Hughes Aircraft Company and Hughes Communications, Ford Aerospace, Aussat, American Satellite Corp., The DGT, France, Societe Europeenne des Satellites, Eutelsat, Britsat, Atlantic Satellites, RCA Americon, Martin Marietta, Geostar, Telesat Canada, ISRO, Orion Satellite Corp., and many others.

Special thanks to the following individuals: Theo Pirard of the Space Information Centre, Belgium, for supply of tables; John Anderson and Lloyd Ludwig of Britsat for understanding of the technical details of satellites and the broadcasting environment in the UK, Sue Faulkner of Interspace for typing much of the manuscript.

CONTENTS

SECTION I

INTRODUCTION

OVERALL SUMMARY

Satellite communications have been regarded as a major part of the explosive growth in communications that has occurred over the last twenty years. But the future of the technology now appears to be uncertain as it has proved to continue to be risky and as other technologies now present competition.

But it is self-evident that satellites have already made a major contribution to the international telecommunications infrastructure and have had a major impact on TV. Satellites are no longer a "new" technology; a vast amount of experience has already been gained in their benefits and shortcomings. Whilst it is notoriously difficult to predict the future of communications, it has now become fairly clear what future we may expect for satellite communications until the end of the twentieth century.

Past optimism that satellites would provide much of the wherewithal for the global village of communications, including multi-channel TV, computer to computer communications, interactive home services, breakdown of PTT monopolies and massive job opportunities have now been tempered with reality that the wired world is at best a very long term target which involves a lot of uncertainties.

On the other hand, recent, almost nihilistic, comments that there is no future for satellite communications are as unrealistic as the global village concept. Whilst there is clear evidence to show that satellites may not be suitable for trunk telephony in the Western world, there is even clearer evidence to show their economic viability in such areas as TV traffic. There are other markets, such as business services, mobile communications and Radio determination where there are stronger elements of uncertainty, but such elements are present in any new area of activities. To dismiss satellites as irrelevant to such new markets, is, at present, a highly premature position to take.

The economies of the Western world are currently undergoing some painful transitions from physical production of goods to information based societies. This is not surprising as technology is easily transferred between countries. Basic production technologies are no longer a monopoly of the dominant West. Newly developing or industrialising countries, particularly in the Far East, have proved remarkably adept at using Western production know-how, coupled with cheaper labour, to become major manufacturing bases. To maintain their high standards of living and business competitiveness, Western countries (including, needless to say, Japan) are increasingly viewing information as a central and critical resource alongside the five traditional factors of production, land, labour, capital and entrepreneurship.

The information concept covers both technology as well as administrative information. The former is easily transferred between countries, but the latter is less well understood in concept. It appears that whilst advanced countries can maintain a competitive lead over other countries by continuously being ahead in the R & D game, in the administrative side, the control over information is likely to prove to be much more significant in maintaining competitiveness than ownership of information. Some current advanced US business thinking takes the viewpoint that only those businesses in the West where information is treated as a critical and strategic in-house resource will be able to survive. Those companies may not necessarily be producing technically advanced products. They include companies in such areas as food retailing, financial services and motor car manufacturing.

Central to this concept is the need to communicate the information, most of which is straightforward day-to-day management information. It is not surprising that companies such as General Motors have incorporated the control and distribution of information as a central feature of the long term corporate strategy. The company illustrates, however, that there is a high degree of uncertainty as to the direction in which such information based companies will develop.

It is often assumed in the USA and Europe that motor car manufacturing will eventually transfer to developing countries where labour costs are cheaper. But recent experience is suggesting that in organisations where information is being treated as a central resource, production is moving back to those locations where labour is most skilled, primarily in the West. This is because labour is no longer the critical cost factor in the location decision. For example, in the production of silicon chips, the USA initially responded to cheaper Japanese imports by shifting production off-shore, mostly to the Far East. The Japanese then responded to falling costs, not through finding cheaper labour, but through automation of production and implementation of strict quality control methods. The Japanese beat the Americans in this battle and the latter eventually were forced back to producing in the USA. The key to this development was information. Automation implies substitution of information based technology for labour. Quality control implies internal information procedures to identify production weaknesses and correct them. The logic is that the production location decision will be based not on the cost of labour but, perhaps, on the cost of robots. This, in turn, implies the need for a small but highly skilled labour force.

In a broader sense, companies like General Motors have a much wider information need which extends well outside of the production function into distribution, marketing, advertising, personnel, finance and general management. US thinking views the information resource as a tool which is giving an increasing amount of power to the administrative (or bureaucratic) function in such large companies and, indeed, within society as a whole. Whilst the reader is almost certainly aware that this trend has been prevalent throughout the twentieth century, it is only in the last decade that companies have begun to become aware of the critical nature of data processing and communications in almost all sectors of business and the need to effectively implement internal policies to develop a competitive edge in applying the technologies involved.

In the field of satellite communications there are a number of lessons to be learnt from this uncertain but important environment. IBM recognised the general trend in the 1970s when it entered into the partnership with Aetna Life and Comsat to form Satellite Business Systems, the US domestic satellite operator which intended to provide specialised satellite communications services. This reflected both the convergence of data processing and communications requirements and a perceived rapid increase in demand. As is well known, Satellite Business Systems failed in its primary mission. It was early into the market which both failed to grow as rapidly as expected and in which growth took a different direction than expected. Companies have found that developing information strategies is a slow affair and in which the amount of information that is needed for transimission is small.

Consequentially, the transmission needs are smaller. Satellite Business Systems concentrated on offering cheap space segment transmission facilities but in conjunction with expensive ground equipment. From a technical viewpoint, it offered time division multiple access (TDMA) techniques best suited to high volume traffic where the trade off between the costs of the space segment favour lower space segment charges. A similar trade offis found in trunk telephony via satellites, which accounts for the large size of the Intelsat Standard A earth stations.

More recently, interest in specialised business services via satellite has shifted towards the use of cheap and small dishes in conjunction with frequency division multiple access techniques. These are known by the term VSATs (Very Small Aperture Terminal systems). Although they are inefficient users of satellite capacity, they are not used for the high volume applications intended in the original Satellite Business Systems concept.

This shift towards interest in lower information transmission requirements reflects that the underlying demand is dependent on changing management techniques needed to accommodate the changing internal information requirements and capabilities created by modern technologies applied within the company to both production and administrative functions.

That Satellite Business Systems should have got it so wrong in its initial estimates of demand is not, upon reflection, surprising. There is a high degree of uncertainty (to the extent that there is uncertainty about uncertainty) in the communications world. We really do not know, even in the relatively short time scale, which direction communications will take. There may well be an explosive growth in demand for business communications in the next few years; it appears equally realistic to assume that the growth may be slow and unsteady as companies struggle with their internal information needs.

An intuitive best guess given here is that it will take two decades for business communications services to mature. There will be much political opposition to the developments which already clearly suggest that the successful companies will rapidly grow. This will lead to further loss of political control as multinational companies take over many of the world's markets or when regulation of such markets as financial services breaks down as such markets become international. The author expects that business communications will be one of the most important markets that will need to be addressed by the satellite communications industry, but that, unlike TV traffic or public switched telephony services, there will be no quick pot of gold.

There are similar parallels in other communications markets outside of specialised business communications. The use of satellites for distribution of TV has proved to have been highly successful in some markets and a non-starter in others. The Direct Broadcast Satellite concept has failed to materialise on a commercial basis, but demand for satellite capacity for networking of terrestrial TV and for distribution of cable TV traffic has facilitated major changes in the structure of US broadcasting as well as offering the consumer a wide variety of choice. But such developments have not been experienced so far in Europe, where new media remains a marginal business with few companies making profits at the moment.

The reasons are detailed later in this introduction. It is sufficient to say here that the political implications of satellite TV are enormous. Whilst in a country such as the USA, the restructuring that has followed on from the introduction of satellite based transmission methods is largely politically acceptable, this is not the case elsewhere in the world. Restructuring elsewhere implies not only transborder broadcasting but the use of English language programming, mostly from the USA, which threatens what some politicians see as cultural sovereignty. It implies a one-way trade in programming material from the USA with, because of language barriers, no opportunity to export programming back.

Some more radical observers see the satellite TV industry as a potential form of US domination over Europe. There are weaknesses in this argument in so far as it is likely that most US programming will prove to be unacceptable to the consumer. For example, even in the United Kingdom, which shares a common language with the USA, popular US comedy programming is frequently just not amusing to UK audiences. Moreover, transborder TV in Europe using satellites has to become a commercial proposition before it becomes a threat. This is still some way off and requires coordinated changes in regulation in the many countries that make up Europe. The European Economic Community is attempting to do this in the face of some opposition from its member countries. The plan involves protection of the European audiovisual market from US programming products. The proponents of the idea that satellite TV is a potential trojan horse for US business, cultural and political interests appear to forget that in programming at least, non-US countries have a considerable amount of ammunition with which to fight back.

Whilst there remains a lot of uncertainty about the future of satellite TV in Europe, it is clear that satellites have proved to be a very effective tool in the development of the television industry and that that role is likely to continue. Unlike specialised business communications, the use of satellites for TV is well understood. At worst, Europe will continue to use satellites for TV news feeds and occasional use traffic.

In the public switched telephony field, one organisation dominates the satellite side of the business, Intelsat. This organisation has become very aggressive in the face of changing circumstances. About two thirds of its revenue comes from international voice circuit traffic. The organisation has, in the first half of the 1980s, seen the rate of growth in this traffic fall well below forecasts. This in turn has led to over capacity in its space segment. It has responded by selling off capacity for domestic communications and cutting back on plans for the number of satellites in its next generation satellite system, Intelsat VI. It has also vigorously pursued development of other satellite services, such as specialised business communications and TV traffic.

Nevertheless, the Intelsat organisation still appears to be facing an uncertain future which is detailed elsewhere in this book. Whilst it is obvious that Intelsat is likely to face considerable competition from fibre optics in its heavy route telephony markets, the position is compounded by the trend towards deregulation of international communications, particularly at the moment in Japan. Intelsat's weakness is that it is a carriers' carrier, unable to market its services directly to the end user. It is suggested here that much of the shortfall in growth in Intelsat voice circuit use in the first half of the 1980s has been caused by its member PTTs (notably KDD of Japan) charging the end user a mark-up greatly in excess of cost. This in turn has led to the pressures which are now liberalising Japanese communications.

A notable feature of the current communications market is that the Pacific Ocean region, including Japan, is seen as the major growth pole, or market opportunity. The interest in that market by potential private satellite opportunities follows declining interest in the Atlantic Ocean region, where a more mature market is faced by increasing competition from fibre optic submarine cables. Like the Atlantic region, the Pacific region involves a number of plans for such fibre optic systems. Whilst there is evidence to show that fibre optic technology has become very cost competitive with satellites, there is perhaps another reason why fibre optics are likely to mop up most of the trunk international telephony market.

Intelsat is an international consortium. Its members each have limited influence on the organisation and are restricted (at least in principle) to the competitive activities they can engage in against the organisation. In effect, they cannot operate international private satellite systems used to carry public switched telephony traffic. The rules of Intelsat membership, however, only apply to satellite communications. The members are free to engage in submarine fibre optic cables in competition with Intelsat. Moreover, they are free to choose whatever partners they want in such ventures. This allows PTTs to enter the most lucrative communications markets, leaving Intelsat to provide services to the less lucrative areas, such as developing countries. Moreover, Intelsat is obliged to provide global communications on a non-discriminatory basis. In other words, it has to serve the less lucrative markets, whereas PTTs are not obliged to do so with new fibre optic cables.

Intelsat is now suggesting that it should be allowed to charge on a discriminatory basis to off-set the competition from fibre optics. During 1986, the organisation cut its telephony charges by 25%. It is therefore responding to the threat from the new technology. Nevertheless, it appears clear that it pays some of the more powerful PTT members to break ranks from the cooperative and offer competitive alternatives to the system.

Overall, these changing circumstances for the organisation reflect growing liberalisation and deregulation of international telecommunications. It is difficult to see how Intelsat can be protected from the changes. The deregulation itself reflects both deeper changes in the structure and operation of western economies and changing attitudes towards both the structure of domestic and international communications.

The Intelsat organisation was formed at a time when it was generally regarded that telecommunications were a natural monopoly. That is to say, competition increased costs to the consumer. Whilst there is still plenty of evidence to show that telecommunications remain a natural monopoly in an unchanging environment, many proponents of liberalisation argue that preservation of the monopoly organisations is in effect a method of "a priori" planning unsuitable for conditions of uncertain technology developments and demand. Put a different way, the argument says that the structure of international communications needs a high degree of flexibility because it is uncertain what is going on. To keep power in the hands of monopoly organisations decreases the level of flexibility needed to meet changing circumstances.

The same logic applies equally to regional and domestic communications satellite systems. Nevertheless, increasing liberalisation of telecommunications in the USA, Japan and the international markets is not being matched at the national level in Europe. Indeed, central to the European satellite communications issue at present is the difficulty Societe Europeenne des Satellites is having with its Astra satellite project, in competition with the PTT owned Eutelsat system.

The reasons why individual countries in Europe are reluctant to deregulate (the UK apart) are, again, complex. It is suggested here that one of the principal reasons is that European telecommunications equipment manufacturers are in deep economic trouble. None has a domestic market large enough to provide the demand level needed to justify large amounts of R & D expenditure in modern telecommunications equipment, whether that be satellite or terrestrial switching technology. They are therefore unable to compete with the products of North American or Japanese manufacturers. Deregulation and liberalisation would open up their domestic markets to foreign competition with which they could not compete. European based companies could, nominally, find a way round this problem by merging themselves and creating a larger, European open market for equipment.

However, this has not happened for several reasons. European companies have found it difficult to cooperate. There are cultural and legal differences between companies in different countries which make the problems worse. Moreover, many of the companies are dependent on their parent governments for preferential procurement policies, R & D subsidies and protective non-tariff policies which both cause conflict between countries and make the individual companies dependent on granting political favours back to their governments, such as the preservation of uneccessary jobs.

It is generally recognised that the European telecommunications manufacturing industry is a disaster waiting to happen. This of course raises questions as to whether the position is true in the manufacture of satellites. It is self-evident that manufacturing of satellites in Europe is a fragmented business, with five countries in the prime contractorship business – France, UK, West Germany, Italy and Sweden. None of these countries has more than a minor domestic market for communications satellites. Evidence also suggests that the European satellites are up to 80% more expensive than their US equivalents and that delivery time is up to 20% longer. From an engineering viewpoint, US satellites are consistently more advanced than those of Europe.

On the other hand, in the manufacture of communications satellites, Europe has had to (successfully) cooperate. There is no really true national manufacturer of satellites; all of the prime contractors are members of European consortia and any individual satellite order requires substantial involvement by other European manufacturers. Moreover, European governments still appear willing to put public money into satellite R & D. Europe also has a regional captive customer in the Eutelsat organisation.

It is therefore concluded for the time being that Europe will remain in the satellite manufacturing business. Whether it is in practice doing itself a favour by doing so remains open to question. Essentially the ASTRA project is a method of saying it is not. In a glib sense, there are arguments to say that the business of satellite communications is about communications, not satellites.

There are many more jobs to be created by allowing pan-European TV via satellite than there is in the construction of a single satellite system capable of providing such a service. For example, it has been estimated by the author that a UK Direct Broadcast Satellite system would create up to 100,000 jobs in programming the TV channels involved, manufacture and distribution of ground receivers, collection of subscription fees and other related activities. However, the manufacture of the three satellites needed for the system would create no more than 2,000 permanent jobs.

Similarly, it may be argued that Japan is misdirecting its efforts in developing communications satellites. It has always been and is now behind the USA in the space segment side of the business. Its satellites remain small and uncompetitive on the international markets. The country has no great need for satellites for internal communications. Liberalisation and deregulation of the domestic market has been paralleled with the purchase of US built satellites to serve domestic needs. Whilst some observers (and presumably the Japanese themselves) believe that the country will catch up with the USA in satellite technology, the evidence is largely missing. On the other hand, it is clear that Japan is now a major force in the manufacture of ground equipment, a market in which European manufacturers are largely unsuccessful except where their markets are protected.

The USA communications satellite manufacturers have had very few new orders over the last two years. However, it is difficult to assess the impact of the down turn in orders on their long term finances. All three manufacturers are subsidiaries of major corporations. Hughes Aircraft is a subsidiary of General Motors, RCA Astro-Electronics is a subsidiary of General Electric and the space activities of the two companies are in the process of being merged. Ford Aerospace is a subsidiary of the Ford Motor Corporation. It is not clear how much defence work the three manufacturers receive from the US Department of Defence, particularly in relation to the Strategic Defence Initiative. It is possible that one or more of the commercial satellite manufacturers will withdraw from the comsat market to concentrate on more lucrative defence work.

This leads on to the long term justification for continued public involvement in satellite technology. Communications technology has always been closely related to defence requirements and there is no reason to suggest that it will not continue to be so. The expertise involved in the development of satellite communications has a direct impact on the militarisation of space through Star Wars. Non-US governments have a strong interest in retaining a seed core industry for the development of space weaponry. Otherwise, they will become even more dependent on the USA for their defence needs.

Whilst it is outside the scope of this book to discuss the military aspects of satellite communications, it is worthwhile to note the connection with commercial communications satellites.

There are also other long term justifications for communications satellites over and above the commercial and military needs. They are largely long-term strategic policy justifications which are, perhaps, not quite so cynical as the military and commercial factors.

No society can justify its activities without some belief or feeling about the objectives or the benefits of its actions. Embodied in the thinking of Western societies since the 18th century has been the concept of progress. Part of the motivation behind the satellite and space industries is that they can significantly contribute to human understanding and welfare.

Whilst most people are healthily sceptical about the rhetoric of politicians, vested interests and space enthusiasts are arguing the case for yet more expenditure, when both the developed and developing worlds have already contributed vast sums of money to the two industries. Whilst it may be argued that the only real interest in the activities should be those justified by commercial criteria, in practice very little of past expenditure has been so justified. This includes much of the expenditure on satellite communications particularly where state owned telecommunications administrations or broader government policies towards industry, R & D and communications have been involved.

The two semi-philosophical beliefs underlying the endeavours of the world's satellite and space activities are widely different.

The engineer's and scientist's viewpoint is that Space represents an escape from mankind's earthly bound narrowness and a method of exploring his boundaries and therefore his inner self. It has some public appeal and politicians, providing the funds, have not been slow to exploit it, both in the USA and the Soviet Union.

Particularly in the USA and to a lesser extent in Europe, there is a large "space movement" lobbying for prestigious space projects. Many senior scientists and engineers belong to the organisation involved in these movements.

The second viewpoint belongs to the social scientist, who sees the development of new communications technology as a method of man overcoming his ignorance of the world. The late Ithiel de Sola Pool described the new communications methods as "Technologies of Freedom". The satellite world contains some of the best minds around and it is no mere coincidence that the term "global village" is so often used in discussing the impact of new communications technology. It is closely linked to de Sola Pool's idea that the dissemination of information and knowledge is a step towards liberating mankind from his ignorance and giving more control to him over his earthly destiny.

The concept is more recent than that of the scientist and engineer and reflects the painful transition in the Western world's economy towards information based activity and the development and application of a glittering array of electronics based technology. Communication has been a central feature of those applications. They have also been paralleled with significant developments in our understanding of how organisations work and both have led to major changes in the way that communications are being organised. In all of these matters the lead has been taken by the USA, with Europe, in particular, a long way behind.

It is never easy to predict or speculate on the future but it is clear that for the time being, say, up to the year 2,000, that the application of information and electronics technology will be the dominant feature of the developed economies. Beyond that, some observers see that materials and life sciences will replace electronics as the driving technology in the West. Like communications, the space industry has a major role to play in this new area and, indeed, has been building up to it over the last decade and a half.

There are a number of conclusions that have been derived in writing this book. These fall into the broad framework outlined above and place some markers to those involved in the two activities as to which direction the world's satellite industry is heading in.

1. The future of satellite communications is interdependent with the future of the space industry. Commercial satellites require cheap and reliable launch vehicles. For the time being these are not available, primarily because Shuttle has not lived up to expectations. The failure of Shuttle appears to have set the US space programme back about ten years and is also impacting on the European space programme.

2. The primary use of communications satellites within the Western world is going to be for broadcasting. New technologies involving fibre optics are already providing more cost effective solutions to point-to-point communications requirements.

3. There are also likely to be a number of smaller, more specialised markets served by satellites. These include international telephony along thinner route pathways currently served by Intelsat, maritime and aeronautical mobile communications, radio determination satellite services and general fixed satellite services in some developing countries.

4. Western governments are getting smarter in their R & D policies, both towards technology in general and (hopefully) space. This is likely to have a negative impact on such proposals as the International Space Station and plans for very advanced launch vehicles. Similarly, it is not expected that large sums will continue to be poured into experimental communications satellites unless they are now tied to broader policy objectives towards communications.

5. The satellite communications industry is not in a desperate state as some observers have commented. The lack of a large number of orders over the last three years is a myth. What has happened is that past growth expectations have failed to materialise as the industry has matured and communications organisations have looked to other technologies to meet changing needs. A steady flow of orders is expected over the next few years as existing operators replace time expired spacecraft.

6. The decline in interest has arisen because Direct Broadcast Satellite (DBS) TV has not proved to be a commercial proposition and because the total demand for communications capacity has not been as great as expected. Fibre optics will take away a lot of the demand for fixed satellite service capacity telephony and other specialised point-to-point traffic.

7. The difference between Direct Broadcast and Fixed Satellite Service systems has already narrowed to the point of disappearance. This has occurred because of developments in ground receiver equipment coupled with the use of 50 Watt transponders on satellites. It is possible that a "second generation" DBS satellite concept, using spacecraft in the 100–120 Watt transponder range, appears.

8. It is still difficult to predict what, if any, future there is for specialised business services via satellite. On the one hand, engineers are presenting systems that appear commercially viable; on the other, past experience shows a long list of commercial failures. Whilst satellites are presented as a step towards Interactive Switched Digital Network (ISDN), the latter is an engineering led concept, rather than a demand led innovation. It is expected that the future of specialised business services via satellite will become clear by the end of 1987. At best, the future appears to lie in a small market for point-to-multipoint traffic.

9. Technological developments in satellites are expected to be limited over the next decade. The major new innovations such as the "switchboard in the sky", intersatellite links and the use of lasers are not expected to be commercially available to satellite operators before 1995. Satellites derived from such programmes as NASA's ACTS spacecraft are expected to show a significant improvement in circuit costs but may be of limited interest to the broadcasting markets.

10. Within the USA, changes in the regulatory and policy environment have largely been beneficial to satellite communications. Europe is largely saddled with policies of the past, reflecting protective national interests and an unwillingness to take the risks of change. In broadcasting particularly, the policies are having a major detrimental effect on satellite communications and, unless changed, will lead to a very weakened broadcasting industry. Europe's failure to grasp the new opportunities will lead to major losses in employment opportunities.

11. Despite the rhetoric of the benefits of satellite communications to developing countries, there

there appears to be little evidence to support the arguments in favour of increased use by such nations.

12. The concern over shortages of orbital slots and frequencies which became major policy issues between 1980 and 1985 has for the most part now become a non-issue. This is because large scale demand for satellite communications has failed to materialise and because new developments in spatial frequency reuse and use of the Ka-band now being developed. The failure of conventional DBS TV is expected to eventually open up the frequency spectrum. There remains some concern in the USA that insufficient frequencies are being allocated to Mobile Satellite Services.

13. The concern expressed by developing countries that the developed world would use all the orbital slots and available frequencies before they have time to develop their own satellite systems has, for similar reasons, become of minor importance. WARC '85 (World Administrative Radio Conference, 1985) also provided compromises to the requirements of developing countries and Intelsat.

14. Many of the plans for satellite systems listed in this book are likely to fail. There is little possibility of entrepreneurial start ups or existing communications companies starting new systems. The risks are just too great for a market that is not particularly promising. Moreover, insurance rates have rocketed over the last three years and there remains concern about the availability of insurance as well.

15. Over the last ten years, the greatest cost saving development in satellite communications has been in the ground rather than the space segment. There is still some way to go in reducing ground segment costs which favour the swing towards broadcasting use of satellites. The reductions will have little impact on telephony use, where the ground capacity has already been installed and where the primary emphasis remains on keeping the space segment as low as possible. The reduction in ground segment costs has kept up the interest in business services after the failures of Satellite Business Systems and expensive Time Division Multiple Access (TDMA) networks.

SUMMARY OF NORTH AMERICA

SUMMARY

The North American business environment for satellites and space is currently in a state of turmoil and presents any company entering the scene with a lot of risks, competition and uncertain demand. There are a number of well established operators, such as RCA Americom, Hughes Communications, Western Union, Satellite Business Systems, GTE, Comsat and American Satellite Company with surplus capacity, particularly in C-band. There is likely to be a shortage of Ku-band capacity over the next two or three years as the demand increases and lack of a launch vehicle prevents extra capacity coming on line. At the time of writing, demand from broadcasters remains buoyant, accounting for half of US domestic satellite capacity.

Most of the major regulatory changes covering both broadcasting and telecommunications are now in place. As interest switches away from satellite communications, pressures for further regulatory reform will come from different directions. The changes in regulations over the last 15 years have placed the US communications business in a much stronger position than is the case for Europe. They have also provided the consumer with a much greater variety of broadcasting and telecommunications services and continues to provide a climate which promotes innovation.

The overall telecommunications business is in the process of restructuring following the break up of AT & T and subsequent regulatory changes. The current process is one of consolidation. The satellite operators who depend heavily on carrying telephony or similar services are being most affected by changes in the market environment. The telecommunications carriers are shifting to fibre optic technology and such new services as Very Small Aperature Terminal (VSAT) systems have an uncertain future.

There remains some residual interest in Direct Broadcast Satellite (DBS) TV which is likely to linger on for some time; however, the difference between DBS TV and Fixed Satellite Service (FSS) TV services continues to narrow as ground equipment technology improves and more medium power satellite capacity becomes available.

Mobile Satellite Services (MSS), both in Canada and the USA continue to attract interest, but this will be a small and specialised market. The most interesting feature has been the development of a new category of MSS services known as Radio Determination Satellite Services (RDSS). These provide a more limited type of service covering location and short text messages. So far, the Geostar Corporation has proved to be the leading proponent of RDSS and is now looking at a global service.

In Canada, the next generation ANIK E satellite system reflects over optimistic demand forecasts of the early 1980s which led to Teleast's current overcapacity. The initial procurement for ANIK E consists of only two satellites. Nevertheless, Canada's geography continues to favour satellite communications for a wide variety of services.

The continuing delays in the Shuttle programme have had a major impact on the US space programme, setting it back about ten years. At the time of writing, US policy appears to be confused and uncertain. It is likely that one commercial Expendable Launch Vehicle (ELV) operator will set up shop and begin operations from 1989. Current favourite is Martin Marietta with the Titan family.

US SPACE POLICY

This is dominated by three major programmes; SHUTTLE, the International Space Station and the Strategic Defense Initiative. The Advanced Communications Test Satellite (ACTS), although perpetually under threats of funding costs, remains a major long term R & D commitment. The three major programmes all involve a degree of future uncertainty. There are doubts that "Star Wars" is a workable proposition and may also be wound down as a result of arms limitation agreements with the Soviet Union. In either case, there is a negative impact on the need for launch vehicle capacity.

The International Space Station could also prove to be unfundable in the current political environment. The need for the Station is under question both in the USA and amongst the international partners in Europe and Japan. Some see that Materials Processing is the only significant activity provided by the Station and that it has limited scientific and commercial appeal. Interest has also been dampened by delays in the current Shuttle programme.

The environment for satellites and space is dominated by four changing political facets. These are summarised as follows:

1. In the wake of the loss of Challenger in January 1986, the Federal government is lukewarm in providing funding for NASA's plans to replace the orbiter and to develop the International space station. NASA has lost a lot of public and political credibility over the Challenger affair and previous considerable delays in the Shuttle programme.

Moreover, public expenditure cutbacks forced on the federal agencies has also impacted on NASA. The current administration, like most Western governments since 1973, has been unable to plan federal expenditure on a long term basis. This is partly because it has been unable to control it and partly because it is under pressure to obtain "better value for money" in its activities.

2. Partially as a result of the loss of Challenger, the USA has become very lukewarm as to the commercial viability of satellite communications. Nearly 25 years ago the first AT & T Telstar satellite was launched with the promise of cheap and reliable satellite communications in the future. Insurance claims on satellites over the last three years have indicated that reliability has yet to be achieved.

Unit costs per circuit via satellite have also levelled out over the last seven years and it is not expected that there will be any further economies of scale until the next generation of satellites come on stream in the late 1990s. In the meanwhile, unit costs of fibre optic cable are falling fast and still have several years to go before they level out. Many in the industry believe that fibre optic technology will be a major threat to satellite communications both within the USA and on international routes currently served by Intelsat.

3. The decline of interest in satellite communi-

cations has an immediate effect on the prospects for launch vehicles. The Americans are still trying to sort out which conventional launch vehicle will replace shuttle for launching of commercial satellites. However, some observers believe that the Strategic Defence Initiative R & D programme is also likely to collapse in the next 18 months. This exercise was a major potential user of launch vehicle capacity.

The implication is that if SDI, communications satellite operators and the International Space Centre are all in trouble, who needs launch vehicles? The policies being pursued by the Reagan administration appear to be covering bets against future launch vehicle capacity requirements, rather than providing a fixed amount of capacity. Funding for the replacement orbiter to Challenger can be stopped at relatively short notice. It is also unclear where future funding will come from.

4. The process of liberalisation of communications (both TV & radio and telephony and business services) has proved popular and is likely to continue. It is having a major impact on the structure of all sectors of communications and creating a lot of new business opportunities which may or may not involve satellite communications. There has consequently been a shift in lobbying pressures away from the aerospace side of satellite communications towards the communications carriers and users sectors. This also partially reflects fundamental shifts in the US economy away from manufacturing to an Information Technology based society.

US BROADCASTING AND SATELLITES

The broadcasting industry in the USA is regulated by the Federal Communications Commission (FCC). Unlike Europe, the Americans have never felt it necessary to establish a national, state sponsored, public service broadcasting system. Nevertheless, it has had to grapple with the problem of limited availability of frequency spectrum for TV. Historically, this resulted in the FCC carving up the frequencies in such a manner that it created three dominant national networks, ABC, NBC and CBS.

The inherent tendency towards monopoly in broadcasting caused by limited availability of the frequency spectrum led to regulations on ownership of the TV stations to which the networks distribute their programming. Until recently any company was limited to owning 7 stations; the networks gained national coverage through having other stations affiliated to them. During the 1970s the networks began to be faced by competition. It came at first from cable TV and slightly later from Video Cassette Recorders (VCRs).

Competition from DBS TV has not materialised. It was not until 1986 that the full impact of competition began to be felt. By that time, the primetime audience share of the big three networks had fallen from 90% in 1978 to less than 70%.

Much of this shift in their fortunes has resulted from satellite technology as well as regulatory changes by the Federal Communications Commission and the Courts. The availability of satellites and cable network distribution has shifted the emphasis in broadcasting economics away from control of the frequency spectrum to control over programming supplies.

The initial impact of satellites was in cable TV. They allowed fledgling cable TV channel providers to distribute their channels nationally without any substantial investment in expensive terrestrial transmission networks. In a chicken and egg situation cable operators, who previously were able only to distribute off-air programming or biked in material suddenly found a whole new source of programming. The advertisers also saw that the new cable TV channels could provide a useful media outlet if they could get on to enough cable networks.

The industry began to take off when the first domestic use satellite became available in 1976. Developments in ground equipment also allowed the price of receive only terminals to fall rapidly. These technical developments involved Gallium Arsenide Field Effect Transistors. A decade later the cable industry had matured with the majority of areas likely to be cabled already having been so.

The use of satellites has already put the broadcasting industry in a position to break the stranglehold of the three national networks over local stations. It is argued that the RCA Satcom K (Ku-band) satellites could destroy the network system altogether, although this appears unlikely. Before this satellite system became available, no broadcaster, except the big three networks, had the means to distribute its programming simultaneously to most other TV stations. Even where the stations had access to satellites, their receive terminals were often located some miles away from their broadcasting studios. This meant that they had to pay for expensive landlines or microwave links.

The RCA system circumvented this by offering medium power transponders requiring only a small receive terminal which could be located on the customer's premises. Moreover, RCA gave the terminals away to the broadcasters on the condition that they could be used only with Satcom K services. Most stations have taken up the offer and are now able to receive any transmission through the satellite system.

It also means that any station which wishes to distribute its programming nationally can do so without going to syndication and without using the network system. It has a third benefit in that it allows a small local TV station to provide live coverage of events or news in its area for onward distribution nationally. The Satcom K system is particularly useful for Satellite News Gathering (SNG) with a direct uplink from the outside broadcasting unit to the satellite. This means that the outside film crew are not required to return footage to the central studios and gives the local station a much wider field for its opertions.

There is a tendency for local stations in an area usng the Satcom K SNG service to form consortia for news gathering, thus placing themselves in a stronger position vis-a-vis the networks.

The process of shifting economic power away from the networks to local stations was evident before the arrival of Satcom K late in 1985. The independent TV stations, which are those not affiliated to the three national networks, had already become a significant force and distribution of programming to collections of independent stations (syndication) a big business. It has, however, meant that programming prices to independents have risen substantially.

Some observers believe that the restructuring

of the industry now apparent will lead to a dozen or so big television groups in addition to the three networks. It will involve increasing quasi-networking operations using satellites and vertical integration of production with broadcasting. Rupert Murdoch, who, in Europe owns the aggressive satellite delivered Sky Channel, is attempting to establish a fourth American network in this changing environment. There is little doubt that he will succeed although it is unlikely that it will be a clone of the existing networks which are in any case changing. Murdoch is one of many who have proposed a fourth US network. He is, however, aided by new rules that allow him to purchase up to 12 TV stations.

However, more importantly, Murdoch has taken control of production facilities and libraries, having bought Twentieth Century Fox film and TV studios. He has been able to finance the costly exercise from movement of his London based newspapers out of Fleet Street to modern facilities at Wapping. The resultant cost reductions have yielded him around \$75–\$100 million a year.

Other factors in the changing structure are the growth of major new urban areas, such as Dallas–Fort Worth, Miami, Atlanta, Houston and Denver and changing Federal Communications Commission regulations. These have created new markets for the independents outside of the control of the New York based networks.

The liberalisation of the TV broadcasting industry has followed the appointment of a pro-liberalisation FCC Chairman, Mark Fowler. Nevertheless, much of the justification for the level of regulation had already become suspect. The traditional monopoly power of the big three network stations had derived from lack of sufficient frequency spectrum.

The open skies Domsat decision by the FCC (1972) had, in effect, opened up the satellite frequencies to broadcasters who used them to develop cable. This allowed, for the first time, a real degree of choice to viewers and provided a market allocation method to reflect consumer choice. About one half of US homes are now passed by cable TV networks. Moreover, the introduction of the VCR has further widened the choice. The regulation changes made under Fowler are as follows:

1. Advertising. Existing voluntary restraints have been removed to allow any amount of time on a TV channel to be given over to advertisements.

2. Content. Similarly, rules that required that a percentage of a TV station's time be given over to information (5%), local (5%) and non-entertainment (10%) programming were dropped.

3. Ownership. Apart from allowing 12 stations to be owned by any one organisation, the FCC also relaxed attribution rules allowing some cross ownership of cable TV companies by broadcasters and allowing an individual to own up to 5% of a media station before being defined as an owner.

4. Franchise renewal procedures were modified to favour existing franchise holders.

5. More stations were franchised.

US TELECOMMUNICATIONS AND SATELLITES

The process of liberalisation of regulations covering telecommunications in the USA has been more complex than has been the case with broadcasting. Telecommunications are regulated by the courts, individual states and the FCC. The process of liberalisation of telecommunications has tended to involve more regulation rather than deregulation (although in specific areas liberalisation requires a degree of deregulation). Generally, the hitherto dominant monopoly, AT & T, has found itself faced with more regulation than its new competitors, such as MCI. The regulatory restrictions placed on the organisation at its break up have subsequently been eased.

The need for regulation under conditions of liberalisation arise simply because "laissez faire" type policies under conditions of natural or organisational monopoly leads to continuation of that monopoly power rather than new competition. The natural monopolies involved in telecommunications are not paralleled in broadcasting where the monopoly power derives from a lack of a price mechanism to allocate the frequency spectrum and the usual organisational features which push companies to defend their markets.

The process of liberalisation of telecommunications in the USA and the break up of AT & T has been widely criticised by the general public who have been used to cheap local telephone services cross-subsidised from long distance and business users. The efforts of the 1980s have attempted to match cost with charges, eliminating cross-subsidies and allowing new services and new organisations into the market. One area of business use which received a considerable cross subsidy was the leased line service.

Increases in rates in this area have made some organisations look to alternative means, such as Private Satellite Networks (PSNs) using, for example, VSAT terminals. Depending on who you believe, the development of PSNs/VSAT systems is either a damp squib of a market opportunity or the great business opportunity of the late 1980's. It is still too early to determine whether the market will materialise.

The FCC has already considerably eased the regulations covering Very Small Aperture Terminal (VSAT) systems by now allowing submissions for authorisation of terminals on a block basis rather than terminal by terminal applications. The future of the VSAT market is closely tied to past and future developments in the overall regulation of US domestic and international telecommunications. The break up of AT & T (which was a regulation issue controlled by the courts rather than the FCC) and general policies to reduce cross-subsidisation in the industry, have helped in turning the market towards Very Small Aperture Terminal (VSAT) systems.

Large users of telecommunications are faced with a situation where leased lines on the public telecommunications network (hitherto cross-subsidised from Message Toll and Wide Area services) have increased in real terms. Moreover the users are now faced with having to deal with several telecommunications carriers for their long distance communications needs. This has created both organisational and billing problems. Suppose for example, a company whose major facilities are outlets spread right across the USA. Each of these would be connected to the local Bell telephone network, which in turn would be connected to the long distance telecommunications carriers (still dominated by AT & T). The long distance carriers are connected to other local Bell operating companies or operators.

To establish a communications network connecting all of the company's premises or outlets requires use of up to several hundred separate telecommunications companies' networks. (Although, in practice most of the capacity involved was part of the old Bell/AT & T network and therefore the number of companies involved is usually not as great). The immediate problem for any organisation wishing to make significant changes to its national communications network is that it now has to deal with many telecommunications carriers instead of just Bell/AT & T as was the case before breakup of the latter. Moreover, if anything technical goes wrong with a long distance line on the system, up to three telecommunications companies could now be involved; the two local companies and the long distance carrier.

Therefore, some companies have been looking at VSAT systems to bypass both the local telecommunications carrier and the long distance carriers, and thereby creating a system which uses only one satellite communications carrier. This method does not avoid use of the existing terrestrial system as the latter is still needed by the user for local communications.

Nevertheless, the general industry feeling is that although Very Small Aperture Terminal (VSAT) systems present a potentially useful option, companies are not rushing to install them. For example, in 1985 General Motors announced a plan to link 13,000 of its dealers to a VSAT system for collection and distribution of Data on sales, inventories and orders. The company still has to begin implementing the system. Some observers believe that there is just far too little data involved to justify the expense. There is certainly no evidence to back early market predictions that there would be an explosion in Very Small Aperture Terminal (VSAT) type traffic.

If it does grow significantly, a learning curve needs to be passed through before the industry knows where and how best to use the technology. According to some industry sources, the VSAT concept is best suited to data networks where the user has locations a long distance from the host computer and where there are upward of at least 100 locations spread nationwide across the USA or 200–300 locations in a region of up to several adjacent states.

The concept is made a little more complex by the situation where the site of a host computer can be linked through existing terrestrial facilities to a local teleport. This would often be the case where the host computer is located in corporate headquarters in a major city such as New York. Under these conditions, the breakeven point for a VSAT network, in terms of numbers of installations may be much lower.

The biggest disappointment in the VSAT market has been Federal Express's withdrawal of its "Zapmail" electronic courier service. The company was seen as the precursor to the VSAT industry in general. Nevertheless, "Zapmail" was a specialised service and the company had very ambitious plans for it. The withdrawal (in September 1986) caused considerable surprise to many in the industry.

The VSAT technology has been seen, like many other business services via satellite, as a step towards Integrated Services Digital Network (ISDN) and also as a threat to it because it allows bypass of the public switched network. Those who say it is a step mention that it allows higher data rates than are often available on the old and yet to be updated public network.

By allowing a market to develop before the switched terrestrial network can allow high speed transmission of data reduces the future financial risks of investment in the terrestrial ISDN network, because there will be a ready market for it to serve. They have a point in that the ISDN concept has tended to be engineering rather than market led. At its most extreme, ISDN is probably politically untenable because it claims that control of computer processing equipment should be in the hands of a telecommunications monopoly and outside the hands of the user.

ISDN is an ideal technology for PTT monopolies because it swings the balance of telecommunications economics towards natural monopoly (ie where any competition will increase rather than reduce costs to the user), but it ignores the obvious fact that technological development is uncertain and that new applications of technology may swing the balance against ISDN. ISDN is at best a very long term concept and if it does justify monopoly power then there is an offsetting cost in terms of organisational bureaucracy which will stifle further developments beyond ISDN.

OTHER OBSERVATIONS ON THE US SATELLITE SCENE

In September 1986, Comsat announced that it was "merging" with Contel. The moves reflects both the changing structure of the US telecommunications industry and the declining interest in point-to-point satellite communications, both on a national and international basis.

Comsat is selling out to Contel (formerly known as Continental Telephone) after experiencing its first quarterly loss in nearly twenty five years. This dramatic move has not surprised some observers we have spoken to; although the company has, in the short term, been affected by losses on the STC spacecraft. With little prospect of now selling them, the telecommunications market is moving against Comsat as telecommunications carriers are investing in fibre optics for trunk telephony.

There is little prospect of rapid growth in new satellite services, such as VSAT systems, to offset both US and international competition from fibre optics in voice traffic. Moreover, as a satellite company, Comsat is not strong in the broadcasting markets, these having been wrapped up by RCA Americom and Hughes Communications.

The official position is that Comsat and Contel are merging. This is a neat piece of Public Relations which masks the true nature of the arrangement. It is in practice a takeover of Comsat by Contel. Because of its special nature, being the US signatory to Intelsat and Inmarsat as well as a monopoly provider on international satellite communications, Comsat cannot hand over those powers to another private company. Contel is three times the size of Comsat and has major satellite interests of its own, including Spacecom and American Satellite Co.

Before the announced "merger", Comsat was in a very weak position as the telecommunications world is rapidly moving away from the satellite services Comsat provides. The move neatly diversifies Comsat out of dependence on a declining market. Comsat has failed to diversify away from its core Intelsat orientated business. It has had two major attempts, both of which have failed. These were its partnership with IBM and Aetna Life in Satellite Business Systems; the other was the STC DBS project. Contel had already successfully expanded and diversified itself out of core markets.

The new company will continue to be known as Comsat. Under the terms of the "merger", this involves an exchange of shares between the two companies with Comsat offering 0·94 of its own shares for each one in Contel.

It is already clear to many in the US market that the future of satellite communications lies in broadcasting. The business services sector is likely to remain small and specialised, with a tendency towards point-to-multipoint services. Whilst the problems of the satellite industry have, over the last year, been largely ascribed to the very high risks involved in launching satellites and the delays caused by grounding of vehicles, this has masked a more fundamental change in the industry. Long distance public switched communications are very price elastic; typically this elasticity is around −1 (unity): any fall in price results in the carrier's revenue remaining the same as users increase their use in response.

But as satellite communications space segment technology has matured, the cost per voice circuit on a satellite has stopped falling. As use is highly price elastic, growth rates in the industry have fallen. Intelsat charges have remained static over the last four years after having fallen continuously since the system began operating. It is not therefore surprising that Intelsat has experienced a decline in the growth rate of demand for its voice circuits. Moreover, this directly relates to Comsat which, in practice, is Intelsat.

The technological developments in satellite communications applied during the 1980's have mostly related to the ground segment. This has had little impact on trunk telephony, which accounts for two-thirds of Intelsat's revenue. The trunk telephony market is best served by keeping space segment charges as low as possible at the express of ground equipment.

The mature satellite technology is now faced by competition from a rapidly emerging technology, fibre optics, where cost per circuit is expected to fall for some years to come. US long distance domestic carriers are moving their trunk voice traffic off satellites onto these systems. 1986 has already seen the move of GTE's Sprint service from Spacenet to fibre optic.

Telecommunications administrations are investing heavily in transoceanic fibre optic systems, both in the Pacific and Atlantic regions. These have yet to come on stream (TAT–8 is not due to until 1988). It is not surprising that PTT administrations such as British Telecom have got themselves a reputation of being hostile towards satellites. For trunk voice services, satellites are now yesterday's technology.

It is perhaps a matter of history repeating itself. Marconi originally envisaged that his radio technology would provide telephone services "without wires", hence the still commonly used term in the UK for a radio receiver, a "wireless". By 1919, it was apparent to some that the great market for radio technology would be in point-to-multipoint services: broadcasting. This is where it is now apparent that satellites are effective. It is not surprising as a satellite is no more than a collection of radio repeaters.

THE PRIVATE US INTERNATIONAL SATELLITE SYSTEM PROPOSALS

Over the last year, the pressures on Intelsat have shifted away from the threat of private international satellite systems to the combined pressures of its own overcapacity and the imminent prospects of fibre optic transoceanic cables.

Most of the current US proposals for private international satellite systems involve entrepreneurial start up companies; the large US domestic satellite operators are either not involved or are involved at the fringes. For example, RCA Americom has dropped its plans to modify a Satcom satellite to provide six transponders over the North Atlantic. Hughes Communications are currently only vaguely looking at use of the Atlantic Satellites system for transatlantic traffic. Companies such as Orion, Cygnus, ISI and Finansat have yet to raise finance to start operations.

The company in the strongest position is PanAmSat. The latter has tackled a market were fibre optics are unlikely to play a role; the Latin American scene.

All of the private international satellite systems require arrangements with overseas PTTs for up and down links. The major PTTs are already investing heavily in fibre optic links and it is difficult to envisage that they would willingly allow private competition in.

The private operators are also restricted by the US government to providing services other than public switched traffic. The satellite business services market has not proved to be a significant opportunity; nor have international TV channels, although there remains a buoyant demand for occasional use TV. Finally, Intelsat's surplus capacity has allowed it to provide cheap capacity to its members for domestic operations, further undercutting the position of the proposed private operators. It is therefore concluded that the private venture satellite proposals are likely to fade away.

Some observers believe that Hughes communications has a global plan for satellite communications involving Atlantic satellites in Europe and its Japanese partnership, JCSAT. Much more likely is that the company has a pragmatic approach and that both JCSAT and Atlantic Satellites are stand alone projects.

SUMMARY OF EUROPE

SUMMARY OF THE EUROPEAN SATELLITE AND SPACE SCENE

The European satellite and space scene remains, at the time of writing, distinctly different from that of North America. Nevertheless, the recent experiences in the USA are expected to impact on Europe sooner or later, leaving the differences largely at regulatory and organisational level.

Europe has come later into the satellite and space field than the USA, only obtaining operational launch vehicles and satellites in the 1980's. There are several consequences from this, not least that all European satellites operate in the Ku-band, rather than C-band which still predominates. The late entry means that Europe has a shortage of satellite capacity which may not begin to ease until 1988–89. It also means that Europe has a well developed terrestrial telecommunications network; coupled with the small geographic area of Europe (and within that area, strong concentration of population around a geographic core within the West) there is limited demand for point-to-point services.

It means that European based satellite operators are all chasing the broadcasting market. The position is made more complex by the fact that governments have been willing to invest in DBS TV (notably the TELE-X, OLYMPUS, TDF-1 and TV-SAT systems). These are likely to prove to be costly mistakes; so far some $1,300 million has been invested in the construction of DBS satellites over 8 years. It is proving very difficult to turn the systems into commercial operations.

Unlike the USA, European communications are dominated by monopoly, State owned, telecommunications administrations which are also used as instruments of industrial and R & D policies. Governments have encouraged their telecommunications administrations to invest in European manufactured satellites, either directly as operators (France and West Germany) or through Eutelsat.

Part of the aim has been to build up the satellite manufacturing capabilities of European industry. Nevertheless, there is little doubt that the manufactures are uncompetitive with US industry for the space segment and uncompetitive with the Far East for the ground segment. Satellite manufacturing is split between several countries and companies, each with limited capacity and small order books. The companies work in Consortia, adding to their overall costs. The small production runs and limited past experience also arises doubts amongst some potential operators as to the reliability of the products.

With a turn down in their domestic order books over the last three years, it is therefore not surprising that US satellite manufacturers have been attempting to sell their satellites to Europe. The first success was with the Luxembourg Astra system using an RCA-4000 platform. Hughes Communications has followed up with the Dublin based Atlantic Satellites. Both systems are proposed by countries with no vested interests in the European satellite manufacturing business and likely to suffer only limited consequences from breaking the telecommunications administration (PTT) stranglehold on satellite communications.

Nevertheless, European governments have a deep seated objective in creating autonomy from the USA in the space field. There are undoubtedly long term strategic, economic, cultural, political and defence reasons lying behind this. One consequence of the problems with Shuttle is an apparently increased determination to push ahead with both the European participation in the International Space Station and a parallel project to have an independent manned launch vehicle capability.

Whether the Europeans can reach agreement amongst themselves on these projects remains open to some doubt. None of the European countries have the experience or funding to go it alone, but truly cooperative projects have a habit of not working. In most of the European satellite and space projects, one country dominates in each. Eutelsat is an exception.

SATELLITE TV IN EUROPE

The European broadcasting scene differs from that of the USA in the following ways:

1. Broadcasting is organised along national lines, with a strong public service ethos in each country. When the radio frequency spectrum is in short supply, the public service broadcasting ethos probably makes economic sense. It is a form of rationing and when done well there is evidence to suggest that it provides better overall consumer satisfaction than the commercialised off-air TV which has predominated until recently in the USA.

2. Nevertheless, the public service broadcasters (which include heavily regulated commercial companies, such as those in the UK), have been protected from competition from such broadcasting methods as transborder broadcasting and cable TV. There are two consequences. European broadcasters and advertisers are generally weak in their abilities to use television to promote goods and services. Italy and the UK are exceptions to this. Secondly, many broadcasters, both in the USA and Europe belief that there is an untapped demand from both the consumer and the advertiser for more television, either in the form of cable TV, DBS TV or more off-air broadcasting.

3. European markets are much more fragmented than is the case in the USA. Language alone creates a major barrier. It can be overcome by such methods as dubbing or sub-titles, but cultural preferences (which are very real and not just a political ploy to keep out US programming) remain. The advertising markets are also largely organised along national lines; few advertising campaigns are organised along European lines. Pan-European broadcasters such as SKY have found that they have had to help produce a European advertising industry.

4. Related to this is the fact that programme copyright ownership is still organised along national lines; just as the legal rules are based along similar lines. The European Commission is pressing to harmonise these rules.

5. European TV transmission standards are not only incompatible with the US NTSC transmission standards but are divided amongst themselves into the alternative PAL and SECAM methods. Europe is also unable to agree on new transmission standards for

satellite television. It must be emphasised that variations in transmission standards are a reflection of deliberate attempts to protect markets, rather than a reflection of different perspectives on the technical performances of each standard.

6. Unlike the USA, most of Europe remains uncabled and this situation is likely to continue for some time. Of the major countries, only Germany is currently cabling on any significant scale. The reasons are complex but three stand out as significant:

A. Planning or zoning regulations. Generally these require that cables are buried under the ground rather than strung up along telephone poles. A typical average cost of a new cable TV drop in the UK is around $1,000. US experience is that the resale value is now around $1,400. The latter figure indicates that the cost of a new drop needs to fall to around $500–$750 to make cable an economic proposition. In cities like London "make good" payments (a form of institutionalised corruption practised by most of the London boroughs) can add further costs.

B. Cable has come late to Europe in comparison with the USA. The consumer has already invested in VCR equipment and some European governments have opened up the terrestrial frequencies in advance of cable TV. Consequently, cable TV is unattractive to a lot of consumers. In the UK, Germany, France and Italy, the new terrestrial channels have provided some popular and very good television which makes cable even less attractive to the consumer.

C. European governments are still very restrictive on the content of cable TV programming. Although the European Commission is trying to establish common rules, such programming as pornography, common in the USA and popular with many consumers, is largely ruled out. It is rare in Europe that governments have conceeded freedom of choice to the viewer.

D. There is still relatively little in the way of popular, mass appeal, cable TVprogramming available in Europe. Many of the cable TV channels are put together by existing public service broadcasters. The truly commercial channels are slowly building up their audiences and the quality of their programming under some very difficult commercial constraints.

E. Finally, some potential commercial interests are put off by conflicting and changing policies which swing between terrestrial, off-air and cable TV. European broadcasting remains both a highly political and uncertain affair.

6. Direct Broadcast Satellite (DBS) TV remains a major issue in Europe, not least because governments are having to justify substantial past investment in the field. The European electronics equipment manufacturers remain a significant lobbying force in DBS TV although most broadcasters appear indifferent. The lack of cable TV networks also favours DBS TV. Europe's fragmented markets mean that there may still be more opportunity for DBS than in the USA. Geography also means that Western Europe can be covered by a single DBS satellite with an EIRP (signal strength) performance of in excess of 62 dBW and without any significant problem of time zones.

7. There is also a common and broader cultural difference between the USA and Europe which both continents often fail to realise. The Europeans are more respectful of authority and institutions and less willing to take the risks of being wrong than Americans. Do not expect Europeans to want change and innovation as rapidly as Americans.

EUROPEAN POLICY AND REGULATION MAKING

European policy and regulation making still largely remain at national level, with a fringe element belonging to a transnational framework. The latter mainly covers areas where there is no significant conflict with national policies or where national policies are unable to operate, either because they are too expensive or because they conflict with international treaty obligations. For the purposes of the satellite and space industry the following European and International bodies either have an organisational role or have some limited influence on policy making:

A. The European Commission and Parliament. In practice this has been limited to meeting the terms of the Treaty of Rome covering broadcasting and a number of R & D initiatives that have sometimes paralleled national initiatives. The latter include RACE (in telecommunications) and ESPRIT (in electronics) R & D programmes. The European Commission has been arguing that broadcasting is an economic activity and therefore open to free competition between EEC members. It is also in the nature of the EEC that policies favour EEC members at the expense of non-EEC countries.

Therefore, high on the list of priorities, has been to keep US interests (in particular, as the USA tends to dominate media) out of the market. Nevertheless, the European Commission's efforts in this area have not been received well by its member states and the organisation appears to remain a political and economic lightweight. The EEC Commissioner of Culture is now taking a growing interest in audiovisual matters. In September 1986, the current Commissioner, Carlo Ripa Di Meana, called for a "global audio-visual perspective in Europe aimed at unifying the European market" and issued an appeal for film and TV interests to imitate agricultural lobbies in battling for finance (The EEC Common Agricultural Policy costs it $225 million per week in addition to national agricultural subsidies and higher than necessary prices paid by consumers by protectionist measures).

B. The European Broadcasting Union (EBU). This organisation has attempted to see that Europe adopts common transmission standards for Direct Broadcast satellite (BDS) TV. Although it is probably not critical that Europe does so for DBS TV (as it appears to have no future) the adoption of common standards has particular relevance to industrial policies towards television in general.

The argument is that Europe needs to adopt a common standard to build up a common market large enough to provide the economies of scale to compete with the Japanese and to provide the patents to prevent the Japanese from copying European technology to re-export it to Europe. In the case of DBS transmission standards, the European Broadcasting Union appears to have failed in that there is no agreement to use its recommended C–Mac standard. It is not surprising in that Europe has considerably (and

EUROPEAN SATELLITES FOR TELEVISION BROADCASTS

SYSTEM NAME/ OPERATOR	*EUTELSAT I*/ EUTELSAT	*TELECOM I*/ D.G.T.	*TV–SAT*/ BUNDESPOST (4)	*ASTRA*/ S.E.S.	*OLYMPUS*/ E.S.A.	*DFS KOPERNIKUS*/ BUNDESPOST	*A.S.*/ATLANTIC SATELLITES	*EUTELSAT II*/ EUTELSAT
Organisational type	Public cooperative	Governmental entity	Governmental entity	Private limited	Governmental/ international	Governmental entity	Private limited	Public cooperative
Total investment/ dollars	About 450 million (2)	About 500 million ?	Around 250 million	Some 150 million	Around 615 million	375 million	Some 400 million	400 million
Specific services developed by system (ITU/IFRB category)	Communications, TV distribution, data transmission (FSS)	Communications, video-conferencing, TV distribution, data (FSS)	Direct TV broadcasts (DBS)	TV distribution in Western Europe (FSS) (5)	Direct TV broadcasts (DBS) (6)	Communications, data transmission, TV distribution (FSS)	Transatlantic communications, TV distribution, direct TV (FSS + DBS) (7)	Communications, TV distribution, data transmission (FSS)
Operational start (scheduled start)	October 1983	January 1985 (summer 87) TDF–1 (end 87) Tele–X (in 1988)	TV–SAT	(1st half 1988)	(2nd half 1988)	(Mid 1988)	(Early 1990)	(Late 1989?)
Downlink frequencies/GHz (1)	10·95–11·2 11·45–11·7	12·5–12·75	11·7–12·5	11·2–11·45	12·1–12·25	11·45–11·7 12·5–12·75	11·7–12·06 10·7–10·85/12·5–12·75	10·9–11·2/11·45–11·7 12·5–12·75
Receive bandwidth/MHz	72	36	27	24–26	27	44 and 90	27 and 54?	31, 36 and 72
Polarisation	Linear	Linear	Circular	Linear	Circular	Linear	Linear	Linear
Shape of beam	Circular	Circular	Elliptical	Shaped eurobeam	Elliptical	Elliptic	Shaped eurobeam	Shaped eurobeam
Number of sattelites	up to 4	up to 3	1 (or 2) for each	1 (+1)	1 (+1?)	Up to 3	Two	Four (up to 8?)
Type of platform	ECS/OTS bus	CS/OTS bus	Spacebus 300	RCA 4000	Olympus	ECS/Spacebus 100	HS–393	Spacebus 100B
Satellite stabilisation	3-axis	3-axis	3-axis	3-axis	3-axis	3-axis	Spin	3-axis
Lift-off weight/kg	1,150	1,185	2,130	1,800	2,537	1,400	2,280	1,625
Weight in orbit/kg	700	700	1,025	1,020	1,600?	840	1,500?	985
Lifetime/years	Seven	Seven	Seven	Ten	5 to 7	Ten	Ten	Ten
Available power/kW	0·8–1	1·1	3·2	2·85	3·6	1·1	3?	3
Eclipse protection	Partial	Complete	Complete	Complete	Complete	Complete	Complete	Complete
Number of transponders × power amplifier (redundant transponders)	10 × 20 W (2 × 20 W)	6 × 20 W) (3 × 20 W) (3)	4 × 230–260 W (2 × 230–160 W)	16 × 47 W (6 × 47 W)	2 × 230 W (1 × 230 W)	10 × 20 W (5 × 20 W)	5 or 10 × 100 W +8 × 50 W? (TBD)	16 × 50 W (4 × 50 W)
Prime contractor (State)	British Aerospace (United Kingdom)	Matra (France)	Eurosatellite (Germany)	RCA Astro-Electronics (USA)	British Aerospace (United Kingdom)	Siemens/GESAT (Germany)	Hughes Aircraft (USA)	Aerospatiale (France)
Launch system	Ariane 3 + SYLDA	Ariane 3 + SYLDA	Ariane 2	Ariane 4 + SPELDA	Ariane 3	Ariane 3 + SYLDA	Ariane 4 or Titan III?	Ariane 4 + SPELDA (8)
Tariff per transponder (a) lease per year (b) sale	(a) some 2·5 million dollars (b) no sale	(a) 3 million dollars? (b) no sale	(a) 8–15 million dollars (b) no sale	(a) 5 million dollars (b) 30 million	(a) TBD/experiments (b) no sale	TBD	TBD	(a) 3·5 million dollars (b) no sale
EIRP in Western Europe/dBW	40–46	44–48	60–65	50–51	62–63	49 (Germany)	60 (British Islands) 50 (Western Europe)	47–51 (Europe)
Geosynchronous locations (scheduled satellites)	13°E, 7°E (10°E)	8°W, 5°W (3°E)	TV–Sat/TDF–1 (19°W) Tele–X (5°E)	(19°E)	(19°W)	(23·5°E, 28·5°E)	(31°W)	36°E, 3°E, 19°E?
Present status	Operational with 2 satellites	Operational with 2 satellites	Delays and unknowns about D–2 MAC standard	First private venture with comsat in Europe	Technological satellite for experiments	Delays due to Ariane launch services	Awaiting for Irish DBS and communications franchise	Depending on Ariane launch services and on modifications
Remarks about performances	Mainly used for TV distribution	Only for domestic services	TV reception with 0·5 m diameter antennas	TV reception with 0·8 m diameter antennas	TV reception with 0·5 m diameter antennas	+ Experimental payload at 20/30 GHz bands	Mainly dedicated to TV broadcasts/ exchanges	To be used with small earth terminals (TV, data)
NOTES: (1) Only Ku-band spacecraft and TV broadcasts	(2) European investment through ESA (European Space Agency) ECS programme	(3) Plus 6/4 GHz and 8/7 GHz communications packages	(4) + *TDF–1*/ TELEDIFFUSION DE FRANCE + *TELE–X*/ NOTELSAT	(5) Only unidirectional TV links with a medium-power communications satellite	(6) Leased by Italian RAI and by EBU + experimental communications payload	First operational demonstration of systems developed by GESAT	(7) Mixing communications and direct TV mission, proposed to UK for DBS channels	(8) Considering the Soviet offer of Proton launch services

deliberately) failed to adopt standardised transmission SECAM in the past. For example, France adopted the SECAM transmission standard for colour TV during the 1960's whereas most of the rest of Europe adopted the PAL system. The whole of Western Europe avoided use of the US NTSC standard. These differences have nothing whatsoever to do with the technical merits of the various systems. They relate directly to protecting markets from competition.

Similarly, the Europeans have argued for and adopted Open Standards Interface (OSI) technical standards for computer communications, against the adoption of the de facto IBM SNA (Systems Network Architecture) communication standards. This is not based on technical merit, even though there are technical merits to OSI. The aim is to protect European industry. In the latter case conflicts of interest between European countries is likely to result in at least one European country breaking ranks and adopting ASN.

C. CEPT (Council for European Posts and Telecommunications Administrations). A rather secretive and cosy cartel of European telecommunications administrations which sets both technical standards between its members and charges. It was out of CEPT that Eutelsat was formed.

It has virtually no interest in making any of its activities public. It remains a dark horse amongst European organisations and therefore its activities are highly suspect.

D. The European Space Agency (ESA). This is an R & D organisation with the emphasis on development rather than research. It is a central feature of European industrial policy towards satellite and space activities but by no means controls it. Only half of European expenditure on space is accounted for by ESA. Where ESA does account for it, it tends strongly to relate projects where one country is supplying a large proportion of the expenditure and therefore retains ultimate control over it. Where the dominant countries in the organisation have felt that it is in their own interests, they have either opted for bilateral research (such as TV–SAT/TDF–1) or national research outside the framework of the organisation. Moreover, the organisation is, and has tended to be, dominated by one country, France.

E. Eutelsat. Conceptually a European clone of Intelsat, this organisation is a consortium of European telecommunications administrations which aims at providing European regional communications via satellite. Like Intelsat, its origins are closely linked to industrial policy rather than communications policies. Unlike Intelsat, it took twenty years for the Eutelsat space segment to become operational by which time many of the potential industrial benefits to Europe had disappeared. Nevertheless, the organisation is an international body and consequently the 26 member countries are tied to it by treaty as well as economic arrangements. Most countries are, by tradition, unwilling to break treaty arrangements. The specific treaty arrangements which, in effect, regulate the satellite activities of its PTT members are those covered in its articles of association. In particular, those arrangements covering economic harm to Eutelsat effectively rule out a PTT operating or using a separate satellite system for regional as distinct to national services.

Eutelsat is fundamentally different from other European organisations in that it is not dominated by one or two countries. Therein probably lies its fundamental weakness. It is based on the Intelsat concept, which although nominally an international organisation, was created by and for the interests of the USA. There is no such dominant interest in Eutelsat, and therefore no such dominant support. Many in Europe believe that Eutelsat will break up under its own internal conflicts. Unlike Intelsat, it was too long in the making and lacks the underlying originality to make a go of high risk opportunities. Moreover, it appeared at a time when it was apparent that the real benefits to Europe would come from providing satellite services rather than manufacturing satellites. It remains tied to European industrial policy, rather than Information technology policy.

OTHER SATELLITE PLANS FOR EUROPE

The dominance of telecommunications administrations is creating problems for a number of proposals and plans, most notable of which is the Luxembourg Astra satellite system. However, the various plans for private international satellite systems in the USA (Orion, ISI and PanAm Sat) are all capable of providing intra European services. They all also require authorisation in Europe but none has been forthcoming so far. Similarly there are other plans for European broadcasting satellites which could have a major impact on Eutelsat's activities. These are from Atlantic Satellites in Dublin, offering both fixed satellite service and DBS capacity, from Britsat London, offering DBS capacity for fixed satellite service type services as well as its primary mission to broadcasters and European Business Satellite in Sweden.

The nature of the potential all encompassing Eutelsat agreement has led to some political inconsistencies. Like Intelsat, Eutelsat is partially a product of industrial policy to provide a captive market for the European satellite manufacturing industry. By ruling out separate satellite systems, it is also reducing the number of sales opportunities for the manufacturers and for the domestic satellite systems operated by its member telecommunications administrations. The French Telecom 1 domestic satellite system appears to encroach as far as possible on Eutelsat territory without actually breaking the rules. Its Ku–band footprint is able to cover all of the populated parts of Europe although, so far, its use outside France has been limited. The French position appears to be one of covering all future potential options.

The PTTs have also had disagreements within the Eutelsat forum about the future of rival communications satellite systems. It appears that British Telecom has seriously considered using the Astra system and pulling out of Eutelsat altogether.

Intelsat. The relationship between European telecommunications administrations and Intelsat is similar to that of the Eutelsat case. The combination of the two arrangements largely prevents European telecommunications administrations from leasing or purchasing Intelsat capacity for European wide, as distinct from national, broadcasting. The US authorisation of private international satellite systems making up competition in Intelsat has not yet been paralleled in Europe, either in allowing the US private companies downlink or uplinks in Europe or allowing

European private international (as distinct from regional) satellite systems.

The reason appears largely to do with lack of interest in private satellite systems. The Europeans were the first to threaten Intelsat with separate satellite systems, notably UNISAT, Telecom 1 and F–Sat/Videosat. The relationship with Intelsat is likely to lead to coordination problems for Astra as and when the issue is raised in the organisation.

Bilateral Arrangements. It could be argued that in practice, the EEC has a strong element of being dominated by a bilateral informal partnership between West Germany and France. Nevertheless, the expansion in membership since 1972 has strongly undermined this. Moreover, the EEC is primarily an economic entity and a number of areas of pan-European activity fall outside its scope. These include R & D and most communications matters. One consequence is that initiatives for European activities have largely come from national or bilateral sources.

For example, in Direct Broadcast Satellites, France and Germany agreed on a bilateral R & D arrangement in the late 1970's (TV–SAT/TDF–1). This has led to further attempts to create a dominant orbital slot for DBS at 19 degrees West, using either French or German built satellites. The slot would be both used by those countries allocated the slot under WARC '77 (World Administrative Radio Conference, 1977) and countries like the UK allocated 31 degrees West. Similarly, the adoption of European transmission standards for DBS has tended to fall along bilateral rather than pan-European terms. In the European answer to the US Strategic Defense Initiative programme, Eureca, the Europeans have tended to group around an initial bilateral framework outlined by France and Germany.

EUROPEAN COMMUNITY POLICY ON BROADCASTING

The current European Commission initiatives in this field are in two areas of which one is the most significant. The latter is a proposal for a European Council directive called "Television Without Frontiers". The other was a proposal for a united satellite TV transmission standard for Europe. This has made no real headway and is probably not nearly as significant as "Television Without Frontiers". Europe can get by without unified technical standards. However, it appears that the Commission's directive is central to the development of the European satellite TV market. It is not altogether clear at the time of writing that the directive will become Community legislation.

"Television Without Frontiers" is the name of both the current Directive (1986) and a past (1984) Green (discussion) Paper of the European Commission intended to provide a European common market for TV programming. The proposals fall into two groups. There are technical and economic recommendations which appear to fall within the scope of the Treaty of Rome. A second set focuses on cultural matters although there is a grey area between economic and cultural affairs. They cover such matters as quotas on European content.

The directive is generally not popular amongst producers broadcasters and governments in Europe although this unpopularity is not a fair indication of the merits of the proposals. The essential problem with any such legislation is that it over-rides national sovereignty in a field in which the EEC has, hitherto, had no influence. For example, the British government in 1986 made it clear that it generally agreed with the view of a House of Lords Select Committee, which rejected virtually all of the proposals of the commission's Green Paper, Television without Frontiers, on which the draft directive is based. The EEC Commission's draft directive on the harmonisation of community broadcasting legislation proposes the following areas which are of greatest concern to European national governments.

1. All member states should have at least one television channel carrying advertising.

2. It includes plans for minimum programme quotas of community production and of production outside the traditional broadcasting organisations. There are major disagreements between EEC member countries over these and other politically sensitive proposals. These have yet to be resolved.

The Proposals were originally set out in the European Commission's Green Paper, "Television Without Frontiers", published in 1984. It covered three areas which the Commission felt harmonisation were immediately feasible to implement: advertising, copyright and youth protection. The Commission's intention was to bring broadcasting within the scope of the Treaty of Rome without having a radical, and politically unacceptable, set of proposals. The Commission also believed that many national controls over broadcasting were being undermind by technical developments, notably satellite and cable TV.

The quota proposals in the Green Paper have been strongly criticised because, it is felt, they would add unnecessarily to programming costs. The commission believes that the quotas are needed to defend the European production industry against the threat posed by cheap imports, mainly from the USA. This has some political acceptability in European countries. Nearly all have regulations which insist on some proportion of domestic or European content in TV channels. The rules do vary considerably from country to country and in the broadest sense are proportional to the size of the population. The larger the size of the population, the lower the proportion of programming allowed from outside the country.

The Commission is considering two types of programme production quota:

1. A minimum of 30% of programmes to be produced within the EEC, rising to 60% after three years.

2. Independent production should account for 5% of programmes, rising to 10% after three years.

The Commission is concerned that developments in the audiovisual market could result in a large proportion of US or other non-EEC material coming into Europe either through EEC based broadcasters, through direct satellite transmission from the USA or through non-EEC European broadcasters. A totally open European market in broadcasting could be flooded with cheap US imports, with a major impact on the European production industry. Therefore the Commission is considering "community preference". Tied to this is the notion of an independent production quota which would encourage the expansion of a production base outside the traditional broadcasting organisations.

The quota proposal was raised during consultation which followed the publication of the Green Paper. The idea was also raised by the European Parliament when it discussed the Green Paper in 1985. Members suggested that the inclusion of such a measure would increase support for the idea of harmonisation.

The concept has been accepted by the Council of Europe, which adopted a resolution saying that a "reasonable" quota should be allowed. If the Commission eventually accepts a quota, there are likely to be further problems regarding its level. Some broadcasters argue that a high percentage would lead to an increase in costs because they would no longer have free access to cheap non-EEC products. This could lead to increases in licence fees, or to further dependence on advertising and a breakdown in the Public Service Broadcasting ethos.

The Commission has also proposed that all EEC member states should have at least one channel carrying advertising. Only Denmark and Belgium do not at present allow TV advertising, although both are introducing them. The Commission has subsequently dropped this requirement. However, under its current proposals, member states would not be allowed to prevent the reception by their citizens of advertising broadcast from other members, provided that these broadcasts met certain community standards. Therefore members would be more or less forced to allow advertising on domestic TV, otherwise their national advertising markets would be undermined by broadcasters from outside their country.

Yet cross-border commercial TV and radio have not only been a feature of the European broadcasting scene for half a century but have also been actively supported by governments of countries to which they were directed. CLT in Luxembourg was originally set up by French commercial interests with support of the French government and is still largely French controlled. Indeed France has allowed Italian interests into its domestic TV scene in the last year (and is now trying to get them out). The position is complicated because expanding cross-border broadcast activities conflict with job creation in the recipient countries.

Acceptance of the Directive is still some time off. It is not, even if it passed by the European Parliament, likely to come into effect until the end of the decade. The Commission does this week discuss, and accept, the draft directive, then it will be published and sent to the Council of Ministers some time after Easter 1987. The Council would then consult with the European Parliament, with the Economic and Social Committee, and with the individual member states. This will probably take until autumn at least. The adoption of an amended directive could take another two years.

The European Commission argues that its directive, as far as advertising is concerned, would have little effect on cable and satellite TV. Concern appears to be arising from proposals covering sponsorship. There is also some doubt as to whether the Commission should be dealing with broadcasting matters.

The Commission has said that only Denmark has questioned its competence to deal with the issue of pan-European TV. Denmark has said that broadcasting is a cultural issue and does not therefore come under the Treaty of Rome. The Commission has replied that broadcasting is still an economic activity and that all economic activities are covered by the Treaty.

Denmark could not by itself prevent the final adoption of a directive by the Council of Ministers. Adoption requires only a majority of member states rather than a unanimous vote. The European Council has brought forward to the middle 1987 the date by which it expects negotiations on the directive to be complete.

Reflecting the discussion nature of the Green Paper, the draft directive has included a lot of current European thinking on the future of audiovisual markets. But the addition of proposals that all channels should include minimum quotas of community produced programming, and of material from independent producers, has created concern amongst almost all broadcasters and from many governments, including the UK, West Germany and Denmark. The provisions are supported by Italy and France. The inclusion of the quotas may also secure support for the Directive from the European Parliament, which suggested the idea in the first place.

THE GREEN PAPER

The Commission wants a free flow of programmes by ensuring that member states cannot keep out the broadcasts of other member states, provided that they compiled with minimum and common standards. The paper was published by the Commission in 1984 and covered the following areas:

COPYRIGHT

This is a controversial area. The commission's proposals are supported by cable operators but opposed by broadcasters. Like the Green Paper, the draft directive proposes to remove the right of copyright holders to prohibit the simultaneous relay of signals by cable in another country. The Commission is aiming, instead, for a policy whereby copyright holders should have the right to receive fair remuneration. Where copyright is invoked, and in the absence of agreement through negotiation, re-transmission should be made possible within two years by the application of a statutory licence.

QUOTAS

The draft directive proposes that a minimum of 30 per cent of non-news programmes should be produced within the EEC rising to 60 per cent over three years and that independent production should account for 5 per cent of programme budgets, rising to 10 per cent after three years.

The quota may be a greater burden for new cable and satellite channels, buying much of their material from outside the EEC, principally from the USA. There are also a number of current and proposed cable TV channels for Europe originating from the USA. These include Cable News Network (not altogether classic news programming), a general entertainment channel from Turner broadcasting, Anglovision (in which NBC has a strong interest), Financial News network and the USIA Worldnet channel. The quota would increase costs and would make certain specialist services, such as an English-language film channel, extremely difficult to run. The quotas are intended to apply to individual channels, and not to national or organisational output in total. The category of programmes excluded from the EEC quota calculation includes sports coverage and game shows.

ADVERTISING

Green Paper proposals that all member states should be required to allow advertising on domestic services have been dropped. This is because the availability of advertising-supported cross-border broadcasts is leading member states in this direction anyway.

Two EEC members do not allow advertising on television, Denmark and Belgium. They are now introducing commercial TV because they are already allowing in transborder advertising supported by cable TV. The minimum advertising standards proposed by the draft directive are as follows:

1. A maximum of 15 per cent of airtime allocated to advertising.

2. A prohibition on the advertising of tobacco and tobacco products.

3. Restrictions on the advertising of alcohol.

4. The advertising airtime limit applies only to cross-border broadcasts, not to domestic ones.

SPONSORSHIP

The draft directive says that an undertaking should not exert "undue influence" over editorial content.

BUSINESS COMMUNICATIONS IN EUROPE

European satellites have tended to aim specifically at the business communications market, although so far success has been limited.

The EUTELSAT I spacecraft (except F1) each carry two SMS (Satellite Multi-Services) transponders. At the time of writing the service is virtually unused. Potential users have complained of the high prices charged by telecommunications administrations for using the system. The service uses Frequency Division Multiple Access (FDMA) techniques which are inefficient users of the satellite capacity but which allow cheap ground equipment.

France opted for TDMA techniques on its Telecom 1 business services, which are also available through Eutelsat. The French experience has followed that of Satellite Business Systems in the USA. The market has proved to be small and users are reluctant to invest up to $1 million in the expensive ground equipment.

Both the Eutelsat and Telecom 1 methods have had their critics; the Nordic TELE–X preoperational satellite was designed with a business payload aimed at keeping the ground segment costs as low as possible. To do so, it opted for DBS power transponders using Frequency Division Multiple Access (FDMA) techniques. Nevertheless, interest in TELE–X is now minimal amongst its backers, primarily because broadcasters are not interested in using its TV capacity.

Deutsche Bundespost's DFS–Copernicus satellite system has been designed with business communications in mind, but the Bundespost is understood to be concerned that the Telecom 1 system has proved unpopular. it has leased capacity on the latter as an interim measure before it is able to offer its own business capacity.

It has claimed that the business services are an interim step towards the Interactive Switched Digital Network (ISDN) concept, in which it is now establishing pilot projects. Similarly, the Italians have claimed that their Intalsat preoperational system will allow them to offer ISDN type services before the terrestrial system is installed.

The European Space Agency's Olympus satellite has been designed to test new business services and the agency has been active in trying to obtain experimental users of the capacity. It has met with limited success.

One major potential market for specialised business communications is the European financial environment. The liberalisation of financial institutions, particularly in the City of London, and the spread of world wide, twenty four hour dealing in currencies, stocks, shares, bonds and other financial instruments, has, and is having a major impact on the telecommunications scene. Indeed, the liberalisation of UK telecommunications and the creation of Mercury Communications has much to do with pressure from the City to improve communication services.

Nevertheless, the UK telecommunications carriers are investing heavily in fibre optics to serve the new opportunities. Cable and Wireless (who own Mercury Communications) are in partnership with KDD of Japan and investing in fibre-optics in that area. British Telecom, along with other European telecommunications administrations, is investing in TAT–8 and TAT–9 for North Atlantic communications and has been selling leases in advance to the City. These have proved to be popular.

The implication is that the City of London (Europe's dominant financial market) requires global, rather than regional capacity. Nevertheless, in certain areas it remains interested in the use of business satellite services, which can offer more immediate and global capacity than is currently available via the terrestrial and submarine network.

The Stock Exchange is developing its own private satellite network, using Intelsat's International Business Service (IBS).

European pressure was a significant factor in the development of Intelsat's IBS service. The now defunct UNISAT project originally envisaged a fixed satellite service payload for North Atlantic business traffic because, at the time, Intelsat had no such service either available or planned. Unisat was, in some ways, the "cat amongst the pigeons" which led to US proposals for private international satellite systems in competition with Intelsat.

Related to the interest in business communications via satellite is the structure of European communications carrier activities. Europe is still dominated by national, satellite owned telecommunications administrations (PTTs); the UK is still the exception amongst the major countries and progress towards liberalisation is slow. There are a number of reasons for this:

1. Although the classic PTT monopoly exists on a national basis, the telecommunications administrations have tended to compete amongst themselves

in strategies towards international communications. For example, British Telecom offers considerably lower rates on international traffic than the Deutsche Bundespost; as a consequence, companies like Ford Motors have established their corporate European headquarters in the UK rather than in Germany.

2. Many of the European national markets are small and do not justify the large scale private communications carriers such as MCI found in the USA.

3. Telecommunications remain a major feature of national industrial policies towards both information technology and equipment manufacturing in general. The liberalisation of the US telecommunications scene has been dominated by demand led rather than the supply led factors still dominant in Europe. European governments have been unwilling to shed inefficient telecommunications equipment manufacturing capacity.

They are locked in a position not experienced by the Americans. Virtually all their domestic markets are too small to justify the high R & D costs in state of the art telecommunications equipment, such as switching systems and satellites. Through ownership of monopoly telecommunications administrations, the governments can create captive customers for their overpriced or poorly performing equipment. If they break that control, it would allow cheaper or better performing equipment in from abroad without corresponding access to other markets to allow the domestic manufacturers either to specialise or gain economies of R & D expenditure.

The UK apart, much of the liberalisation of telecommunications experienced so far in Europe has been about tinkering with the edges of the PTT monopolies, such as allowing non-PTT supplied equipment to interface with the PTT network.

4. The European Commission has tended to look at R & D rather than liberalisation policies. Although it is within the scope of the Treaty of Rome for it to insist that there is a free market in telecommunications equipment, there appears that little can be done until such time as the state sponsored telecommunications administrations are taken away from state control.

5. European based companies have limited experience of working with each other, either in R & D (which such EEC projects as Esprit and RACE are meant to help overcome) or in production and marketing of complementary product lines. Governments are therefore even more reluctant to allow liberalisation and the consequential competitive threat to their domestic industries.

6. Europe has been slow in comparison with Japan and the USA to shed its manufacturing base and move towards an information and services based economy. This is partly because it has been unable to create the job opportunities in the latter. It has consequently tried to hang on to what it already has, rather than release resources to new fields. The telecommunications administrations are major employers in Western Europe and their trade unions have considerable political clout.

It is therefore concluded that progress towards liberalisation will be slow and patchy in Europe. Even in those areas where there is no apparent need to protect PTT monopolies, such as in cable TV network operations, telecommunications administrations are likely to continue to retain their power, as is the case in West Germany. The cost is that the European telecommunications equipment manufacturing industry will continue to lose out in its export markets and the European consumer will pay too much for telecommunications services.

SUMMARY OF THE FAR EAST

FAR EASTERN REVIEW

Developments in Japan and the Pacific Ocean region have been attracting the interests of both the telecommunications and aerospace industries in the USA and Europe. Japan has largely been responsible for the creation of a Far Eastern growth pole, requiring considerable investment in an international telecommunications infrastructure. Japan is also aiming to be a major force in the satellite and space manufacturing sectors. It is already a major force in the commercial ground equipment sector. The Far East is also a major opportunity for domestic and regional satellite operators, with the following satellite systems already under construction or in operation:

CS–2 (Japanese domestic)
SC–3 (Japanese domestic)
BS–2 (Japanese domestic)
BS–3 (Japanese domestic)
JC SAT (Japanese domestic)
SCS (Japanese domestic)
AUSSAT
(Australian domestic with regional ambitions)
PALAPA (Indonesian domestic/regional system)
INSAT (Indian domestic)

In addition to these the Pacific Ocean region is now attracting the interest of international private satellite operators in competition to Intelsat. The region is experiencing traffic growth rates well above those of the Atlantic Ocean Region. Within the ASEAN sub-region and the Pacific Island nations, there is a perceived need for improved basic telecommunications infrastructure communications as well as growing demand.

China has also been opening up and is a possible market for US satellite manufacturers. It has already entered the international space business by offering its Long March 2 and 3 vehicles for launch of commercial satellites. From 1992, Japan will also be able to offer its advanced H–II vehicle, the development of which will place itself in the position of a major space hardware manufacturer. Moreover, despite allowing sales of US manufactured satellites to domestic telecommunications companies, Japan continues with a major R & D programme in communications satellite manufacturing.

India has successfully implemented its INSAT I domestic communications system and has ambitions to become an indigenous manufacturer of commercial communications satellites. Along with Australia and Indonesia, it has a ground equipment manufacturing capability as well but in all three cases, the industries survive because of preferential procurement policies or tariff barriers. All three are dependent on the import of critical components from the USA, Europe and Japan.

Nevertheless, developments in the region are largely along national lines, there is no equivalent to Eutelsat or ESA; such matters as transborder broadcasting and regional communications are handled through bilateral arrangements or through organisations such as the ITU. The factors that bind the region together for broadcasting, telecommunications and R & D activities are much weaker than they are in Europe and, apart from a generally higher economic growth rate than the rest of the world, it is difficult to distinguish this as a coherent region. Rather, the picture is one of three major nations, Japan (tied closely in direction to South Korea and Taiwan), China and India, plus two sub-regions consisting of the Pacific Islands nations and ASEAN, and a number of countries out on a limb, notably Australia and New Zealand.

Bearing this in mind, there are still several features of the region's satellite communications scene which are of special interest.

DBS TV

There are four satellite systems of interest. Only one is a true DBS system, the Japanese BS–2/3 system. The technical experience of BS–2 has been disappointing, even though the Japanese had prior experience with the preoperational BSE satellite launched in 1977. The failure of BS–2a raised questions about the overall reliability of high power TWTAs needed for true DBS services. The tubes were only 100 watt products, considerably below the 240 watts envisaged in most of the European DBS proposals. 100 watts has been considered for UK and US services but the general feeling in the satellite world is that new TWTAs need to be developed. NEC was expected to provide new 100 watt transponders for the BS–3 DBS system now under development for Japan. However, such were the risks that the option was dropped in favour of the existing Thomson tubes. The BS–2 system now appears to have been downgraded to a preoperational project, leaving commercial DBS operations to the BS–3 system.

A second lesson learned from the BS–2a failure was the problem of project management when a satellite is procured using a wide variety of imported systems and components. The Japanese response has been to reorganise its domestic satellite manufacturing arrangements.

The problems have also impacted on the Japanese Television Receive Only (TVRO) manufacturing industry which saw BS–2 as a method of building up domestic demand from which it would launch an assault on the international markets, already being in a strong position through supply of ground equipment for the full range of FSS services. BS–2a left a lot of equipment and wasted development and tooling-up costs on the hands of the Japanese manufacturers. The Japanese DBS industry is now sceptical about the underlying reliability of DBS spacecraft and the long term commercial viability of DBS services. Nevertheless, DBS TV is part of a much broader set of policies toward TV and industry. It is unlikely that Japan will abandon its governmental corporatist policies towards DBS, particularly as it is seen as a major step towards High Definition TV.

Both India and Australia have opted to use FSS spacecraft to provide DBS TV. Both are relatively modest attempts to provide TV to rural areas outside the range of terrestrial transmitters, rather than as alternatives to off-air TV in populated areas. Nevertheless, Australians are planning to develop specialised Private Satellite Networks for distribution of specialised TV channels to clubs and hotels. Australian broadcasting interests involved (Robert Holmes à Court, Rupert Murdoch) have interests in satellite TV in the UK. Murdoch has also set up the fourth national US network and appears to be aiming at a global TV network.

Both the Insat I and Aussat systems are used primarily for networking TV. The Insat I system has been used to create India's national terrestrial TV network, whereas Aussat has been used to alter the structure of the existing national networks.

China has taken advantage of the Intelsat SHARE project to set up an educational TV programme in the country. This is a major project in its own right and the country is purchasing Intelsat Transponders to develop the project. The country appears to have dropped plans for its own domestic direct broadcast satellite system, which was being handled by an organisation known as the China Broadcasting Satellite Corporation. Nevertheless, the country retains a strong interest in developing its own communications satellite technology and there remains a possibility that it will use its own fixed satellite system for some forms of TV services.

FIXED SATELLITE SERVICES

The most critical development in the region in this field is the break up of NTT's and KDD's monopolies on Japan's national and international telecommunications. As explained elsewhere in this introduction, this is leading to a major increase in demand for international communications capacity in the region. Much of the increase in demand will not, however, go to satellites. Telecommunications carriers world-wide are busy investing in transoceanic fibre optic cables to meet the growing demand.

Within Japan, the liberalised telecommunications market has allowed Hughes Aircraft Company and Ford Aerospace to sell satellites to domestic operators in which they have a financial investment. Japan Communications satellite and the Satellite communications company both have class one telecommunications licenses but face competition from new companies using fibre optics. The Mercury Communications concept is being used by the national railway operator, Japan National Railways (JNR), to install a national fibre optic system alongside its railway tracks.

KDD is now losing its monopoly on international traffic coming in or out of Japan, with new operators looking to use Intelsat capacity as well as those planning new fibre optic links. So far, there have been no substantive plans for alternative satellite systems to Intelsat to provide international communications for Japan.

However, there are a number of proposals for alternative satellite systems to serve the Pacific region in general. These include Finansat, Columbia Communications Corporation, Pan Am Pacific Satellite, McCaw Space Technologies, Pacific Satellite Corporation and an expanded Aussat system. Most of these proposals envisage long distance telecommunications links between the Pacific region and distant continents although Pan Am Pacific envisages links between the Pacific island nations.

AT & T, Teleglobe Canada, KDD, Hawaiian Telephone Company and the Philippine Long Distance Telephone Company are expected, in 1988, to complete a trans-Pacific fibre optic cable. This would provide 40,000 voice circuits and involve a link from the USA to Hawaii with branch to Japan and Guam.

FIBRE OPTIC COMPETITION

FIBRE OPTIC COMPETITION TO SATELLITES

The impact of fibre optics on satellite communications have been a strong topic of interest in the trade press since Tony Rutkowski of the Federal Communications Commission published an article in Telecommunications magazine in early 1983. Since then, the US satellite manufacturers have received only a handful of orders from commercial concerns and telecommunications operators have increasingly turned to investing in fibre optics for long distance and international communications. Most recently it has become apparent that the technical risks in launching satellites is much higher than was apparent in 1983. This has, in turn, meant that commercial satellite operators are faced with uncertainty about scheduling their launches and about availability of launch insurance.

Much of the past comment on the battle between fibre optics and satellite communications has concentrated on the relative technical merits of the two technologies. They have tended to ignore the much more important commercial and operational factors influencing telecommunications carriers in their choice between the two.

The approach here has been to concentrate on the commercial, cost and economic factors which are influencing the decisions. It is concluded that there is clear evidence to show that fibre optics will generally replace satellites for trunk telephony purposes. Satellites are significantly more effective for point-to-multipoint TV distribution services. There may be a number of business services for which satellites remain more cost effective than terrestrial based telecommunications systems. For obvious reasons, fibre optics cannot be used for mobile communications. There are certain mobile communications markets, such as maritime communications, where satellites have advantages over all other means of communications, but these are generally specialised services with limited total demand.

It is not clear how policy structure and changes are affecting the relative merits of each of the two technologies. Nevertheless, the nature of fibre optics appear to favour arguments that telecommunications is a natural monopoly. Therefore, it is possible that existing monopoly PTTs are investing in fibre optics to preserve that monopoly.

LIMITATIONS OF FIBRE OPTIC TECHNOLOGY

For communications purposes, developments in fibre optic technology have, so far, been limited to the actual transmission segment of the network. The developments in optical switching technology have remained in the laboratory. Thus, in its application to telecommunications, fibre optics technology is a substitute for one part of the telecoms network, the transmission link. It is dependent on the existing (but rapidly changing) switching technologies which also serve other terrestrial transmission systems. Put another way, fibre optics are limited entirely to being part of the hierarchal telecommunications network. This is not the case with satellite communications which can provide a complete by-pass of the terrestrial network. Particularly in broadcasting, where the terrestrial network does not have the cost effective capacity to provide point-to-multipoint broadband capacity, fibre optics have little to offer at present.

Fibre optics are primarily of merit because of their ability to increase transmission capacity between the switching centres in a terrestrial telecommunications network.

THE FIBRE OPTIC SWITCHING PROBLEM

In the United Kingdom, the present Conservative government, in establishing its policy towards cable TV originally envisaged that the country would be cabled with fibre optics, with local TV systems based on switched star technology and with a national fibre optic network provided by Mercury Communications (and whatever British Telecom wanted to install). The Mercury/BT systems have developed but what little investment there has been in cable has been entirely dependent on co-axial technology.

The local fibre optic cable concept failed immediately because no suitable switches were available. Although optical or photonic switches are available in the laboratory, fibre optic switching (and repeaters) require that the optical signal be converted back into an electronic signal before switching (or amplification). The state of the art (1986) optical switching technology only allows about 4 channels to be switched. The development of optical switches cable of switching several thousand circuits appears to be many years away.

Therefore fibre optics are necessarily dependent on conventional switching technology and are in practise only a transmission technology. Nevertheless, in comparing them with satellite technology, the dependence on conventional switching technology (and therefore, in practice, existing switching facilities) gives fibre optics a competitive advantage. Modern electronic switching has considerably reduced the cost of purchase and maintenance of switching centres and has allowed digital communications to take an increasing proportion of the overall telecommunications traffic. Fibre optics are a digital transmission method and fit into the growing trend. As at 1986, 27% of US switching centres are digital. By 1990, about half are expected to be digital.

THE FINAL MILE PROBLEM

In the link between the end user and the first switching centre in the telecommunications network, the use of high capacity fibre optics still remains limited. Most of the traffic in this part of the system is low volume and therefore does not require the new technology. Moreover, there remains the extra cost of converting the electrical/electronic signals used by the end users equipment (whether it be a TV set, telephone or data processing equipment) to and from an optical form. For local cable TV networks it is theoretically possible to distribute TV channels to households without switching. This is the "tree and branch" method where the main set of signals is tapped outside the household, with the full set of signals going into the household. In conventional cable TV networks there are many taps. It can be done with fibre optics but each tap results in a very large degradation of the main set of signals. This in turn requires the use of a repeater after a few taps to amplify the weakened signal. Fibre optic repeaters are expensive because they require that the optical signal be converted back into an electronic form and, once amplified, back into an optical form.

Despite these limitations on the use of optical communications, there remain special conditions where fibre optics are being applied to the local link. These are in large urban areas such as New York where the national and international satellite gateway (teleport) is connected to large business users in the centre of the city. Some industry observers see such conditions as making satellite communications technology complementary with fibre optics. This may be the case in the short run but the evidence appears to suggest that in the long run, such local fibre links will increase the swing away from satellites. Once the local link is interconnected with national and international fibre optic networks, there are both cost and operational arguments for abandoning the satellite segment altogether. This is primarily because such local links are serving the point-to-point market.

DECREASING FIBRE OPTIC COSTS

Despite the limitations of optical communications technology, it is a technology that is subject to rapidly decreasing costs of application. These stem from two basic developments:

1. Monomode transmission techniques which reduce either the need for repeaters or increase the transmission capacity rate or both in a fibre optic link. The repeaters are an expensive factor in the overall costs of the link, particularly over long distances and in submarine links. Monomode fibres were first field tested in 1983 and developments in the technology are expected to produce significant cost reductions per circuit for at least the next decade. Dispersion-shifted optical fibre technology is also understood to be leading to significant cost per circuit reductions. Some communications engineers argue that repeaters will eventually be needed at intervals measured in 1,000s of kilometres, which, for most links, means not at all.

2. The transmission capacity (or bit rate capability) of fibre optics is now exceeding the capabilities of the electronic equipment which the fibre optic technologies interface with. Increasing capacity of interfacing equipment is overcoming this problem.

For example, the 12,000 voice circuit TAT–8 transatlantic link, due to come on stream in 1988 is expected to require a capital investment equal to a 144,000 voice circuit capacity transatlantic link available in 1995.

Another example is Mercury Communications UK fibre optic network for the UK. This uses monomode technology with an initial bit rate of 140 Mbit/s and with regenerators placed at 25 kilometre intervals. Mercury says that this rate can now be increased to 560 Mbit/s without changing the actual cables and without more regenerators.

LIMITATIONS OF SATELLITE TECHNOLOGY

A number of commentators have argued that there are significant developments in satellite technology which are expected to influence the balance between satellites and fibre optics. But many of the so called developments cover the use of multiple and rapidly switchable spot beams and on-board satellite switching. The major R & D project involved in this field is NASA's Advanced Communications Test Satellite, due to fly in 1992. That satellite is aimed at providing the space segment technology for (basically) telephony traffic and, perhaps, specialised business communications using the VSAT system. Operational versions of the satellite are not expected to become available until at least the mid-1990s, nearly a decade away. It is not reasonable to expect telecommunications companies to wait that long for the satellite technology needed to compete with fibre optic technology. It appears that fibre optic technology is already competitive with satellites for long distance telephony and traffic on a point-to-point basis. The figures quoted above on TAT–8 and its successors indicate that by 1995, the cost per circuit via fibre optic across the North Atlantic will fall by a factor of 12 before advanced satellites become available. In the meanwhile, there will be a continuous fall in the costs of fibre optics although there will be a substantial reduction in the cost per circuit on the Intelsat system.

An alternative perspective on this is that satellite R & D economics are suffering from diminishing marginal returns and are now therefore a mature technology. It is no longer worthwhile investing large sums of money in them, even though, from an engineering viewpoint, major improvements in performance are possible.

INCREASING SATELLITE COSTS

The Intelsat organisation is currently launching the last satellites of its fifth generation design (INTELSAT V). At 1976 costs, each of the fifteen Intelsat V spacecraft cost $34 million. So far, 2 of the satellites have been lost on launch. Therefore the cost of each in orbit satellite is $39 million. Adjusting for inflation, it is estimated here that the cost at 1986 prices is $60 million.

SATELLITE LAUNCH AND RISK MANAGEMENT

The Intelsat satellite cost figures are meant to illustrate the limited impact of technology applications to the space segment of satellite communications, rather than overall cost changes in satellite communications. The experience of the last five years points to a substantial increase in the actual commercial costs of the space segment resulting from increases in insurance and launch charges.

An illustrative example is the cost of an industry standard Hughes HS-376 spacecraft in orbit, comparing 1982 with 1988 prices, when the last of this design is expected to be launched. The cost per circuit has more than doubled in 8 years (with all other factors remaining equal). Rather than a technology offering significant cost reductions, the current trend is one of major cost increases as it becomes apparent what the actual risks are in satellite communications. Put another way, it is now clearly apparent that improvements in payload capacity on satellites have been off-set, in cost terms, by the engineering side not being able to match the increased risks of more advanced technology – either in launch vehicles or the satellites themselves.

Moreover, satellite communications tend to be commercially more risky than terrestrial network facilities. The specialised, almost bespoke, nature of commercial satellites, means that once they are purchased by an operator for a designated service, they represent sunk costs. For example, the very substantial losses experienced by Satellite Business Systems since it began operations in 1980 do not reflect the technical risks of the technology. SBS has lost in excess of $400 million purely because no one wanted its original services and they are not suitable for other applications. Therefore satellite operators need to look for a higher rate of return than, say, local telephone operating companies. There is currently a market for satellites which are "surplus to the requirements" of commercial satellite operators. Amongst the satellites on offer in this market are the Palapa B2 spacecraft, two Comsat DBS, one SBS, one Telesat Canada and (unconfirmed) one RCA spacecraft. These satellites have been notoriously difficult to sell, even though all except the Comsat spacecraft are basically similar to those of other operators.

Events in 1984 to 1986 showed that the risks of launching satellites were considerably greater than had been previously expected, leading to insurance rates (where insurance is available) rising to 30%. At the time of writing it is by no means certain that a satellite insurance market exists in any substantive form. The commercial satellite operators are also faced with an additional risk arising from the grounding of all the Western world's launch vehicles. It is suggested here that under these circumstances, the financing of a commercial satellite system is no longer a realistic proposition. The satellite operator is no longer able to show that insurance will be available on launch and can therefore no longer provide reasonable guarantees for loan finance. The satellite operator can no longer plan his future operations with any degree of uncertainty. It is now difficult to predict when a satellite will be launched and even more difficult to predict when launch vehicle capacity will be available to launch a replacement in the event of a failure of the first. Nor are future launch charges particularly certain. There is virtually no free launch capacity available up until 1990. The shortage of capacity, which has created a seller's market, may continue for some years after 1990.

These changes in market circumstances over the last three years suggest that only the financially stronger satellite operators will survive. More particularly, they will be the operators such as Intelsat which are big enough to carry some launch risks internally or those satellite operators such as Hughes communications and RCA Americon which are owned and backed by large corporations able to self insure against the risks.

It has been argued that satellites, once in orbit, are less prone to damage by natural catastrophes than terrestrial fibre optics links. Undersea cables are subject to breakage from ships' anchors and turbidity currents whilst land lines are subject to storm damage, it is argued. However, these arguments appear to be largely irrelevant in assessing the economics of the two technologies. For voice traffic, satellites are dependent on the public switched network and therefore part of the point-to-point links involved are subject to the same type of interruptions as fibre optic landlines. With all satellites there remains a risk of total in-orbit failure and a considerably higher risk of partial failure. This is particularly the case with new designs of spacecraft. The satellite manufacturers are under continuous pressure to improve their designs and therefore the risk is likely to remain as long as there is a market for satellites. Moreover, in the comparative analysis of failure, it appears dishonest to ignore the very high risks of launch.

VSATs AND FIBRE OPTICS

It is argued by some US proponents of fibre optics that they can now be used to provide feeder links to large office buildings from the local switching centre. If this is the case, it means that many businesses may gain direct access to national fibre optic networks. Proponents of Very Small Aperture Terminal (VSAT) business services via satellite have argued that there will be considerable demand for their systems because most businesses will not be directly connected to fibre optics which are essentially limited to trunk links. On the other hand, it has been argued that one of the most attractive features of VSATs is that they provide a complete by-pass of the terrestrial system. This has become attractive because, since the break up of AT & T, it has become increasingly difficult to deal with numerous carriers involved in a long distance private network when previously AT & T would have handled the whole exercise. Counter to this are claims that the growth in interest in fibre has largely been created by the break up of AT & T. It has allowed competitors to invest more rapidly in fibre than AT & T which has had to postpone plans for fibre optics because of a detailed review by the Federal Communications Commission. The Commission has not subjected the fibre optic plans of AT & T's competitors to such detailed scrutiny allowing them to move ahead more quickly.

PRICE ELASTICITY AND THE IMPACT ON FIBRE OPTICS AND SATELLITES

From 1967 through to 1981, Intelsat cost per circuit charges to its members were reduced each year. But by the end of 1981, they had levelled out and have remained static since (although they are expected to fall by 25% next year). Again this illustrates that satellite technology has matured. However, it appears to have had a major impact on the rate of growth in demand for international (i.e. long distance) communications.

At the time of writing, about 67% of Intelsat's revenue is accounted for by voice circuits. In the US domestic scene, the percentage is much lower. Voice traffic accounts for about 27% of all US domestic satellite use. In Europe, the figure is even lower; as at September 1986, the Eutelsat system was providing 1,000 voice circuits. This is estimated to currently account for no more than 7% of intra-European satellite traffic. In general, the proportion of satellite traffic worldwide accounted for by voice services has been declining. It is only over the last year that fibre optics have begun to make an impact. The full impact of the newer technology began to be felt in the autumn of 1986 and, in the international scene, is not likely to have an impact until 1988.

The past decline in proportion of traffic accounted for by voice is closely related to satellite economics. As costs of the space segment fell up to around 1983–4, the satellite operators have been able to move into new markets which were hitherto uneconomic to serve. In particular, they have moved into existing and new TV markets. As a result, even though telephony use has been increasing, it has been declining as a percentage of overall satellite use.

COMPARATIVE COSTS OF INTELSAT SPACECRAFT COSTS PER CIRCUIT, IN REAL TERMS

	INTELSAT VA	*INTELSAT VI*
COST OF SATELLITE (1)	$60 million	$140 million (2)
COST OF 1989 LAUNCH (3)	$66 million	$73 million
COST OF INSURANCE	$38 million	$64 million
TOTAL BASIC COST OF SATELLITE	$164 million	$297 million
OPERATIONAL DESIGN LIFE	7 years	10 years
PAYLOAD WEIGHT	260 Kgs	600 Kgs
COST PER Kg OF PAYLOAD (5)	$631,000	$495,000 (22%)
COST PER Kg PER YEAR (5)	$90,100	$49,500 (50%)
VOICE CIRCUIT CAPACITY	15,000	30,000
COST PER CIRCUIT (5)	$11,000	$9,900 (10%)
COST PER CIRCUIT PER YEAR (5)	$1,571	$990 (37%)

Notes (1) The cost comparison is meant to show what it would cost now to purchase the satellites assuming that no satellites had been purchased before, rather than Intelsat's actual costs.

(2) The contract for the five satellites in the Intelsat VI series was for $700 million at 1982 prices, including options which have not been taken up.

(3) The launch cost figures are based on the quoted 1989 cost of launch of an Ariane launch for Intelsat VI. This is $84 million but is assumed to include insurance cover on the cost of the launch vehicle only and excludes the launch insurance of the satellite and its in-orbit life insurance. It is assumed here that the Intelsat organisation could not obtain block insurance for two or more satellites with Intelsat covering the first loss internally. The assumption is made because there is insufficient insurance capacity available.

(4) For comparative purposes, it is assumed here that each satellite is separately and fully insured at a rate of 30% of the cost of the satellite and its launch. In practice, the cost of insuring the Intelsat V series is believed to be below 10% because cover was arranged before underwriters became aware of the actual risks involved.

(5) Percentages given at the end of each line show the percentage saving of Intelsat VI over Intelsat V. In practice, the figures are lower because insurance costs have increased and because discounting factors have not been used to discount the extra three years of life in the Intelsat VI spacecraft.

COMPARATIVE COSTS IN THE INCREASE IN THE PRICE OF SATELLITES, 1982–1988 HS–376 SPACECRAFT

	1982 COST	1988 COST
COST OF SATELLITE FROM MANUFACTURER	$30 million	$30 million
COST OF LAUNCH BY SHUTTLE	$15 million	not available
COST OF LAUNCH BY ARIANE (EST)	not available	$55 million
LAUNCH INSURANCE	$4·5 million	$26 million
COST OF FINANCING TWO YEAR DELAY IN LAUNCH AT 14% PER ANNUM	not applicable	$5 million
TOTAL COST	$49·5 million	$116 million

Reductions in cost per telephony circuit have allowed satellites to move into telephony markets already well served by landlines. Here they have faced more competition than on the early transoceanic fields and have not been able to take as big a market share.

Their operators are often owners of terrestrial facilities which they do not wish to completely replace. Moreover, there appears to be a cut off point at about 800 miles under which satellites are unable to compete with copper landlines on trunk telephony routes. This is one reason why so little internal European telephony traffic is routed via satellite.

COST PER CIRCUIT COMPARISON BETWEEN FIBRE OPTICS AND SATELLITES

Intelsat's own figures have suggested that the cost per circuit on an Intelsat VI spacecraft with multiple-access configuration is around $1133 per circuit and with maximum capacity around $504. The organisation estimates that the cost per circuit on TAT-8 is $1,596 per circuit. The costs of longer distance fibre optics links are higher; Intelsat claim that the cost of a transpacific circuit via fibre is $2,237. Satellites are of course distance insensitive when it comes to cost per circuit. The Intelsat figures indicate that for TAT-8/Intelsat VI technology, the break even point, in terms of distance, is around 3,800 kms, above which the advantage goes to satellites. Past estimates of satellites versus copper terrestrial links have shown a break even point of around 1,200 kms. As has been established earlier, the costs per circuit of fibre optics could fall by a factor of 12.

According to some estimates, each additional fibre optic pair on a transatlantic cable adds an additional 10% to the overall captial costs. Assuming that there is no technical improvement in basic fibre optics capabilities, a trebling of capacity on the next generation transatlantic cable would indicate a cost per circuit of $644. Similarly, the cost per circuit falls to $531, with a fourfold increase to $437 and with a twelve fold increase to $379. This is well below the $504 minimum offered by Intelsat VI. At $644 per circuit, the break even between fibre and satellite (assuming a cost per satellite circuit of $1133) increases to 13,500 kms.

The implication of the above analysis is that the marginal cost of providing capacity on a fibre optic cable is low. However, the analysis implies two other essentially economic features. These are as follows:—

1. For thin route long distance voice communications, satellite communications may remain more cost effective than fibre optics. Typically such routes would be the Intelsat paths connecting the developed world with the less developed countries. For these services, telecommunications organisations are unlikely to invest in large capacity fibre optics.

2. On the heavy routes, for example, across the North Atlantic, current forecasts of future demand indicate figures considerably less than a single fibre optic link could provide. Intelsat has quoted a total market demand of 87,000 voice circuits across the Atlantic by 1995. A single fibre optic cable could provide 144,000 by that time. This is in addition to the capacity on TAT-8 and TAT-9 and that provided by Intelsat which could bring supply up to 342,000 circuits. From the point of view of the fibre optic cable and satellite operators, they will either have to arrange for a per circuit charge several fold higher than the actual cost per circuit or lose a lot of money.

3. Even in the hypothetical case where supply is matched to demand, a large amount of unused back-up capacity is required in case of failure of a cable. This back-up capacity may either be in the form of satellites or other cables.

IMPACT OF COST PER CIRCUIT REDUCTIONS ON FUTURE DEMAND

Intelsat has based its forecasts of future demand by aggregating forecasts supplied by individual members, rather than by its own internal models. No doubt the organisation cross checks the supplied figures against internal forecasts. However, the aggregate forecast method has not proved reliable in the first half of the 1980s. It is suggested here that the reason primarily lies in the fact that international long distance communications are price sensitive and Intelsat's failure to reduce charges in the 1980s has led to a slow down in the growth of traffic. The point is very important because it is also saying that increases in demand are not primarily coming from independent factors such as an increase in trading in the international financial markets. The increase in demand is following and is a response to decreases in international charges.

From this it may therefore be concluded that further falls in charges will lead to a substantial increase in demand for capacity. With a price elasticity of -1, a reduction in charge per circuit will double total demand. Moreover, traditionally international charges levied on end users by PTTs have been considerably higher than charges from Intelsat to the PTTs. This relationship has meant that in most countries apart from the USA, the UK and a few other European countries, demand has been dampened to well below levels of potential demand.

There is further evidence pointing to this argument in the percentage shareholdings of countries in the Intelsat organisation. These shareholdings are a close measure of the use of the system by each member. All other things being equal (including PTT charges), one would expect those countries with the largest international trade activities and located far from their markets to have the largest use of the system. This would indicate that Japan would at least be the second largest user behind the USA or, possibly, the largest user. In practice, however, NTT of Japan has been claimed to charge up to 40 times actual costs for international voice calls. This would suggest that demand is considerably lower than underlying potential. The actual percentage figures of Intelsat ownership confirm this.

As at the end of 1984 the five largest shareholders were as follows:

USA	22·46%
UNITED KINGDOM	13·07%
FRANCE	5·64%
WEST GERMANY	4·09%
JAPAN	3·30%

Japan's lowly position in comparison with that of the UK (whose major trading partner is close at hand (Europe) may well also reflect British Telecom's relatively low international call charges. It may also be argued that the UK has a disproportionate need for

NET PRICE ELASTICITY IN THE USA FOR CUSTOMER DIALLED STATION TO STATION CALLS BY MILEAGE BAND

Mileage Band (miles)	Business Callers	Residential Callers
1–30 miles	−0·28 to −0·44	−0·26 to −0·51
31–85 miles	+0·02 to −0·66	−0·10 to −0·36
86–196 miles	−0·28 to −0·60	−0·33 to −0·87
197–430 miles	−0·48 to −0·94	−0·38 to −1·18
431–3000 miles	−0·47 to −0·94	−0·66 to −1·12

(The net price elasticity for a given market sector is the sum of the own-price and cross-price coefficients. *Source:* Robert R. Auray, 1978)

Price elasticity is an economist's term used to show the sensitivity of changes in the price of a product or service to the total purchases of that good or service. It is normally expressed as a ratio. For example, a price elasticity ratio of −1 means that if telephone charges are reduced by 50%, consumers will double their use of the telephone and therefore spend the same as before on the service. A ratio of −2 means that they will spend twice as much on telephones if the price is halved. A ratio of −0·5 means that if the charges are halved they will spend only 25% less on telephones calls, because, at the same time, they will increase the amount of calls they make by 50%.

Measuring price elasticities in telecommunications is a difficult exercise in which errors are likely to occur. However, the ratios are extremely important because they show how demand for services is likely to change as prices are changed. The above ratios are particularly important because they show that for long distance (and therefore International) services, use of telecommunications is very sensitive to price. For example, if, as it can now be reasonably expected, trans-Atlantic fibre optic cables can result in call charges being reduced by a factor of 4, a price elasticity of −1 indicates that telecommunications operators will need to provide 4 times the capacity that is currently needed to meet demand.

The price elasticity measurements can also be used to help to show where the most lucrative markets are for telecommunications carriers. It has been estimated that long distance telephone charges in Japan and international charges out of that country are up to 40 times what they need to be to make NTT and KDD (the domestic and international telecommunications operators of that country) a reasonable profit on the services. This is one major reason why the Japanese are not heavy users of the Intelsat system despite the country's large and advanced economy. The break-up of NTT's and KDD's monopoly and the expected substantial falls in long distance and international telecommunications charges is expected to lead to a massive increase in demand for telecommunications services.

For business communications one of the reasons why the use of telecommunications is so price sensitive is that relatively small price changes fundamentally change the economics of certain businesses and administrative control procedures within large corporations. The development of international financial markets has been dependent on cost reductions in long distance communications. Similarly, the economics of electronic data bases allow considerable economies of scale on an international basis, but this requires that consumers have cheap telecommunications charges to gain access to the data bases. Within corporations, the convergence of telecommunications and computer technology is facilitating the development of new managerial control practices which are increasingly seen as giving major competitive, strategic, advantages to those companies which are able to develop them first, or most effectively. This is one of the reasons why Japanese companies have been so keen to see NTT and KDD opened up to new competition.

A final point about the price elasticity measures is that they indicate that in most Western countries there is considerable total demand potential which can be exploited by liberalisation of telecommunications and the consequential matching of costs with charges. Traditionally, long distance telecommunications charges have been well above cost and used to cross-subsidise local and short distance charges. But because price elasticity for the latter is well below −1, it means that any increase in charges will result in an overall increase in spending on such services. Similarly, and as is pointed out above, the corresponding decrease in long distance charges is over −1, price reductions in the latter will also lead to an increase in spending. The broad conclusion is that old fashioned monopoly PTTs, which are still the norm in Europe, considerably hamper the growth in telecommunications.

international traffic because London is still the largest financial centre in the world. This is a chicken and egg argument. It could equally be argued that low international call charges reflect that position (and indeed have helped to maintain London's preeminence) at the expense of Tokyo.

IMPACT OF LIBERALISATION

The Pacific Ocean region is now attracting considerable attention from telecommunications organisations. It is argued that the reason primarily relates to the break up of the NTT/KDD telecommunications monopoly in Japan and the consequential realignment of international call charges with actual and potential costs. The position is not matched on the North Atlantic where most European PTT administrations are not expected to significantly reduce transatlantic charges and those that are in a position to do so (the USA and the UK) have already largely done so. Moreover, the transatlantic market is already well served by current and planned cable and satellite systems, leaving an over capacity problem, at least in the short to medium term.

From the point of view of international communications, the Pacific region experience suggests that liberalisation increases the demand for fibre optics. However, there is evidence in the USA that liberalisation is leading to increasing demand for domestic satellite communications for business use. Whilst the telecommunications carriers are currently busy transferring their long distance services from satellite to fibre optic (for example GTE is transferring Sprint this year) there appears to be growing interest in use of satellite VSAT (Very Small Aperture Terminals) to provide private communications systems which bypass the public switched network. The reason lies in the following three factors:

1. Technical developments in earth station technology and transmission techniques (such as spread spectrum) which allow cheap earth terminals. Typically these may be as low as $1,000 for a receive only terminal.

2. The break up of AT&T means that telecommunications users may now have to deal with several instead of one telecoms company. This can create more logistical problems in organising the communications of a company with a large number of sites located across the USA.

3. The realignment of costs with charges that has followed liberalisation has meant that some business charges for use of the terrestrial network have increased considerably, such as leased lines. Whilst fibre optics appear to offer the prospects of lower communications charges in the future, they are part of the overall public switched network.

Moreover, installation of a national fibre optic network is a slow and patchy process, with most investment going into heavy traffic routes. Moreover, VSATs are able to offer digital communications almost immediately, whereas it will be some years before the public switched network will be largely digital. Although interest in VSAT technology remains uncertain, some proponents argue that the technology is complementary to fibre optics in that it allows Integrated Services Digital Network (ISDN) type services to develop before the terrestrial network has developed sufficiently to handle them. Fibre optics provide the transmission capacity for ISDN but also require considerable investment in switching facilities. The same arguments have been applied to the type of business services offered by SBS but the latter failed to stimulate growth in such communications services.

FUTURE ROLES OF SATELLITE COMMUNICATIONS IN THE FIBRE OPTIC ERA

The analysis so far points to the fact that there is limited potential for satellites in the international and domestic trunk route carriage of telephony and associated traffic. As this accounts for some 67% of Intelsat traffic, it is not surprising that the organisation appears to be reducing its capacity by selling off transponders and deciding not to purchase further satellites in the Intelsat VI series over and above the five already under construction. Yet, the Intelsat satellites are perhaps atypical in their design, intended primarily to serve international telephony through use of low power C-band transponders. They are essentially point-to-point communication satellites.

In the US domestic satellite market, common carriers and "other" common carriers were using 12·4% of the total domestic satellite capacity available at the beginning of 1986. Of that 12·4%, AT&T accounted for nearly one-half. GTE's Sprint service accounts for one quarter. The latter, as has been pointed out, will move to fibre optic this year. AT&T's satellite system is new and the company is not expected to immediately move all of its traffic off them until they become time expired. MCI is understood to be moving traffic off its terrestrial microwave system onto the SBS satellites but this appears to be a matter of short term expedience rather than underlying economics. MCI obtained the SBS satellites after most of their costs were written off. They were not therefore obtained at a price which reflects their costs. The immediate prospect, therefore, is that the US satellite operators will lose about 10% of their traffic because of competition from fibre optics.

The conclusion from this is that there remains a lot of traffic in which satellites are a more effective medium than fibres. Video traffic accounts for half of US satellite use, with scheduled programming accounting for half of this and occasional transmissions the other half. TV programming traffic is fundamentally different from voice telephony for the following reasons:

1. It is traffic which does not need switching and there is therefore no reason to use the terrestrial communications network.

2. Broadcasting is a one way point-to-multipoint transmission exercise. There is no equivalent of a break-even point over the distance found in comparing telephony services via satellite and fibre optics. The economics of satellite broadcasting are largely distance insensitive. Instead, the break-even point between satellites and terrestrial transmissions depends on the number of earth terminals which pick up a programming transmission.

3. Broadcasting requires a lot of transmission capacity. Typically, a communications satellite used for distributing TV channels could carry between 12 and 36 TV channels, equivalent to anything between 9,000 and 72,000 multiplexed voice circuits. This would require very substantial investments in fibre optic

communications, particularly as TV requires point-to-multipoint capacity. The three US national TV net works have abandoned landlines in favour of satellites for distribution of TV traffic.

The European experience so far has been that its satellites are largely used for TV traffic. This is far from what was originally expected. The Eutelsat system was intended primarily for voice telephony. The French Telecom 1 system's domestic capacity was intended for Business communications. Neither have proved to be more than marginal sources of traffic.

A very recent indication of the level of interest in fibre optics is illustrated by TAT 9, the trans-atlantic cable planned for operation in 1991 and backed by AT & T, DGT (France), Telefonica (Spain), Teleglobe (Canada and British Telecom). This is expected to have a voice capacity of 75,000 circuits on its main transatlantic segment. At an October 1986 conference of telecommunications operators interested in using the cable, 30 organisations from 21 countries attended and British Telecom claimed that the cable would be sold out.

FIXED SATELLITE SERVICES (FSS)

FIXED SATELLITE SERVICES

These cover three main markets; telephony, broadcasting and a variety of business communications. Within these three main markets there are some distinct differences. Moreover, the markets can be divided up on a geographical basis. The best understanding of current trends and developments may probably come from an initial look at the service types rather than the geographical markets.

1. *TELEPHONY* – Satellites were originally developed to provide telephony capacity for trunk routes, with broadcasting as the secondary market. Satellite technology developments have over the last twenty years reduced the cost of a satellite circuit continuously. During the 1980's, the fall in costs has levelled out. Initially, the high cost per circuit meant that the use of satellites could only be justified over long distances, such as transoceanic routes. It is not therefore surprising that the Intelsat system was the first operational and commercial application of satellites.

It was not until the mid 1970's that satellites began to be used for domestic satellite services. Even then they were confined to countries with large geographic areas, such as Canada and the USA and this broad picture still continues outside of Europe. At the time of writing, about 67% of Intelsat's revenue is accounted for by voice circuits. In the US domestic scene, the percentage is much lower. Voice traffic accounts for about 27% of all domestic satellite use. In Europe, the figure is even lower. As at September 1986, the Eutelsat system was providing 1,000 voice circuits.

The reason is closely related to satellite economics. As costs of the space segment have fallen, the satellite operators have been able to move into new markets which were hitherto uneconomical to serve. In particular, they have moved into existing and new TV markets. As a result, even though telephony use has been increasing, it has been declining as a percentage of overall satellite use. Reductions in cost per telephony circuit have allowed satellites to move into telephony markets already well served by the landlines. Here they have faced more competition than on the early transoceanic fields and have not been able to take as big a market share. Their operators are often owners of terrestrial facilities which they do not wish to completely replace. Moreover, there appears to be a cut off point of about 800 miles under which satellites are unable to compete on trunk telephony routes. This is one reason why so little internal European telephony traffic is routed via satellite.

The reduction in cost per circuit has had two effects. Firstly it has made it more economical to serve thinner routes and, secondly, it has led to substantial increases in the number of long distance and international calls. The price elasticity of demand for long distance and international calls is high, increasing with distance. For example, it has been estimated that the elasticity for international calls is as high as $-1{\cdot}25$. For Japan, a figure as high as $-2{\cdot}28$ has been calculated. The significance and importance of the price elasticity concept to telecommunications in general and long distance telephony in particular has been explained in a previous table.

Nevertheless, two developments in the 1980's have shifted telecommunications economics against the use of satellites.

Firstly, unit costs per satellite circuit have not been falling fast. There is therefore not the substantial growth in use experienced during the 1960's and 1970's. The above analysis indicates that a large proportion of that growth came from falling charges and not from other changing demand factors such as the growth in international trade or the spread of multinational businesses. For long distance and international telephony or data services to continue to grow substantially requires a technology that can offer considerable reductions in charges. Satellites are no longer able to offer the prospect of falling charges, although, it must be admitted, there is considerable scope for PTTs to reduce long distance and international charges by realigning local and long distance charges with actual costs.

Secondly, the rapid application of fibre optics in the US domestic and the trunk international scene is expected to result in further falls in costs per circuit which satellites cannot match, at least for trunk telephony. An analysis of the competitive threat of fibre is given elsewhere in this book. In the overall risk and return comparisons between satellite and fibre, the latter now has a commanding lead for telephony.

Nevertheless, the threat of fibre optics to satellite communications is not as great as some people think. A satellite is essentially a collection of radio repeaters. Any study of the early history of radio communications (say, until 1918) shows that the early thinking on their use revolved around telephony without wires (hence the still commonly used word in the UK for radio receivers is "wireless"). Yet, the great application of radio turned out to be entertainment broadcasting, ie point-to-multipoint rather than point-to-point services. It is not surprising that the same situation is appearing in satellite communications. It appears that satellites may be increasingly used for specialised point-to-multipoint services for business.

2. *BROADCASTING SERVICES* – These are likely to prove the main long term market for satellite services as telephony declines in importance and if, as remains likely, business services grow slowly. The Intelsat system was the first method by which live coverage of events could be provided on an international basis. Nevertheless, the world's broadcasting industry is still divided up largely on national lines, although production has always been international. The primary service markets for broadcasting are as follows:

CABLE TV PROGRAMMING – Satellites were the main ingredient in the formula which led to the growth of the modern multi channel cable system away from its use for relay of terrestrial channels. The satellites allowed a channel to be distributed simultaneously to many cable TV systems with a suitable dish. It made cable both attractive to the consumer and cost effective to produce. There is still no good alternative to satellites for distribution of TV channels to a large number of cable headends. The market has matured in the USA, with the majority of homes likely to be passed, having already been done so.

The explosive growth in US cable which occurred

between 1976 and 1984 has slowed down as the market has become saturated. This is not the case in Europe where cabling continues. Europe is still largely uncabled and presents a major long term opportunity for broadcasters. It is unlikely to see the rapid growth rates seen in the USA. This is because the markets are fragmented across national boundaries and each country has different regulations and policies covering cable, and also different off-air broadcasting competition to cable.

The cable markets tend to be very national in character. Of the US cable TV channels only an edited version of Cable News Network is currently available in Europe. Most of the European satellite delivered cable TV channels are not really available on a European basis. There are three reasons for this state of affairs.

Firstly, copyright is largely carved up on a national basis, with different companies owning copyright in different countries. For example, Cable News Network is unable to broadcast European Broadcasting Union programming material on its European service as it would compete with national broadcasters who generated the material in the first place. It is able to do so in the USA where it is not competing with the European Broadcasting Union members. Similarly, copyright for films is carved up so that a film channel can only be broadcast in the country or countries where the copyright has been cleared. Copyright rules have not been standardised between countries.

Secondly, language and cultural barriers between countries make what is popular programming in one country dull in another. It is also expensive to dub or subtitle programming from one language to another.

Thirdly, there are very few products or services that can be advertised on an international basis. There is therefore not the advertising revenue to support international cable TV channels.

It would appear from this that the Intelsat system has only a limited role in the cable TV market. At least for the time being, this is not the case. Intelsat domestic transponder lease and sales are popular with cable TV channels in Europe. The lack of capacity on the Eutelsat system and the preemptible nature of current Eutelsat leases means that the UK and West Germany are relying heavily on Intelsat capacity. For such services there are shortcomings to the Intelsat capacity in that it is not normally available for international distribution of cable TV and that each satellite has a maximum of only six Ku-band transponders available.

Cable channel providers and cable network operators naturally like as many channels as is possible to be on one satellite to reduce the investment needed in receiving equipment. Intelsat's primary broadcasting market, however, is likely to remain in international occasional use video. The essentially still national character of cable TV means that most of the traffic is passing through national or regional satellite systems.

NETWORKING – The use of satellites by TV networks to distribute their programming to affiliates has become the norm in the USA with all three of the big networking companies abandoning their terrestrial systems. This is an expanding sector of satellite communications as the American broadcasting scene is restructuring and new mini-networks emerge. It is also a growing satellite opportunity elsewhere in the world, for example, in India and Australia.

It has yet to become a feature of the European broadcasting system.

OCCASIONAL USE – This area covers broadcasting of special events and news. It tends to have a very strong International element. Its development is closely linked to the progress of the Satellite News Gathering business. It is the primary video use of the Intelsat system.

3. *BUSINESS COMMUNICATIONS* – This is a mish-mash of services ranging from plain old telephony through to data communications, facsimile transmissions and video and audio conferencing. Developments in this market reflect both the introduction of new telecommunications technology, the application of information technology within the office and on the factory floor and changes in telecommunications regulation. The growth in specialised business services as distinct from business use of telephony has taken place from a very small base. These specialised services are subject to a great deal of uncertainty as to demand and appropriate technology. The principal forms of such services are as follows:

Data communications from computer to computer. It is already clear that most companies do not have a need to transmit vast amounts of such data and that past experience has shown growth to be erratic and uncertain. Witness the Satellite Business Systems saga in which IBM took a stake in the late 1970's in the face of the then perceived convergence of computer and communications technology. The concept involved a massive investment in both the space and ground segments and demand failed to materialise. Instead, the US market has looked to less ambitious satellite data communications services through VSAT systems.

Videoconferencing. At one time some observers believed that there would be a massive growth in these services as companies realised that they were a very cost effective method of avoiding travel. In practice the technology has not generally proved to be a substitute for travel and growth has remained low. On the other hand, some companies have found the technology useful for specialised services such as in-house education (IBM) or internal corporate communications (Ford). It therefore continues to expand, albeit modestly.

Facsimile transmission. Federal Express's experience in this field has proved to be a disappointment and it remains an interesting but limited opportunity for the satellite world.

Value Added Networks (VANS). There has been great difficulty in defining this as a service but it is expected to grow considerably, despite pressure to restrict trans-border data flows. Typically such services include electronic mail and distribution of data base material. Other such services rapidly growing in importance and overlapping with computer to computer communications are being found in financial services, partly prompted by the deregulation of the international financial markets. Such services as Reuters require international telecommunications links either via satellite or through submarine cable. Under certain circumstances, the Intelsat or other satellite systems have distinct advantages over other methods in that they can already provide global links and can be used to rapidly install regional systems. VANS have a tendency to be point-to-multipoint services and therefore are inherently suitable for satellite technology where long distances are involved.

DIRECT BROADCAST SATELLITE SERVICES

DIRECT BROADCAST SATELLITE MARKET

At one time this appeared to be a promising market for the satellite manufacturers, but, by 1984, there were considerable doubts being raised both in the USA and in the UK about the commercial viability of Direct Broadcast Satellite (DBS) TV. US thinking was moving towards using medium power Fixed Satellite Service (FSS) satellites for DBS type services. Three principal factors appear to underline the decline in interest in DBS TV.

1. The economics of conventional Direct Broadcast Satellite TV require that a considerable amount of money be spent up front in procuring a space segment, in working capital and opening up a production line to mass produce ground receivers. It is estimated that at least $500 million is required to establish a fully operational and commercially viable Direct Broadcast Satellite TV system. Yet there are no revenues to be derived until the system starts transmission. Thereafter, it requires a rapid build up in audiences.

2. The Direct Broadcast Satellite concept was originally conceived in 1977 and since that time there have been unanticipated developments in ground receiver equipment which have rendered the original concept as outdated. It was envisaged that a 200–230 watt Direct Broadcast Satellite transponder would allow reception with a Television Receive Only terminal dish 90 cms in diameter. With current ground equipment, this has now fallen to 30 cms. A 90 cm dish can now be used to pick up quality TV pictures from a 50 watt Fixed Satellite Service (FSS) transponder. The lower power transponder means that more transponders can be packed onto a satellite providing either the viewer with a lot more variety or the satellite operator with extra capacity to offer other services and diversify risks. With more transponders, the cost per transponder falls making the proposition more attractive to broadcasters. The availability of say 16 medium power transponders on a Fixed Satellite Service System, compared with five on a Direct Satellite system also reduces the risks of DBS type services in two other ways. Firstly, the broadcaster can use the system to serve several markets at once. For example, these could be the cable TV market, the Television Receive Only (TVRO) terminal, satellite master antenna TV system markets and the occasional video markets, none of which are capable of being served by expensive Direct Broadcast Satellite transponders. Secondly, if the broadcaster fails to make a go of his service, the capacity can be switched to other Fixed Satellite Service services, instead of wasting unwanted capacity.

3. Conventional Direct Broadcast Satellite is a new product aiming at a market where there is no clear evidence of sufficient unfulfilled consumer demand. Most western countries already have good quality terrestrial TV and in the large urban areas have, or are installing, cable TV. The Video Cassette Recorder has given the consumer further choice, both in the variety of programming and the ability to time shift. In Europe and the USA changing policies are leading to more off-air television and a closer match of TV programming with consumer demand, for example, France has introduced three terrestrial national TV channels over the last three years in addition to its three public service channels. West Germany is introducing low power TV. The USA has been allowing new TV stations into a previously restricted environment.

There are also other objections to Direct Broadcast Satellite TV as a commercial proposition. Firstly, the cost of the ground receivers is still an unknown factor. Various industrial sources have suggested that the unit price will be around $450–$700. This may be too great for consumers already obtaining an increasing amount and variety of programming from other sources.

The amount of potential advertising revenue available may be too small to allow advertising based programming, leaving the alternative of pay-TV. Thus the consumer would have to pay twice for Direct Broadcast Satellite; once for the equipment and then again for the subscription charges. The experience in cable pay TV ventures is that there is a very high churn rate which, amongst other factors, reflects a large degree of consumer indifference towards the product. But with cable, the consumer does not have to also pay an upfront cost of $450–$700.

The traditional Direct Broadcast Satellite concept also involves a large degree of still unproven technology. The Japanese experience has been that the travelling wave tube amplifiers and power supplies have not met operational requirements. Similarly, the French have had to abandon plans to use domestically produced Thomson travelling wave tube amplifers on the TDF–1 satellite. A second, unrelated problem, is that technical standards for transmission have not been agreed on an international basis. Each country has adopted its own preferences (partially to protect their equipment manufacturers from foreign, mostly Far Eastern, competition). The lack of agreement in Europe also impacts on those proposals involving transborder broadcasts such as ASTRA, TDF–1/2 and possibly, Britsat and Atlantic satellites. It puts the ground equipment manufacturers in a difficult position in that they still do not know which standard to tool up to.

The current position in Direct Broadcast Satellite TV is that there remain a number of proponents of the technology, both in the USA and Europe but that all of the current satellites under construction or completed (TV–SAT, TDF–1, STC, BS–2 and TELE–X), have no finalised proposals for their use by broadcasters. The situation could perhaps be described as one where even if the satellites are constructed and given to the broadcasters at substantially below cost, it is not possible to turn them into commercial propositions. This does not auger well for those planning DBS satellite projects from scratch.

Most of the support for Direct Broadcast Satellite TV in Europe and Japan is now coming from TV and ground receiver manufacturers or governments interested in broader policies towards such areas as High Definition TV. Broadcasters are largely disinterested in the technology. The French TDF–1/2 satellite project has been temporarily bailed out by the equipment markets, for example. The current right wing government has said that no more public money is to go into the project over and above that which is already committed. Therefore, it is requiring that commercial interests pay for the rest of the construction and launch costs of the second satellite (TDF–2) and fund the programming of the system. At the time of writing, the future of the project remains highly uncertain, although, at worst, it is likely that TDF–1 will be launched as a demonstration satellite.

Similarly, West Germany has stopped funding of the construction of the second TV–SAT spacecraft until such time that it becomes clearer who, if anyone, will use the operational system. The Nordic countries have totally failed to come up with a proposition for use of the TELE–X Direct Broadcast Satellite payload and it is highly unlikely that a second spacecraft, needed to turn the system operational, will be ordered.

Indeed, a complete, two satellite Direct Broadcast Satellite system is currently available on the market. This is Comsat's unlaunched and unused Satellite Television Corporation's (STC) system, abandoned by the company in 1984 when it became apparent that it was not a feasible commercial proposition for the US market. So far, there have been no takers for the STC satellites, even though they have been up for sale for two years.

Despite some 11 US proposals for Direct Broadcast Satellite being listed in this book, so far only STC has ordered any such satellites and the remainder are paper satellites only. Christian Broadcasting network has been mentioned as a potential user of the STC satellites under a plan in which Dominion Video Satellite would operate them. Some observers believe that the Hubbard organisation's plan for Direct Broadcast Satellite, called USSBCI, is the best thought-out concept, but this also remains a paper plan after four years of effort.

There has been a shift in interest in the USA towards so called second generation DBS TV systems which use satellites with transponders in the 100–120 power output range. The interest is paralleled in Europe with Eutelsat's plans for a "Europesat" DBS system for the 1990's. A US company, Advanced Communications Corporation, has gone as far as ordering a "second generation" DBS satellite system. It is not clear to what extent such satellite systems will be used for services otherwise carried on Fixed Satellite Systems, such as distribution of cable TV channels and networking of terrestrial off-air TV. Whilst some in the industry were saying in 1984 and 1985 that DBS and FSS concepts were merging, the appearance of some interest in these second generation systems has not, so far, led to suggestions that merging is continuing.

The Americans have already put together the RCA SATCOM K system, using 45 watt KU-band transponders and capable of quasi-Direct Broadcast Satellite services. Even the most optimistic supporters of Direct Broadcast Satellites in the USA now envisage only a specialised market for DBS TV, serving, perhaps, rural communities outside the reach of cable or specialised demand for certain types of religious belief.

Yet despite this gloomy US environment, interest in Direct Broadcast Satellite remains strong in Europe. There are five main reasons why this is so:

1. There is still strong pressure from industry and government departments interested in the broader impacts on the electronics industry. Governments have sunk an estimated $1,300 million into R & D on four DBS satellites projects (TDF–1, TV–SAT, TELE–X and Olympus). Total abandonment is politically unacceptable, particularly when the level of total national R & D expenditure is a national and public political issue, as is the case in France. The adoption of new transmission standards for Direct Broadcast Satellite TV is also aimed at winning back a large part of the consumer TV reception market lost to the Far East. In projects such as OLYMPUS and TV–SAT/TDF–1, the countries involved are tied together in bilateral and multilateral arrangements which make it difficult for any one partner to pull out.

2. European countries remain worried about the cultural impact of transborder TV, particularly that originating from the USA. TV in Europe is still strongly influenced by some high minded public service ideals. By developing their own satellite systems the Europeans hope that they may mop up the surplus demand for extra and new TV before it is absorbed by commercial and rather banal US programming. There is still a widely held belief in Europe, weakly supported by research, that increasing crime rates are a result of TV, particularly US TV. It is also a way of protecting European programme producers and broadcasting organisations. There is some hypocrisy in this position in that it is very clear that those European countries that have or are planning to, invest in Direct Broadcast Satellites, are looking to export their programming across European borders, thus creating possible cultural strains within the continent.

3. Some European supporters of high power Direct Broadcast Satellite TV claim that the technical developments in ground receivers, far from making high power obsolete, have actually increased their consumer attractiveness. Indeed, some argue that high power Direct Broadcast Satellite technology is suitable for Very Small Aperture Terminal system business services using 30 cm dishes as an alternative to the so far largely unsuccessful European attempts at Fixed Satellite Service (FSS) business communications. They argue that, unlike the USA, Europe has not invested heavily in satellites and it can take advantage of the new technology without the dead hand of the past. The TELE–X satellite, for example, uses two DBS power transponders for business services and the original planners of the system implicitly criticised the use of low and medium power transponders for business services.

4. European proponents of Direct Broadcast Satellite TV have also argued that there is a greater unmet European demand for TV than is the case in the USA. Associated with this is a greater potential advertising income base. The argument is that broadcasting in Europe has been carved up on a national basis, often with restrictions on advertising and foreign programming content. The existing broadcasters are therefore unable to afford to generate enough domestic production of the quality to fully satisfy audiences or they are unable to import enough quality programming.

It is still an essential feature of the European programming market that imported material is a lot cheaper than producing the same quality domestically. The public service broadcasting ideal is also said to lead to untapped demand because it is twisting what is essentially a mass media market into a hybrid specialist and mass market. Moreover, the proponents also argue that Europe is still largely uncabled and is likely to remain so for many years. Less than 15% of European homes are passed by cable TV networks compared with 50% in the USA. This means that most Europeans are denied the choices available to Americans, who, despite a high churn rate on cable appear to be willing to pay for it.

5. There is also a physical difference between Europe and the USA which predicates in favour of European high power Direct Broadcast Satellite TV. The large geographic area of the USA requires that a DBS satellite system either uses large footprints (and therefore require large dishes) because the EIRP (signal strength) performance is poor, or more satellites be employed to provide smaller beams and higher EIRP. It is not possible to cover the whole of continental USA with a single high power DBS satellite with EIRP performance in the 60+ dBW range. Moreover, there are major time zone differences between the East and the West coast of the USA, which means that the same channel has to be broadcast at different times on separate transponders to provide a complete US service. These time zone differences are much less significant in European (one hour against four). Moreover, it is possible to provide a footprint of in excess of 60 dBW for virtually all of western Europe from one satellite. The problem is that Europe carved up its Direct Broadcast Satellite frequencies at World Administrative Conference '77 on a national basis and it appears highly unlikely that agreement between all European countries on frequencies for such a pan-European service could be found easily.

Yet the prospects for conventional high power Direct Broadcast Satellite TV in Europe appear to be going the same way as those in the USA. This does not mean that there is no future for DBS TV either in Europe or elsewhere in the world. The future appears to be in using medium power Fixed Satellite Service (FSS) systems. Such systems are already being used in Australia and India, so far primarily to serve rural populations outside the range of terrestrial transmitters. Although on a prima facie basis the implication is that the market is a specialised one quite different from the markets envisaged in Europe, Australian interests are looking at serving specialised urban markets using the quasi-Direct Broadcast Satellite capacity on AUSSAT. These initially cover services to clubs and other such organisations but with the eventual aim of providing the service to any private individual willing to pay for a Television Receive Only terminal and, if necessary, a subscription fee. Similarly in Europe, Societe Europeenne des Satellites is planning to market direct reception Television Receive Only terminal equipment for Direct Broadcast Satellite services. The implication is that DBS TV has converged with the other broadcasting satellite services; cable TV distribution, networking and occasional use video.

The Europeans have already committed themselves to the EUTELSAT II satellite system, planned to be available from 1989. This is, in terms of performance, very much an ASTRA clone aimed largely at the broadcasting market. Beyond that system Eutelsat also has plans for a pan-European second generation Direct Broadcast Satellite system to follow on from the TDF–1 type projects called EUROPESAT, current thinking is that it will be available from around 1994. The organisation is proposing the use of 120 watt transponders providing a European EIRP (signal strength) performance to allow use of dishes of around 60–90 cm. This type of Direct Broadcast Satellite performance is being proposed by some UK DBS interests and Atlantic Satellites in Dublin, but in the latter case, the performance will only be available for the UK. All three proposals remain on paper only.

COMPARATIVE TABLE OF SATELLITES FOR TELEVISION BROADCASTS

CHARACTERISTICS	*LOW-POWER SATELLITE*	*MEDIUM-POWER SATELLITE*	*HIGH-POWER SATELLITE*
Transponder power	From 10 W to 30 W	From 30 W to 60 W	From 100 W to 250 W
Satellite type	Communications (FSS)	Communications (FSS)	Direct broadcast (DBS)
EIRP per channel	35 dBW–45 dBW	45 dBW–55 dBW	55 dBW–65 dBW
Transponder lease	1·5–3 million dollars/year	3–6 million dollars/year	6–12 million dollars/year
TVRO parabola	1·6 to 3·2 m in diameter	0·8 to 1·6 m in diameter	0·4 to 0·8 m in diameter
Receiver price	More than 2,000 dollars	Around 1,000 dollars	less than 1,000 dollars?
OPERATIONAL SATELLITES Starting year: Name of satellite (frequencies) *System covering Europe*	1974: Westar (C-band) 1975: RCA Satcom (C-band) 1975: Intelsat IVA (C-band) 1975: Raduga (C-band) 1976: Comstar (C-band) 1976: Palapa (C-band) 1980: *Intelsat V* (C-/Ku-bands) 1980: SBS (Ku-band) 1982: Anik D (C-band) 1983: Galaxy (C-band) 1983: *Eutelsat I* (Ku-band) 1983: Telstar 3 (C-band) 1984: Spacenet (C-/Ku-bands) 1984: *Telecom I* (Ku-band) 1985: Morelos (C-band) 1985: Gstar (Ku-band) 1985: American Satellite (C-/Ku-bands) 1985: Brazilsat/SBTA (C-band) 1986: STW/China (C-band)	1978: *Gorizont* (C-band) 1978: Anik B (Ku-band) 1982: Anik C (Ku-band with spotbeams 1982: Insat I (S-band) 1985: Arabsat I (S-band) 1985: Aussat I (Ku-band) 1985: RCA Satcom (Ku-band)	1976: Ekran (UHF) 1984: BS–2 (Ku-band)
PROJECTED SATELLITES Planned launch Name of satellite (frequencies) *System covering Europe*	1987: *PamAmSat* (C-band) 1988: *Kopernikus* (Ku-band) 1988: *Intelsat VI* (Ku-band) 1988: JCSat (Ku-band) 1989: Spacecom (Ku-/Ka-bands) 1989: Comsat (Ku-band) 1989: Fordsat (C-band) 1990: *Italsat* (Ka-band) 1991: *AMS–Israel* (Ku-band)	1987: *Lutch*? (Ku-band) 1987: *GDL–Astra* (Ku-band) 1988: Galaxy (Ku-band) 1989: Fordsat (Ku-band) 1989: *Eutelsat II* (Ku-band) 1990: Insat II (S-band) 1991: Anik E (Ku-band)	1987: *TV–Sat* (Ku-band) 1987: *TDF–1* (Ku-band) 1988: *Tele–X* (Ku-band) 1988: *Olympus* (Ku-band) 1988: Dominion Video/STC (Ku-band) 1990: BS–3 (Ku-band) 1990: *Atlantic Satellites* (ku-band) 1990: USSB (Ku-band) 1990: Galaxy DBS (Ku-band) 1992: *Europesat* (Ku-band)
ADVANTAGES of the system	– low-cost and high-reliability of satellite repeaters – large number of available satellite repeaters for television broadcasts – highly useful with cable TV systems	– medium-cost and high-reliability of satellite repeaters – inexpensive TVRO stations for easy installation – large number of available satellite repeaters	– low-cost and easily operational TVRO terminals for individual reception – powerful TV signals with no severe interference, for high-definition TV broadcasts
DISADVANTAGES of the system	– expensive and complicated TVRO systems – coordination requirements, compatibility with other communications systems – risks of severe interferences during broadcasts	– coordination requirements with other communications satellite systems – utilisation restricted by national monopolies of PTT administrations – some risks of interference	– very expensive space segment for a limited number of repeaters – yet uncertain reliability for the high-power repeaters (tubes) – use of new TV standards

MOBILE SATELLITE SERVICES (MSS) AND RADIO DETERMINATION SATELLITE SERVICES (RDSS)

MOBILE SATELLITE SERVICES

Mobile Satellite Services (MSS) are in practice a group of different services aimed at different markets. All are specialised. The individual services that are usually grouped under the term MSS are as follows:

MARITIME MOBILE SATELLITE SERVICES – The International Maritime Satellite organisation (Inmarsat) was the first civil satellite operator to offer these services. The organisation is able to offer both voice and telex services to shipping operators.

By the nature of maritime activities, the service lends itself to both global coverage and international operation. The experience of Inmarsat has been one of steady growth in demand since it took over the Comsat owned Marisat spacecraft. Generally speaking, the demand experienced by Inmarsat has exceeded initial expectations. However, the system so far has largely been confined to use by ships above 1,600 tons gross weight. There are about 25,000 such vessels worldwide; Inmarsat has about a 15% penetration of that total market and current indications are that demand continues to grow steadily.

The organisation has therefore been promoting the development and application of equipment and services suitable in terms of weight, transmission capacity and cost for smaller vessels. The organisation admits to there being uncertainty about the demand for the newer services and is therefore taking a gradualist approach to developing the market.

LAND MOBILE SATELLITE SERVICES – This is defined here as "true" land mobile communications covering one and two way voice and telex services. The position is somewhat more complex as so far there have only been proposals from the USA and Canada for domestic systems and these have tended to include other MSS services as well, such as RDSS, fixed telephony and some inshore maritime mobile communications. The division between the various MSS services is somewhat blurred in the various proposals that exist. Canada is making its MSAT system for domestic use but to be operated in conjunction with a US system. There are twelve proposals in the USA still being considered by the Federal Communications Commission. It is expected that only one applicant will eventually emerge as the market is limited and risky. Included within the overall concept of land MSS systems is RDSS and aeronautical communications.

Much of the demand for land mobile services is better served by cellular radio than satellites. Line of sight to geostationary satellites remains a major problem and ground equipment and satellite user charges are likely to be considerably higher than cellular radio, whether for voice or text. The MSS concept is largely aimed at remote or rural requirements, which by definition limits the overall market. The markets are also characterised by low communications capacity requirements from each user and the need for wide area coverage.

There is still only limited interest in Europe for "true" land MSS systems. However, the UK is currently working on proposals for a Molniya type satellite (T–SAT) which could provide much better line of sight than geostationary MSS satellites. This is because the Molniya type orbit virtually places the satellite vertically overhead for about eight hours per day.

VERY THIN ROUTE FIXED SATELLITE SERVICES – One application of MSS voice services is for temporary or permanent rural telephony to and from a fixed point. These can also be provided by Fixed Satellite Systems using VSAT terminals. Many of the US and the Canadian MSAT proposals envisage the use of MSS for such fixed voice services. Technically there is no reason why the Inmarsat system cannot be used for the same purpose, although the organisation is not marketing equipment for such applicants.

RADIO DETERMINATION SATELLITE SERVICES (RDSS)

Over the last two years, this service has attracted considerable attention. So far, the principal proponent has been the GEOSTAR Corporation in the USA, which appears to be aiming a global system in conjunction with other partners. Essentially, RDSS differs from "true" MSS in that it does not offer voice services. It is limited to navigation and positioning services and one or two way digital text messages. Whilst some MSS operators are including RDSS services amongst their plans, RDSS is also an alternative to MSS. Both the user ground equipment and the space segment charges are cheap in comparison with MSS voice services. With the initial RDSS services a completely separate satellite system is not necessary. Initial Geostar operations are using "piggyback" payloads on conventional FSS spacecraft.

The major development during 1986 in this field was the decision by France to enter the European field in conjunction with Geostar through a newly created company, LOCSTAR. It appears that Locstar will begin operations using a piggyback payload on the ARABSAT 1C satellite, to be launched in 1989. Dedicated Geostar satellites are not expected to be available until the early 1990's. Geostar is also looking for potential partners in other parts of the world.

AERONAUTICAL SATELLITE SERVICES

Again, there is some overlap between RDSS and MSS services in this field. The Geostar concept envisages use of RDSS for aircraft navigation, traffic control and distress services complementary to existing such systems. Nevertheless, the major current development in aeronautical satellite communications is the decision by Inmarsat to enter the field. This has followed on from an over-ambitious AEROSAT proposal of the 1970's. Again, Inmarsat is taking a cautious approach to the new services. These have followed on from current military air-to-air and air-to-ground satellite communications.

The current prospects for all types of civil MSS satellite systems appear limited to an expanded Inmarsat system, Geostar and probably one combined US/Canadian system. All the services are aimed at specialised markets with often widely differing needs and largely incompatible user ground equipment. There also appears to be a considerable long term potential in certain MSS areas; others are not likely to prove commercially viable.

ORIGINS OF MOBILE SATELLITE SERVICES

MSS services have been around for a long time, but the primary application has been in military communications. Apart from the Inmarsat system, which started with Marisat capacity built for lease to the US navy, MSS services have yet to see civil application on a commercial basis. Part of the reason is that demand is uncertain, and, in total, fairly limited in comparison with FSS services. For example, it is envisaged that there is room for only one MSS system for the USA. There are no plans for true land MSS services anywhere in the world outside North America, although GEOSTAR and LOCSTAR have ambitions for RDSS services.

The milestone event in establishing land MSS services dates back to 1979 when the Mobile World Administrative Radio Conference (WARC '79) allocated the 806–890 MHz band to the services in the Americas. The major driving forces behind the WARC '79 decision were the USA, Canada, Sweden and Japan. Canada and the USA had been discussing the land MSS concept since the late 1960's. In 1982, an agreement between the Canadian and US governments, allocating shared spectrum in the 800 MHz range together with coordination procedures between the two countries began to get the ball rolling, although subsequent development has remained slow. In 1983, Canada notified the IFRB of plans for its MSAT MSS system, including use of the L-band (1·544–1·660 GHz) and the 806–890 MHz band. The Canadian Department of Communications also signed an agreement with NASA covering cooperation in development of a North American MSS system. Subsequent private US proposals have also envisaged coverage of Mexico.

It was not until early 1985 that the US Federal Communications Commission (FCC) issued a notice of proposed rule making and set the scene for US domestic MSS operations. The notice resulted in 12 applicants to operate a land MSS system; most had what were described as "big ambitions backed with little money". it was and is expected that the FCC will grant only one license amongst the twelve because the market, in total, is too small to support more. The US proposals fall into two groups. One proposes use of small satellites capable of providing limited services; the other proposes larger spacecraft with higher capacity capable of providing "true" MSS services. The different approaches reflect different perceptions of the economics of MSS services. One critical policy issue was involved in these applications. This was the allocation of UHF frequencies to MSS. The applications mostly concentrated on providing UHF services with a limited amount of L-band capacity. However, the FCC was under pressure to allocate more UHF spectrum to cellular radio which was proving to be very popular.

In 1986 the Commission allocated only L-band frequencies to US MSS (it allocated 4 MHz in UHF to MSS out of deference to Canada which still wanted to use both bands). The decision appears to have weakened the prospects for a joint Canadian and US approach to the operation of an MSS system. Moreover, the Canadian MSAT proposals also envisage "true" MSS services which may prove incompatible with a limited US service that may emerge from the 12 applicants. The Canadians also claim that allocation of L-band only will considerably increase the cost of MSAT. Canada is considering the possibility of a first generation MSAT system using only L-band, although its Department of Communications has said that it would be prepared to absorb the cost increases of using UHF because of the strategic importance of the band to federal government communications requirements. Unlike the US position, there is considerable public funding (directly and indirectly) proposed for MSS operations. The Federal Government allocated CAN$50 million in May 1986 to fund development of MSAT hardware and has made a commitment of CAN$126 million covering lease of MSAT capacity for its own commercial requirements. At the time of writing, it appears that a full Canadian MSAT system will not be available until the early 1990's. The country is studying the possibility of a limited interim service.

GEOSTAR AND INMARSAT

The slow progress and uncertain progress in North America has to some extent been offset by innovations from the GEOSTAR corporation, an entrepreneurial start up venture with big ambitions but also with a realistic and modest approach to developing the market. Geostar has limited itself to RDSS services, initially using satellite payloads carried piggyback on FSS communications satellites; it does not plan to start using a dedicated satellite system until the 1990's. The approach has been to keep the cost of the ground equipment (transceivers) as low as possible (with unit costs expected to fall below $1,000) as well as low transmission charges to the user. The capital investment required by Geostar is also low. As and when it moves to a dedicated satellite system, it will have already built up a user base to transfer to the satellites. A third significant factor in the Geostar approach is that it is global. To achieve this objective it is entering into partnerships with non-US organisations. The first such partnership is with the French national space agency. A new company, Locstar, backed by European public and private funds is expected to be formed in 1987 and will provide RDSS services to Europe and the Middle East in conjunction with Geostar. At the time of writing it is likely that a piggyback RDSS payload will be carried on the ARABSAT 1C satellite, for which CNES appears to have "arranged" a launch in May 1989 atop ARIANE V37. Geostar says that it is looking for a similar partnership in the Far East.

Although the Geostar RDSS services are limited in comparison with full voice MSS services, the approach makes a lot of commercial sense and undermines the market for true MSS services, either planned for North America or through the Inmarsat system. It is now fairly certain that Geostar will be a major force in MSS, probably running as a specialised second to Inmarsat. Indeed, it is possible that a market will develop where MSS services are all provided on a global basis, with Inmarsat concentrating on a voice centered package of services aimed at maritime and aeronautical markets. Geostar would concentrate on text and positioning services for land based aeronautical and maritime users with limited communications capacity requirements.

Inmarsat's proposed aeronautical services, currently under development, cover voice and telex traffic for air traffic control (ATC), passenger messages and internal communications for the airline operator. The Geostar system offers navigation and distress services and "mid-air collision avoidance" capabilities rather than full ATC services. What ATC capabilities it does offer are envisaged for use in developing countries.

The Inmarsat approach to MSS has some parallels with that of Geostar. It initially began services without a heavy investment in the space segment, leasing capacity from Comsat, Intelsat and ESA. It also tackled a single market, maritime communications for large ocean going vessels, where there was an obvious need for both better and price competitive communications. Once it had established itself in these markets, it opted for its own dedicated satellite system (INMARSAT II). It is only now beginning to move into the aeronautical services market. For the latter, it has not opted for dedicated satellites; the Inmarsat II system will be used. When the possibility of international satellite aeronautical services was considered during the 1970's, a dedicated Aerosat system was envisaged. It proved to be too ambitious for commercial operations.

VERY SMALL APERTURE TERMINAL (VSAT) MARKETS

VSAT MARKETS

Within the USA and Canada, and to a lesser extent in Europe, there is currently a lot of interest in the application of small dish satellite terminals to provide specialised business communications. The small dish systems are now commonly known by the acronym "VSAT" and are intended for such services as provision of private telecommunications systems for collection and distribution of data, either on a one way or two way basis. The technology is seen by some as a major opportunity for satellite communications and by others as yet another attempt at specialised satellite business communications which is likely to go the same way as past attempts.

Very Small Aperture Terminal (VSAT) systems typically use frequency inefficient FDMA (frequency division multiple access) techniques which keep the cost of the ground segment low but require a lot of satellite capacity. The individual earth terminals are between 1·2 and 1·8 m in diameter and are therefore small enough to be located on the customer premises. Typically, they may be used to provide links directly to an earth station located at corporate head offices or to a teleport on the outskirts of the city in which the corporate headquarters are located. A microwave or landline link would connect the corporate headquarters with the teleport.

The system has so far been primarily applied in the USA using both C-band and Ku-band satellites. The data may typically be accounting, sales or inventory data between personal computers at point of sale terminals located in regional offices or retail outlets and a central host computer located in corporate head offices. The system might, for example, be used by a major retail chain for stock control, ordering and distribution, credit card verification, cash management and so forth. The small dish terminal concept allows high data rates, and many believe that it is only viable when such rates are used. This implies that the economics are fundamentally dependent on the need to collect and distribute large amounts of data. Some observers believe that despite the convergence of computer and communications technology, such large volumes of data only exist in specialised environments. For most large businesses, there is just not sufficient data to justify VSAT systems.

The market for very small apetture terminal systems was spurred on by Federal Express (Fedex), who decided that it could be used for electronic courier services as an alternative to its overnight package delivery service. FEDEX offered facsimile services using VSATs with the possibility of having terminals located on customer premises as well as its own. The company planned to eventually have its own satellite system (Expresstar) to support its services were marketed under the name Zapmail. However, much to the surprise of the communications industry, Fedex announced in September 1986 that it was abandoning Zapmail.

Fedex's initial interests were followed up by a proposal from General Motors (with a major corporate policy to combine communications, computer CAD/CAM and automated manufacturing technologies) to link 13,000 of its dealer outlets in the USA with a small dish system. General Motors has yet to implement the plan. However, US retail chain K-Mart has signed an agreement with GTE to install a VSAT system to connect its point of sale systems on a real time basis with corporate facilities. This further stimulated interest in the market.

However, the satellite manufacturers and operators have been looking at providing other small dish business services for some years before Fedex's lead. The Intelsat International Business Service (IBS), introduced in 1983, brought it to the international market place. The technology is marketed in the UK, by British Telecom International, under the name SATSTREAM. The experience of the UK is that there is limited demand from specialised users, such as remote printing, but the potential large users, such as the City of London financial institutions, have yet to take it up.

Principal manufacturers of very small aperture terminal systems are Equatorial Communications (USA), M/A Com (USA), NEC (Japan) and Comsat (USA) in partnership with Mitsubishi (Japan).

VSATs are seen by some industry participants (Comsat, for example) as a step towards the Interactive Switched Digital Network (ISDN) concept: others see it as a threat to ISDN because they allow a complete bypass of the public telecommunications network. VSATs also allow introduction of fully digital services before the public switched network is upgraded to carry such communications links. In the USA, for example, it is estimated that only half the switching centres will be fully digital by 1990. A fully digital terrestrial network is a necessary element in the full ISDN concept. However, both of these arguments can lead to the suggestion that VSAT systems are only interim technologies before the full development of advanced terrestrial communications.

There is also some concern that fibre optics may compete with VSAT systems. This appears to be a more complex issue. The VSAT systems are competing with the public network and it is the public network which will benefit most from fibre optic, allowing it to reduce long distance costs on the trunk routes. Although VSAT systems allow data services off the trunk fibre optic systems, most business premises are located in urban areas or conglomerates connected by the trunk networks. This suggests that the proponents of VSAT systems are arguing that they are best applied to the specialist markets where links are required to remote locations. Within the USA and Canada there is a considerable such market. Some 20% of the population of the USA lives in rural areas (compared with less than 10% for Europe. Distances are considerably greater in the USA than they are in Europe. Although satellite communications are (except on a global basis) distance insensitive, terrestrial communications generally are not. But the balance in the USA has been towards reducing the costs of long distance communications (by a combination of regulatory and industry structure reform and technological developments). This implies that the distance insensitivity of satellite communications is becoming less important.

Furthermore, satellite technology has generally matured over the last six years whereas fibre optic technology is still subject to significant cost and performance improvements. These relate to developments in switching and reductions in the number of repeaters required in a given length of fibre optic cable. The actual fibres themselves have been significantly improved, thus further reducing the need for repeaters and improving the capacity for each $ worth of investment. Some in the satellite industry are arguing that it pays to link communities with populations of less than 10,000 by fibre optic. If that is the case, only a short term or specialised use for VSAT systems can be envisaged.

Nevertheless, VSAT systems have a significant potential for broadcasting data and information. There is little doubt both in the short and the long term that satellite communications have a strong future in broadcasting type point-to-multipoint type communications. This is what they are naturally best at. The major doubt as regards to VSAT systems is whether there is any demand for such services by business.

Whilst it is readily apparent that Western businesses are increasingly regarding telecommunications as a major corporate resource and that computer and communications technology is also becoming interdependent within organisations, the volume of transmission capacity needed to handle most intra-company data transmissions is limited. There is simply not much information needed to be transmitted each day between, say, a supermarket and its corporate headquarters and/or vice versa. Therefore, in terms of market size, VSAT systems are aimed at a small total market. On the other hand, it is a rapidly growing underlying segment of the total telecommunications market.

Within the UK a VSAT system is being developed to distribute video and text on horse and dog racing to betting shops. This is one of the few examples known to the author of a point-to-multipoint VSAT system under development in Europe and applies to a very specialised market niche. It is at times difficult to distinguish it, on a technical and operational basis, from a TV channel using Television Receive Only terminals. The difference appears to be largely legal.

Agence Presse France is developing a VSAT type network for distribution of news information to various sites in France using the TELECOM 1 satellite system. It is not an alternative to the organisation's or the news industry's news communications needs which are essentially point-to-point or two way and often without the need to distribute news simultaneously to a lot of points.

The Intelsat International Business Service (IBS) is intended for use by VSAT systems on an international basis. So far there have been few takers. The case is similar to that of the EUTELSAT SMS service using FDMA. Outside the USA the VSAT concept has less attraction. It tends not to provide an alternative to the public telecommunications network because the latter, and the VSAT terminals and the satellites which they are used in conjunction with, are under the control of the same monopoly PTT. For the same reason, there is not the advantage that it avoids the organisational problems of dealing with several telecommunications companies. In Europe, it creates additional problems in that the PTT administrations operate on a national basis, leaving those companies which want to operate small dish systems across national borders with problems of coordinating parts of their system with different PTTs, all of whom have some control over the system. Europe has yet to create a common market for VSAT systems.

An alternative market for VSAT systems may be in developing countries where the terrestrial communications system may be inadequate, inefficient or non-existent. India has been looking at various such options for use with the C-band Fixed Satellite Service (FSS) capacity on its INSAT I satellite. So far it has yet to install an operational system although a demonstration system is due to come on stream in 1986–87. Australia is applying VSAT systems for use with its Aussat Ku-band capacity.

SATELLITE NEWS GATHERING (SNG)

This is a registered name of Conus Communications' service and has been applied as a generic name to all such services. Conus Communications objects to such use and although the trade may use the term in private, publications have been forced to use the term "News Gathering via Satellite" (NGS). In this section the term Satellite News Gathering is used throughout.

Satellite News Gathering is a satellite business opportunity which has largely been developed in the USA over the last five years. It is part of the ongoing trend towards electronic news gathering allowing footage to be uplinked directly from the site of origination to a satellite for forward transmission to one or more TV stations or facilities. Through this method, which uses portable truck mounted uplink stations, live coverage is facilitated for one or more TV stations or networks.

It is essentially a cost reducing exercise in that live outside broadcasts have been available to viewers for years. It is just that they are, using conventional land lines, expensive and therefore impractical for many purposes. This is particularly the case when the need is for live or rapid coverage of a sudden event.

The Satellite News Gathering market has been of particular benefit to US independent TV stations in the USA. These companies have been increasing their share of prime time audience viewing at the expense of the three national networks and SNG has been able to further the trend. Forty percent of the average local television station's advertising revenue derives from its news programming.

The networks have developed their own satellite news gathering services to support their affiliates, but there are a number of very successful Satellite News Gathering services which have been developed outside the network system. Most notable of these have been continental USA communications, which are owned by Hubbard Broadcasting, an independent TV station owner.

The US Satellite News Gathering market is still somewhat fragmented, with differing technical standards between different services. This may be deliberate in that it protects the networks' declining hold on the broadcasting market. There are two main transmission systems used for Satellite News Gathering in the USA. These are Single Carrier Per Channel (SCPC) and Demand Assignment Multiple Access (DAMA). It is expected that if any industry standard is not adopted there may be some reorganisation within the sector. There are currently ten Satellite News Gathering services available in the USA.

LAUNCH VEHICLES

THE LAUNCH VEHICLE MARKET

Understanding of this market is critical to understanding of the alleged threats to satellite communications and future prospects for space activities in general.

By any standards the launch vehicle industry is in a mess. At the time of writing there is not one single commercial operational launch vehicle backed by an open production line which is currently operational any where in the Western World. Of the launch vehicles that are in existence, only the European ARIANE can currently take orders for launches and there is no available launch slot on this until 1990. Moreover it is grounded until at least 1987 by a fundamental design flaw in the third stage. Shuttle has been withdrawn from the commercial satellite launch market although some 14–17 commercial payloads (not all comsats) already booked on the system are expected to use the system.

There is still no firm date when Shuttle will recommence launches. Current planning indicates no earlier than February 1988. The production lines for both Delta and Atlas Centaur have been closed and no decision to reopen them is expected before the Spring of 1987. The TITAN vehicle is currently available only for US military launches and no final decision to commercialise it is expected before the Spring of 1987. If it is commercialised, it will be unable to fly payloads before 1989.

The Chinese are offering commercial launches on their Long March 3 (CZ–3) vehicle, but no launch slot is available until the end of 1987. Many US satellite operators and launch vehicle engineers are sceptical about the reliability of Long March, which has been developed in isolation from Western launcher technology.

The US Expendable Launch Vehicle market is suffering from the combined problems of lack of firm federal policy and uncertainty as to the total demand for launch vehicles from commercial satellite operators. Four vehicles are lined up for commercial operations:

DELTA, through either McDonnell Douglas or Transpace Carriers Incorporated. A commercialised Delta is not considered likely. It is possible that NASA will order some further vehicles to be constructed from long lead items already in stock, but it is unlikely that the production line will reopen.

ATLAS–CENTAUR/ATLAS K, through General Dynamics. It is generally thought that this launcher will have to win the contract for the Medium Expendable Launch Vehicle (MELV) under study for the USAF. The production line is now closed.

TITAN III, through Martin Marietta. This is considered by many to be the most likely launch vehicle to be commercialised. The TITAN IV version is already under development for the USAF.

JARVIS, through Hughes Aircraft Company and Boeing. This is considered to be somewhat of a crank proposal for commercialised operations.

The Japanese have a very advanced expendable launcher, H–II, under development which should be available from 1993, but this has no impact on the immediate shortage of capacity.

Finally, the Soviets have been offering the PROTON launch vehicle but it is generally considered that the USA will not allow any satellite containing US components, systems or designs (i.e. all Western built satellites) to fly on the vehicle.

This sad state of affairs affects the satellite and space scene in the following ways:

1. Commercial communications organisations investing in new hardware now perceive that they could be faced with years of delay in both the launch of the current satellites they have on order and any future systems.

2. Launch insurance is now very expensive (with rates of at least 30%) or not available at all. The launch vehicle operators and satellite manufacturers are circumventing part of the problem by offering insurance directly to the satellite owner. However, this has so far proved to be only partial coverage, requiring the owner to go into insurance markets for the remainder. The costs of satellites, including launch, are too great for independent satellite operators to self insure. Only the very largest organisations, such as Intelsat, have the wherewithall to do so. It therefore appears to rule out entrepreneurial operators from the satellite market. The lack of suitable insurance to cover the risks means that financial institutions would not be prepared to put up the finance for the overall satellite project. The very high insurance rates and lack of space insurance capacity meant that several satellites were launched in 1985 either partially or not insured.

3. The launch charges for satellites have "skyrocketed" because it is now a sellers market. Satellite operators who have committed to purchasing satellites are willing to pay to launch as soon as possible to avoid interruption in services or because the extra launch charges are less than the costs of financing an unused and completed satellite.

4. The lack of any prospect of a return to cheap Shuttle launches undermines the fundamental economics of satellite communications at a time when fibre optic costs are falling rapidly.

5. Some of the satellite users are also threatened with insolvency if the right satellite capacity is no longer available. In Europe, some satellite TV broadcasters are concerned that there is no in-orbit spare capacity on the Eutelsat system and if the Eutelsat I F1 satellite fails, they will have to transfer to Intelsat which means that some cable networks will be unable to receive the channels.

6. Other potential satellite users may be unwilling to opt to use the technology because of uncertainty that there will be satellite capacity in the future, or, perhaps, because a shortage will lead to an increase in satellite user charges.

7. Ground equipment manufacturers may be unwilling to invest in Research and Development and tooling up for new product lines if they perceive that there is an unavoidable risk from lack of suitable satellite capacity for their customers.

SECTION II

INDIVIDUAL SATELLITE SYSTEMS

COMMERCIAL DETAILS

OPERATOR	Advanced Business Communications Incorporated (ABCI)
ADDRESS	8729 Belmart Road, Potomac, MD 20854, USA
TELEPHONE NO.	(703) 827-8282
TELEX NO.	
FAX NO.	
SENIOR STAFF	Dr Matt C. Nilson, President Samuel W. Fordyce, Executive Vice President
PURPOSE OF PROJECT	Provision of medium power Ku-band capacity for US domestic operations
PROJECT STATUS	Suspended
PROJECT VALUE	Not available

SATELLITE USERS

Company envisaged providing the following Ku-band 5 services to CONUS.

1) Voice and Data Communications
2) Video Teleconferencing
3) Satellite Master Antenna Television
4) Cable TV distribution
5) TV programme networking

PROJECT OVERVIEW

ABCI lost its licence in 1985 to construct and launch its satellite system revoked by the Federal Communications Commission. However, the company advises that it intends to fight on. The future of the two Hughes HS–393 satellites under construction for the company is not clear. Industrial sources suggest that the FCC revoked ABCI's licence because the company was over dependent on Hughes Aircraft Company for its finance. The ABCI order "received" by Hughes was the first for a Hughes' new HS–393 satellite. In order to clinch the deal and make the breakthrough into the market place, Hughes provided substantial finance for ABCI to the extent that the FCC considered ABCI to be "a front" for Hughes' ambitions.

The ABCI system offered two operational Ku-band satellites, each with 16 50 Watt transmitters. These would be available for sale or lease in what ABCI call a "shopping centre" mode whereby buyers or leasors would utilise them for their own communications networks, or would be "common carriers". Satellites are designed to enable customers to select from three different geographical coverage areas which cover either the entire CONUS or Eastern or Western half of CONUS. The Company claims that the high power levels of the ABCI satellites make them particularly suited to provide voice and data communications for networks employing small terminals located in downtown areas and to distribute TV signals to small receiver dishes. Systems applications include voice and data communications, video teleconferencing, satellite master antenna TV, Cable TV and multi-channel cable systems and TV network interchanges.

ABCI TYPICAL TRAFFIC MODES

TRAFFIC APPLICATION	CHANNELS PER TRANSPONDER	EARTH STATION ANTENNA DIA. TRANS./RCV. (METERS)	LINK AVAILABLE (%)	SYSTEM PERFORMANCE
FM TV (SMATV)	1	6.0/1·0	99·9	Video S/N = 42·8 dB (45 dB in clear weather)
FM TV (CABLE TV)	2	7·0/7·0	99·95	Video S/N = 51 dB
FDM FM (4000 voice channels)	1	8·0/8·0	99·95	TTNR = 51·0 dB
DIGITAL (80 MBPS)	1	4·5/4·5	99·95	BER 1 × 10 exp. −7
DIGITAL (50 MBPS)	1	2·0/2·0	99·95	BER 1 × 10 exp. −7
DIGITAL SCPC T–1 (1·544 MBPS)	24	3·5/3·5	99·95	BER 1 × 10 exp. −7
SMALL TERMINAL T–1 SERVICE	10	2·0/2·0	99·95	BER 1 × 10 exp. −7
DIGITAL SCPC (50 KBPS)	600	3·0/3·0	99·95	BER 1 × 10 exp. −7
COMPANDED SSB	6700	8·0/8·0	99·95	TTNR = 51·0 dB
ONBOARD PROCESSER T–1 SERVICE (FDM UP & TDM DOWN)	32	2·0/2·0	99·95	BER 1 × 10 exp. −7

ABCI SATELLITES' COVERAGE PATTERNS ("FOOT PRINTS")
WITH POWER FLUX DENSITY CONTOURS

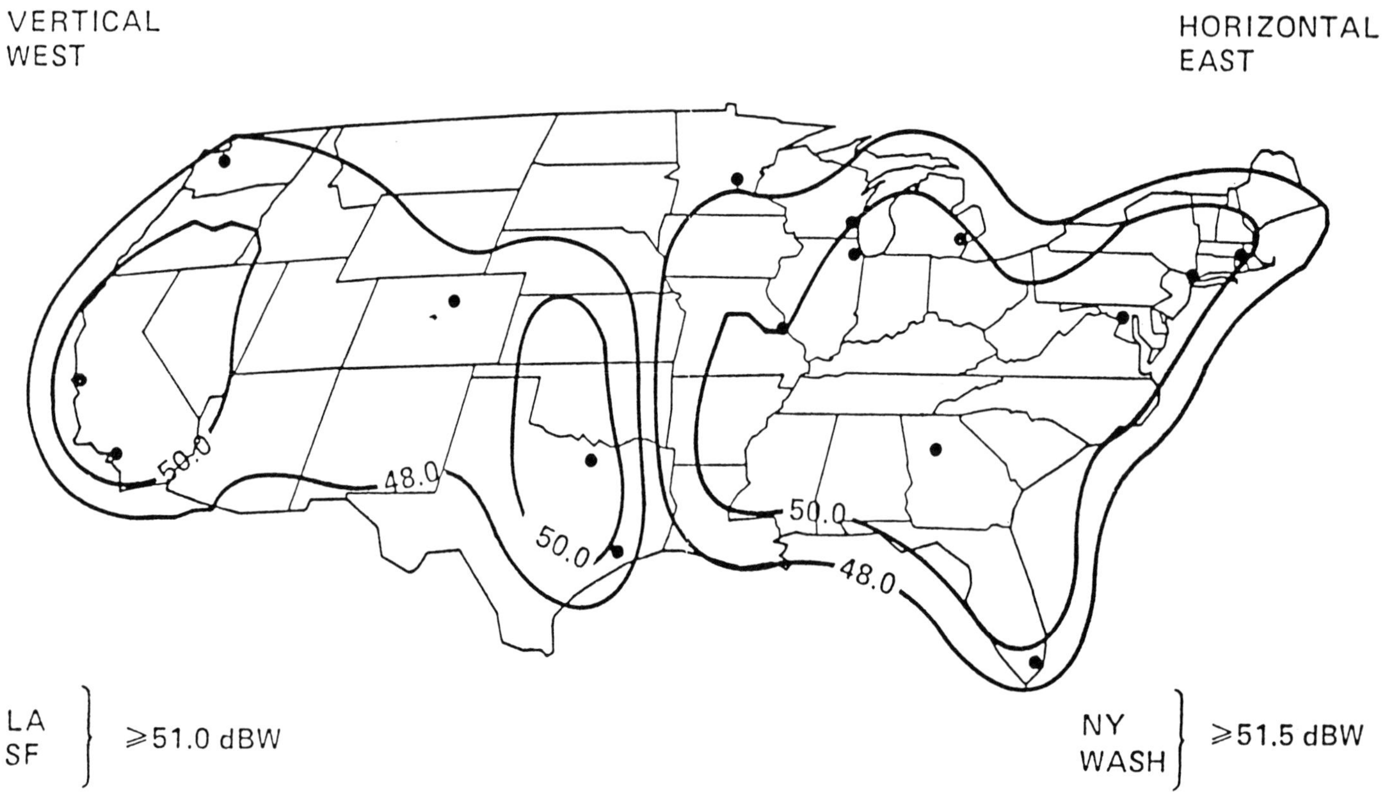

TECHNICAL SUMMARY

PLATFORM	HS–393
PRIME CONTRACTOR	Hughes Aircraft Company
OTHER MAJOR CONTRACTORS	not applicable
STABILISATION	Spin
LAUNCH MASS	not applicable
MASS IN ORBIT	1,350 kg
POWER SUPPLY, END OF LIFE	2,375 Watts
DESIGN LIFE	10 years
ORBITAL POSITIONS	ABCI–1, 83° West ABCI–2, 130° West
STATION KEEPING ACCURACY	+/−0·05° (E/W & N/S)
ECLIPSE PROTECTION	Full
SPACECRAFT DIMENSIONS	2·5 m diameter 3·0 m height at launch 10·0 m in orbit

NO. OF ACTIVE TRANSPONDERS	16
NO. OF SPARE TRANSPONDERS	8
TWTA POWER OUTPUT	50 Watts
FREQUENCY RANGE:	
UPLINK	14·0–14·5 GHz
DOWNLINK	11·7–12·2 GHz
TRANSPONDER BANDWIDTH	52 MHz
EIRP MINIMUM, BEAM EDGE	48 dBW
CHANNEL POLARISATION	dual linear, horizontal & vertical
RECEIVE GAIN	0 dB/K
RECEIVE SATURATION FLUX DENSITY	−87 to −78 dBW/m^2

COMMERCIAL DETAILS

OPERATOR	Advanced Communications Corporation
ADDRESS	1000 Savers Federal Building, Little Rock, Arkansas 72201, USA Also at c/o Shea and Gould, 1627 K Street NW, Washington DC 20006, USA
TELEPHONE NO.	(501) 663-2720 and (202) 833-9850
TELEX NO.	
FAX NO.	
SENIOR STAFF	Daniel H. Garner, Richard Rodin, Legal Counsel to ACC, Donald Diment
PURPOSE OF PROJECT	Provision of "second generation" DBS capacity for US domestic TV channels
PROJECT STATUS	Initial Order for the satellites placed in October 1986 but full funding has yet to be raised
PROJECT VALUE	Contract for first two satellites and associated services but not launch. Insurance has a potential worth of $200 million plus, according to RCA
LAUNCH DATES	First launch is projected for late 1989 or early 1990. Launch vehicle has not yet been decided but satellites are compatible with Ariane, Atlas Super-G and Titan III

SATELLITE USERS

Advanced Communications Corporation has not yet announced customers for the system which, at the time of writing, appears to remain a speculative venture.

The company says that it is to open a national marketing information centre in Washington DC in the Autumn of 1986.

It also says that the satellites will be used to transmit entertainment, educational and informational programming to homes and commercial locations equipped with dish antennas of 65 cms or less depending on the service involved.

PROJECT OVERVIEW

Advanced Communications Corporation was one of the so called "second round applicants" to operate a US DBS satellite system. It subsequently modified its plans for the space segment and on 26th September 1986 received authority from the Federal Communications Commission to operate a two satellite system with each satellite carrying 16 125 Watt Travelling Wave Tube Amplifiers. This represents what is seen in the USA as a "second generation" Direct Broadcast Satellite concept, able to provide as many channels as were available on fixed satellite service spacecraft available in the early 1980s. The new concept from Advanced Communications Corporation takes advantage of both developments in receiver technology and in the availability of the larger RCA–5000 platform from the late 1980s. Nevertheless, interest in the second generation concept has proved to be rather limited in contrast to the position in DBS TV between 1980 and 1984. It is not clear where Advanced Communications Corporation's funding for the system is coming from. The company appears to be backed by private interests without the support of major corporations.

ACC's original authorisation from the Federal Communications Commission covered two satellites (and a ground spare), each equipped with six 200 Watt transponders and located at 101 degrees West. The plan envisaged that three channels would be provided by ACC and three other transponders leased out to other TV channels. The three channels provided by ACC would have consisted of entertainment, information and educational material. It was granted FCC authorization on 4th December 1984 and announced its intentions to file for the amended system in November 1985.

TECHNICAL SUMMARY

PLATFORM	RCA–5000
PRIME CONTRACTOR	RCA Astro-Electonics
OTHER MAJOR CONTRACTORS	not available
STABILISATION	3-axis
LAUNCH MASS	not available
MASS IN ORBIT	not available
POWER SUPPLY, END OF LIFE	not available
DESIGN LIFE	not available
ORBITAL POSITIONS	110° and 119° West
STATION KEEPING ACCURACY	not available
ECLIPSE PROTECTION	not available
SPACECRAFT DIMENSIONS	not available
SPAN, TIP-TO-TIP	not available

ACTIVE TRANSPONDERS	16
SPARE TRANSPONDERS	not available
TWTA POWER OUTPUT	125 Watts
FREQUENCY RANGE:	
UPLINK	not available
DOWNLINK	not available
TRANSPONDER BANDWIDTH	not available
EIRP, MINIMUM BEAM EDGE	not available
POLARISATION	not available
RECEIVE CIVERAGE (G/T)	not available
SINGLE CARRIER SATURATION FLUX DENSITY	not available

COMMERCIAL DETAILS

OPERATOR	National Aeronautics and Space Administration
ADDRESS	Lewis Research Centre, 2100 Brookpark Road, Cleveland, Ohio 44135, USA
TELEPHONE NO.	(216) 433-4000
SENIOR STAFF	Andrew J. Stofan, Director, Cleveland Research Centre Dr John M. Klineberg, Deputy Director John Taylor, PR Officer Robert S. Lawton, Manager, ACTS programme, RCA Astro-Electronics
PURPOSE OF PROJECT	Research and Development into advanced satellite communications technology (see text)
PROJECT STATUS	Satellite under construction but subject to possible delays due to budgetary problems
PROJECT VALUE	Satellite has been ordered from RCA Astro-Electronics under a contract worth $260 m including $70 m for ground segment, contract excludes launch costs
LAUNCH DATES	1989 at the earliest, using Shuttle

SATELLITE USERS

The ACTS satellite is essentially a Research and Development exercise. Following past experience of NASA R&D satellite projects, use of the satellite will eventually be opened up to "experimental" users to test and develop new services. These users may not necessarily be US organisations.

NASA envisages that potential users of the technology demonstrated by ACTS include organisations interested in packet switching, electronic mail, mobile satellite services, interactive home video, restrictive networks and more conventional services such as wideband data services, public telephony, CATY distribution, teleconferencing, video distribution etc. The satellite will allow two basic types of services.

1) Trunking. This will accommodate high volume users in large cities, using the satellites IF matrix switch.
2) Customer Premises Services will employ smaller and cheaper earth stations located on user premises, such services will use either scanning spot beams (see text) or fixed beams and the satellites on-board processing capabilities.

Federal Express has already declared that it intends to sponsor an experiment on ACTS.

PROJECT OVERVIEW

Research and Development expenditure on US satellites comes from four major sources. There are spin-offs from Department of Defence work, Intelsat R&D contracts, expenditure provided by the satellite manufacturers and NASA projects. The latter has a long history of developing experimental communications satellites dating back to the earliest days of satellite communications.

However, the late 1970's and early 1980's saw no major dedicated experimental project until 1985 when Advanced Communications Technology Satellite was finally funded by the US Congress after pressure from the House Science and Technology space science and technology subcommittee. The ACTS project had been sitting on the shelf for some years whilst NASA lobbied hard to obtain funding. Central to the change of heart in the House was the gearing up of R&D in Japan, with authorisation of the large CS–4 experimental communications satellite and a launch vehicle, H–II, capable of launching satellites which would compete with US products. The CS–4 project will give Japan expertise in moving away from the traditional "dumb" communications satellite, which is essentially no more than a sophisticated radio relay, towards the "intelligent" satellite as well as providing it with further expertise in the Ka-band. During the early 1980's it was generally considered that engineers would have to move up into the 30/20 GHz Ka-band as the Ku-band frequencies became progressively more congested. Subsequently, more moderate growth in demand for satellite communications, including the near collapse of DBS, has made this development much less pressing, particularly as rain attenuation remains a serious problem in the Ka-band.

The central feature of the ACTS project therefore is its capability as an "intelligent bird" (The phrase coined by Professor Barry Evans at the University of Surrey, England which is involved in the similar T–SAT project.) Evans argues that the major problem now facing satellite engineers is meeting the demand for business services using small fixed earth stations and mobile communications. Current "dumb" satellites are limited in their ability to meet such requirements despite evidence to the contrary in the explosive growth of the V–SAT market in the USA. In the small earth station sector, TDMA techniques provide low satellite tariffs but require very expensive earth stations. For example, Telecom 2 earth stations can cost upwards of $1 m. FDMA techniques, as used on the ECS–SMS, have much lower earth station costs but "crippling" space segment tariffs because the satellites are used so inefficiently with this method. The problem is to match the satellite service with the access technique.

The ACTS system addresses this problem by two basic innovations which are interdependent. Firstly the use of narrow, multiple spot beams each illuminating an area of 400 km diameter. These narrow spot beams are able to scan the entire coverage area of the satellite, which, in the ACTS case, is the whole of CONUS. There are in addition two fixed beams on the satellite. The use of small spot beams allows considerable spacial frequency reuse, and a high EIRP which reduces the size of the earth terminals. The scanning process also allows the satellite to match its capacity with fluctuating user demand between geographical areas at any one time. Each scanning spot beam needs only to cover areas that have a demand for satellite capacity at any one time and needs only to stay there whilst that demand lasts. The scanning spot beam concept is comparable to packet switched data networks where users share the use of a transmission path. In essence, it overcomes the problems experienced by a conventional satellite, which typically covers a large area and handles traffic which originates or is destined between a limited number of geographical areas within that region. Much of the spectrum is wasted in providing coverage to areas where it is not needed either because there are no users or because demand fluctuates throughout the day. The scanning technique on ACTS enables the user to determine use of the spacecraft according to demand. It provides a five fold increase in the efficiency of satellite communications links. NASA quotes that the capacity of ACTS is up to 200,000 voice circuits – the Intelsat VI satellite has a capacity of around 30,000 voice circuits. Quoted cost saving to the customer is up to 50%. The access technique is TDMA (see next paragraph); the small coverage cells reduce the length of the back haul circuits needed to connect smaller user premises.

However the use of multiple spot beams and scanning techniques also requires substantial development of on-board processing techniques, creating the "switchboard in the sky" description of ACTS. Such switching has to be with minimal control from the ground. ACTS uses on board switching as follows: Intermediate frequency matrix switching is used to interconnect high volume wideband traffic over the fixed beams. This traffic can be routed by ground station control. Its shortcoming is that it is not suitable for the low volume traffic on the satellite using the scanning spot beams. These require that the up-link and down-link traffic be unbundled on the satellite. On a conventional satellite the information in an uplink signal cannot be separated out on the satellite and is downlinked in the same bundle as it is uplinked. To unbundle the information on-board the satellite so that individual streams of information can be diverted to different downlinks requires baseband processing. The uplinked traffic is demodulated and decoded before entering the baseband switching processor. This switches the individual signals according to information in each signal, as well as providing error detection and correction through Forward Error Connection (FEC) techniques. The signals are then encoded and modulated before downlinking either on fixed or scan beams. Coupled to the baseband processor are input and output memories. The system is the "brain" of the satellite which allows the scanning beams to meet the spacial and time variations in capacity demand.

The TDMA access system is therefore controlled by both the baseband processor for the customer premises service and from a master ground station at the NASA Lewis Research Centre for the Trunking services on the satellite. The combination of multiple spot beams requires the development of an advanced multiple beam antenna on the satellite to provide highly focused beam patterns. These in turn are required to eliminate interbeam interference. The use of Ka-band assists in the design of the antenna. Beam concentration helps in providing transmission margins to overcome the high rain attenuation in the Ka-band. Power output of repeaters can also be switched from 8 Watts to 40 Watts to overcome rain attenuation when it occurs. The communications payload, together with the baseband processing equipment, on the ACTS satellite, will weigh around 20% more than a conventional payload with the same nominal performance. In total the entire payload accounts for about 30% of the in-orbit weight of the satellite.

TECHNICAL SUMMARY

PLATFORM	
PRIME CONTRACTOR	RCA Astro-Electronics
MAJOR SUBCONTRACTORS	TRW Defense and Electronics Group, Multibeam communications payload ($100 m) Motorola Aerospace Electronics Office, Baseband processor ($25 m) Electromagnetic Services, Antenna beam forming network ($5 m) Hughes Aircraft, TWTA's ($3 m) Comsat Laboratories, Ground Segment ($70 m)
STABILISATION	
LAUNCH MASS	
MASS IN ORBIT	
POWER SUPPLY, END OF LIFE	
DESIGN LIFE	
ORBITAL POSITIONS	
STATION KEEPING ACCURACY	
ECLIPSE PROTECTION	
SPACECRAFT DIMENSIONS	
SPAN, TIP-TO-TIP	
NO. OF ACTIVE TRANSPONDERS	
NO. OF SPARE TRANSPONDERS	
TWTA POWER OUTPUT	
FREQUENCY RANGE:	
UPLINK	
DOWNLINK	
TRANSPONDER BANDWIDTH	
EIRP MINIMUM, BEAM EDGE	
CHANNEL POLARISATION	

Artists impression of NASA's Advanced Communications Technology Satellite (ACTS) under construction by RCA Astro-Electronics

COMMERCIAL DETAILS

OPERATOR	Telesat Canada
ADDRESS	333 River Road, Ottawa, Ontario, K1L 8B9 Canada
TELEPHONE NO.	(613) 746-5920
TELEX NO.	(053) 4184/Telesatott
FAX NO.	(613) 748-8712
SENIOR STAFF	Eldon D. Thomson, President & CEO John Almond, VP, Corporate Developer Frederick M. Bartlett, VP, Administration Harry Kowalik, VP, Space Systems Linda M. Rankin, VP, Telecoms Ronald W. Wieleba, Secretary Donna Wishart, Public Affairs
PURPOSE OF PROJECT	Provision of Ku-band capacity for telesat Canada's domestic satellite services (Anik A and D systems are C-band and Anik B is hybrid C/Ku-band)
PROJECT STATUS	All three satellites have been launched but Anik C1 is "stored" in orbit
PROJECT VALUE	Telesat quotes the cost of an Anik C satellite at "C$45 m" although the Anik C3 satellite was insured for US$65 m
LAUNCH DATE	Anik C3, 12th November 1982, STS-5 + PAM D Anik C2, 18th June 1983, STS-7 + PAM D Anik C1, 12th April 1985, STS-51D + PAM D

SATELLITE USERS

Four members of Telecom Canada, a consortium of telephone operators, has a commitment to lease four Ku-band transponders from Telesat Canada up until 31st December 1998. According to Telesat "this obligation has a revenue value in excess of $100 m".

Charges in the regulatory environment in Canada in 1985 now allow Telesat to "directly market" partial channel services – prior to 1985 the organization had operated as a "carrier's carrier" dependent upon terrestrial carriers for the development and marketing of voice, data and image telecommunications services.

The Anikast 1400 SNG service is available to broadcasters and only uses the Anik C satellites.

As at February 1986, the following TV channels were carried on Anik C3, the primary satellite TV spacecraft in the Anik C series.

Trans-ponder	Channel	Company
8	Knowledge Network–Western	Knowledge Network
9	Alberta Access–Western	Access Network
12	Superchannel/First Choice–Western	Superchannel and First Choice
18	Atlantic Sat. Net.–Eastern	Atlantic TV Network
19	Premier Choix/TVEC–Eastern	Premier Choix/TVEC
24	LaSette (TVFQ–99)–Eastern	La Sette (TVFQ–99)
26	TV Ontario–Eastern	TV Ontario
31	Superchannel/First Choice–Eastern	Superchannel and First Choice
32	Radio Quebec–Eastern	Radio Quebec

From 6th September 1986, Television Quatre Saisons started using the satellite. Basic annual lease charge for a full transponder is C$2 m.

PROJECT OVERVIEW

Telesat Canada's 3 satellite Anik C series was one of the leading North American Ku-band projects of the early 1980's but demand for its capacity fell well short of the expectations of the late 1970's. Telesat has tried to sell the Anik C1 spacecraft as a consequence – it had been launched over a year late because of problems with the PAM–D Apogee kick motors which failed on the Palapa B2 and Westar VI satellites, which also use a Hughes HS–376 platform. After launch in April 1985 Anik C1 was placed in in-orbit storage until such time as a buyer could be found. As at mid-1986, no buyer had been found. It is expected to remain in storage for up to another three years (unless a buyer is found) whereupon it will be brought into service for Telesat. The problem of overcapacity in the Canadian Ku-band services is compounded by the Anik B satellite exceeding its design lifetime. Nevertheless, the North American market is likely to see a loss in supply of Ku-band capacity resulting from the considerable shortfalls in launch vehicle capacity resulting from the "grounding" of Shuttle and Ariane in 1986. The Telesat Canada capacity is available to US satellite operators and, indeed, the Anik C2 satellite has been used for quasi-DBS operations in the USA.

The use of the Anik C2 satellite by a company called United Satellite Communications Inc. from late 1983 to 1985 was followed by the DBS industry with great interest. USCI, a US company, used transponders on the satellite, leased through GTE Satellite Corporation to provide a direct-to-home scrambled pay TV service to the North East of the USA. USCI opted for the "low power" Anik C system to gain early entry into the DBS market. The service used 2·5–4 foot diameter dishes and provided three channels of films and sports and two channels of UPI news and programme listings mixed with music videos. However USCI failed to attract customers; it appeared that the maximum number of subscribers was not much more than 10,000. Part of the problem lay in the high cost of installing the dishes which helped to rapidly drain the company of resources needed to market the service. It attempted to find up to $200 m from private investors and later, to merge with Comsat's STC but eventually the company went bankrupt, leaving considerable doubts as to the long-term viability of any form of DBS service. Telesat Canada, during 1984 and 1985, put a lot of planning effort into using the Anik C system for a similar quasi-DBS service for rural Canada. To quote Telesat's 1985 Annual report (published in April 1986) "Telesat (has) prepared a comprehensive business plan addressing demonstrated market interest in a direct-to-home television service using the Anik C satellites ... an attractive package of licensed program channels would be marketed on a subscription basis to underserved households beyond the economic reach of cable ...."

Nevertheless, it is difficult to see why the Canadian quasi-DBS service would suceed where USCI failed. Canada has a smaller population than the area served by USCI. Distribution and installation costs of Ku-band TVRO's may have fallen as the industry has gone through its learning curve (and shifted production off-shore to the Far East) but the start-up costs of the exercise could mean it is equally as risky as the USCI venture. Although the "quasi-DBS" market looks suspect, particularly as the VCR continues to penetrate the market, Telesat has been very successful in attracting broadcasters to the Ku-band capacity on the Anik C system. These broadcasters fall into three groups. Those broadcasters, involved in terrestrial transmissions, looking to use the Anik system to provide a distribution network to serve terrestrial transmitters, can overcome many of the problems involved in terrestrial links. A second group are using the system to distribute TV channels dedicated to cable, to cable head ends. The third group belong to the "occasional group" users. Telesat supports the latter with transportable uplink facilities (in both C and Ku-band). According to Telesat's PR machinery, the occasional use RV traffic has, since 1981, been growing at an annual rate of 35%.

Each of the Anik C satellites has 16 operational transponders (plus four spares), each capable of up to two colour TV programmes and 1,344 telephone circuits. Signals are downlinked from each satellite. Signals use four spot beams for downlink, which are spread across Canada in an East–West fashion, and provide a minimum signal strength of about 46·3 dBW for most of Canada and down to around 31 dBW within the USA, within 500 miles of the Canadian border, and within the largely unpopulated North of Canada. Given that the vast majority of Canadians live within 300 miles of the Canadian border, the system can provide strong coverage for nearly all Canadian broadcasters, telephony operators and businesses. According to Telesat "Anik C's antenna coverage includes virtually all of populated Canada, with four contiguous spot beams serving the Western, Western–Central, Eastern–Central and Eastern regions of the country. Telesat's customers are (thus) able to choose regional, half, or whole country coverage…." The high power of the Anik C satellites, has enabled Telesat to introduce a satellite News gathering service which it markets under the name Anikast 1400. A 2·2 metre dish, mounted on the back of a van equipped with a generator, transmitter, receiver, test and other electronics was made available in 1983 and demonstrated that the equipment could allow a broadcaster to be on air within 20 minutes.

Initially Telesat made available its own equipment to broadcasters but now leases or sells customer-dedicated terminals. The Anikast 1400 service is defined by Telesat as a "transportable earth station television transmit service, providing one video and two audio channels, and furnished for an eight-hour availability period which includes one hour of TV transmission time." This basic service costs the customer C$1,550 per hour and additional periods of transmit time are available for C$35 per 10 minutes or C$210 per hour.

TECHNICAL SUMMARY

PLATFORM	HS–376
PRIME CONTRACTOR	Hughes Aircraft Company
OTHER MAJOR CONTRACTORS	Spar Aerospace, Fleet Industries, SED, Com Dev (45% by value to Canadian subcontractors)
STABILISATION	Spin
LAUNCH MASS	1160 kg
MASS IN ORBIT	567 kg (B.O.L.)
POWER SUPPLY, END OF LIFE	900 Watts
DESIGN LIFE	10 years
ORBITAL POSITIONS	Anik C1, 108·5° W Anik C2, 112·5° W Anik C3, 117·5° W
STATION KEEPING ACCURACY	
ECLIPSE PROTECTION	
PROPULSION SYSTEM	
SPACECRAFT DIMENSIONS	6·43 m high, 2·16 m diameter
NO. OF ACTIVE TRANSPONDERS	16
NO. OF SPARE TRANSPONDERS	4
TWTA POWER OUTPUT	15 Watts
FREQUENCY RANGE:	
UPLINK	14·030–14·457 GHz
DOWNLINK	17·730–12·157 GHz
TRANSPONDER BANDWIDTH	54 MHz
EIRP MINIMUM, BEAM EDGE	47 dBW, Canada, 32 dBW USA
CHANNEL POLARISATION	Linear, both vertical & horizontal
RECEIVER (UPLINK) COVERAGE (G/T)	2 dB/K
SINGLE CARRIER SATURATION FLUX DENSITY	−80 dBW/sq. m.

Artists impression of the launch of the first Anik C satellite (Anik C3) by Shuttle in November 1982. All three Anik Satellites were launched via Shuttle and this launch was on the first commercial flight of STS. The Anik C3 was one of two Hughes HS–393 missions launched on this flight; the other was for SBS (SBS–3). Both used the PAM–D upper stage. All three Anik C satellites were launched via Shuttle, although the launch of Anik C1 was delayed by problems with the PAM D upper stage

COMMERCIAL DETAILS

OPERATOR	Telesat Canada
ADDRESS	333 River Road, Ottawa, Ontario, K1L 8B9 Canada
TELEPHONE NO.	(613) 746-5920
TELEX NO.	(053) 4184/Telesatott
FAX NO.	(613) 748-8712
SENIOR STAFF	Eldon D. Thomson, President & CEO John Almond, VP, Corporate Development Frederick M. Bartlett, VP, Administration Harry Kowalik, VP, Space Systems Linda M. Rankin, VP, Telecoms Services Ronald W. Wieleba, Secretary Donna Wishart, Public Affairs Bernie D'Lima, Broadcast Sales
PURPOSE OF PROJECT	Provision of C-band capacity for Canadian domestic services
PROJECT STATUS	Both satellites are now in orbit but Anik D2 is stored at 111·5° West until it is needed to replace the Anik B satellite, between the end of 1986 and the Spring of 1987
PROJECT VALUE	The two satellites were procured from Spur Aerospace under a 1979 contract worth C$80·8 m. Anik D2 satellite was insured for US$75 m
LAUNCH DATES	Anik D1, 26th August 1982, Delta 3920 flight 164 plus PAM D Anik D2, 8th November 1984, STS 51A plus PAM D

SATELLITE USERS

The satellites are designed to carry voice, data and video traffic.

As at February 1986, the following video traffic was carried on Anik D1.

Trans-ponder	Channel	Company
2	The Sports Network	TSN
4	Global TV Network	Global
6	MuchMusic	MuchMusic
8	CHCH–Hamilton	CANCOM
9	WDIV–Detroit	CANCOM
10	WXYZ–Detroit	CANCOM
11	CBC North	CBC
14	TCTV–Sherbrooke	CANCOM
15	CBC North	CBC
16	CITV–Edmonton	CBC
17	CBC North	CBC
18	CITV–Edmonton	CANCOM
19	CBC North	CBC
20	CBMT–Montreal	CBC
21	WTVS–Detroit	CANCOM
22	BCTV–Vancouver	CANCOM
23	WJBK–Detroit	CANCOM
24	Parliamentary Network	CBC
24	CBC North	CBC

The satellite also carried 14 radio services as a subcarrier to the above. Basic annual lease charge for a full transponder is C$1·3 m. The Anikast RadioNet 2 provides 7·5 kHz audio channels on a 24 hour-per-day and occasional use basis and is separate from the subcarrier radio services using the above video channels. Rate for the Anikast RadioNet 1 service is $40,000 per annum including uplink service. The occasional rate service costs between $100 and $200 per hour, it uses SCPC technology.

PROJECT OVERVIEW

Like the Anik C system, the C-band Anik D system is a product of over optimistic forecasts of traffic made in the late 1970's. Consequently, the second satellite, Anik D2, is in an in-orbit store at 111·5° West until such time it is needed to replace the ageing Anik B satellite. Although an early provider of Ku-band capacity, Telesat Canada started out in the C-band business and the Anik D satellites, which use the same Hughes HS–376 platform as the Ku-band Anik C satellites, represent Telesat's current generation of C-band capacity.

The Anik D satellites were the first series of Telesat spacecraft to be "primed" from a domestic Canadian company Spar Aerospace. Prior to placement of the order in 1979, Spar had been building up its expertise by acting as a sub-contractor to US commercial satellite manufacturers. The approach was markedly different from the European one which had been, and is, to build up its satellite manufacturing capabilities through prime contractorships being supported by a series of consortia and a protected market place. Consequently, the Europeans have had to "reinvent the wheel". The Canadian approach has allowed its industry to make use of technology already developed in the USA without the need for expensive R&D programmes. It also means that the Canadian products are generally proven reliable before flight. Because of this approach, Spar has been able to enter the export market for satellites. It was prime contractor for the Brazilian SBTS system, which also used a Hughes HS–376 platform.

Telesat's Data distribution service uses the Anik D satellite. This provides a transmission rate from 75 bits/s to 19·2 kbits/s. Receive stations use 1·2 m diameter dishes. Service is one-way only. The first major user of the Anikom 100 service was Dataline Inc of Toronto which uses it to distribute its Canquote financial information service. Anikom combines packet switching and spread spectrum modulation techniques. The Anik D2 satellite was subject to a successful insurance claim for C$6 m in 1985. This followed a sudden "spin-up" on the 8th March which resulted in loss of station keeping fuel and failure of a redundant electronic unit. Telesat says that there was strong evidence to suggest that these problems were caused by a large electrostatic discharge: the space environment near the satellite was conducive to such an event.

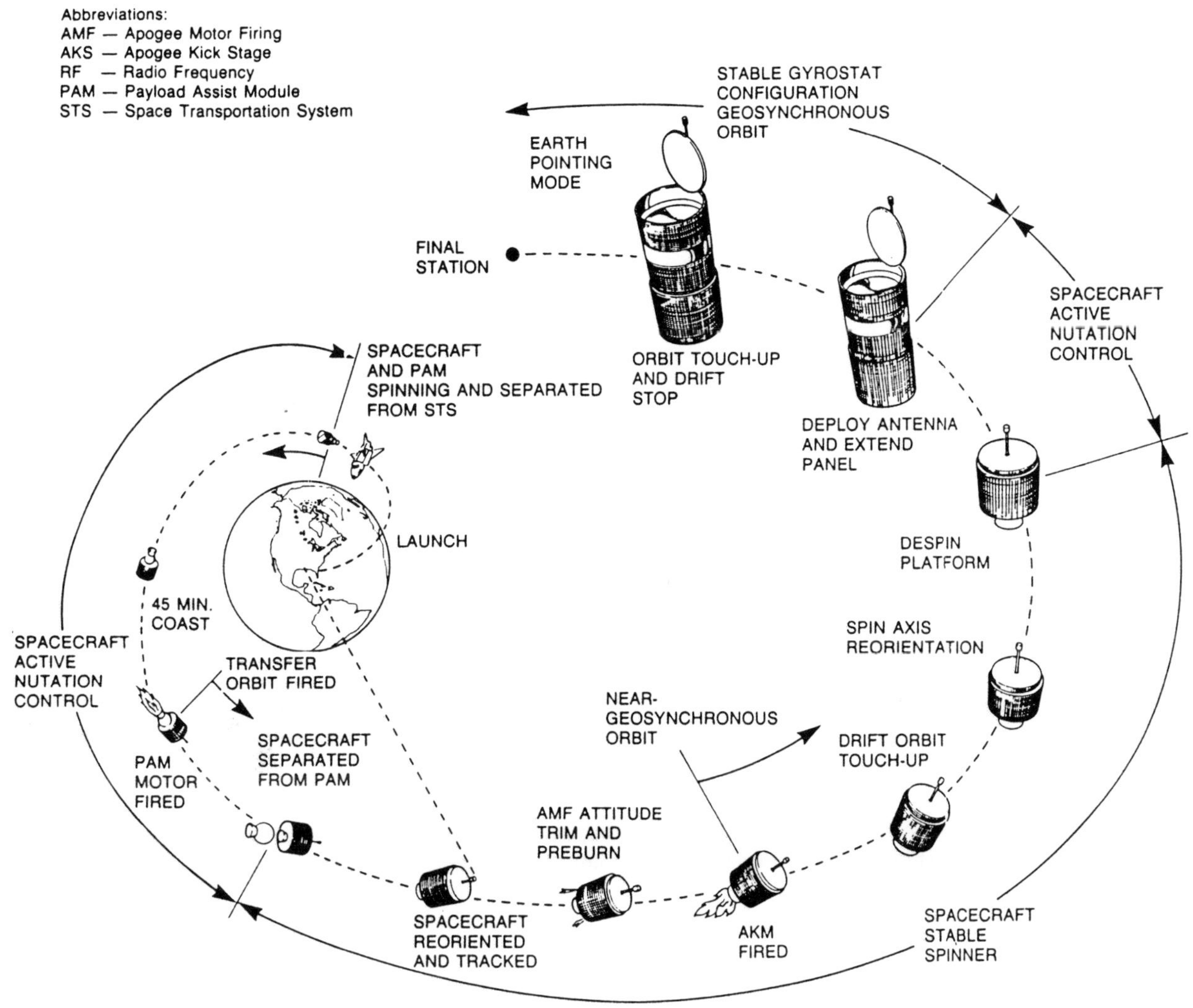

Anik D2/STS 51A mission scenario

TECHNICAL SUMMARY

PLATFORM	HS–376
PRIME CONTRACTOR	Spar Aerospace
OTHER MAJOR CONTRACTORS	Hughes Aircraft Company, SED, Fleet Industries, COM DEV
STABILISATION	Spin
LAUNCH MASS	1240 kg
MASS IN ORBIT	660 kg (beginning of life)
POWER SUPPLY, END OF LIFE	1000 Watts
DESIGN LIFE	8 years
ORBITAL POSITIONS	Anik D1, 104·5° West Anik D2, 111·5° West
STATION KEEPING ACCURACY	
ECLIPSE PROTECTION	Full
SPACECRAFT DIMENSIONS	6·57 m high, 2·16 m diameter

NO. OF ACTIVE TRANSPONDERS	24
NO. OF SPARE TRANSPONDERS	
TWTA POWER OUTPUT	11·5 Watts
FREQUENCY RANGE:	
UPLINK	5·925–6·425 GHz
DOWNLINK	3·700–4·200 GHz
TRANSPONDER BANDWIDTH	36 MHz
EIRP MINIMUM, BEAM EDGE	36 dBW
CHANNEL POLARISATION	Linear, both vertical & horizontal orthogonal
RECEIVER (UPLINK) COVERAGE (G/T)	−3 dB/K
SINGLE CARRIER SATURATION FLUX DENSITY	−81 dBW/sq. m.

TELESAT CANADA — ANIK E

COMMERCIAL DETAILS

OPERATOR	Telesat Canada
ADDRESS	333 River Road, Ottawa, Ontario, H3A 2S4 Canada
TELEPHONE NO.	(613) 746-5920
TELEX NO.	053 4184 telesatott
FAX NO.	(613) 748-8712
SENIOR STAFF	R. W. Jestin, Director, Anik Systems E. H. Martin, Director of Operations and Business Development B. L. Olsen, Director, Broadcasting Services Division
PURPOSE OF PROJECT	Provision of domestic C-band and domestic and transborder Ku-band capacity to replace Anik C & D
PROJECT STATUS	Final order for the system was due in October 1986
PROJECT VALUE	CAN$450 m including launch and insurance
LAUNCH DATE	Both due for launch in 1990 but launch vehicle not selected. Ariane 4, Long March, Titan III and Atlas-Centaur are under consideration. It had been to use Shuttle.

SATELLITE USERS

The two satellites in this series are intended to replace both the Anik C and Anik D satellites from 1990. Both C- and Ku-band capacity users will transfer to Anik E. The principle differences for users between the Anik C and D systems and Anik E are that Anik offers enhanced transborder services aimed at Canadian companies with communication needs in both Canada and the USA and enhanced Ku-band capacity providing better EIRP performance within Canada. It also provides national as well as spot beam Ku-band capacity within Canada.

PROJECT OVERVIEW

The two satellite Anik E system is Telesat Canada's next generation satellite system intended to provide capacity to replace the Anik C and D satellite systems during the 1990's. Anik E is a hybrid C and Ku-band satellite system and will use the new RCA 5000 platform, although the prime contractor will be Spar Aerospace. Both satellites are due for launch in 1990 with services commencing in the third quarter of that year. If the two satellites are successfully launched and placed in-orbit, the second satellite will not be needed until 1993. Under those circumstances, it will be placed in storage orbit and the communications subsystems turned off. If it proves not possible to place the satellite in storage orbit (because the launch is shared with another satellite), extra fuel may be carried to provide for two years of extra station keeping requirements. If one of the satellites is lost on launch, a third spacecraft will be ordered for launch in 1993.

Payload Performance: The C-band performance is similar to that of Anik D, providing national coverage with an EIRP of at least 35·5 dBW. Solid State Power Amplifiers, each with a power output of 11·5 Watts, will be used The Ku-band capacity consists of 16 50 Watt TWTA's. Of these, eight are intended for East (horizontal polarisation) and eight for West (vertical polarisation) beam services. A national beam will also provide services in both polarisations. There is also intended an "enhanced cross-border capabilities" beam giving an EIRP performance in the USA of 42–45 dBW. There is a fairly complex switching procedure on-board the satellite which allows transponders to switch from one beam to one or more beams, depending on which transponder is involved.

TECHNICAL SUMMARY

PLATFORM	RCA 5000
PRIME CONTRACTOR	Spar Aerospace
OTHER MAJOR CONTRACTORS	RCA Astro-Electronics (for platform)
STABILISATION	3-axis
LAUNCH MASS	2500 kg
MASS IN ORBIT	1210 kg (empty)
POWER SUPPLY, END OF LIFE	2900 Watts
DESIGN LIFE	12 years
ORBITAL POSITIONS	104·5 to 117·5° West
STATION KEEPING ACCURACY	
ECLIPSE PROTECTION	Full
SPACECRAFT DIMENSIONS	
SPAN, TIP-TO-TIP	24 metres
NO. OF ACTIVE TRANSPONDERS	24 C and 16 Ku-band
NO. OF SPARE TRANSPONDERS	6 C
TWTA/SSPA POWER OUTPUT	11·5 (C) and 50 (Ku) Watts*
FREQUENCY RANGE:	
UPLINK	
DOWNLINK	
TRANSPONDER BANDWIDTH	36 MHz (C), 54 MHz (Ku)
EIRP MINIMUM, BEAM EDGE	35·5 dBW (C-band) 48 dBW (Ku-band E & W beam) 44·5 dBW (Ku-band national)
POLARISATION	Linear orthogonal
RECEIVER COVERAGE (G/T)	−30 dB/K EOC (C-band) −1 dB/K EOC (Ku-band)
SINGLE CARRIER SATURATION FLUX DENSITY	

*C-band uses Solid State Power Amplifiers, Ku-band uses TWTA's. Apparently Telesat Canada were worried about the uproven nature of 50 Watt SSPA's.

TELESAT CANADA, REVIEW OF ORGANISATION

History: Telesat Canada is the domestic satellite operator for Canada. Legislation to create an organisation to plan, design, build and operate a Canadian satellite communications system was introduced into the country's Parliament in 1968. The statute, the Telesat Canada Act, was passed and received Royal Assent in the summer of 1969, it was to become the world's first domestic satellite operator when the Anik A1 satellite was successfully launched by Delta on 9th November 1972. Canada's peculiar geography, being a country considerably larger than continental USA but with a population one tenth of the US total was the major reason why Canada entered the domestic communications satellite field so early, before US operators were able to dominate the field.

Function of Telesat: Prior to 1985, Telesat Canada was a "carriers carrier" unable to retail its major services directly to the end user. In the 1969 legislation it was intended that the company would be complementary to, and not competitive with, the telecommunications services offered by other Canadian telecommunications companies. It was given a monopoly of Canadian domestic satellite communications and was (and is) subject to regulation by the Canadian Radio-television and Telecommunications Commission (CRTC). Telesat Canada is a separate organisation from Teleglobe Canada which is responsible for most of Canada's international telecommunications. Teleglobe, for example, is the Canadian signatory to Intelsat and Inmarsat. However, subject to intergovernmental approval, Telesat Canada does provide cross border and US domestic services.

Telesat is owned by the major domestic common carrier telecommunications operators in Canada and the Federal government, with the latter having the largest, but not the majority, shareholding. It is a profit making organisation but its pricing structure is regulated by the CRTC. This regulation works two ways. Firstly it places an upper price limit which Telesat Canada can charge its customers for capacity. Secondly, up to 1985, it provided Telesat with "unlimited" transfer payments under its connecting Agreements with (other) Telecom Canada members. These payments had been unlimited to guarantee Telesat a return on equity equal to the weighted average of those of two Telecom Canada members, Bell Canada and the British Columbia Telephone Company. In return for this arrangement, Telesat was obliged to return to Telecom Canada half of any earnings above the Bell/B.C. Tel weighted average. These arrangements are now being phased out. The transfer payments are capped at \$20 m for 1985, 1986 and 1987 and will cease thereafter. The obligation to return "excess" earnings to Telecom Canada is also being dropped.

These adjustments reflect the more general moves in Canada towards liberalising telecommunications. Up to 1985 Telesat had been able to market only whole channel services directly to broadcasting and cable TV organisations for carriage of radio and television programmes. These same broadband broadcasting channel services were, and are, also available to the "regulated telecommunications carriers (RTCC) such as Bell Canada. All other satellite services offered by Telesat were retailed by the RTCC's. However, from 1985 Telesat was allowed to directly market partial channel services, taking it one step out of the role of carrier's carrier and reducing its dependence on terrestrial carriers for development and marketing voice, data and video services.

1985 also saw other areas of liberalisation of Canadian telecommunications. The CRTC opened up resale and sharing and interconnection of infra-exchange systems. According to Telesat, these decisions, coupled with terms of Telesat's Telecom Canada connecting agreement, "will facilitate the company's provision of end-to-end private line services throughout Canada." The CRTC also began licensing of privately owned transmit earth stations from April 1986.

View of Telesat's main communications control earth station at Allan Park, Ontario, north of Toronto. The station controls all of the Anik series of satellites. (photograph courtesy of Telesat Canada)

Specialist Services offered by Telesat: Telesat is best known outside North America for its consulting services. These in particular include feasibility studies for new satellite systems and services. Contracts have included feasibility studies for the South Korean government on its aborted Domsat plan, the "21 Club" looking at providing a UK DBS system, Swiss DBS, US DBS systems, and the Canadian government for the Radarsat earth observation satellite and MSAT. Telesat also supplies satellite control systems and related software and provides launch mission tracking services. For example it provided tracking services from Allen Park for Hughes Aircraft Company for two Aussat launches and mission support and launch director services for the two Brazilian SBTS satellite launches with related satellite operations training. The company is currently providing consultancy services to Granada TV of the UK who are bidding for a DBS franchise (see UK DBS).

The "Anikast" series of services offered by Telesat cover both C and Ku-band television and radio distribution services including full period (24-hour-per-day) and occasional use transmissions. Included in this family are SNG (Satellite News Gathering) services. The Anikast Radio Net 2 service provides distribution of radio channels (see Anik C system). During 1984, Telesat proposed to the Federal government that a 14 channel subscription TV service (eight channels in English, six in French) be introduced on the Anik C system. This would be aimed at areas "underserved" by terrestrial TV. This radical proposal has to be seen in the light of Canada's (largely unsuccessful) broadcasting policies (as well as Telesat's desperate need to sell capacity on Anik C). The whole issue was under review at the time of writing. The quasi DBS service would use dishes of around 1·2 metres. The Anikast 1400 Ku-band service is Telesat's SNG service.

The Data Distribution Service started in 1985 and provides one-way data in real time, in a point-to-multipoint configuration, using C-band capacity.

Size and Structure of Telesat Canada: As at 31st December 1985, Telesat Canada employed 573 people, up 10·2% on 1984. The relatively high staffing of this satellite operator reflects its quasi-public sector position. The company has been actively seeking work outside its main sphere of operations to mop up some of its personnel expertise. Notable amongst its efforts has been outside consulting work, but in 1985 its consulting services outside Canada earned only C$3·8 m out of a total revenue from all operations of C$101,233,000. Of that revenue figure 45·8% was accounted for by Domestic Broadcast services, 41·4% by voice, data and image services and 4·6% by US-to-US services. As at 31st December 1985, Telesat's operational in-orbit satellites were valued at C$252 m, before depreciation. In addition the Anik D2 and Anik C1 satellites, held as in-orbit spares were valued at C$133 m, before depreciation. Earth station and other property assets were valued at C$166 m (before depreciation). Total assets, after depreciation were valued at C$355 m on which Telesat earned profits of C$23 m before tax.

Future Developments at Telesat Canada: The expected shortage of Ku-band capacity in North America between 1987 and 1990, caused by the lack of launch vehicles to launch new US Ku-band satellites, may provide Telesat with a significant opportunity to use its surplus capacity on the Anik C system. The company is already adjusting to liberalisation of the North American telecoms/satellite markets. Its 10·2% staff growth in 1985 occurred mostly in sales and product development areas, which were expanded so that the company could capitalise upon its new role as a service retailer.

Telesat does not plan to get involved in high-power DBS TV, opting instead to push for low power Ku-band FSS quasi-DBS services. However, outside its conventional FSS services, it is looking to operate the M–SAT mobile satellite system in conjunction with US partner(s). It is possible that this system could become operational in 1990.

Telesat's Experimental Services Programme has made available two C-band and two Ku-band transponders until 30th June 1987 at 75% discount rate. According to Telesat "its purpose is to encourage the development of new applications for satellite telecommunications and foster their evolution to commercial services." Some 25 applications for experimentation had been received or were pending by the middle of 1986 and included tests of new earth station technology, video teleconferencing, private business networks, tele-education and tele-health services, market trials for broadcasters and live video news releases.

COMMERCIAL DETAILS

OPERATOR	Antares Satellite Corporation
ADDRESS	4700 Syracuse Parkway, 11th Floor, Denver, Colorado 80237, USA
TELEPHONE NO.	(303) 695-3715
SENIOR STAFF	David Murray Drucker
FINANCIAL BACKERS	Equity in Antares Satellite Corporation is currently held by 1) United Paging Corporation (60% of equity) – a wholly owned subsidiary of United Cable TV Corporation which operates 42 cable nets serving 680,000 subscribers in 15 states. It says it has sufficient internal financial resources to proceed with DBS TV 2) Echosphere Corporation, a distributor of satellite earth stations owns 30% 3) AB Hirschfeld Press owns 10%. It is a lithographic printer and owns Colorado Carphone, a mobile radio company
PURPOSE OF PROJECT	Antares application with the FCC primarily covers provision of DBS capacity but the company is interested in other services – see Project Overview
PROJECT STATUS	The Company was given conditional authorisation by the FCC in September 1985 to construct two satellites for in-orbit operation but has to meet the Commission's due diligence requirements by September 1986
PROJECT VALUE	not available
LAUNCH DATE	First launch in 1990

SATELLITE USERS

Antares intends to initially lease all its channels to programmers on a non-common carrier basis but may eventually programme one or more channels. Leased channels would primarily be used for subscription television and pay-per-view video entertainment.

However other options are being looked at which are largely described as services intended for direct reception by the general public and one and two way business services including data transmission. Amongst these services are:

1) a "meeting channel"
2) downloading of video and audio entertainment, computer software and games, stock market information and news topics. Information would be selected by the consumer by telephone
3) Teleshopping
4) Ku-band data services (both one and two way) for business using microterminals
4) Land mobile satellite services
5) International DBS and business services

PROJECT OVERVIEW

The proposals put by Antares to the Federal Communications Commission are both innovative and relatively controversial. The Company is primarily proposing a Direct Broadcast Satellite system to provide coverage to CONUS. However the satellites are both planned to include steerable spot beams for Alaska, Hawaii, Puerto Rico and areas outside the USA. Spot beams are available to offer additional channels to metropolitan areas. Requested orbital slots at 61°5′ West and 157° West could provide services to Western Europe, South America and Asia through the spot beams. Antares–2 would carry two fixed spot beams for Hawaii and Alaska.

Design of spacecraft chosen at this stage is Eurosatellites "Spacebus 300" platform, used for TDF–2/TV–SAT. The first Antares satellite will carry eight 100 Watt transponders connected to a steerable spot beam to deliver signals to discrete geographical areas. The two spot beams on Antares–2 will be fixed. Each half-CONUS or spot beam transponder will be capable of delivering an EIRP signal level of 51 dBW at beam edge. The Company says it does not plan a ground or in-orbit spare at this stage but will keep long-lead items in store to construct a replacement satellite within one year of failure of an in-orbit operational satellite.

In addition to leasing capacity to DBS channel programmers on a non-common carrier basis, Antares says it "is exploring the possibility of making available both one and two way business services utilising small antennas... such services are possible in the Ku-band using spread spectrum techniques such as are currently being employed in C-band data services. Antares anticipates that data rates up to 19,200 kbit/s can be achieved using 24 inch receive only antennas, or 48 inch transmit/receive antennas installed at customer sites in view of either Antares–1 or Antares–2." One of Antares shareholders, United Cable, plans to market terrestrial data services. The Antares proposals would complement these.

Antares also says that "preliminary analysis indicates that a land mobile satellite service would be compatible with the spacecraft design... Because the primary beneficiaries of direct-broadcast and land mobile satellite service are persons residing in remote and rural areas of the USA, the marriage of these services on a single satellite system should be given serious consideration."

In respect of its thinking about teleshopping Antares proposes "A video channel (which) will carry extended product demonstrations or commercial information. Ordering can be as easy as entering a product number, a method of payment (credit card information may be stored in the system) and a confirmation number for security... Through automatic downloading of viewing credits, the system will make impulse viewing and purchasing convenient." Other services considered include an encrypted digital transmission format, capacity for high definition television, delivery of films to cinemas, teletext distribution of written material, distribution of weather satellite images, weathermaps and weather radar scans. "To the extent permissible by the (Federal Communications) Commission, one channel could be allotted as a meeting channel... If the organisation (using the channel) wished, they could recoup their cost by charging the viewers a very small fee; addressability means that their members can be charged a different price than the general public."

TECHNICAL SUMMARY

PROPOSED PLATFORM	Eurosatellite Spacebus 300
PRIME CONTRACTOR	Either MBB or Aerospatialé
STABILISATION	3-axis
LAUNCH MASS	not available
MASS IN ORBIT	not available
POWER SUPPLY, END OF LIFE	not available
ORBITAL POSITIONS	61·5° West, 157° West
STATION KEEPING ACCURACY	not available
NUMBER OF CHANNELS PER SATELLITE	8 for ½-CONUS coverage 2 for spot beams
TWTA POWER OUTPUT	100 Watts (8 transponders) 20 Watts (2 transponders)
EIRP MINIMUM, BEAM EDGE	51 dBW

COMMERCIAL DETAILS

OPERATOR	Arab Satellite Communications Organisation (Arabsat)
ADDRESS	PO Box 1038, Riyadh, Saudi Arabia
TELEPHONE NO.	1 464 66 66
TELEX NO.	201300 arbsat sj
FINANCIAL BACKING	The Arab Satellite Communications Organisation is funded by the 22 members of the Arab League but of these, Saudi Arabia, contributing 20%, is the largest backer. The Arabsat organisation has a record of not being prompt in paying its bills and apparently has had difficulty in getting money out of some of the 22 members of the Arab League
PURPOSE OF PROJECT	Provision of regional telecommunications satellite capacity to members of the Arab League. Also provides quasi-DBS capacity
PROJECT STATUS	System became operational in 1985 although it still faces technical and financial problems
PROJECT VALUE	The three satellites were purchased from Aerospatialé for \$134 m. Arabsat says it paid \$23·5 m for its Ariane V12 launch and \$20·5 m for its STS 51G launch, including cost of PAM–D motor
LAUNCH DATES	Arabsat 1, 8th February 1985, Ariane V12 Arabsat 2, 18th June 1985, STS 51G + PAM D Arabsat 3, ground spare

SATELLITE USERS

Arabsat spacecraft are multi-mission satellites providing quasi-DBS capacity and telecommunications capacity. Primary users are expected to be PTT's using the spacecraft for international and domestic telephony. The System offers domestic leases. It is being used for distribution of TV channels between Arab States. Use of the system has been handicapped by problems in member states of the Arab Satellite Communications Organisation, and the fact that only twelve countries had established major earth stations by the end of 1985. Bahrain and Jordan were the first two countries to establish earth stations. The Arabsat Communications Organisation charges rates roughly equal to those charged by Intelsat which has led to suggestions that Intelsat users are unlikely to switch to Arabsat. Services offered by Arabsat include telephony and TV distribution, quasi-DBS operations, as well as specialised services such as video-conferencing, electronic mail and data collection and distribution. It is not clear at this stage to what use the DBS capacity is to be put.

ARABSAT MEMBER NATION	% OF MEMBERSHIP SHARES
Saudi Arabia	26·6
Bahrain	4·0
Egypt	10·4
United Arab Emirates	6·6
Iraq	2·6
Jordan	3·3
Kuwait	8·2
Lebanon	6·3
Libya	18·2
Morocco	0·5
Mauritania	0·2
Oman	0·6
Qatar	5·0
Syria	1·7
Somalia	0·3
Sudan	2·1
Tunisia	0·6
Yemen Arab Republic	0·7
Democratic Republic of Yemen	0·6
Palestine	0·2

PROJECT OVERVIEW

The Arabsat project has been described in one trade magazine as a complete mess, mostly because of Organisational rather than technical problems. The description appears to over-exaggerate the problems of organising a \$500 m exercise amongst 22 varied and less developed countries. The problems have been compounded by changes in the economic environment since the Arab Satellite Organisation was formed in 1976 and because of political differences between the 22 members of the Arab League who finance the outfit. The Western press also concludes that the project is in a mess. Its problems have been described as a "huge financial crisis" – rising costs have led to a budget deficit estimated in excess of \$50 m caused by the inability or unwillingness of some members of the Arab League to finance the equity needed by the organisation and over-runs on cost. There have been reports that Arabsat has, to put it tactfully, been tardy in paying Aerospatiale for the satellites, although Aerospatiale has denied them. The Director General of Arabsat, Ali al Mashat, was dismissed from his post in 1986 because of poor management performance.

Origins of the Arabsat project date back to the end of the 1960's and was first put forward as a concept to the conference of Arab Information Ministers in Bizerte, Tunisia. The Arab Satellite Communications Organisation was formed some six years later, in 1976, by the Arab League. The Arabsat Organisation consists of the General Assembly of Signatories comprised of the PTT or Communications Ministers of the 22 participating countries, the Board of Directors, a Council of nine members and an Executive Secretariat headed by the Director General. The stated objectives of the organisation are to "serve the economic and technical independence of the Arab region, augment the sense of national prestige, play an important role in the organisation of economic and social structures by strengthening the linguistic and cultural homogeneity, by facilitating decentralisation of industry, by slowing the migration of population to urban centres, by maximising the use of available centralised facilities and expertise, etc., and lead, finally, to the realisation of the Arab's aspirations in prosperity and unity."

In plain English, this means that the project is aimed at 1) providing prestige, 2) providing "welfare" by bringing radio and TV to all areas of the Arab world and providing educational programming, 3) increasing Arab unity, 4) providing independence from the Western dominated Intelsat system, 5) stimulate establishment of space communications industries in the Arab world, 6) providing improved telecommunications needed since Arab countries began benefiting from the oil revenue boom after 1973. The project was originally envisaged as a "welfare" exercise but the 1973 oil revenue boom led to growing awareness for the need to provide an improved telecommunications infrastructure to service expanded industrial bases and bigger economies. Thus the primary use of Arabsat was eventually set as telecommunication. The Arab Satellite Communications Organisation also assists member states in design of their earth stations and "conducts studies" in the field of space science and technology.

However, since the early 1980's the project has had a somewhat colourful record. In 1981 it selected Aerospatiale of France as prime contractor for the space segment. However over half the value of the contract was sub-contracted to Ford Aerospace and Communications Corporation (FACC) in the USA. The US government objected to exporting high technology to an organisation whose membership included the two bogy-men of the Arab world, Libya and Palestine (i.e. the PLO). It took some months, and a corresponding delay in the project, before export authority was permitted by the US government. Egypt, potentially the largest user of the satellite, was expelled from the Arab Telecommunications Union after the Camp David agreement in 1979. It decided to boycott Arabsat as a result of its expulsion from the Union. Iraq and Iran decided to go to war.

Oil revenues began to decline in the face of declining prices and falling output. Moreover from the formation of the Arab Satellite Communications Organisation in 1976 to the satellite system becoming operational in 1985 was a period of nine years in which national PTT's had to find alternative capacity. Consequently, the organisation lost a lot of potential revenue to Intelsat and alternative terrestial systems. By 1985 it became apparent that there was a deficit of around \$50 m on Arabsat's books and pleas were being sent out to member countries not to abandon the project. As the first satellite became operational it was also apparent that only two countries, Bahrain and Jordan had established local earth stations to use the system, so the organisation was unlikely to obtain an early and significant revenue stream. Confidence in the project was further shaken by technical problems with the first spacecraft which resulted in the loss of a significant proportion of its station keeping fuel.

The project also caused a reaction from Israel who decided they wanted a hybrid C/Ku-band spacecraft at 15° East on which the Saudi Minister of Communications, Dr Alawi Darwish Kayyal has commented "They have chosen an orbit and frequency for their satellite that would negatively affect Arabsat and which is against all international law." The Arabs have formally complained to the International Frequency Registration Board. The Arab Satellite Communications Organisation has also been concerned that "an unfriendly power" might try to jam the spacecraft command systems. Thus the TTC system has a security encoding system believed to be at least equal to a conventional US military system.

The Arabsat spacecraft are relatively technically advanced; the design is now being marketed by Aerospatiale under the Spacebus 200 banner. They employ a graphite epoxy structure and a unified propulsion system. The three Arabsat spacecraft are all identical. Arabsat 1A is the operational in-orbit spacecraft. 1B is the in-orbit spare and 1C is designated the ground spare. In order to maximise the operational capacity on each spacecraft, there are no spare C-band TWTA's to support the 25 operational transponders, although there is a spare TWTA for the quasi-DBS S-band service, which in total requires two operational TWTA's to provide one channel.

The 25 C-band transponders, each using an 8·5 Watt TWTA manufactured by Hughes Aircraft to (249–H designation) are provided for telephony, low and medium data transmission, multiplexed telex/telegraphy and radio and TV distribution. Bandwidth per transponder is 33 MHz, lower than the US norm of 36 MHz. EIRP at beam edge is 31 dBW (see Technical Summary). Lease rate per transponder per annum is around \$800,000 a year. There are no spot beams on the satellite. There is a single coverage beam for the whole of the Arab League region. Primary users of the C-band capacity are expected to be PTT's. C-band payload has a capacity of 8,000 telephone circuits. Each satellite has one channel for quasi-DBS services which uplinks at 6 GHz along with the 25 C-band transponders but downlinks at 2·5 GHz (S-band). The S-band channel uses two 50 Watts (similar to Hughes tubes carried on Insat–2). Each TWTA providing a total power output of 100 Watts. EIRP at beam edge of 41 dBW. Reception with 3 metre antennas is expected (or at least hoped) that there will be around 200 TVRO's using the system. However, so far only experimental use of the quasi-DBS channel has been made. The Arab Satellite Communication Organisation has said that it hopes that the antenna dishes of TVRO's can be produced locally in Saudi Arabia or other member countries.

Arabsat Satellite Ground Control Network, providing telemetry, tracking and command (TT&C) and monitoring (M) functions consists of 1) A primary TTC&M earth station with a minimum SSC capability, located in Saudi Arabia, 2) A secondary TT&C earth station with minimum SCC capability, located in Tunisia and 3) A primary Satellite Control Centre, (SCC) in Saudi Arabia. The limited SSC capability included in the primary earth station provides back-up for the primary SCC in the event of its failure. The secondary TT&C earth station provides back-up capability for spacecraft TT&C and/or SCC in the event of catastrophic failure of the primary station or the communications links. The secondary TT&C in Tunisia also falls within the footprint of Eutelsat's satellite multi service (sms) business capacity, it will therefore provide access from/to the Arabsat system for the European data network, Euronet and its database, DIANE.

Five types of ground station are available for the Arabsat system: 1) The largest, 11 metres, for use with the 6/4 GHz capacity, are intended for large metropolitan use. These will handle heavy trunk-route telephony traffic and TV channels on both PAL and SECAM colour standards, 2) A smaller, 8 metre, earth station, suitable for medium to low-density SCPC telephony traffic and TV receive only (but not TV transmit) are intended for urban areas with lesser requirements. 3) S-band quasi-DBS services utilise 3 metre TVRO terminals. 4) C-band television transmissions can be picked up with a 4·5 inch terminal, particularly intended for use in remote locations. 5) Small 1·6 metre transportable C-band terminals for emergency communications. The 4·5 metre terminals are also capable of providing services like remote printing of newspapers and one-way data flow. The variety of earth terminals indicate that the Arabsat spacecraft are multi-mission birds although it is not entirely "user friendly" in the way that, say, Tele–X is. Despite its colourful record to date, the Arab Satellite Communications Organisation has achieved an operational satellite system and this has been described by Hamdi Kandeel, Director of UNESCO's Free Information and Communications Policies Department as the only strategic Arab accomplishment in communications. During 1986, it is expected that serious work will begin on defining the requirements of the Arabsat II system.

TECHNICAL SUMMARY

PLATFORM	
PRIME CONTRACTOR	Aerospatiale
OTHER MAJOR CONTRACTORS	Ford Aerospace (70% of contract), Selenia, AEG Telefunken, MDAC, MBB, Bertin
STABILISATION	3-axis
LAUNCH MASS	1195 kg (Ariane), 1270 kg (STS)
MASS IN ORBIT	592 kg
POWER SUPPLY, END OF LIFE	1300 Watts
DESIGN LIFE	7 years
STATION KEEPING ACCURACY	$-1/+0{\cdot}12°$
ORBITAL LOCATIONS	Arabsat 1, 19° East Arabsat 2, 26° East

ECLIPSE CAPACITY	All 25 C-band channels fully operational in eclipse
NO. OF ACTIVE TRANSPONDERS	25 in C-band, 1 in C/S-bands
TWTA POWER OUTPUT	8·5 Watts (25 C-band) 50 Watts (3 C/S-band) TWTA's of which two operate in parallel and one is spare
TRANSPONDER BANDWIDTH	(All) 33 MHz
FREQUENCY RANGE:	
UPLINK (C-band)	5·925–6·435 GHz
DOWNLINK (C-band)	3·7–4·2 GHz
(S-band)	2·54–2·655 GHz
EIRP MINIMUM, BEAM EDGE	C-band, 31 dBW S-band, 41 dBW

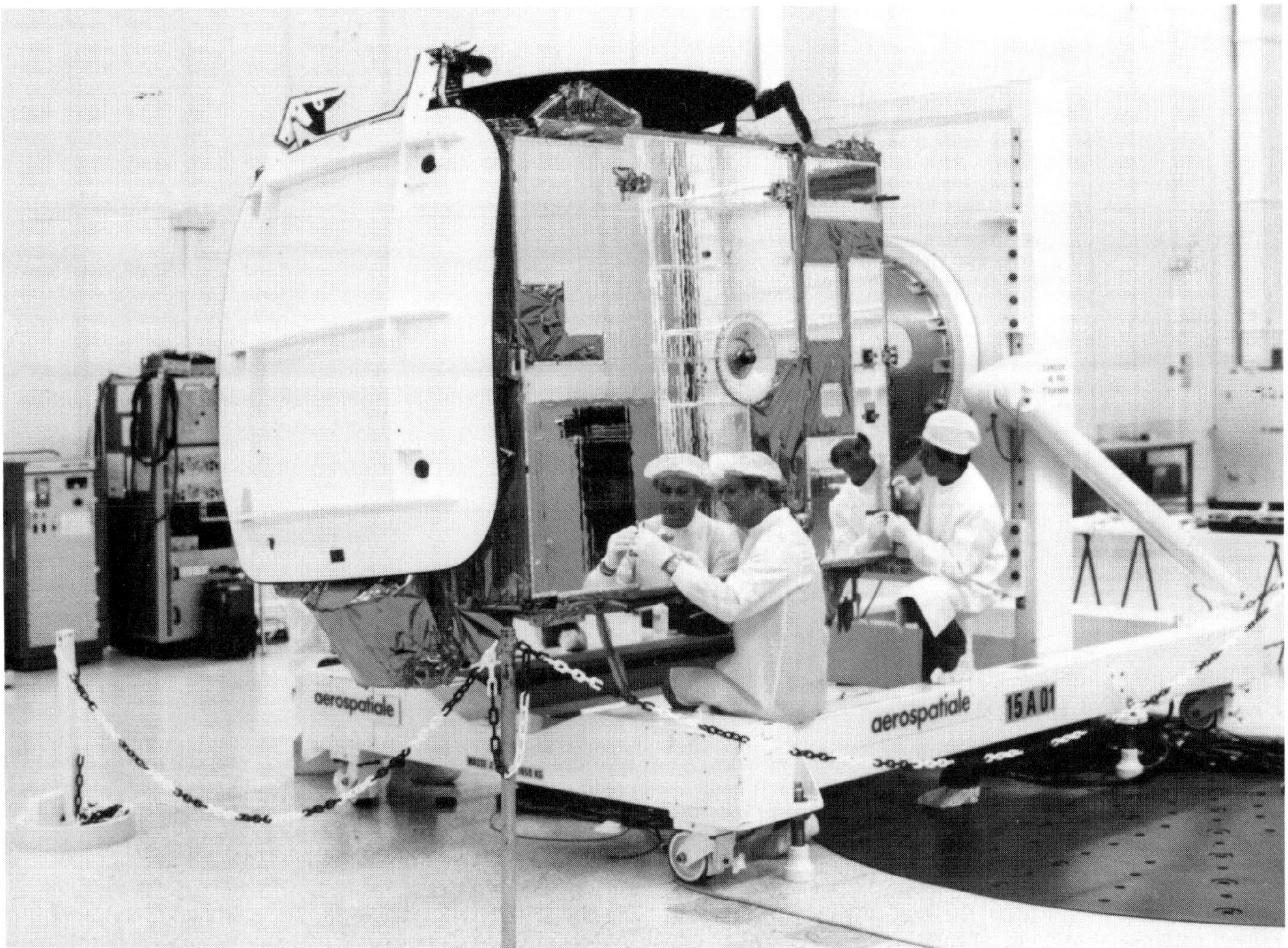

Second flight model of Arabsat shown here at Aerospatiale's facility at Cannes, France (photo courtesy of Aerospatiale)

COMMERCIAL DETAILS

OPERATOR	American Satellite Company (a wholly owned subsidiary of Continental Telecom)
ADDRESS	1801 Research Boulevard, Rockville Maryland, 20850-3186, USA
TELEPHONE NO.	(302) 251-8300
SENIOR STAFF	John N. Lemasters, President Stuart Meister, VP & General Counsel James Becker, VP, Finance & Treasury Arthur E. Parsons, Senior VP, Sales and Marketing George P. Roberts, Senior VP, Operations and Engineering Deborah Arrington, Director, Marketing and Communications Atol Tucker, Assistant VP, Customer Service Oho W. Hoernig, P&PR
PURPOSE OF PROJECT	Provision of C- and Ku-band capacity for US domestic operations
PROJECT STATUS	One of two satellites launched and fully operational. Second launch delayed by non-availability of launch vehicle
PROJECT VALUE	The two satellites were procured under a contract worth $120 m
LAUNCH DATES	ASC–1, 27th August 1985, STS 51–I plus PAM–D ASC–2, was due 27th January 1987, STS 71–C plus PAM–D

SATELLITE USERS

Prior to the launch of its own satellite system ASC was already a well established satellite communications company serving some 450 customers through capacity purchased on the Western Union Westar system. The ASC 1&2 satellites were procured to provide capacity for this expanding customer base. For its customers ASC provides three major types of communications services.

National Network Services is operated as a shared public network and is therefore accessed by several hundred major corporations. It provides private lines for voice, low-speed data, facsimile and video teleconferencing through 39 city pairs.

Commercial Systems uses dedicated earth stations located directly on the customers premises. It by-passes both local and trunk telecommunications networks and can be used to integrate voice, high-speed data, facsimile and video teleconferencing over the same network.

Government Systems provides dedicated communications for the US government and military. Earth stations are located directly on the customers facilities and services include voice, data, facsimile and video teleconferencing facilities. ASC's first major contract, in 1974, was in this sector.

PROJECT OVERVIEW

American Satellite Company has been operating since 1974 but its first satellite was not launched until 1985. It had originally been set up with the intention of operating its own satellite system. Its early history is perhaps illustrative of why so many start-up satellite operators "failed" in the 1980's in their attempts to become satellite operators. ASC is somewhere between being an entreprenurial start-up venture made good and a major corporation with the "internal" resources to venture into becoming a satellite operator. ASC is (now) wholly owned by a major US Telecoms operator, Continental Telecom, but was previously a 50:50 partnership with Fairchild Industries. Back in the mid-1970's when it was originally looking at purchasing its own satellite system, it found a lower risk route to profitability in leasing capacity on someone else's satellite (Western Union to be precise). This allowed it to build up its market and expertise before ordering its own satellite system in 1983.

By the time its first satellite was launched, ASC had "over 450 of the nation's largest businesses and government agencies" using its private and shared network services via the Westar satellite system for voice, data facsimile and video communications. This not only provided a rapidly growing cash income whilst the costly spacecraft were being built but also provided a basis on which it raised a $350 m syndicated line of credit to finance its growth through to 1992. It owns 20% of the Westar III, IV and V satellites. The new ASC satellites were purchased to meet increased demand and provide new services. Essentially, ASC grew by finding traffic in areas that other companies ignored or served poorly. By 1985, it had over 170 earth stations for use with the Westar system and the ASC–1 satellite. Most of these stations were two-way. ASC also has a track record for innovation in providing new products and services in satellite communications. It was the originator of the first satellite transmitted computer-to-computer service and the first domestic US satellite communications carrier to offer all digital transmission technology.

When ASC placed its initial order for two RCA satellites in 1983 it also had an option on a third which included an order for long-lead items. ASC decided not to take up the option and the third satellite was eventually constructed for PamAmSat. During 1983, it had been intended that the ASC–3 satellite be the ground spare in the system.

According to ASC, capacity on ASC–1 was "sold-out" before launch and, by the end of 1985, 50% of the capacity on

ASC–2 had been sold. Prior to the loss of the Shuttle orbiter Challenger on 29th January 1986, ASC–2 had been due to be launched on 27th January 1987. With a two year delay in the Shuttle programme and lack of capacity on other launch vehicles, it appears that ASC–2 will not fly before 1989.
ASC also has another problem with its first two satellites. The Federal Communications Commission was unhappy with ASC's design because, in the Ku-band, dual polarisation was omitted. The Commission preferred the dual polarisation to preserve the Ku-band frequency spectrum. In the event, it granted ASC authority to operate the Ku-band capacity on a temporary basis, and it can order ASC to stop using them.

ASC opted for the single polarisation to keep the size and weight of the satellite down. Dual polarisation would have required a larger satellite and an Apogee Kick motor and a larger share of the Shuttle payload bay. In arriving at the final satellite specification, ASC dropped several features included in its RFP. This included on-board signal processing, switching and cross-band coupling. Again this was partly the result of an engineering trade-off between mass, power, lifetime and payload capabilities. One novel feature introduced on the system was encryption of the satellites' command links. The first commercial satellite system to include this protection was the Arabsat system. The ASC system carries a considerable amount of US governmental traffic.
The payload capabilities of the ASC satellite design is summarised at the end of this section. The predominance of C-band capacity on the spacecraft reflects ASC's heavy past investment in its earth station system which it began developing before Ku-band capacity was commercially available in the USA.

There is a single antenna reflector on each satellite providing Ku-band coverage to the 48 CONUS states and a dual reflector providing C-band coverage to the 50 states and Puerto Rico. The C-band capacity uses a mixture of Solid State Power Amplifiers (SSPA's) each with 8·5 Watts RF output and Travelling Wave Tube Amplifiers each with 16·2 Watts RF output. Ku-band capacity uses only TWTA's. At the time the satellites were ordered, state-of-the-art SSPA technology meant that RF power was limited to about 10 Watts maximum. The SSPA's were designed and built by RCA Astro-Electronics who are now proposing 50 Watt SSPA's, although some satellite operators appear reluctant to use such unproven technology. The 8·5 Watt SSPA's were already space proven when ASC–1&2 were ordered, having already flown on the Satcom–V satellite launched in October 1982. The TWTA's on ASC–1&2 are manufactured by Hughes Electronics Dynamics.

National Network Services: ASC's National Network supplies private line communications to companies in major metropolitan areas in the USA. The service is a shared public network and provides businesses which have small to medium-sized traffic requirements with a variety of communication services ranging from voice and data to facsimile and video teleconferencing. ASC claims that some of the advantages of the National Network include:

- Cost savings of up to 50% over terrestrial services;
- Compatability with existing telephone systems so conversion to ASC circuits is transparent to the user;
- ASC provides complete technical and service support from installation through operation and maintenance;
- A lower bit error rate than leased telephone lines.

Once a private line becomes operational, signals travel from user locations over local telephone lines to an American Satellite Central Office (CO). The CO signals are then multiplexed into a single stream of information and routed to a nearby ASC earth station which uplinks the signal to the satellite. The satellite

In a scene which has temporarily disappeared into history, a satellite is launched from the cargo bay of the Shuttle orbiter, Discovery. The satellite is ASC–1 and the date is 27th August 1985. This was the first RCA built satellite to be launched via Shuttle. The Perigee motor which is clearly visible underneath the satellite is a McDonnell Douglas PAM–D motor. (photograph courtesy of RCA Astro-Electronics)

retransmits it to the receiving station where the process is reversed.

To increase the efficiency of its National Network offerings, ASC has upgraded the network to include digital as well as analog facilities. Through the use of Time Division Multiple Access (TDMA) technology, the improved digital system can now transmit information at high speeds up to 64 Mbps. It was the first such commercial system of its kind in the United States.

Commercial Systems: Commercial Systems are end-to-end private communication networks which use on-premises earth stations and completely by-pass terrestrial systems. Commercial Systems are designed for corporations with large volumes of information to transmit to regional sites, or those with a specialised communications requirement which can only be satisfied with a private system. The first private corporate network using on-premises earth stations was implemented by American Satellite in 1977 for Sperry Univac.

Since commercial systems employ digital transmission technology, voice, high-speed data, facsimile and video teleconferencing communications can be simultaneously integrated over each system. Commercial Systems provide several advantages over terrestrial networks. They offer greater accuracy. Because a single satellite transponder replaces hundreds of terrestrial long-distance facilities, switches, and single boosters, the digital service yields a lower transmission error rate than that of typical terrestrial communication networks. ASC's commercial systems can also be upgraded or expanded within a much faster time frame than conventional telephone lines.

To reduce space segment costs further, American Satellite has introduced advanced switching technology which enables corporations with very large networks to use their space segment more efficiently. This technology dynamically reallocates satellite capacity throughout the network in response to changes in network demand. When channels between any two sites within the network become idle, they can be reassigned to other sites which wish to communicate.

Unlike terrestrial networks, Commercial Systems are also capable of operating in the broadcast mode. In essence, once the satellite receives a signal, it retransmits it so every earth station within a two- to three-million square mile radius can receive the signal. This is advantageous for audio, video and facsimile communications.

A transportable earth station was introduced by ASC in 1981, as part of its Commercial Systems service. This service includes an integrated solid-state electronics package and a mobile 4·6 metre antenna. Once moved, the antenna can be ready for use within 48 hours after its arrival.

In the last five years, the demand for ASC's Commercial Systems product line has doubled. Because of the company's ability to custom-tailor services for its customers, American Satellite currently provides private networks for many of the nation's largest Fortune 1000 companies including Abbott Laboratories, Federal Express, E. F. Hutton, *USA Today* and Sears Communication Company.

Government Systems: ASC provides private line, 56 Kbps data, wideband data, and digitally secure voice channels to the military and key government agencies. Service is provided through ASC's government dedicated earth station network which permits direct communications between military bases and other government installations located throughout the nation. It is the largest network of its kind in the domestic USA.

The company's first government dedicated network was established in 1974 for the Department of Defense Naval Oceans Systems Command (Acoustic Research Net). ASC's government networks are now used to transmit such diverse information as global weather maps, Space Shuttle operational support data, and nuclear research data. The company provides high performance communications channels that operate at data ranges exceeding 3 million bits of information per second. This speed of transmission represents more than 100 single-spaced typewritten pages printed per second, with a bit error rate of less than one error in every 100 million bits of transmitted information.

Artists impression of the RCA Astro-Electronics built ASC–1 satellite as deployed in geostationary orbit. The hybrid design (providing both C- and Ku-band capacity) is based on the RCA Satcom design

Government agencies and military bases can access the ASC network anywhere in the Continental US, Midway Island, Puerto Rico, Hawaii and Guam by adding on-site earth stations and/or satellite capacity. Earth stations can be installed within a minimal time frame and can be placed directly on site. Users can also access the nationwide network by sharing earth stations at a nearby government facility.

To date, ASC furnishes communications service to the Defense Advanced Research Projects Agency (DARPA), Defense Communications Agency (DCA), Department of Energy (DOE), and the National Aeronautics and Space Administration (NASA). Some of the other government agencies communicating via ASC networks include strategical and technical missions of the armed forces, bulk-encrypted Automatic Voice Network (AUTOVON), and the World Wide Military Command and Control System (WWMCCS).

TECHNICAL SUMMARY

PLATFORM	Satcom 3000
PRIME CONTRACTOR	RCA Astro-Electronics
OTHER MAJOR CONTRACTORS	See notes below
STABILISATION	3-axis
LAUNCH MASS	2802 lb
MASS IN ORBIT	1483 lb
POWER SUPPLY, END OF LIFE	1215 Watts
DESIGN LIFE	10 years
ORBITAL POSITIONS	ASC–1, 128° W ASC–2, 81° W ASC–3 (allocated), 101° W
STATION KEEPING ACCURACY	+/−0·05° (E/W and N/S)
ECLIPSE PROTECTION	Full
PROPULSION SYSTEM	STS + PAM–D, Delta 3920 + PAM–D or Ariane
SPACECRAFT DIMENSIONS	4·3 × 5·3 × 10·5 ft
SPAN, TIP-TO-TIP	46·7 ft
SATELLITE CONTROL CENTRE	Atlanta, Georgia
NO. OF ACTIVE TRANSPONDERS	J8 at C-band, 6 at Ku-band
NO. OF SPARE TRANSPONDERS	4 at C-band, 1 at Ku-band
TWTA POWER OUTPUT	16·2 W + 8·5 W SSPA's
FREQUENCY RANGE:	
UPLINK	14 GHz and 6 GHz
DOWNLINK	12 GHz and 4 GHz
TRANSPONDER BANDWIDTH	36 MHz and 72 MHz
EIRP MINIMUM, BEAM EDGE	34 dBW – 42 dBW
CHANNEL POLARISATION	Horizontal and vertical frequency reuse except in Ku-band which is horizontal only on the downlink

TRANSPONDER TRANSMISSION CHARACTERISTICS

	Transponder Bandwidth	*Nominal CONUS EIRP*	*Amplifier Power*	*Redundancy*	*Coverage*
6 C-band (H)	36 MHz	34 dBW	8·5 W SSPA	7 for 6	50 States & Puerto Rico
6 C-band (V)	36 MHz	34 dBW	8·5 W SSPA	7 for 6	49 States & Puerto Rico
3 C-band (H)	72 MHz	37 dBW	16·2 W TWTA	4 for 3	50 States & Puerto Rico
3 C-band (V)	72 MHz	37 dBW	16·2 W TWTA	4 for 3	49 States & Puerto Rico
6 Ku-band (H)	72 MHz	42 dBW	16·2 W TWTA	7 for 6	48 States

MAJOR SUBCONTRACTORS
to RCA Astro-Electronics for the ASC–1&2 satellites

TWTA s Division	Hughes Electronics Dynamics PO Box 2999 Torrance, CA 90509
Multiplexers and Switches	Com Dev Ltd 155 Sheldon Drive, Cambridge, Ontario, Canada NIR 7H6
Tower (C-band)	Heath Tecna 19819 – 84th Avenue South, Kent, Washington 98031
Tower (Ku-band)	Exxon Corporation US 276 and Old Laurens Road, Fountain Inn, SC 29644
Reflectors	Composite Optics 9623 Distribution Avenue, San Diego, CA 92121
Attitude Control	
Sun Sensor	Adcole Corporation 330 Bear Hill Road, Waltham, MA 02154
Earth & Horizon Sensors	Lockheed Space Systems Division, PO Box 504, Sunnyvale, CA 94088
Rate Measurement Assembly	Northrop Corporation 100 Morse Street, Norwood, MA 02062
Electric Power	
Battery Cells	Eagle Picher PO Box 47, Joplin, MO 64801
Solar Cells	ASEC 15251 East Donjulian Road, City of Industry, CA 91746
Solar Array Substrate	Heath Tecna 19819 – 84th Avenue South, Kent, Washington 98031
Propulsion	
Thrusters	Rocket Research Yorke Center, Redmond, Washington 98052
Propellant Tanks	PSI 2017 Camfield Avenue, Los Angeles, CA 90004
AKM	Morton–Thiokol PO Box 241, Elkton, MD 21921
Structure	Parsons of California 3437 Airport Way, PO Box 6189, Stockton, CA 95206

COMMERCIAL DETAILS

OPERATOR	American Satellite Company
ADDRESS	1801 Research Boulevard, Rockville, Maryland, 20850-3186 USA
TELEPHONE NO.	(301) 251-8300
SENIOR STAFF	John N. Lemasters, President Stuart Meister, VP & General Council James Becker, VP, Finance & Treasury Arthur E. Parsons, Senior VP, Sales George P. Roberts, Senior VP, Operations Deborah Arrington, Director, Marketing Atol Tucker, Assistant VP, Customer Service
PURPOSE OF PROJECT	Provision of C, Ku and Ka-band capacity for ASC's domestic US satellite services
PROJECT STATUS	FCC construct, launch and operate, authority given for ASC–3. Order has yet to be placed
PROJECT VALUE	not available
LAUNCH DATES	ASC–3, September 1989 (if launch services available) ASC–4, ground spare

SATELLITE USERS

See section on ASC 1&2. The ASC 3&4 satellites are intended to meet growing capacity requirements of ASC's satellite services and provide back-up ground spare capacity. ASC–4 is intended as a ground spare satellite.

PROJECT OVERVIEW

These two satellites are ASC's next generation of spacecraft. The company has yet to place an order for them but received FCC construct, launch and operate authority for ASC–3 in 1985. ASC–4 is a ground spare. ASC has considered including Ka-band capacity on ASC 3&4, but their primary purpose is the provision of C- and Ku-band capacity. In the C-band, the ASC 3&4 satellite would carry 24 C-band transponders of 9 Watt power output each and a bandwidth of 36 MHz. Up to 12 of these transponders could be paired to create six 72 MHz units at 17 Watts power output. In the Ku-band there are 19 54 MHz 21 Watt transponders. Eight of these would be available as East and West spot beams. The Ka-band capacity considered by ASC is also aimed at commercial users. It would include three spot beams, each with 500 MHz bandwidth and using 8 Watt TWTA's. A beacon transmitter would cover CONUS with 100 MHz bandwidth.

The original ASC–3 satellite was an option plus an order for long lead items placed with RCA Astro-Electronics as part of the order for ASC 1&2. That ASC–3 was designated a ground spare but ASC did not take the option up fully. The satellite was completed by RCA Astro-Electronics for PanAmSat.

TECHNICAL SUMMARY

PLATFORM	not decided
PRIME CONTRACTOR	not decided
OTHER MAJOR CONTRACTORS	not decided
STABILISATION	not decided
LAUNCH MASS	not decided
MASS IN ORBIT	not decided
POWER SUPPLY, END OF LIFE	not decided
DESIGN LIFE	Minimum 8·5 years
ORBITAL POSITIONS	ASC–3, 64° West ASC–4, ground spare
STATION KEEPING ACCURACY	not decided
ECLIPSE PROTECTION	Full protection required
PROPULSION SYSTEM	not decided

SPACECRAFT DIMENSIONS	not decided
SPAN, TIP-TO-TIP	not decided
NO. OF ACTIVE TRANSPONDERS	24 at 9 Watt, 19 at 21 Watt
NO. OF SPARE TRANSPONDERS	not decided
TWTA POWER OUTPUT	9 Watts and 21 Watts
FREQUENCY RANGE:	
UPLINK	5·925–6·425, 14·0–14·5, 30 GHz
DOWNLINK	3·7–4·2, 11·7–12·2, 20 GHz
TRANSPONDER BANDWIDTH	36 MHz (9 W TWT), 54 MHz (other)
EIRP MINIMUM, BEAM EDGE	not available
CHANNEL POLARISATION	as per ASC 1–2

COMMERCIAL DETAILS

OPERATOR	Société Européenne des Satellites (SES)
ADDRESS	63, avenue de la Liberte, Luxembourg City, Boite postale 1781-L1017, Luxembourg
TELEPHONE NO.	(352) 499471-1
TELEX NO.	60229 sesat lu
FAX NO.	(352) 471219
SENIOR STAFF	Jean Dupong, Chairman, Board of Directors Viscount Etienne Davignon, Executive Committee Pierre Meyrat, General Manager Marcus Bicknell, Commercial Director Carlo Rock, Press & PR Aldis Grinbergs, Engineering and Operations Director Donald Angliss, Head of Satellite Engineering Celso Azevedo, Head of Communications Engineering Chris Schram, Operations & Maintenance Manager Daniel Showalter, Satellite Control Engineer Alex Crofts, Station Maintenance Head
PURPOSE OF PROJECT	Provision of medium power Ku-band capacity for European broadcasting needs, including cable TV, networking and quasi-DBS services
PROJECT STATUS	First satellite financed and due for launch in 1988. Operator trying to come to an agreement with Eutelsat
PROJECT VALUE	Full two satellite system including launch and insurance is expected to cost \$190 m. Astra–1 cost \$45 m plus launch (\$37 m), some insurance, TT&C and working capital. SES's funding is now \$96 m
LAUNCH DATES	Astra–1, February 1988, Ariane 3, V27 Astra–2, 1990, Ariane or others

SATELLITE USERS

At the time of writing, the only customer that has been announced for the Astra–1 satellite is SCANSAT (Sweden). It intends to use three transponders for quasi-DBS services. Three transponders are reserved for CLT but the company has yet to confirm that it wants them.

There have been numerous reports of other organisations planning to take capacity on the system. The most persistent is that British Telecom will take up to six or more for its customers currently using Intelsat or Eutelsat or potential customers looking for satellite capacity. Robert Maxwell is understood to have put in a bid for three transponders and Thames TV is reported to be looking at the possibility of taking one transponder.

Some observers believe that British Telecom will pull out of Eutelsat altogether and use SES instead. At the time of writing SES says that it is negotiating with 4–5 PTT's over their use of the system. It appears unlikely that British Telecom will pull out of Eutelsat in favour of SES. Moreover, SES has yet to announce any firm arrangements for uplink facilities with BT or other European PTTs. However, it is expected that SES will announce firm customers for the ASTRA satellite during 1987.

In a submission to Eutelsat in February 1986, the Luxembourg government claimed that the market for Astra was about 56 new broadcasting channels looking for capacity in the near future. As at November 1986, it is understood that there were letters of intent to take 23 transponders on ASTRA although it is not known if this includes CLT. Whether these letters of intent include the condition that SES orders a second satellite remains an unknown factor.

SES and the Luxembourg government have been trying to negotiate a lease/lease back arrangement with Eutelsat. This appears to have been abandoned but a number of existing satellite TV broadcasters have expressed distinct reservations about committing to ASTRA unless SES comes to an arrangement. They fear that unless SES does so, European PTTs will either block downlinks or stop distribution of their channels on PTT owned cable networks.

PROJECT OVERVIEW

Introduction: The Astra project aims at providing medium power Ku-band European satellite communications capacity for the broadcasting industry and is in competition with the PTT owned Eutelsat organisation. It is a product of Luxembourg's fairly unique position in European's changing transborder broadcasting market and the government's heavy reliance on income from that source. It has little to lose from breaking its European arrangements with other Eutelsat members: a lot to lose from failing to keep up with new media technologies and a lot to gain from creating for itself an expanded market niche. Société Européene des Satellites and the government of Luxembourg have taken some time to get the Astra act together but, at the present time, the organisation appears to be in a strong position to challenge both the changing telecommunications environment in Europe and the entrenched PTT monopoly over satellite communications. To assess the SES project requires, at the time of writing, a degree of informed speculation.

As at September 1986, the company is in the process of attempting to reach an "accommodation" with a reluctant Eutelsat. It also appears to have the support of at least one major European PTT (British Telecom). Although the company has only one confirmed customer, it has the finance needed to complete and launch its first satellite. Hitherto, a number of observers had commented that the company was a speculative venture. Nevertheless, it has raised some \$96 million, presumably with the shareholders knowledge that it is likely to suceed in its ambitions. It appears that the company and its users are not releasing details of their agreements and that the company does have a secure customer base.

Background to the Astra Project: Luxembourg established itself as a major force in the European transborder broadcasting market in the 1930's. By that time most European countries had established their broadcasting systems along the lines of the national public service broadcasting model, with limited opportunity for commercial radio. Nevertheless, a number of countries were concerned that commercial radio would be broadcast in from other countries. France took a logical step and allowed French interests to own the transborder broadcasting coming into the country. Those interests opened shop in Luxembourg under the name Radio Luxembourg. After World War II the company moved into television and is now known as

CLT. Luxembourg has no significant domestic market of its own to serve, having a population of a quarter of a million. Broadcasting has become its largest industry and the State's principal source of revenue.

In 1977, the World Administrative Radio Conference allocated five DBS channels to each European country, including Luxembourg. The latter country had no need of these for its domestic market but saw that its traditional outlets in other countries could be dominated by the new media, threatening the viability of CLT. It was therefore one of the first European countries to develop a plan for using its DBS capacity. The LUXSAT project covered a high power DBS system for transborder TV in which CLT would have played a major role. It was competitive with both the French TDF–1 and German TV–SAT systems. Eventually France offered CLT capacity on the TDF–1 system at the cost of abandoning LUXSAT. Luxembourg did abandon the project in 1983, although by that time it was clear that it was not a viable proposition under any circumstances. At the same time some US thinking favoured the use of medium power FSS satellites for broadcasting, arguing that they could be used for both quasi-DBS services and other broadcasting markets such as cable TV. This was brought to the attention of the then Luxembourg Prime Minister, Pierre Werner.

The country then began to take medium power seriously and in 1983 announced that it was looking at a concept known as "GDL". It took in US expertise to set the project up. This mostly came in the form of Clay Whitehead, who had had considerable success in setting up the Hughes Communications Galaxy system. After some procrastination, the Luxembourg government set up a fully Luxembourg controlled and owned company, Société Luxembougeoise de Satellites. This was awarded a franchise for the space segment in May 1984 and appointed Coronet to study, develop and operate the system. Coronet was headed up by Clay Whitehead and was backed by European private investors and cable TV channel company HBO, from the USA (this was before HBO found out that it could not keep on expanding indefinitely). The project came under immediate criticism from PTT's, European governments and Eutelsat, being variously described as a Coca-Cola satellite (by Louis Mexandeau in France), anti-European (by Andrea Caruso at Eutelsat) and so forth. It was also not particularly popular with CLT as is explained later. Coronet eventually ordered a satellite but fell out of favour with the Luxembourg government.

By the start of 1985, it was clear that Coronet was failing in its dealings with the Luxembourg government who were in the process of sponsoring a new company to take over the project. In late December 1984, the goverment had submitted proposals to Coronet for the development of GDL. Whitehead rejected them. Whitehead eventually failed in the Coronet exercise, partly, it appears, because it was too American and partly because the Luxembourg government disagreed with some of the financial aspects of Coronet's practices. Others have commented that the Luxembourg government's handling of the Coronet affair was weak and that it made major errors in re-establishing GDL under the control of the fully European backed Société Européene des Satellites. Coronet had aimed to get its satellite launched in 1986 at a time when the lease contracts on the Eutelsat–I F1 satellite were up for renewal. The problems over Coronet and the delay in setting up SES lost a year on the project and the failure of Ariane has added another year so that the first Luxembourg satellite will not go up until at least February 1988, by which time Eutelsat will have another satellite in orbit and will be close to having its Eutelsat–II system operational. The latter is very much based on the Astra concept. There have also been criticisms of the early management of SES although the company has gradually got this problem sorted out.

SES was formally created in the Spring of 1985 and, like Coronet, opted for a US manufactured satellite. In fact it purchased the partly completed satellite ordered by Coronet, which is virtually the same as the RCA Americom SATCOM K birds.

Chronology of the GDL Project

August 1983	Luxembourg makes public its plans for a medium power satellite system called GDL, having abandoned the LUXSAT DBS plan.
26th September 1983	The Luxembourg PTT submits its plan to the IFRB of the ITU.
8th November 1983	Establishment of the initial commercial entity to develop and operate the GDL system under the management of Clay Whitehead.
February/March 1984	Attempts to set up a US backed company fail but Clay Whitehead retained to manage the project.
25th May 1984	SLS formed and granted a licence to Clay Whitehead and Coronet which is planning only minority US shareholding in the venture.
15th June 1984	Luxembourg PTT administration begins coordination process with the IFRB.
23rd July 1984	New Luxembourg government under Jaques Santer gives its support to GDL.
26th October 1984	Joint declaration of the French and Luxembourg government allowing CLT use of the TDF–1 DBS satellite.
21st December 1984	After slow progress in finding investors in Coronet, the Luxembourg government "makes new proposals for a new and solid start to the GDL project". These reject the American type structure and reduce the role of Whitehead. He rejects them.
5th February 1985	The government reaffirms its support for the GDL project and the ITU publishes request for coordination.
1st March 1985	Société Européenne des Satellites formed with only European shareholders.
15th May 1985	Coronet finally closes down shop and its management team is disbanded.
3rd September 1985	SES board decide to go ahead with development of the GDL system. The Luxembourg government is concerned about the delays in organising users for the TDF–1 satellite.

CLT, DBS and Astra: Despite the importance of Luxembourg's efforts to gain a stake in the new transborder TV technology, CLT has been remarkably cool towards Coronet and Astra. This is partly because the company has to court the favours of the countries into which it broadcasts, notably France and Germany, both of which have rival interests to Astra. Most notable of these is the Franco-German TDF–1 and TV–SAT DBS system. CLT has plans to use both the satellites. In the French case, it is also trying to get a major stake in the national terrestrial channel, La Cinq. It appears that its willingness to use TDF–1 is a political move aimed to gain approval for its proposals for La Cinq. The TDF–1 satellite no longer has a French domestic market as the terrestrial frequencies have been opened up for further channels.

CLT was ruled out from the new terrestrial market by the last socialist government in France. Two reasons have been given. One was that its news coverage was too critical of the socialists. The second was that control of the organisation has shifted away from French interests towards a broader North European, and particularly Belgian, base.

Preliminary footprint of GDL-6 Satellite
Orbital position: 19° E

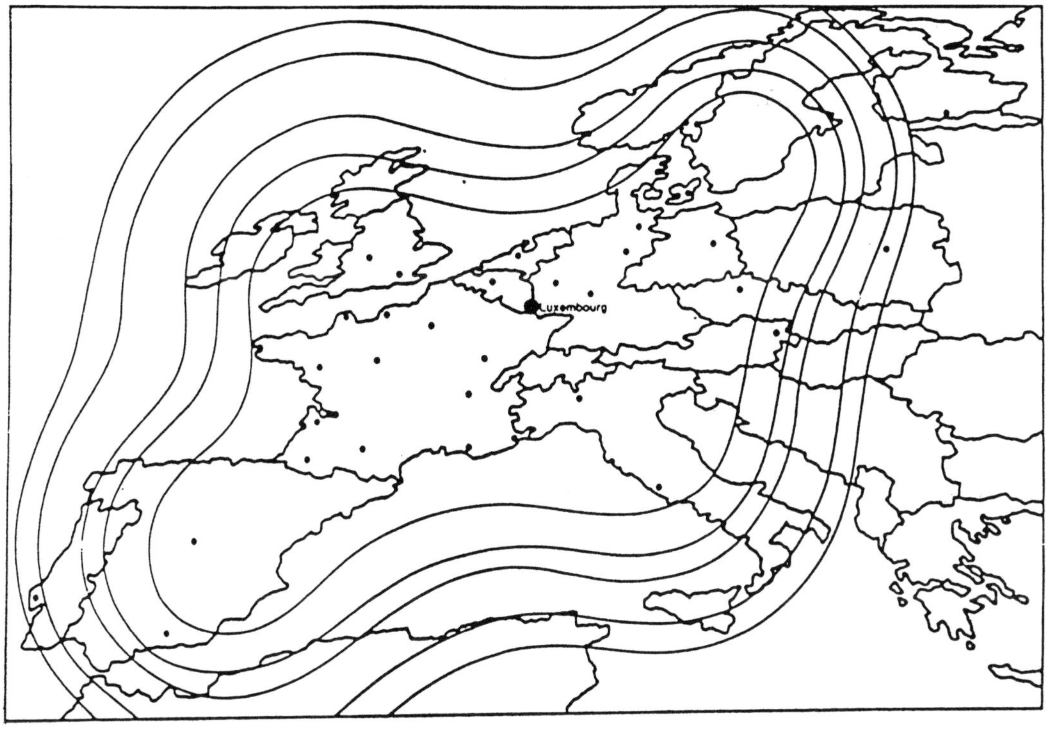

CLT has three transponders reserved for it on Astra. It has yet to announce its intention to take them up. Nevertheless, it appears to be a logical option to do so. The TDF–1 satellite has yet to be commercialised and there is some doubt about its future. The TV–SAT system is in a similar position, with a final order for the second satellite in the system. At the same time, CLT is expanding its activities into new areas, such as Belgian commercial TV and, possibly, La Cinq, and may need both networking and distribution capacity.

SES's relationship with Eutelsat and European PTT's: SES and the Luxembourg government argue that the Astra system will not cause Eutelsat economic harm for the following reasons:

1) Astra is a broadcasting satellite system and Eutelsat is a telecommunications satellite system. Therefore, it is a matter of comparing two unlikes.

2) Astra will carry traffic which otherwise would not be carried on the Eutelsat system. As a dedicated broadcasting system designed for its particular market, Astra will be creating new demand. It would also create a large enough market for Eutelsat to expand its activities over and above the level that would exist if Astra did not go ahead.

3) Astra is not aimed at the markets for which Eutelsat is intended to serve. This is point-to-point telephony, business communications and the EBU. The Luxembourg government has restricted SES to serving the broadcasting industry.

The points are largely rhetorical. The Eutelsat executive remains hostile to the project, which threatens both its standing with the European PTT's and its commercial viability. It is in direct competition with the Eutelsat–II system due to come on stream in 1989–90. As a consequence, it has appeared unlikely that Astra would pass the tests of economic harm to Eutelsat. This in turn would undermine the Astra project. Many PTT's would be unlikely either to allow uplinks or downlinks on the system. It means that SES would be unable to sign up customers for the system and would have to finance the second Astra satellite as a speculative venture.

The result is that Luxembourg has attempted to come to an accommodation with Eutelsat. The current proposals are that it leases its capacity to Eutelsat who, in turn, lease it back to SES for forward lease on to PTT's or broadcasters. One major problem in this is the issue of the second Astra satellite. As it currently stands, only one satellite is on order and SES refuse to discuss if and when it plans to order a second. It has an option with RCA which expired at the end of 1986. Eutelsat appear to be arguing that if Astra–2 is ordered it will add too much capacity to the European market and also place Eutelsat in a position where it would be underwriting part of the financial risks of SES.

Nevertheless, the members of Eutelsat are taking a different position to its executive. The latter has wanted to abandon the negotiations over the "arrangement" and, incidentally, abandon the coordination over economic harm. The PTT members have rejected the Executive's proposal and told it to find a solution with Luxembourg.

Eutelsat asked its members, in the Summer of 1986, to submit requests for modifications to the Eutelsat–II system. Requests came from the UK, Portugal, Spain, France, Italy and Germany and also a request from Denmark to provide communications with Greenland. Other Scandinavian countries, surprisingly, did not submit any requests. Those that did, apparently, are those most interested in leasing capacity on the Eutelsat–II system. The UK and Holland objected to the final form of the modifications. The modifications appear to be in direct response to the competition from Astra. The new Eutelsat–II modifications are expected to add some two to three months to the construction time of the spacecraft, but as Arianespace cannot launch the first until July 1989 (at least according to its latest launch manifest) this will have no impact on the quoted launch date. The three satellites in the new series are to be located at 36°, 3° and 12° East. The possibility of locating one satellite at 19° East has also been evaluated.

Strained relationship between Eutelsat and Luxembourg over the Astra system led, in August 1986, to what is being described as a diplomatic incident. In reply to questions from a Deputy in the Luxembourg lower house (Chamber of Deputies), Prime Minister Jacques Santer told the House that everything was going well with the Astra project and that negotiations with Eutelsat were positive and not only progressing but doing so well. Santer added that a rapid solution to the negotiations was expected. Eutelsat apparently took offence at this on the grounds that they had found the proposals from SES unacceptable and, at the time, the negotiations had come to a complete standstill. Eutelsat spoke to Luxembourg journalist Mario Hirsch, who had previously been involved in setting up Coronet and has subsequently been a strong critic of the Luxembourg governments handling of the satellite project and revealed the position, thus, in effect, undermining the integrity of PM Jacques Santer.

Santer had replied to the Deputy because the Deputy had questioned as to why Astra was not on a copy of a launch manifest he had received and as to whether SES was in financial difficulties. It was also tactless of Eutelsat to have written to Hirsch who is highly unpopular with the Luxembourg government over his criticism of the SES exercise.

SES's Finances: SES, in the meanwhile, surprised some industrial observers by announcing at the end of August 1986, that it had completed the doubling of its share capital and now had all of the funds for the first satellite in the system, including launch, insurance and working capital needs. The surprise came because the company has not announced any customers (apart from Scansat), has not announced any agreements for uplinks and downlinks outside of Luxembourg and has still to come to an arrangement with Eutelsat. Thus, on a prima facie basis, the shareholders are still investing funds on a speculative basis. But as there is a degree of secrecy about the company's current activities, this has led to speculation that it had finalised unannounced agreements with one or more European PTT's directly. Some of Eutelsat's members have been hostile towards the executive over the last year and this hostility may reflect a back door deal with SES.

SES's share capital now stands at LUX Frcs2025 million ($48 million). The equity capital is also matched by debt finance of the same value giving the company a total funding of LUX Frcs4000 million ($96 million). The Luxembourg government has provided guarantees on debt finance to the value of equity invested in the company. This guarantee covers reimbursement of the debt if the project fails. It only lasts up until the end of 1988. When SES was set up in March 1985, the Luxembourg government leaned heavily on locally based subsidiaries of foreign companies to invest in the company. The supporting loan guarantees are now a major political threat to the government if the project fails.
SES will need further funding to procure and launch the second Astra satellite but the company has not revealed the option price for the satellite. The option price expired at the end of 1986 unless the option was taken up by that date. SES's debt finance is guaranteed up until the end of 1988 by the Luxembourg government.

As expected, the August 1986 increase in capital has involved changes in the percentage shareholdings in the company. A twelfth shareholder has entered the scene. The Belgian financial group Société Nationale d'Investtissements (SNI) has agreed to invest up to LUX Frcs50 million ($1·2 million) in SES and will take a 2·5% shareholding in return. It is understood that the Kirkbe organisation has not increased its equity investment in this round. The inclusion of SNI puts SES in a stronger position to obtain Belgian PTT authorisation for distribution of channels transmitted through Astra onto Belgian cable networks. SNI is a publicly owned investment organisation with close links with the similarly named and similarly functioning SNI company in Luxembourg. However, the Belgian SNI still has to obtain approval of its planned investment in SES shares.
Prior to the August 1986 increase in capital, there were 11 shareholders in the company, holding 90% of the equity and 99% of the voting power. These were as follows:

Artists impression of SES's Ku-band spacecraft which uses the large RCA 4000 platform (courtesty RCA Astro-Electronics)

Banque Generale du Luxembourg SA
Banque Internationale a Luxembourg SA
Compagnie Luxembourgeoise de la Dresdner Bank AG
Deutsche Bank Compagnie Financiere SA
Kinnevik International AB
Kirkbe AS
Natinvest SAH
Realisations et Investissements en Technologies Avancees
Societe Generale de Belgique
Caisse d'Epargne de l'Etat
Societe National de Credit et d'Investissment
Thames Television

SES's LEGAL POSITION: In a submission to all the Eutelsat members prior to the November 1986 board meeting of the organisation, the Luxembourg government outlined its opinion on the legal position of its ASTRA system vis a vis Eutelsat. The position is essentially the same as that taken in February 1986 when the Luxembourg government submitted documentation to Eutelsat claiming that economic coordination was not necessary.

Under the title "Does Luxembourg Need to Co-ordinate with Eutelsat?" and with a covering letter from Jacques Santer (Prime Minister of the country), opinions from 8 legal experts (mostly in the academic sphere) were included. The covering letter from Santer stated "Let me make it quite clear that my government is under no circumstances ready to compromise on its sovereign rights and is not to renounce the project at its final stage."

The Luxembourg government concluded that the eight legal opinions "all ... reach the same conclusion. This is that SES will be a broadcast satellite service, which in the view of those giving opinions falls squarely within the definition in the Treaty (Eutelsat Convention) of 'specialised telecommunications services' and not '(international) public telecommunications services'. Accordingly the issue of 'significant economic harm' which applies only to 'international public telecommunications services' is irrelevant to both the Grand-Duchy of Luxembourg and SES."

According to the Luxembourg Government, the Eutelsat Convention provides a mechanism for assessing whether utilisation of space segment equipment is technically compatible with the Eutelsat space segment. The Convention also provides, but only in relation to "international public telecommunications services", that the various parties and signatories to the Convention shall make an assessment as to whether utilisation of space segment equipment by persons other than Intelsat will cause Eutelsat "significant economic harm". This point has "given rise to lengthy and difficult discussions between Eutelsat and Luxembourg".

"Eutelsat contends that the service to be operated by SES is an 'international public telecommunications service' and should therefore be examined in relation to it causing significant economic harm. The effect of this would be to delay or hinder the effective use of SES's satellite," says the Luxembourg Government. This presumably is circumstantial evidence to suggest that Eutelsat is playing for time. SES contends that the service it will offer is not an "international telecommunications service", and that therefore a consideration of "significant economic harm" is not only time wasting but unnecessary and inappropriate under the Eutelsat Convention.

MARKETING THE ASTRA SYSTEM A notable difference between Eutelsat and Societe Europeenne des Satellites is in the approach taken to marketing satellite TV. As a carriers' carrier, Eutelsat is largely passive in this exercise, leaving it to its PTT members. Under its Commercial Director, Marcus Bicknell, SES is, at the time of writing, developing a plan to support both the channel providers and the TVRO industry involved directly or indirectly with ASTRA.

SES is now placing considerable emphasis on the quasi-DBS capability of its satellite system. The Company is saying that an 85 cm dish is all that is necessary for direct reception in most of Europe. The Company will work closely with the TVRO industry "to ensure that attractively priced reception equipment is easily available to a potential buyer asking how to get ASTRA". From early 1987, SES will introduce a sticker or label which retail TVRO outlets will be encouraged to attach to suitable equipment. The label will read "Compatible With ASTRA". SES will also provide further marketing support to TVRO manufacturers. Seminars for manufacturers and trade press advertisements will be used to introduce the SES marketing campaign during 1987.

The Company will also develop a promotional campaign for ASTRA services "to ensure that television viewers in Europe are aware of the name ASTRA and its attractive proposition". Although individual channel providers will be responsible for the promotion of their own channel, SES will be providing a "supplementary" and "coordinating" campaign. The Company will commit funds to those parts of Europe where the most prominent of the ASTRA channel operators "are in favour" of the ASTRA campaign.

SES is also to offer optional support services for TV channel operators, either prior to launch of a channel on ASTRA or afterwards in analysing results. Perhaps necessary is advice on regulatory affairs. Also on offer is technical and market research/analysis support.

COST EFFECTIVENESS OF TYPICAL EUROPEAN SATELLITES

Power type	*Low power*	*Medium power*	*High power*
% European homes within 50 dBW contour	0%	70%	85%
Accessible homes 1995	22 million	60 million	99 million
Transponder lease cost per year	\$1·5–4 m	\$4·5–5·5 m	\$7·5–30 m
Average cost per year	\$2·75 m	\$5 m	\$18·75 m
Cost in 1995 per million homes per year	\$125,000 m	\$83,000 m	\$189,000 m

(Table courtesy of Societe Europeenne des Satellites, December 1986)

The figure above 22 million accessible homes for low power satellites refers to cable TV homes. The corresponding figures for medium and high power relate to direct to home satellite TV reception. Estimates of Direct to Home reception are based on the number of homes with sufficient consumer purchasing power to buy a video cassette recorder ... from this it is deduced that the purchase of satellite receiving equipment is possible.

Baseline transponder lease charge per annum is \$5 million, compared with \$3.2 million for Eutelsat II non-preemptible rates. SES suggests that its higher charge reflects a superior level of support for the user. Moreover, its rates in the early years will be lower than the baseline figure for those users taking long term leases. This, SES suggests, will allow new channels to cut down on their cash flow requirements until such time as their services achieve profitability. The Company is also offering outright sale of transponders at a baseline price of \$30 million. It is believed that these baseline prices are based on the assumption that a second satellite is in orbit and that non-preemptibility terms are included.

The Second Astra Satellite: Potential users of the Astra system are, at the time of writing, concerned that only one satellite has been ordered by Société Européenne des Satellites, leaving any user of the Astra 1 satellite with the problem of finding alternative capacity in the event of its partial or total failure. SES's option to purchase a second satellite from RCA at a fixed price expired at the end of 1986 without such an order being placed. The company refused to discuss the issue of the second satellite with the press during 1986. However, it is clear that the ordering of this satellite could have created major problems with the negotiations with Eutelsat for lease/lease back of Astra capacity. There are also indications that SES discussed with the French PTT and Communications Ministry the possibility of purchasing an Aerospatiale Spacebus 100B satellite (the same platform as used for Eutelsat II). Under the arrangement, SES would have provided the Ku-band capacity considered for Telecom 2, leaving the latter providing C-band and military capacity. It is not known how far this proposal has been accepted by the French. As one trade source put it, "there is a big difference between (SES) flirting with and copulating with European PTTs".

POTENTIAL MARKETS FOR THE ASTRA SATELLITE

Country	*Cable homes*	*SMATV homes*	*DBS homes*	*Total TV homes*	*Within ASTRA 48 dBW*
	*(1)	*(2)	*(3)	*(4)	*(5)
Austria	0·3 m	0·6 m	0·5 m	2·83 m	91%
Belgium/Lux	2·0 m	0·1 m	0·3 m	3·62 m	100%
Denmark	0·2 m	0·9 m	0·3 m	1·89 m	100%
Finland	0·1 m	0·4 m	0·3 m	1·62 m	0%
France	0·4 m	7·6 m	7·2 m	20·5 m	100%
Ireland	0·3 m	—	0·5 m	0·8 m	94%
Italy	—	5·8 m	11·8 m	18·4 m	76%
Netherlands	3·0 m	0·8 m	0·3 m	4·6 m	100%
Norway	0·2 m	0·25 m	0·4 m	1·3 m	67%
Spain	—	2·5 m	6·9 m	12·3 m	87%
Sweden	0·2 m	1·6 m	1·0 m	3·6 m	84%
Switzerland	1·2 m	0·25 m	0·15 m	2·3 m	100%
UK	0·15 m	2·5 m	11·5 m	20·8 m	100%
West Germany	1·7 m	10·4 m	8·6 m	24·5 m	100%
Total	10·65 m	33·7 m	49·75 m	119·0 m	60–70 m *(6)

NOTES: *(1) Shows actual number of homes connected to cable TV networks as at 1986.
*(2) SMATV systems are those actual community antenna TV systems installed in block of flats which can be upgraded with a satellite dish to receive satellite TV.
*(3) DBS homes are those households with the consumer purchasing power to buy a VCR and therefore are financial affluent enough to purchase a TVRO terminal for use with ASTRA.
*(4) All TV households in Western Europe.
*(5) Percentage of TV homes in each country within the ASTRA 48 dBW zone.
*(6) SES calculates that some 60–70 million households constitutes the effective universe of households for ASTRA services at present.

Source: SES, December 1986, derived from "New Media in Europe", J. Tydeman/E. J. Kelm, 1986.

It is also known that SES has been considering 100–120 Watt transponders for the Astra 2 satellite thus creating a concept similar to that of Eutelsat's Europesat study. The Spacebus 100 platform would be too small to carry 16 such transponders, the only available option being the RCA 5000 platform. SES has not publicly confirmed that it is considering such an option for Astra 2.

However, SES suggested at the end of 1986 that there was little point in proceeding with the then current negotiations with Eutelsat which would have effectively transferred some $70 million from SES to Eutelsat. The apparent abandonment of negotiations suggested that the way was clear for SES to order the second satellite, although, as at December 1986, the company still needed to raise the equity finance to do so.

SES has suggested, euphemistically, that its shareholders should commit to the purchase when it has confirmed orders for the transponders on the first satellite. This suggests that the order will at least be subject to major negotiations from January 1987. The company plans to co-locate the satellite at 19·2 degrees East with the Astra 1 satellite.

Luxembourg has requested orbital slots at 1 degree East and 20 degrees West. The co-location of the Astra 2 satellite with Astra 1 at 19 degrees East may create problems in obtaining a sufficient frequency spectrum, particularly if Astra 2 has 16 operational transponders. A supplementary 250 MHz band would be required.

SES has a launch reservation with Arianespace for launch of Astra 2 in 1990 on Ariane 4. However, the company has been looking at other launch options and admits that amongst these are Titan III or IV.

SES and UK DBS: There is little doubt that Société Européenne des Satellites sees the UK broadcasting scene as a major opportunity and that its Astra system is in competition with the medium to high power satellite options proposed to the Independent Broadcasting Authority. Indeed SES argued to UK broadcasters considering participating in UK DBS as well as the Authority that its system could provide the space segment for UK DBS as well as the bonus of European coverage. The Independent Broadcasting Authority (regulatory authority for commercial TV and radio in the UK) announced on 11th December 1986 that the Granada led British Satellite Broadcasting consortium had won the bidding for UK DBS. SES were poised to take advantage of the inconsistencies and weaknesses in UK DBS thinking to promote its own space segment. The subsequent story is intriguing.

Société Européenne des Satellites immediately placed advertisements in the British national press suggesting that there were considerable advantages to the UK broacasting industry in using Astra for both distribution of channels to cable networks and for quasi-DBS services. It claimed that it was offering six to eight channels to UK broadcasters. The UK press, authorities and most (but not all) broadcasters appeared unaware of the significance of the offer which would give the country unregulated quasi-DBS services and transborder broadcasting at least two years before the planned British Satellite Broadcasting (BSB) services and rendered any plans for the latter commercially unavailable.

Britsat, offering a high power DBS satellite system with European coverage, gave the Independent Broadcasting Authority's Director General, John Whitney, in the closing stages of selection of a UK DBS contractor "the clearest possible warning" of the impending commercial impact of Astra on UK DBS to the effect that if the Authority granted an award without taking it into consideration, "the long run damage to British Television interest would be incalculable". Britsat made it clear to the Authority that there was a clear battle between UK DBS and the Astra system.

It is clear that in choosing British Satellite Broadcasting (BSB), the IBA has selected a consortium which neither has arranged the finance to carry the concept through to final fruition nor has produced a final satellite plan. The process of allocation of only three channels is an admission that the IBA believes that BSB cannot raise the finance if other DBS channels are available before BSB's system is operational and established. It is implicit in the allocation of only three channels that BSB could not survive competition from other DBS channels. SES was publicly stating that it was offering 6–8 channels.

The failure of the IBA to put together a tenable plan for UK DBS follows on from its unsuccessful attempts which led to the 21 consortium. Whilst SES told us that the IBA has done the best possible job under the circumstances, the events question both the competence of the civil service in allocating responsibility to the Authority (twice) and the Authority's ability to recognise that it neither has the ability to regulate new media nor the commercial judgement to recognise that its plans would be rendered irrelevant by an aggressive Luxembourg.

It rapidly became known to the public that a number of UK broadcasters were already moving towards using the Astra system. Within days of the Independent Broadcasting Authority's announcement that BSB had won the contract to supply UK DBS TV, it became known that Thames TV, a commercial TV company regulated by the Authority, was investing in SES and looking at taking a transponder. Before the end of the month, it also became known that publisher Robert Maxwell had put in an

SUMMARY OF CURRENT, PLANNED OR POTENTIAL SATELLITE TV CHANNELS IN EUROPE

	CHANNEL	OPERATIONAL DATE	CATEGORY OF PROGRAMMING	TARGET MARKET AND LANGUAGE	COUNTRY OF ORIGIN	SATELLITE USED OR PLANNED
1	Sky Channel	Now	General Entertainment	Europe (E)	UK	Eutelsat I
2	Superchannel	Now	General Entertainment	Europe (E)	UK	Eutelsat I
3	Premiere	Now	Subscribed Film	UK/Europe (E)	UK	Intelsat V
4	Children's Channel	Now	Children's	Europe (E)	UK	Intelsat V
5	Screen Sport	Now	Sport	Europe (E)	UK	Intelsat V
6	New World Channel	Now	Religion	Europe (E)	Norway	Eutelsat I
7	Worldnet	Now	News	Europe (E)	USA	Eutelsat I
8	Lifestyle	Now	Woman's	UK (E)	UK	Intelsat V
9	Arts Channel	Now	Up Market Arts	UK (E)	UK	Intelsat V
10	Bravo	Now	Pay Film	UK (E)	UK	None
11	Europa TV	Suspended	General Entertainment	Europe (EFGPS)	Holland	None
12	Home Video	Now	Subscribed Film	UK (E)	UK	None
13	CNN Europe	Now	24 hour News	Europe (E)	USA	Intelsat V
14	MTV Europe	1987	24 hour Music	Europe (E)	USA/UK	Intelsat V
15	(Unnamed, Maxwell)	1987/1988	General Entertainment	Europe (E)	UK	Astra?
16	Nashville Network	1987/1988	C&W Music	Europe (E)	USA	n/a
17	Financial News Network	Now	Business News	Europe (E)	USA	Intelsat V
18	Anglovision	1987/1988	General Entertainment	Europe (E)	USA/Ireland	n/a
19	Visnews/World News	1987/1988	24 hour News	Europe (E)	UK/Europe	n/a
20	Carlton Communications (?)	1987/1988	n/a	Europe (E)	UK	n/a
21	Thames TV (?)	1988	n/a	Europe (E)	UK	Astra
22	BBC Overseas Services	1987/1988	News & General Entertainment	International	UK	Intelsat
23	BBC One	Now	Rebroadcast	Europe (E)	UK	None
24	BBC Two	Now	Rebroadcast	Europe (E)	UK	None
25	ITN	1988	24 hour News	Europe (E)	UK	n/a
26	Now (BSB)	1990	News & Events	UK (E)	UK	UK DBS
27	Galaxy (BSB)	1990	General Entertainment	UK (E)	UK	UK DBS
28	Screen (BSB)	1990	Pay Film	UK (E)	UK	UK DBS
29	Yorkshire TV (?)	1988	n/a	UK/Europe (?)	UK	n/a
30	Health Channel	1987/1988	Health	UK (E)	UK	None
31	West Nally	1987/1988	n/a	Europe (?)	UK (?)	n/a
32	Samuel Broadcasting	1987/1988	Business News	Europe (E)	USA	n/a
33	EBN	1987/1988	Business News	Europe (E)	UK/Norway	n/a
34	RTL-Plus	Now	General Entertainment	Germany (G)	Luxembourg	Eutelsat I
35	SAT-1	Now	General Entertainment	Germany (G)	Germany	Eutelsat I
36	Teleclub	Now	Subscribed Film	Swiss/German (G)	Switzerland	Eutelsat I
37	3-SAT	Now	General Entertainment	Germany (G)	Germany	Intelsat V
38	Music Box	Now	Pop Music	Germany (G)	Germany	Intelsat V
39	Bayern III	Now	Rebroadcast	Germany (G)	Germany	Intelsat V
40	WDR III	Now	Rebroadcast	Germany (G)	Germany	Intelsat V
41	Eins Plus	Now	General Entertainment	Germany (G)	Germany	Intelsat V
42	Eureca TV	Now	n/a	Germany (G)	Germany	Intelsat V
43	Impulse TV	1987/1988	n/a	Germany (G)	Germany	n/a
44	TV-5	Now	General Entertainment	Europe (F)	France	Eutelsat I
45	TV-6	Now	Music	France (F)	France	Telecom I
46	La Cinq	Now	General Entertainment	France (F)	France	Telecom I
47	Cristal TV	1987/1988	Music	France (F)	France	Telecom I
48	STV	1987/1988	Music	France (F)	France	Telecom I
49	La Sept	1988	Culture	Europe (F)	France	TDF–1/2
50	TVE (R. Hersant)	1987/1988	General Entertainment	Europe (F)	France	n/a
51	RTL-Television	1988	Rebroadcast	France (F)	Luxembourg	TDF–1/2
52	Canal-Plus	Now	General Entertainment	France (F)	France	None
53	Canal J	Now	Children's	France (F)	France	None
54	RAI 1	Now	Rebroadcast	Europe (I)	Italy	Eutelsat I
55	RAI 2	Now	Rebroadcast	Europe (I)	Italy	n/a
56	Canale 5	Now	Rebroadcast	Italy (I)	Italy	None
57	Italia 1	Now	Rebroadcast	Italy (I)	Italy	None
58	Rete 4	Now	Rebroadcast	Italy (I)	Italy	None
59	Fininvest	1987/1988	n/a	Europe (?)	Italy	n/a

SUMMARY OF CURRENT, PLANNED OR POTENTIAL SATELLITE TV CHANNELS IN EUROPE

	CHANNEL	OPERATIONAL DATE	CATEGORY OF PROGRAMMING	TARGET MARKET AND LANGUAGE	COUNTRY OF ORIGIN	SATELLITE USED OR PLANNED
60	Scansat 1	1988	General Entertainment	Scandinavia	Sweden/UK	Astra (?)
61	Scansat 2	1988/1989	n/a	Scandinavia	Sweden/UK	Astra (?)
62	Scansat 3	1988/1989	n/a	Scandinavia	Sweden/UK	Astra (?)
63	NRK	Now	Rebroadcast	Norway (N)	Norway	Eutelsat I
64	El Pais	1987/1988	General Entertainment	Spain (?)	Spain	n/a
65	ABC	1987/1988	General Entertainment	Spain (?)	Spain	n/a
66	Grupo 16	1987/1988	General Entertainment	Spain (?)	Spain	n/a
67	RTVE	Now	Rebroadcast	Spain (S)	Spain	None
68	Filmnet ATN	Now	Subscriber Film	Low Countries	Netherlands	Eutelsat I
69	Euro TV	1987/1988	General Entertainment	Europe	Netherlands	n/a
70	Telecine Romandie	Now	Subscriber Film	Switzerland	Switzerland	n/a
71	EBU/Eurovision	Now	Networking	Europe	Europe	Eutelsat I
72	EBU/Eurovision	Now	Networking	Europe	Europe	Eutelsat I

NOTES: This table shows current or potential satellite TV channels in Europe including terrestrial channels using satellites for transmission outside their country of origin. It is a rough guide to the potential market for Astra (and Eutelsat) transponders. Omitted from the list are radio channels and local TV channels currently limited to off-air or cable transmissions. These channels may make use of satellites to distribute their own programming or to receive incoming programming from other broadcasters.

At the end of 1986, SES said that it had identified 170 companies in Europe which have projects to launch TV channels or who already run new media services of various sorts. It believed that 55 of these were likely to launch a service and 23 had already signed letters of intent with SES.

offer to SES for three transponders to carry a film channel (Premiere), a music channel (MTV) and a general entertainment channel. Thus Maxwell is in a position to get three (quasi-DBS) channels without the need to finance an expensive and risky satellite. The consumer, purchasing a TVRO terminal would also have more choice than with the BSB proposals, with up to eight UK channels available on the first spacecraft and up to a further 16 on a second Astra satellite. The news also suggested that Maxwell had abandoned plans to use the French TDF–1/2 DBS system.

In November 1986 SES stated that it had received letters of intent from 23 companies out of 55 it considers likely to launch satellite TV services. The company planned to "manage" the selection of channels on the satellite to include roughly three channels for each of the German and French speaking markets as well as 6–8 for the UK market. Included in the proposed selection would be certain thematic channels, such as film channels and a news channel.

SES is also offering the broadcasters a comprehensive range of support services including the establishment of a TVRO standard for use with the system which would mean that dishes would be marketed under the name Astra. This approach is significantly different from Eutelsat which does not liaise directly with its users.

SES had initially been careful in promoting the potential DBS use of the Astra system. However, after the loss of Ariane V18 in the spring of 1986 and consequential delays in the launch of the first Astra satellite until 1988, SES stated that it was using the delay to consider modifications to the first Astra satellite to provide a higher EIRP performance to "targeted markets" in Western Europe. These modifications will improve the downlinked signal strength for those markets by 2 dBW, allowing direct reception with 65 cm dishes in the UK, France, Germany, Switzerland, Northern Italy and the Low Countries. (Outside this area, the satellite will be able to distribute to cable TV headends.) For England, this would allow dishes in the 60 cm size. The dish size compares with 45 cm dishes being quoted by BSB for use with their system.

TECHNICAL SUMMARY, ASTRA 1

PLATFORM	RCA 4000
PRIME CONTRACTOR	RCA Astro-Electronics
STABILISATION	3-axis
LAUNCH MASS	1793 kg
MASS IN ORBIT	
POWER SUPPLY, END OF LIFE	2850 Watts
DESIGN LIFE	10 years
ORBITAL POSITIONS	Astra–1, 19° East Astra–2, 1° East
STATION KEEPING ACCURACY	+/−0·05 dBW (E/W & N/S)
ECLIPSE PROTECTION	Full
NO. OF ACTIVE TRANSPONDERS	16
NO. OF SPARE TRANSPONDERS	6

TWTA POWER OUTPUT	45 Watts* (Astra 2500 Watts)
FREQUENCY RANGE: UPLINK	14·250–14·500 GHz* 13·000–13·250 GHz (Astra–2)
DOWNLINK	11·200–11·450 GHz* 10·700–10·950 GHz (Astra–2)
TRANSPONDER BANDWIDTH	24–26 MHz*
EIRP MINIMUM, BEAM EDGE	50 dBW*
CHANNEL POLARISATION	Dual linear
RECEIVE COVERAGE (G/T)	6·6 dBW/sq.m.*
SATURATION FLUX DENSITY	

*Under review

COMMERCIAL DETAILS

OPERATOR	RCA Americon for Alascom Incorporated
ADDRESS	949 East 36th Avenue, Pouch 6607, Anchorage, Alaska 99502 USA
TELEPHONE NO.	(907) 264-7388
SENIOR STAFF	Charles E. Robinson, President Lee Wareham, VP, Operations Mike Nichols, Director, Marketing Tom Jensen, Public Affairs
PURPOSE OF PROJECT	Provision of C-band capacity for Alaska, used by Alascom for message toll services (long distance telephony) and distribution of radio and TV
PROJECT STATUS	Aurora 1 satellite fully operational Aurora 2–4 yet to be ordered but planned for 1990's
PROJECT VALUE	Aurora 2 was purchased, in-orbit, from RCA Americom for $88 m. In 1984, Alascom said that it intended to spend $221 m on the Aurora 2–4 satellites
LAUNCH DATES	Aurora 1, 27th October 1982, Delta Aurora 2–3, not before 1990 Aurora 4, ground spare

SATELLITE USERS

Alascom Incorporated is the long distance telecommunications operator for Alaska. In 1982 it purchased the Satcom V Spacecraft from RCA Americom to provide capacity for Alaska. This satellite was renamed Aurora 1.

Its primary user as such is Alascom. Any inhabited site in Alaska with a population total exceeding 25 is equipped with an earth station connecting into Aurora. These stations also fed into or are fed by the terrestrial network for the surrounding countryside. The primary service is telephony but the combination is also used for telex, data, telegram, radio and TV services.

RCA Americom also leases four transponders on the satellite under a contract due to expire in 1988. Lease rate is $1 6 m per year.

The Aurora 2–4 satellites are intended to replace Aurora 1 and provide intra as well as interstate capacity. Services will include telephony, private network and video and new services such as packet switching, video-conferencing and videotex.

It is intended that capacity on the in-orbit spare at 137° West will be offered to other satellite operators as "in-orbit spare capacity" so that they can turn preemptible services into fully protected services.

As at February 1986, two transponders on Aurora 1 were being used for TV channel traffic. Transponder 20 carried the Learn/Alaska channel and 22 carried the Alaska satellite TV project.

PROJECT OVERVIEW

The first of the Aurora satellites (Aurora 1) was purchased from RCA Americom by Alascom in 1982 for $88 m. RCA Americom continues to operate the satellite on behalf of Alascom and leases six transponders on it. In 1984 Alascom filed with the FCC for authority to construct launch and operate two further C-band satellites and for construct authority for a ground spare. It requested orbit slots at 141° West and 137° West. The latter was for the in-orbit spare. The FCC has given construct, launch and operation authority for one new satellite, Aurora 2, to be located at 146° West.

TECHNICAL SUMMARY

	Aurora 1	*Aurora 2–4*
PLATFORM	Satcom 3000	
PRIME CONTRACTOR	RCA Astro–El	
OTHER MAJOR CONTRACTORS		
STABILISATION	3-axis	
LAUNCH MASS	1084 kg	
MASS IN ORBIT	599 kg	
POWER SUPPLY, END OF LIFE	1100 Watts	
DESIGN LIFE	10 years	10 years
ORBITAL POSITIONS	142° W	146° W 141° W* plus ground spare
STATION KEEPING ACCURACY		
ECLIPSE PROTECTION		
PROPULSION SYSTEM		
SPACECRAFT DIMENSIONS		
SPAN, TIP-TO-TIP		
NO. OF ACTIVE TRANSPONDERS	24	24
NO. OF SPARE TRANSPONDERS		
SSPA POWER OUTPUT	8·5 Watts	10 Watts
FREQUENCY RANGE:		
UPLINK	5·925–6·425GHz	5·925–6·435MHz
DOWNLINK	3·7–4·2 GHz	3·7–4·2 MHz
TRANSPONDER BANDWIDTH	36 MHz	36 MHz
EIRP MINIMUM, BEAM EDGE	38, 34 & 28 dBW	
CHANNEL POLARISATION		

*requested but not granted

COMMERCIAL DETAILS

OPERATOR	Aussat Pty Ltd
ADDRESS	MLC Centre, 19 Martin Place, Sydney, NSW 2000, Australia
TELEPHONE NO.	(02) 238 7800
TELEX NO.	
FAX NO.	
SENIOR STAFF	W. G. Gosewinckel, CEO and GM W. L. Nowland, Spacecraft Manager R. C. Johnson, Deputy General Manager Mark D. Harwood, Communications Systems Manager
PURPOSE OF PROJECT	Provision of Ku-band capacity for Australia and Papua New Guinea for voice, data and video traffic including quasi-DBS
PROJECT STATUS	Satellite system is now operational. Aussat 3 satellite has yet to be launched
PROJECT VALUE	Initial capital of Aussat Pty Ltd is A$100 m. A further A$350 m was raised through loans to get the project off the ground. The three satellites cost A$300 m including two launches, two TT&C centres and major city earth stations
LAUNCH DATES	Aussat 1A, 27th August 1985, STS 51–I plus PAM–D Aussat 1B, 28th November 1985, STS 61–B plus PAM–D Aussat 1C, March 1987 by Ariane V19 or August 1987 by Ariane V22

SATELLITE USERS

1) The Homestead and Community Broadcast Satellite Service (HACBSS) is used by the Australian Broadcasting Corporation to distribute to the rural outback areas.

2) The Australian Broadcasting Corporation also used the normal satellite services for relay of programmes between studios, distribution of programmes to provincial transmitters and distribution of Radio Australia programmes from studio to transmitter. Commercial TV and radio networks also use the system for similar services including transmission of news.

3) Telecom Australia uses the system for telephony, telex, fax and data transmissions particularly to remote communities.

4) The Department of Aviation is using the low power transponders to link air traffic control and flight service centres to aircraft.

5) The Department of Defence uses the system for "internal administrative communications".

6) The system is being developed to provide educational services to remote areas through agencies such as the School of the Air.

7) Capacity is available for business services such as electronic funds transfer, voice, video and data transmission services for mining operations in remote locations.

8) Commercial TV interests are expected to use the HACBSS capacity on the system for their own quasi-DBS services.

9) The system also provides domestic telecommunications capacity with Papua New Guinea through that country's Post and Telecommunications Corporation.

10) The Aussat 1C spacecraft has been modified to provide one 30 W and two 12 W transponders for use by up to 20 Pacific Island countries.

PROJECT OVERVIEW

The Australian domestic satellite system, Aussat, has proved to be one of the bigger successes amongst satellite projects of the 1980's, with demand exceeding expectations to the extent that the operator decided to launch the third satellite in the system (originally designated a ground spare) before the first two satellites became fully operational.

The origin of the Aussat system dates back to 1977 when the Commonwealth (Central) government established a Task Force to examine the feasibility of establishing an Australian satellite system. The Task Force made its recommendations in July 1978 and in October 1979 the government announced its decision to proceed with the planning and design of an Australian satellite system. The project was then placed in the hands of a team drawn from the Overseas Telecommunications Commission (Australia), Telecom Australia and the Australian Broadcasting Corporation with consultancy services from Comsat and the European Space Agency. A call for tenders was issued in October 1980 and contracts were signed with Hughes Communications International in May 1982 for supply of the three identical satellites and associated ground-based satellite control facilities. Meanwhile Aussat Pty Ltd was formed on the 6th November 1981 to develop and operate the system. Subsequently Telesat Canada was used as a consultant in both the technical and commercial fields. The first two satellites were launched in 1985 by Shuttle and the launch of the third was brought forward from a possible 1988 to 1986 launch. However, the Aussat 1C satellite had been booked on Ariane and the technical problems with the third stage motor of that vehicle means that Aussat 1C will probably fly in 1987.

Aussat Pty Ltd is a wholly publically owned company with 75% of its A$100 m equity capital held by the Commonwealth government, the remaining 25% by the Australian Telecommunications Commission. A further A$350 m in loan capital was raised from two syndicates of banks. The first syndicate, lead by the Commonwealth Trading Bank of Australia, raised a domestic loan of A$175 m. The second, also raising A$175 m was an overseas syndicate lead by Manufacturers Hanover Trust Company. These were considered as bridging loans which Aussat intends to refinance "at the appropriate time with permanent long-term finance." The legal structure of Aussat Pty Ltd is one of a limited liability company operating as a private company. However it is required to provide capacity for services that are in the national interest and provide services "at reasonable non-discriminatory prices". It currently employs some 200 staff. In addition to operating the satellite system, Aussat owns eight "major city earth stations" purchased from Mitsubishi

Australia Ltd. These are located in the six state capitals and Canberra and Darwin. They are used to provide satellite access for city based customers. The Australian Broadcasting Corporation is the major user. Aussat also owns single voice channel "minor" earth stations for use in field demonstrations and marketing to prospective low capacity users. However, most other earth stations using the Aussat system are owned by the users, not by Aussat Pty.

The technical characteristics of the Aussat satellites provide an interesting insight into the types of services offered by Aussat Pty. Excluding the communications payload, the Aussat spacecraft are very similar to the standard HS–376 satellites. This allowed the first satellite to be completed within 30 months of contract start-up. Outwardly, the Aussat spacecraft are similar to other HS–376 satellites, except that the large single reflector common on HS–376 satellites is replaced by a complex arrangement of three smaller reflectors. The smallest of the three, at 0·6 metres, gives the widest coverage, providing a national beam. The other two, at 1·0 and 1·1 metres, are used to form the spot beams for four areas of Australia. Each reflector works in the vertical and the horizontal polarisation mode. There are seven transmit (downlink) beams available. Each satellite has an on-board microwave switch matrix which provides "flexible" connectivity between the 15 operational transponders and transmit antenna beams.

Each satellite has two types of Ku-band TWTA's. The 12 Watt TWTA's are intended for conventional Fixed Satellite Services providing a minimum EIRP for Australia of 36 dBW. These TWTA's were based upon the 16 Watt and 20 Watt transponders used on the GTE Spacenet (RCA Satcom) and SBS (HS–376 satellites). The HACBSS service, which is in effect a quasi-DBS service required a nominal spot beam EIRP of 47 dBW. This required use of 30 Watt TWTA's. The TWTA's came from Hughes Electron Dynamics Division and the power supply for them from Hughes Space and Communications Group. The on-board microwave switch matrix, which is reconfigured by ground command, provides connectivity between the 11 × 12 Watt transponders, the 4 × 30 Watt transponders and the seven transmit antenna. This capability was incorporated to provide flexibility in meeting uncertain market demand.

Amongst the most interesting of services on the Aussat system is the quasi-DBS Homestead and Community Broadcasting Satellite Service (HACBSS). Under WARC–77 arrangements the 12·5–12·75 GHz band was allocated for downlinks shared between FSS and DBS systems. The Aussat downlinks are in the 12·25–12·75 GHz range and the HACBSS service lies in the upper end of this range. WARC–77 allocated Australia orbital slots at 98° East and 128° East, providing 63–64 dBW EIRP, for true DBS services. The 30 Watt transponders on Aussat gives an EIRP in the 47 dBW range allowing use of TVRO's in the 1·2–1·8 metre range. The provision of the quasi-DBS capacity on Aussat closely parallel's Telesat Canada's attempts to have a quasi-DBS service (of up to 14 channels) introduced on its Anik C Ku-band system.

The latter would provide an EIRP minimum of 47 dBW for most of Canada except the far north where it would fall to around 31 dBW. The provision of a quasi-DBS service on Telesat also parallels that of the Aussat provisions in that both are aimed at rural markets, either "underserved" in the number of terrestrial channels available or without any service. There are some 300,000 people in the remote areas of Australia who can receive no terrestrial television or radio broadcasts and some one million whose reception is "technically inadequate". The public service broadcasting, the Australian Broadcasting Corporation (ABC) provides a national TV channel as well as several radio channels. In some areas ABC is the only service available.

Nevertheless, Australia is, surprisingly, one of the most urbanised countries in the world and there are some 50 commercial TV stations, grouped into three large networks, which serve the more populated areas. It is these services, and similar radio services, which are least represented in the outback. Not surprisingly some commercial TV operators, notably Rupert Murdoch and Kerry Packer, have been looking to use the Aussat system for quasi-DBS services. Nevertheless, the initial quasi-DBS service was given to ABC and commercial interests blocked

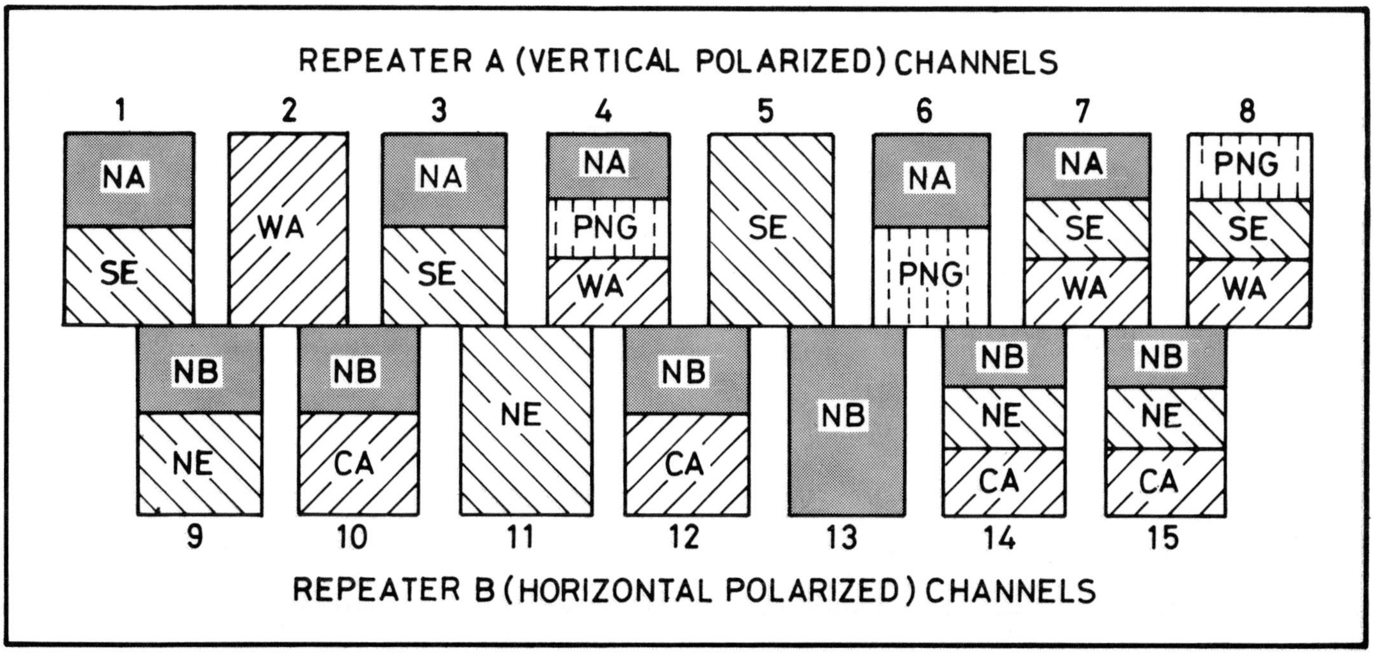

Aussat transmit beam switching capability. Diagram shows alternative beams which can be selected for each repeater. Note that repeaters 7, 8, 14 and 15 use 30 Watt TWTA's intended for the quasi-DBS HACBSS service

Key to transmit beams is as follows:

NA	National beam A	NB	National beam B	WA	Western Australia
SE	South East	PNG	Papua New Guinea	NE	North East
CA	Central Australia				

out. The dominant Australia culture is European and broadcasting has a "special significance" for Australians. "Broadcasting has assumed major responsibility as a primary conservator of Western culture." Like Europe's Australian broadcasting policy places a high emphasis on the need to maintain public service broadcasting. The HACBSS concept was a product of ABC to show that as a good public service broadcaster, it would provide coverage to 100% of Australia. Two of the commercial networks, after some delay, have subsequently joined the HACBSS system. The HACBSS service is unusual in that it abandoned use of the PAL transmission standard in favour of Scientific Atlanta's B–MAC standard.

ABC's use of HACBSS involves four 30 Watt transponders with four spot beams to provide regional variations on its national HACBSS service. The organisation also uses the HACBSS service to by-pass sections of the National Broadcasting Service terrestrial transmission network. The ABC channel is downlinked to existing rural NBS transmitter stations for rebroadcasting terrestrially to local communities. The 12 Watt transponders are also available to public and commercial broadcasters for relay of programmes from major studios to other centres for further distribution and interconnection of production studios in major cities for programme interchange. Charges for use of the space segment of Aussat are detailed at the end of this section of the book. However, the earth station charges for one-way services, using Aussat facilities are as follows (prices as set in July 1984).

One-way Services	Annual Price A$	Monthly Price A$
TV Uplink	160,000	13,333
TV Downlink	40,000	3,333
Radio Program Channel Uplink	20,000	1,667
Radio Program Channel Downlink	5,000	417

Like the Australian Broadcasting Corporation, the Department of Aviation is a "preferred" user of the Aussat system. That is to say it was identified as a customer in the Aussat Memorandum of Association, having been singled out at an early stage as a major user. The Department is using the system to improve voice and data links between 46 major manned air traffic control and flight service centres. The links are used for dissemination of flight service data and coordination of air traffic control for aircraft entering and leaving each manned centre's airspace.

More significantly, the Aussat system is being used to provide full VHF air-to-ground coverage on all domestic commercial flight routes above 20,000 feet. Prior to the new system, VHF coverage was limited to airspace surrounding each manned centre. HF communications were used elsewhere. 55 new unmanned VHF air-to-ground masts are being linked via Aussat to designated manned centres. The Department of Aviation has purchased 202 aeronautical earth stations from NEC Australia. They are installed in identical pairs at 101 separate locations. The service uses four 12 Watt transponders, two on each of the currently in-orbit satellites, and national beams.

Similarly the Remote Telephony Satellite Service also used 12 Watt transponders and national beams. It also uses Demand Assignment Multiple Access Facilities and SCPC/compressed FM transmissions. Initial service extended telephony to approximately 400 remote area subscribers. The transmit/receive earth stations for this service fall into two categories. Low capacity stations able to support only a small number of voice channels; these may for example, be used on isolated homesteads and use dishes between 3·7–4·5 m. Higher capacity terminals, with 3·7–6·4 m dishes are capable of supporting 12 voice channels.

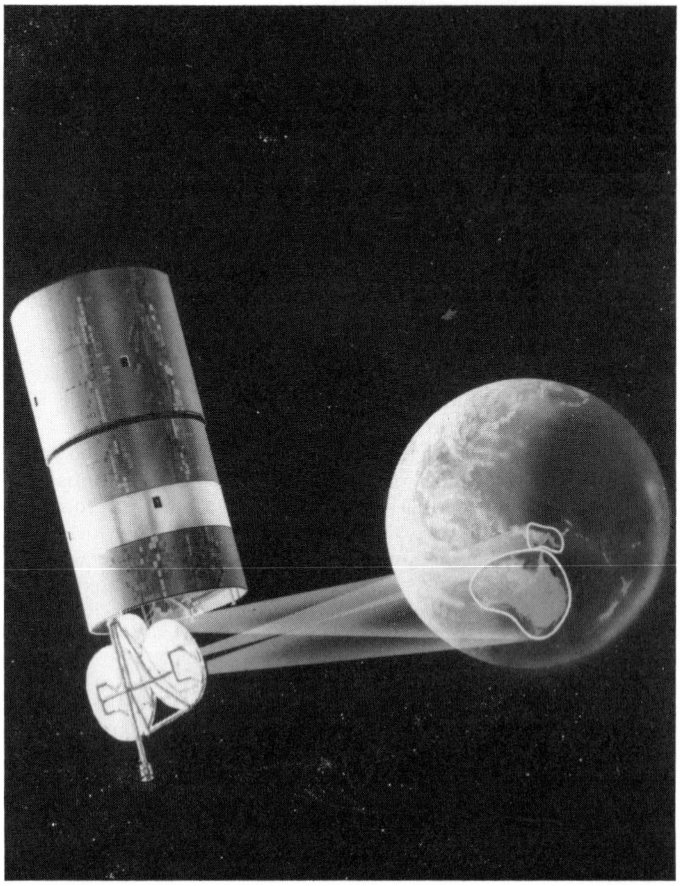

Artist's impression of the Aussat–1A satellite over Australia. The platform is a HS–376 design. Although the picture appears to be upside-down, a little bit of thought shows that it is the right way up. The Aussat satellites were the first HS–376 designs to use an orbit normal south configuration (the on-station spin axis is directed southwards). (Courtesy of Aussat)

Voice Channels

The eight Major City Earth Stations (MCES's) of the Aussat system are owned by Aussat Pty. These are the major gateway stations for broadcasters, business users, Telecom Australia and others. The earth stations were procured from Mitsubishi Australia Ltd and, according to Aussat, about 50% of the contract, worth A$32 m, was accounted for by Australian content. The stations in Sydney, Perth, Adelaide and Darwin are equipped with two antennas, one for each operational satellite; the others at Brisbane, Melbourne, Hobart and Canberra each have one antenna. The Darwin and Brisbane antennas are 18 metres; all the others are 13 metres. Aussat earth station charges are listed below. These include space segment charges and are full time.

Two-way Services (includes space segment)	Annual Price A$	Monthly Price A$
Voice Channel	8,000	667
9·6 Kilobits/sec Digital Channel	10,000	833
48 Kilobits/sec Digital Channel	20,000	1,667
2·048 Megabits/sec Digital Channel	195,000	16,250

The development of the Aussat system created a dilemma for policy makers in that country. It was the first major Australian space project and by far the bulk of the expenditure on the system was placed outside the country. This has lead to a serious consideration of building up domestic Australian satellite manufacturing capabilities. At one-time it was suggested that Australia should develop a prime contractor capability for the second generation Aussat–2 system. Although this was ruled out

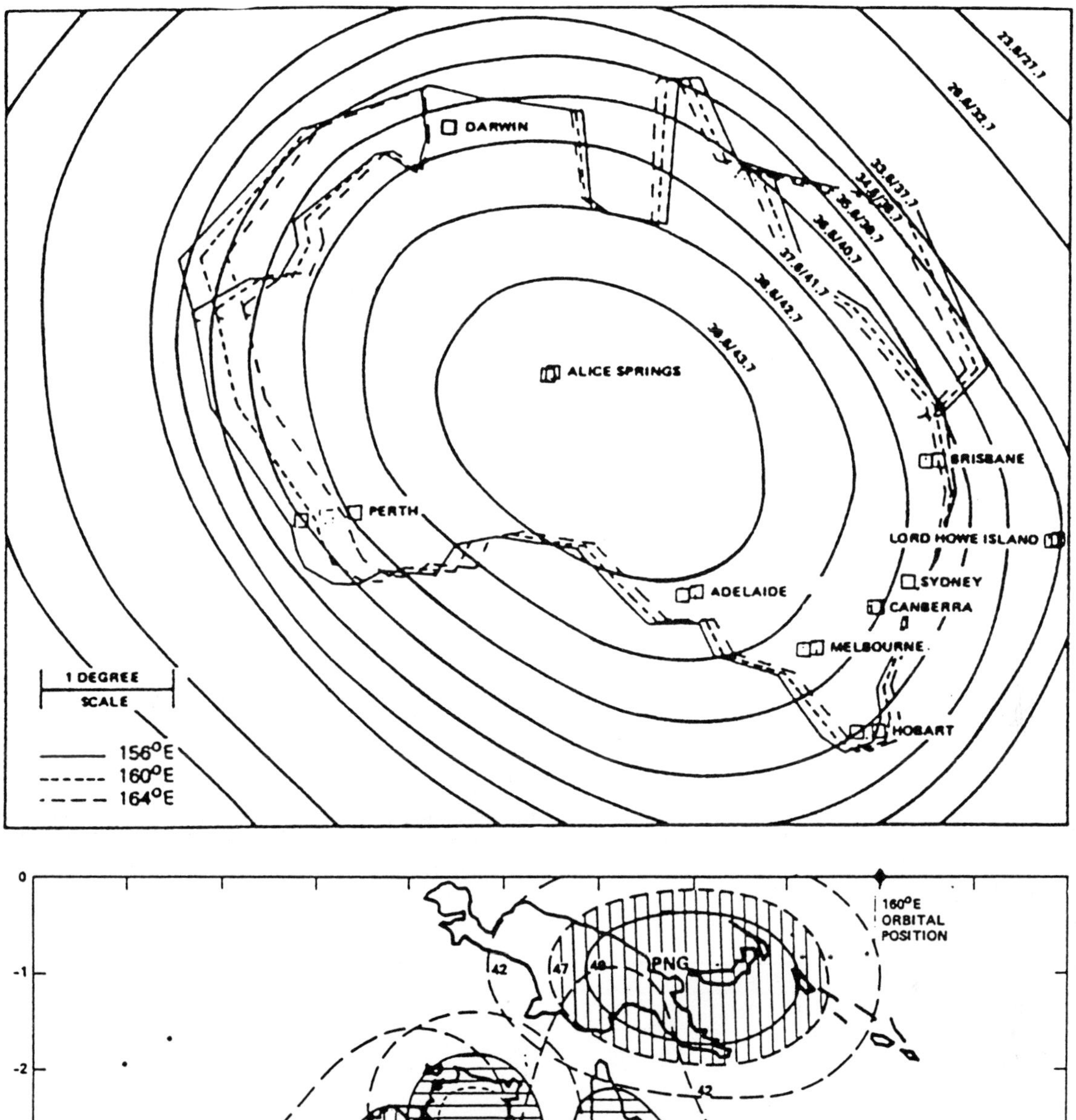

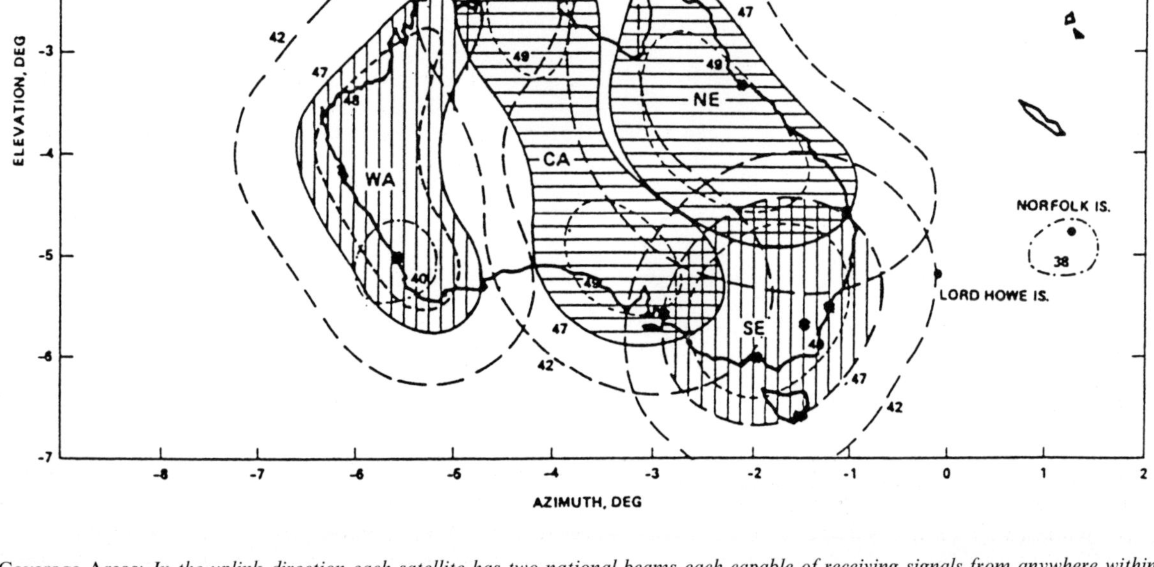

Aussat Coverage Areas: *In the uplink direction each satellite has two national beams each capable of receiving signals from anywhere within Australia. In the downlink direction, each has two national beams, designated NA and NB. The above diagram gives typical NA Beam EIRP performance in dBW for 12 Watt and 30 Watt channels. Coverages are shown in true view projection for each orbital location and include normal mode pointing errors. The downlink direction also provides 4 spot beams on each satellite covering: 1) Western Australia, 2) Queensland, 3) South Australia, 4) New South Wales, Victoria and Tasmania. Small spot beams also cover Lord Howe and Norfolk Islands. In addition there is separate uplink and downlink capacity available in a separate Papua New Guinea spot beam. The above diagram shows typical EIRP performance in dBW for 30 Watt channels in each spot beam at 160° East*

as not feasible, partly because there was insufficient time to develop the capability before the order for Aussat–2 system is placed, the option is apparently still open for the Aussat–3 system. It is expected that Australia will seek to contribute "significantly" to the manufacture of the Aussat–2 system, despite a still very weak production capability. As one Australian put it, the country's technology capabilities are still largely confined to building high-tech sheep-dips.

Nevertheless, it is estimated that Australia will be spending between A$370 m and A$500 m by 1995 on satellite services, including use of the Intelsat system. The country also received a small amount of sub-contract work on the Aussat–1 system. The offset agreements with Hughes for the latter gave Australian companies A$50 m of work for supply of spacecraft and ground control equipment for other Hughes satellites. Direct involvement of Australian firms in the Aussat–1 space segment was a mere A$5·4 m. This involved contracts with AWA and STC for contributions to the TT&C facilities and wiring harnesses for satellite operations. Such offset and subcontracting arrangements are common in the bidding proposals for satellites, although, in general, the US satellite manufacturers insist on retaining tight control over the contracts. Aussat Pty has been careful to point out that Australian industry has "participated" in the Aussat–2 system. It may reasonably be expected that Aussat Pty comes under increasing pressure to source more of its equipment in Australia. This is because, by mid-1986, it was becoming clear that the Australian economy was faced with a structural problem, being over-dependent on the export of raw materials and commodities suffering from long-term price depressions.

Aussat Pty is now in the process of obtaining its second generation Aussat–2 system. It has already asked for expressions of interest, and is winding up that stage ready to go to tender. Phase A feasibility studies started early in 1985. Phase B (design study) is due to be completed by 1988. Phases C and D (fabrication through to launch) would begin in early 1989, leading to launch in 1992. The system is likely to be geared towards regional as well as national requirements and be in competition with the Pacstar system.

According to Mark D. Harward of Aussat Pty, the modifications to the Aussat–1C could lead to "a dedicated Southwest Pacific regional network designed to provide to all island nations the type and qualities of service that truly reflect the region's needs." Aussat–1C has been modified to include one 30 Watt transponder and two 12 Watt transponders for possible use by up to 20 Pacific island nations.

TT&C&M (14 metre) and two smaller transmit/receive antenna at Aussat's Major City Earth Station, Belrose, Sydney. (Photograph courtesy of Aussat Pty)

PRICING OVERVIEW

The National satellite system has a range of transponders with footprints as described in the *Network Designer Guide* published by Aussat Pty Ltd. The prices for a whole transponder for different use possibilities can be calculated by using a base price and a number of factors.

The base price as at 1st July 1985 for:

- 12 Watt Protected Transponder
- National Beam Usage
- 12 Month Contract
- Unlimited Use,

is: Annual Payment A$2,100,000
Monthly Payment in advance A$175,000

This price will be multiplied by the factors outlined below depending upon the exact beam, power level and level of protection required.

Beam		Power		Protection Status	
National	1·0	12 Watt	1·0	Preferred*	1·1
SE Spot	1·0	30 Watt	1·6	Protected	1·0
NE, WA, CA Spot	0·9			Pre-emptible**	0·9
PNG Spot	0·9				

*Preferred Customers are those identified in the Aussat Memorandum of Association and are:
1) ABC – for HACBSS service
2) Department of Aviation – for air navigation and related safety services
3) Telecom – for remote telephony/emergency services.
**Pre-emptible capacity can be taken out of service to provide facilities to restore capacity to protected or preferred customers.

Volume Discount

A discount will be applied when an order is placed for the simultaneous use of two or more whole transponders on a full-time basis:

Discount for 2 transponders	2%
Discount for 3 transponders	3%
Discount for 4 transponders	4%
Discount for 5 or more transponders	5%

CONTRACT OVERVIEW

Contracts between Aussat and its customers will generally allow unrestricted use of the satellite capacity, and will include the following features:

- Variable length contracts.
- Equal monthly payments in advance as the standard arrangement.
- Renewable at the end of the contract period at the then published terms.
- Option to expand capacity.
- Annual price reviews.
- Variations possible by negotiation.

Contractual restrictions will be minimal, namely:

- Transponder capacity used by Aussat's customers may not be resold to third parties.
- Customers of Aussat will not be permitted to carry third party traffic.

PRODUCT OVERVIEW

Satellite Transponder Service

This service will be available to all organisations. Whole transponders or fractions of a transponder will be available on a full or part-time basis. This transponder service may be accessed through Aussat's Major City Earth Stations or through the customer's own earth stations.

When a whole transponder is used to provide a Satellite Program Service (SPS), the user will be restricted to the transmission of one usable television picture.

Full Transponder Full-time

The basic price, as modified by the factors described in the Pricing Overview, is applicable.

Part Transponder Full-time

The charges applying to part use of a transponder are dependent on the percentage of transponder resources used by the customer (this percentage will be determined by Aussat on the basis of power or bandwidth used). The relationship between the price for transponder resources used and the base price is calculated by a formula, as illustrated in the following table:

Transponder Fraction	Annual Price A$	Monthly Price A$	Approximate Fraction of whole Transponder Price
0·1	161,700	13,475	0·08
0·2	352,800	29,400	0·17
0·3	573,300	47,775	0·27
0·4	823,200	68,600	0·39
0·5	1,102,500	91,875	0·53
0·6	1,411,200	117,600	0·67
0·7	1,749,300	145,775	0·83
0·8 & above	2,100,000	175,000	1·00

Part-time Full Transponder Use

Transponder capacity will be available for occasional use applications. The charge per hour will depend on the time of day and the annual committed hours for this service.

The following price applies for: 12 Watt Protected Transponder; National Beam; Price as at 1st July 1985; Valid for fiscal year. Base price: A$1000 per year.

Prices will be varied (time of day):

Prime Time (4pm – 12pm)	1·0
Day Time (8am – 4pm)	0·8
Night Time (12pm – 8am)	0·5

Committed Annual Usage

Up to 250 hours per year	No discount
250–1000 hours per year	10% reduction
Greater than 1000 hours per year	20% reduction

Committed Annual Usage	Hourly Price		
	Prime time A$	Day time A$	Night time A$
Up to 250 hours	1000	800	500
250 to 1000 hours	900	720	450
More than 1000 hours	800	640	400

This basic pricing structure of the Aussat space segment (published February 1984) excludes earth station charges. The pricing structure was considered by Société Européenne des Satellites as a basis for the Astra pricing system but was rejected by SES as unsuitable. The basic Earth station service charges, as at July 1984 are shown in the main text of the Aussat description.

Projected Transponder Demand on the Aussat system by 1988 (1984 forecast)

	High power	Low power
ABC	4	5
SBS		2
Major TV Networks		7
Second set of high power transponders by TV interests	4	
Sound Broadcasting		0·5
Department of Aviation		4
Department of Defence		1
Telecom Australia		1
Papua New Guinea		1
Data and Other Business Communications		2·5
Occasional Use		1
TOTAL	8·0	25·0

TECHNICAL SUMMARY

PLATFORM	HS–376
PRIME CONTRACTOR	Hughes Aircraft Company
OTHER MAJOR CONTRACTORS	Hughes Electron Dynamics for TWT's. Hughes Space and Communications for associated flight power supplies
STABILISATION	Spin
LAUNCH MASS	1250 kg
MASS IN ORBIT	655 kg (beginning of life)
POWER SUPPLY, END OF LIFE	860 Watts, miminimum
DESIGN LIFE	8 years
ORBITAL POSITIONS	Aussat–1A, 156° E Aussat–1B, 164° E Aussat–1C, 160° E
STATION KEEPING ACCURACY	+/−0·05° (E/W & N/S)
ECLIPSE PROTECTION	Full
SPACECRAFT DIMENSIONS	2·17 m diameter, 6·6 m high
NO. OF ACTIVE TRANSPONDERS	15 (4 at 30 Watts, 11 at 12 Watts)
NO. OF SPARE TRANSPONDERS	4 (2 at 30 Watts, 2 at 12 Watts)
TWTA POWER OUTPUT	30 Watts and 12 Watts
FREQUENCY RANGE: UPLINK	14·0–14·5 GHz
DOWNLINK	12·25–12·75 GHz
TRANSPONDER BANDWIDTH	45 MHz
EIRP MINIMUM, BEAM EDGE	47 dBW, spot, 36 dBW, national
CHANNEL POLARISATION	orthogonal, horizontal and vertical polarisation

COMMERCIAL DETAILS

OPERATOR	Telecommunications Satellite Corporation of Japan (TSCJ)
ADDRESS	2-12-18 Shibadaimon, Minato-Ku, Tokyo 105, Japan (see text)
TELEPHONE NO.	432-28-01
SENIOR STAFF	Hiroshi Hirose, President
PURPOSE OF PROJECT	Provision of DBS capacity to provide reception in areas not covered adequately by terrestrial TV
PROJECT STATUS	Due to become operational during 1986
PROJECT VALUE	Two satellites, plus launch, insurance and operations cost $255 m of which NASDA provided one-third and NHK two-thirds
LAUNCH DATES	BSE, 7th April 1978, Delta BS–2A, 23th January 1984, N–II BS–2B, 12th February 1986, N–II

SATELLITE USERS

It was intended that public service broadcasters NHK would use BS–2A to transmit two TV channels to areas where reception of its existing terrestrial TV channels was poor or non-existent. However, when two transponders on the satellite failed its use was limited to experimental transmission of one channel – this use was subsequently shelved indefinitely. It was also intended to use the satellite systematically for experiments in advanced broadcasting systems such as high definition television, PCM stereophonic sound broadcasting and facsimile broadcasting during periods when the two main TV channels were not transmitting.

The two NHK (Nippon Hoso Kyokai) channels which originally used BS–2A were an educational and general interest channel. At the time of writing the two channels have not been transferred to BS–2B, although this was the intention. If the channels are transferred to BS–2B, then BS–2A will have to be turned off, as there would otherwise be interference between the two satellites.

The BS–2 system was due to be transferred from NASDA to NHK ownership as and when the satellites were proved successful in-orbit. The actual operations of the in-orbit satellites is handled by a company called the Telecommunications Satellite Corporation of Japan.

PROJECT OVERVIEW

The origins of Japan's efforts to develop a domestic Direct Broadcast Satellite system is bound up in Japanese industrial policy and the structure of existing terrestrial television. It is well known that for many years the Japanese have had a coherent industrial strategy toward building up capabilities in the television industry, most recent impacts including its domination of the VCR markets in the Western World and its increasing dominance of the satellite earth terminal market. But, as far back as 1963, Japanese industry, with the approval of MITI (Ministry for International Trade & Industry), were actively working towards an aggressive attack on Western markets. In that year, the Japanese TV receiver manufacturing companies met publicly to create a cartel to regulate the sales of their TV sets in the USA. One consequence of the complicated intrigues that involved this cartel was a sharp rise in prices of TV sets in Japan and much lower prices on exports to the USA. It is of interest that Japanese public service broadcaster NHK became interested in DBS TV not long after the 1963 TV set cartel, which eventually wiped out much of the European and US TV set industry. However, the Japanese interest in DBS also reflects the strong position of public service broadcasting in that country.

The structure of broadcasting in Japan is split roughly 50:50 between NHK and private television stations. But NHK has a considerable degree of autonomy from the government – it is virtually "owned" by the licence fee paying public. It is also the richest public service broadcaster in the world. But where government control has affected the organisation it has tended to push it towards developing DBS. It has a legal obligation, dating back to 1950, to make broadcasting services available to the entire population. This is a major task. In 1983 it calculated that there were still 400,000 households in Japan unable to receive terrestrial TV broadcasts.

The legal obligations, its wealth, and perhaps its autonomy, allowed NHK to take a front seat in the development of a Japanese DBS system. It contributed nearly two-thirds of the cost of the BS–2 programme. However, despite some favourable conditions in Japan for DBS, it has so far created major problems for NHK and the Japanese experience of DBS to date is not an inspiration to other countries looking at establishing domestic DBS systems. Part of the problem lies in "bad luck", the other parties in R&D and industrial policy. Until recently, Japanese policy towards satellites has been to develop its own domestic satellite manufacturing capability by ensuring that all domestic communications satellites are procured from Japanese manufacturers and, if possible, launched on Japanese launch vehicles. Unfortunately, Japanese satellite manufacturers are not good at manufacturing satellites, being heavily dependent on imported systems and sub-systems which has created problems in project management and quality control.

Moreover, Japanese launch vehicles have been too small to launch the size of satellite needed for economic communications. As a consequence, Japanese satellites have proved to be extremely expensive, somewhat unreliable and with limited payload capacity.
The first Japanese DBS satellite, BSE (also called Yuri) failed totally after 27 months when the power supply to all three 100 Watt TWTA's went down. BSE was an experimental, demonstration satellite but was the world's first dedicated DBS spacecraft. The BS–2A satellite, launched in 1984, was intended

for operational use, with NHK transmitting two existing terrestrial channels through the satellite. However, two of its three transponders quickly failed, leaving NHK with the option of running an experimental service through the remaining transponder. Consequently consumers refused to purchase DBS TVRO's, which were retailing for around $850 each.

In the two years after launch of BS–2A, only 5000–6000 such units were sold. The failure of BS–2A was eventually established as being caused by the unexpected overheating of Travelling Wave Tubes procured from Thomson-CSF of France through General Electric. One consequence was that the Japanese placed the Prime Contractorship for the next generation DBS satellites, BS–3, with Nippon Electric Corporation. It also resulted in a delay in the launch of BS–2B whilst NASDA attempted to find the cause of failure of BS–2A. The Japanese also, as a consequence, questioned both the viability and future of domestic DBS TV.

NHK was strongly criticised for its involvement in the project, which was seen as costly and high-risk and to help save face it did not increase its 1985 licence fee. Even before the BS–2B satellite was launched in 1986 there were considerable doubts about its reliability based on the experience of BSE and BS–2A. Launch insurance for BS–2B was at a rate of 31·5%, then a Lloyds record. Nevertheless, despite setbacks with the project it is unlikely that the Japanese will jeopardise their ambitions in HDTV and other transmission/broadcasting technologies by abandonment of their DBS TV efforts. It has already committed to the follow-on DBS project BS–3 which will eventually replace BS–2B However, NHK has become wiser about the technical risks involved in DBS.

Aviation Week and *Space Technology* reported in July 1986 that NHK was reluctant to assume ownership of BS–2B and wanted NASDA to guarantee the reliability of the satellite. The satellite had suffered from a malfunction in its in-orbit testing. The failure involved an integrated circuit in the satellite's primary altitude control system but a redundant system was available to correct the spacecraft's altitude. It was reported that the probability of BS–2B surviving beyond five years had fallen from 84% to 74%. Similarly, in 1985 NASDA stipulated in the contract for the BS–3 system that there would be provision for penalty payments. The negotiations for this penalty clause pushed the launch of BS–3 back by a year.

However, in its grand industrial strategy, Japan has been caught out twice over its DBS plans. Firstly, Europe has opted for MAC transmission standards on which it holds patents. MAC is intended as a precursor to the introduction of HDTV. Similarly Europe has insisted that it will not comply with Japanese/US plans for a world standard for HDTV. European countries have proved fairly unanimous in adopting MAC although agreement has not been reached on what form of MAC to introduced. NHK has been working on 1125 line HDTV for over a decade and the consequential exclusive from European markets of this standard is a serious blow to Japanese DBS intentions.

TECHNICAL SUMMARY

PLATFORM	
PRIME CONTRACTOR	Toshiba Corporation
OTHER MAJOR CONTRACTORS	General Electric
STABILISATION	3-axis
LAUNCH MASS	670 kg
MASS IN ORBIT	350 kg
POWER SUPPLY, END OF LIFE	890 Watts
DESIGN LIFE	5 years
ORBITAL POSITIONS	BSE, BS–2A, BS–2B, 110° E
STATION KEEPING ACCURACY	+/−0·1° (N/S & E/W)
ECLIPSE PROTECTION	
PROPULSION SYSTEM	
SPACECRAFT DIMENSIONS	1·2 × 1·32 × 2·89 m
SPAN, TIP-TO-TIP	9 m
CONTROL CENTRE	Kimitsu Satellite Control
NO. OF ACTIVE TRANSPONDERS	2
NO. OF SPARE TRANSPONDERS	1
TWTA POWER OUTPUT	100 Watts
FREQUENCY RANGE:	
UPLINK	
DOWNLINK	11·7067–12·031 GHz
TRANSPONDER BANDWIDTH	70 MHz
EIRP MINIMUM, BEAM EDGE	55 dBW
CHANNEL POLARISATION	−8·2 dB/K

COMMERCIAL DETAILS

OPERATOR	Telecommunications Satellite Corporation of Japan (TSCJ)
ADDRESS	2-12-18 Shibadaimon, Minato-Ku, Tokyo 105, Japan
TELEPHONE NO.	432-28-01
TELEX NO.	
FAX NO.	
SENIOR STAFF	Hiroshi Hirose, President
PURPOSE OF PROJECT	Provision of domestic DBS capacity to replace and add to capacity provided by BS–2
PROJECT STATUS	Satellites under construction
PROJECT VALUE	Space segment is budgeted to cost $168 m
LAUNCH DATES	BS–3A, August/September 1990 by H–I BS–3B, mid-1991 by H–I

SATELLITE USERS

It is intended that two of the three operational channels on this satellite will be used by NHK. The plan is for the latter to transfer its two channels from the BS–2B satellite.

A private sector company, Japan Satellite Broadcasting Corporation, will use the remaining channel. The corporation is a consortium of publishers, electronic and other industrial companies and has committed 27 billion Yen ($115 m) to the project over seven years. NHK and JSBC will be providing 65% of the development cost funds of BS–3, the remainder being provided by NASDA.

Nevertheless, there is considerable doubt about the economics of the BS–3 project, particularly as other countries have failed, so far, to establish their own domestic DBS systems. This has dampened down the arguement that BS–2 and BS–3 will provide a basis to enter the export markets. Similarly JSBC has revised downwards it revenue forecast through use of BS–3. Some estimates indicate that BS–3 will only make a profit after seven years of operation, others believe it will never make a profit.

PROJECT OVERVIEW

This project has been considerably affected by the technical problems with the BS–2 satellite system. In the light of the failure of BS–2A, prime contractorship was awarded to NEC, it had been expected that Toshiba would win the prime contractorship as it had done with BS–2A. NEC's bid to build the satellite was at a slightly higher price than Toshiba and it had fewer Japanese components. However NEC has experience in Travelling Wave Tube Amplifiers and BS–3 will use NEC TWTA's instead of the Thomson tubes on BS–2. Because of the problems with BS–2, NASDA, who are procuring the satellite on behalf of the broadcasters, insisted on incorporating penalty clauses in the contract for BS–3. This also delayed the project. Originally intended for launch in 1988 to replace BS–2, launch of BS–3 is now scheduled for 1990 and 1991.

A follow on system to BS–3, BS–4 is currently under consideration. This would require first launch in 1997. As Japan is already developing larger satellite platforms and will have the large H–II vehicle available from 1992, the BS–4 design is expected to be able to handle up to eight operational DBS channels.

TECHNICAL SUMMARY

PLATFORM	Satcom 3000
PRIME CONTRACTOR	Nippon Electric Company
OTHER MAJOR CONTRACTORS	RCA Astro-Electronics
STABILISATION	3-axis
LAUNCH MASS	
MASS IN ORBIT	550 kg
POWER SUPPLY, END OF LIFE	
DESIGN LIFE	
ORBITAL POSITIONS	BS–3A, BS–3B, 110° E
STATION KEEPING ACCURACY	
ECLIPSE PROTECTION	
PROPULSION SYSTEM	

SPACECRAFT DIMENSIONS	
SPAN, TIP-TO-TIP	
NO. OF ACTIVE TRANSPONDERS	3
NO. OF SPARE TRANSPONDERS	2
TWTA POWER OUTPUT	100 Watts
FREQUENCY RANGE:	
UPLINK	
DOWNLINK	
TRANSPONDER BANDWIDTH	
EIRP MINIMUM, BEAM EDGE	
CHANNEL POLARISATION	

COLUMBIA COMMUNICATIONS CORPORATION

COMMERCIAL DETAILS

OPERATOR	Columbia Communications Corporation
ADDRESS	8th Floor, 1136 Union Mall, Honolulu, HI 968134, USA
TELEPHONE NO.	(808) 523 1770
SENIOR STAFF	Clifford Laughton, President Deke Slayton
PURPOSE OF PROJECT	Provision of Ku-band capacity for the Pacific Ocean region
PROJECT STATUS	Conditional authorisation has been granted by the Federal Communications Commission
PROJECT VALUE	Not available
LAUNCH DATES	Not available

SATELLITE USERS

Details of potential users of this proposed system are not available at this stage.

PROJECT OVERVIEW

One of many new venture companies that sprang up the first half of the 1980's to develop and operate a private telecommunications satellite system, Columbia Communications Corporation has survived with radically altered plans to offer capacity in the Pacific Ocean region. Columbia Communications Corporation is a subsidiary of Columbia Astronautics Corporation.

The company originally requested authorisation from the Federal Communications Commission for a hybrid Ku-band and C-band two satellite system for US domestic communications. This plan envisaged each satellite having 24 Ku-band and 24 C-band transponders on HS–393C platforms. The satellites would have been located at 62·5 and 147·5 degrees West. The C-band capacity would have covered most of continental USA but the Ku-band capacity on each satellite would have provided only half-CONUS coverage. The FCC dismissed the company's application for this system in 1985 and the company presented an application for a satellite system for the Pacific Ocean Region. Conditional authorisation for this has now been granted by the Federal Communications Commission. This is for a single satellite in orbit with a ground spare. Each satellite would carry 44 Ku-band transponders.

COMMERCIAL DETAILS

OPERATOR	McCaw Space Technologies (satellite to be operated under the name Celestar International Satellite System)
ADDRESS	2000 116th Avenue NE, PO Box 8367, Bellevue, WA 98009, USA
TELEPHONE NO.	(206) 453-1115
SENIOR STAFF	Lon Levin, Attorney Joe Walter, Vice President
FUNDING OF PROJECT	McCaw expects "no trouble" in meeting financing needs through transponder contracts, debt and equity. McCaw is using Paine Weber to assist in raising finance
PURPOSE OF PROJECT	Provision of Ku-band capacity for International Telecommunications services in the Indian and Pacific Ocean regions
LAUNCH DATES	Celestar 1, 1991 by Shuttle, Ariane or comparable vehicle
PROJECT VALUE	June 1986 filing with the Federal Communications Commission shows cost of the entire programme as $382·1 m. This covers three satellites ($110 m), launch ($120 m) and insurance ($56 m) – insurance rate budgeted is 24%

SATELLITE USERS

McCaw Space Technologies envisage that services on the satellites would be offered to multi-national businesses, financial institutions, international organisations, video programmers and distributors. The services offered McCaw claims are not offered by Intelsat and will not be interconnected with the public switched network.

McCaw will offer transponders for sale and long-term lease on a non-common carrier basis. It sees its primary customers as financial institutions.

PROJECT OVERVIEW

McCaw Space Technologies filed the application in June 1986 with the US Federal Communications Commission for its Celestar International Satellite System. McCaw was already an applicant for domestic US mobile and radio determination satellite services but this new application put McCaw in the field of international satellite operations.

The plan involves two in-orbit operational satellites, plus one ground spare. The first satellite at 170° East would provide coverage of West Coast of North America, the South Pacific and the Asian Pacific rim. The second satellite, would be located at 70° East and cover the Indian Ocean region, Europe, Africa and Asia including India, China, Eastern and Southern Africa and the Middle East. Both satellites would be accessed from Guam in the Pacific Ocean which would allow double hop transmissions for the entire service area.

Each satellite would use 102 active transponders with 17 spare transponders for its 17 spot beams. On-board matrix switching would allow any uplink channel in any beam to be switched to any downlink channel in any beam on the satellite at 17° East. Twenty four transponders would be directed to North America and 12 towards Guam. Eleven spot beams with six transponders each would serve other regions. On the satellite at 70° East, 24 transponders would similarly be directed towards Guam but 13 spot beams with six transponders each would service other areas. Use of Guam as a hub for the system would allow inter-connection with the TPC–3 undersea cable.

Design of the satellites uses the RCA 5000 platform which is also planned for the Telesat Anik E system and Federal Express's Expresstar. At the time of writing RCA Astro-Electronics had not been selected for the construction contract.

TECHNICAL SUMMARY

PLATFORM	RCA 5000 (base configuration)
PRIME CONTRACTOR	RCA Astro-Electronics (expected)
OTHER MAJOR CONTRACTORS	
STABILISATION	3-axis
LAUNCH MASS	
MASS IN ORBIT	
POWER SUPPLY, END OF LIFE	
DESIGN LIFE	
ORBITAL POSITIONS	170° East (Celestar 1) 70° East (Celestar 2) plus one ground spare
STATION KEEPING ACCURACY	
ECLIPSE PROTECTION	
PROPULSION SYSTEM	
SPACECRAFT DIMENSIONS	
SPAN, TIP-TO-TIP	
NO. OF ACTIVE TRANSPONDERS	102 (per satellite)
NO. OF SPARE TRANSPONDERS	17 (per satellite)
SSPA POWER OUTPUT	5 Watts
FREQUENCY RANGE:	
UPLINK DOWNLINK	Ku-band
TRANSPONDER BANDWIDTH	36 MHz
EIRP MINIMUM, BEAM EDGE	
CHANNEL POLARISATION	
NO. OF SPOTBEAMS	17 (per satellite)

COMMERCIAL DETAILS

OPERATOR	CableSat General Corporation
ADDRESS	1150 West King Street, Cocoa, Florida 32922, USA
TELEPHONE NO.	(305) 632-1000
TELEX NO.	
FAX NO.	
SENIOR STAFF	Ray Kassis, President Robbie Gray, Assistant to the President
FINANCIAL BACKING	Entrepreneurial start up
PURPOSE OF PROJECT	Provision of C-band capacity for domestic US operations to be used by cable television channels
PROJECT STATUS	Has construction request filed with the Federal Communications Commission
PROJECT VALUE	not known
LAUNCH DATES	Unlikely to have a first launch before 1988

SATELLITE USERS

System is intended for use by cable network programmers but could also carry audio, data and other traffic.

Company says its goal is to encourage development of satellite cable networks that may be programmed through videoplexing to permit individual time zone feeds across the USA.

PROJECT OVERVIEW

This is an entrepreneurial start-up company aiming to provide C-band capacity for cable TV programmes in the USA. It is a project with major problems to overcome. Firstly, at the time of writing, there is a lot of surplus C-band capacity in the USA, owned by established operators who are not likely to view another C-band system with passive indifference. Secondly, the start-up type of satellite operation is not likely to obtain finance. This is especially significant at a time when satellite operators are having difficulty in obtaining insurance cover, a necessary pre-requisite. Moreover, it has to obtain FCC construction, launch and operating authorisation.

The company proposes two in-orbit operational satellites, between 60° West and 70° West and between 140° West and 150° West. The first would provide coverage of the Eastern and central time zones and be known as CableSat East. CableSat West is intended to provide coverage of the mountain and Pacific time zones, Hawaii and Alaska. Each satellite is planned to carry 24 FSS C-band transponders. One extra C-band transponder is planned for use by amateur radio operators and would be known as the Amateur Radio Network Experiment Transponder (ARNET). A segment of each of the 24 transponders would provide addressable emergency broadcast system message for weather or other emergency services.

TECHNICAL SUMMARY

PLATFORM
PRIME CONTRACTOR
OTHER MAJOR CONTRACTORS
STABILISATION
LAUNCH MASS
MASS IN ORBIT
POWER SUPPLY, END OF LIFE
DESIGN LIFE
ORBITAL POSITIONS
STATION KEEPING ACCURACY
ECLIPSE PROTECTION
PROPULSION SYSTEM

SPACECRAFT DIMENSIONS
SPAN, TIP-TO-TIP
NO. OF ACTIVE TRANSPONDERS
NO. OF SPARE TRANSPONDERS
TWTA POWER OUTPUT
FREQUENCY RANGE:
UPLINK
DOWNLINK
TRANSPONDER BANDWIDTH
EIRP MINIMUM, BEAM EDGE
CHANNEL POLARISATION

COMMERCIAL DETAILS

OPERATOR	Comsat General Corporation
ADDRESS	950 L'Enfant Plaza SW, Washington, DC 20024, USA
TELEPHONE NO.	(202) 863-7739
SENIOR STAFF	Elizabeth Young, VP, Marketing Don Flora, VP, Sales Robert W. Kinzie, President
PURPOSE OF PROJECT	Provision of C-band capacity for US domestic telecoms operators, notably AT&T, for CONUS, Alaska, Hawaii and Puerto Rico
PROJECT STATUS	System is nearing the end of its design life and is being replaced by Telstar 3 and GSTAR
PROJECT VALUE	Comsat invested $197 m in full for this system
LAUNCH DATES	Comstar D1, June 1976, Atlas-Centaur Comstar D2, July 1976, Atlas-Centaur Comstar D3, July 1978, Atlas-Centaur Comstar D4, February 1981, Atlas Centaur

SATELLITE USERS

The Comstar D system was leased to AT&T for the AT&T–GTE Domestic Communications Satellite System which carried voice, data and video traffic, for CONUS, Alaska, Puerto Rico and Hawaii. AT&T now has its own Telstar 3 system, in C-band, and GTE now has its own C-band capacity in Spacenet and Ku-band capacity in Spacenet and GSTAR.

Comstar D2 was used to provide replacement transponders to RCA Americom.

The capacity on the Comstar satellites was essentially used for long distance telecommunications traffic between major earth stations with 30–32 metre dishes. Each transponder has a capacity of about 1500 half circuits. It is estimated that by 1982, Comsat had recovered its initial investment through $200 m of revenue received by that date.

PROJECT OVERVIEW

Comsat General Corporation is one of six major subsidiaries of the Communications Satellite Corporation, commonly referred to as Comsat. Comsat general operates and owns the Comstar D system, along with the Marisat system and is planning its own Ku-band domestic satellite system for the USA (see Inmarsat I and Comstar K sections for further details of those two systems). Comsat General's approach has been to lease out its satellite capacity to major communications operators. The Comstar D system capacity was leased primarily to AT&T although that company now uses its own Telstar satellites. At one time two Comstar D satellites were leased to AT&T. By 1985 there was only one operational Comstar satellite left and Comsat lost its $1 m a month contract with AT&T. The Comstar D4 satellite is expected to be retired from 1988.

Comsat General's role as a domestic satellite operator dates back to a decision in 1972 by the Federal Government in favour of an "open skies policy". Prior to 1972, Comsat had been prohibited under the 1962 Communications Satellite Act from operating US domestic satellites. Although that act was responsible for the creation of Comsat and gave it a monopoly on the operation and supply of satellite services in the USA, it stipulated that those services could only be used for international transmissions. Indeed, despite developing the original Telstar 2 and two satellites, AT&T was also barred under the 1962 act from being a domestic satellite operator. It must be remembered that in 1962 it was not altogether certain that satellites would be a feasible commercial proposition, it was highly unproven technology and the US government felt that its most important role would be in trunk international telecommunications which could support the heavy financial commitment needed to develop a satellite system.

It also had foreign policy and trade considerations in the forefront of its thinking. The creation of Intelsat was a result of US efforts and it provided the country with an effective captive customer for its then new satellite technology. Comsat was formed as the US signatory to Intelsat. However, the 1962 Act, if read carefully, suggested that the Federal Government only intended a temporary monopoly for Comsat. The 1972 open skies policy was part of a much broader US trend towards liberalisation of telecommunications which climaxed in the break-up of AT&T in the 1980's.

The 1972 policy lead to a series of three domestic satellite systems. Western Union launched its first Westar satellite in 1974, RCA launched its first Satcom satellite the following year and Comsat launched Comstar for AT&T in 1976. A further three Comstar satellites were launched by 1981. Three have now been retired.

TECHNICAL SUMMARY

PLATFORM	Intelsat IV–A
PRIME CONTRACTOR	Hughes Aircraft Company
OTHER MAJOR CONTRACTORS	
STABILISATION	Spin
LAUNCH MASS	1520 kg
MASS IN ORBIT	864 kg (beginning of life)
POWER SUPPLY, END OF LIFE	610 Watts
DESIGN LIFE	7 years
ORBITAL POSITIONS	Comstar D4, 127·5° W Comstar D1–D3 now retired
STATION KEEPING ACCURACY	
ECLIPSE PROTECTION	
PROPULSION SYSTEM	
SPACECRAFT DIMENSIONS	2·4 m diameter, 6·1 m height
SPAN, TIP-TO-TIP	
CONTROL CENTRE	Washington DC

TT&C LOCATIONS	Southbury, Conn., Santa Paula, California
NO. OF ACTIVE TRANSPONDERS	24
NO. OF SPARE TRANSPONDERS	
TWTA POWER OUTPUT	5 Watts and 5·5 Watts
FREQUENCY RANGE:	
UPLINK	5·925–6·425 GHz
DOWNLINK	3·700–4·200 GHz
TRANSPONDER BANDWIDTH	34 MHz
EIRP MINIMUM, BEAM EDGE	31 dBW and 33 dBW
CHANNEL POLARISATION	Linear, frequency reused by horizontal and vertical cross polarisation
RECEIVE COVERAGE (G/T)	−9 dB/K
SINGLE CARRIER SATURATION FLUX DENSITY	−66·7, −72·7, −75·7 and −81·7 dBW/sq. m

COMMERCIAL DETAILS

OPERATOR	Comsat General Corporation
ADDRESS	950 L'Enfant Plaza SW, Washington, DC 20024, USA
TELEPHONE NO.	(202) 863-7739
SENIOR STAFF	Robert W. Kinzie, President Don Flora, VP, Sales Elizabeth Young, VP, Marketing
PURPOSE OF PROJECT	Provision of domestic US Ku-band capacity
PROJECT STATUS	Comsat has FCC construct, launch and operating authority but has yet to order the two satellites involved
PROJECT VALUE	The Company said in 1984 that it planned to spend $317·5 m on the Comstar K space and ground segment
LAUNCH DATES	First launch not before 1989

SATELLITE USERS

The Company has said that it intends to use the satellites for common-carrier and non-common carrier services. These include sales and long-term leases of capacity, occasional use, private network and end-to-end services.

PROJECT OVERVIEW

Strictly speaking, this system is Comsat's third attempt at providing Ku-band capacity for US domestic requirements. Its first was through its one-third shareholding in Satellite Business Systems, which it off-loaded to IBM and Aetna Life at the end of 1984. The second was its STC DBS system. It is still looking to sell these two satellites and has no intention of operating them. The Corporation has been under considerable financial pressure during the 1980's, partly as a consequence of these two aborted forays into satellite operations and partly because the organisation grew up as a rather overweight monopoly and has had difficulty in adjusting to the competition of the 1980's.

During 1985, it received construct, launch and operating authority for two new Ku-band satellites and construct authority for a ground spare. It was intended that launch would not take place before 1988. At the time of writing (Summer 1986) no firm order had been placed with a satellite manufacturer, and Comsat had asked for a six month extension of its deadline from the Federal Communications Commission to begin construction of the satellites.

TECHNICAL SUMMARY

PLATFORM	
PRIME CONTRACTOR	
OTHER MAJOR CONTRACTORS	
STABILISATION	
LAUNCH MASS	
MASS IN ORBIT	
POWER SUPPLY, END OF LIFE	
DESIGN LIFE	
ORBITAL POSITIONS	Comstar K1, 76° W Comstar K2, 134° W Comstar K3, ground spare
STATION KEEPING ACCURACY	
ECLIPSE PROTECTION	

PROPULSION SYSTEM	
SPACECRAFT DIMENSIONS	
SPAN, TIP-TO-TIP	
NO. OF ACTIVE TRANSPONDERS	16
NO. OF SPARE TRANSPONDERS	
TWTA POWER OUTPUT	40 Watts
FREQUENCY RANGE:	
UPLINK	14·0–14·5 GHz
DOWNLINK	11·7–12·2 GHz
TRANSPONDER BANDWIDTH	54 MHz
EIRP MINIMUM, BEAM EDGE	
CHANNEL POLARISATION	

Note: Use of spot beams has been considered.

COMMERCIAL DETAILS

OPERATOR	Telecommunications Satellite Corporation of Japan
ADDRESS	2-12-18 Shibadaimon, Minato-Ku, Tokyo 105, Japan
TELEPHONE NO.	432-28-01
SENIOR STAFF	Hiroshi Hirose, President
PURPOSE OF PROJECT	Provision of Ka- and C-band domestic satellite communications capacity
PROJECT STATUS	Both satellites are now in orbit and system operational
PROJECT VALUE	System, including launch, insurance and TT&C facilities cost $220 m
LAUNCH DATES	CS–2A, 4th February 1983, NII CS–2B, 6th August 1983, NII CSE, 15th December 1977, NI

SATELLITE USERS

Principal users of the CS–2A satellite are:

1) The Telecommunications Administration, (NTT, 2),
2) National Police Agency (NPA),
3) Ministry of Construction,
4) Japan National Railways (JNR),
5) Electric Power Companies Group,
6) Fire Department,
7) Ministry of Posts and Telecommunications (MPT).

The list of users masks the fact that the CS–2A satellite is underutilised (CS–2B is the in-orbit spare). The satellites were very expensive to construct and launch and each has limited capacity. Much of the use so far has been on a quasi-experimental basis.

PROJECT OVERVIEW

The CS–2 system's origins date back to 1973 when the Japanese government agreed to funding of what was then called an "Experimental medium-capacity Communications Satellite" (CSE). This was an R&D project to build up Japanese experience in the Ka-band frequency range. At the same time Japan committed itself to an experimental DBS satellite, BSE. Both were intended to be "test-beds" for operational domestic satellite systems, and both had to be small because the only operational Japanese launch vehicle in the period was the Delta derived NI. The CSE satellite was successful and is still used for experimental work. On the basis of its successful in-orbit operation, Nippon Telegraph and Telephone, the then state owned monopoly PTT, in conjunction with the Ministry of Posts and Telecommunications, finalised the go-ahead in 1979 for an operational communications satellite system, CS–2. This used a design very close to that of CSE.

The two CS–2 satellites were launched in 1983 to become Japan's first operational domestic satellite system. However the small satellites were expensive, both in unit cost, and, more importantly, in cost per transponder or cost per telephone circuit using the satellite system. NTT has had difficulty in justifying their use and even more difficulty in convincing customers to use them. Moreover, the satellites were heavily dependent on US technology; Japanese industry accounted for only about 30% of the work on the satellites. The USA saw the Japanese as blatantly protecting their satellite communications market by insisting that such expensive satellites be used. Also in the first half of the 1980's, Japan was moving towards liberalising its telecommunications operations, allowing new competition against NTT and part privatisation of the latter. Japanese business also had a vested interest in seeing new services develop. Consequently, and after US pressure, Japan, in 1985, opened up its domestic satellite communications business. The result is that

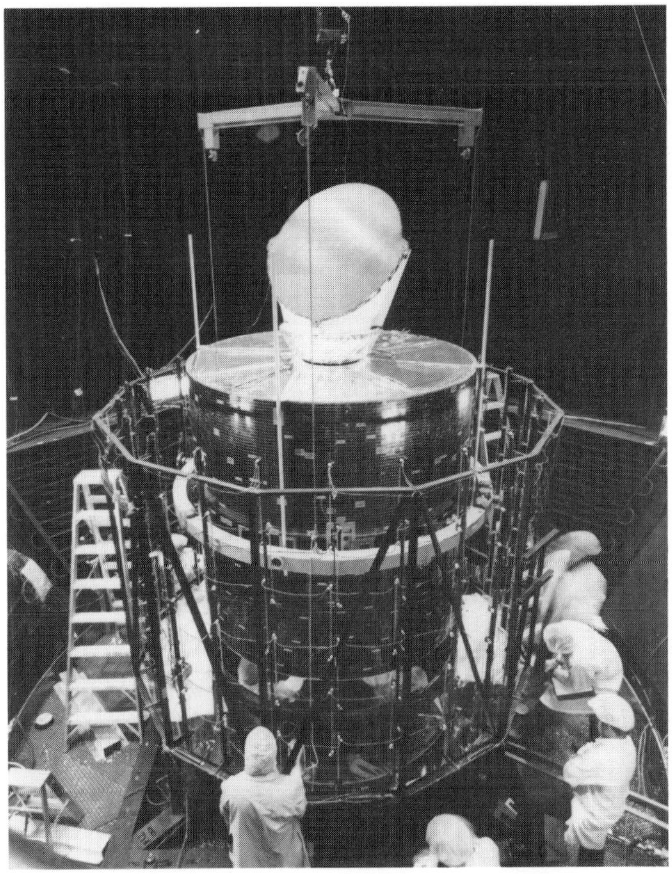

CS–2A spacecraft being prepared for pre-launch space environment tests at Ford Aerospace and Communications Corporation's Western Development Laboratories at Palo Alto, California

two joint Japanese/US ventures have obtained Class I telecommunications licences (the same licence as held by the privatised NTT) to provide domestic satellite communications services using satellites wholly manufactured in the USA. These joint ventures are Japan Communications Satellite, using Hughes built spacecraft, and Satellite Communications Corporation using Ford build satellites.

Nevertheless, the Japanese appear not to have the slightest intention of abandoning the development of their own domestic communications satellite programme. The follow-on replacements to CS–2, the CS–3 satellites, have already been ordered. Moreover, a fourth generation satellite system, CS–4 is being planned. This would involve much larger satellites than the small CS–2 and CS–3 satellites and provide capacity competitive in volume with US manufacturers. To support the CS–4 system, the Japanese government is already committed to funding a pre-operational version, ETS–VI, which like the CS–4 spacecraft, will be launched on H–II. Moreover, because of their uniqueness, the CS–4 satellites will not be cheap. Projected cost of a two satellite CS–4 systems is estimated at between $400 m and $600 m.

Thus the CS–2 satellites remain part of a broad strategy. One major element in this strategy has been not so much creating an ability for Japan to compete with US aerospace manufacturers but creation of a highly competitive ground equipment market. In reaching the latter objective, the Japanese have been spectacularly successful. This is despite the fact that the CSE, CS–2 and CS–3 satellites operate primarily in the Ka-band, a frequency range that has not, so far, been adopted for operational satellite use elsewhere in the world.

TECHNICAL SUMMARY

PLATFORM	NATO–III
PRIME CONTRACTOR	Mitsubishi Electric Corporation
OTHER MAJOR CONTRACTORS	Ford Aerospace
STABILISATION	Spin
LAUNCH MASS	670 kg
MASS IN ORBIT	350 kg (beginning of life)
POWER SUPPLY, END OF LIFE	375 Watts
DESIGN LIFE	5 years
ORBITAL POSITIONS	CS–2A, 132° E CS–2B, 136° E
STATION KEEPING ACCURACY	
ECLIPSE PROTECTION	
PROPULSION SYSTEM	
SPACECRAFT DIMENSIONS	2·184 m diameter, 2·045 m long 3·288 m deployed
NO. OF ACTIVE TRANSPONDERS	2 (C-band), 6 (Ka-band)
NO. OF SPARE TRANSPONDERS	
TWTA POWER OUTPUT	
FREQUENCY RANGE:	
UPLINK	
DOWNLINK	
TRANSPONDER BANDWIDTH	C-band, 180 MHz; Ka-band 130 MHz
EIRP MINIMUM, BEAM EDGE	30 dBW (C-band); 37 dBW (Ka)
CHANNEL POLARISATION	

COMMERCIAL DETAILS

OPERATOR	Telecommunications Satellite Corporation of Japan
ADDRESS	2-12-18 Shibadaimon, Minato-Ku, Tokyo 105, Japan
TELEPHONE NO.	432-28-01
PURPOSE OF PROJECT	To replace the CS–2 system in providing domestic C- and Ku-band satellite communications for Japan
PROJECT STATUS	Both satellites currently under construction
PROJECT VALUE	Not available
LAUNCH DATES	CS–3A, February 1988, H–I CS–3B, August 1988, H–I

SATELLITE USERS

The CS–3 system is the follow-on system to CS–2 and therefore its primary users are likely, at least initially, to be those using CS–2, notably NTT and its customers who are tied in to using the Ka-band frequencies. For new customers, the CS–3 system is likely to face competition from the inherently more economic JCSAT and SCS domestic satellite systems. The CS–3 satellites have a competitive advantage over use of the new private systems. CS–3 operates primarily in the Ka-band instead of Ku. Although Ka-band communications suffer from rain attenuation, the Ku-band is used heavily in Japan for terrestrial microwave communications. It is therefore difficult to coordinate Ku-band uplinks. It is possible that the Ku-band services may be limited to services that require few uplinks such as distribution of TV programming, leaving the Ka-band open to business services which may require many uplinks. The Ka-band may be more amenable to businesses than major carriers because the latter would find the rain attenuation problem even more unacceptable than business.

PROJECT OVERVIEW

Like the CSE and CS–2 satellites, the CS–3 continues with the use of a Nato–III Ford Aerospace derived spin stabilised platform, although CS–3 is larger. It also continues with the use of the Ka-band section of the frequency spectrum. The CS–3 satellites are almost certainly the last of the family. For the CS–4 and ETS–VI programme, Japan is switching to much larger three-axis stabilised platforms. It already uses three-axis stabilised platforms for its BS–2 and BS–3 DBS spacecraft. For operational purposes the two CS–3 satellites are straight replacements for the two CS–2 satellites now in orbit. From both a technical and commercial viewpoint, the CS–3 system is of relatively little interest, except for its use of the Ka-band. Like the CS–2 satellites, CS–3 satellites have limited payload and are expensive. The main feature of interest about CS–3 is its place in Japanese industrial policy towards satellite communications. This is briefly explained in the section on CS–2.

TECHNICAL SUMMARY

PLATFORM	NATO–III
PRIME CONTRACTOR	Mitsubishi Electric Corporation
OTHER MAJOR CONTRACTORS	Ford Aerospace & Comm. Corporation NEC (transponders)
STABILISATION	Spin
LAUNCH MASS	
MASS IN ORBIT	550 kg
POWER SUPPLY, END OF LIFE	800 Watts
DESIGN LIFE	7 years
ORBITAL POSITIONS	
STATION KEEPING ACCURACY	
ECLIPSE PROTECTION	
PROPULSION SYSTEM	
SPACECRAFT DIMENSIONS	
NO. OF ACTIVE TRANSPONDERS	2 (C-band), 10 (Ka-band)
NO. OF SPARE TRANSPONDERS	
TWTA POWER OUTPUT	
FREQUENCY RANGE:	
UPLINK	
DOWNLINK	
TRANSPONDER BANDWIDTH	
EIRP MINIMUM, BEAM EDGE	
CHANNEL POLARISATION	

CS–4 (SUPER CS)

COMMERCIAL DETAILS

OPERATOR	Telecommunications Satellite Corporation of Japan
ADDRESS	2-12-18 Shibadaimon, Minato-Ku, Tokyo 105, Japan
TELEPHONE NO.	432-28-01
PURPOSE OF PROJECT	To provide domestic communications capacity
PROJECT STATUS	Development work began in 1987
PROJECT VALUE	Total system including development, launch, insurance and control facilities is expected to cost between \$408 m and \$612 m
LAUNCH DATES	1992, by H–II 1994, by H–II

SATELLITE USERS

PROJECT OVERVIEW

Intended to follow on from the ETS–VI experimental satellite, CS–4 will employ a large platform and payload to match. Development work was not expected to start until at least 1987, with launch of a first satellite in 1992 and a follow on in 1994. The design is expected to carry between 10 and 20 beams and 60–70 operational transponders providing a capacity for 100,000 simultaneous telephone circuits. CS–2 satellites, in contrast each have a capacity capable of carrying no more than 6000 telephone circuits.

TECHNICAL SUMMARY

PLATFORM	ETS–VI
PRIME CONTRACTOR	Mitsubishi Electric Corporation (expected)
OTHER MAJOR CONTRACTORS	Ford Aerospace (expected)
STABILISATION	3-axis
LAUNCH MASS	
MASS IN ORBIT	2000 kg
POWER SUPPLY, END OF LIFE	
DESIGN LIFE	
ORBITAL POSITIONS	
STATION KEEPING ACCURACY	
ECLIPSE PROTECTION	
PROPULSION SYSTEM	
SPACECRAFT DIMENSIONS	
NO. OF ACTIVE TRANSPONDERS	60–70
NO. OF SPARE TRANSPONDERS	
TWTA POWER OUTPUT	
FREQUENCY RANGE:	
UPLINK	
DOWNLINK	
TRANSPONDER BANDWIDTH	
EIRP MINIMUM, BEAM EDGE	
CHANNEL POLARISATION	

COMMERCIAL DETAILS

OPERATOR	Direct Broadcast Satellite Corporation
ADDRESS	7315 Wisconsin Avenue, Suite 518E, Bethesda, Maryland 20814, USA
TELEPHONE NO.	(301) 652-4663
TELEX NO.	710 824 0096
SENIOR STAFF	Wilbur Pritchard, President Harley Radin, Vice President
PURPOSE OF PROJECT	Provision of DBS capacity for the USA
PROJECT STATUS	The Company gave up its construction permit in July 1985 in order to increase its requested number of transponders per satellite to 16 and decrease power output to 100 Watts.
PROJECT VALUE	DBSC signed a contract in April 1984 for \$177m for two satellites supplied by Ford Aerospace.
LAUNCH DATES	Not available, although DBSC had contracts with Arianespace for launch of the two satellites in late 1987 and early 1988

SATELLITE USERS

DBSC originally planned to offer all of the high power capacity on its DBS system to broadcasters and limiting its own activities to operating the satellite. It had signed an agreement with world communications for the latter to market the capacity as well as the spot-beam capacity on the satellite.

Original plans envisaged serving continental USA was two service areas – the eastern half and the western half, with six channels for each area. Spot beams were available to serve Alaska, Hawaii, Puerto Rico and the Virgin Islands and also allot extra channels to major population areas or targeted rural areas.

Users envisaged by DBSC were subscription and advertiser-supported television programmers and audio channels. It also envisaged business use for internal video, voice, data and facsimile transmissions.

PROJECT OVERVIEW

Direct Broadcast Satellite Corporation was a first round DBS applicant which received Federal Communications Commission authority on 9th October 1984 to construct and launch a ten-channel DBS system, having met the commission's "due diligence" requirements. It was apparently a close decision by the Commission as the company's President, Wilbur Pritchard, had to argue hard that his start up venture could raise the finance for the project. The commission's authorisation covered two in-orbit satellites each with six 200 Watt channels and 12 45 Watt spot beams. It became apparent in 1985 that there was little chance, at least immediately, of raising the finance needed for the project and Wilbur Pritchard was forced to sell the major part of his shareholding in DBSC just to continue operations.

In July 1985, the company gave up its construction permit, nominally to change the design of its space segment by increasing the number of channels per satellite to 16 but with 100 Watt TWTA's instead of 200 Watts. Whether this revised plan will attract the substantial funding needed remains an open question and is of course heavily dependent on the general future of DBS and the ability of start-up satellite operators to attract funds. DBSC received authorisation in September 1986 from the FCC for its revised satellite plan. It envisages two satellites, each serving half continental USA. There are also four channels on spot beams to provide regional coverage and increased power for possible High Definition TV.

COMMERCIAL DETAILS

OPERATOR	Deutsche Bundespost
ADDRESS	Telecommunications Central Headquarters, Am Kavalleriesand 3, 6100 Darmstadt, West Germany
TELEPHONE NO.	(6151) 83 49 08
TELEX NO.	4 19 511 FTZD
FAX NO.	(6151) 83 49 08
SENIOR STAFF	Otto Schmeller, Manager for Domestic Satellite Projects
PURPOSE OF SYSTEM	Provision of Ku and Ka-band domestic communications capacity for West Germany. Capacity is intended for point-to-point telephony traffic, business communications, data and cable TV channel distribution
STATUS OF PROJECT	Satellites under construction with launches in 1988 and 1989. However, the project has been delayed by technical and organisational problems as well as delays in the Ariane launch schedule. The system was due to become operational in 1986
PROJECT VALUE	Satellites were ordered under a fixed price contract worth DMks 815 million ($400 million) at 1982 prices. Prices includes two launches, ground equipment and project management support. It excludes satellite TT & C facilities. Existing facilities near Munich will be used in conjunction with two new Ku-band TT & C stations. The main TT & C facilities will be owned and operated by DFVLR. Split of the DMks 815 million contract was as follows: Space segment, 70% Earth segment, 25% System studies and project management, 5% The Contract with Arianespace for launch of the two satellites was worth approximately DMks 200 million ($100 million).
LAUNCH DATES	DFS-COPERNICUS 1, Ariane 4, V31 July 1988 DFS-COPERNICUS 2, Ariane 4, V35 March 1989 or V36, April 1989 DFS-COPERNICUS 3, ground spare

SATELLITE USERS

The DFS-Copernicus satellite system is intended for West Germany's domestic use only, including links to Berlin. It will provide capacity to follow on from West Germany's unsuccessful use of Telecom 1 capacity for TDMA business services, and a domestic alternative to Eutelsat SMS capacity. Additionally provided will be capacity for cable TV channels and terrestrial TV broadcasters seeking networking capacity, capacity for Deutsche Bundespost's trunk voice traffic needs and capacity for development of new business communications services, partially as a step towards ISDN services.

However, due to delays in the project, Deutsche Bundespost has purchased six Ku-band transponders on the Intelsat V satellite at 60 degrees East. These transponders are now allocated to TV channels which may not wish to move to DFS-Copernicus. At one time it was expected that the German terrestrial broadcasters, ARD and ZDF, would use the system for networking purposes. Other broadcasters using the Eutelsat I system for German language TV broadcasts may also not want to move to DFS-Copernicus for fear of losing their markets outside that country. The system is also likely to face competition from ASTRA and the German DBS satellite, TV–SAT.

DFS-Copernicus is intended to provide Ku-band capacity within the Federal territory of West Germany (excluding Berlin) for point-to-point trunk transmissions (voice and data), voice and telematic services for business, and capacity for TV programming.

Between the Federal Republic and Berlin it is intended to provide both Ka and Ku-band capacity for high capacity voice and data communications and TV transmissions. Services in the Ka-band will initially be offered on an experimental/preoperational basis but are expected to become fully operational in due course.

In all cases, Deutsche Bundespost is likely to have strong control over the use of the satellite system, including ownership of the earth stations at cable head ends. Liberalisation of telecommunications is proceeding slowly in Germany although the position is under review.

PROJECT OVERVIEW

Introduction: The DFS-Copernicus system (also spelled DFS-Kopernicus) is a West German domestic communications satellite series currently under construction and due for launch in 1988 and 1989. The need for such a domestic satellite system has been questioned both in the light of substantial capacity available from the late 1980s on the Eutelsat, Intelsat, ASTRA and TV–Sat satellites and the moderate success of the French DGT in finding business users and broadcasters for its Telecom 1 Ku-band capacity. The DFS-Copernicus system is seen by many in Europe as an engineering led project rather than commercially led exercise. This is perhaps not a surprising viewpoint as the satellite has been ordered by the PTT, Deutsche Bundespost. Like AT & T before divestiture and European PTTs in general, The West German PTT has been criticised for being engineering led, with only weak attention being paid to marketing. DFS-Copernicus has also been criticised as a product of West German industrial policy (as has TV–SAT), in that it was ordered to provide West German aerospace and communications equipment manufacturers with a domestic market on which to build an export base. According to the West German business daily newspaper *Handlesblatt* of 14th June 1982, the then West German Minister for Posts and Telecommunications believed "that the construction of a West German satellite is essentially for future export prospects in West German industry, since only companies possessing complete networks of ground based digital communications and transmission systems and satellite systems will remain competitive on the internationl market." This is perhaps unfortunate in that European satellite manufacturers have, without exception, been unsuccessful in selling satellites to organisations outside Europe. As well, the European market is too small to support a healthy industry which includes several prime contractors.

The project is now running over a year late, having been intended to become operational in 1986: the delays have been caused by both technical and organisational problems and. more recently, delays in the Ariane launch schedule. The system had been ordered in June 1983 (through a letter of intent) after long preliminary discussions within Germany. Development work started in July 1983. Delivery contract was signed in December 1983. The country had established its expertise in satellite communications with the Franco-German preoperational Ku-band Symphonie satellites of the early 1970s. At the time DFS-Copernicus was ordered, the country was already developing the TV–SAT DBS spacecraft as part of a Franco-German cooperative programme.

Planned services using DFS-Copernicus: Planned use of the DFS-Copernicus system falls into three sectors, TV traffic, business communications and point-to-point telephony for the West German public switched network.

1. TV services are intended to provide nationwide distribution of regional TV channels and distribution of cable TV channels to cable head ends. The intention was to allocate five 44 MHz Ku-band transponders to this service with five feeder earth stations located near to the studios of regional transmitters. However, regional broadcasters are already using the Intelsat V transponders owned by the PTT. However, the 44 MHz transponders (of which there are seven operational on each satellite) were also designed for point-to-point traffic. All four operational 90 MHz Ku transponders can each carry two high quality TV channels. The German broadcasting scene is currently changing rapidly with the advent of low power terrestrial TV stations, rapid extension of cable TV and potential use of the TV–SAT DBS system. The country is regarded as the most promising in Europe for new media developments. Therefore, it may be assumed that there will be demand for satellite capacity for TV traffic but that the future direction is uncertain.

TV services using the 44 MHz transponders will use receive terminals of about 3·5 metres diameter for professional activities. Uplink stations of 4·5 metres are planned as are portable TV transmit stations for use with the Ka-band 90 MHz transponder on each satellite. These are 3·5 metres in diameter.

2. A nationwide demand assigned TDMA network for business communications was originally envisaged. Capacity for this would include two of the seven 44 MHz transponders for "telematic" services. The network is seen as a step towards a West German ISDN network. An initial 30 small earth stations (3·5 metres) is included in the DFS contract together with TDMA references stations. This service was intended to follow on from Deutsche Bundespost use of Telecom 1 Ku-band capacity. Like Telecom 1, a digital transmission rate is 60 Mbit/s.

In the TDMA/DA system, demand assignment of capacity will use a two stage concept. Each traffic station will be assigned a variable basic capacity for use in the outgoing direction. If the traffic to be handled exceeds this capacity, the respective traffic station will receive an additional demand-assigned capacity from a centrally managed capacity pool. For two transponders and 30 traffic stations envisaged initially, a capacity of 880 equivalent 64 kbit/s duplex channels (without forward error correction) can be reached. Set up times are relatively short. The system is optimised for a network with 30 to 100 stations using 2–4 transponders. This corresponds to an average traffic of 20 to 40 Erlang with a bit rate of 64 kbit/s. More traffic can be handled through single channel per burst (SCPB) techniques. This is aimed at stations with very low traffic rates, whereas highly loaded stations will have up to 95% of their traffic using the variable, locally administered main traffic bursts. For these stations, SCPB will only handle traffic peaks.

System Configuration

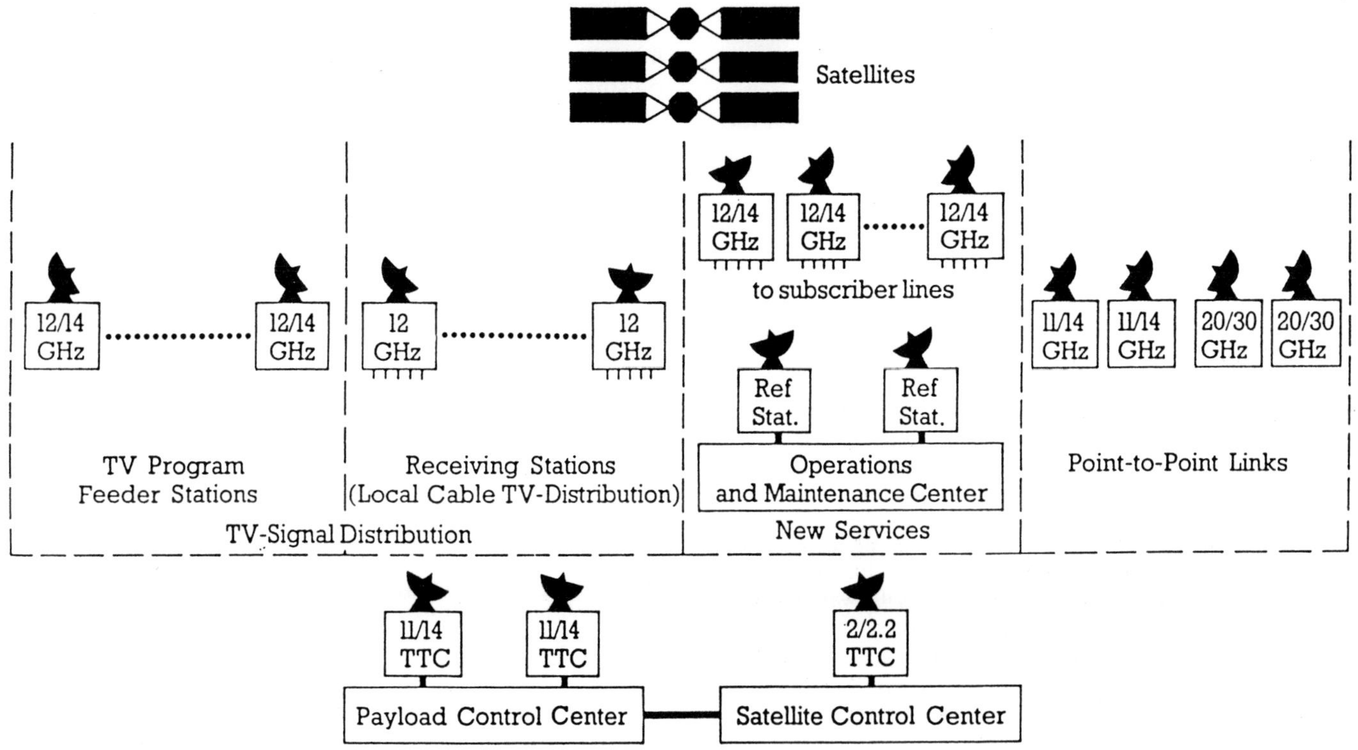

It is possible to connect terrestrial links directly to the satellite based TDMA ground segment. However, it is envisaged that the terrestrial links will go through switching equipment associated with the traffic stations where switches, traffic concentration and signalling will be adapted to handle the traffic. Additional digital access modules (30 channel multiplexers) will extend the coverage area within which users can connect their traffic to the traffic station. One of the principal roles of the switching system is to distinguish between traffic to be routed outside of the geographic area covered by the local end of the system and that which is to be routed to areas within it. In the latter case, there would not normally be a need to use the satellite capacity as the message would be routed terrestrially. Up to 100 users could be connected through the switching system to the traffic station.

All single channel subscribers to the satellite TDMA service, that is, subscribers with 72 kbit/s requirements, and PABX tielines will be connected to digital access modules before access to switching. Wideband users will be connected directly to switching equipment via 2·048 Mbit/s links. Up to 17 terrestrial links of 2·048 Mbit/s could be handled by the switching system which is connected via six 2·048 Mbit/s trunk links to the TDMA system. Where terrestrial links are connected directly to the TDMA system, a 2·048 Mbit/s channel is envisaged.

As a further step towards ISDN, it is also envisaged that at a later date, it will be possible to distribute the traffic flow by adding the switching equipment as an interface to the terrestrial network. This would allow the end user to have only one line for transmission of information. The switching equipment would identify whether a given piece of traffic would be best suited for onward transmission via satellite or the terrestrial network.

The TDMA service involves two types of call; dialled calls and reserved calls. The reserved calls differ from dialled calls with respect to the set-up procedure, the grade of service during set up and security factors. A reserved connection is set up automatically by the traffic sation's switching equipment at the requested time and using reserved capacity in areas where blocking could occur. With this service, point-to-point, point-to-multipoint and conference connections (video and voice) are possible.

It is not clear if Deutsche Bundespost intends to offer FDMA based VSAT services through DFS-Copernicus. Such services, with terminals located on customer premises, would provide a complete by-pass of the terrestrial network and, perhaps, call into question the organisation's ambitions to provide a national ISDN network on which initial work is underway.

3. Point-to-point communications through DFS-Copernicus are envisaged to bridge temporary bottlenecks in the terrestrial network and to provide a high quality independent transmission link. The point-to-point services are envisaged to use three 11/14 GHz (90 MHz bandwidth) transponders and one 20/30 MHz transponder (90 MHz bandwidth). A maximum of one duplex 140 Mbit/s connection (two transponders), two TV channels (one transponder) and one 70 Mbit/s duplex connection (via the Ka-band transponder) can be implemented. The 44 MHz Ku-band transponders can also be used for point-to-point links, allowing a bit rate of 70 Mbit/s per transponder. A bit/rate of 140 Mbit/s allows transmission of 1920 simplex channels per transponder. This number can be doubled by using digital speech interpolation techniques. For date transmissions 1920 lines with 64 kbit/s can be used. This falls to 1440 lines if forward error correction is required.

The Ku-band earth stations for point-to-point services will be Intelsat Standard C (18 metre) systems for high capacity trunk links and high quality TV transmissions on 90 MHz transponders or a smaller design (8 metres) for medium to high capacity requirements, including TV point-to-point transmissions through 44 MHz transponders. The Standard C stations are not included in the contract price for DFS-Copernicus. For the Usingen-Berlin services, stations currently used with the Eutelsat system will be taken over. For the Ka-band point-to-point services, new 11 metre stations were included in the original DFS-Copernicus contract. One station will be located at Usingen and one in Berlin.

For telephone and data transmissions Quadrature Phase Shift Keying (4 PSK) techniques with coherent modulation will be used. If two broadband transponders are available, the forward and the backward directions will be transmitted over 140 Mbit/s PSK modems. In the case of only one transponder being used for duplex operation, two PSK-carriers will have to be transmitted.

The DFS-Copernicus space segment: The system will consist of three satellites of which one is designated as a ground spare. The in-orbit operational satellite, DFS-Copernicus 1, will be located at 23·5 degrees East and the in-orbit spare at 28·5 degrees East. These conventional spacecraft are small, having a launch mass of 1400 Kgs. However, they will be the first operational satellites outside the Far East to use the Ka-band frequency. However, most of the capacity will be in the Ku-band, thus following the European norm. Bipropellant propulsion is employed in this 3-axis stabilised design.

The design carries a payload of two antennas, eleven active and six redundant transponders. Use is made of a combined off-set Gregorian antenna – one operating at 20/30 MHz and the Other for Ku-band capacity. Linear polarisation of signals is employed with frequency reuse in the Ku-band. The two 14 GHz FET receivers will have a redundancy of two out of four. The redundant 30 GHz receiver will be equipped with a parametric amplifier. In the 11/14 GHz band there will be five transponders for three active ones. In the 12/14 GHz band ten transponders for seven active ones and in the Ka-band two transponders for one active one. TWTA power output is 20 watts per TWTA. Twenty seven watt TWTAs had been considered for the Ka-band service but this option appears to have been dropped.

The platform is body stabilised using a momentum wheel and the bipropellant propulsion system. Both the thrusters for stabilisation and positioning and the apogee motor will be fed from the same capillary tanks. Pointing accuracy of the satellite is between $+/-0{\cdot}16$ degrees. The power supply is designed for full eclipse protection with two redundant nickel cadmium batteries. Thermal control is passive but with heater support. Service life of the satellite is 10 years. All components, including the solar generator and fuel system, are designed for 10 years of operation.

The tracking, telementry and command system operates in two frequencies. For the transfer phase and in case of an emergency during operation, S-band will be used. During normal operations TT & C signals will use the Ku-band. The S-band TT & C station is owned by DFVLR and has been used for previous missions. It will be connected to the satellite control centre, which, during the transfer stage, will make use of NASA and ESA S-band stations. Two new Ku-band TT & C stations are being installed and connected to the payload control centre.

Soon after the order for the DFS-Copernicus system was placed, the contractors involved established a formal consortium, Gesat, to market the design on a worldwide basis as well as ground equipment, and also provide consultancy services. Since it was established in October 1984, the marketing company has not registered an order for a satellite but, at the end of 1986 said that it was having "serious discussions in South America, Africa and Asia" and was looking at adapting the DFS-Copernicus design to the needs of developing countries. Plans did not require a "complete redesign" of the satellite.

TECHNICAL SUMMARY

PLATFORM	DFS-Copernicus
PRIME CONTRACTOR	DFS Consortium led by Siemens and including ANT Nachrichtentechnik, MBB and Standard Elektrik Lorenz
STABILISATION	3-axis
LAUNCH MASS	1,400 Kgs
MASS IN ORBIT	815 Kgs
POWER SUPPLY, END OF LIFE	1,600 Watts
DESIGN LIFE	10 years
ORBITAL POSITIONS	DFS-Copernicus 1, 23·5 degrees East DFS-Copernicus 2, 28·5 degrees East DFS-Copernicus 3, ground spare
STATION KEEPING ACCURACY	+/−0·07° (E/W, N/S)
ECLIPSE PROTECTION	Full
SPACECRAFT DIMENSIONS	
SPAN, TIP-TO-TIP	
ACTIVE TRANSPONDERS	Ku-band, 3 @ 90 MHz, 7 @ 44 MHz, Ka-band, 1 @ 90 MHz
SPARE TRANSPONDERS	Ku-band, 2 @ 90 MHz, 3 @ 44 MHz, Ka-band, 1 @ 90 MHz
TWTA POWER OUTPUT	All at 20 watts
UPLINK FREQUENCIES	14·000–14·250 GHz 14·250–14·500 GHz 29·130–30·300 GHz
DOWNLINK FREQUENCIES	12·500–12·750 GHz 11·450–11·700 GHz 19·330–20·230 GHz
TRANSPONDER BANDWIDTH	see Active transponders
EIRP, MINIMUM, BEAM EDGE	49·1 dBW (12·5–12·75 GHz) 49·2 dBW (11·45–11·7 GHz) 48·0 dBW (19·33–20·23 GHz)
POLARISATION	Linear in both Ku and Ka with frequency reuse in Ku
RECEIVE COVERAGE (G/T)	9·6 dB/K (12/14 GHz), 8·9 dB/K (11/14 GHz), 7·7 dB/K (20/30 GHz)

Transponder Frequency Plan

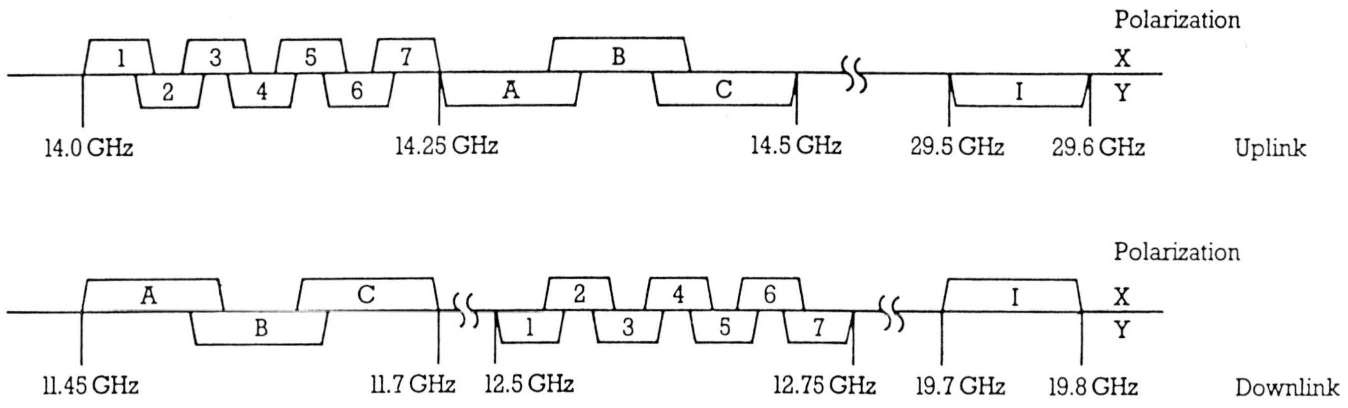

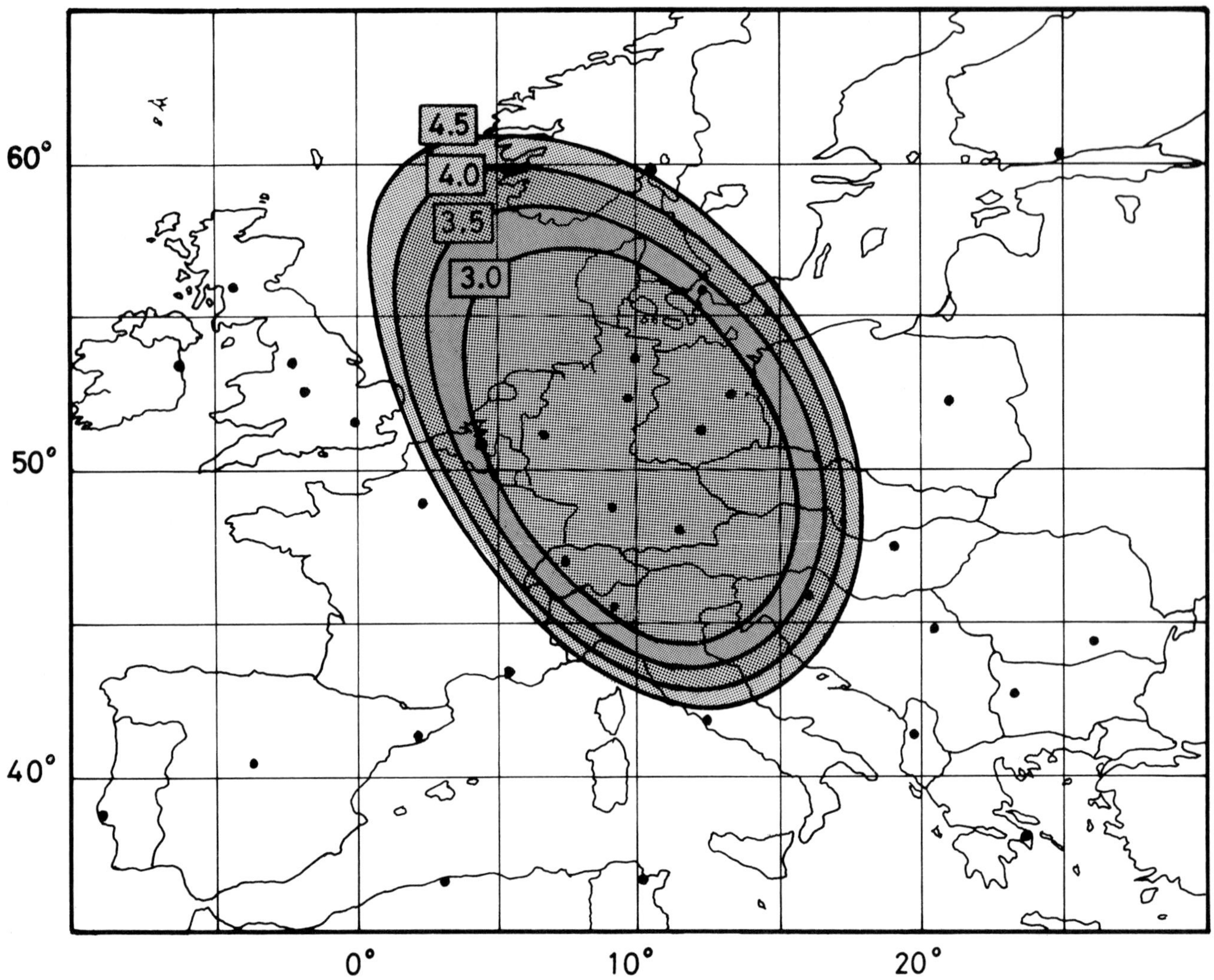

Footprint of TV capacity available on DFS-Copernicus. Figures quoted are size of Antennas required to pick channels up.

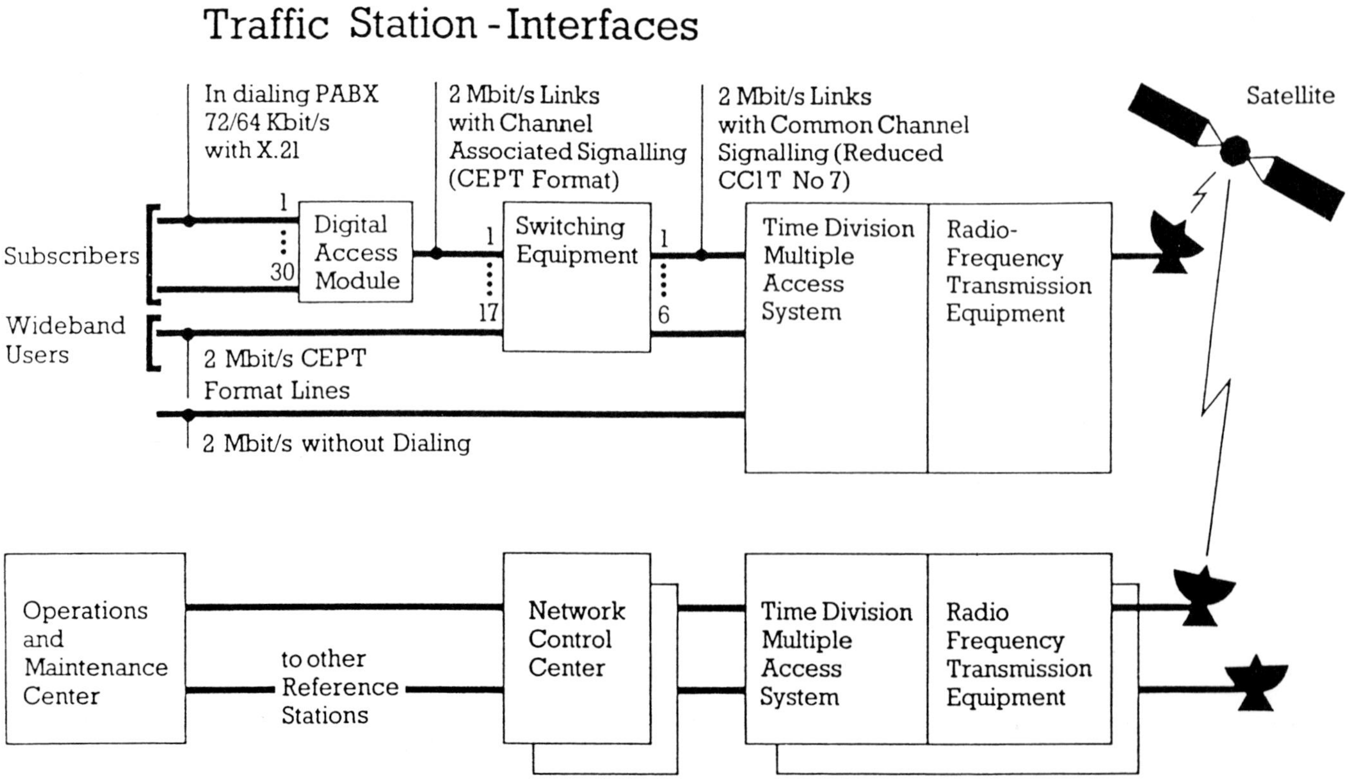

Interfaces to the Switching Network

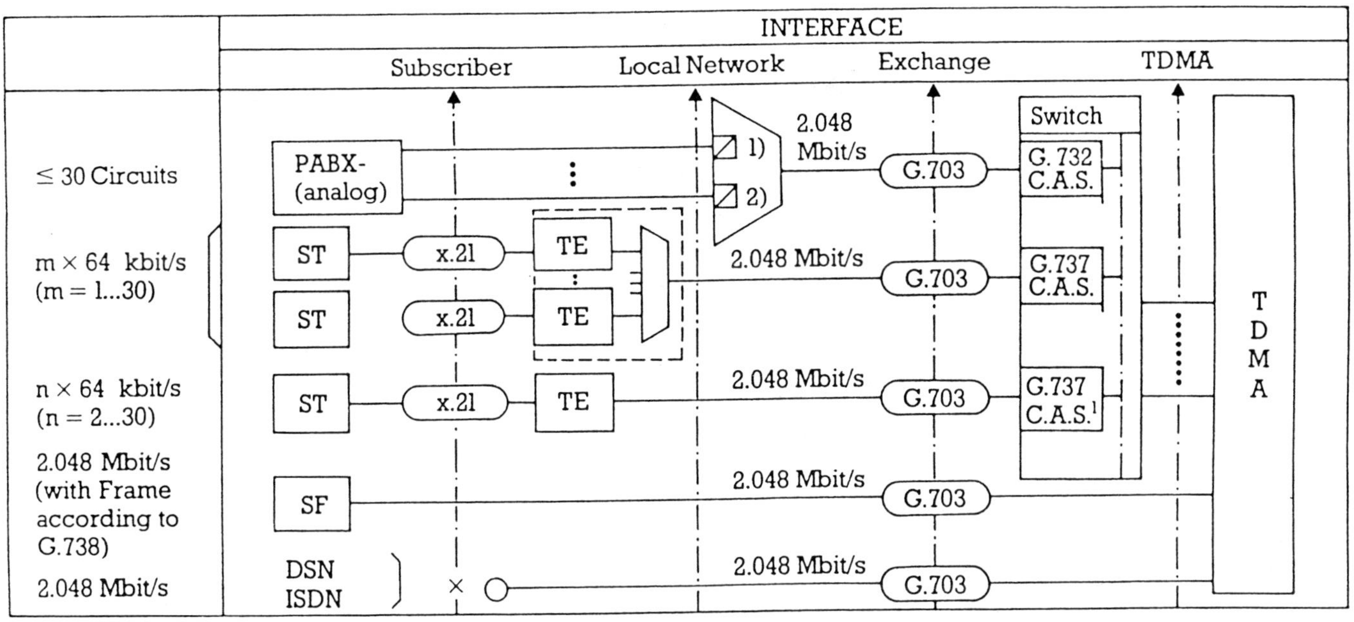

ST:	Subscriber Terminal	ISDN:	Integrated Services Digital Network	1)	A/D Conversion and Adaptation
TE:	Transmission Equipment	DSN:	Digital Switching Network		
TDMA:	Time Division Multiple Access	X:	Exchange	2)	Direct Connection to TDMA System possible
PABX:	Private Automatic Branch Exchange	C.A.S.:	Channel Associated Signalling		

Earth Station 12/14 GHz Specifications

		Traffic Station				Reference Station	
		Internal Zone		External Zone			
		Transmitter	Receiver	Transmitter	Receiver	Transmitter	Receiver
Frequency/GHz		14.0 – 14.25	12.5 – 12.75	14.0 – 14.25	12.5 – 12.75	14.0 – 14.25	12.5 – 12.75
Min. power flux density		– 116.6 dB (W/m²)		– 117.8 dB (W/m²)			
Antenna:	Reflector diameter	3.5 m		4.5 m		4.5 m	
	Polarization	vertical	horizontal	vertical	horizontal	vertical	horizontal
	3 dB Beamwidth	0.36°	0.4°	0.28°	0.31°	0.28°	0.31°
	Gain	51.7 dB*	50.7 dB**	54.3 dB*	53.2 dB**	54.3 dB*	53.2 dB**
Receiver:	Type		GAS FET, uncooled		GAS FET, uncooled		GAS FET, uncooled
	Noise temperature		270 K		270 K		270 K
	Sensitivity**		25.5 dB/K		28.0 dB/K		28.0 dB/K
Transmitter:	Type	TWT, air cooled		TWT, air cooled		TWT, air cooled	
	Power	250 W		250 W		600 W	
	Output power level	22.6 dB (W)		22.6 dB (W)		27.7 dB (W)	
	EIRP	72 dB (W)		74.5 dB (W)		75 dB (W)	

*at 14 GHz and input of ortho mode transducer   **at 12.5 GHz and input of low noise amplifier

Earth Station 20/30 GHz-Specifications

	Transmitter	Receiver
Antenna reflector diameter	11 m	
Polarization	linear	
Tx/Rx	vertical/horizontal	
Min. power flux density	−115 dB (W/m²)	
Tracking system	monopulse	
Redundancy	step track	
Beacon frequency	19.7 to 19.705 GHz	
	Transmitter	Receiver
Frequency range	29.5 to 30.0 GHz	19.7 to 20.2 GHz
Antenna:		
3 dB Beamwidth	≈ 0.06°	≈ 0.09°
Gain	≥ 68.5 dB*	≥ 64.95 dB**
Receiver:		FET, helium cooled
Noise temperature		100 K
Sensitivity		42.2 dB**
Transmitter:	TWT, air cooled	
Power	1 KW	
Output level	27.9 dB (W)	
EIRP	≥ 92.5 dB (W)*	

*at 30 GHz ** at 20.2 GHz

COMMERCIAL DETAILS

OPERATOR	Dominion Video Satellite Inc
ADDRESS	801 Laurel Oak Drive, Suite 710, Naples, Florida 33940, USA
TELEPHONE NO.	(813) 262-1433
TELEX NO.	
FAX NO.	
SENIOR STAFF	Robert W. Johnson Pelle Karlsson (Director)
FINANCIAL BACKERS	
PURPOSE OF PROJECT	Domestic DBS operator providing capacity and some programming for an alleged market of 8–12 million evangelical christians in the USA
PROJECT STATUS	Company has received construct, launch and operating authorisation from the Federal Communications Commission for its two satellite system. It has also met FCC "due diligence" requirements. However the company is understood to be considering switching from 230 Watt transponders to 100 Watts.
PROJECT VALUE	Cost of two satellites, TT & C facilities, launch and insurance costs are estimated at $250m
LAUNCH DATES	Not available Not available

SATELLITE USERS

Dominion expects to programme one channel on its DBS system. Christian Broadcasting Network says it plans a family programming channel. An educational channel providing tuition on a pay-per-view basis is planned. Dominion says it expects to provide an all new channel as a joint venture with other companies.
Target market is some 8–12 million evangelical christians which Dominion says are underserved by religious and family programming. Programming is expected to be provided on a 24-hour-per-day basis. Channel uses are as follows:

RARC Channel EAST	WEST	Service
02	04	Religious
03	08	Educational
10	12	General Video Entertainment Films
14	16	News, Public Affairs
20	18	General Video Entertainment Films
24	22	Narrowcasting, speciality programming
28	26	Sports, cultural pre-emptible channel for High definition TV
30	32	Films, Professional Business, Pre-emptible channel for High Definition TV

PROJECT OVERVIEW

An odd proposal for a DBS TV system but a survivor from eight original applicants for DBS operations and now holding Federal Communications Commission construct, launch and operating authorisation, having met the commission's "due diligence" requirements. The commission requires that the potential operator meet these due diligence requirements before they assign orbital positions and channels (i.e. frequencies). The requirements mean that the operator has to have signed up with a manufacturer for supply of the satellites and a similar agreement with a launch vehicle operator. Both satellites are planned to be co-located at 119° West, with one covering the Western half of the USA, the other the Eastern. Both satellites are capable of transmitting by use of either quarter or half-CONUS beams, each with a beam edge EIRP of 54 dBW.

The Company has also had discussions with Comsat regarding the purchase of the two STC spacecraft but no firm commitment to use them has so far been forthcoming.

TECHNICAL SUMMARY

PLATFORM	HS-394
PRIME CONTRACTOR	Hughes Aircraft Company
STABILISATION	Tri-spin
LAUNCH MASS	
MASS IN ORBIT	
POWER SUPPLY, END OF LIFE	5000 Watts
ORBITAL POSITIONS	DCS-1, 119° West DVS-2, 119° West
STATION KEEPING ACCURACY	
DESIGN LIFE	10 years
PROPULSION SYSTEM	
NO. OF CHANNELS PER SATELLITE	8 (operational)*
TWTA POWER OUTPUT	230 Watts*
EIRP MINIMUM, BEAM EDGE	54 dBW*

CHANNELS ASSIGNED	*No*	*Pol.*	*frequency centres GHz*
DVS-1	2	LHC	17·38858/12·23858
	6	LHC	17·3969/12·2969
	10	LHC	17·45522/12·35522
	14	LHC	17·51354/12·41354
	20	LHC	17·60102/12·50102
	24	LHC	17·65934/12·55934
	28	LHC	17·71766/12·61766
	30	LHC	17·74682/12·64682
DVS-2	4	LHC	17·36774/12·26774
	8	LHC	17·42606/12·32606
	12	LHC	17·48438/12·38438
	16	LHC	17·5427/12·4427
	18	LHC	17·57186/12·47186
	22	LHC	17·63018/12·53018
	26	LHC	17·6885/12·5885
	32	LHC	17·77598/12·67598

*Under review

COMMERCIAL DETAILS

OPERATOR	European Business Satellite
ADDRESS	Apelvagen 42 S-182 75 Stockshund, SWEDEN
TELEPHONE NO.	(8) 856204
TELEX NO.	14036 ibscom s
PRINCIPAL OFFICERS	Leif Lundquist, Managing Director (Lundquist estimates he will need a staff of around 75 to run EBS)
FINANCIAL BACKERS	EBS is essentially a venture capital backed company looking for the main body of its finance from partners in Sweden, in Norway and the UK where it has established subsidiaries into which partners are expected to invest (see overview for current shareholders)
PURPOSE OF PROJECT	Provision of capacity dedicated to business use via earth terminals with dishes ranging from 1 to 5 metres. Coverage would be European
PROJECT STATUS	Currently seeking operating licences and subsequent funding
LAUNCH DATES	Not available
PROJECT COSTS	Not known, possibly in excess of $200m is required for a 2 satellite system

SATELLITE USERS

The Company says that the main categories of user to whom EBS will initially offer either the purchase or lease of transponders will include:
—national and international corporations spread over several sites and/or countries
—distributors of news and data bases
—suppliers of various types of value-added network services
—government and public agencies, including embassies and consulates or networks of regional offices
—distributors and broadcasters of radio and television programming
—national PTTS (including deregulated bodies) and international groups of PTTS.

Examples of specialised services EBS intends to provide are
—private corporate networks
—video teleconferencing
—corporate training, education and conference services
—data base dissemination
—remote printing
—tele-courier services
—radio and television programme relay
—general broadband network applications.

PROJECT OVERVIEW

This is a venture capital backed scheme aiming to provide a business satellite system for Europe completely outside the PTT framework and in competition with Eutelsat, Intelsat and domestic systems (DFS-Kopernikus and Telecom 1) although its head, Leif Lundquist, says it will not compete with the pre-operational TELE-Y satellite. To get this project off the ground, Lundquist needs 70 licences for the system from national governments in Europe, virtually all of which have international obligations to Eutelsat and Intelsat which they may, in effect, have to breach in giving EBS a licence. Lundquist has concentrated on obtaining licences in three countries, Sweden, the UK and Norway, although none so far have granted licences. Licences are a pre-requisite to obtaining the funding for this system.

European Business Satellite is headed by a parent company in Sweden EBS AB, which is 95·2% owned by management venture capital AB of Sweden and 4·8% by Lundquist. MVC in turn is owned in equal shares by four Swedish companies, Enator AB, Praktikertjanst AB, Forvaltnings AB Sekretaren and Skandiu Investment AB. EBS AB has a joint venture in Norway, EBS A/S in which shipping company Klosters Reden A/S has a 50% shareholding. EBS AB has a 100% owned subsidiary in the UK, EBS European Business Satellite Ltd. However, total funding is currently no more than seed corn money and Eutelsat appears unconcerned about the prospect of competition from EBS. EBS says it will form national companies in other European countries. The national companies will together own the satellites and the controlling earth stations. The national share will be matched to the market share of each national company.

The growth of business services via satellite has so far been somewhat erratic in both North America and Europe, partly because it has been exploited on a trial and error basis. SBS, for example, has yet to make money on its North American system and the French DGT appears to have been grossly over ambitious in choosing advanced TDMA techniques for Telecom 1 which involve very expensive ground equipment. Part of the problem is that the introduction of new services has been handled by satellite operators rather than specialist service companies. The development of ground terminal technology currently coming on steam, and expected to continue for the foreseeable future, and, in the 1990's the introduction of on-board satellite switching, are expected to lead to substantial growth in the business communications market. Currently active in the industry, VSAT, is already resulting in a log of new US developments. Thus, despite the apparent slow progress of EBS it is probably aiming at the most promising future market in Europe.

EBS claims that the technical and operational characteristics of the proposed EBUS spacecraft are substantially more advanced than those of any other proposed or operational European satellite system. Primarily this is because of the flexible access techniques which will be used in the EBS system. It will allow customers to use a number of different types of advanced coding, modulation and access techniques. Small and inexpensive "micro" (VSAT) earth stations with antenna diameters ranging from 1–2 metres can be used to transmit and receive data at rates up to about 200 kbit/s. Higher volumes of traffic could use dishes of between 2 and 5 metres.

TECHNICAL SUMMARY

PLATFORM	To be chosen from amongst tried and tested options
ORBITAL POSITIONS	EBUS-1, 4° West EBUS-2, 7° West
NO. OF ACTIVE TRANSPONDERS	16
FREQUENCY BAND	11/14 GHz

COMMERCIAL DETAILS

OPERATOR	Atlantic Satellites Limited
ADDRESS	Shell House, 20 Lower Hatch Street, Dublin, Ireland
TELEPHONE NO.	0001 785722
TELEX NO.	90696 pdf ei
FAX NO.	0001 606762
SENIOR STAFF	James J. Stafford, Chairman Jerry Farrell, Managing Director Donald Flynn, Finance & Administration R. F. Alden, Director Albert (Bud) Wheelon, Director Steven Dorfman, Director J. E. Koehler, Director
PURPOSE OF PROJECT	Provision of DBS capacity for the UK. Provision of FSS capacity for Europe and across the North Atlantic
PROJECT STATUS	Still at the planning stage
PROJECT VALUE	Full system (two satellites) including launch and insurance costs is expected to cost \$300 m
LAUNCH DATES	First launch not before 1989

SATELLITE USERS

Irish interest in providing DBS capacity has been aimed at the UK market rather than its own small and well provided domestic market. Atlantic Satellites and the Irish government have had extensive discussions with the British government and the UK Independent Broadcasting Authority about use of the satellite by those companies seeking franchises for DBS from the Authority. Atlantic Satellites has also talked directly with the bidders. At the time of writing none has expressed a strong interest. There is one Irish based consortium interested in using the DBS capacity for a TV channel backed by US interests. This is the rump of the Westsat consortium, which failed in its bid to obtain a satellite franchise from the Irish government.

Atlantic Satellites was expected to begin strong marketing of its capacity in the Autumn of 1986 when it opens its London office. It has already had discussions with British Telecom about the latter's use of the FSS payload. BT has apparently rejected the idea. AS says that it is initially concentrating on marketing the FSS capacity for European services.

PROJECT OVERVIEW

Introduction: Although nominally a Dublin based company, Atlantic Satellites is 80% owned by Hughes Communications of the USA. It is aiming to use the 31° West orbital slot allocated to the US and Ireland at WARC '77 to provide a two satellite hybrid system for US DBS and FSS services for Europe and across the North Atlantic. It currently envisages that one of the two satellites would be operated with control facilities in the UK and the other with control facilities based in Shannon in the Republic of Ireland. Each satellite would be providing in-orbit back up capacity to the other. The entire project is in direct competition with the Eutelsat–I and II systems and the Luxembourg Astra system. It is also a route for Hughes Aircraft Company to enter the European market to sell its satellites.

Background to the Project: Like most of the countries in Western Europe, Ireland was allocated five channels at WARC '77 for DBS TV. It was also allocated the 31° West orbital position along with the UK, Spain, Portugal and Iceland. As a small country (population 3·5 m), already heavily cabled and with a good variety of TV (two RTE Irish channels and the four UK national channels), it was one of the least likely of European countries to develop its own national DBS system. Therefore, Ireland has looked to other markets to exploit the DBS capacity allocated to it. In practice this has meant that its thinking has been closely tied to events in the UK. It has had discussions with Spain and Portugal about share use of a DBS system but this idea was abandoned because of lack of interest. Much of the thinking behind the current plan is based on the idea that if the UK fails (again) to get a national DBS project together, the Irish would step in and provide the capacity to any broadcaster interested in it.

The Republic was a little late in getting its act together on DBS TV. Strong moves were started in the UK in 1979 to get a DBS project together, after it was apparent that West Germany and France would develop their own domestic DBS systems (TV–SAT and TDF–1). The European Space Agency had been looking at an experimental DBS platform (H–SAT) as long ago as 1972 and this eventually, in 1979, came out as Olympus. It was not until 1983 that the Irish government advertised for franchises to operate a national DBS system, by which time the UK was deeply involved in the Unisat project to provide two DBS channels operated by the BBC and six FSS transponders for North Atlantic services.

The closing date for the Irish franchise bids was August 1984 and there was some interest. Four consortia submitted plans to the Department of Communications in Dublin. These were Atlantic Satellites, then a new venture backed by Irish businessman James Stafford, Westsat (a consortium of Bord Telecom (the national telecoms administration), RTE (Radio Telefis Eirann, the state owned public service broadcaster), with Guiness Peat Aviation and Allied Irish Investments, Dublin based Independent Newspaper Group and United Satellites of the UK. It appears that United Satellites was never in the running and it closed up shop (in June 1985) before the "winner" was announced. At that time it appeared that UK DBS prospects had totally disappeared as the government sponsored "21 consortium" had broken up. The 21 consortium had found that it could not produce a viable financial plan for DBS and also been consistently told by the British government that it had to procure satellites from the UK.

The Irish government also ruled out the Westsat consortium as it included two state owned companies, each with their own financial pressures. State ownership meant that ultimately the government would be underwriting a large proportion of the cost of DBS. The Independent Newspaper group was primarily interested in the programming side of DBS rather than the hardware. The government therefore announced in September 1984 that it had selected Atlantic Satellites to negotiate with over the award of the franchise.

Atlantic Satellites considered various options for procuring a space segment. Amongst its early thinking was use of the STC birds or purchase of a small satellite with a cheap launch. After having approached Hughes in mid-1985. Atlantic Satellites announced in March 1986 that Hughes Communications were taking an 80% shareholding in the company. The move had to be approved by the Board of General Motors, who by then had taken control of the company. It was a move that gave two advantages to Hughes. Firstly, it gave them a potential captive and first European customer for a satellite and placed Hughes communications with a foothold with which to expand its communications activities into Europe. Hughes Communications already has a foothold in Japan with its stake in the JCSAT partnership. Prior to the Atlantic Satellites deal, Hughes Aircraft Company has been somewhat overshadowed by RCA Astro-Electronics sales success with the Astra system. With the Intelsat system suffering from overcapacity and some strong doubts about the future of satellite use by telephony carriers, the satellite manufacturing industry has problems in finding sales. The move into Europe, like the move into Japan, reflects the better quality of US satellite systems.

The takeover of Atlantic Satellites by Hughes also involved appointment of a board consisting entirely of Hughes employees, although James Stafford retained chairmanship. As a sop to European politics, British Aerospace have been offered a "substantial" amount of subcontracting work on the satellites. The latter company is already a subcontractor to Hughes on the Intelsat–VI satellites.

Atlantic Satellites is expected to open an office in London, late in 1986, as part of its efforts to begin serious marketing of its satellite capacity. Recent developments in the company show a concerted effort to move forward with the project. It is understood that a marketing team will be built up in London to serve the whole of Europe. Although AS is not believed to be a preferred option to any of the current bidders for UK DBS franchises, the underlying weaknesses in the bidder's plans suggest that AS may be looking to mop up on the scene when the plans, as many believe, collapse. It was awarded its franchise by the Irish government at the end of 1986.

The Atlantic Satellites Proposal: The company is proposing two hybrid FSS/DBS satellites each with 10 DBS transponders of 120 Watts (a maximum of eight which could be operated simultaneously) on a beam switchable between coverage of Ireland and coverage of the UK. The transponders would provide an EIRP minimum at UK beam edge of 59 dBW, allowing TVRO dishes of 90 cm. The gradual fall off in performance around the beam edge means that cable headends in Europe would also be able to pick up the DBS channels. In practice, the DBS channels appear to be equivalent of European wide FSS channels. Each satellite would also have 16 FSS channels each using a 50 Watt transponder. These are for both European services and services across the North Atlantic. The European FSS coverage beam footprint is virtually identical to that of the Eutesat II satellite located at 36° East, both in terms of geographic coverage and EIRP performance. The FSS transponders would provide a minimum EIRP performance of 45 dBW over all of Western Europe and CONUS and 49 dBW for most of the two areas.

A maximum of eight DBS transponders could be operated at any one time, with one FSS transponder operating. Alternatively, 16 FSS transponders could be operational at any one time but under such circumstances no DBS transponders could be operational. Typically, the first satellite could offer three fully protected DBS channels and three back-up DBS channels in conjunction with five protected FSS channels and seven preemptible FSS channels ... "After the first spacecraft is filled, the second satellite can be launched to provide an additional 16 channels of FSS service as well as on-orbit back-up to the DBS channels. The following operational alternatives for the design:

	DBS CAPACITY	FSS CAPACITY
1.	No DBS transponders	and 16 50 Watt FSS transponders
2.	3 DBS transponders	and 12 50 Watt FSS transponders
3.	5 DBS transponders	and 8 50 Watt FSS transponders
4.	6 DBS transponders (on one satellite)	and 6 50 Watt FSS transponders on the same satellite with 16 50 Watt transponders on a back up satellite
5.	8 DBS transponders	and 1 50 Watt FSS transponder on each of two satellites

Option 4 is available for a back-up satellite system for a UK DBS satellite operated by another company.

The satellites are intended to use an "extended life" (15 year design life) Hughes HS–393 platform rather than Hughes' HS–394 platform designed specifically for DBS. The HS–393 platform is a derivative of the Intelsat VI design. It is essentially a telecommunications satellite platform not well suited to conventional DBS requirements. This is because it is spin stabilised and is unable to generate the power supplies of up to 4000 Watts needed for a large DBS payload. The technical limit of the HS–393 platform is around 2700 Watts. On the other hand, the Atlantic Satellites proposal implicitly involves a diversification away from the risks of total dependence on the DBS market. In any case, the HS–394 platform would be an expensive option as it requires a considerable amount of development work before the first operational satellite could be finished.

One interesting feature of the Atlantic Satellite design is the use of two dishes on each satellite. This is required because the DBS transmissions will use circular polarisation techniques and the FSS transmissions will use linear techniques. So far Hughes has not developed a way of combining the two techniques on the same dish on a satellite.

Users of the System: This is far from clear at the time of writing. The company says that it has had discussions with the Independent Broadcasting Authority in the UK who have told them that it has no objections if the current UK franchise bidders want to use the system. The company has also talked with both the franchise bidders and other organisations involved in satellite broadcasting. It has, for example, talked with the Mirror Group who have variously planned to use the French TDF–1 DBS satellite and the Luxembourg Astra system. Only one UK company has expressed a public interest in using the system. This is National Broadcasting Limited (James Lee) which is bidding for a UK DBS franchise. It says that it may consider use of Atlantic Satellites if it loses out on its UK application.

The rump of the Irish Westsat consortium, under Fred O'Donovan, is planning one DBS channel primarily using US backing and programming. RTE is unlikely to take any major stake in the software side of the venture. There has been past talk of it taking two channels but it is a small broadcaster (with a turnover of around $90 m per annum) and is already committed to DBS through the Europa TV venture. That channel, which has been losing money heavily, is due to use the Olympus satellite when the latter is launched. Nevertheless, the European satellite TV scene is expected to change considerably in the future and Atlantic Satellites lack of current customers is not an indication of the prospects for the system.

The whole project has, at various times, been suggested as attempting to ride on the tailcoats of US DBS, offering capacity to those broadcasters unable to get on a successful UK DBS project and, alternatively, as an option if UK DBS fails. It is not clear which is the current position. Nevertheless, there is also a lot of capacity on the satellite for cable TV channels and other organisations wanting FSS capacity. At the moment the company says that it is concentrating its mind on providing capacity for traffic within Europe. Hughes Communications parent company, General Motors has, as part of its corporate strategy, a major plan to develop an integrated computer, automated manufacturing and corporate communications system on a world wide scale. To what extent this includes substantial use of satellite communications remains unclear, particularly to take a significant amount of business from this source.

At the moment, however, it appears that Atlantic Satellites is aiming to compete directly with Eutelsat and Société Européenne de Satellites. These organisations are, for the time being, primarily in the broadcasting business.

TECHNICAL SUMMARY

PLATFORM	HS–393
PRIME CONTRACTOR	Hughes Aircraft Company
OTHER MAJOR CONTRACTORS	Possibly British Aerospace
STABILISATION	Spin
LAUNCH MASS	Not available
MASS IN ORBIT	Not available
POWER SUPPLY, END OF LIFE	Not available
DESIGN LIFE	15 years
ORBITAL POSITIONS	Both satellites are planned to be located at 31° W
STATION KEEPING ACCURACY	+/−0·05° (E/W & N/S)
ECLIPSE PROTECTION	Full
SPACECRAFT DIMENSIONS	Not available
TRANSPONDER OPTIONS PER SATELLITE	16 FSS & 0 DBS or 12 FSS & 3 DBS or 8 FSS & 5 DBS or 6 FSS & 6 DBS or 1 FSS & 8 DBS
NO. OF ACTIVE TRANSPONDERS	10 DBS and 16 FSS
NO. OF SPARE TRANSPONDERS	A maximum of 8 DBS or a maximum of 16 FSS transponders available
TWTA POWER OUTPUT	120 (DBS) & 50 (FSS) Watts
FREQUENCY RANGE:	
UPLINK	
DOWNLINK	11·700–12·100 GHz (DBS)
TRANSPONDER BANDWIDTH	27 MHz for both DBS and FSS transponders
EIRP MINIMUM, BEAM EDGE	59 dBW (UK, DBS) 45–49 dBW for FSS in Europe
CHANNEL POLARISATION	Circular for DBS and linear for FSS
RECEIVE COVERAGE (G/T)	Not available
SINGLE CARRIER SATURATION FLUX DENSITY	Not available

COMMERCIAL DETAILS

OPERATOR	NASDA (National Space Development Agency of Japan)
ADDRESS	World Trade Centre Building, 4–1 Hamamatsu-cho 2–Chome, Minato-ku, Tokyo 105, Japan
TELEPHONE NO.	81-3-435-6111
TELEX NO.	J28424 (Abb: NASDA J28424)
FAX NO.	81-3-433-0796
PURPOSE OF PROJECT	Experimental satellite to test maritime and aeronautical communications techniques
PROJECT STATUS	ETS–V was undergoing final checkout prior to launch in 1987
PROJECT VALUE	Not available
LAUNCH DATES	August 1987 by H–I

SATELLITE USERS

The ETS–V project is being handled by the Radio Research Laboratory of the Ministry of Posts and Telecommunications, the Electrical Navigation Research Institute of the Ministry of Transportation and NASDA. The satellite only has a design life of 1·5 years and is intended for use by maritime and aeronautical communications experiments. Mobile communications experiments include oceanic flight control over the Pacific area, navigational aid and search and rescue.

PROJECT OVERVIEW

The Engineering Test Satellite–V (ETS–V) is a test bed for maritime and aeronautical communications, to establish Japanese capabilities in 3-axis stabilisation and to confirm the use of the three stage version of the H–I vehicle for launching Japan's still small satellites. It is also "intended to accumulate domestic technology for the development of the next generation application satellites". ETS–V has required considerable attention to the problems of thermal and attitude control systems and the use of lightweight structures. It also employs a new Japanese data system and solar array technology. The BS–2 spacecraft employed 3-axis technology but this was largely imported from the USA. Three-axis technology is intended for the next generation Japanese satellites – ETS–VI and CS–4.

TECHNICAL SUMMARY

PLATFORM	ETS
PRIME CONTRACTOR	
OTHER MAJOR CONTRACTORS	
STABILISATION	3-axis
LAUNCH MASS	550 kg
MASS IN ORBIT	
POWER SUPPLY, END OF LIFE	
DESIGN LIFE	1·5 years
ORBITAL POSITIONS	
STATION KEEPING ACCURACY	
ECLIPSE PROTECTION	
PROPULSION SYSTEM	

SPACECRAFT DIMENSIONS	
SPAN, TIP-TO-TIP	
NO. OF ACTIVE TRANSPONDERS	
NO. OF SPARE TRANSPONDERS	
TWTA POWER OUTPUT	
FREQUENCY RANGE:	
UPLINK	1·644–1·6485 GHz
DOWNLINK	1·542–1·5464 GHz
TRANSPONDER BANDWIDTH	
EIRP MINIMUM, BEAM EDGE	
CHANNEL POLARISATION	

COMMERCIAL DETAILS

OPERATOR	NASDA (National Space Development Agency of Japan)
ADDRESS	World Trade Centre Building, 4–1 Hamamatsu-cho 2–Chome, Minato-ku, Tokyo 105, Japan
TELEPHONE NO.	81-3-435-6111
TELEX NO.	J28424 (Abb: NASDA J28424)
FAX NO.	81-3-433-0796
SENIOR STAFF	
PURPOSE OF PROJECT	Preoperational experimental satellite
PROJECT STATUS	Under development
PROJECT VALUE	Not available
LAUNCH DATES	1992 by H–II

SATELLITE USERS

PROJECT OVERVIEW

ETS–VI is intended as the pre-operational model of the yet to be ordered CS–4 domestic communications satellite system. In terms of size ETS–VI is a major step forward for Japanese satellite manufacturing capabilities giving it a platform large enough to begin competing with US industry. The move to the large platform design was facilitated by the availability of the H–II launch vehicle and ETS–VI is expected to be the first payload of that launcher.

TECHNICAL SUMMARY

PLATFORM	
PRIME CONTRACTOR	
OTHER MAJOR CONTRACTORS	Toshiba
STABILISATION	3-axis
LAUNCH MASS	2000 kg
MASS IN ORBIT	
POWER SUPPLY, END OF LIFE	
DESIGN LIFE	
ORBITAL POSITIONS	
STATION KEEPING ACCURACY	
ECLIPSE PROTECTION	
PROPULSION SYSTEM	

SPACECRAFT DIMENSIONS	2 × 3 × 2·8 m
SPAN, TIP-TO-TIP	
NO. OF ACTIVE TRANSPONDERS	
NO. OF SPARE TRANSPONDERS	
TWTA POWER OUTPUT	
FREQUENCY RANGE:	
UPLINK	
DOWNLINK	
TRANSPONDER BANDWIDTH	
EIRP MINIMUM, BEAM EDGE	
CHANNEL POLARISATION	

EUROPESAT

COMMERCIAL DETAILS

OPERATOR	European Telecommunications Satellite Organisation (Eutelsat)
ADDRESS	Tour Maine-Montparnasse, 33 Ave du Maine, 75755 Paris, Cedex 15, France
TELEPHONE NO.	33 14 538 4747
TELEX NO.	203823 eusat f
FAX NO.	33 14 538 3700
SENIOR STAFF	Dr. Andrea Caruso, Director General G. Payet, Director, Technical Department W. Shroeder, Director, Operations Department D. G. Hardman, Director, Administration and Finance Charles Morrow, Director, Legal Affairs H. Billig, Head of Space Segment Division Philip Binet, Head of Documentation and PR Martine Papo, Associate Strategic Planner
PURPOSE OF PROJECT	Provision of pan-European DBS capacity for the 1990's
PROJECT STATUS	At planning stage. First launch not envisaged before 1993
PROJECT VALUE	During 1985, Eutelsat quoted a cost of $310 million for a fully operational Europesat system. This includes construction, launch and insurance and TT&C facilities
LAUNCH DATES	First launch not envisaged before 1993

SATELLITE USERS

Europesat is Eutelsat's long term thinking on the future of European satellite television capacity. the concept is highly regarded by a number of European broadcasters despite it being no more than a paper plan.

Current Eutelsat thinking is that the system would not be launched before 1993/1994 and would represent second generation DBS capacity to follow on from such systems as TV–Sat and TDF–1 which are unable to provide pan-European coverage.

It is not clear who the customers for the system would be. Eutelsat has envisaged a separate operating company for the system which would be able to market capacity directly to the end user. It is also not clear that users would be limted to those based in countries which would be members either of Eutelsat or the operating company. Similarly, the Eutelsat II system will provide capacity which is aimed at the broadcasting markets and could provide capacity for quasi-DBS services. It is possible that some users of Eutelsat II may not want to transfer to Europesat.

PROJECT OVERVIEW

Introduction: Europesat is a concept for DBS capacity for Europe for the 1990's under a study by Eutelsat and intended to provide follow-on capacity to the TDF–1/2 and TV–Sat national concepts developed during the 1980's within the framework of WARC '77. The concept was developed by Eutelsat so that the organisation could retain its position as the "leader" in providing satellite capacity for European broadcasters. The concept appears to be in direct competition with the Astra concept being developed by Société Européenne des Satellites in Luxembourg. Like Astra, Europesat does not comply with the WARC '77 allocation of channels and frequencies for DBS TV. However, the WARC '77 arrangements will have expired by the time Europesat becomes available. Studies of the Europesat concept are still in progress and therefore the project is still far from being defined.

Background to the Project: The Europesat concept is the result of a $300,000 study initiated by Eutelsat in 1985. It reflected a change of sentiment amongst the satellite operators of the Western world away from the high power DBS concept envisaged in WARC '77 and adopted for TDF–1, TV–Sat, Tele–X and Olympus. Demand for satellite capacity from European broadcasters has indicated the need for European rather than national coverage. Technical developments in ground receiver equipment through the 1980's and into the 1990's means that much smaller TVRO dishes can be used with such medium power satellite concepts as Europesat.

The preliminary study by Eutelsat came under some criticism from PTT members and was therefore widened to cover a greater variety of options. Initial proposals were for a satellite

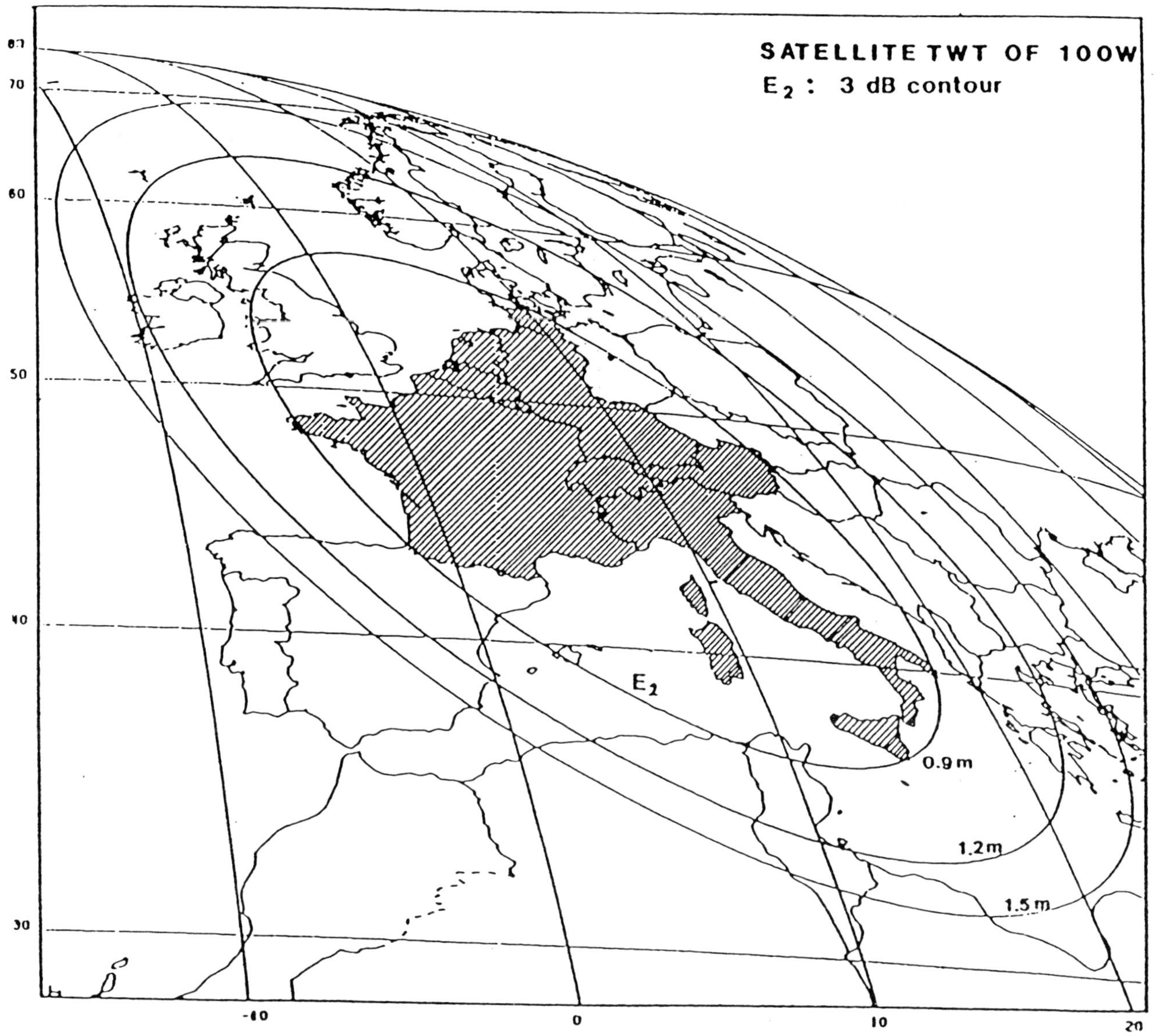

Above and over pages: some options for downlinks on Europesat

with either 12, 16 or 20 transponders in the 100–125 Watt range (and possibly as low as 60 Watts). This would have been available for launch in 1991 or 1992 and provided capacity to replace the TDF–1/TV–Sat spacecraft. The Europesat spacecraft would have been located at 19° West, the same orbital position allocated to TV–Sat and TDF–1 but would have provided wider European coverage. Germany and France would have had to give up their rights to use the slot. To some extent this plan was eclipsed by the decision to order a Eutelsat II system with medium power transponders suitable for quasi-DBS channels and available to broadcasters from 1989.

It now appears that the Eutelsat executive favours a two satellite Europesat system with each satellite having ten 120 Watt transponders and enabling reception with 60 cm dishes. Both satellites would be located at 19° West. This Europesat concept is very close to that of the second Astra satellite under consideration by Société Européenne des Satellites. However, the latter company is understood to be planning for launch of Astra 2 in 1990, some years ahead of the envisaged launch of Europesat. Indeed, the Eutelsat II spacecraft design is seen as a competitive response to the first Astra spacecraft. Astra 2 is expected to have 16 fully eclipse-protected transponders in the 100 Watt range and will be co-located with Astra 1 at 19° East. It is understood that current thinking in Eutelsat envisages full eclipse protection for the Europesat design.

The footprint envisaged for Europesat appears to remain a major problem. During 1985 Eutelsat looked at five options for service areas ranging from complete coverage of Europe to only the most commercially concentrated areas in Western Europe. The European Space Agency is looking at the problem of finding

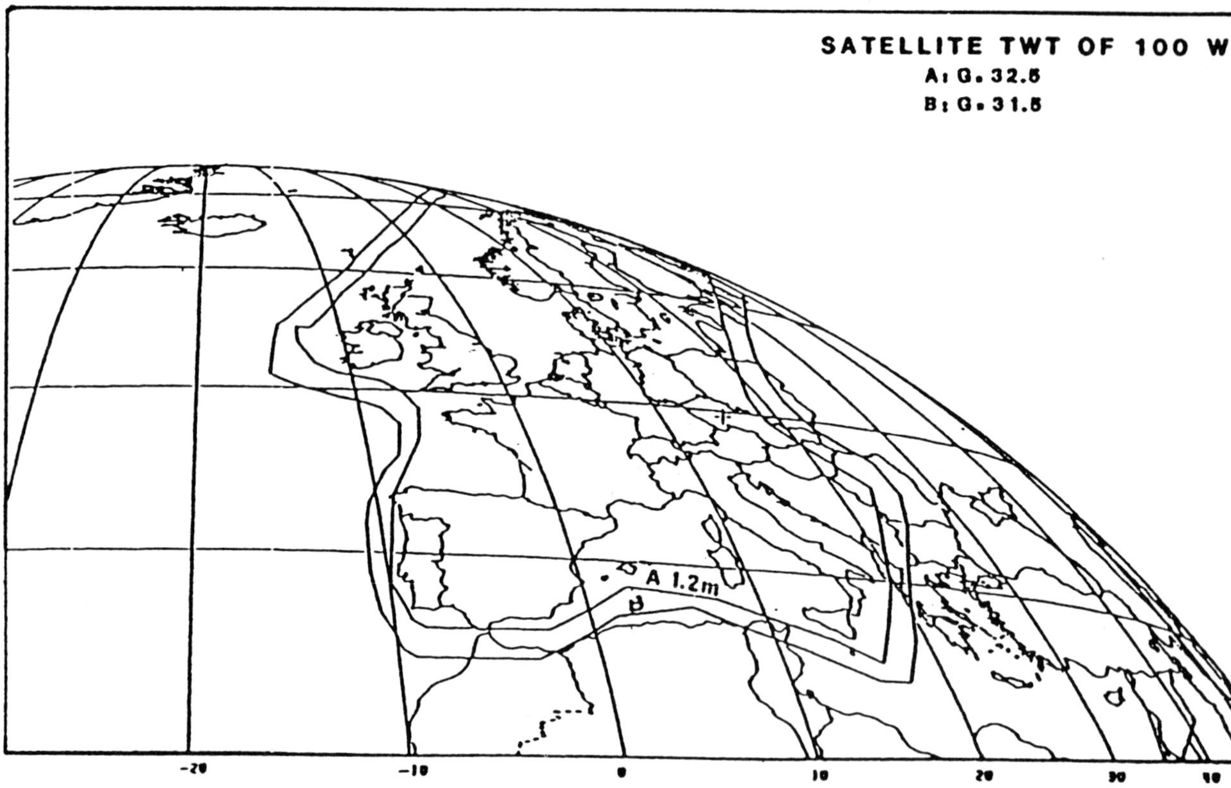

suitable frequencies for the concept. This requires that at least one and probably more countries give up the frequencies allocated to them at WARC '77 and that any finalised frequency plan does not cause interference with national used frequencies either used terrestrially or for other satellite systems. It will also require agreements between countires. However, it appears that the Europesat concept will not require a new WARC to be called. Article 4 of Appendix 30 of the WARC '77 agreement states that assignments that are different from those of the plan must be subject to the agreement of the Administrations affected before assignments are recorded, and enjoy the same status as the plan.

The organisation claims that some 15 of its members are interested in the concept and cost per transponder would be about one fifth that of a high power DBS transponder on TDF–1/TV–Sat type satellites.

OPTIONS FOR THE EUROPESAT DESIGN

TYPE OF PLATFORM	TRANSMIT POWER			
	60 WATTS		100 WATTS	
	FULL ECLIPSE CAPABILITY	NO ECLIPSE CAPABILITY	FULL ECLIPSE CAPABILITY	NO ECLIPSE CAPABILITY
Half Ariane 4 Class (Spacebus 100B – 2100 Kg)	16 Transponders	18 Transponders	10 Transponders	12 Transponders*
Ariane 44P Class (TV–Sat/TDF–1 – 3000 Kg)	22 Transponders	34 Transponders	14 Transponders	22 Transponders
Ariane 44LP Class (Large Olympus – 4000 Kg)	32 Transponders	38 Transponders*	20 Transponders	22 Transponders*

*Limitation due to solar generator.

Source: Eutelsat, 1985.

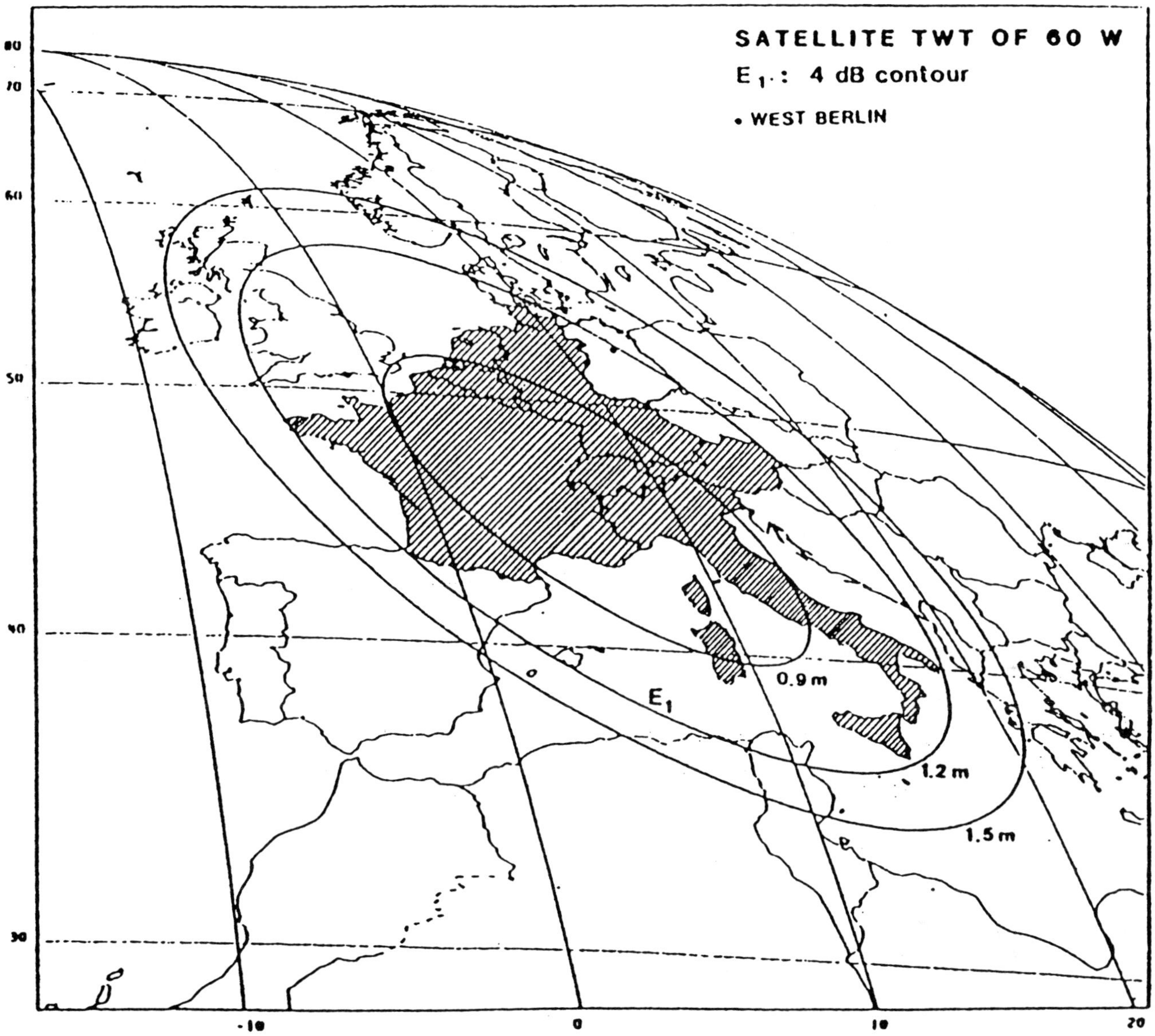

TECHNICAL SUMMARY

PLATFORM	Not decided. Could be a TDF–1/TV-Sat platform or Olympus or Spacebus 100B
PRIME CONTRACTOR	A European manufacturer is envisaged
STABILISATION	Probably 3-axis
LAUNCH MASS	1500–2250 Kg or 4000 Kg (Olympus class)
POWER SUPPLY, END OF LIFE	Not available
DESIGN LIFE	Not available
ORBITAL POSITIONS	Between 19° W and 29° E 19° West appears to be preferred
STATION KEEPING ACCURACY	Not available
ECLIPSE PROTECTION	Probably

SPACECRAFT DIMENSIONS	Not available
SPAN, TIP-TO-TIP	Not available
NO. OF ACTIVE TRANSPONDERS	15–20 per satellite
NO. OF SPARE TRANSPONDERS	Not available
TWTA POWER OUTPUT	60–120 Watts
FREQUENCY RANGE:	
UPLINK	17·3–18·1 GHz
DOWNLINK	11·7–12·5 GHz
TRANSPONDER BANDWIDTH	Not available
EIRP MINIMUM, BEAM EDGE	55 dBW
CHANNEL POLARISATION	RHCP & LHCP
RECEIVE COVERAGE (G/T)	Not available
SINGLE CARRIER SATURATION FLUX DENSITY	Not available

COMMERCIAL DETAILS

OPERATOR	European Telecommunications Satellite Organisation (Eutelsat)
ADDRESS	Tour Maine-Montparnasse 33, F-75755 Paris Cedex 15, France
TELEPHONE NO.	33 14 538 4747
TELEX NO.	203 823 EUSAT
FAX NO.	33 14 538 3700
SENIOR STAFF	Andrea Caruso, Secretary General G. Payet, Director, Technical Department W. Schroeder, Director, Operations Department D. G. Hardman, Director, Admin. and Finance Philippe Binet, Head, Documentation and PR
PURPOSE OF PROJECT	Provision of Ku-band regional satellite system for European telecommunications administrations to provide capacity for telephony, TV and business services
PROJECT STATUS	System operational but with two further satellites still due for launch. Their launch is delayed
PROJECT VALUE	Average cost of each of the five satellites to the European Space Agency was $49 m. The satellites are leased by ESA to Eutelsat
LAUNCH DATES	Eutelsat I–F2, 16th June 1983, Ariane 1, L6 Eutelsat I–F2, 4th August 1984, Ariane 3, V10 ECS–4, March 1987, Ariane 3, V19 ECS–5, January 1988, Ariane 3, V26

SATELLITE USERS

The Eutelsat system was originally envisaged as providing capacity for telephony traffic and for the distribution of Eurovision programming for the European Broadcasting Union. The latter leases two transponders on Eutelsat I–F2 directly from Eutelsat for the Eurovision service.

Apart from the EBU, Eutelsat is a carriers' carrier. Its capacity is used by the European PTT's who are members of the organisation. These are listed in the project overview. For telephony traffic, Eutelsat I–F2 is currently used and PTT's pay Eutelsat in accordance with the amount of telephony traffic they pass through the system. Use of the satellite for telephony commenced in 1986 and it is expected that 20 PTT's will use the system for this purpose.

For cable TV traffic transponders are leased by Eutelsat to individual PTT's who then sub-lease them to broadcasters. The terms and conditions of the lease between Eutelsat and the PTT depends on the number of countries to which the TV channel is downlinked. Indeed, the entire pricing structure of Eutelsat capacity is related to European Telecoms tariffs organised by the CEPT. The TV broadcasters using Eutelsat are listed in the project overview. One transponder on Telecom I is available for sub-lease by Eutelsat for TV traffic.

The Eutelsat I–F2 and ECS–4 and 5 satellites each have two specialised transponders and an antenna for business services offered by member PTT's. This is known as the satellite multiservice systems and operates in SCPC/FDMA mode. TDMA business capacity is available through Eutelsat on Telecom I. SMS users include Reuters and EDS.

Current television channel providers using the Eutelsat system are detailed in the Project Overview below.

PROJECT OVERVIEW

History of Eutelsat: The origins of Eutelsat date back to the early 1960's when Europe was working on its involvement in the US dominated Intelsat proposal. Various Europeans considered a similar organisation to provide satellite communications within Europe as well as a European dominated alternative to Intelsat to provide global communications. France, in particular, was concerned about US domination of Intelsat. In 1964 the European Conference on Satellite Communications (CETS), formed to coordinate Europe's position in the negotiations which lead to the formation of Intelsat, started to look at the possibility of a European satellite system. Like the US thinking behind Intelsat, which gave that country a captive export market for its satellite industry, the European thinking focused on providing an industry and technical capability. The related origins of Eutelsat to Intelsat perhaps partly explain both the organisational similarities between the two organisations – both are essentially carriers' carriers, have a similar legal foundation and monopoly power over satellite communications – as well as their close relationships with industrial policies.

In 1966 the European Conference on Satellite Communications asked the European Space Research Organisation (ESRO) and the European Launcher Development Organisation (ELDO), both precursors to the current European Space Agency, to undertake the first technical study of a European Satellite system. This request covered a system which was limited to providing capacity for the European Broadcasting Union (EBU) for Eurovision programming in Europe and the Mediterranean basin. It became apparent that this could not financially justify a satellite system and in 1969 emphasis was placed on primarily providing telephony capacity but with secondary capacity for the EBU. It was not until the early 1980's that it became apparent that Eutelsat's primary initial market would be in distribution of cable TV channels. The design of the current series of Eutelsat spacecraft reflect the early emphasis on EBU requirements. For example, each satellite has four transponders (numbers 1, 3, 7 and 9) which can downlink through a Eurobeam; all other transponders (except SMS) can only downlink through spot beams.

In 1969 an ad hoc working group was formed to further define the programme to satisfy the needs of the Conference of Post and Telecommunications Administrations (CEPT – a rather shadowy organisation which coordinates tariffs and technical matters of European PTT's) and the EBU. This working group consisted of CEPT representatives, senior governmental officials, the EBU, ELSO and ESRO, and was within the framework of the European Space Conference (ESC). In July 1970 a decision was taken at the fourth European Space Conference to start a programme to implement a regional European satellite system and develop European industrial capacity to manufacture the system.

In July 1971, a satellite working group under the coordinating committee for Satellite Telecommunications of CEPT, concluded in favour of an organisational concept which was subsequently called Eutelsat. This organisation would own, operate and manage the space segment on behalf of European PTT's wishing to participate in the project. The group also concluded that extending terrestrial systems in Europe would be cheaper than providing a regional satellite system. This later became a critical factor in coordination with Intelsat. European PTT's are signatories to Intelsat and are therefore required to ensure that their involvement in other satellite systems do not cause economic harm to Intelsat. As it was envisaged that Eutelsat was not an economic proposition, it would not divert traffic away from Intelsat as the latter would not have provided uneconomic capacity for that traffic. As it is today, it appears that Eutelsat's principal traffic, cable TV channels, is an economic proposition and the organisation competes with Intelsat in providing capacity for this service in Europe.

Nevertheless, the real economics underlying the Eutelsat system are not so clear. Development of the Eutelsat 1 system involved a publicly funded R&D programme through the European Space Agency. This involved two Orbital Test Satellites (OTS) and further R&D expenditure on the operational Eutelsat 1 spacecraft. Construction of the Eutelsat 1 spacecraft was funded by the European Space Agency with the largest single source of funds coming from the British government. It is reported that each of the five Eutelsat 1 spacecraft cost ESA $49 m. The satellites are leased by ESA to Eutelsat. It is difficult to distinguish which part of this cost is accounted for by R&D or by an indirected subsidy to the European satellite manufacturing industry or to Eutelsat. The contract price for Société Européenne des Satellites first Astra satellite, a considerably larger design than Eutelsat 1, was $45 m and that contract was placed in 1985, some six years after the Eutelsat 1 contract. In addition, the long delay between placing the order for the Eutelsat 1 spacecraft and final delivery into orbit, has tied up funds which otherwise would be earning interest.

The OTS/MAROTS/MARECS systems and Eutelsat: In late 1971 ESRO/ELDO proposed a new programme for development of the European space segment. This envisaged an experimental phase to develop a test satellite (the OTS project), followed by a technological phase after the launch of OTS. This was intended to develop the final spacecraft design based on the OTS experience and the creation of Eutelsat. In 1973, when the ESRO/ELSO organisations were in their final death throes, the European Space Conference confirmed its intention to implement this programme and develop a maritime version of OTS called MAROTS which would provide an alternative to Comsat's Marisat (Inmarsat I) mobile maritime communications satellite system. But the collapse of ELDO because of the inability to successfully launch its Europa launch vehicle, left Europe without a launch vehicle. The USA was refusing to launch European communications satellites in order to protect its own satellite industry.

Moreover, ESRO collapsed with ELDO leaving Europe without a coordinated R&D programme. However, particularly after pressure from France, a replacement organisation, the European Space Agency, was formed in 1975 and charged with the responsibility of R&D in three major areas – communications satellites (largely dominated and funded by the UK), Spacelab (largely dominated and funded by West Germany) and the Ariane launch vehicle (dominated and mostly funded by France). This tripartate arrangement provided a workable, if not modest, framework for the development of the European Space Industry. It was also able to build upon Europe's experience in Ku-band satellite communications initiated with the Franco-German Symphonie satellites. However ESA was and is a Research and Development organisation and was not charged with the responsibility of operating a commercial satellite system, or, indeed, operating Ariane on a commercial basis.

In January 1976, the coordinating committee for Satellite Telecommunications of CEPT approved a new report with recommendations for a European telecommunications satellite system and CCTS proceeded to draw up draft agreements to create an interim organisation, along the lines of the interim Intelsat organisation, to manage the system. CCTS also negotiated with ESA to provide the space segment for the yet to be formed interim Eutelsat organisation. The interim Eutelsat organisation was intended to establish and operate the space segment, on a monopoly basis, of the European telecommunications system pending the establishment of a definitive organisation to operate on a commercial basis. However, because of delays in establishing Inmarsat, it was agreed that the MAROTS spacecraft, being developed by ESA using the same platform as OTS, would also be managed by the interim organisation.

In March 1977 a conference of European PTT's resulted in agreement on the final form of interim Eutelsat and the agreement entered into force on 30th June 1977. The agreement was signed by 20 CEPT members. There were subsequent supplementary agreements in 1977 and 1978 on MAROTS and ECS. The interim Eutelsat organisation initially consisted of an Assembly of Signatory Parties, two Councils (ECS and MAROTS) and a permanent executive called the General Secretariat. The MAROTS and ECS Councils were responsible for development of "reliable space segments" adapted to their respective requirements. In 1978, the MAROTS project was modified to incorporate an ECS platform rather than an OTS platform and the satellites became known as MARECS – an acronym for MARitime European Communications Satellite. Once the Inmarsat organisation was formed, it was decided that it would be responsible for procurement of the satellites from ESA rather than through Eutelsat. Therefore the MAROTS Council ceased to exist in 1981, leaving Interim Eutelsat to concentrate on its ECS satellites.

The executive of interim Eutelsat began to function in 1978 with staff drawn from the satellite working group of the coordinating committee for Satellite Telecommunications of CEPT and staff seconded from European PTT's. First Secretary General was Francois Job who continued in his position until 1981 when Andrea Caruso took over the position. Whilst Interim Eutelsat was involved in MARECS, it had permanent staff in London to handle that activity, but most of the staff were in Paris.

Agreements on provision of the ECS satellites and use of OTS were finalised in 1979. ESA was then responsible for providing, launching and maintaining in orbit the ECS satellites (which are known as Eutelsat 1 satellites after launch) including provision of control facilities at Redu in Belgium. ESA procured five satellites, but Eutelsat initially took only two of these. Its third was taken up by Eutelsat when it became apparent that demand from the cable TV channel industry was substantial. When this was lost on launch, Eutelsat took a fourth, as replacement, and finally, a fifth in the light of technical problems with the first, and further demand for capacity. The first ECS satellite was not quite as intended, not providing full eclipse capability and having, nominally, one too few operational transponders. The ECS/MARECS series was Europe's first production run of operational commercial communications satellites and it appears that the manufacturing industry suffered from lack of experience.

In the late 1970's Interim Eutelsat had been studying the potential market for satellite business services and at the end of 1982 it was decided to modify the ECS-2, 3, 4 and 5 satellites to carry two extra Ku-band transponders and a new antenna for a new service designated Satellite Multiservice System (SMS). At the same time, it was agreed to the equivalent of one Ku-band transponder on Telecom 1 for the service. The agreement for the latter was signed on 31st March 1982. As not all Eutelsat members are able to fully benefit from the SMS service they received some financial compensation as a consequence.

Agreement with the European Broadcasting Union for lease of two transponders was finalised in January 1982. Under this arrangement, the EBU leases two transponders on Eutelsat I–F2 directly from Eutelsat. These are transponder numbers 3 and 9 and are non-preemptible. Two further transponders on Eutelsat I–F2, numbers 1 and 7 are designated reserves to numbers 3 and 9, in case of failure of the latter.

The first Orbital Test Satellite, OTS–1, was lost on launch, by Delta, in September 1977 but a back-up, OTS–2, was successfully launched by Delta on 11th May 1978. This was used for experimental services and was initially intended for only three years of operation. However, it surpassed this expectation and was also effectively used for commercial operations. Most notable of these was a cable TV channel established by a British entrepreneur, Brian Haynes. Satellite Television PLC subsequently was taken over by News International and operated as Sky Channel. It is now on Eutelsat I–F1. This was a precursor to a rush of applicants to European PTT's wanting to use the Eutelsat system for distribution of cable TV channels. To accommodate the demand Interim Eutelsat arranged to lease transponders on a preemptible basis on Eutelsat I–F1, which was designated an in-orbit spare to the Eutelsat I–F1 satellite.

Demand, however, far exceeded capacity leading to the need to launch ECS–3, 4 and 5. Eutelsat eventually decided it wanted a sixth ECS satellite but its signatories, notably the UK and France, objected to this, perhaps because, by the time it is available, in 1988, there will only be a one year gap before the second generation, Eutelsat II, system becomes operational. When the first Eutelsat I–F1 satellite became operational in 1983, the OTS satellite, which had suffered a transponder failure, was no longer required and its telecommunications payload was turned off. However ESA continued other tests with OTS until the end of 1985.

Formation of the Definitive Eutelsat Organisation: The definitive Eutelsat Organisation came into effect on the 1st September 1985. However, the interim Eutelsat Organisation began to consider arrangements for a definitive organisation in 1979. The progress in establishing a definitive organisation proved to be slow, certainly in comparison with the need to establish a commercial organisation. This was partly accounted for by the fact that Eutelsat is, in practice, an intergovernmental organisation – its members are largely government owned and controlled PTT's which historically have been civil service departments. Moreover, establishing pan-European organisations has never proved easy as there are inevitable conflicts of interest, differences in outlook, long chains in the decision taking process and no country capable of dominating the environment sufficiently to push through a decision quickly. Finally, the European PTT's are notorious for being engineering, rather than commercially led.

In May 1982 an Intergovernmental Conference in Paris agreed on the texts of a Convention and Operating Agreement for the definitive Eutelsat Organisation. By the end of 1983, 24 European countries had signed the Convention and Operating Agreement. But they were slow to ratify them and a second Intergovermental Conference was held in December 1983 to extend the ratification deadline by 19 months to 15th July 1985.

Competition and Eutelsat: Eutelsat is very much in the mould of European Telecommunications operators, providing services complementary to, rather than competitive with, terrestrial services, and based on the assumption that telecommunications is a natural monopoly. Eutelsat has argued that its monopoly power over provision of regional European satellite communications capacity is necessary because of the very high commercial risks of operating in that market and its inherent limited opportunity to make profits from satellite communications. Nevertheless, Eutelsat is both game-keeper and poacher when it comes to competition for satellite capacity.

As poacher, it competes with Intelsat, although that competition is currently masked by the shortage of capacity on the Eutelsat system. Eutelsat's relationship with Intelsat has, in the past, been stormy. The distinction between a regional satellite system such as Eutelsat providing capacity for many countries and Intelsat for most countries is somewhat arbitrary and it is difficult to see how Eutelsat does not inflict economic harm on Intelsat. This has particularly been the case over provision of capacity for cable TV traffic, a situation not originally envisaged. When Intelsat originally coordinated with Eutelsat on the economic harm issue, it only gave limited period clearance of Eutelsat. This was in 1981 and was for five years. Eutelsat objected strongly and was eventually given full clearance. Back in the early 1960's when the original Intelsat agreement was being hammered out, the USA wanted conditions that would have ruled out Eutelsat but, primarily on French insistence, provision was made for such organisations so that they did not provide "significant" economic harm, rather than any harm. The reason was partly to provide an outlet for the European aerospace industry rather than any belief that competition in telecommunications was desirable.

Nevertheless, once Eutelsat overcame its problems with Intelsat, Luxembourg decided it wanted to be in the FSS broadcasting satellite business. Luxembourg, a member of Eutelsat, not only proposed a system which carried cable TV channels but also decided it would procure the satellite in the USA, thus breaking the monopoly European satellite manufacturers have over their domestic market.

Moreover, Luxembourg opted for a satellite payload, using 45–50 Watt transponders which many European broadcasters found attractive. Luxembourg also intended to launch its first satellite just before the leases on Eutelsat I–F1 came up for renewal, thus directly aiming to "lift" Eutelsat's customers. Yet, Luxembourg's ambitious plans began to face problems and it now appears that it will not launch its first satellite until early 1988. By 1989, if it can find launch capacity, Eutelsat will have its second generation system available which will provide payload performance as good as Luxembourg's. Moreover, the Luxembourg satellite owner and operator, Société Européenne des Satellites, is, at the time of writing, caught in a trap. European governments, who are members of Eutelsat and have a financial interest in protecting that organisation, will have to grant authorisation for uplinks and downlinks to the US built system. They are under obligation not to do so if the Luxembourg system is deemed to cause Eutelsat economic harm.

Therefore the Luxembourg government may have to coordinate with Eutelsat on the issue of economic harm before it can tell its potential customers that they can use the satellite. But because it can't do so, it has not been able to announce new customers. The Luxembourg government has argued that its satellite system will not do Eutelsat any economic harm because: 1) The Luxembourg system will have new customers, rather than customers taken from Eutelsat, 2) It will stimulate the European broadcasting market and create greater overall demand for satellite transponders such that there will be more demand for Eutelsat capacity than would be the case if the Luxembourg system did not go ahead. Luxembourg also argues that as the Eutelsat I system is not designed for broadcasting traffic. Broadcasters are having difficulty making money because they are losing out on quasi-DBS options and the SMATV market.

Unless a satellite is built to address their needs, they will pull out of the business, and 3) Luxembourg has also said it does not need to coordinate with Eutelsat because the Eutelsat I system was built for telephony and European Broadcasting Union traffic. Because Luxembourg's satellite is intended solely for use by broadcasters, it will not touch that traffic and therefore cannot cause Eutelsat economic harm.

All Luxembourg's arguments have, in a de facto sense, turned out to be fairly spurious. Eutelsat has already responded by designing its Eutelsat II system with virtually the same performance as the Luxembourg system. Eutelsat has also argued that it cannot coordinate with Luxembourg on the grounds that as Société Européenne des Satellites has no customers, it cannot determine which customers it will take away from Eutelsat. However, despite serious concern within the Eutelsat executive about the impact of the Luxembourg system on Eutelsat, the Eutelsat members told the executive in 1986 to come to a cooperative arrangement with Société Européenne des Satellites, if possible. The Luxembourg government, concerned by the continuing speculative nature of its satellite project, the inability of SES to raise further equity capital, the uncertainty over uplinks and downlinks and its own hefty financial commitment to loan guarantees, offered in July 1986, terms to Eutelsat for a potential arrangement. This is reported to involve a lease/lease back condition.

At the time of writing, the Eutelsat executive was most unhappy about the terms offered by the Luxembourg government – particularly as they allowed SES to purchase further satellites in its current generation system and to go in for cooperative ventures, outside the Eutelsat framework, for a second generation system. There also appear to be problems in the terms meeting Eutelsat's Articles of Association. Nevertheless, the final decisions are down to Eutelsat's members, the PTT's, rather than its Executive. Some PTT's, notably British Telecom, have taken a keen interest in taking capacity on the Luxembourg system. Moreover, it appears that France has long been considering offering capacity in competition to Eutelsat and Intelsat. The French domestic satellite, Telecom 1, can provide C-band coverage of the Eastern seaboard of the USA and the Caribbean region providing potentially direct competition with Intelsat. In 1983 France notified the International Frequency Registration Board of its intention to use orbital slots at 11° West and 7° East for a system known as F–SAT and 37·5° West and 32° East for a system called Videosat. The latter in particular could provide virtually international, rather than regional, coverage. The F–SAT system would operate in the Ku-band using 50 Watt transponders. It is remarkably similar to proposals emerging for the Telecom II system. During the Summer of 1986, French State Secretary for Communications, Gerard Longuer suggested that capacity on the Telecom II system (which, like F–SAT, is expected to use 50 Watt transponders) will be made available to non-French customers in Europe.

In practice, Eutelsat already faces some competition from the Telecom I domestic satellite system. When originally launched it was intended that the Telecom I system would not be used for cable television traffic. However, the lack of demand for Telecom I's TDMA based business services, left it under utilized and the satellite was opened up for broadcasters and radio. The first broadcasters to use it were La Cinq and TV–6. As French cable television is beginning to expand, more cable TV channels are expected on it. In West Germany, the yet to be launched DFS–Kopernicus domestic satellite system was designed to carry cable television traffic. It is technically possible that one or more German language TV channels using Eutelsat will move to that system, although, so far, none have disclosed their intentions to do so.

Eutelsat's current position and its impact on competitors: Eutelsat is a very politically powerful organisation because of its backing from Europe's largely state owned PTT's and its role in industrial policy. For example, Brussels based Media International, looking at satellite options for carriage of TV channels, warned its shareholders (News International and Group Brussels Lambert) to be very careful over looking at Luxembourg's Société Européenne de Satellites. It advised that News International should not transfer Sky Channel from Eutelsat to SES until the latter had got its problems sorted out. In effect it said that if Sky Channel came off Eutelsat, vindictive European PTT's could stop distribution of the channel on cable networks.

Sky Channel has had a good reason to want to move onto another satellite. It has been using Eutelsat I–F1 since the latter became operational in 1983. But the leases on Eutelsat I–F1 are all preemptible. This means that if Eutelsat I–F2 fails most of the traffic on that satellite will move to Eutelsat I–F1 and some users of the latter will have to move off the satellite altogether. Eutelsat does not have a third satellite in-orbit to provide back up capacity. There are two reasons for this. One is that it was never originally envisaged that Eutelsat would carry a substantial amount of cable TV traffic. The other reason is that the Ariane vehicle has proved to be unreliable. The third Eutelsat spacecraft, ECS–3, was lost on launch when the third stage of Ariane VI–5 failed in September 1985. The European Space Agency arranged for a replacement to be launched on Ariane VI–9 in May 1986. Unfortunately Ariane VI–8's third stage failed, revealing a generic design fault in the ignition sequence of the third stage. As a result, the third stage had to be redesigned and tested before the next Ariane (VI–9) could fly and launch ECS–4. This was expected in the Spring of 1987.

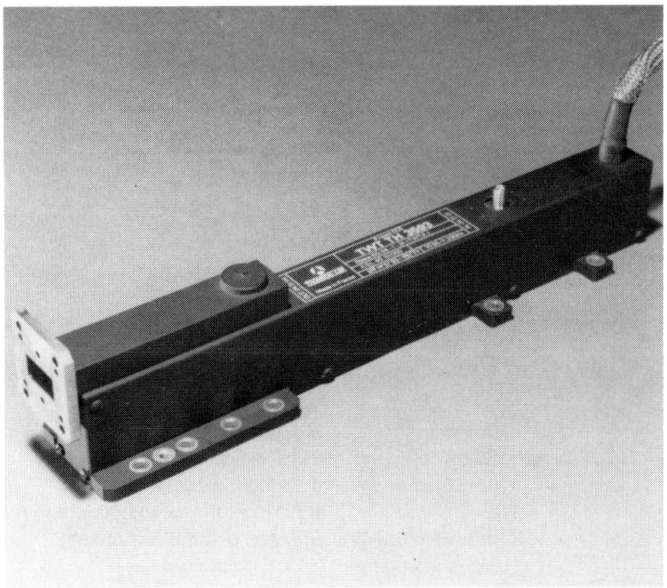

Thomson–CSF travelling wave tube (model TH 3593) used on the Eutelsat–I satellites. The tube delivers a power output of 20 Watts in the Ku-band frequency range

Sky's desire to secure its position was compounded by a serious problem in the travelling wave tube of the channel leased by it on Eutelsat I–F1. During 1985, this suffered a degredation in performance such that the uplink power had to be doubled to maintain the required downlink signal strength. A similar problem was experienced on the channel used by Filmnet on the same satellite. It is unlikely that further degradation of the Sky Channel transponder can be offset by increasing the uplink power. Both of the travelling wave tubes involved were manufactured by AEG Telefunken.

In November 1986, transponder 2 on Eutelsat I–F1, used by the West German cable TV channel 3-Sat, suffered a total failure. 3–Sat transferred to the transponder previously used by Europa TV, which had ceased operations a few days previously. 3–Sat was moved to Intelsat V at 60° East the following month.

Each Eutelsat I satellite carries six Thomson–CSF transponders and six AEG–Telefunken transponders. On the Eutelsat I–F2 satellite, two travelling wave tubes built by Thomson–CSF, have similarly caused concern. Europe has been a pioneer in the development of Ku-band payloads and ground equipment but, prior to the order for the Eutelsat I system there had been little experience in using the frequency on an operational basis. The 20 Watt travelling wave tubes built by AEG–Telefunken were selected for use on the Hughes built SBS and Anik–C satellites. The experience of Hughes with these tubes suggest that the European manufacturers suffered from lack of experience in manufacturing tubes. Hughes found that the

Eutelsat–I services: Current primary use of the Eutelsat system is for TV traffic. The entire capacity on Eutelsat I–F1 is used by cable TV channels. Two transponders on Eutelsat I–F2 are used by the European Broadcasting Union on a full-time basis for Eurovision programming distribution. Norway leases one transponder on Eutelsat I–F2 to distribute its NRK public service broadcast channel. Allocation of channels on the Eutelsat–I system as at 31st December 1986, was as follows:

Transponder Number	*Beams available on transponder*	*TV channel and Country of Origin or other use of transponder*	*Spot Beam Used*	*Polarisation*	*Centre Frequency*	*Beam Edge EIRP*	*TWT Manufacturer*
EUTELSAT I–F1, 13° EAST							
1	Eurobeam or SW	Radio Televisore Italian, Italy (see notes)	SW	X	10·991667 GHz	41 dBW	AEG
2	SE	3-SAT, Germany	SE	X	11·075 GHz	31 dBW	Thomson
3	Eurobeam or SW	Vacant	SW	X	11·158333 GHz	41 dBW	Thomson
4	SW	TV–5, France	SW	X	11·491667 GHz	41 dBW	AEG
		New World Channel, Norway	SW	X			
		Worldnet, USA	SW				
5	SA	not allocated	SA	X	11·575 GHz	41 dBW	Thomson
6	SW	Sky Channel, UK	SW	X	11·658333 GHz	41 dBW	AEG
7	Eurobeam or SW	Teleclub, Switzerland	SW	Y	10·991667 GHz	41 dBW	Thomson
8	SE	RTL Plus, Luxemboug (see notes)	SE	Y	11·075 GHz	31 dBW	AEG
9	Eurobeam or SW	Filmnet, Holland (see notes)	SW	Y	11·158333 GHz	41 dBW	AEG
10	SW	SAT–1, Germany	SW	Y	11·491667 GHz	41 dBW	Thomson
11	SA	not allocated	SA	Y	11·575 GHz	41 dBW	AEG
12	SW	Music Box/Superchannel (UK)	SW	Y	11·658333 GHz	41 dBW	Thomson
EUTELSAT I–F2, 7° EAST							
1	Eurobeam or SW	Reserve EBU transponder		X	10·991667 GHz	35 or 41 dBW	AEG
2	SE		SE	X	11·075 GHz	31 dBW	Thomson
3	Eurobeam or SW	European Broadcasting Union	Eurobeam	X	11·158333 GHz	35 dBW	Thomson
4	SW		SW	X	11·491667 GHz	41 dBW	AEG
5	SA or SE			X	11·575 GHz		Thomson
6	SW	NRK, Norway (preemptible lease)	SW	X	11·658333 GHz	41 dBW	AEG
7	Eurobean or SW	Reserve EBU transponder		Y	10·991667 GHz	35 or 41 dBW	Thomson
8	SE		SE	Y	11·075 GHz	31 dBW	AEG
9	Eurobeam or SW	European Broadcasting Union	Eurobeam	Y	11·158333 GHz	35 dBW	AEG
10	SW		SW	Y	11·491667 GHz	41 dBW	Thomson
11	SA		SA	Y	11·575 GHz		AEG
12	SW		SW	Y	11·658333 GHz	41 dBW	Thomson
SMS–1	SMS Eurobeam	Eutelsat for PTT's	Eurobeam	X	12·541667 GHz	37 dBW	not available
SMS–2	SMS Eurobeam	Eutelsat for PTT's	Eurobeam	Y	12·541667 GHz	37 dBW	not available
ECS–4 (to be launched), 10° EAST							
1	Eurobeam or SW	See notes below. Channels allocated	tbd	X	10·991667 GHz	35 or 41 dBW	AEG
2	SE	On the third Eutelsat spacecraft are	SE	X	11·075 GHz	31 dBW	Thomson
3	Eurobeam or SW	as follows	tbd	X	11·158333 GHz	35 or 41 dBW	Thomson
4	SW	NP, Turkey, one, Spot East beam	SW	X	11·491667 GHz	41 dBW	AEG
5	SA or SE	NP, Spain, one, Spot Atlantic beam	tbd	X	11·575 GHz		Thomson
6	SW	NP, UK, two, Spot West beam	SW	X	11·658333 GHz	41 dBW	AEG
7	Eurobeam or SW	P, Italy, two, Spot West beam	tbd	Y	10·991667 GHz	35 or 41 dBW	Thomson
8	SE	P, Denmark, one, Spot West beam	SE	Y	11·075 GHz	31 dBW	AEG
9	Eurobeam or SW	P, Sweden, one, Spot West beam	tbd	Y	11·158333 GHz	35 or 41 dBW	AEG
10	SW	P, Norway, one, Spot West beam	SW	Y	11·491667 GHz	41 dBW	Thomson
11	SA	Italy is expected to use one transponder	SA	Y	11·575 GHz		AEG
12	SW	for RAI and possibly one for Canal Cinq	SW	Y	11·658333 GHz	41 dBW	Thomson
SMS–1	SMS Eurobeam		Eurobeam	X	12·541667 GHz	37 dBW	not available
SMS–2	SMS Eurobeam		Eurobeam	Y	12·541667 GHz	37 dBW	not available

Notes: Eutelsat I–F1 is intended to carry only nine operational transponders. Ten are currently in use. Radiotelevisione Italiana is due to transfer to ECS–4 when the latter becomes operational. RTL plus will then take the vacated transponder and use the Spot West beam. Transponder 12 on F1 is not eclipse protected; all the other Eutelsat transponders are. The transponders are allocated to the PTT's on a pre-arranged basis. Transponder 9 used by the Dutch based Filmnet is allocated to Belgium and sub-leased to Filmnet through the Belgium PTT, RTT. All transponders on F1 are preemptible. Lease rates are set for fixed periods of three years. The rate depends on the number of countries in which the downlink is (officially) received. The scale ranges from a minimum of one country to a maximum of five. Over five the lease rate is the same as five. The transponders used by Italy and Luxembourg on F1 are under "provisional" leases.

ECS–4. In 1985, nine transponders on Eutelsat I–F3 were allocated to national PTT's. F3 was lost on launch in September 1985. It is assumed here that the allocation remains valid for ECS–4 but, in 1986, Turkey purchased two transponders from Intelsat for TV distribution and therefore may not require its lease on ECS–4. NP = non-preemptible lease, P = preemptible lease. The allocation of transponders suggests that ECS–4 will be a "cable hot bird". However, given the technical problems with Eutelsat I–F1, it appears likely that it will act as a replacement for F1 rather than fully as a source of extra capacity. The latter will be available when ECS–5 is launched in 1988.

KEY: SA = Spot Atlantic beam, SE = Spot East beam, SW = Spot West beam.

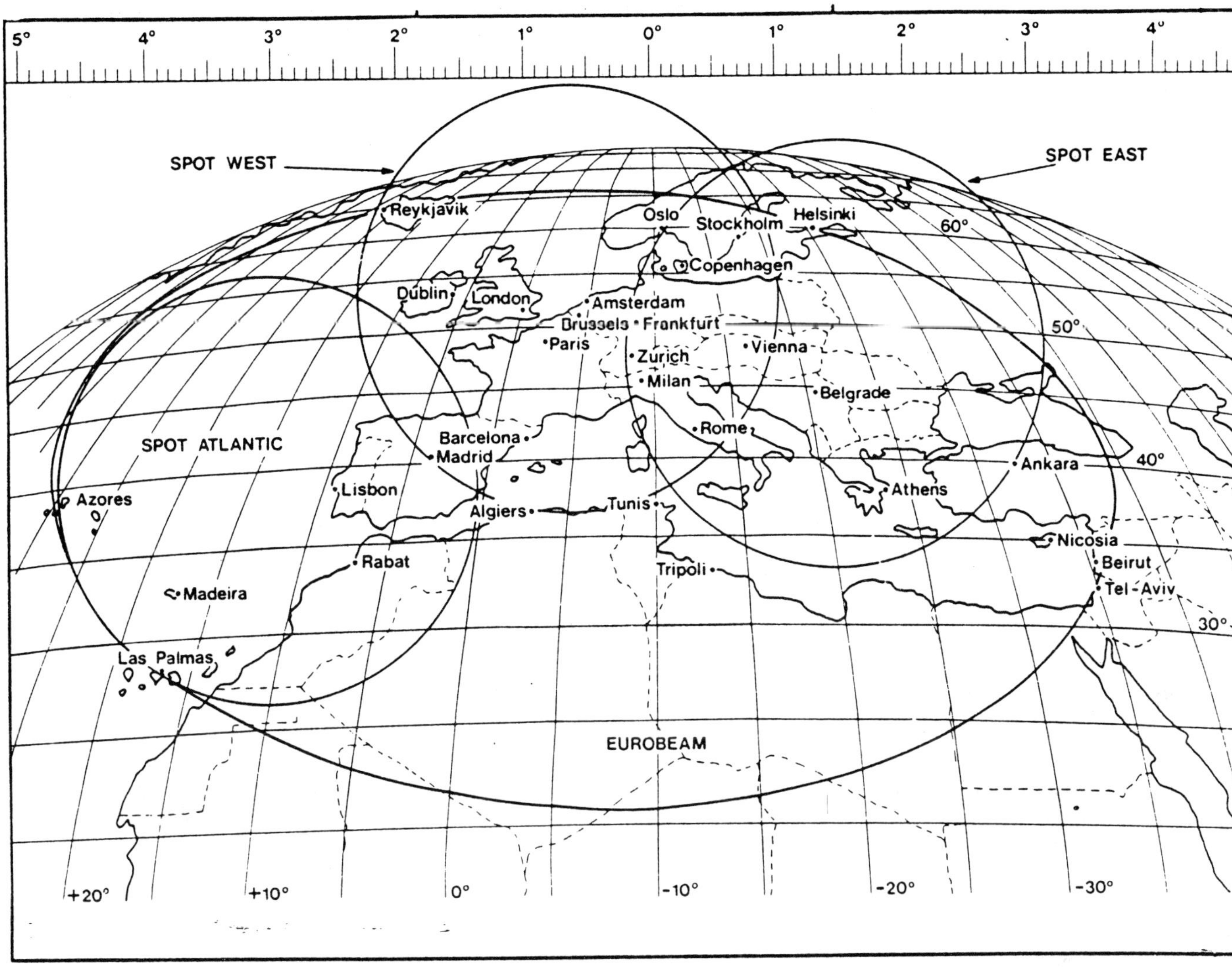

Basic footprint coverage of the Eutelsat–I satellites. This is for the primary payloads of the satellite. The downlink footprint of the SMS service is shown below. The Eurobeam downlink for primary services is principally used by the EBU and for telephony traffic. The Spot Atlantic beam has, so far, only been used on an experimental basis by Catalan TV in Spain. Spot East and West are used for cable TV traffic

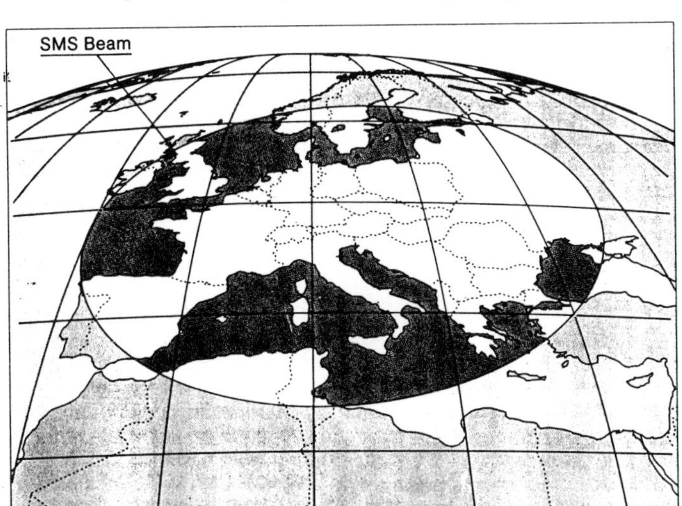

rejection rate for the tubes was many times greater than had been envisaged. It had expected to reject some 20% of the tubes manufactured by AEG–Telefunken – in practice it turned out to be some four times this figure. However, of those accepted, the tubes proved to be very reliable. It is understood that the AEG–Telefunken tubes on Eutelsat I–F1 did not go through the same rigorous inspection as applied to those delivered to Hughes.

On the other hand, when the Eutelsat I system was ordered, it was not expected that they would be subjected to the heavy loading of two TV channels which can run for 24 hours per day every day of the year. The technical problems with Eutelsat I–F1, the demand for more capacity and more security and the loss of ECS–3 put Eutelsat under pressure to put more capacity up quickly. At one time there was concern that Eutelsat I–F1 would not last beyond mid-1986. British Telecom looked at taking capacity on Luxembourg's Astra satellite as one method of getting round the problem and, at the time of writing, is renewing its leases to UK customers with clauses that mean it will go to a higher power satellite if available. Eutelsat has taken two extra ECS satellites to provide extra and back-up capacity, ECS–4 and 5, but the PTT's refused its request for a sixth (ECS–6). ECS–5 does not look like being available in-orbit until 1988 – it had been intended to launch it by Ariane V24 in March 1987.

However, the greatest impact of the combination of competition from Société Européenne des Satellites and the shortcomings in the Eutelsat I design, has been on the design for Eutelsat II. Originally intended to use 30 Watt travelling wave tubes, the concept was similar to that of the original satellite system. However, the design now uses 50 Watt tubes and is remarkably close to the Astra concept. The Eutelsat II system and its origin is described in the next section. Eutelsat has also been working on a concept for a broadcasting satellite system for

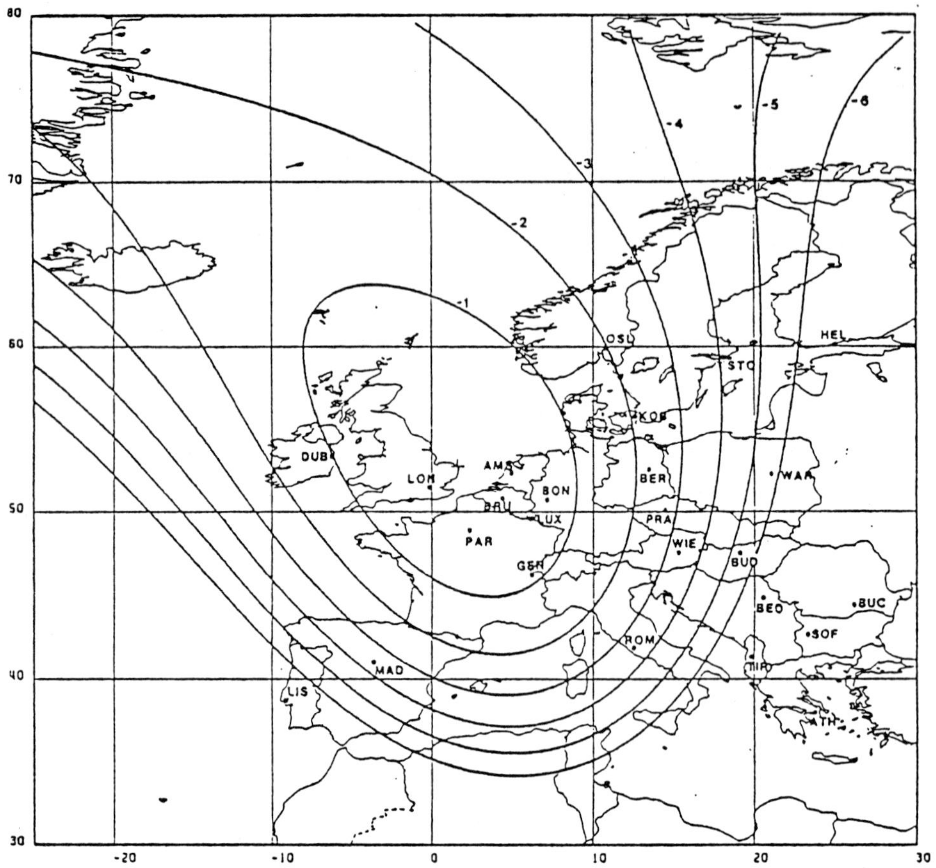

Spot West beam footprint on Eutelsat I–F1 at 13° East. Inner-most contour shows a signal strength of 46 dBW; the outermost is 41 dBW. This spotbeam is used by RAI, Europa TV, TV5, Worldnet, WPN, Sky Channel, Teleclub, Filmnet, SAT–1 and Music Box/Superchannel. Typically, within the 46 dBW contour a 1·6 metre TVRO dish can provide acceptable quality pictures

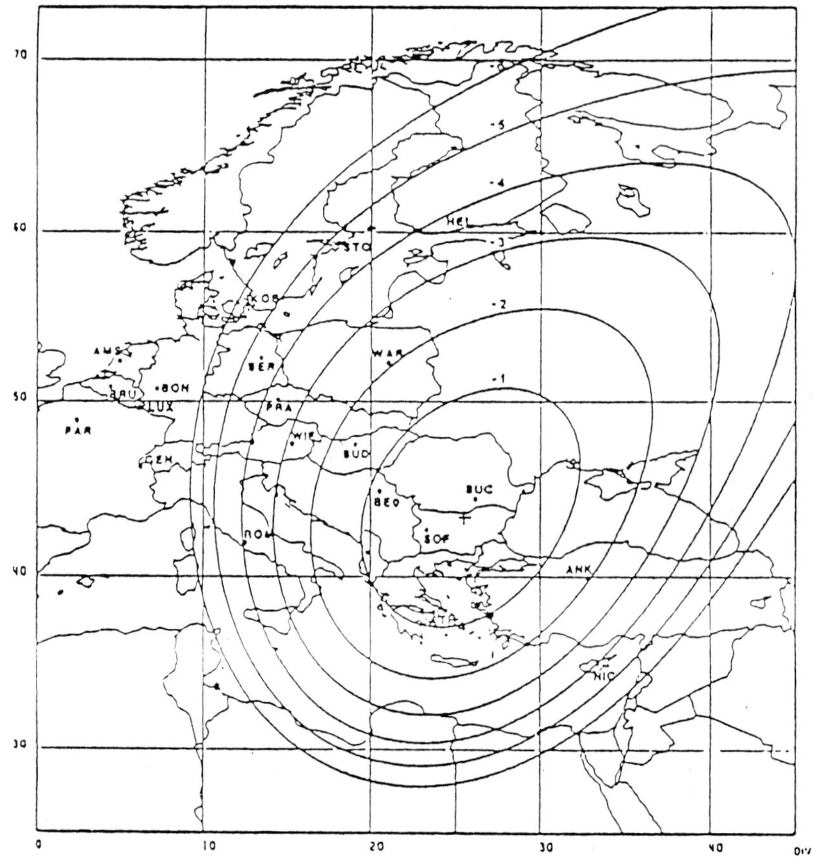

Spot East beam footprint on Eutelsat I–F1 at 13° East. Inner-most contour is the 46 dBW threshold, outermost is 41 dBW. The 31 dBW contour cuts through the UK and Spain. Channels using this beam are 3–SAT and RTL–Plus

Artist's impression of the Eutelsat I spacecraft (courtesy British Aerospace)

the mid-1990's and beyond known as Europesat. This appears to address the needs of DBS TV as well as cable/SMATV services and is similarly described in a following section of this book.

The Eutelsat Space Segment: It was originally envisaged that the Eutelsat space segment would consist of three satellites:

1) Eutelsat I–F1 would be the in-orbit back-up satellite,

2) Eutelsat I–F2 would be the in-orbit operational satellite,

3) Eutelsat I–F3 would be the ground spare.

However rapid growth of interest in cable television in Europe in the first half of the 1980's led to a substantial demand for capacity to distribute specialised TV channels to cable networks. RTT's were able to fill the entire capacity on Eutelsat I–F1 with cable TV channels. It was therefore decided that the slightly larger Eutelsat I–F3 would be designated as the principal satellite for television channels. However this satellite was lost on launch in September 1985 and the decision was taken to procure, through ESA, the ECS–4 satellite, now due for launch in May/June 1986. Decision was also taken to procure a fifth satellite of the ECS series. It is not clear what the roles of these extra satellites are, but it is understood that ECS–5 is to replace the troubled Eutelsat I–F1 satellite. The Eutelsat I–F1 capacity is available only on a preemptible basis. The satellite is not quite as advanced as subsequent satellites in the series in that it is neither equipped to carry SMS traffic nor has full eclipse protection capability.

The Eutelsat I–F1 and subsequent satellites each have a Eurobeam footprint and three Spotbeam coverages. The Eurobeam coverage is used for downlinking of telephony and television and for the uplinking of television. The Eurobeam covers much of North Africa and extends out into the Atlantic to provide services to the Canaries and Azores and up to Iceland and Eastwards to include Turkey, much of European Russia, and Israel. The three spotbeams, Spot Atlantic, Spot West and Spot East are used for downlinking of telephony. The Eutelsat I–F2 and subsequent satellites carry two extra transponders for satellite multi-services (SMS). The footprint for this service is somewhat smaller than the Eurobeam and is centred on the Western part of Europe where most potential customers are concentrated. On the uplink side all transponders on the satellite are connected to an antenna providing Eurobeam coverage. On the downlinking side they are connected to three spot beam antennas as well; a number of switches provide some flexibility in the allocation of capacity to the different beams, but the interconnection of some transponders is permanent.

The 500 MHz of frequency spectrum available for each satellite is divided into six channels of 83·33 MHz spacing and a useful bandwidth of 72 MHz. Frequency re-use allows a total capacity of 12 channels. However on Eutelsat I–F1 only nine can be used simultaneously in sunlight, and five in eclipse. On subsequent spacecraft, nine transponders can be operated simultaneously in both sunlight and eclipse.

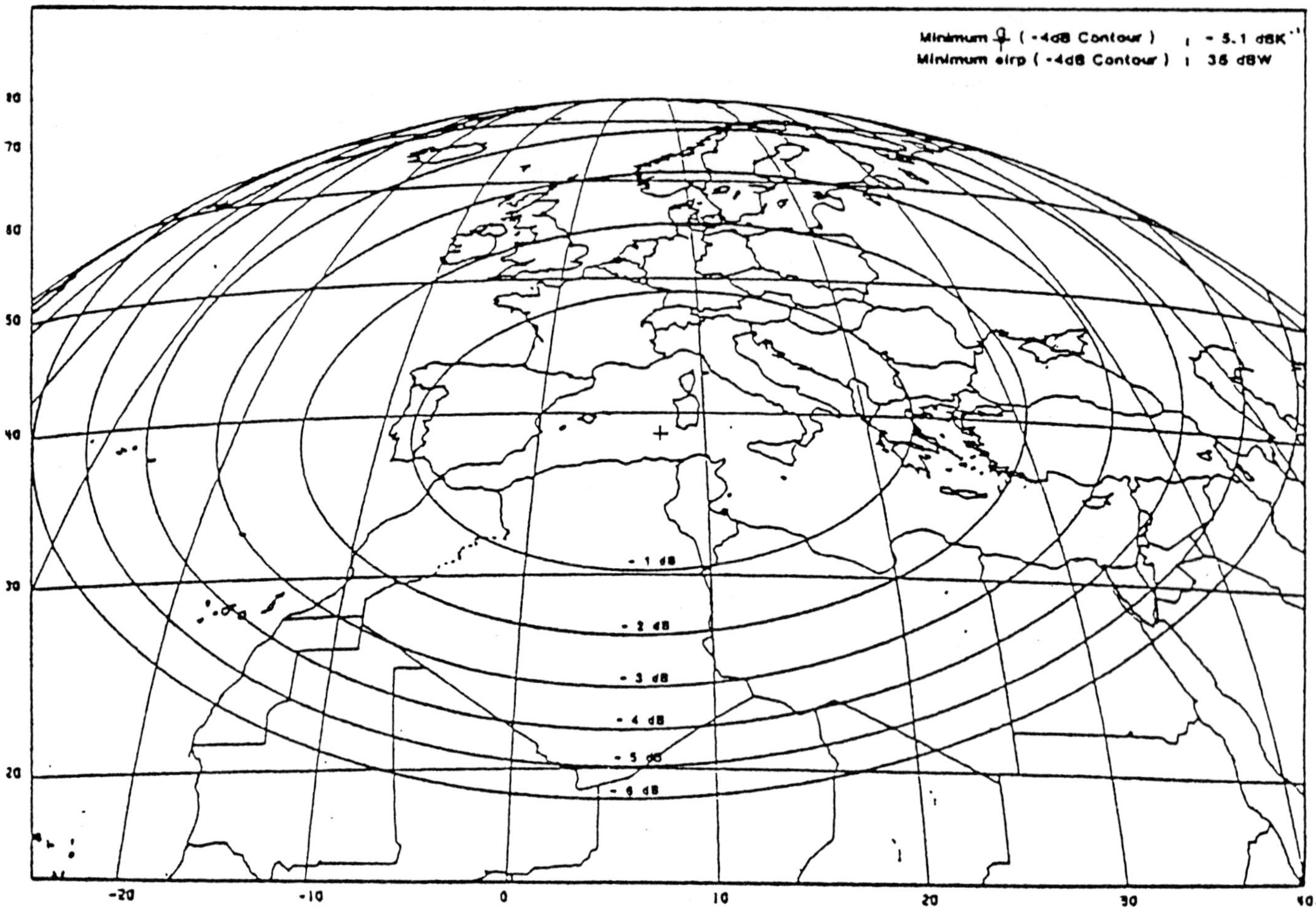

Eurobeam coverage of Eutelsat I–F2 at 7° East. This beam is used for all uplinks and downlinks for EBU broadcast transmissions. The spotbeams are not used for uplinks. Innermost contour on the Eurobeam is 39 dBW; outermost is 34 dBW

Spacecraft design: The ECS spacecraft are medium sized, three axis, stabilised satellites and externally similar to the OTS satellites. All are Ku-band birds. Eutelsat I–F2 and subsequent satellites in the series have more batteries to provide eclipse protection and a new design of antenna for Eurobeam coverage. The basic platform design has been used for the Telecom 1, Skynet–IV and Marecs satellites and has been developed into the Eurostar design being offered by Satcom International.

EUTELSAT SERVICES

1) *Digital telephony*; In theory, if not in practice, the principal mission of Eutelsat is to provide high-capacity digital links for routing telephone traffic between the international switching centres of its member countries. Each transponder has sufficient bandwidth to carry a digital transmission rate of 120 Mbits/s or the equivalent of 1800 telephone circuits. The transmission system is TDMA. The telephone traffic service uses 17 metre earth stations of which there are expected to be no more than 20 using the Eutelsat system. All these stations have access to all satellite transponders via the same Eurobeam antenna. However they can only receive their own signals if these are sent through transponders connected to a spotbeam providing coverage for their geographical location. The Eutelsat telephony mission is integrated with the terrestrial network in Europe and therefore interconnected with international telecommunications systems, including Intelsat and transatlantic submarine cables. However it is not normal practice to double hop telephone transmissions through two satellites if at all avoidable. The first of the Eutelsat spacecraft to carry telephone traffic was the Eutelsat I–F2 satellite whose capacity is also shared with the satellite multi service requirements and TV traffic. The DMA reference and monitoring stations for the system are located at Fucino in Italy and Guadalajara in Spain.

2) *Distribution of Cable TV Channels*: Capacity on Eutelsat I–F2 was effectively sold out for use for this traffic before the satellite was launched. The transponders were allocated to member countries in a crude process roughly in proportion to each members investment in Eutelsat. Eutelsat believes that it may now be the largest carrier of such satellite TV channels in the World. One channel on Eutelsat I–F2 is also used by Norway for TV distribution. It is expected that ECS–4, due for launch in mid-1986, will be the main cable TV satellite, not least because there are technical problems with Eutelsat I–F1 and in theory because Eutelsat I–F1 is the in-orbit back-up to Eutelsat I–F2, its capacity is preemptible. Lease rates for ECS–4 will be higher than Eutelsat I–F2 because the capacity is non-preemptible. However, demand for capacity for TV distribution already exceeds Eutelsat's capacity, so many European broadcasters are using Intelsat as an alternative. More capacity in the form of ECS–5 or 6 is expected. One transponder on Eutelsat I–F2 is also used for occasional TV transmissions.

3) *Eurovision Services*: Two transponders on Eutelsat I–F2 are allocated to the European Broadcasting Union for distribution of Eurovision programmes to their members. System

uses frequency modulation although digital transmission techniques are expected to be introduced.

4) *Satellite Multi-services System (SMS)*: This service is in fact carried on two separate satellite systems, Eutelsat I–F2 et seq. and Telecom–1. The decision to include an SMS service was taken after development work on the ECS satellites was well underway. Eutelsat asked ESA in 1980 to study inclusion of two extra transponders on the satellites for this service. It was not feasible on ECS–1 but was possible on ECS–2 to 5. The SMS service uses SCPC/FDMA techniques. However Eutelsat also leases one transponder on Telecom–1 which uses TDMA techniques. The ECS/SMS service is philosophically different from the Telecom–1/SMS system in that the ECS/SMS service was proposed as a quick response to a growing demand for wide-band digital services across international borders within Europe. The Telecom–1 system is explicitly part of a plan to integrate terrestrial services with satellite communications in pursuit of an ISDN network. Eutelsat has left much freedom to national PTT's in developing ECS/SMS services.

The SMS service allows uni-directional or bi-directional digital links to be established between users in a point-to-point or point-to-multipoint configurations. Data rates offered to subscribers are initially limited to 64, 128, 256, 1920 Kbits/s although other multiples of 64 Kbits/s may be provided. The system allows video-conferencing using a European standard based on a bit rate of 2048 Kbits/s. SMS satellite channels can be interconnected with terrestrial networks at each end, but can also be linked directly to the customer terminal. Many of the new services that the ECS/SMS system will provide, require the transmission of signals in one direction only (transfer of data files, pictures, drawings, electronic mail) or two-way communications of an asymmetric nature-database consultation, remote use of computers, video-transmissions etc. Eutelsat is offering channels for SMS services on the basis of full-time lease, regular part-time lease, reservation by subscription or reservation for one-off use.

INVESTMENT SHARES OF EUTELSAT MEMBERS

Austria	1·96%	Malta	0·05%
Belgium	4·90%	Monaco	0·05%
Cyprus	0·97%	Netherlands	5·45%
Denmark	3·27%	Norway	2·50%
Finland	2·72%	Portugal	3·05%
France	16·35%	San Marino	0·05%
Germany (Wast)	10·79%	Spain	4·63%
Greece	3·18%	Sweden	5·45%
Iceland	0·05%	Switzerland	4·35%
Ireland	0·22%	Turkey	0·93%
Italy	11·45%	United Kingdom	16·35%
Liechenstein	0·05%	Vatican City	0·05%
Luxembourg	0·22%	Yugoslavia	0·96%

Source, Eutelsat, December 1986

TECHNICAL SUMMARY

	ECS–1	*ECS–2 to 5*	*OTS*
PLATFORM	ECS	ECS	ECS
PRIME CONTRACTOR	British Aerospace heading up MESH consortium	British Aerospace heading up MESH consortium	British Aerospace heading up MESH consortium
OTHER MAJOR CONTRACTORS	Matra, ERNO, Aeritalia, ATN, Selenia Spazio, Saab, SEP	Matra, ERNO, Aeritalia, ATN, Selenia Spazio, Saab, SEP	Matra, ERNO, Aeritalia, Selenia Spazio, Saab, SEP
STABILISATION	3-axis	3-axis	3-axis
LAUNCH MASS	1030 kg	1140 kg	837 kg
MASS IN ORBIT (Beginning of Life)	610 kg	680 kg	440 kg
POWER SUPPLY, END OF LIFE	900 Watts	900 Watts	600 Watts
DESIGN LIFE	7 years	7 years	3 years
ORBITAL POSITIONS	10° East	Eutelsat I–F2, 7° East Eutelsat I–F3, lost ECS–4 (expected), 13° East ECS–5, not known ECS–6, not known	5° East (finally but 10° East initially)
STATION KEEPING ACCURACY			
ECLIPSE PROTECTION	Partial (five transponders)	Partial (nine transponders)	Partial (two transponders)
TOTAL NO. OF TELECOMS TRANSPONDERS	12	12	6
NO. OF SIMULTANEOUSLY OPERABLE TRANSPONDERS	9	9	6
NO. OF SMS TRANSPONDERS	–	2	–
TWTA POWER OUTPUT	20 Watts	20 Watts	–
FREQUENCY RANGE:			
UPLINK	14–14·5 GHz	14–14·5 GHz	–
DOWNLINK (TELECOMS)	10·95–11·2 and 11·45–11·7 GHz	10·95–11·2 and 11·45–11·75 GHz	–
DOWNLINK (SMS)	–	12·50–12·75 GHz	
TRANSPONDER BANDWIDTH	80 MHz	80 MHz	40 MHz (2), 120 MHz (2), 5 MHz (2)
EIRP MINIMUM, BEAM EDGE:			
EUROBEAM	34·8 dBW	34·8 dBW	–
SPOTBEAM	40·8 dBW	40·8 dBW	–
SMS BEAM		39·8 dBW	–

COMMERCIAL DETAILS

OPERATOR	European Telecommunications Satellite Organisation (Eutelsat)
ADDRESS	Tour Maine-Montparnasse, 33 Avenue du Maine F-75755, Paris Cedex 15, France
TELEPHONE NO.	+33 14 538 4747
TELEX NO.	203 823 EUSAT
FAX NO.	+33 14 538 3700
SENIOR STAFF	Andrea Caruso, Director General W. Schroeter, Director, Operations G. Payet, Director, Technical Department D. Hardmann, Director, Admin & Finance Philippe Binet, Head, Doc & PR
PURPOSE OF PROJECT	Provision of regional Ku-band capacity for use by member European PTT's and the EBU for TV, telephony and business coms. Replaces Eutelsat–I system
PROJECT STATUS	First three satellites already ordered and delivery due to begin in 1989
PROJECT VALUE	First three satellites were ordered under a contract worth $225 m. Launch, control and insurance costs are extra
LAUNCH DATES	Eutelsat II–F1, July 1989, Ariane V39 Eutelsat II–F2, January 1990, Ariane V43 Eutelsat II–F3, date to be announced, Ariane

SATELLITE USERS

Like the Eutelsat–I system, Eutelsat–II is intended for use by the members of the Eutelsat organisation, with the latter acting as a carriers' carrier. Membership of Eutelsat is given in the section on Eutelsat–I. For TV traffic individual transponders are leased to member PTT's and sub-leased on to individual broadcasters. The EBU has a separate arrangement with Eutelsat for its capacity. These broad arrangements are expected to continue for Eutelsat–II. However, changes amongst the TV broadcasters are likely as new entrants are expected and some existing uses of the Eutelsat–I may transfer to other satellite systems.

Similarly, it is possible that the new system will prove more attractive to business users and Private Satellite Networks (PSN) than the current low power satellites available in Europe although there is some doubt if Europe's PTT structure will facilitate such developments. Eutelsat says that the new spacecraft will be used for SMS business services, "data transmission services" and telephony and TV services. The satellites have a potential for quasi-DBS TV. By the time the system becomes operational, TVRO technology may allow reception within the 51 dBW contour with dishes as small as 40 cm and 50 cm within 49 dBW. At this stage there are no definitive plans in Europe to use the satellites for DBS and Eutelsat is still looking at its own next generation DBS system, Europesat.

PROJECT OVERVIEW

Introduction: The Eutelsat–II system is the second generation satellite segment for the Eutelsat organisation and is due to come on stream in 1989–90. The project has been heavily influenced by the potential competition from the Astra project of Société Européenne des Satellites and the dominating need to meet the requirements of broadcasting organisations. The new system may face considerable competition from the domestic satellite systems of West Germany (DFS–Kopernicus) and France (Telecom–II) and possibly from Atlantic Satellites who are now offering both DBS and FSS capacity with European coverage. The project is also of major importance to the European aerospace industry, now faced with competition from US manufacturers in its domestic market. The Eutelsat organisation itself is also likely to face pressures from changing national policies towards telecommunications. Increasing liberalisation and the introduction of competition are inherent threats to Europe's hitherto cosy PTT cartels of which Eutelsat is an off-shoot. Some Telecoms administrations may either pull out of Eutelsat or back competitive systems to Eutelsat–II (or both).

Background to the Eutelsat–II system: When Eutelsat originally looked at its requirements for the 1990's, it decided that it needed about 30 transponders. At the time it looked as if advanced technologies such as intersatellite links and onboard switching using large platforms would be available in the period. It soon became apparent that they would not be ready in time and an "interim" generation system, employing conventional technology with a smallish platform would be needed. Eutelsat issued the RFP (Request for Proposals) for Eutelsat–II in February 1985. This was for an unambitious design similar in performance to the existing Eutelsat–I spacecraft. The organisation wanted low power with TWTA's in the 30 Watt power output range with three footprints essentially matching existing services. A Eurobeam was intended to service European Broadcasting Union needs. A wider continental beam was able to provide services for all European PTT's including those behind the iron curtain who have not decided (yet) to join the organisation. Finally a Western beam roughly corresponded to the West spotbeam on the Eutelsat–I system which is used by cable TV broadcasters. The concept lead to some criticism of Eutelsat. For example, in 1985 trade sources were suggesting that "the design of the (Eutelsat–II) satellites is not radically different from the present series and they will have little more capacity for television... they are the product of planning which experience of cable and satellite has shown to be obsolete...". At the time US thinking was strongly in favour of medium power (50–60 Watt) for both FSS services and possible quasi-DBS services.

The possibility of such medium power services, particularly for the distribution of TV traffic, had already been seized upon by Luxembourg which was threatening to undermine Eutelsat's position as well as the European satellite manufacturing sector. The Coronet company, later to be replaced by SES, was proposing to use US built spacecraft which are significantly cheaper than those from Europe and are available within 30 months (against 36–42 months from Europe). The Astra satellite has cost SES $45 m whereas each Eutelsat–II satellite, with an almost identical payload performance, costs $75 m.

By 1986, when it was announced that Eutelsat had modified the payload performance to match that of Astra, it was apparent that there was a strong sense of urgency to get the new spacecraft launched as quickly as possible. The problems in the launch vehicle market gave Eutelsat a competitive boost in that it delayed the launch of Astra to at least February 1988, not much more than a year ahead of the first Eutelsat–II satellite. The extent of the threat from Astra can be judged by the fact that during 1986 sub-leases between British Telecom and UK broadcasters for capacity on Eutelsat–I and an implicit clause that BT transfer the customer to Astra if it were available before Eutelsat–II. Latest potential competitor in the Eutelsat market place is the Dublin (Ireland) based Atlantic Satellites whose FSS

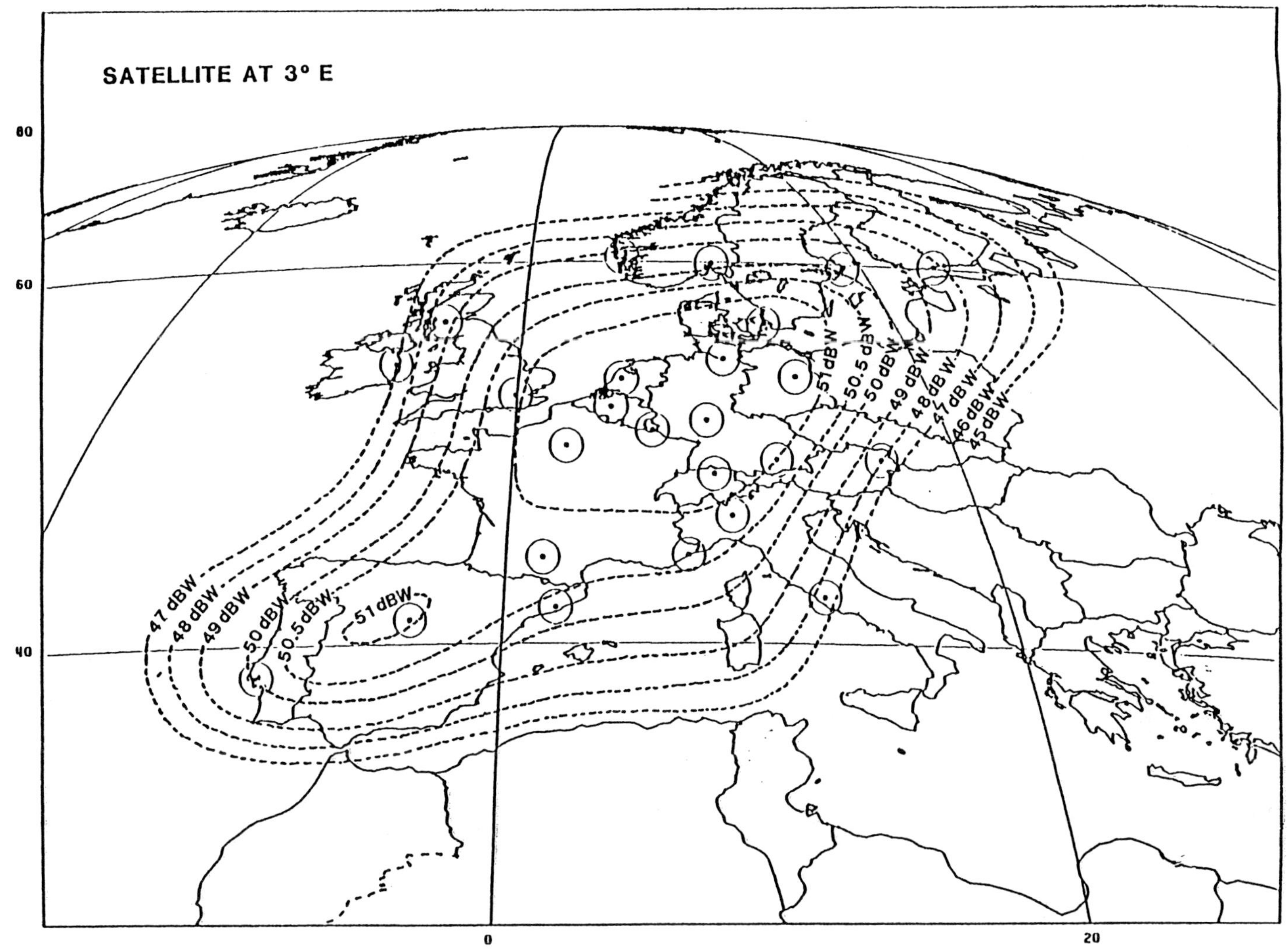

footprint is almost identical to that of the Eutelsat–II satellite at 3° East. AS can also provide overspill of its proposed UK/Ireland DBS capacity into Europe. The proposal is a direct competitor to Astra as well as Eutelsat.

Technical developments in ground receiver equipment meant that some sectors in the satellite communications industry saw a potential market for quasi-DBS services and VSAT business communications systems in Europe using medium power satellites. This thinking led to approaches to the Luxembourg government to allow a competitive system to Eutelsat. The result was the Coronet project which eventually metamorphised into SES's Astra satellite system. The Eutelsat member PTT's recognised this threat and British Telecom, in particular, was formative in pushing for changes to the Eutelsat–II system.

British Telecom's Influence: Unlike other European PTT administrations, British Telecom had by 1985 become a private sector (regulated) organisation facing long term competition in its domestic and international markets from Mercury Communications. The exact nature of the relationship between British Telecom and the conventional PTT dominated Eutelsat organisation still remains unclear but it appears that the former has been willing to throw in its lot with Société Européene des Satellites. This would have broken the regional monopoly of Eutelsat, who, at the best of times, work in a difficult market. Such a move would have seriously threatened the existence and raison d'etre of Eutelsat. During 1985 British Telecom requested that Eutelsat provide a modified satellite system within the Eutelsat–II programme. This would have involved a dedicated operational satellite with 50 Watt transponders. BT also talked with SES about options to use that company's space segment as an alternative to Eutelsats'. SES was then due to launch its first satellite in 1987. But in May 1986 Eutelsat ordered the Eutelsat–II system with specifications very close to those requested by British Telecom for its "dedicated satellite".

The about turn by Eutelsat surprised some in the industry but in retrospect it reflected the obvious fact that the regional satellite market in Europe is dominated by the needs of the broadcasting industry. There are now numerous satellite systems looking for a share of this market. These include the domestic Telecom–1 and 2 and DFS–Kopernicus systems, SES's Astra system, the TDF 1/2 and TV-SAT DBS projects, Intelsat, Panamsat, and also proposals from Atlantic Satellites and the various options open to other European countries for DBS TV. Eutelsat is hampered in serving the broadcasting market because it is a carriers' carrier with individual broadcasters negotiating with their PTT's for capacity. Those PTT's, of course, prefer to see broadcasters use their own domestic satellite systems and the broadcasters prefer to negotiate directly with competitive satellite operators. Eutelsat's problems have been compounded by technical problems with the Eutelsat–I spacecraft, a shortage of capacity on that system (leaving some broadcasters to use the Intelsat system), the loss of ECS–3 on launch and delays in the launch of ECS–4 and 5.

Moreover, it appears that British Telecom have strongly considered pulling out of the Eutelsat organisation. It has resisted attempts by the Eutelsat executive to strengthen the organisation's competitive position against new-comers, particularly Société Européenne des Satellites. Eutelsat prepared an internal report in 1986 outlining proposals to sink the Astra project. But British Telecom supported moves which forced the Eutelsat executive to negotiate with Luxembourg to come to an

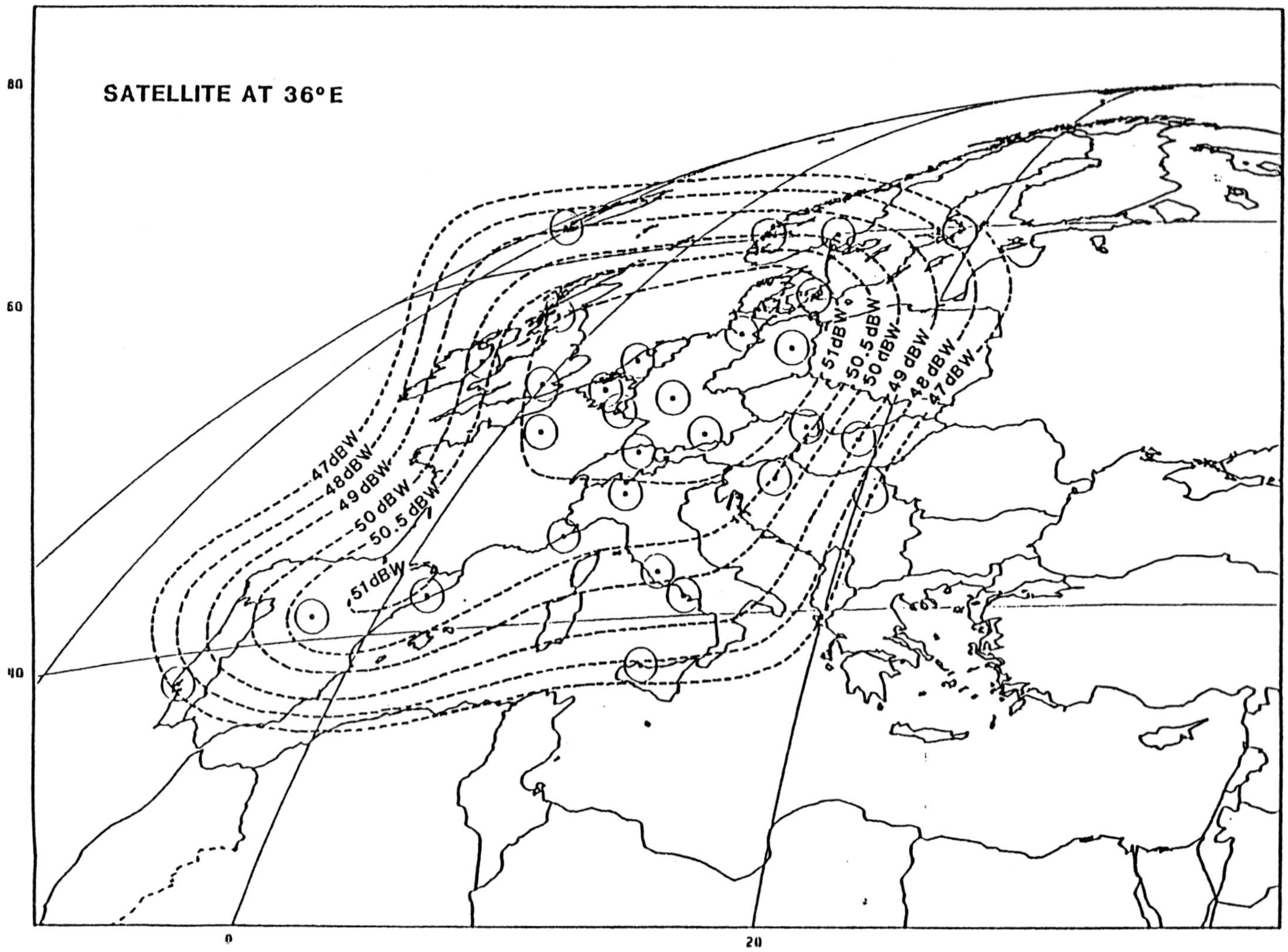

accommodation with SES. It also objected to Eutelsat plans for a sixth Eutelsat–I satellite and a fourth satellite in the initial order for the Eutelsat–II system.

This has led to some speculation that Eutelsat placed the order for the Eutelsat–II system with Aerospatiale, rather than the rival Satcom International consortium, because France is much less likely to pull out of Eutelsat. Prior to the award of the contract in May 1986, British Aerospace had primed all European (as distinct from national) communications satellites. British Aerospace, as a partner with Matra and Selenia Spazio in the Satcom International bid, would have primed the satellites. The official Eutesat version of the decision was that there was very little to choose between the two bids but some observers remain sceptical as to why the less experienced Aerospatiale team got the contract. Similarly, it was decided to finance the Eutelsat–II system by Eutelsat raising loan finance rather than by the more traditional route of member PTT's providing capital contributions. This would make it easier for Eutelsat's other members if BT pulled out of the organisation. There is also concern that the increasing (but still slow) moves toward telecommunications liberalisation in Europe would result in other PTT administrations taking an aggressive attitude towards Eutelsat. The latter remains a product of the conventional communications policies of the 1960's and 1970's, with a cumbersome and slow (ish) decision taking process and structure which requires constant referral to member PTT's who often have conflicting interests. Eutelsat is far from being a shadowy organisation but its members have been used to cooperating in secretive organisations such as CEPT which are largely outside the gaze of either the press or the public.

The Eutelsat–II contract: The contract for the new system was placed, in May 1985, with the Aerospatiale consortium which included Marconi Space systems (UK), MBB–ERNO (West Germany), Alcatel Thomson (France), Aeritalia (Italy), ETCA (Belgium) and L. M. Ericsson (Sweden). The initial order was for three spacecraft with options for five more. The contract value was for ECU225 m ($225 m). One or more options could have been taken up if British Telecom wanted its own dedicated system, but it appears that BT had provisionally requested lease of only six transponders. Eutelsat publicly placed a lot of emphasis on the tight delivery schedule for the spacecraft and said that there were significant performance guarantees relating to on-time delivery. Nevertheless, Aerospatiale was given 36 months to deliver the first satellite from the time of the order, contrasting to typical US performance now standing at 28 months. Nominally Eutelsat was required to go out to international tender for procurement but in Europe's politically charged telecommunications environment this means to European suppliers only.

Unlike the Eutelsat–I system, Eutelsat is responsible for procurement and operation of the new system. Eutelsat–I was procured and is operated by the European Space Agency with the satellites leased to Eutelsat. ESA has no involvement in Eutelsat–II which will be owned and operated by Eutelsat which now has its own project management team. Eutelsat is constructing its own control facilities for the system.
The platform is a modified Spacebus 100 (100B) design derived from that used for Arabsat but with a significant increase in weight (1625 kg at launched compared with Arabsat's 1200 kg). Each satellite will carry 16 operational 50 Watt TWTA's. Eight transponders have a bandwidth of 72 MHz and eight have 36 MHz for narrow-band operation. Power supply is 3000 Watts, compared with 1300 Watts for the Arabsat spacecraft. The failed

Satcom International bid offered a Eurostar platform with eight transponders at 60 Watts and eight at 35 Watts although it appears that the specification could have been reconfigured to 16 at 50 Watts. The Spacebus 100B configuration is a conventional three-axis design whose most novel feature is the use of multi-feed antenna reflectors for double polarisation and reconfigurable shaped beams.

Eutelsat's Board of Signatories, meeting in an extraordinary session on the 4th and 5th of September 1986, decided to "implement certain modifications to the technical characteristics of the (Eutelsat–II) satellites in order to provide a better response to the requirements of a number of signatories". The most significant of these is an enlargement of the service zone of the high gain beam allowing direct reception of TV channels with TVRO's of around one metre. The footprint from 3° East now shows a minimum EIRP of 47 dBW for all of Western Europe except Ireland, Southern Italy and the North of Sweden and Norway. A second major modification is that all the 16 operational transponders on each satellite will be able to operate on the main beam; previously only 12 transponders were able to do so.

Eutelsat asked its members to submit requests for modifications in time for the 4th September meeting. Requests came from the UK, Portugal, Spain, France, Italy and Germany and also a request from Denmark to provide communications with Greenland. Other Scandinavian countries, surprisingly, did not submit any requests. Those that did, apparently, are those most interested in leasing capacity on the Eutelsat–II system. The UK and Holland objected to the final form of the modifications. The new Eutelsat–II modifications are expected to add some two to three months to the construction time of the spacecraft, but as Arianespace cannot launch the first until July 1989 (at least according to its latest launch manifest) this will have no impact on the quoted launch date. The three satellites on order are to be located at 36°, 3° and 12° East. The possibility of locating one satellite at 19° East has also been evaluated.

Eutelsat–II and the Astra Project: The Eutelsat–II system is clearly a competitive response to SES's Astra project and there have been differences of opinion between the Eutelsat executive, opposing Astra, and the Eutelsat member PTT's looking for suitable satellite capacity for their customers. Eutelsat has tried to bring forward the launch of the Eutelsat–II system to meet the needs of still unprofitable cable TV channels looking for quasi-DBS services. The original CORONET proposals envisaged taking away traffic from the Eutelsat–I system but subsequent problems in Luxembourg and the unavailability of launch vehicles means that Astra is now competing directly with the Eutelsat–II system.

Luxembourg, which had signed the definitive Eutelsat agreement but whose parliament has yet to ratify it, has invested a considerable amount of money and political effort in the Astra project but has also tried to come to some kind of "accommodation" with Eutelsat. It appears that it has been proposing a lease/leaseback arrangement for Astra capacity whereby Eutelsat would have nominal control over the Astra space segment but leaving SES free to market it on a commercial basis. There also appear to be institutional reasons why this is unacceptable to Eutelsat. These relate to Eutelsat's Articles of Association which nominally require it to go out to "international tender" for procurement of its space segment.

In practice this means European sources but the Astra satellite is being procured from the USA. SES is also continuing to suggest that it will procure a second satellite which will add further capacity to the market for the Eutelsat–I and II systems. The strained relationship between Eutelsat and Luxembourg over the Astra system led, in August 1986, to what is being described as a diplomatic incident. In reply to questions from a Deputy in the Luxembourg lower house (Chamber of Deputies), Prime Minister Jacques Santer told the House that everything was going well with the Astra project and that negotiations with Eutelsat were positive and not only progressing but progressing well. Santer added that a rapid solution to the negotiations was expected.

Eutelsat apparently took offence at this on the grounds that they had found the proposals from SES unacceptable and that, at the time, the negotiations had come to a complete standstill. Eutelsat wrote to Luxembourg journalist Mario Hirsch, who had previously been involved in setting up Coronet and has subsequently been a strong critic of the Luxembourg government's handling of the satellite project, and revealed the position, thus, in effect, undermining the integrity of PM Jacques Santer.

Santer had replied to the Deputy because the Deputy had questioned as to why Astra was not on a copy of a launch manifest he had received and as to whether SES was in financial difficulties. It was also tactless of Eutelsat to have written to Hirsch who is highly unpopular with the Luxembourg government over his criticism of the SES exercise.
SES submitted new proposals to Eutelsat in August 1986 in an attempt to reach an agreement. Eutelsat then said that there was insufficient time to evaluate them before the Council meeting in Late September. Eutelsat also said that it also wanted to know the outcome of the coordination of Astra with the Intelsat system.

TECHNICAL SUMMARY

PLATFORM	Spacebus 100B
PRIME CONTRACTOR	Aerospatiale
OTHER MAJOR CONTRACTORS	MBB-ERNO (WG), Alcatel Thomson (FRC), Marconi SS (UK), Aeritalia (ITLY), ETCA (BLGM), L. M. Ericson (SWEDEN), CASA (SPAIN)
STABILISATION	3-axis
LAUNCH MASS	1625 kg
MASS IN ORBIT	985 kg
POWER SUPPLY, END OF LIFE	3,000 Watts
DESIGN LIFE	7 years
ORBITAL POSITIONS	36° East, 12° East, 3° East, 19° East (possible)
STATION KEEPING ACCURACY	
ECLIPSE PROTECTION	Full

SPACECRAFT DIMENSIONS	2·5 metre by 2·0 metre
SPAN, TIP-TO-TIP	22·4 metre
NO. OF ANTENNAS	2
RELIABILITY	0·74 for 16 chnls, 7 years
NO. OF ACTIVE TRANSPONDERS	16
NO. OF SPARE TRANSPONDERS	8
TWTA POWER OUTPUT	50 watts
FREQUENCY RANGE:	
UPLINK	14·0–14·5 GHz
DOWNLINK	10·95–11·2, 11·45–11·7 and 12·5–12·75 GHz
TRANSPONDER BANDWIDTH	36 MHz (8) and 72 MHz (8)
EIRP MINIMUM, BEAM EDGE	47 dBW
CHANNEL POLARISATION	Linear with frequency reuse through orthogonal polarisation
RECEIVE COVERAGE (G/T)	−3·5 dB/K at beam edge
NO. OF COVERAGE AREAS	Uplink 1, downlink 3

COMMERCIAL DETAILS

OPERATOR	Federal Express
ADDRESS	889 Ridgelake Boulevard, Memphis, TN 38119, USA
TELEPHONE NO.	(901) 766-7700
SENIOR STAFF	Allan McArtor, Senior VP, Telecoms Dick Fellows, Project Manager, Satellite Development Gilbert Nook, VP, Satellite Systems
PURPOSE OF PROJECT	Provision of Ku-band capacity for Federal Express's voice, video, image and data circuits
PROJECT STATUS	The company has construct, launch and operate authority from the FCC but has yet to order satellites
PROJECT VALUE	In 1984 the company said that it intended to invest $1224 m in satellite distribution by 1995. This is mostly for ground equipment
LAUNCH DATES	Expresstar B, 1989, launch vehicle undecided. Expresstar A launch date undecided. Expresstar C is ground spare

SATELLITE USERS

The Expresstar system was primarily intended to provide Federal Express with Ku-band capacity for its "Zapmail" electronic courier service. This was being developed as a means of replacing the physical distribution of documents, a field in which Federal Express plays a dominant role, through its overnight/same day courier services and a large fleet of aircraft. However, Fedex has now abandoned Zapmail.

The "Zapmail" service used two way VSAT terminals on the customer premises for point-to-point, multipoint to multipoint and point to multipoint electronic distribution of documents at data rates of 56 kb/s and above. The company was hoping to have 200 VSAT units installed on customer premises by the end of 1986. The satellite capacity required for this is provided by GSTAR and Westar 3. This will be replaced by Expresstar once the satellite is launched.

Federal Express says that its main current thinking about the use of Expresstar is dominated by the Zapmail requirements. However, it has considered use of the system for internal voice, video and data communications. It is also possible that the company offers file transfer and other transaction services through the satellite system.

It has also said that it may lease surplus capacity on the first satellite before its own traffic demands build up to fill it.

Given that Federal Express has abandoned Zapmail, future use of Expresstar is uncertain, as are the company's intentions to construct it. A formal order for two satellites has been placed with RCA Astro-Electronics.

PROJECT OVERVIEW

Federal Express's Expresstar satellite project is, in itself, a relatively straightforward project to provide the company with Ku-band capacity. However, its real significance lies in the major service which Federal Express was developing for use by its customers. This is the Zapmail electronic courier services. Federal Express's major line of business is a physical document courier service which is aimed at the high value sector of the postal market, where speed and reliability cannot be matched by the US postal service and the international postal service. However, the physical distribution of documentation is, by any means, slower than the possibilities offered by electronic means. The carriage of information by electric, and therefore virtually instantaneous means, dates back to the 1850's with the development of the electric telegraph. Although this service was largely undermined in the 20th century by the spread of the telephone network and increased numbers of connections, the arrival of electronics in communications equipment and the rapid reduction in computer power costs, has meant that the costs of carriage and distribution of information by electronic means is falling in real terms. In the trade-off between cost and speed of carriage, the cost factor is favouring faster delivery. In the case of Zapmail, this means instantaneous delivery. The changing balance between cost and speed threatens Federal Express's core business, at least in the long-term.

In developing "Zapmail" Federal Express opted for use of satellites for its main transmission requirements. In doing so it limited its market size in the broader electronic mail market. Moreover, other new technology applications in terrestrial communications, such as fibre optic and broadband facilities, would have eventually swung the market away from Federal Express. Nevertheless, the company argued that use of Ku-band satellite capacity for Zapmail at 56 kb/s or above shows savings of 60% compared with terrestrial US telecoms capacity. The overall cost saving, including purchase of VSAT terminals may currently be around 30 to 50%. Key to this development has been the falling cost of VSAT terminals (now below $10,000 for a 1·8 metre dish) and the capability of handling higher speed data rates (56 kb/s and above). Most US terrestrial telecoms infrastructure cannot handle speeds above 19·2 kb/s. The VSAT terminals are also capable of a high level of data transmission reliability, with bit error rates of between 10^{-5} and 10^{-20}. Federal Express had been very optimistic and aggressive in developing its Zapmail and it was therefore all the more surprising when it abandoned its plans in the Autumn of 1986. In 1985 it surprised the US satellite industry with an announcement that it had contracted with M/A Com Telecommunications for the supply of up to 25,000 VSAT earth terminals which it designated as on-premise terminals (OPT). It was expected that 100 to 200 of these would be operational by the end of 1986. The large order with M/A Com was aimed at reducing unit costs of the terminals. The company's plans envisaged "at least 50,000 OPT's" by the end of the 1990's connecting some 250,000 users. Under the then current arrangements, Federal Express would have owned the OPT's. The OPT's are linked by co-axial cable to one or more facsimile machines on the customer's premises. These are called "Zapmailers". Transmission costs would have been 25 cents per page in addition to the lease/rental purchase costs of the Zapmailers. Also under consideration by Federal

Express were larger terminals linked to local distribution networks and customers, through terrestrial microwave, fibre optic or cable television networks.

The Zapmail market sector is also distanced from the US C-band VSAT data distribution services. These services have, to some large extent been aiming at the leased line market. Here AT&T has dominated the service with terrestrial services. Such phone lines have a maximum bit rate often no more than 19·2 kb/s. The voice grade leased lines for data is very substantial, providing a large part of the computer to terminal network in the USA. Traditionally, terrestrial leased lines have been cross-subsidised from conventional message toll services which, in turn, have also cross-subsidised local telephone traffic. Liberalisation of the US telecoms market and the break up of AT&T has meant that such cross-subsidies are being phased out, hence terrestrial leased line charges have increased at a time when alternative C- and Ku-band services are falling in cost. Equatorial Communications, a California based manufacturer of C-based small aperture terminals, estimates that, current, two-way satellite C-band services using such equipment can produce savings of around 30%.

Federal Express began looking at satellite communications for electronic courier services in 1980. It filed for FCC construct, launch and operate authority for its own satellite system, on 7th November 1983. As an interim measure, in case the application failed, it made a preliminary agreement with a company called United States Satellite Systems Incorporated. This was a new venture company aiming to provide a domestic US Ku-band satellite system (called USAT) but it folded in 1985 when the FCC rejected its application for authority to construct, launch and operate the system. The preliminary agreement with Federal Express involved the latter using capacity up to the two full satellites in the four satellite USAT system. However, Federal Express gained authority from the FCC in 1985 for its own system. This covered two in-orbit satellites and a ground spare. First launch was due in 1989, to place a satellite at 124° West. This launch date would have been delayed by the non-availability of launch vehicles. The second satellite was assigned the 77° West orbital slot but Federal Express had not yet set a firm date for launch. In the meanwhile, it had been talking with the FCC and GTE about the possibility of the latter placing a GSTAR satellite at 124° West so that when the first Expresstar satellite was launched, customers would not have to realign their dishes. Federal Express's agreement with GTE called for up to eight transponders on GSTAR.

For the Expresstar system, Federal Express planned a "telehub" near its headquarters in Memphis. The system required a double hop. Each satellite had a series of 27 MHz transponders for quarter CONUS coverage, switchable to full CONUS coverage, and 54 MHz transponders for full CONUS coverage. The latter would have used the uplink from the customer's premises to the satellite and down to the telehub. The 27 MHz transponders would have used the uplink from the telehub to the satellite and down to the customer's premises. The telehub served as a network control point. The 27 MHz transponders provided an EIRP of 50 dBW in quarter CONUS coverage and 44 dBW for full CONUS coverage. The 54 MHz transponders provided 47 dBW into Memphis. The on-premise customer earth stations would only have required 1·2 metre dishes for Zapmail services via Expresstar, compared with 1·8 metres at present.

Federal Express was also looking at the mobile satellite services industry to provide communications with its fleet of 8000 road vehicles and 80 aircraft.

TECHNICAL SUMMARY

PLATFORM	RCA 5000
PRIME CONTRACTOR	RCA Astro Electronics
OTHER MAJOR CONTRACTORS	
STABILISATION	3-axis
LAUNCH MASS	
MASS IN ORBIT	
POWER SUPPLY, END OF LIFE	
DESIGN LIFE	
ORBITAL POSITIONS	Expresstar B, 124° West Expresstar A, 77° West Expresstar C, ground spare
STATION KEEPING ACCURACY	not available
ECLIPSE PROTECTION	Full
PROPULSION SYSTEM	not available
SPACECRAFT DIMENSIONS	not available
SPAN, TIP-TO-TIP	not available
NO. OF ACTIVE TRANSPONDERS	34
NO. OF SPARE TRANSPONDERS	
TWTA POWER OUTPUT	50 Watts
FREQUENCY RANGE:	
UPLINK	
DOWNLINK	
TRANSPONDER BANDWIDTH	27 MHz and 54 MHz
EIRP MINIMUM, BEAM EDGE	44 dBW, 47 dBW, 50 dBW
CHANNEL POLARISATION	

FINANSAT

COMMERCIAL DETAILS

OPERATOR	Finansat
ADDRESS	
TELEPHONE NO.	
TELEX NO.	
FAX NO.	
SENIOR STAFF	Ken Susskind, Senior VP, Legal Regulatory and Technical Planning Myer Feldman, CEO and Chairman Louis Beck Stanley Vaughn, VP for UK
FINANCIAL BACKING	The company is looking for funding from three sources: 1) a wealthy joint venture partner to provide up to $300 m 2) or foreign partners with possible US public offering 3) (last resort) a combination of supplier, user and insurance finance
PURPOSE OF PROJECT	Provision of C-band capacity for financial institutions world-wide
PROJECT STATUS	At planning stage although an FCC authorisation request was submitted in 1985
PROJECT VALUE	System will require funding of around $300 m according to Finansat
LAUNCH DATES	not available

SATELLITE USERS

This project was originally envisaged as a system to provide satellite capacity in conjunction with small earth stations to the international financial community. The company is said to have had discussions with Citibank and Bank of America in the USA, banks in Tokyo, Shanghai and Hong Kong and various "European financial institutions". It has appointed a Vice-President for the UK, Stanley Vaughn although the company is understood to have looked for UK equity finance as well.

An original plan envisaged an "exclusive" service for financial institutions but the company subsequently discovered that the demand from this sector was insufficient to justify the system and would leave the system with "enormous excess capacity". The company has subsequently looked at other sources of demand. However the system envisages use of C-band capacity only, which may limit its value outside the USA.

The type of services envisaged are C-band equivalents of Ku-band VSAT networks being installed in the USA.

Finansat says it is neither likely to cause economic harm to the Intelsat system nor does it want to. Company says it is not "at present" interested in video or in competing with Intelsat for switched voice "traffic". Its emphasis is on data transmission which Intelsat cannot fulfill under present or planned conditions.

Overall objective is to link financial institutions in the USA with branches or ancillary offices in Europe and Asia.

PROJECT OVERVIEW

Finansat is one of a number of proposals to provide international private satellite communications. It is a start-up venture and given the problems involved in raising large sums of money for such satellite projects, problems in the insurance sector and competition from Intelsat and new private international satellite operators, Finansat's future appears uncertain.

TECHNICAL SUMMARY

PLATFORM	
PRIME CONTRACTOR	
OTHER MAJOR CONTRACTORS	
STABILISATION	
LAUNCH MASS	
MASS IN ORBIT	
POWER SUPPLY, END OF LIFE	
DESIGN LIFE	
ORBITAL POSITIONS	47° West, 178° West
STATION KEEPING ACCURACY	
ECLIPSE PROTECTION	
PROPULSION SYSTEM	

SPACECRAFT DIMENSIONS	
SPAN, TIP-TO-TIP	
NO. OF ACTIVE TRANSPONDERS	36 (per satellite)
NO. OF SPARE TRANSPONDERS	
TWTA POWER OUTPUT	
FREQUENCY RANGE:	
UPLINK	} C-band
DOWNLINK	
TRANSPONDER BANDWIDTH	72 MHz
EIRP MINIMUM, BEAM EDGE	
CHANNEL POLARISATION	

COMMERCIAL DETAILS

OPERATOR	Ford Aerospace Satellite Services Corporation/ AT&T,
ADDRESS	1140 Connecticut Avenue NW, Washington DC 20036, USA (company merging into AT&T Long Lines Satellite Communications, Bedminster, New Jersey, NJ 07921 USA)
TELEPHONE NO.	FASSC (202) 785-4000 AT&T (201) 234-4000
SENIOR STAFF	Alan Parker, Chairman* Rex Hollis, President* John Fedak, VP Engineering* Donald Shroeder, Chief Financial Officer* Susan Pearce, Public Affairs Director* Karl Savatiel, Director of Satellite Communications, AT&T *Ford staff; not expected to transfer to AT&T
PURPOSE OF PROJECT	Originally planned to provide hybrid Ku/C-band US domestic capacity for a variety of services
PROJECT STATUS	Company is being sold to AT&T who may opt for different satellies than those envisaged by FASSC. FASSC has FCC construct, launch and operating authority and construction of satellites commenced before sale of company to AT&T
PROJECT VALUE	FASSC's plans envisaged expenditure of $400 million on construction, launch and operation of the two satellites and a ground spare
LAUNCH DATES	Fordstar 1, 1992 Fordstar 2, 1993 Fordstar 3, ground spare (FASSC had launch reservations on Shuttle but also was looking at Ariane 4 in 1986)

SATELLITE USERS

The Fordstar proposals of Ford Aerospace Satellite Services Corporation are, in effect, now defunct as the project has been taken over by AT&T. The latter company has not revealed its plans to use the orbital slots involved in the takeover beyond suggesting that it will be using them for C-band capacity to replace the Telstar 3 satellites which will become time expired in the 1990s. It is possible that AT&T will also want to provide Ku-band capacity.

Under the deal which resulted in FASSC being taken over by AT&T, FASSC's parent company will get use of three Ku-band Conus transponders on the first satellite launched by AT&T under the purchased authorisation. Ford will not be allowed to re-sell the capacity. It is not clear what will happen if AT&T fail to order the satellites or concentrate on C-band only.

Despite its past and current reputation as a terrestrial telecommunications company, AT&T is a lightweight amongst US domestic satellite operators and should it go ahead with modified FASSC plans, it is likely to have only a small share of the market. AT&T has had ambitions in the VSAT market and private satellite networks.

FASSC had envisaged that its system would be used for a variety of "existing and new telecommunications and video services". C/Ku-band interconnectivity would have allowed established users with C-band systems to access the Ku-band ground without the need for new uplink facilities. Amongst such potential users could have been Satellite News Gathering operators and VSAT system operators perhaps wanting some C-band capacity for areas of high rainfall and Ku-band for other areas. FASSC also planned to "provide a variety of voice, video and data services via satellite serving thin route and wideband applications for commercial and governmental customers."

FASSC had also expected that three Ford companies would use the system. These were Starnet, a San Diego based long distance telephone company purchased by Ford Aerospace & Communications Company in 1984, the Ford Motor Company and the Ford Motor Credit Corporation. The three would have accounted for some 10–15% of Fordstar capacity.

PROJECT OVERVIEW

Introduction: Ford Aerospace Satellite Services corporation had intended to launch a two satellite hybrid C/Ku-band system for US domestic purposes. In December 1986 it was announced that the company was being sold to AT&T for $2·7 million. Whilst the trade press suggested that this indicated a renewed interest in satellite communications by AT&T and therefore a boost for the industry in general, many observers had believed that Ford would have abandoned the Fordstar plan anyway. Construction of the Fordstar satellites had already begun before the sale but FASSC has requested FCC permission to delay launch from 1989 to 1992–93. AT&T says that it intends to put the contract for the space segment out to tender. The purchase price offered for FASSC did not include the initial work done on construction of the satellites for FASSC. AT&T's purchase price covered only intellectual property, a business plan, a detailed technical plan and a statement of work. The five staff of FASSC are not expected to transfer to AT&T. The conclusion is that AT&T purchased FASSC to obtain the two orbital slots allocated by the Federal Communication Commission and the company's licence. AT&T appears to want the slots from the early 1990's for the follow on satellites to its current Telstar 3 system. The satellites in that system were launched in 1983–85. There are suggestions that AT&T is also looking to operate Ku-band capacity in the 1990s. AT&T has said that the Ford based satellites would allow AT&T to continue services to its C-band tenants on currently active Telstar satellites as well as expand its Ku-band Skynet Very Small Aperture terminal (VSAT) services. However, there appears to be no pressing need for the latter requirement as that satellites would not be launched until the early 1990s.

AT&T intends to fold FASSC, along with schemes for the two large satellites planned by the company, into its own satellite organisation in Bedford, New Jersey. AT&T and FASSC have both filed petitions with the FCC to allow the FASSC licence to be transferred. AT&T has refused to reveal when it intends to issue RFP for the satellites which will fill the orbital slots involved in the purchase, but has confirmed that

Ku-BAND (50-STATE) COVERAGE AREAS

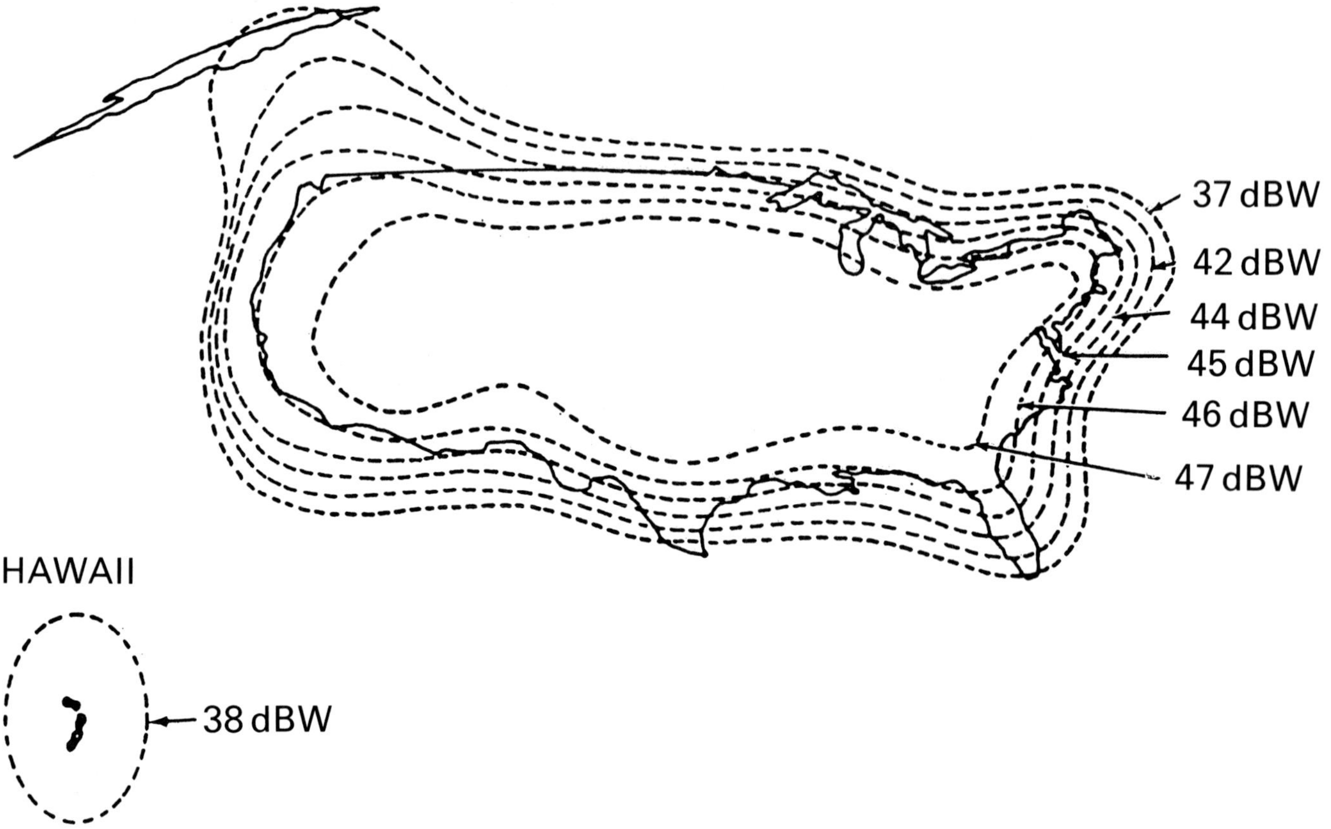

REGIONAL SPOT BEAM COVERAGE AREAS

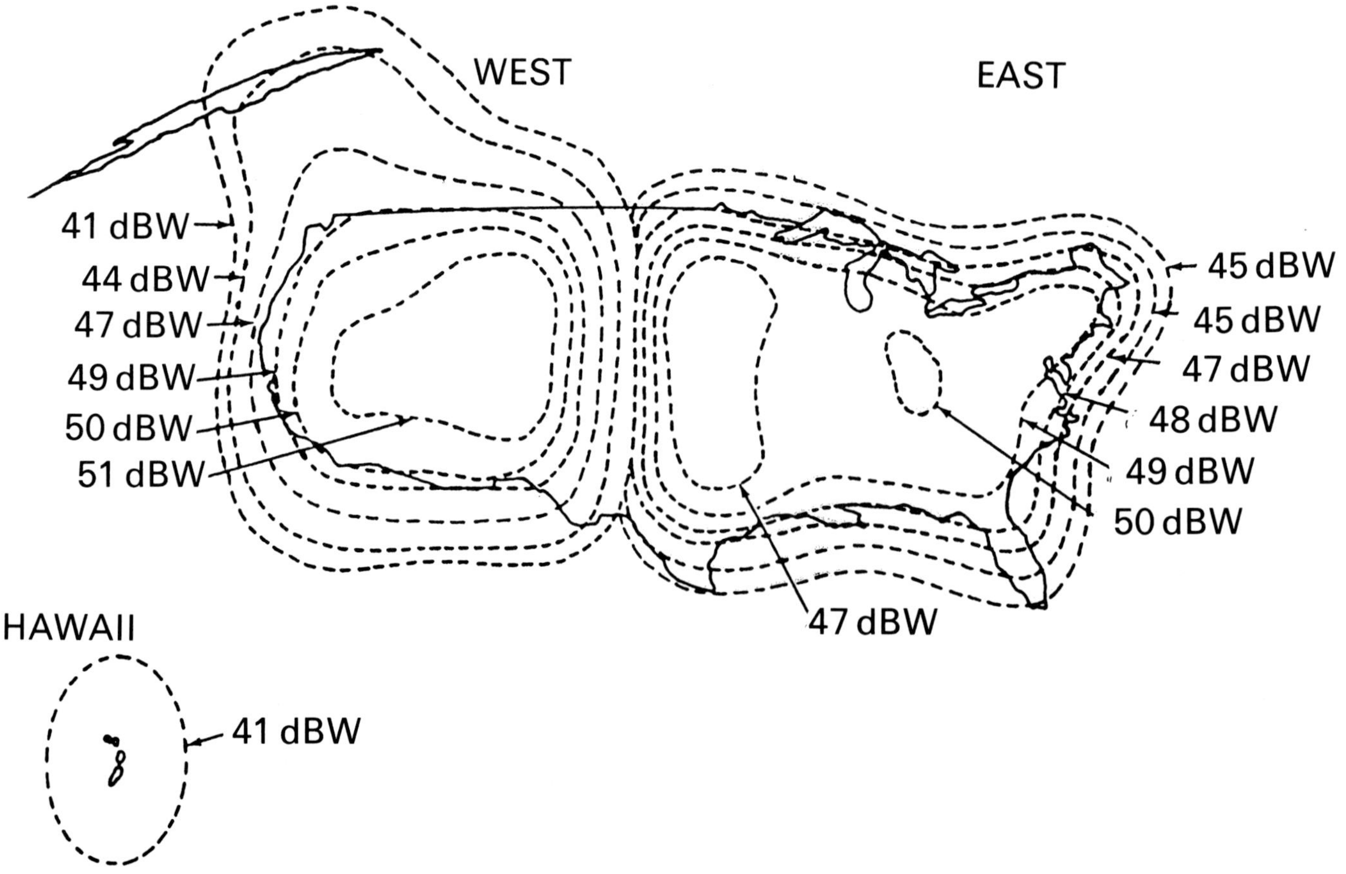

procurement will go out to competitive tender. In commenting in the trade press on the takeover, FASSC Chairman Alan Parker said that the deal was "very, very favourable to the satellite business in this country (the USA)", since AT & T's "bold statement about their plans on satellites ... goes a long way to dispel the notion that satellites are passé, that they've had it, and that fibre is the only way to go." Similarly he commented that satellite operators should rejoice that a company like AT & T, "which carries the Good Housekeeping sign of approval" among communications companies, would endorse the economics of hybrid satellites.

Prior to the takeover, FASSC had arranged for Ford Aerospace & Communication to order long-lead items for the Fordstar satellites. This work "would be available to AT & T" and Ford (FACC) "should be in a very good position (and will) diligently pursue the contract with AT & T." There was some trade talk prior to the take over of FASSC that AT & T was negotiating to take over FASSC's parent company, Ford Aerospace and Communications Corporation. However, AT & T has denied that and confirmed that the contracts for the satellites planned by FASSC will be put out to open bid.

FASSC plans: Although the future of this project is now open to a great degree of uncertainty, it is worthwhile reviewing the FASSC plans not only because they may influence AT & T but also because they reflect many of the problems in the US domestic satellite communications market in the mid-1980s.

Ford Aerospace and Communications was the last of the three US manufacturers of commercial communications satellites to set up its own satellite operating company. It created Ford Aerospace Satellite Services Corporation. in 1983 after both Hughes and RCA had had several years of experience in operating their own systems. The availability of captive customers for their satellites allowed RCA and Hughes to both demonstrate their spacecraft in orbit to potential customers and extend production lines to gain economies of scale in production and R & D. The failure to establish its own operating company is seen by some observers as a prime reason why Ford has been so unsuccessful in selling commercial communications satellites in the 1980s. However, Ford had the contract for the 15 satellite Intelsat V series which, perhaps, meant that it was overspecialised in large spacecraft. The hybrid Fordstar satellites were based on the Intelsat design.

FASSC filed for authorisation with the Federal Communications Commission in 1983 and then expected to launch its satellites in 1987. It received authorisation in August 1985 at which time it planned to launch in 1989, either by Shuttle or Ariane. Following the loss of Shuttle in early 1986, it became apparent from the trade press that the project was not moving rapidly, although the company met its August 1986 deadline to begin construction. It applied to the FCC in the Summer of 1986 to delay launch until 1992–93 on the grounds of uncertainty in the launch vehicle sector, although a number of observers

Artist's impression of the Fordstar satellite which uses the Supersat platform. Design is derived from the Intelsat–V series but power systems, structure and platform are totally new

commented that the company was stalling for time before making a decision on the project's future. The company admitted that the project could have been cancelled because of the combination of lack of advanced orders for its capacity and uncertainties over availability of launch vehicles and high insurance costs ... it was considering a "range of options ... that could result in phasedown or curtailment". By November 1986 it became clear that negotiations between AT & T and Ford were underway and which led to the December 1986 announcement that AT & T was taking over FASSC for its orbital slots.

The Fordstar satellites proposed by FASSC would, had they been launched in 1987, have been the largest and most powerful satellites in operation offering twice the number of communications channels and twice the power of any satellite in operation. RCA now has a three axis stabilised design in the same class, the RCA–500 platform available for launch from 1989. Hughes has the HS–393 platform derived from Intelsat VI. The design envisaged on-board switching between C- and Ku-band transponders using techniques developed by Ford for the Intelsat V series. Each satellite would have had 24 C-band and 24 Ku-band transponders, each of 36 MHz bandwidth. At one time Ford Aerospace and Communications Corp proposed a new upper stage, HPPM (High Performance Propulsion Module) to launch the spacecraft and similar designs sold to other companies. However, the collapse of the Shuttle commercial launch programme from 1986 means that this project is now dead.

C-BAND (50-STATE) COVERAGE AREAS

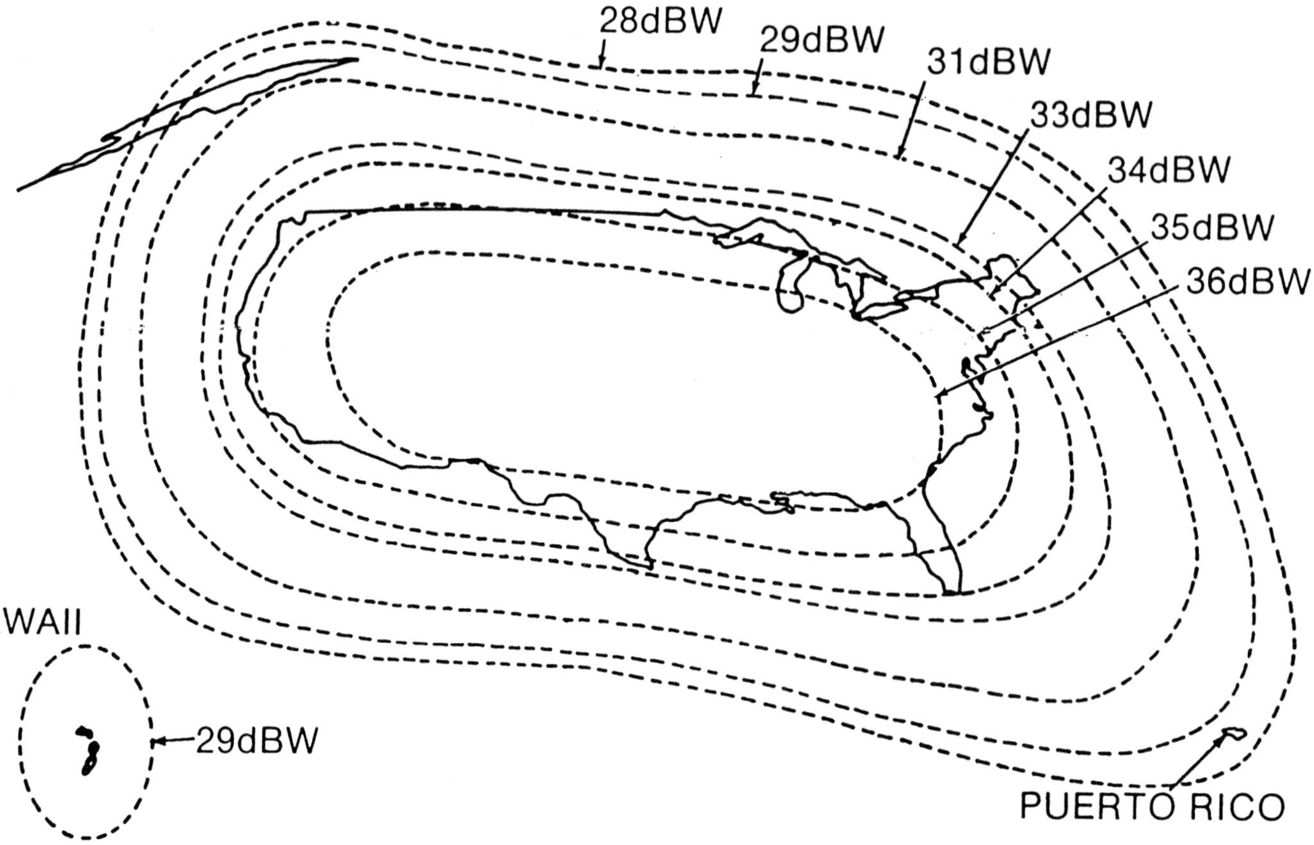

TECHNICAL SUMMARY

PLATFORM	Intelsat–V (see note below)*
PRIME CONTRACTOR	Ford Aerospace Communications
STABILISATION	3-axis body stabilised
LAUNCH MASS	7691 kg
MASS IN ORBIT	1360 kg (BOL)
POWER SUPPLY, END OF LIFE	3875 Watts (4500 Watts BOL)
DESIGN LIFE	10 years
ORBITAL POSITIONS	Fordstar 1, 101° West Fordstar 2, 93° West Fordstar 3, ground spare
STATION KEEPING ACCURACY	+/−0·03° (N/S & E/W)
ECLIPSE PROTECTION	Full
SPACECRAFT DIMENSIONS	2·7 × 2·5 × 1·65 m
SPAN, TIP-TO-TIP	20·4 m
NO. OF ACTIVE TRANSPONDERS	48 (24 at Ku, 24 at C-band)
NO. OF SPARE TRANSPONDERS	12 (Ku) and 4 (C)
TWTA POWER OUTPUT	45 Watts (Ku), 10 Watts (C)**
FREQUENCY RANGE:	
UPLINK	6 GHz and 14 GHz
DOWNLINK	4 GHz and 12 GHz
TRANSPONDER BANDWIDTH	36 MHz (all 48 TWTA's)
EIRP MINIMUM, BEAM EDGE (101° West satellite)	CONUS, C-band 35 dBW CONUS, Ku-band 46 dBW Ku-band East beam 49 dBW Ku-band West beam 49 dBW
SATELLITE G/T (101° West satellite)	−3·0 db/°K, C-band CONUS −1·0 db/°K, Ku-band CONUS
CHANNEL POLARISATION	

*The basic Intelsat V design is being marketed to commercial customers under the name Supersat.

**C-band will use solid state Power Amplifiers at 10 Watts instead of TWTA's

COMMERCIAL DETAILS

OPERATOR	Hughes Communications Galaxy Incorporated
ADDRESS	Worldway Postal Center, PO Box 92424, Los Angeles, California 90009, USA
TELEPHONE NO.	(213) 607-4000
SENIOR STAFF	Robert Waldron, Senior Vice President Jim Ramo
FINANCIAL BACKERS	Hughes Communications is a wholly owned subsidiary of Hughes Aircraft Company which is owned by General Motors
PURPOSE OF PROJECT	Provision of medium power (100 W) DBS capacity for CONUS
PROJECT STATUS	
PROJECT VALUE	$81 m in expenses before system becomes operational. $396 m total for 10 years of operation
LAUNCH DATES	Galaxy DBS–1, late 1988/early 1989, by Shuttle Galaxy DBS–2, early 1989, by STS

SATELLITE USERS

If this satellite becomes operational it will offer 32 television channels (16 on each spacecraft) for sale or lease on a non-common carrier basis directly to programmers. Coverage will either beam quarter CONUS, half CONUS or full CONUS. EIRP at beam edge, depending on beam, ranges from 48 dBW to 54 dBW allowing reception by dishes of under 1 metre diameter.

In addition to DBS services, satellite is intended for use for distributing TV channels and programming for TV networks and to cable head ends and SMA TV markets.

Hughes Communications also envisage a variety of secondary applications of the satellite including videoconference, data transmission and digital telephony.

PROJECT OVERVIEW

Hughes Communications moved into the DBS TV scene as a second round applicant for DBS construction permits (submitting its application on 12th January 1984) and was granted construction, launch and operating authority. It then took the necessary steps to meet the due diligence requirements of the Federal Communications Commission. As a result Hughes was assigned all 32 channels available at the 101° West orbital slot. The 101° West orbital slot is generally regarded as the best slot available to US DBS operators and Hughes' monopoly of the available channels left a lot of potential DBS operators looking elsewhere.

Both satellites in the system are capable of providing full continental US coverage beams with an EIRP at beam edge of 48 dBW. Each can also provide half CONUS coverage (51 dBW at beam edge) and quarter CONUS (EIRP of 54 dBW). East of West half CONUS beams can be used for transmissions on a channel-by-channel basis and each East or West half CONUS beam can, in turn, be split into one of two possible quarter CONUS beams. The East half CONUS beam can be divided into either an East or East central quarter CONUS beam. The coverage area of each quarter CONUS beam is approximately one time zone. The satellites will use shaped antenna patterns to distribute concentrated signals to highly populated areas of the USA. The beam-shaping techniques improve EIRP by 3–4 dBW.

The spacecraft itself, which uses a Hughes HS–394 platform, is novel in that it combines spin stabilisation with three axis stabilisation. This is necessary to provide high electrical power levels needed for the DBS payload. Sun-tracking solar panels have a tip-to-tip span of 115 feet; the spin-stabilised centre body also carries solar cells.

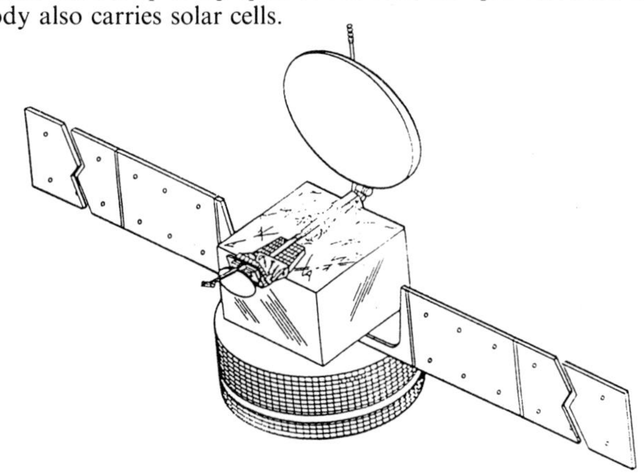

TECHNICAL SUMMARY

PLATFORM	HS–394
PRIME CONTRACTOR	Hughes Aircraft Company
OTHER MAJOR CONTRACTORS	
STABILISATION	Tri-spin
LAUNCH MASS	
MASS IN ORBIT	
POWER SUPPLY, END OF LIFE	4000 Watt
DESIGN LIFE	
ORBITAL POSITIONS	Both have FCC assignments at 101° West
STATION KEEPING ACCURACY	10 years
NO. OF ACTIVE TRANSPONDERS	16
NO. OF SPARE TRANSPONDERS	
TWTA POWER OUTPUT	100 Watt
FREQUENCY RANGE:	
UPLINK	17·300–17·800 GHz
DOWNLINK	12·200–12·700 GHz
TRANSPONDER BANDWIDTH	24 mHz
EIRP MINIMUM, BEAM EDGE	Full CONUS 54 dBW Half CONUS 51 dBW Quarter CONUS 48 dBW

COMMERCIAL DETAILS

OPERATOR	Hughes Communications Galaxy Incorporated
ADDRESS	Worldway Postal Center, PO Box 92424, Los Angeles, California 90009, USA
TELEPHONE NO.	(213) 607-4117
SENIOR STAFF	John Koehler, President & CEO Steven Petrucci, Executive VP Robert Waldron, Senior VP Jack Hayden, Chief Financial Officer Bruce Lederman, General Counsel Cindi S. Whalen, Director of Public Affairs
PURPOSE OF PROJECT	Provision of C-band domestic capacity for voice, video and data traffic in the USA
PROJECT STATUS	System fully operational
PROJECT VALUE	Galaxy 1–3 cost $112 m excluding launch, insurance and TT&C costs. Launch costs for the three satellites was $78 m, including insurance
LAUNCH DATES	Galaxy 1, 26th August 1983, Delta 170 + PAM D Galaxy 2, 22nd September 1983, Delta 173 + PAM D Galaxy 3, 21st September 1984, Delta 176 + PAM D Galaxy 4

SATELLITE USERS

The Galaxy 1 transponders were sold outright to carry cable TV channels. Purchasers were Time Incorporated (6), Group W Broadcasting (4), Times Mirror Satellite Programming (2), Turner Broadcasting (2), Viacom International (2), Spanish International Network (2), C–SPAN (1), Entertainment and Sports Programming Network, Christian Broadcasting Network, Eastern Microwave and The Disney Channel. Most of these channels were sold in 1981–83 before the satellite was launched.

The two biggest owners of capacity on the Galaxy 2 and 3 satellites, which are dedicated to business use, are MCI, with 24 transponders and Equatorial Communications with 12. Both use the capacity for common carrier services.

Other major users of the Galaxy C-band system are IBM, using Galaxy 2 for its Interactive Satellite Education Network, NHK of Japan for a full time television service from New York to California for upward link to Intelsat. NHK also uses Galaxy 2. Hughes Communications also offers video timesharing for non-fulltime users and occasional use needs. Users include CNN, ESPN, Hughes Television Network, News Link, Post News, Turner Broadcasting and TYSC–Group W.

Computer Distribution Network uses Galaxy 2, transponder 5 for an advertising, promotion and special programming channel for the micro-computer industry.

PROJECT OVERVIEW

The Hughes Galaxy System was one of the bigger successes of the first half of the 1980's. Although nominally a multi-purpose C-band system, carrying television, voice, video and data communications traffic, the Galaxy 1–3 system scored a notable success in providing a "cable hot bird" with Galaxy 2. Under the Presidency of Clay Whitehead, Hughes Communications developed the concept of a "shopping centre" cable TV satellite. This analogy compared the sale of individual transponders on the satellite for cable TV channels to a shopping precinct in which retailers had to locate their premises because other retailers had done so and were attracting the shoppers. The Hughes group, although soundly profitable, was, until its takeover by General Motors in 1985, traditionally short of cash to expand into new ventures. Whitehead was able to both reduce the cash demands and the speculative nature of the Galaxy project by pre-selling the capacity on Galaxy 1. The transponders on that satellite were sold outright to the major cable TV channel operators. It also undertook a market exercise in which cable operators were offered a free antenna or retrofit kit to pick up Galaxy 1 transmissions. Two thousand cable operators took this up.

The second Galaxy satellite was located to provide business coverage. At 74° West it could not provide coverage to Alaska and Hawaii. With Galaxy 3 it was aimed at providing voice, video and data communications. Hughes also pulled off a major deal when it sold 24 transponders on the two satellites to long distance telecommunications operator, MCI. The latter now also owns SBS 1–3. In January 1985 Equatorial Communications purchased 11 transponders on Galaxy 3 having already purchased one transponder on it in 1984. The undoubted success of the Galaxy system lead Clay Whitehead to set up his own company, National Exchange, to operate the proposed Spotnet satellite system. Whitehead also came to Europe to set up Luxembourg's GDL satellite system (now called Astra) through a company called Coronet. For this he envisaged pre-sale of transponders to create a "hot cable bird". Whitehead eventually fell out with the Luxembourg government and Coronet was folded. However, the project now continues close to its original concept under Société Européenne des Satellites.

It is not altogether clear as to what Hughes Communications ambitions are in satellite communications. It has plans to extend the Galaxy FSS system into the Ku-band sector (through the purchase of the Galaxy V–VII satellites) and in the 1990's into the Ka-band. Hughes Aircraft Company was, however, closely involved in the financing of the now defunct ABCI company which, until 1985, was looking to provide a Ku-band satellite system for US domestic needs. Hughes Communications Incorporated also has international ambitions. It is a partner in the JCSAT domestic Japanese communications satellite programme and has an 80% shareholding in the start-up company, Atlantic Satellites, in Ireland. The latter is looking at providing FSS capacity across the North Atlantic and in Europe and DBS capacity for the UK. The company also operates the Leasat system for the US Navy and has its own paper DBS project for domestic US requirements. It also offers International Business Services through the Intelsat system as a method of extending its domestic satellite and terrestrial companies. Whether Hughes Communications has a long term objective to become a global communications satellite operator remains unclear; for the present the approach appears to be on a case-by-case appraisal of commercial opportunities.

When General Motors bought Hughes Aircraft Company, it was clear that the former was interested in Hughes Communications as part of a strategy of developing its own corporate communications, both domestically and world wide. The purchase also relieved Hughes of some of its internal financing problems. Whilst for many years Hughes Aircraft Company has been a vocal supporter of Intelsat's world wide monopoly (it has the contract to supply the Intelsat VI series of satellites) it is noticeable since 1984 that that support has not been forthcoming.

As at February 1986, the following TV channels were carried on Galaxy 1

Transponder Number	*Channel*	*Operator*
2	The Nashville Network	Group W
3	WGN-Chicago	United Video
4	Disney Channel-East	Disney Television
5	Showtime-East	Showtime/TMC
6	Spanish International Network	SN
7	Cable News Network	Turner Broadcasting
8	CNN Headline News	Turner Broadcasting
9	ESPN	ESPN
10	The Movie Channel-East	Showtime/TMC
11	CBN Cable Network	CBN
12	Reques TV	Reques TV
13	C-SPAN	Showtime/TMC
14	The Movie Channel-West	Showtime/TMC
15	WOR-New York	Eastern Microwave
16	PTL–The Inspirational Network	PTL Center
17	WTBS-Atlanta	Turner Broadcasting
18	Cinemax-East	HBO/Cinemax
19	GalaVision	SIN/GalaVision
20	USA Network-East	USA
21	Alternative View Network	AVN
22	The Discovery Channel	The Discovery Channel
23	Home Box Office	HBO/Cinemax
24	Disney Channel-West	Disney Television

In addition, transponder 3 carried 13 radio channels, transponder 13 and transponder 15 each carried 3 radio channels and transponder 7 and 18 each carried 3 radio channels

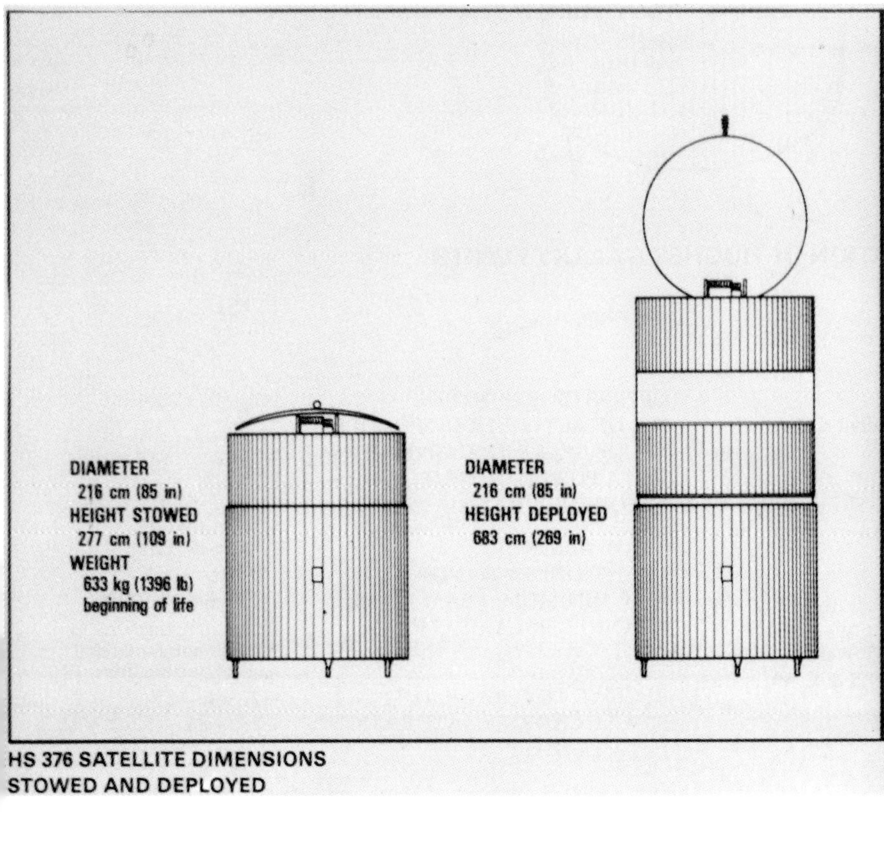

HS 376 SATELLITE DIMENSIONS STOWED AND DEPLOYED

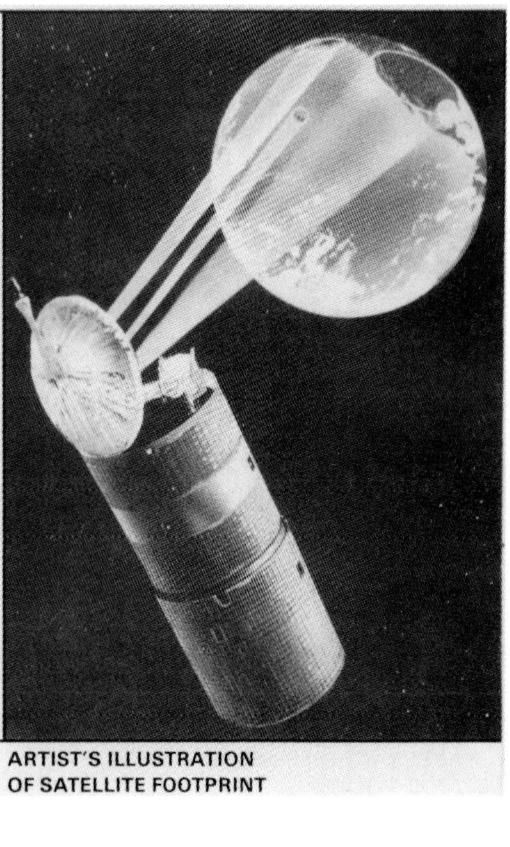

ARTIST'S ILLUSTRATION OF SATELLITE FOOTPRINT

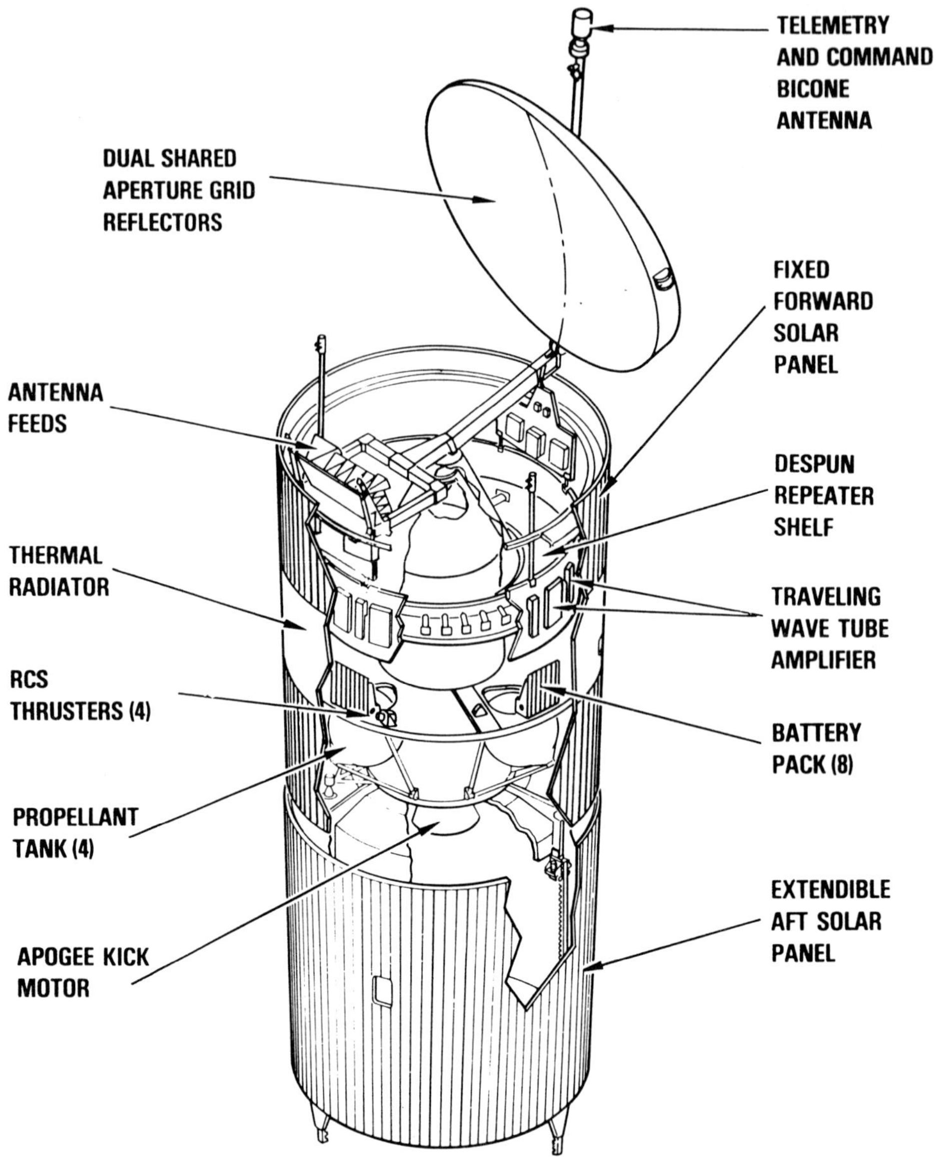

LOCATION OF HUGHES GALAXY I UNITS

TECHNICAL SUMMARY

PLATFORM	HS–376
PRIME CONTRACTOR	Hughes Aircraft Company
OTHER MAJOR CONTRACTORS	
STABILISATION	Spin
LAUNCH MASS	
MASS IN ORBIT	637 kg (BOL)
POWER SUPPLY, END OF LIFE	670 Watts
DESIGN LIFE	10 years
ORBITAL POSITIONS	Galaxy 1, 132° West Galaxy 2, 74° West Galaxy 3, 95° West
STATION KEEPING ACCURACY	+/−0·1° (N/S & E/W)
ECLIPSE PROTECTION	Full
PROPULSION SYSTEM	
SPACECRAFT DIMENSIONS	
SPAN, TIP-TO-TIP	
TELEMETRY & COMMAND CENTRES	Brooklyn, NY and Fillmore, CA

OPERATIONS CONTROL	Fillmore, California
NO. OF ACTIVE TRANSPONDERS	24
NO. OF SPARE TRANSPONDERS	6
TWTA POWER OUTPUT	8·5 Watts
FREQUENCY RANGE:	
UPLINK	5·945–6.405 GHz
DOWNLINK	3·720–4·180 GHz
TRANSPONDER BANDWIDTH	36 MHz
EIRP MINIMUM, BEAM EDGE	34, 30 and 38 dBW*
CHANNEL POLARISATION	Linear, frequency re-use through horizontal and vertical polarisation
RECEIVE COVERAGE (G/T)	−2, −4, −8*
SINGLE CARRIER SATURATION FLUX DENSITY	−88, −86, −82 dBW/sq.m

*For CONUS, Alaska and Hawaii respectively. Note that the Galaxy IV satellite is intended to use solid state power amplifiers rather than TWTA's

COMMERCIAL DETAILS

OPERATOR	Hughes Communications Galaxy Incorporated
ADDRESS	Worldway Postal Center, PO Box 92424, Los Angeles, California 90009, USA
TELEPHONE NO.	(213) 607-4000
SENIOR STAFF	John Koehler, President and CEO Steven Petrucci, Executive VP Robert Waldron, Senior VP Jack Hayden, Chief Financial Officer Bruce Leaderman, General Counsel Cindi S. Whalen, Director, Public Affairs
PURPOSE OF PROJECT	Provision of domestic US Ku-band capacity for data, videoconferencing, private networks and TV traffic
PROJECT STATUS	FCC authorisation given in 1985 but construction yet to start
PROJECT VALUE	Full three satellite system, including launch and insurance, estimated at $297 m
LAUNCH DATES	Galaxy 5, 1989 Galaxy 6, 1989/90 Galaxy 7, ground spare

SATELLITE USERS

Hughes Communications marketing publicity suggests that the company is aiming this system at the business community. A 1985 brochure, for example, states that potential uses of this Ku-band system include the transmission of "corporate voice and data, broadcasting and television signals, financial information services and videoconferencing". The system will allow the company's customers "to use satellite communications technology for a broad variety of purposes directly from their own business locations".

PROJECT OVERVIEW

These three satellites are intended to take Hughes Communications Galaxy Incorporated into the US domestic Ku-band market in the late 1980's. It received FCC construct, launch and operating authority for the two in-orbit satellites from the FCC in 1985, with the intention of launching the satellites between 1988 and 1990. The company appears to be placing strong emphasis on the ability of these satellites to provide links directly to the customer premises using 1–2 metre VSAT terminals. However, the US long distance communications market is, at the time of writing, undergoing a shakeout and restructuring. This has already affected Satellite Business Systems.

Moreover, the crisis in the launch vehicle market which occurred in 1986 after the loss of Challenger and Ariane V18, has set back the launch of satellites by up to two years, which may take launch of the Ku-band Galaxy system into the early 1990's. Unlike the Galaxy 1 system, Hughes Communications does not appear to have a clearly defined a certain market for this Ku-band system.

TECHNICAL SUMMARY

PLATFORM	HS–393
PRIME CONTRACTOR	Hughes Aircraft Company
OTHER MAJOR CONTRACTORS	
STABILISATION	3-axis
LAUNCH MASS	
MASS IN ORBIT	1150 kg (estimate)
POWER SUPPLY, END OF LIFE	2250 Watts (estimate)
DESIGN LIFE	10 years
ORBITAL POSITIONS	Galaxy V, 71° West Galaxy VI, 130° West Galaxy VII, ground spare
STATION KEEPING ACCURACY	
ECLIPSE PROTECTION	
PROPULSION SYSTEM	
SPACECRAFT DIMENSIONS	10 metres high, 3·6 metres diameter (estimate)
SPAN, TIP-TO-TIP	
NO. OF ACTIVE TRANSPONDERS	16
NO. OF SPARE TRANSPONDERS	8 (estimate)
TWTA POWER OUTPUT	50 Watts (estimate)
FREQUENCY RANGE:	
UPLINK	14·0–14·5 GHz
DOWNLINK	11·7–12·2 GHz
TRANSPONDER BANDWIDTH	54 MHz
EIRP MINIMUM, BEAM EDGE	48 dBW
CHANNEL POLARISATION	
RECEIVE G/T	0 dB/K

COMMERCIAL DETAILS

OPERATOR	Hughes Communications Galaxy Incorporated
ADDRESS	Worldway Postal Center, PO Box 92424, Los Angeles, California 90009, USA
TELEPHONE NO.	(213) 607-4000
SENIOR STAFF	John Koehler, President and CEO Steven Petrucci, Executive VP Robert Waldron, Senior VP Jack Hayden, Chief Financial Officer Bruce Leaderman, General Counsel Cindi S. Whalen, Director, Public Affairs
PURPOSE OF PROJECT	Provision of Ku-band capacity for the 1990's
PROJECT STATUS	FCC authorisation given in 1985 but construction yet to start. This is a long term plan for the 1990's
PROJECT VALUE	Not available
LAUNCH DATES	Not available

PROJECT OVERVIEW

This is a Hughes Communications plan for the 1990's in anticipation of the C and Ku-bands being fully committed. The option to use the Ku-band is not without its critics and only time will tell how this Hughes plan develops. It says that its Ku-band satellites will each have approximately six times the capacity of one of its Galaxy C-band spacecraft. The system is envisaged to provide business communications capacity to 16 major metropolitan cities via 16 high power spot beams. The system "will be designed with the possibility of adding inter-satellite communications between Ku- and Ku-band spacecraft for much greater coverage. Principal services provided will be teleconferencing, high speed document transfer, remote printing and other data services.

COMMERCIAL DETAILS

OPERATOR	Geostar Corporation
ADDRESS	101 Carnegie Center, Suite 302, Princeton, New Jersey 08540, USA
TELEPHONE NO.	(609) 452-1130
TELEX NO.	130702 geostar prin
FAX NO.	(609) 452-1171
EUROPEAN ADDRESS	17 rue de la Baume, 75008 Paris, France
EUROPEAN TELEPHONE NO.	(14) 227 78 47
EUROPEAN TELEX NO.	64 21 87 F actbaum
EUROPEAN FAX NO.	(14) 359 77 55
SENIOR STAFF	Dr Gerard O'Neill, President Peter Watkins, Director of Development, Europe
FINANCIAL BACKING	Privately funded venture
PURPOSE OF PROJECT	Provision of Radio Determination Satellite Services and two-way digital message service for the USA (see Overview for further details). Also FGMDSS services
PROJECT STATUS	RDSS service expected to become operational in 1987, but an operational dedicated satellite system is not expected before 1992
PROJECT VALUE	not available
LAUNCH DATES	(fully dedicated satellites expected from 1992)

SATELLITE USERS

Services offered by the Geostar system are as follows:

- navigational positioning
- radiolocation (return of position information to a central site)
- emergency location
- terrain warning to pilots and hazard warning to mariners
- warnings of potential collisions
- approach guidance for aircraft
- two-way digital message service

The system is designed to provide a large number of different services to users in many different transportation modes. These include:

- central control capability for trucking industry
- just-in-Time inventory control coordination for manufacturers, retailers and shippers
- Pig in Transit Location – locating a trailer or container in transit and being carried by one or more modes of transport
- Over The Road Dispatching. Also useful in Europe for TIR lorries
- Railways – for train control purposes amongst other services, rolling stock control, container control, wayside fixed facilities reporting, personnel reporting. Railways are seen as increasingly significant to Geostar
- Air – navigation and distress, air traffic control and approach guidance
- Sea – navigation and distress, particularly for pleasure traffic and small vessels

System allows interconnection to the telephone system for all database services and other services that can be provided through a modem

PROJECT OVERVIEW

The Geostar Corporation's proposals are amongst the most interesting to have come out of the satellite industry for some years. It involves use of a Radio Determination Satellite Service (RDSS) system invented by Gerard O'Neill, a former astronaut who founded the Geostar Corporation in 1982. The Geostar Corporation is an entrepreneurial company currently believed to have around $7 m in capital. Long term plans for the company envisage a domestic US satellite system which provides both RDSS and two-way digital message services. This is envisaged as having the potential of being developed into a global system. However the RDSS system can be implemented without a dedicated satellite system by the addition of low-cost relays to satellites being constructed for conventional FSS systems. The first GEOSTAR relay was carried on the GSTAR–2 satellite launched in 1986. The relay was developed by RCA Astro-Electronics. Geostar has already built a computer control centre at Princeton which calculates positioning and provides message switching.

A second computer centre is planned as back-up and will be located on the Western seaboard. The user accesses the system either by a hand-held or vehicle mounted transceivers and advance orders for over 6000 units had been received prior to the launch of GSTAR–2. Use of GSTAR will become fully operational in mid-1987. The GSTAR–2 payload failed soon after the satellite was launched. Replacement payloads are due for launch on Satcom–F3R and GSTAR–3. A full two way payload is due for launch on GSTAR–4 in 1989. This opportunity places the Geostar Corporation in a stronger position than other entrepreneurial companies proposing to offer new services, whether of an FSS or MSS nature. As and when its own satellite system starts up, it will have already developed a customer base to take up capacity immediately. Just as importantly, the structure for manufacturing transceivers and retailing transceivers and services will be in place.

The economics of the Geostar system are related to those of Mobile Satellite Services. Three of the twelve applications submitted to the Federal Communications Commission in April 1985 for Mobile Satellite Services included plans for Radiodetermination Satellite Services. These were those of the Omninet Corporation, McCaw Space Technologies and MCCA American Satellite Service Corporation. Some in the market believe that Geostar's two-way digital message service provides a more commercially viable method of mobile communications than voice services. Not surprisingly, therefore, Geostar has attracted critics – it is expected that the service charge for use of the Geostar system will be around $20 per month and the cost of

a transceiver, manufactured in volume, would be around $450. Geostar has been criticised for being spectrum wasteful and some competitor companies have also suggested that it is of doubtful economic viability. Moreover, Geostar's early claims for a high level of accuracy for position location (to within a few metres for three-dimensional positioning, about one metre for surface vehicles), has taken a hammering – it is claimed by competitors that the real figures will work out at around 100 metres.
The Federal Communications Commission issued a Report and Order on 13th September 1985 "creating" the new Radio Determination Satellite Service and allocated frequencies in the 1·610–1·6265 GHz and 2·4835 GHz bands, plus feeder links in the 5–7 GHz bands. The company is now initiating the prime system concept in Europe. Comsat General Corporation is under contract to Geostar for satellite design specification and construction supervision. RCA Astro-Electronics, under contract to Geostar, undertook a Preliminary Design Study of the proposed satellite in 1983.

In the first-generation Geostar system (coverage of the continental United States) there are three satellites, all in orbit above the equator at fixed longitudes. Satellite S1 is over the Atlantic, S2 on a longitude corresponding to the central United States, and S3 over the Pacific. The ground station, which can be at any convenient secure location, sends an interrogation radio signal up to one of the satellites many times per second. The satellite relays the signal in a broad beam covering the United States. When the communicator, which can be on a car, truck, boat, aircraft or building or can be hand-carried, receives the signal it responds with a binary sequence ("fingerprint") of pulses which uniquely identify that communicator. Additional digital data or a message may be added to the same signal. The communicator then ignores all further interrogations for a time depending on the class of service selected: a fraction of a second for aircraft in the landing phase of flight; continuously until overridden by the user in the case of message service. That dead-time prevents saturation of the single Geostar channel (nominal 16 MHz bandwidth).

The communicator's pulse code is received at each of the three satellites and relayed by them to the ground station over tight beams, arriving as three identical pulse sequences at three different times. The computer at the ground station identifies the fingerprint, measures the three times, and from that timing information computes the longitude, latitude, altitude and time of response of the communicator. From its memory it retrieves the corresponding data from the last previous response and so calculates the communicator's velocity in all three coordinates (ground track, ground speed and rate of climb). The computer also updates the service charges associated with that communicator for later billing.

The ground station then sends all the navigational information to the user, as a pulse sequence which goes by a tight beam relayed through one of the satellites. The user's receiver identifies the information by its unique binary fingerprint, and presents the data, most simply in alphanumeric form on a calculator-like liquid-crystal display. The delay-time between the user's request and his receipt of the data is approximately 0·6 second. Each Geostar satellite requires an antenna dish of ten metres diameter similar to the antenna of the ATS–6 satellite of 1974. The antenna dish is used to define spot beams both for transmission and reception of signals. The weight and power required for the Geostar satellites is relatively low.

The Cray Computer Corporation ran benchmark tests of prototype Geostar software in the Spring of 1983, and concluded that computer speed and capacity will be more than adequate to handle a large Geostar user population. Each Geostar vehicle communicator uses an output tube of a type manufactured since 1964, or alternatively can be all-solid-state. Pedestrian communicators are solid-state. The 1983 study by RCA Astro-Electronics indicates that the Geostar system can provide in the order of 100 million transactions per hour, divided between position fixes and messages according to customer needs. Messages are in blocks of 32 characters, and any number of blocks can be combined for longer communications.

Global coverage requires a minimum of eight satellites, or ten with dual on-orbit spares. Conventional satellite communications could link several ground stations into one global network. Three Geostar ground stations are sufficient for global coverage. For the temperate zones of the Americas and Europe which have the greatest commercial potential, the precisions in altitude and longitude are good, and latitude is measured slightly less precisely. Close to the equator the precision in latitude deteriorates. Geostar does not cover the extreme polar regions, but the Alaskan North Slope is within its potential coverage area.

The user communicator will be purchasable (from mass-market electronics dealers like Radio Shack or equivalent) in either of two forms. The single-unit "Personal" form will be approximately 1″ × 2″ high × 8″ wide, a little thicker but otherwise similar to a pocket computer. It will weigh a pound or less, and be powered by calculator-type button batteries, which will last about one year. It will have a liquid-crystal display, of 20 to 60 characters, and a keyboard with the 26 letters, the nine numbers, and punctuation marks. In use, it will be set on a desk, a windowsill, the dashboard of a car or boat, or held in the hand. Its antenna will be internal, and will be correctly oriented when the display is right side up. When carried in a vehicle, the personal communicator can be placed at a window (for example on the glare-shield just under the windshield) to avoid signal blockage by the metal roof.

The "vehicle" communicator will be for users who prefer to build the unit into a car, truck, boat or airplane, rather than using the personal version. The vehicle communicator will have a display and keyboard as one unit, connected by a cable (length not critical) to the radio unit. The radio unit (about 1″ × 2″ × 6″) will be located just under the trunk lid or rear deck of a car. Installation will require only the drilling of a small hole ($\frac{1}{4}$″ or less) through which a 2″ long antenna rod will extend upward through a seal from the microwave unit underneath. The advantages of the vehicle communicator will be ease of connection to a vehicle-theft alarm, and its ability to transmit and receive under all conditions, independent of vehicle orientation. ("Personal" communicator users may have to move their units to a window to avoid signal blockage by the metal roof of a vehicle).

For high-security installations, like truck trailers or expensive automobiles with high risk of theft, the microwave unit can be secured against accidental damage or vandalism by location in a one-piece circular disk, flush mounted in the vehicle roof or deck. The disk will be roughly 12″ diameter × 2″ deep, and may be painted.

(See also Locstar for further details.)

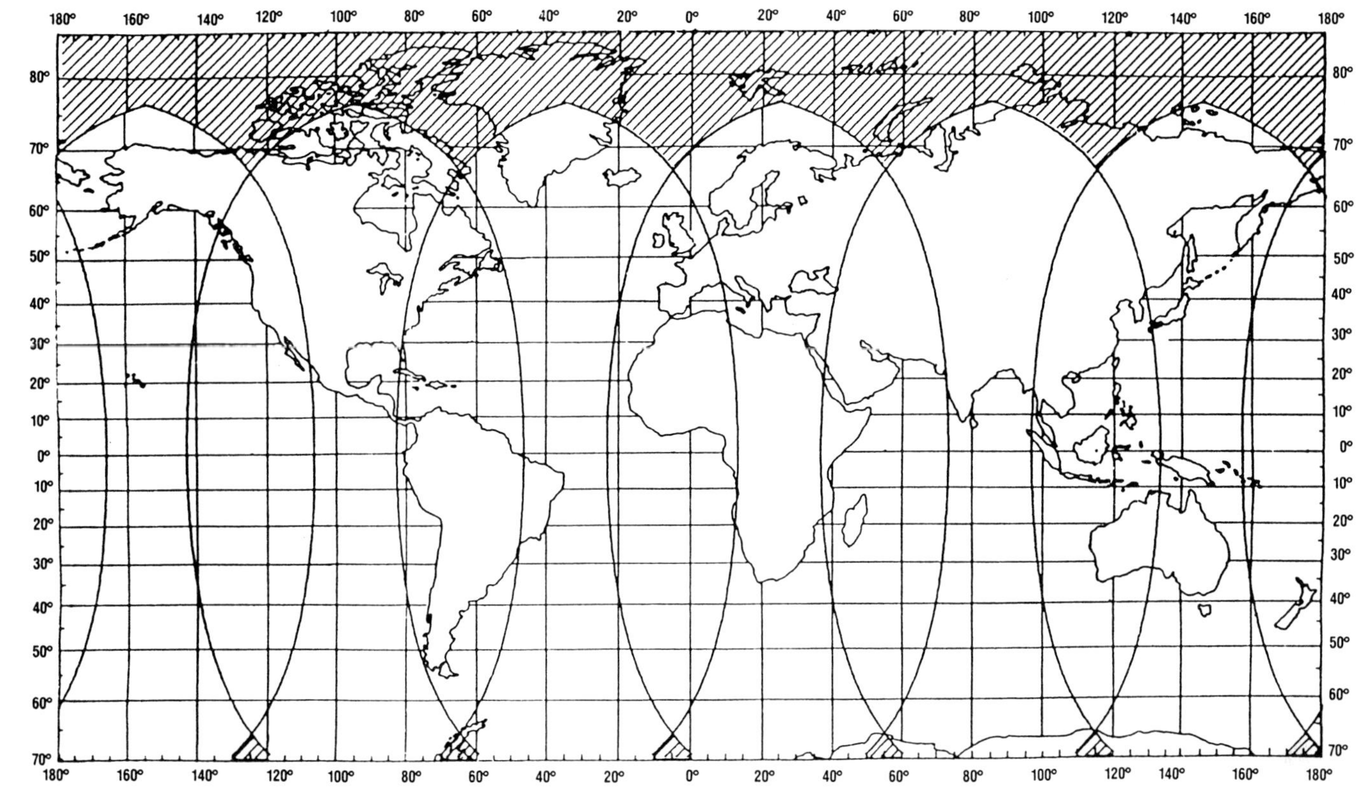

Geostar service coverage for six satellites at 60° orbital spacing. Positioning, radio location, emergency location, two-way communications and data-base services are provided throughout the non-shaded region. The coverage for other than position-related services extends into a portion of the shaded region. Three ground stations are required to support the six-satellite world-wide system. Contour lines indicate coverage limits for individual satellites

TECHNICAL SUMMARY

PLATFORM	PAM DII Class
PRIME CONTRACTOR	Likely to be RCA Astro-Electronics
OTHER MAJOR CONTRACTORS	not available
STABILISATION	3-axis
LAUNCH MASS	2750 lbs
MASS IN ORBIT	1468 lbs (BOL)
POWER SUPPLY, END OF LIFE	2207 Watts
DESIGN LIFE	10 years
ORBITAL POSITIONS	S1, 70° West S2, 100° West S3, 130° West
ECLIPSE PROTECTION	Full
PROPULSION SYSTEM	Delta/PAM, STS/PAM and Ariane III plus another payload

SPACECRAFT DIMENSIONS	not available
SPAN, TIP-TO-TIP	not available
NO. OF ACTIVE TRANSPONDERS	8
NO. OF SPARE TRANSPONDERS	not available
TWTA POWER OUTPUT	not available
FREQUENCY RANGE:	
UPLINK	1·610–1626·5 GHz (LHC)
DOWNLINK	2·4835–2·5 GHz (RHC)
GEOSTAR CENTRAL UPLINK	6·533 GHz
GEOSTAR CENTRAL DOWNLINK	5·117–5·183 GHz
OUTBOUND DATA RATE	32–512 KBPS. Geostar central to users, for beam edge
INBOUND DATA RATE	8–16 KBPS. User to Geostar central data
TRANSPONDER BANDWIDTH	16·5 MHz per channel

COMMERCIAL DETAILS

OPERATOR	Graphic Scanning Corporation
ADDRESS	PO Box 19768, Orlando, Florida 33940
TELEPHONE NO.	(301) 437-7000
SENIOR STAFF	Richard Vega
PURPOSE OF PROJECT	Domestic US DBS TV services
PROJECT STATUS	Remains a paper plan
PROJECT VALUE	not available
LAUNCH DATES	not available

PROJECT OVERVIEW

Graphic Scanning was one of the first round applicants for DBS assignments but lost its Federal Communications Commission authorisation in October 1984 because it failed to complete negotiations with a spacecraft manufacturer (i.e. it hadn't got enough money). The original application with the FCC was a bit odd anyway, proposing two satellites each with only two channels. It also originally proposed to begin covering the West Coast before the East coast. Notwithstanding the set-back at the FCC, it received subsequent conditional approval from the FCC to begin construction of two spacecraft. It had until September 1986 to meet the FCC's due diligence requirements. Like many potential US DBS operators it is considering filing an amendment with the FCC to increase the number of channels on each spacecraft. In this case, the increase would be from two to eight.

TECHNICAL SUMMARY

PLATFORM	Advanced Satcom
PRIME CONTRACTOR	RCA Astro-Electronics
OTHER MAJOR CONTRACTORS	
STABILISATION	3-axis
LAUNCH MASS	not available
MASS IN ORBIT	not available
POWER SUPPLY, END OF LIFE	not available
DESIGN LIFE	not available
ORBITAL POSITIONS	119° West (requested) 148° West (requested)
STATION KEEPING ACCURACY	not available
ECLIPSE PROTECTION	not available
PROPULSION SYSTEM	not available
SPACECRAFT DIMENSIONS	not available
SPAN, TIP-TO-TIP	not available
NO. OF ACTIVE TRANSPONDERS	2*
NO. OF SPARE TRANSPONDERS	not available
TWTA POWER OUTPUT	200 Watts*
EIRP MINIMUM, BEAM EDGE	54 dBW

*under review

COMMERCIAL DETAILS

OPERATOR	GTE Satellite Corporation
ADDRESS	One Stamford Forum, Stamford, Connecticut 06904, USA
TELEPHONE NO.	(203) 965-3464
TELEX NO.	
FAX NO.	
SENIOR STAFF	Dr C. J. (Jerry) Waylan, President Gene Tamplin, Account Manager James McKeeft, VP, Marketing Martin Hyman, Director of Marketing
PURPOSE OF PROJECT	Provision of Ku-band capacity for domestic US communications and replacement of capacity provided by Comsat's Comstar system
PROJECT STATUS	One satellite now operational but project delayed by non-availability of launch vehicles
PROJECT VALUE	The first three GSTAR satellites were constructed under a contract which GTE says was worth \$100 m
LAUNCH DATES	GSTAR 1, 8th May 1985, Ariane 3 GSTAR 2, 28th Mar 1986, Ariane 3, V17 GSTAR 3, 1987, Ariane 3, V22 GSTAR 4, expected mid-1989 by Ariane V37

SATELLITE USERS

Primary use of the GSTAR system is to provide transmission services for individual customers and for groups of customers. These include common carriers, specialised common carriers, resale and value added carriers, companies and government bodies. Transmission services include voice, data and video. GSAT began on private line transmission services for voice and data in 1982, customers using this system migrated to the operational GSTAR system. GTE also leased capacity on the Anik–C2 satellite which was used by USCI for quasi-DBS services. This company went bankrupt but it was intended that the service should migrate to GSTAR.

Principal user of the GSTAR 1 satellite is Hi–Net Communications. Hi–Net uses transponder numbers 19, 20, 23, 24, 31 and 32 to carry ESPN, CNN Headline News, Showtime (East and West), Biznet, WTBS and Satellite Cinema. Also using GSTAR 1 for video are the National Technical University, AMCEE and the Campus Network. The Hi–Net use is for providing the TV channels to Hotels and each channel uses a half-transponder. Hi–Net is a joint venture between Comsat and Holiday Inns.

Federal Express leases one transponder for its Zapmail service and options for a further seven transponders. Other users of GSTAR 1 are Bonneville Telecommunications, ISACOMM, Vitalink and Xerox Computer Services. GSTAR 2 also carries a payload for Geostar but this has failed. GSTAR 3 will also carry a Link 1 payload (or Geostar and GSTAR 4 a two way Geostar Prime payload.

PROJECT OVERVIEW

The GSTAR system is operated by GTE Satellite (GSAT), part of the major US telecommunications operator and equipment manufacturer, GTE Corporation. The latter also owns the GTE Spacenet satellite system (see Spacenet) which was purchased as part of GTE's takeover in 1983 of the Southern Pacific Communications Corporation. The two satellite systems are being combined under the management of GTE Spacenet. GTE Satellite Corporation's involvement in US domestic satellite communications dates back to 1972 when it was formed to operate, in conjunction with AT&T, three of seven earth stations integrated into the US National telecommunications network. From 1976 it used Comsat's Comstar satellite system in conjunction with AT&T. It was granted construct, launch and operate authority in 1980 by the Federal Communications Commission for its own domestic communications satellite system, which, unlike Comstar, operates in the Ku-band. GTE had been authorised in 1973 for a domestic satellite system but opted to sublease capacity on Comstar instead.

The financial prospects for the GSTAR system has been somewhat varied. The original FCC authorisation was for two in-orbit satellites and a ground spare. GTE then requested authorisation to launch and operate the third satellite and authority to construct a fourth, for which long lead items were already ordered. However due to limited demand for the GSAT services, GTE considered selling one of the GSTAR satellites and amongst the potential buyers was Société Européenne des Satellites for its first Astra satellite.

However the GSTAR satellite did not meet SES's specifications. The original GSTAR concept envisaged the first two satellites virtually co-located at 103° West and 105° West; the third satellite has been assigned the 136° West by the FCC. Each of the GSTAR satellites has 16 operational Ku-band 54 MHz transponders. The coverage pattern of the satellites is quite complex. Each satellite can provide coverage of the full 50 States through a CONUS beam and spot beams for Hawaii and Alaska. The antenna coverage patterns can be altered by ground command as are transponder combinations using the spot, regional and CONUS beams. There are two regional beams, East and West. Fourteen transponders can be switched to East regional, West regional or CONUS beam coverage. Two transponders provide 50 State or spot beam (East) coverage. The CONUS and regional beams use 20 Watt TWTA's, while the 50 State transponders utilise 30 Watt transponders. Unlike other US domestic Ku-band satellites, polarisation axes are tilted 26° to reduce rainfall depolarisation in the South East of the USA. GSTAR 1 was also the first Ku-band satellite to use cross

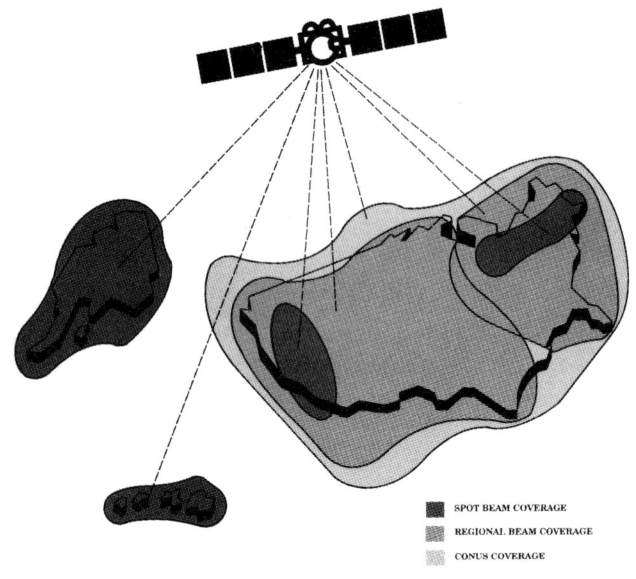

Illustrative view of the GSTAR coverage. GSTAR 1 was the first US domestic satellite to provide full US Ku-band coverage. The ability to use either spot, regional or CONUS coverage is intended to provide high EIRP levels where capacity on the satellite is demanded. Nominal EIRP per transponder is 42 dBW for CONUS, 40 dBW, for Hawaii and Alaska

polarisation. Eight transponders have horizontal linear polarisation, the other eight have vertical polarisation. The eight horizontally polarised channels can be group configurated for East spot beams, the eight vertically polarised channels can similarly be configurated for West spot beams.

Capacity on the GSTAR system is designed for data, video and voice traffic. Customers are able to lease or purchase transponders for full time services. At one time it was expected that United States Communications Incorporated (USCI) would transfer its quasi-DBS service from Anik–C2 to GSTAR 1 but the company folded before it was able to do so. This contract would have been worth $240 m to GSAT. However, the GSTAR system has not proved to be popular with broadcasters and has been "eclipsed" by the RCA Satcom Ku-band system for this service. Principal video user is Hi–Net which provides a package of TV channels for hotels using half transponders for each of 14 TV channels. The GSTAR system became operational after the growth in TV traffic via satellite had levelled out.

The Florida News Network, a consortium of four Florida TV stations (WJXT-TV, Jacksonville, WFTV-Orlando, WTSP-Tampa/St Petersburg and WPLG, Miami) uses the occasional use service on GSTAR for satellite news gathering (SNG). This is under the News Express service offered by GTE. Florida News Network provides a statewide SNG service using truck-mounted portable uplinks with 4·5 metre antenna. System is also used for daily news trading amongst the stations. The 54 MHz transponders are also available for teleconferencing. They are also marketed under GTE's Visual Video Services (VSS) for high-speed data transmissions.

The US telecommunications market is expected to see a growing shortage of Ku-band satellite capacity during the second half of the 1980's as demand increases but capacity is constrained by considerable delays in the launching of satellites and in the Ariane and Shuttle programmes. At the time of writing, commercial forecasts indicate that the increase in demand will come from business use. A full transponder on GSTAR is capable of handling a maximum of either 60 Mb/s or 90 Mb/s. The latter QPSK service uses the spot beam transponder. For voice and data traffic, the GSTAR system uses Time Division Multiple Access and Demand Assigned Multiple Access. A 90 Mb/s transmit/receive earth station requires a 7·7–8·9 m antenna, a 1000 Watt transmitter and a 125–225 K LNA. At 60 Mb/s a 5·5—7·7 m antenna, with a 500 Watt transmitter and a 125 L LNA is needed.

GTE also offers much lower rate channels routed through the GSTAR system according to customer requirements. These are marketed at three bit rates; GDD–1 at 4·8 Kbps, GDD–2 at 9·6 Kbps and GDD–3 at 56 Kbps. These are aimed at remote job entry and realtime computer to computer links, Fax transmissions and Electronic Mail. GTE argues that its digital satellite services offer lower bit-error rates than similar terrestrial services. Largest user of GSTAR for data transmission is likely to be Federal Express, which may take up to eight full-time transponders for Fax transmissions and internal corporate communications. However for the lower rates customers may go through GTE's own earth station system. GTE has a multi-city private channel service capable of transmitting up to 1544 Mbps. This currently operates among city pairs in Chicago, Houston, Los Angeles and New York and customers may use shared network facilities to transmit high and low volumes of information in those cities.

Federal Express has its own dedicated earth stations, either on its customer premises or at its central hub, and is in effect a VSAT system. GTE also offers voice and hybrid voice/data services on GSTAR. Its three Link Line voice services can be combined with its Data Services so that capacity may be used to transmit digital voice and data alternately. For example, the line can transmit voice during the data and batch processing, electronic mail, facsimile, teletype and other data during the night. The GSTAR private line service, which is aimed at these low bit rates for data and voice, is aimed at interstate traffic. Channels are leased to customers by the month and are routed through transponders on GSTAR dedicated to the leased line service. GTE also provides partial transponder leases and occasional use of one or more transponders for periods as short as 15 minutes. These are available for large bit rate requirements as an alternative to the private line channels.

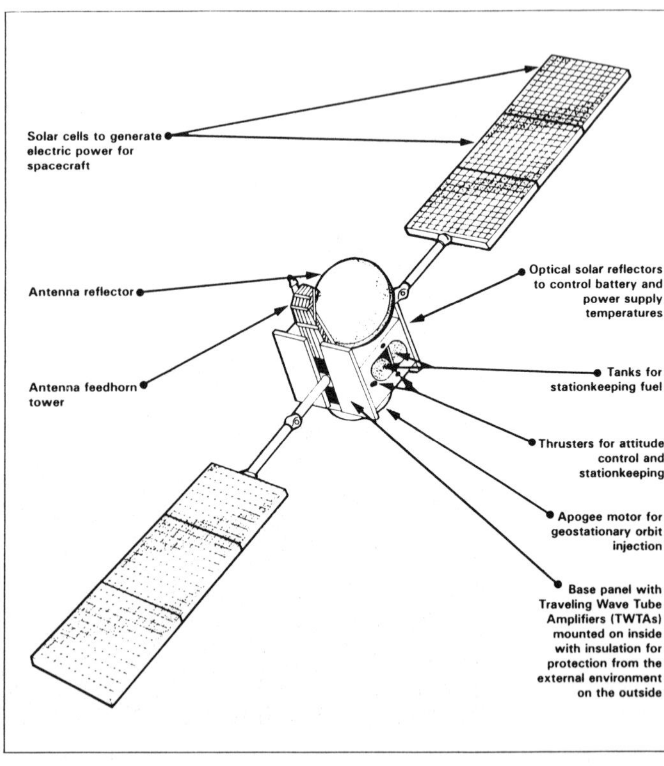

Xerox Computer Services has developed its own GSTAR VSAT service for its customers requiring data transmission services of up to 56 Kb/s. In 1985 it placed an order for 2000 terminals with Telcom General. The service provides a rate of 56 Kb/s outbound from Xerox's Hawthorne Data Centre and 9·6 Kb/s inbound from the VSAT's. The system provides a complete by-pass of the terrestrial telecoms network. By May 1986 20% of the company's customers had been converted to the satellite system. It says that the system provides a 40% saving in telecoms costs over terrestrial lines and a bit error rate improvement from 10^{-4} to 10^{-8}. Xerox is now moving to provide the VSAT service for customers outside its own customerbase, but is not yet extending the service to video or voice traffic. The service is not available for T1 (1·544 Mb/s) data rates which the company sees as a much smaller market in terms of number of users.

The National Technical University provides post-graduate level education to working engineers within companies. The NTU is a consortium of 21 US engineering University departments. NTU has a complete 54 MHz GSTAR transponder which carries two 22 MHz channels. Association for Media-based continuing Education for Engineers (AMCEE) also used the GSTAR system for distribution of short video technical courses.

TECHNICAL SUMMARY

PLATFORM	Advanced Satcom (3000)
PRIME CONTRACTOR	RCA Astro-Electronics
STABILISATION	3-axis
LAUNCH MASS	1250 kg
MASS IN ORBIT	760 kg
POWER SUPPLY, (BOL)	1400 Watts
DESIGN LIFE	10 years
ORBITAL POSITIONS	GSTAR 1, 103° W GSTAR 3, 136° W GSTAR 2, 105° W
STATION KEEPING ACCURACY	+/−0·04° (N/S, E/W)
ECLIPSE PROTECTION	Full
CONTROL CENTRE	McLean, Virginia
TT&C CENTRES	Woodbine, MD, San Ramon, CA

NO. OF ACTIVE TRANSPONDERS	16 (14 at 20 W, 2 at 30 W)
NO. OF SPARE TRANSPONDERS	6 (5 at 20 W, 1 at 30 W)
TWTA POWER OUTPUT	20 Watts (14), 30 Watts (2)
FREQUENCY RANGE:	
UPLINK	14·0–14·5 GHz
DOWNLINK	11·4–12·2 GHz
TRANSPONDER BANDWIDTH	54 MHz
EIRP MINIMUM, BEAM EDGE	40 dBW, min., 47 dBW max.
CHANNEL POLARISATION	Linear (frequency reuse by horizontal and vertical polarisation
RECEIVE COVERAGE, G/T	−1·0 to +2·0; −5·0; −6·0 dB/K
SINGLE CARRIER SATURATION	−92 to −95; −88; −87

Artists impression of the RCA built GSTAR satellite in orbit

COMMERCIAL DETAILS

OPERATOR	Not available. Project under study for the Swiss PTT. Direction Generale de l'Enterprise des Postes, Telephones et Telegraphes
ADDRESS	Viktoriastrasse 21, Berne CH-3030, Switzerland
TELEPHONE NO.	62-11-11
TELEX NO.	32011 ptt ch
SENIOR STAFF	not available
PURPOSE OF PROJECT	Provision of either national or European DBS capacity, possibly in conjunction with Germany and France
PROJECT STATUS	Under study by the Swiss PTT. Broadcasters appear not to be involved at this stage
PROJECT VALUE	not available
LAUNCH DATES	not available

SATELLITE USERS

No information is available at this time on potential users of the Helvesat DBS proposition. It is currently no more than a feasibility study. The Swiss broadcasting company, SSR, is not financially affluent and is unlikely to be able to contribute more than a minimal amount of funding for a DBS TV project.

PROJECT OVERVIEW

Switzerland has been looking at DBS opportunities since the beginning of the 1980's. The current activities revolved around what is essentially a feasibility plan handled by the Swiss PTT, although using outside consultants. It must be considered as an outsider in the attempts to establish European commercial satellite systems.

The project dates back to the allocation of an orbital slot at 19° West at WARC '77, a position shared with France, Germany, Italy and Austria. The Swiss saw this as an opportunity to expand trans-border broadcasting out of the small and linguistically fragmented domestic market for TV. The original proposals came from the Swiss Broadcasting Corporation (SSR) and a now defunct private venture company, Tel–Sat. SSR has, for some years, been under financial pressure and acknowledged that the Swiss market was too small to support a DBS system. It saw the market opportunity with foreign coverage from the satellite but, like the Swiss government, was also concerned that if the country did not get involved in DBS, the country could become swamped by foreign culture coming through French and German DBS satellites. SSR's initial plans for DBS TV included a considerable amount of news and current affairs as well as possible business information. SSR's approach was essentially political and strategic rather than commercial.

Tel–Sat had rival proposals. It was formed in mid-1980 at the instigation of UK entrepreneurs (who also went on to found Sky Channel), Swiss news papers and UK electronics and entertainment company, Thorn-EMI. It originally planned to operate five channels but in February 1981 applied for authority to operate three, one each in French, German and Italian. There was a lot of opposition to the company, particularly because of foreign links. The company struggled on to submit further plans to the government in 1983, but the Swiss government was not particularly supportive of the idea or able to enact legislation to allow the project to proceed.

Eventually, the federal government decided to reject all requests for a concession. The Department of Transportation and Communications then drafted new regulations covering DBS which were subject to voting in a franchise. In the meanwhile, Tel–Sat was dissolved. The new regulations require that DBS should be self-supporting and that any operator be controlled by Swiss interests. In early 1986, the Swiss PTT issued a Request for Proposals to satellite manufacturers in Europe and the USA. This was not part of a serious attempt to develop a DBS system, rather, a part of a feasibility study which included looking at what space hardware was available. The Swiss government had asked the PTT to carry out the study which has remained cloaked in secrecy. Eurosatellite and Deutsche Bundespost (manufacturer and operator of the TV–SAT 1 and 2 satellite systems respectively) were lobbying the PTT hard to purchase a Eurosatellite spacecraft and operate it in conjunction with the French and German DBS spacecraft, sharing in-orbit and ground spare capacity. The German proposals would have helped create a "dominant" orbital slot at 19° West, in competition with the 31° slot allocated to the UK and a potential English language "threat" to European culture.

The Swiss PTT has apparently also looked at Fixed Satellite Service type spacecraft as one option to provide quasi-DBS capacity.

COMMERCIAL DETAILS

OPERATOR	International Maritime Satellite Organisation (Inmarsat)
ADDRESS	40 Melton Street, London NW1 2EQ, UK
TELEPHONE NO.	01-387 9089
TELEX NO.	297201 INMSAT G
FAX NO.	01-387 2115 (G2–3)
SENIOR STAFF	Olof Lundberg, Director General David Wright, Special Assistant to the DG Dr Ahmed Ghias, Technical Director Dr Ramin Khadem, Director of Finance Noel Isotta, Director of Administration Alex da Silva Curiel, Operations Manager Peter Berlin, Space Segment Engineering Manager Capt Jim Fear, Service Development Officer Peter Wood, Aeronautical Project Manager Jim Clarke, Aeronautical Service Development Officer Dr Michael Quigley, Manager Ground Stations Gavin Trevitt, Press & Public Relations Officer Stephanie Collins, Press & Public Relations John Feneley, Inmarsat Council Chairman Geoffrey Hall, Inmarsat Council Vice-Chairman
PURPOSE OF PROJECT	Provision of International civil maritime communications and, from 1987, aeronautical services
PROJECT STATUS	System fully operational on a global basis. Aeronautical services have yet to become operational
PROJECT VALUE	The Marec's satellites are leased from ESA under a contract worth $31 million per annum. The three Marisat spacecraft are currently leased from Comsat under a contract worth $6 million for three years
LAUNCH DATES	Marisat F1, 19th February 1976, Thor Delta Marisat F2I F2, Ariane 4, V37 September 1989 Marisat F3I F3, 14th October 1976, Thor Delta Marecs A, December 1981 Marecs B, 10th September 1982, Ariane V5 Marecs B2, 10th November 1984, Ariane V11 Intelsat MCS A, 28th September 1982, Atlas Centaur Intelsat MCS B, 19th May 1983, Atlas Centaur Intelsat MCS C, 18th October 1983, Atlas Centaur Intelsat MCS D, 4th March 1984, Atlas Centaur

SATELLITE USERS

At the time of writing, some 4700 ships, mostly over 1600 gross tons, are equipped by their owners or operators with Inmarsat standard A terminals for maritime communications purposes. This is Inmarsat's core market and its initial raison d'etre. There are some 25,000 vessels in this class and Inmarsat is expected to increase its market penetration with the standard A terminal and its successors. Ship owners and operators are located worldwide and include those behind the Iron Curtain as well as those in the West. Operators purchase their own ship terminals. Inmarsat is responsible for setting space segment charges but the end user also has to carry PTT charges, both on the mark up on the space segment and for the coast earth terminals and the terrestrial links involved.

Having been successful in its initial mission, Inmarsat is now expanding into new services and market segments through developments of new ground terminals and transmission techniques and obtaining more space segment capacity through the Inmarsat II system.

The Standard A terminal currently costs around $25,000–$30,000 per unit and thus can only be justified for larger vessels. The Standard C terminal has subsequently been developed for less sophisticated requirements of vessels under 1600 tons. In terms of numbers of units, this represents a potentially much larger market. Ex-works price of the Standard C terminal is expected to be around $5,000 per unit.

From 1987, Inmarsat is expected to be offering operational aeronautical communications services. Services will initially use the current Inmarsat space segment and provide public voice and air traffic control services. Primary users will be the major airline operators. In conjunction with these services and new maritime communications, Inmarsat is also aiming to pride positioning and location services although it already can provide a basic service along these lines.

On a case-by-case basis, Inmarsat also offers space segment capacity for terrestrial communications. This is primarily aimed at use in major emergencies where a small transportable (as distinct from mobile) terminal can be erected. Typically, the terminal involved is a stripped-down version of the Standard A ship terminal.

Inmarsat also provides the Secretariat services for the Cospas-Sarsat emergency beacon system. This is primarily an administrative service for the countries participating in the programme, the USA, USSR, Canada and France. The Inmarsat space segment does not carry Cospas-Sarsat payloads although Inmarsat argues that it is complementary with the Cospas-Sarsat service.

Inmarsat has also been looking at the possibility of providing global positioning and location services as an alternative to the nationally owned military US Navstar and Soviet Glonass systems. It has looked at options for polar orbiting satellites in support of this service. At the time of writing it appears that such an Inmarsat service may be many years away.

PROJECT OVERVIEW

Introduction: The International Maritime Satellite Organisation, commonly known as Inmarsat, leases capacity on three satellite systems, Marecs, Marisat and Intelsat V to provide international mobile communications via satellite to ship operators and owners. The organisation is relatively new and has achieved considerable success in its initial mission. It is now in the process of expanding into the provision of aeronautical communications. The organisation is owned by 47 countries and its structure is based on that of Intelsat, although the very different nature of the market it serves means that it operates in a fundamentally different way.

History of Inmarsat: Like Intelsat, Eutelsat, Arabsat and Intersputnik, the origins of Inmarsat date back to the 1960s. In 1966, the Inter-governmental Maritime Consultative Organisation (IMCO), an agency of the United Nations began studying the operational requirements for a maritime satellite system. The 1971 World Administrative Radio Conference allocated a frequency in the L-band part of the spectrum for such services. In April 1975 the IMCO convened the first of three sessions to establish the Inmarsat organisation, the last of which was held in September 1976. Some 26 countries then signed the operating agreement and convention. These came into force on 16th July

INMARSAT MEMBER COUNTRIES, SIGNATORIES AND INVESTMENT SHARES AS AT DECEMBER 1986

COUNTRY		*SIGNATORY*	*INVESTMENT %*
1	United States	Comsat	28·77248
2	United Kingdom	British Telecom PLC	19·18878
3	Norway	Norwegian Telecommunications Administration	14·17552
4	Japan	Kokusai Denshin Denwa (KDD)	7·40414
5	Canada	Teleglobe Canada	4·35796
6	USSR	Morsviazsputnik	3·66942
7	Netherlands	Netherlands PTT Administration	2·54766
8	France	Direction Generale des Telecommunications	2·50230
9	Singapore	Telecommunications Authority of Singapore	2·26260
10	Denmark	Posts and Telegraph Administration	1·65334
11	West Germany	Budesministerium fur das Posts und Fernmeld'n	1·63992
12	Greece	Hellenic Telecommunications Organisation	1·40455
13	Australia	Overseas Telecommunications Commission	1·30604
14	Spain	Compania Telefonica Nacional de Espana	1·30208
15	Italy	Telespazio	1·03226
16	Brazil	Embratel	0·84551
17	Sweden	Swedish Telecommunications Administration	0·64926
18	Kuwait	Ministry of Communications	0·61937
19	Saudi Arabia	Ministry of Posts, Telegraphs & Telephones	0·55397
20	India	Videsh Sanchar Nigam Limited	0·51617
21	Poland	Office of Maritime Economy	0·51617
22	China	Beijing Marine Communications & Navigation Company	0·38870
23	Indonesia	PT Indosat	0·38495
24	South Korea	Korea Telecommunications Authority	0·34892
25	Belgium	Regie des Telegraphes et des Posts	0·33015
26	Argentina	Entel	0·14679
27	Finland	General Directorate of Posts & Telecoms	0·14679
28	New Zealand	Post Office Headquarters	0·11161
29	Bulgaria	Shipping Corporation Voden Transport	0·10834
30	Liberia	Republic of Liberia	0·10621
31	Egypt	Arento	0·09912
32	Philipplines	Philcomsat	0·09912
33	Portugal	Companhia Portuguesa Radio Marconi	0·09572
34	Gabon	Telecommunications Internationales	0·07459
35	United Arab Emirates	Ministry of Communications	0·07424
36	Sri Lanka	Oversea Telecommunications Service	0·06525
37	Iran	Telecommunications Company of Iran	0·05000
38	Algeria	Ministere des Postes et Telecommunications	0·05000
39	Chile	Entel-Chile	0·05000
40	Iraq	Republic of Iraq	0·05000
41	Oman	Sultanate of Oman	0·05000
42	Pakistan	Pakistan Telegraph & Telephone Department	0·05000
43	Tunisia	Republic of Tunisia	0·05000
44	Bahrain	Ministry of Transportation	0·05000
45	East Germany	Ministerium fur Posts und Fernmeldsehen	0·05000
46	Malaysia	Government of Malaysia	0·05000

(Source: *Aeronautical Satellite News*, December 1986)

1979 on which date the Inmarsat organisation officially came into being. It assumed operational responsibility for the Standard A maritime communications system from 1st February 1982. Meanwhile, Comsat had launched and was operating three Marisat spacecraft for the US Navy which were also being used for civil maritime communications. Global services were available on this system from the end of 1979 when some 400 vessels were equipped with terminals for use with the system.

In 1968 the International Civil Aviation Organisation (ICAO) formed a panel on the Application of Space Techniques Relating to Aviation (ASTRA). From this, in 1972, the USA, Canada and ESRO (Later to metamorphise into the European Space Agency) set up a project to establish a dedicated aeronautical communications satellite system called Aerosat. However, it soon became apparent that there was insufficient demand to justify the expensive system envisaged and work on the project was suspended in 1974 and the project finally abandoned in 1978. However, the intergovernmental organisation which led to the formation of Inmarsat also recommended that the latter organisation study the possibility of providing aeronautical communications. In 1981 when the International Civil Aviation Organisation invited satellite operators to consider allocating capacity for such services. As an organisation using an international and global network of satellites, Inmarsat naturally was in a position to respond to the invitation. In October 1985, the Inmarsat Convention was amended to allow the organisation to develop and offer aeronautical services on a commercial basis. This move followed a recommendation to the effect from the Inmarsat Council in 1983. The RFP for the second generation Inmarsat II spacecraft, issued in August 1983, included a requirement to provide an aeronautical communications capability. It is working with the International Civil Aviation

Authority, the International Air Transport Association (IATA), Societe Internationale de Telecommunications (SITA) and R & D agencies, such as the European Space Agency, to develop the new services.

Meanwhile, from an initial base of 400 installed Standard A ship borne analogue terminals, growth in use in the Inmarsat maritime service exceeded expectations by some margin, reaching 4700 units by September 1986, with installations running at about 1000 per year.

At the present time Inmarsat's growth and expansion into new mobile communications services has resulted in some opposition. For example, US commercial interests have argued that they should be able to offer aeronautical services using the Inmarsat space segment. The International Air Transport Association has recommended to its members that all future aeronautical satcoms should be provided by the two airline owned organisations, SITA and Arinc. IATA has claimed that Inmarsat charges for the services will be too high.

Structure of the Inmarsat organisation: The structure and control of the Inmarsat organisation is based on that of Intelsat. It is an international organisation and its staff of around 140 are international civil servants recruited, mostly, from member countries. It is owned by its signatories which represent some 47 countries. The actual signatories are mostly government departments of those countries or PTTs. The US signatory is Comsat which is also the US signatory to Intelsat. Each country's investment share in the organisation is based on their actual use of the Inmarsat system with a proviso that the signatory has a minimum share of 0·05%. Unlike Intelsat, Inmarsat has a large number of communist countries amongst its membership, including the Soviet Union. Membership of Inmarsat is open to all countries and all of them may use the system.

Inmarsat is headed by an assembly of parties which consists of representatives of the member states. It meets once every two years and considers and reviews general policy, long term objectives and activities and purposes. Each country has one vote in the assembly. The assembly expresses its views to the Inmarsat Council which is, in effect, a more powerful body. The Council consists of the representatives of the 18 signatories with the largest investment shares and four signatories representing elected by the Assembly to represent the smaller signatories. The Council meets three times per year and each signatory in the Council has voting rights according to its investment share. The roll of the Council has formally been described as having responsibility "for providing the space segment necessary for carrying out the purposes of Inmarsat in the most economic, effective and efficient manner". The Inmarsat Directorate is the executive arm of the organisation and is responsible for day-to-day running of the organisation.

Like Intelsat, Inmarsat is described as a non-profit making organisation, although this description appears to relate more to accounting practices than economics. Its space segment revenues are expected to cover operating, maintenance and administrative costs and also "repay capital with compensation". In that sense it is profit making but, unlike private commercial companies, is not necessarily profit maximising.

The Inmarsat Space Segment: The Inmarsat system initially relied on the Marisat spacecraft owned by Comsat, which is the US signatory to the organisation. The three satellites in the series were capable of providing near global communications but have limited capacity (approximately 12 voice circuits per satellite). On their own, they were also unable to provide in-orbit spare capacity at the same time as providing near global capacity. Inmarsat has never owned these three. The civil capacity is leased to Inmarsat by Comsat who in turn are also responsible for operating the spacecraft. All three continue to provide in-orbit back-up capacity for Inmarsat having been replaced by Marecs and Intelsat spacecraft for primary use. The leases on the Marisat spacecraft were renewed for three years in the middle of 1986. This was done as an insurance policy against delays in launching of new satellites caused by problems with Shuttle and the Ariane launch vehicle. All three Marisat spacecraft are well beyond the end of the design lives of five years. However, the new lease is for three years with options for extensions.

Inmarsat leases two Marecs satellites from the European Space Agency. These two spacecraft are based on the ECS design developed for Eutelsat but are wholly dedicated to Inmarsat requirements. They are operated by the European Space Agency. However, before they came into operation it was found that the combination of Marecs and Marisat was insufficient to meet Inmarsat operational requirements. Five Intelsat V spacecraft were therefore modified to carry Maritime Communications Services Payloads in addition to their normal payloads for Intelsat services. One of these satellites was lost on launch. Again they are operated by Intelsat, not Inmarsat. All three satellite systems are, however, coordinated through the Control centre at Inmarsat headquarters in London. At the moment, each of the Intelsat V MCS satellites can handle 30 voice channels and the Marecs satellites 40 each. The first Marecs satellite has suffered from a number of technical problems and has been relegated to the status of an in-orbit spare. The second Marecs satellite was lost on launch when the Ariane flight 5 vehicle failed in September 1982. Its replacement, Marecs B2 was not launched until 1984.

The Inmarsat II system, detailed in the next section of this book, will be the first to be both owned and operated by Inmarsat.

CURRENT IMARSAT SPACECRAFT AVAILABILITY

OCEAN REGION	*SPACECRAFT*	*LAUNCH DATES*	*STATUS*
Atlantic Ocean (AOR)	Marecs B2	9th November 1984	Operational
Atlantic Ocean (AOR)	Intelsat V–MCS B	19th May 1983	Spare
Atlantic Ocean (AOR)	Marisat–F1	19th February 1976	Spare
Indian Ocean (IOR)	Intelsat V–MCS A	28th September 1982	Operational
Indian Ocean (IOR)	Intelsat V–MCS C	19th October 1983	Spare
Indian Ocean (IOR)	Marisat–F2	14th October 1976	Spare
Pacific Ocean (POR)	Intelsat V–MCS D	4th March 1984	Operational
Pacific Ocean (POR)	Marecs A	20th December 1981	Spare
Pacific Ocean (POR)	Marisat–F3	9th June 1976	Spare

(Source: *Aeronautical Satellite News*, December 1986)

INMARSAT EXISTING AND PLANNED EARTH STATIONS (CURRENTLY MARITIME ONLY)

	COUNTRY	LOCATION	COVERAGE REGION	OPERATIONAL STATUS
1	Scandinavia	Eik	Indian Ocean	In operation
2	Japan	Ibaraki	Pacific Ocean	In operation
3	Japan	Yamaguchi	Indian Ocean	In operation
4	Singapore	Singapore	Pacific Ocean	In operation
5	UK	Goonhilly	Atlantic Ocean	In operation
6	USA	Santa Paula	Pacific Ocean	In operation
7	USA	Southbury	Atlantic Ocean	In operation
8	Kuwait	Umm-al-Aish	Atlantic Ocean	In operation
9	France	Pleumeur Bodou	Atlantic Ocean	In operation
10	Brazil	Tangua	Atlantic Ocean	In operation
11	USSR	Odessa	Atlantic Ocean	In operation
12	USSR	Odessa	Indian Ocean	In operation
13	Italy	Fucino	Atlantic Ocean	In operation
14	Greece	Thermopylae	Indian Ocean	In operation
15	USSR	Nakhodka	Indian Ocean	In operation
16	USSR	Nakhodka	Pacific Ocean	In operation
17	Saudi Arabia	Jeddah	Indian Ocean	In operation
18	Poland	Psary	Atlantic Ocean	January 1987
19	Poland	Psary	Indian Ocean	January 1987
20	Cuba	N/A	Atlantic Ocean	1987–90
21	India	Belapur	Indian Ocean	1987
22	Iran	Near Tehran	Indian Ocean	1987
23	Egypt	Maadi	Atlantic Ocean	Late 1987
24	Bulgaria	Varna	Atlantic Ocean	After 1987
25	Bulgaria	Varna	Indian Ocean	After 1987
26	Spain	Guadalajara	Indian or Atlantic Ocean	1987–88
27	China	Beijing	Pacific Ocean	1988
28	China	Beijing	Indian Ocean	1988
29	Saudi Arabia	Jeddah	Atlantic Ocean	1988–89
30	Chile	N/A	Atlantic Ocean	1991
31	Iran	Near Tehran	Atlantic Ocean	N/A
32	Singapore	Singapore	Indian Ocean	N/A
33	Argentina	Mar Del Plata	Atlantic Ocean	N/A
34	Hong Kong (UK)	Hong Kong	Pacific Ocean	N/A
35	Pakistan	Karachi	Indian Ocean	N/A

The coast earth station segment of the Inmarsat system: Ownership and control of the ship terminals in the Inmarsat system are in the hands of the end user. However, the system requires the use of land based terminals which uplink messages to the satellites for onward transmission to ships and receive the return messages, via the satellites, which have originated from the ships. The land based terminals, known as Coast Earth Stations (CES). The Coast Earth Stations are owned by signatories rather than Inmarsat. As at the end of 1986 there were 20 such stations for maritime communications use in operation, with a further 16 under construction or planned. As the number of such stations grows, average total charges to ship users will tend to fall (all other factors remaining equal) because terrestrial links will tend to shorten.

Inmarsat also sub-contracts operation of the three Network Coordination Stations for its system. Comsat operates the Atlantic Region Station at Southbury, Connecticut. The two stations for the Indian and Pacific Ocean Regions are both located in Japan (at Yamaguchi and Ibaraki). The Network Coordination Stations regulate the distribution of available telephone channels to ship and Coast Earth Stations. Telex channels are directly assigned to ships by individual Coast Earth Stations.

The existence of 16 Coast Earth Stations provides an element of competition within the Inmarsat system because each is owned and operated by different telecommunications administrations and the Inmarsat user can choose between using stations. Factors in such a choice would include the length of the terrestrial links involved with a certain station and local telecoms administration charges for those links and use of the station.

Inmarsat services, Maritime: The basic service, aimed at large vessels equipped with the Standard A terminal, includes telex, telephone, facsimile medium (2400 bit/s) and high speed (56 Kbit/s) data links, telegram, distress and information and assistance services including weather forecasts. The Standard A terminal has a one voice/one telex circuit capacity. Dish size of the latest generation of such terminals is around 85 cm diameter and cost around \$25,000–\$30,000. This compares with around \$60,000 for a 1·2 metre terminal available in the early 1980s. Weight has fallen from around 270 Kg to 50 Kg in the lightest versions. Inmarsat space segment charges are now \$4·50 per minute for a duplex telephone call.

Some 4700 vessels, including ships and oil rigs were equipped with Standard A terminals at the end of September 1986. New installations appear to be running at about 1000 units per year. Penetration of the target market is about 12%. The Standard A system (which includes all of the ship terminals, the space segment and coast earth stations) is limited in its coverage to latitudes between 75 degrees North and South because the Inmarsat system uses geostationary satellites only. There have also been some prohibitions on use of Inmarsat terminals in some ports and territorial waters. The prohibitions relate to the early days of spark radio transmitters which could interfere with local transmitters on shore or emit sparks hazardous to fuel depots. Inmarsat has worked to have these prohibitions removed for its system.

Under development to serve this and new markets are the Standard B and Standard C systems. Standard B is a digital replacement for Standard A, whereas Standard C is a new type of service using a very small terminal (capable of being covered

Soviet Liquid Petroleum Gas carrier equipped with an Inmarsat Standard A terminal

by a large yogurt carton) for a more limited range of activities, including emergency back up capacity to Standard A and B terminals. A standard C terminal would typically cost around $5,000 and would be attractive to operators of small vessels which would not justify the cost of a Standard A or B system. Similarly, Standard C or its derivatives could also be used for aeronautical services. Standard C maritime terminals could be used with the current Inmarsat space segment and Cost Earth Stations. Standard C is basically a store and forward message transfer system using a low-gain near omni-directional antenna and capable of two-way data transmission at rates up to 600 bit/s. The service will cover telex, teletext and low speed data. Sea trials and initial operations are expected in 1987.

A new service being introduced for maritime users is the Enhanced Group Call system which will probably be better known under two names, "Safetynet" and "Fleetnet". These involve a single message sent to more than one vessel but not to vessels not needing the message. For example, an operator of a fleet of vessels may want to send information to all his ships and would therefore use the "Fleetnet" service. Weather information could be sent to all ships in an area in the "safetynet" service. Trials of the services are expected to begin in the Spring of 1987.

Inmarsat services, aeronautical: These are expected to become operational from 1988, with demonstrations taking place in 1987. The services are aimed at providing both operational requirements for the airline operators and in-cabin passenger services, including voice and text. Operational requirements include air traffic control such as air to ground data links and for transmission of flight data and weather reports.

A number of new aircraft-carried terminals are under development for use with the Inmarsat system. In 1986 Inmarsat issued contracts worth $3·3 million for development of a number of terminal designs. Flight testing of the antennas is due to begin in 1987. These follow on from operational, but non-commercial use, of modified maritime Standard A terminal aboard a Boeing 747 aircraft.

The Contracts are with four manufacturers. Ball Aerospace of the USA is developing a patch antenna which will fit nearly flush with the skin of an aircraft and uses electronically steered phased array antennas. Collins, also of the USA, is developing an avionics unit which may be used with the Ball antenna on a demonstration aboard a US cargo aircraft.

Racal Electronics of the UK (a company with a major strategic involvement in mobile communications) has received a contract for its Avionics division, covering both an avionics system and an antenna (work on the latter has been sub-contracted to the company's antenna division). The antenna is an electronically steered blade containing several small elements and is capable of being mounted atop of any aircraft. The company is participating in a joint programme with British Telecom and British Airways which is expected to lead to an operational system in 1988. E-systems is working on a mechanically steerable terminal. This could fit, for example, within the vertical stabiliser of a Boeing 747 or under a small radome. Potential users of Inmarsat aeronautical services will not be limited in choice to these four supliers or designs. Although all designs will have to meet certain technical standards, not all of which have been finalised. The objective is to give users some degree of choice to interface with their particular avionics equipment, service use and aircraft design. The four contracts require that each company delivers a prototype unit for flight testing in 1987. There is some indication that operational versions may prove cheaper to purchase than HF systems.

Amongst the first airline operators to use the Inmarsat aeronautical services is expected to be Northwest Airlines which has ordered 10 Boeing 747s equipped with terminals and due to be delivered from 1988. The decision to have the equipment installed relates to the fact that the version of aircraft ordered (747–400) dispenses with the need for a flight engineer. The two crew cockpit is seen to require improved communications systems for safety purposes.

User segment charges for the aeronautical services have yet to be finalised by Inmarsat's signatories. A number of them have, however announced plans to either upgrade maritime Coast Earth Stations or construct new facilities for the aeronautical services. These include the United Kingdom, Argentina, Japan, Kuwait, Singapore, the USA and the USSR.

The Inmarsat II spacecraft have been modified to provide an extra 2 MHz of bandwidth in the L-band for aeronautical services, in addition to the 1 MHz initially incorporated in the design.

Arinc, which is owned by airlines and manages airline communications in North America and the Pacific, has criticised the Inmarsat plans for aeronautical communications and is proposing to operate its own system using payloads carried piggyback on other satellite operators' spacecraft. Arinc's proposals involve the use of TDMA techniques in contrast to Inmarsat which intends Single Carrier per Channel (SCPC) techniques. The Arinc approach is to aim for a fully integrated aeronautical communications system covering both airline operation and passenger needs. Inmarsat's approach tends to be much more evolutionary, leaving the airlines with several service level options.

Inmarsat services, navigation: Inmarsat is currently looking hard at providing satellite navigation services and one reason why it opted to renew leases on its Marisat spacecraft in 1986 was to provide the capacity to do so. Navigation services will be aimed at both aeronautical and maritime users and demonstrations are expected to begin 1987. Inmarsat's interest stems from work done by the International Civil Aviation authority's (ICAO) FANS (Future Air Navigation Systems) study committee. It has been looking at the future of communications, navigation and surveillance for air traffic services. Inmarsat is also looking at providing an "integrity" service which will enable users to verify automatically the availability and correct operation of other navigation systems such as Navstar and Glonass. It is studying the use of Standard C terminals for the radiodetermination activities.

As a minor point, the existing Inmarsat Maritime service has a limited radiodetermination service. Automatic positioning reports on vessels can be received by ship owners whose Inmarsat terminals are linked to their navigation equipment.

TECHNICAL SUMMARY

	MARISAT	*INTELSAT MCS*	*MARECS*	*INMARSAT II*
PLATFORM				
PLATFORM TYPE		Intelsat V	ECS	Eurostar
PRIME CONTRACTOR	Hughes Aircraft Co.	Ford Aerospace	British Aerospace	British Aerospace with
OTHER MAJOR CONTRACTORS			Satcom/Mesh consortium	Hughes responsible for payload
STABILISATION	Spin	3-axis	3-axis	3-axis
LAUNCH MASS		1870 kg	1006 kg	1160 kg (Ariane)
MASS IN ORBIT		1072 kg	563 kg	690 kg
POWER SUPPLY, END OF LIFE	300 Watts	297 Watts (MCS only)	723 Watts (solstice)	1130 Watts
DESIGN LIFE	5 years	10 years	7 years	10 years
ECLIPSE PROTECTION	Full	Full	Full	
C- to L-REPEATER				
RECEIVE BAND	6·42–6·434 GHz	6·4175–6·425 GHz	6·420–6·435 GHz	6·435–6·441 GHz
TRANSMIT BAND	1·537–1·541 GHz	1·535–1·5425 GHz	1·5375–1·5425 GHz	1·53–1·546 GHz
RECEIVE G/T	–19·6 dBK	–18·6 dBK	–17·0 dBK	
L-BAND EIRP:				
LOWER POWER MODE	–	27·6 dBW	–	–
MEDIUM POWER MODE	27·0 dBW	–	–	–
HIGH POWER MODE	–	32·6 dBW	34·2 dBW	39·0 dBW
C-BAND ANTENNA:				
RECEIVE GAIN	16·0 dBi	16·1 dBi	16·0 dBi	–
POLARISATION	RHC	RHC	RHC	RHC
L-BAND ANTENNA:				
TRANSMIT GAIN	14·4 dBi	14·7 dBi	18·0 dBi	–
POLARISATION	RHC	RHC	RHC	RHC
CAPACITY	12 channels	30 channels	40 channels	125 channels
L- to C-REPEATER				
RECEIVE BAND	1·6385–1·6425 GHz	1·6365–1·644 GHz	1·6385–1·644 GHz	1·6265–1·6475 GHz
TRANSMIT BAND	4·195–4·199 GHz	4·1925–4·200 GHz	4·1945–4·200 GHz	3·600–3·621 GHz
RECEIVE G/T	–17·0 dBK	–14·9 dBK	–13·2 dBK	–
C-BAND EIRP	18·8 dBW	20·0 dBW	15·4 dBW	24 dBW
L-BAND ANTENNA:				
RECEIVE GAIN	14·4 dBi	14·7 dBi	17·5 dBi	–
POLARISATION	RHC	RHC	RHC	RHC
C-BAND ANTENNA:				
TRANSMIT GAIN	16·0 dBi	16·1 dBi	16·0 dBi	–
POLARISATION	LHC	LHC	LHC	LHC
CAPACITY	20 channels	> 50 channels	> 50 channels	250 channels

COMMERCIAL DETAILS

OPERATOR	International Maritime Satellite Organisation (Inmarsat)
ADDRESS	40 Melton Street, London NW1 2EQ, UK
TELEPHONE NO.	01-387 9089
TELEX NO.	297201 INMSAT G
FAX NO.	01-387 2115 (G2–3)
SENIOR STAFF	Olof Lundberg, Director General David Wright, Special Assistant to the DG Dr Ahmed Ghias, Technical Director Dr Ramin, Khadem, Director of Finance Alex da Silva Curiel, Operations Manager Peter Berlin, Space Segment Engineering Manager Capt James Fear, Service Development Officer Dr Michael Quigley, Manager Ground Stations Gavin Trevitt, Press & Public Relations Officer Stephanie Collins, Press & Public Relations John Feneley, Inmarsat Council Chairman Geoffrey Hall, Inmarsat Council Vice-Chairman
PURPOSE OF PROJECT	Provision of replacement and extra capacity for current and new capacity for new Inmarsat mobile services, notably aeronautical and distress activities
PROJECT STATUS	Three satellites ordered in April 1985 for launch from 1988. Inmarsat has options for six more spacecraft. Some of the options are likely to be taken up
PROJECT VALUE	The three satellites and options were under a contract worth $153 million excluding launch, insurance and TT & C facilities. Options for six further spacecraft could be worth $250 million. Contract for the TT & C facilities is worth $24 million
LAUNCH DATES	INMARSAT–II F1, Ariane 4, V33 November 1988 or V35 March 1989 INMARSAT–II F2, Ariane 4, V37 May 1989 or V40 September 1989 INMARSAT–II F3, Ariane 4, 1990–91

SATELLITE USERS

The Inmarsat II system will be used by both current Inmarsat services and soon to be introduced aeronautical services. The system has been procured to supply both replacement and extra capacity for existing and new services.

At the end of 1986 some 4800 vessels were equipped with Inmarsat Standard A terminals. This figure had not originally been expected until the end of the 1980s. Inmarsat has been developing Standard B and Standard C ship terminals which are expected to further increase use of the system.

The Inmarsat II system is also needed to provide capacity for aeronautical services which will become operational from 1987. The extent of demand for these services is still subject to a degree of uncertainty. The primary market is seen as public correspondence and airline operations communications with air traffic control bundled together under secondary markets. Inmarsat believes that voice traffic will predominate in the two market sectors, although there will be a trend towards data communications which are both more cost effective for the user and more spectrum efficient for the supplier.

The Inmarsat II satellites also incorporate L-band capacity for Emergency Position Indicating Radio Beacons (EPIRBs) for the Future Global Maritime Distress and Safety System (FGMDSS) which is under development for the International Maritime Organisation and which is expected to become operational in 1990. Users of EPIRBs are expected to include maritime, terrestrial and aeronautical operators.

Although the Inmarsat organisation is modelled on Intelsat, by nature of the services it offers, it is closer to the end user and consequently is perhaps closer attuned to market requirements and trends insofar as it is not really a carriers carrier. The end users (ship owners and airliner operators are responsible for purchase of their own mobile terminals with the PTTs responsible for providing land based terminals and terrestrial links and setting end user rates.

Perhaps consequentially, Inmarsat has been keen to develop new services, having a substantial R&D programme covering new earth terminals and transmission techniques. Unlike Intelsat, it has not concentrated heavily on R&D in the space segment. This reflects in the relatively conservative design of the Inmarsat II system.

Moreover, the nature of mobile communications means that the market involved is much more fragmented and uncertain than that faced by Intelsat, thus requiring more attention to end user needs.

So far, there has been little attention paid by Inmarsat to terrestrial mobile communications markets. This has been left to other organisations, although there may be an overlap with future Inmarsat navigation and positioning capabilities.

PROJECT OVERVIEW

Introduction: The Inmarsat II system, currently under construction and due for launch from 1988, is Inmarsat's so called second generation satellite system, although the organisation currently uses three separate satellite systems. Inmarsat II represents a significant increase in Inmarsat capacity as well as allowing the organisation to have its own satellite system under single control operated by itself. The satellites are being delivered at a time when the Inmarsat organisation will have exceeded its initial expectations of demand for maritime communications as well as developing new services for the aeronautical industry.

Bidders for the Inmarsat contract: Inmarsat issued the RFP for the Inmarsat II system in August 1983. Although nominally procurement was open to international tender, the politics of such International organisations as Inmarsat meant that any bids would have to reflect national participation in the Inmarsat organisation. Therefore, it is not surprising that the only two bids received were from consortia led by European companies. These bids came from British Aerospace, with Hughes and Matra as principle co-contractors and Marconi Space Systems with Aerospatiale and Ford as principal co-contractors. Nevertheless, Inmarsat argued that the Inmarsat II contract was the first to be

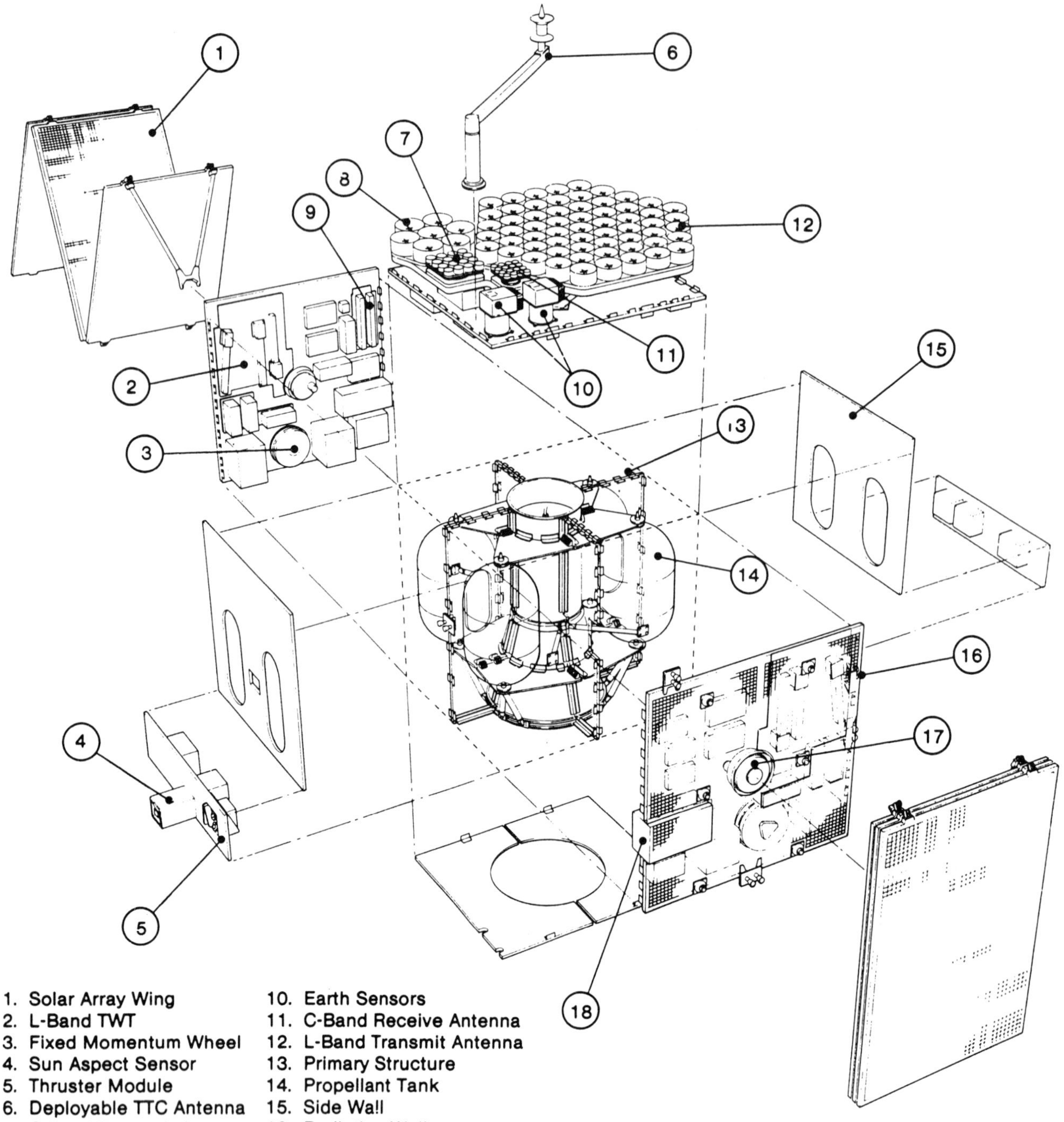

Inmarsat II spacecraft geometry

placed by a European based satellite operator which involved genuine competition in the European satellite construction business.

The British Aerospace team were eventually awarded the contract in April 1985. It appears that their bid was preferred because it involved a more conservative design. Both bidding designs were generally similar, involving medium sized three axis stabilised platforms with bi-propellant propulsion and phased array antennas. However, the Marconi design involved a single high power TWTA for the satellite to ship link against use of four modified Marisat TWTAs combined in parallel on the BAe design. Marconi also opted for solid state power amplifiers for satellite to shore links. The BAe design involves TWTAs for this sector. The BAe design uses conventional paints, blankets, optical reflectors and electric heaters to achieve thermal control whereas Marconi opted to add a network of heatpipes. Similarly, BAe opted for proven nickel cadmium batteries rather that nickel hydrogen envisaged in the Marconi design. In opting for the BAe design, Inmarsat employed a considerable amount of outside consultancy expertise, both from Comsat and the European Space Agency.

The Inmarsat II contract: The Inmarsat II contract

Artist's impression of the Inmarsat II spacecraft which uses the Sakom International Eurostar platform (courtesy BAC Space & Communications Division)

involves the Inmarsat organisation in the purchase of the satellites, although leasing arrangements had been considered. The satellites will be the first satellites owned outright by the organisation. The Marecs series is leased from the European Space Agency as are the Marisat capacity from Comsat and the Intelsat MCS payloads. The initial contract for the Inmarsat II system consisted of three spacecraft with options for six more. Inmarsat argued that the reason for not ordering more than an initial three was that it needed more time to study the growth of its existing maritime services and plan for the development of aeronautical services.

Launch of the three satellites was originally booked for Shuttle (two satellites) and Ariane. The Soviets lobbied hard right through until 1986 to have launches booked on Proton, arguing that the organisation was obliged to use the vehicle because it was the cheapest available and reliable to boot. Nevertheless, CoCom restrictions on transfer of Western technology meant that the Soviet offer was unacceptable. Problems with the Shuttle means that the first two Inmarsat II satellites will be launched on Ariane. The organisation has been looking at other launch options including the Chinese Long March 3 and US expendables such as Titan III for launch of the third and subsequent satellites in the series.

Inmarst third generation plans: The Inmarsat II system is intended to provide capacity for the 1990s. However, the Inmarsat organisation has been looking at options for a third generation system for several years and it is possible that such a system could be awarded before 1990. This could involve the use of on-board message switching and frequencies in the Ka-band (20–30 GHz) as well as the current L-band. Multiple spot beams are also envisaged. Inmarsat is working with the European Space Agency on the Aramis project which may pave the way for the final configuration of the third generation Inmarsat system.

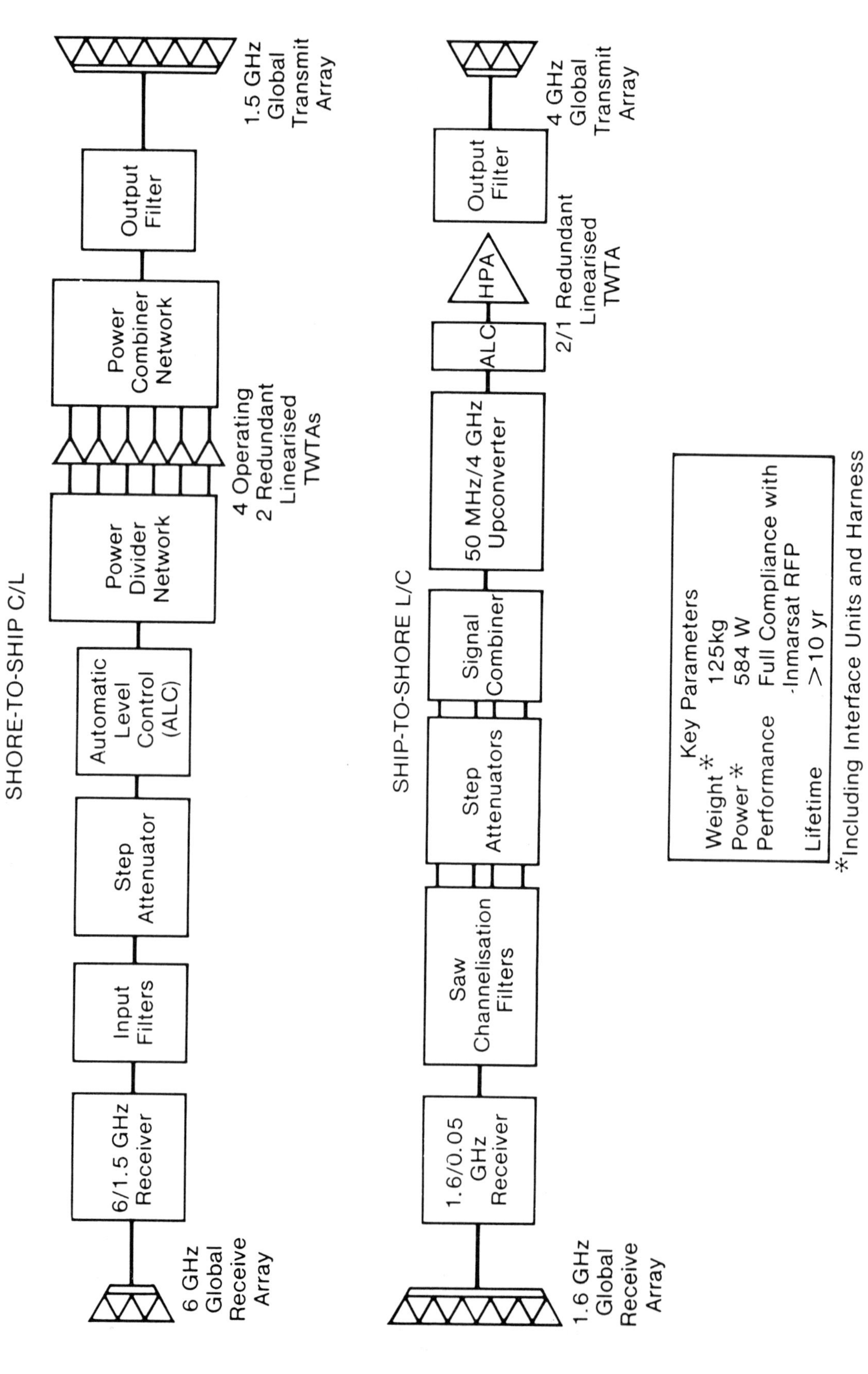

COMMUNICATIONS SUB-SYSTEM BLOCK DIAGRAM

FREQUENCY PLAN

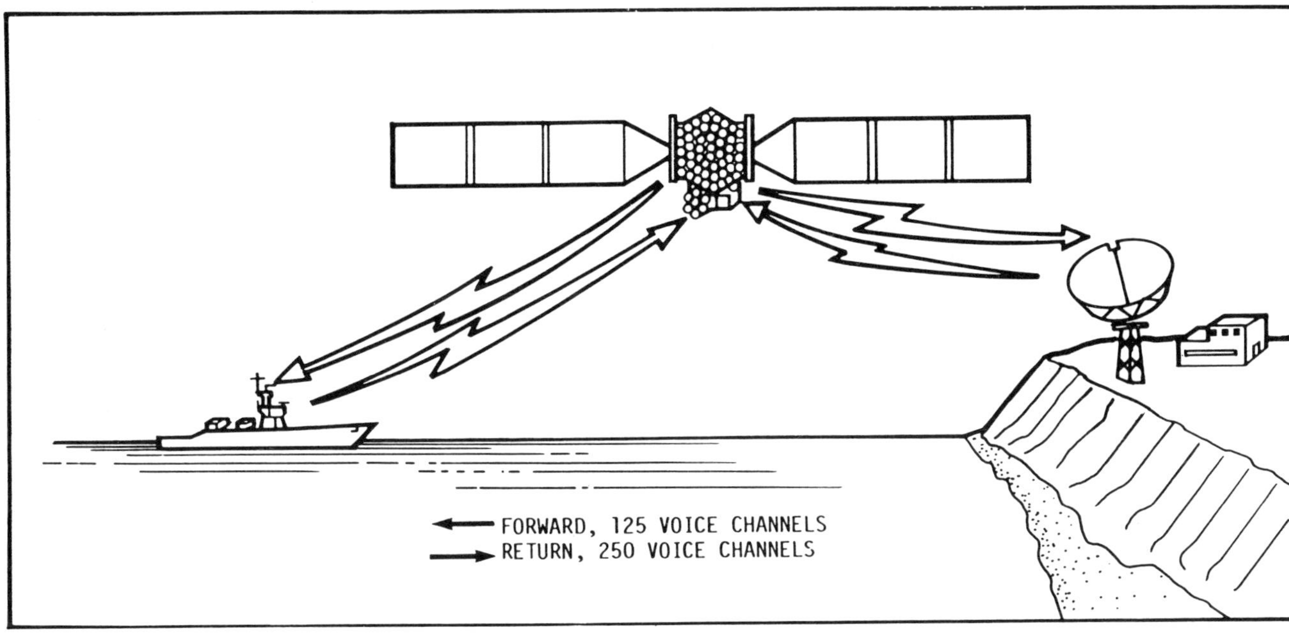

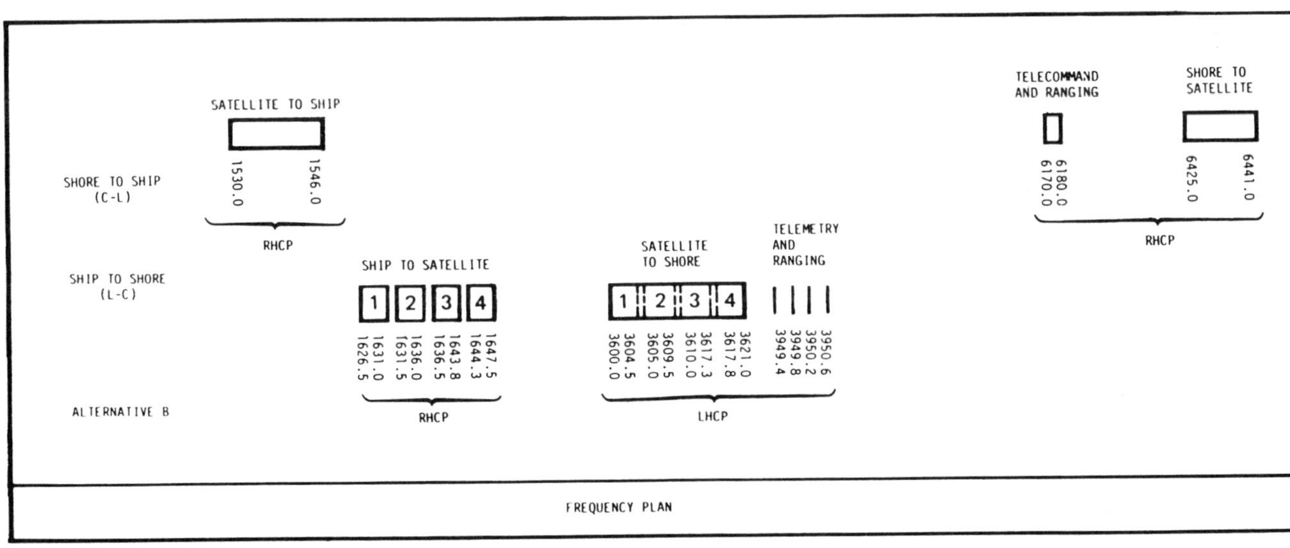

TECHNICAL SUMMARY

PLATFORM	Eurostar
PRIME CONTRACTOR	British Aerospace
OTHER MAJOR CONTRACTORS	Hughes Aircraft Company, Matra, Fokker, Spar NEC, MBB, Teldix, Galileo and McDonnell Douglas
STABILISATION	3-axis
LAUNCH MASS	1162 Kgs (Ariane launch)
MASS IN ORBIT	690 Kg
POWER SUPPLY, END OF LIFE	1130 Watts
DESIGN LIFE	10 years
ORBITAL POSITIONS	Not available at time of writing
STATION KEEPING ACCURACY	
ECLIPSE PROTECTION	Full

SPACECRAFT DIMENSIONS	2·557 × 1·586 × 1·48 m
SPAN, TIP-TO-TIP	15·23 metres
NO. OF ACTIVE TRANSPONDERS	6 (4 for shore to ship, 2 for ship to shore)
NO. OF SPARE TRANSPONDERS	3 (2 for shore to ship, 1 for ship to shore)
FREQUENCIES	
SATELLITE TO SHIP	1·530–1·546 GHz
SHIP TO SATELLITE	1·6265–1·6475 GHz
SATELLITE TO SHORE	3·6000–3·621 GHz
SHORE TO SATELLITE	6·425–6·441 GHz
POLARISATION	
SATELLITE TO SHIP	RHCP
SHIP TO SATELLITE	RHCP
SATELLITE TO SHORE	LHCP
SHORE TO SATELLITE	RHCP

COMMERCIAL DETAILS

OPERATOR	Department of Space, Insat Programme Office
ADDRESS	F-Block, Cauvery Bhavan, Kempegowda Road, Bangalore, Karnataka, 560 009 India
TELEPHONE NO.	27371
TELEX NO.	0845-499
SENIOR STAFF	Shri P. P. Kale, Project Director, Insat I Programme Shri P. Ramachandran, Project Director, Insat II Programme Professor Jai P. Singh, Programme Director, Insat Shri N. Pant, Chairman, Insat Space Segment Board
PURPOSE OF PROJECT	Provision of domestic satellite capacity for quasi-DBS, telephony and earth observation/meteorology
PROJECT STATUS	Fully operational but requires launch of second satellite. Second generation Insat II system now under preparation
PROJECT VALUE	The Insat IA and IB satellites cost $130 m including launch and TT&C facilities. Insat IC cost $58 m excluding launch
LAUNCH DATES	Insat IA, 10th April 1982, Delta F161 + PAM–D Insat IB, 31st August 1983, STS 8 + PAM–D Insat IC, January 1988, Ariane 3 V26 Insat ID, to be decided

SATELLITE USERS

The Insat system is a multipurpose communications and meteorology satellite system. Principal users are as follows:

1) *TV Traffic*: At the time of writing, three transponders are used for this. Two are the S-band quasi-DBS transponders used by TV India in conjunction with small TVRO's in rural and remote locations. TV India also uses a C-band FSS transponder for distribution of its programming to regional terrestrial transmitters. The same transponder is also used for occasional video traffic.

2) *Telephony*: System is used for trunk telephony in the country by the Director General of Posts and Telecommunications, using the C-band low power transponders.

3) *Meteorology*: The Indian Office of Meteorology uses the observation and data collection payloads on the satellite in support of its weather activities.

4) *Radio*: The system is used by All India Radio for networking of five channels as well as distribution through the quasi-DBS capacity.

5) *Private Satellite Networks*: A number of public sector organisations use the system for internal transmission. These include the Oil Natural Gas Commission, the National Thermal Power Corporation, the Central Water Commission and the Press Trust of India. (See project overview for further details)

PROJECT OVERVIEW

Introduction: The Insat I programme is one of the most ambitious satellite programmes in the developing world. Unlike other such projects, the Indians have not opted for off-the-shelf hardware but have, instead, had developed a satellite design tailored to its perceived requirements. This proved to have created problems. The satellite design is novel in that it uses a specialised platform and three distinct payloads. One is for DBS TV to serve rural communities, most of which are outside the range of terrestrial TV transmitters. The second is a convential C-band system used for internal trunk telephony and data services and replacing capacity leased on the Intelsat system. The third is a meteorological observation payload used in conjunction with data collection facilities which collect meterorological data from remote and unattended platforms.

The consequential complex design of the satellite is said by some engineers to be the reason why the first satellite in the series suffered from a total in-orbit failure. Related to this are criticisms that the Indian Department of Space provided project management that was insufficient for such a project. However, the subsequent Insat IB satellite has performed flawlessly although it included a number of modifications over the Insat IA satellite.

Background to the Project: The Insat I system is a joint venture of the Department of Space, the Department of Telecommunications, the India Meteorological Department, All India Radio and Doordarshan. The overall coordination and the management of the Insat system rests with the secretaries-level Insat Coordination Committee. The Department of Space has direct responsibility for establishment and operation of the Insat space segment. The order for the first two satellites in the series was placed with Ford Aerospace on 19th July 1978. It followed on from past Indian use of satellites on a demonstration basis. It had used NASA's ATS–6 satellite for a rural education programme known as SITE (Satellite Instructional Television Experiment). This was a quasi-DBS experiment to provide educational TV programming to remote and rural areas using simple TVRO terminals.

The project ran for some nine months and was successful enough for the Indians to include the quasi-DBS payload on Insat I. India followed up SITE with a demonstration use of the Franco-German Symphonie satellite. This covered telecommunications and was known as STEP (Satellite Telecommunications Experiment Project).
The Insat I system was initially envisaged as a twin satellite system consisting of a primary satellite backed up by an active in-orbit spacecraft. The Insat IA satellite was lost in September 1982 and a replacement IC, ordered. A fourth satellite ID, was ordered in 1985 to cover the transition from the Insat I system to

the domestically produced and ambitious Insat II system.

The Insat IA satellite was launched on 10th April 1982 but failed totally on 4th September of that year. The total failure was related to difficulties in deploying one of the antenna reflectors and the counter balancing solar sail was not deployed as a consequence. At the beginning of September the satellite experienced its first solar eclipse. The sensors on the satellite misread the moon as the sun because the satellite was not in its proper attitude. As a result and because of various other problems, the satellite attempted to correct its attitude errors and consumed all of its station keeping fuel. This left India without any capacity and it turned to Intelsat (two transponders) and the Soviet Union (one Statsionar–6 transponder for TV distribution) for temporary capacity until the Insat IB satellite became available.

Telecommunications Services on Insat I: The telecommunications payload is used for telephony, data and television point-to-point services. Radio is also distributed on this payload as part of the Radio Networking Service (RNS) designed to provide national and regional feeds for retransmission. A typical radio receive-only station is capable of handling the five audio channels of All India Radio (AIR). At the end of 1985 92 AIR stations were in the Insat I radio network. In addition to the normal entertainment functions, there are channels for standard time and frequency signals, disaster weather signals, etc. As at the end of 1985, 3956 two-way voice or equivalent circuits were loaded onto the Insat system and were operational. The Department of Telecommunications was then using 29 fixed and four transportable terminals. The Oil and Natural Gas Commission also has two fixed terminals for the system. A six station satellite-based digital communications network for the National Thermal Power Corporation is also being installed. Indian Telephone Industries is planning a satellite based communications network for its own communications needs. TV stations, 173 of them (out of a total of 179 in India) were also receiving programming through the system. The Indians also say that new initiatives, such as rural telegraphy and business communications are being progressively introduced on the system.

A Satellite-Based Rural Telegraphy Network pilot project in the North Eastern Region is due to start in 1988–89. A demonstration scheme for the Press Trust of India for news and facsimile dissemination through low-cost S-band terminals located on customer premises is due to begin operation in 1986–87.

Quasi-DBS Service: The medium power DBS TV is used for educational and agricultural programming to village receivers in rural areas, for national TV networking and for transmitting UGC-sponsored higher Educational Television. The rural service includes a school Educational Television (ETV) programme feed in selected states and for area-specific DBS TV to the rural TVRO's. The transponders are also used for the five channels of radio. A village TVRO terminal is typically around 3·6 metres diameter and system noise temperature of 610 Kelvin. The G/T is 8·2 dBi/K. At the end of 1985 1500 DBS TVRO terminals had been installed for Insat I services. The educational nature of the programming using these service is seen by some observers as a major feature of Indian economic development.

Meteorological Services on Insat I: The meteorological radiometer takes 30 minutes to scan the entire earth with 4000 picture elements. The ground resolution of each picture element in the visible spectrum is 2·75 × 2·75 km. The infrared resolution is 11 × 11 km. There is also provision for scanning less than the full earth in both East–West and North–South directions. Data Collection Platforms (DCP's) are placed on both land and sea areas for relay through Insat I to the Meteorological Data Utilisation Center at New Delhi. The system is capable of handling up to 400 DCP's but the first phase requires only 100. The Sensory and DCP characteristics of the service are summarised below.

CHARACTERISTICS OF VHRR

PARAMETERS	*VISIBLE*	*INFRA-RED*
Spatial resolution	2·75 km	11·0 km
Scanning lines	4548	1137
Spectral wavelength	0·55–0·75 μm	10·5–12·5 μm
Quantisation level	1024 (10 bits)	1024 (10 bits)
Field of view	76·8 mrad	307·0 mrad
Detectors	Silicon photodiodes	HgCdTe
Radiometer performance	S/N 78 for 2·5% albedo	NE (ΔT) 0·2 K at 300 K

Frequency of rad data (VHRR) transmission:	4034·55 MHz
Data rate:	4000 kbit/s

CHARACTERISTICS OF DCP's

Frequency of transmission	402·75 MHz (uplink)
Downlink frequency	4038·1 MHz
Bit rate	4·8 kbit/s
EIRP (Uplink)	16·5 dB
Mode of transmission	Burst mode (pseudo random)
Burst length	87 ms
No. of samples	10:7 analogue :3 digital
No. of bits on one frame	422 bits

Design of the Spacecraft: The spacecraft is a novel and unusual design having only one solar panel and solar sail to balance it. The reasoning behind this has been well described by Walter Morgan of the Communications Centre of Clarksburg, "the scanning radiometer detectors must operate at a cold temperature. Radiation cooling (using the near-absolute zero of deep space) provides the low temperature. If the sun, earth or moon were to look down the cooling cone, the detector temperature would increase, resulting in either degraded performance or detector destruction. The geometry dictates that the cooling cone looks either North or South. On the South end of the spacecraft is the solar array with solar cells that would emit and reflect the heat from the sun into any cooling cone. To place a second array on the North side would not work because the cooler side looks toward the North. A single solar array has a center of solar pressure different than the centre of gravity. The sun will attempt to turn the spacecraft over the same way the wind turns a windmill. To avoid this problem, a counterbalancing solar sail is used. The five section solar array is closer to the body than a typical body stabilised satellite design. This provides a short lever arm to the sun. A 3·35 metre high solar sail cap is located at the end of a long (12·19 metre or 40 ft) boom to provide a compensating lever arm. There is little likelihood of it being seen by the radiation cooler. The base of the cap has a diameter of only 92 cm. Without the solar sail, the spacecraft is inherently solar pressure unbalanced. Fuel must be used to remove the attitude errors due to the unbalance torque. Magnetic torque coils are used for fine controls."

The Insat ID satellite incorporates several modifications. It has a larger propulsion tank to provide a longer lifetime and additional redundant Travelling Wave Tube Amplifiers. It has larger batteries which give the satellite full eclipse protection. The first three only had partial eclipse protection.

Future of the Insat Programme: At the time of writing, the completed Insat IC satellite is sitting around awaiting launch. This is expected in January 1988, some 10 years after the initial two satellites in the system were ordered. The Indians are now looking to develop their own design of domestic communications satellite and construction of two test Insat II satellites was sanctioned in April 1985. They are due for launch in 1989 and 1990, about the time that the Insat ID satellite will be launched. The Insat II TS (Test Satellite) spacecraft are similar to the Insat I basic configuration but are heavier (by about 50%). They have a considerably enhanced payload capability. Prime Contractor will be Hindustan Aeronautics. They will be followed by three operational Insat II satellites some time during the 1990's. No operational dependence on the Insat II TS satellites is necessary but it is expected that the second of the satellites will be used on an operational basis.

TECHNICAL SUMMARY

PLATFORM	Insat
PRIME CONTRACTOR	Ford Aerospace & Communications Corporation
OTHER MAJOR CONTRACTORS	
STABILISATION	3-axis
LAUNCH MASS	1180 kg (Ariane launch)
MASS IN ORBIT	600 kg
POWER SUPPLY, END OF LIFE	900 Watt
DESIGN LIFE	7 years
ORBITAL POSITIONS	Insat IA, 74° East Insat IB, 74° East Insat IC, 94° East Insat ID,
STATION KEEPING ACCURACY	
ECLIPSE PROTECTION	Partial (Insat ID, full)
PROPULSION SYSTEM	
SPACECRAFT DIMENSIONS	2·18 × 1·55 × 1·42 metres
SPAN, TIP-TO-TIP	19·4 metres
TT&C CENTRE	Hassan, India
COMMUNICATIONS PAYLOAD	
NO. OF ACTIVE TRANSPONDERS	12 Watt
NO. OF SPARE TRANSPONDERS	
TWTA POWER OUTPUT	4·5 Watt
FREQUENCY RANGE:	
UPLINK	5·935–6·425 GHz
DOWNLINK	3·710–4·200 GHz
TRANSPONDER BANDWIDTH	36 MHz
EIRP MINIMUM, BEAM EDGE	32 dBW (EOL)
CHANNEL POLARISATION	
RECEIVE COVERAGE (G/T)	−4·5 dB/K
QUASI-DBS PAYLOAD	
NO. OF ACTIVE TRANSPONDERS	2
NO. OF SPARE TRANSPONDERS	
TWTA POWER OUTPUT	50 Watt
FREQUENCY RANGE:	
UPLINK	5·855–5·935 GHz
DOWNLINK	2.555–2·635 GHz
TRANSPONDER BANDWIDTH	36 MHz
EIRP MINIMUM, BEAM EDGE	42 dBW
POLARISATION	
RECEIVE COVERAGE (G/T)	
DATA COLLECTION PAYLOAD	
NO. OF ACTIVE TRANSPONDERS	1
NO. OF SPARE TRANSPONDERS	1
TWTA POWER OUTPUT	0·3 Watt
FREQUENCY RANGE:	
UPLINK	402 MHz
DOWNLINK	4038 MHz
TRANSPONDER BANDWIDTH	200 MHz
VHRR RADIOMETER	
DOWNLINK FREQUENCY	4034 MHz
EIRP MINIMUM, BEAM EDGE	8·5 dBW

DISASSEMBLED VIEW OF INSAT-II (PRELIMINARY)

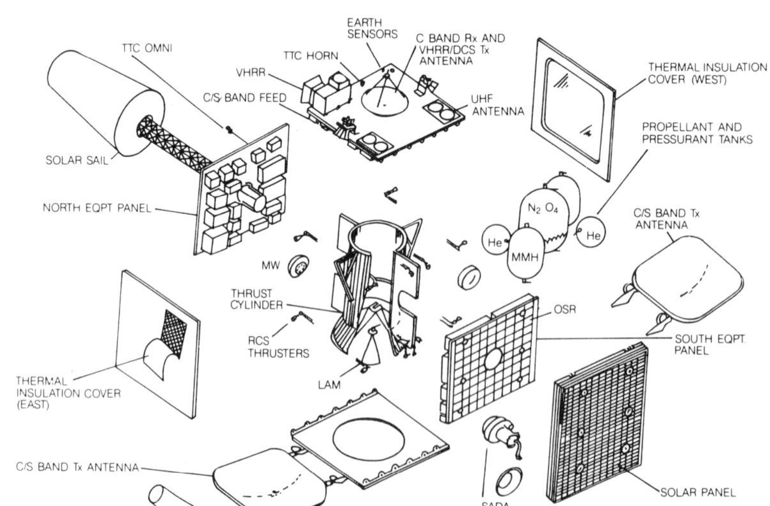

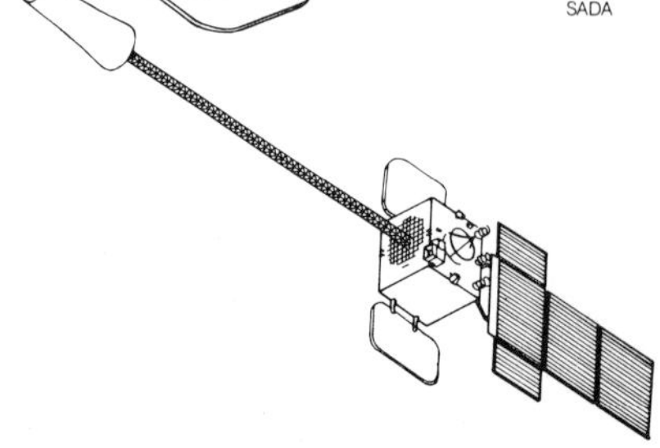

Not to scale

Artists impression of the Insat I spacecraft in orbit

COMMERCIAL DETAILS

OPERATOR	International Telecommunications Satellite Organisation (Intelsat)
ADDRESS	3400 International Drive, NW Washington DC 20008–3098, USA
TELEPHONE NO.	(202) 944-6800
TELEX NO.	(WUT) 892707, (WUI) 64290
SENIOR STAFF	John Hampton, Acting Director General David T. Tudge, Deputy Director General, Finance & Staff David M. Leive, Legal Advisor Alberto Cruzat, Director of Finance and Accounting Maarshall Sneiderman, Director of Procurement Ramamswami Parthasarathy, Director of Technical Programs Robert Leahy, Director, Public and Media Relations Tony A. Trujillo, Senior Executive, Public and Media Relations
PURPOSE OF PROJECT	Provision of capacity for international public telecoms services and lease of such capacity for domestic services. Acts on behalf of its members only (is a carrier's carrier)
PROJECT STATUS	Fully operational but suffering from changing political and commercial environment
PROJECT VALUE	As at 31st December 1984, the investment in the Intelsat space segment was $1535 m net and $2500 m gross. The five Intelsat VI spacecraft are costing $700 m. The first seven Intelsat V birds $235 m and the full 15 $34 m each on average (1976 prices)
LAUNCH DATES	See launch summary

SATELLITE USERS

Intelsat provides satellite capacity for the PTT administrations of the 112 countries which are members of the organisation (see Project Overview for the special case of Comsat). It is in effect a carrier's carrier and it's direct users are those PTT administrations. Its primary role is to provide international capacity for the world's public telecommunications system. This means that it is dominated by telephony requirements. The system is also used, through PTT's, by broadcasters for international television links.

Four Intelsat V satellites carry maritime communications payloads used by Inmarsat.

Lease and outright sale of capacity is available for use by members for domestic communications.

In addition non-member countries can use the Intelsat system at higher rates. Some 60 countries do so.

Primary identifiable users of the Intelsat system are as follows:
1. INTERNATIONAL TELEPHONY: PTT's of 111 member countries, Comsat (USA) on behalf of US carriers and 60 non-member countries' PTT's. Mercury Communications (UK), KDD (Japan) on behalf of Class 1 Licence holders, Telespazio (Italy), OTC (Australia) and various other overseas telecoms administrations on behalf of domestic carriers.
2. INTERNATIONAL TELEVISION: National public service broadcasters, news agencies and international TV channels for both regular broadcasts and coverage of special occasions.
3. IBS: Companies and large organisations for private networks carrying voice, data, telex, video-conferencing and facsimile.
4. INTELNET: Private satellite network operators for remote printing and information dissemination and gathering.
5. VISTA: Developing countries and countries with requirements for remote location communications.
6. CABLE RESTORATION SERVICES: National PTT's and overseas telecoms administrations.
7. DOMESTIC TELECOMMUNICATIONS SERVICES: See text.

PROJECT OVERVIEW

Introduction: Intelsat was the first and is still the largest commercial satellite operator. It was a product of US space and foreign policies of the early 1960's which also led to men on the moon. Now nearing a quarter of a century old, the international PTT cooperative has passed its period of substantial growth and falling charges to its users. It is now faced with overcapacity in its system and the prospect of competition in an environment where it has hitherto had monopoly power. It now looks as if that competition will come from fibre optic submarine telephony cables rather than alternative private sector international satellite systems. The organisation has responded vigourously over the last three years to these changing circumstances but still comes under considerable criticism. This is partly because it is a carrier's carrier and the end user usually pays a considerable mark-up to the PTT for use of Intelsat capacity. This is because most PTT's usually cross-subsidise their local and private consumer charges from long distance and international services. Moreover, it is in the nature of Intelsat that it takes a long time to respond to changing circumstances. Intelsat is also a victim of a distinct cooling of interest in the West towards satellite communications. Intelsat is also a fairly difficult organisation to understand.

There has been little in the way of sound academic study of the institution and some observers believe it is misleading in the information that it gives out. Yet as the organisation which created commercial and global satellite communications, there is no doubt that it has been of major benefit and has had a significant impact on the Western world's communications.

Intelsat was formed in 1964 to provide global satellite communications for any country wishing to join the organisation and to provide an outlet for the new satellite technology then being developed in the USA. At that time communications were seen as a natural monopoly and Intelsat was given a monopoly in the international satellite communications field. The Soviets were the first to break this when they set up a competitive organisation in 1971, Intersputnik. In the USA the federal administration established the Comsat (Communications Satellite Corporation) in conjunction with private funding, to act as the US signatory to Intelsat and provide the US based ground segment of the Intelsat system. In other countries the ground segment is owned (normally) by the local PTT administration, Intelsat being left to operate only the space segment. The Intelsat organisation is owned by 111 of the world's PTT administrations and, in the case of the USA, Comsat. The percentage ownership of each PTT is in direct proportion to their utilisation of the system and is adjusted annually to reflect changes in the utilisation.

Despite repeated claims from Intelsat that it is a non-profit organisation, it is required to make a return of 14% on members investment. Its investments are funded either from capital contributions from members or from revenue. Its pricing structure is based on cost plus but with the proviso that its charges are non-discriminatory. That is to say, the charge from Intelsat to a member PTT for a circuit is the same whether the PTT takes one or several hundred circuits. It is the same between all member PTT's and the same no matter how much total volume the PTT takes on a route or on the system in total. A single hop circuit charge is also distance insensitive. In practice it is difficult to conclude that this non-discriminatory policy is a valid description of the organisation's total package of charges. It now offers a much wider variety of services and terms and conditions than it did ten years ago. Some observers believe that it is dumping surplus capacity on the international market to obtain a marginal rather than a cost based return and to protect itself from competitive satellite proposals.

Competition to the Intelsat System: Optimism about the commercial prospects of satellite communications, some valid criticisms of the Intelsat organisation's services and a general climate of telecommunications liberalisation in the USA led, during the early 1980's to a number of proposals for international satellite systems which threatened the organisation and its objectives of providing global and non-discriminatory satellite services. The first serious proposal was the French Telecom 1 satellite, nominally aimed at the French domestic communications market but with a lot of C-band capacity for serving the small French overseas departments in the Caribbean.

The footprint for these conveniently covered a swath down the Eastern seaboard of the USA and thus could provide transatlantic communications. It was quickly followed up by a proposal from the UK to include telecommunications on a DBS spacecraft, Unisat, located at 31° West. The Americans appeared to have been slow on the uptake of the possibilities for transatlantic communications and, for once, followed Europe. In 1983, the first US proposal was announced. This was for Orion and was quickly followed by proposals from ISI, RCA Americom, Cygnus, PanAmSat, Finansat, Systematics General and others still in the course of development. All of these proposals would require agreements with the USA and with at least one non-US country.

To give this agreement the countries involved would have to break their agreements with the Intelsat organisation not to allow competitive satellite systems. So far only the USA, which has proceeded a lot further than any other country in liberalising its domestic telecommunications, has given such authority. Moreover that authority is conditional in that the new satellite operators should not offer capacity for public communications services (which account for the bulk of Intelsat's revenue). The relationship between Intelsat and its member countries is close to that of international treaty arrangements. It is the norm in international relations that countries are unwilling to be seen to break treaty obligations and the Intelsat case appears to be no different. The new proposals for international satellite systems also came at the time when growth in international telecommunications began to slow down. So serious was the slowdown, which had not been expected, that Intelsat had to ask its members to substantially improve their traffic forecasts. The organisation had been dependent on these from its member PTT's in order to produce global forecasts with which it planned its capacity.

The slow down has left the organisation with surplus capacity and it eventually responded to the mis-match between capacity and demand by dropping the options on the bulk of the Intelsat VI spacecraft. There were options for 11 such vehicles in addition to the initial firm order for five. One of the major arguments put forward by the proponents of private international satellite systems was that they would be providing capacity for services (then) unserved by the Intelsat organisation. This particularly related to business services, such as private corporate networks using terminals located directly on the customer premises. It was a valid criticism that Intelsat was slow in introducing the capacity for these services. They were seen by some telecommunications sectors as a major growth area for the mid-1980's onwards. The private satellite operators claimed that as Intelsat was not providing these services they would not be diverting traffic away from the organisation if they did so. Intelsat responded in 1983 with its International Business Service (IBS), which had been planned over the previous 18 months.

By 1986, it had become apparent that Intelsat had seen off the serious contenders in the international satellite scene apart from PanAmSat. Some telecommunications companies and experts had studied the possibilities during the 1981–85 period and had dismissed the idea as commercially not viable. For example, Cable & Wireless, with almost a world-wide captive market through its Hong Kong, Chinese, Caribbean, UK and Middle East activities, studied the possibility of its own satellite system but abandoned the idea without ever making its plans public. UK telecommunications consultant John Anderson found as early as 1982 that money could not be made with a system for the North Atlantic. Intelsat has been very aggressive in defending its monopoly position by both offering new services and terms and by lobbying both in Washington DC and amongst individual member nations. (Strictly speaking, Intelsat is not allowed to lobby but overcomes the problem by describing its lobbying as "education"). It has a number of major strengths to its advantage in defending its market. It is capable of self-insuring against the high risks of launch, unlike the private satellite systems which need outside insurance to raise the finance for their satellites. The current lack of insurance capacity means that any new venture is unlikely to succeed in raising its required finance.

However, Intelsat now faces competition from new fibre optic technologies. Even though on a nominal cost per circuit, satellite communications may well be highly competitive with the best future offerings of fibre optic trunk systems across the Atlantic, the high risks involved in satellites make fibre optics an attractive investment for PTT's. Intelsat is currently arguing that a fully loaded Intelsat spacecraft over the North Atlantic can provide a cost per circuit one-third of that of fibre optic. Nevertheless telecommunications organisations, who are often members of Intelsat and who have a strong financial interest in it, are now investing heavily in transoceanic trunk fibre optic cables, or, in Europe between the UK and the continent. Some indication of Intelsat's concern can be seen from a "best case" forecast presented by it at the end of 1985. This predicts a market demand for around 87,000 circuits in 1995 for the North Atlantic, with a supply of 342,000 circuits.

The ratio of supply to demand under slightly different assumptions increases from 4:1 to 8:1 onto 10:1 under the worst case (supply of 868,000 circuits). For customised services, demand is 13,050 circuits, indicating that there is no great market growth expected in this field. On the other hand Intelsat has always been interdependent with the international submarine cable system. It is normal practice, for example, to only use a single hop for a voice circuit using the Intelsat system. Where the distances are so great that a double hop would be needed, cable is substituted for one of the hops. This is done for technical reasons as the length of the transmission path on a double hop would cause too much delay for a quality duplex voice circuit. Similarly the cable and satellite systems provide a degree of emergency back-up for each other in the event of failure or demand temporarily exceeding supply. Moreover there are institutional arrangements which make the systems further interdependent. On the North Atlantic, a fixed proportion of traffic is routed through each system.

Intelsat has also faced competition from regional satellite systems. Here the politics are more complex than that of competitive international systems. Intelsat is itself somewhat of a product of US industrial policies and its non-US members have, since the earliest days of satellite communications, developed similar industrial policies which favour domestic satellite manufacturing capabilities at the expense of the USA. European countries have been particularly interested in developing European captive markets for their satellites. This has been done by the creation of a European regional satellite system, Eutelsat, and the development of a number of domestic satellite systems (DFS-Kopernicus, Telecom 1 & 2, TELE-X and Italsat).

The Eutelsat organisation is closely modelled on Intelsat and there has been some acrimony between the two organisations over economic coordination of Eutelsat. The Intelsat organisation originally only wanted to coordinate Eutelsat for a period of five years leaving Eutelsat with a degree of uncertainty about its future. It has been argued that Eutelsat has not created significant economic harm to Intelsat because it addresses the needs of a market where satellite communications are inherently unprofitable and would therefore not be served by the Intelsat system. Nevertheless, Intelsat's attitude towards the organisation indicates that it has taken the potential competition seriously. Eutelsat was originally envisaged to serve the European telephony market, with a limited amount of capacity reserved for the European Broadcasting Union.

In practice its use has been dominated by use for cable television channel distribution. Its perceived shortcomings in serving this market have led to the Astra project from Luxembourg, proposing a European TV satellite system using satellites procured from the USA. Société Européenne des Satellites, who own the Astra satellite has yet to coordinate with Intelsat but it is expected that there will be a considerable battle to obtain the necessary coordination. The Intelsat system is also much used by European cable TV programmers in the absence of sufficient capacity on the Eutelsat system. The same thinking as is held by SES has led to a proposal, backed by Hughes Communications, for a hybrid DBS/FSS satellite system (Atlantic Satellites) operated out of Ireland and (possibly) the UK to provide national, regional and international satellite communications. Unlike the Eutelsat system, those of Ireland and Luxembourg are backed by small and uninfluential governments. Eutelsat is backed by 26 countries which in total have a considerable influence in Intelsat.

Proponents of competition to Intelsat have argued that such regional systems as Eutelsat are necessary because Intelsat's global and standardised series of satellites are not capable of providing the types of specialised payloads that the increasingly sophisticated telecommunications market requires to develop new

INTELSAT STANDARD EARTH STATIONS

STANDARD* NAME	MINIMUM G/T dB/°K	BAND (GHz)	SERVICES	SIZE IN METRES
A	40·7	6/4	All	26·0–32·5
B	31·7	6/4	All	10·0–13·0
C	39·0 14/12	14/11	All	14·0–20·0
D–1 D–2	22·7 31·7	6/4	VISTA	4·5– 5·5 10·0–13·0
E–1 E–2 E–3	25·0 29·0 34·0	14/11 14/12	IBS IBS, INTELNET	3·5– 4·5 5·5– 6.5 8·0–10·0
F–1 F–2 F–3	22·7 27·0 29·0	6/4	IBS	4·5– 5·5 7·0– 8·0 9·0–10·0
G	—	6/4 or 14/11 14/12	All INTELNET	0·8– 2·5 and up
Z	—	6/4 or 14/11 14/12	All	4·5–18·3
Non-Standard	—	6/4 or 14/11 14/12	All	—

*All earth stations can be used to access the international and domestic networks. However, the Standard Z earth station can have access only to domestic networks

services. Nevertheless, the Eutelsat I series of satellites are not at all innovative and the system clearly remains a product of telecommunications thinking and planning of the 1970's. The Astra system is somewhat more innovative, based on the same thinking as the US RCA Satcom K system. Unlike Intelsat it is aimed at a specific market (TV) in a specific region. It may take traffic away from Intelsat but the backers argue that it will generate a lot of traffic that would not otherwise exist.

Similar in concept to the Eutelsat system has been the Arabsat regional system owned by the 22 members of the Arab League. This is used primarily for telephony traffic and, indeed, some members of the League have opted to use the Intelsat system instead because of the long period it took to get the Arabsat system going. Moreover the members of the Arab League have a limited argument that the system is intended as part of their industrial policies towards satellite communications. However, it passed the then somewhat ambiguous tests for economic harm to Intelsat.

There are a whole group of satellite systems which "compete" with Intelsat but which do not come within the scope of its global monopoly. These are the domestic communication satellite systems which only have to be coordinated with Intelsat for technical purposes. In some circumstances they have directly led to loss of revenue to Intelsat. For example the Brazilian SBTS system replaced capacity leased by that country from Intelsat for domestic purposes only. On the other hand, Intelsat is finding a useful market in sale or lease of capacity to countries who neither have the capacity requirements for their own dedicated satellite systems or who cannot afford them. Unfortunately there appears to be a grey area between what is meant to be a domestic satellite system and one that overspills into neighbouring countries. The Indonesian satellite system is used by ASEAN countries as a regional system. Similarly, the Brazilians have been trying to sell capacity to neighbours. Canadian, US and Mexican domestic satellite systems are used for transborder traffic. Intelsat has rules as to whether these infringements are significant or not and has, since 1983, tried to define the rules more tightly. But it remains a difficult (and often politically unrealistic) exercise to try and rule out every infringement.

Services offered by Intelsat: The organisation's public relations machinery presents a glittering array of services on offer through which it is often difficult to see what the primary missions of the organisation are. A breakdown of its revenues presents a much clearer picture. For 1984 its revenue came from the following sources:

1. Full time International Voice Services	72·5%
2. Domestic Leases	9·1%
3. International Data	6·8%
4. Occasional Use TV	4·6%
5. Inmarsat Lease	2·5%
6. International TV Lease	1·4%
7. Other	2·9%
Total Revenue	$411 m

(source Intelsat 1984–5 Annual Report)

The system provides 1700 earth station–to earth station pathways for this traffic. Of these, 150 account for over 50% of Intelsat's revenue. The specific service offerings are summarised as follows:

1) *International Telephony*: This is Intelsat's primary mission, although as a percentage of its total revenue, it is declining and in 1986 stood at about 67%. The mark-up between what Intelsat charges the PTT's for this service and the final charge to the end user varies considerably between countries, even though the Intelsat charges, per circuit, are the same for each country. The service requires PTT procurement of Standard A, B or C terminals although, for more specialised requirements, Standard G and Z are available. The earth station standards are summarised on page 173.

2) *International TV Programme Distribution*: This was one of Intelsat's earliest services. Users can either lease capacity on a full time or occasional use basis. The system tends to use the same international gateways as are used for telephony traffic, although technical developments in both the space and the ground segments now allows bypass of such facilities. Capacity is available in both C-band and Ku-band. Cable News Network of the USA uses C-band uplink from Atlanta, Georgia, to Intelsat V at 27·5° West and a Ku-band downlink from the satellite to provide coverage to Europe. Occasional use services predominate in this field being used for coverage of such events as the Olympic games. The system is also used for international TV news feeds with capacity either available on a full time basis (partial or full transponder) or occasional use. There are some 10 full time leases at the moment. The pricing structure for these services is somewhat complex with over 100 different preemptible and non-preemptible offerings available. The charging structure is based on parameters such as satellite role, type of transponder connectivity, direction of transmission, EIRP and bandwidth. In 1984, a digital video service was offered. The services are in addition to domestic use of Intelsat capacity for distribution of cable TV or national TV channels.

3) *International Business Service (IBS)*: This was introduced somewhat tardily and as a response to outside competition rather than in anticipation of demand. This service became available on 1st October 1983. It can either use the normal Intelsat international gateways with terrestrial links to and from the customer premises or specialised Standard E terminals on the customers premises or shared between several customers. In the USA, the service is also available through non-Comsat owned International gateways. The service is fully digital and is aimed at providing telex, voice, facsimile, data and videoconferencing. The Intelsat V F13, F14 and F15 satellites have been modified so that the IBS service can use multiple beam interconnection. This allows transmission in one hemispheric or spot beam, East or West, which can be received simultaneously in East and West hemispheres and spot beams at C-band and Ku-band. Modulation and access techniques are QPSK/FDMA using either SCPC or multi-channel per carrier. TDMA is also available on a case-by-case basis. The service can offer capacity from a single 64 Kbps channel to a full 72 MHz Ku-band transponder. Charges for a full time 64 Kbps channel, full time, per month vary from $312·5 to $500, although the user will have to pay the usual PTT mark-up.

4) *Intelnet*: This is a specialised service for one-way low speed date distribution using VSAT's of between 0·8 and 2·5 metres (Standard G). Standard E–3 terminals are also available. Service uses either C-band or Ku-band frequencies. Applications range from networks for broadcasting financial news and weather to networks for gathering environmental and oil or gas exploration data and commercial accounting data such as for inventory control and analysis. Preemptible or non-preemptible terms on a full-time or occasional use basis are available. Fractional or full transponders are available. Fractions are in increments of 1 MHz (Ku-band) and 9 MHz (C-band). Cost of receive only earth stations start at $3000 per unit.

5) *Vista*: Introduced in December 1983, this is a thin route service for rural and remote communities providing 2–4 voice circuits and low speed data applications. It is available on either a domestic or international basis. It uses Ku-band frequencies only.

6) *Cable Restoration Services*: This is to provide back up capacity in the event of failure of a trunk telecommunications cable or if a cable is out of use for repairs. The service is an occasional use offering but from 1985 temporary lease of non-preemptible capacity in increments of 800 circuits has been offered. The

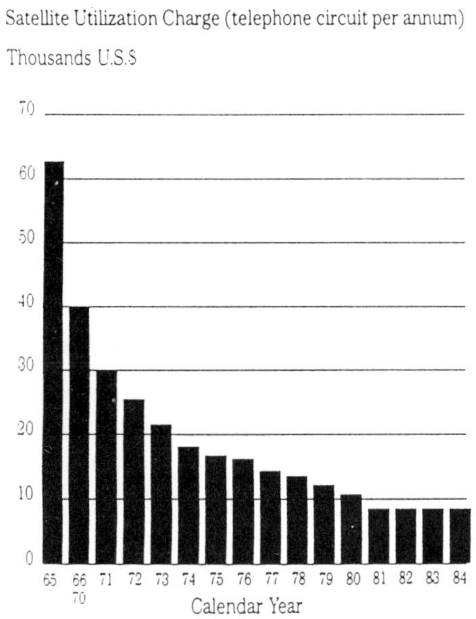

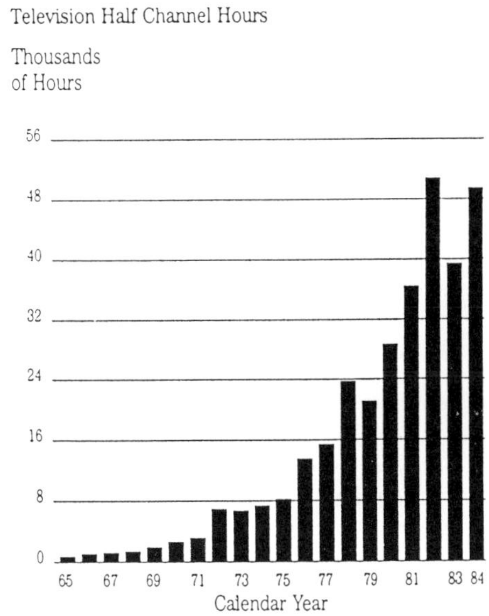

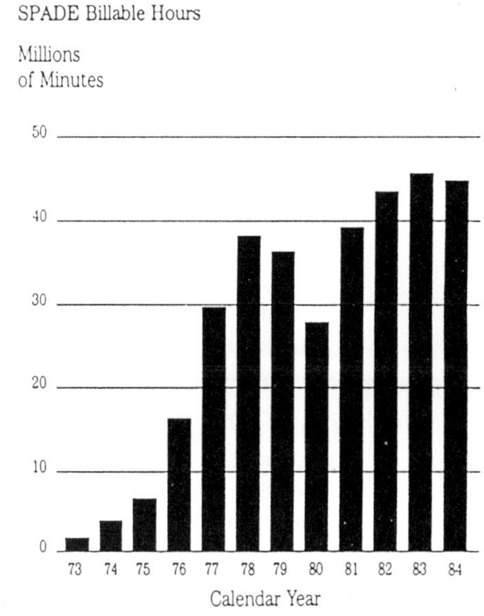

(source Intelsat 1984–85 Report)

revenue contribution to Intelsat from this service tends to be small ($1·7 m in 1984) but it provides a major resource, even if unused, for the international public telecommunications system.

7) *Domestic Telecommunications Services*: The Intelsat organisation has been leasing capacity for domestic communications requirements for a decade. The capacity is used for internal telephony and television programming distribution. From 1986 it has also been selling transponders for these services. In 1985 it asked its members to deem 150 transponders surplus to requirements and it is these that are up for sale. The leased or purchased capacity is not used for regional or international traffic. The prices and terms are dependent on a variety of features such as the age of the satellite, frequencies and so forth. As at the 31st December 1984, the following countries leased capacity for domestic purposes: Algeria, Argentina, Australia, Brazil, Chile, Columbia, Denmark, France, West Germany, India, Libya, Malaysia, Mexico, Morocco, Niger, Nigeria, Norway, Oman, Peru, Portugal, Saudi Arabia, Spain, Sudan, Thailand, United Kingdom, Venezuela and Zaire. Some of these have now returned the capacity to Intelsat as their domestic satellites have come on stream and others have opted to purchase capacity outright. The prices of transponders for outright sale vary between $3 m and $5·5 m and are available on a first come first served basis. According to the organisation, the sale is likely to raise between $65 and $70 m per year.

The Overcapacity Problem: The Intelsat V series was planned and ordered at a time when use of the Intelsat system had been growing at a rate of 15% per annum. This rate began to slow down in the early 1980's at a time when it was apparent that the economies of scale and unit costs per circuit were levelling out. After some 15 years of continuous reductions in Intelsat rates, these too levelled out in the 1980's. There is evidence to suggest that the failure to continue to revise down charges led to the decline in growth rates in use of the system. This suggested that Intelsat should have known not to order and launch so many Intelsat V satellites. International telephone use has a very high price elasticity of demand. Figures for the UK, for example, show that the elasticity is around $-1{\cdot}1$, indicating that a 10% fall in real prices to the end user leads to an 11% increase in spending and a 20% increase in use. Levelling off in Intelsat charges is indicated in the diagram above.

By 1983 it was becoming apparent that the growth rates in use were slowing down. Use of satellite capacity for full-time voice circuits grew at only 10·1% in that year and the organisation was faced with a shortfall in demand over forecasts of 19·8%. A region by region breakdown shadowed the shortfall to be greatest in the Indian Ocean Region (23·5%). In the area where the organisation was then most likely to face competition from private international satellite networks, the Atlantic Ocean Region, the shortfall was a substantial 19·5%. For the Pacific Ocean Region the shortfall was a more modest 12·7%, perhaps indicating better prospects for communications in that area. Similarly, the 1984 overall growth rate was 10·7% (full-time voice circuits only, Atlantic Ocean Region was 7·4% but the Indian and Pacific regions grew at 16·3% and 17·7% respectively – the shortfalls were 16·4%, 24·1% and 4·7% respectively).

Intelsat, Satellite Communications and the Third World: In this section the significance of the Intelsat system to the communications needs of developing countries is reviewed. The issue is of major importance to the organisation in that it has become very active in marketing its surplus capacity for domestic use by those countries and because, in presenting its case against the introduction of competitive international satellite systems Intelsat has argued that thin route service charges may have to increase as a consequence. It is concluded here that satellite communications are primarily of benefit to developed countries and do not address the needs of developing countries. The Intelsat system was created by the Western World for the benefit of the Western World. It is not designed, nor is suitable to meet the needs of the developing world. The implication is that alternative systems or technologies may address those needs more accurately.

There is little doubt that satellite communications are primarily used by developed countries even though most major developing countries are members and users of the Intelsat system and, in many cases, users of regional or domestic satellite systems. At the end of 1985 there were just over 1200 satellite transponders available for civilian use in the non-communist world. Of these 550 were on domestic North American satellites, 48 on European spacecraft and 48 on the Australian Aussat system. Therefore there were available to the developing world,

CHARACTERISTICS OF INTELSAT SERVICES

Service category	*Service Identification*	*Type of traffic carried*	*Modulation*	*Earth Stations (Typical)**	*Band*	*Preemptibility/ Protection*	*Availability*
Voice	Conventional Telephony	Voice, Telex, Facsimile	FDM/FDMA CFM/FDMA	A, B and C	C and Ku	Non-Preemptible, Fully Protected	Full-time
	Digital SCPC	Voice	SCPC/PCM/ PSK	A, B and C	C and Ku	Non-Preemptible, Fully Protected	Full-time
	Demand Assigned SPADE	Voice	SCPC/PCM/ PSK/DAMA	A and B	C	Non-Preemptible, Fully Protected	Full-time
	Thin Route Telephony VISTA	Voice, Telex and Data	SCPC/CFM	D1, D2, A, B	C	Non-Preemptible, Fully Protected. Preemptible also available	Full-time
	Time Division Multiple Access/DSI	Voice	QPSK/PCM/ TDMA	A, B and C	C and Ku	Non-Preemptible, Fully Protected	Full-time
Data	PSK data channels (IDR)	Data	SCPC/PSK or multichannel per carrier	A, B and C	C and Ku	Non-Preemptible Fully Protected	Full-time
	Time Division Multiple Access DNI	Data	QPSK/TDMA/ DNI	A, B and C	C and Ku	Non-Preemptible, Fully Protected	Full-time
	Intelsat Business Service (IBS)	Non-Public switched digital services voice, telex, etc.	QPSK/FDMA uses SCPC or Multichannel per carrier can use TDMA on a case by case basis	A, B, C, E–1, E–2, E–3, F–1, F–2, F–3	C and Ku	Non-Preemptible, Fully Protected	Full-time Part-time Occasional use
	Intelnet	Data	Spread Spectrum (CDMA). Also BPSK/FDMA	Standard G	C and Ku	Non-Preemptible, Fully Protected Preemptible also available	Full-time Occasional use
Video	International leased TV	Video, non-public switched voice	Analog/FDMA	A, B and C	C and Ku	Non-Preemptible, various degrees of protection. Preemptible also available	Long-term lease Short-term lease Occasional
	International Video	Video, non-public switched voice	Analog/FDMA	According to transmission plan approved by Intelsat	C and Ku	Non-Preemptible, various degrees of protection. Preemptible also available	Long-term lease Part-time Occasional
	Improved Video	Video, non-public switched voice	Three 24 MHz carriers per transponder	According to transmission plan approved by Intelsat	C and Ku	Preemptible only available	Long-term lease Occasional
	Digital TV	Video	Digital/FDMA	According to transmission plan approved by Intelsat	C and Ku	Non-Preemptible, Fully Protected Preemptible also available	Long-term lease Occasional
	Integrated Video/ data/and, non-public switched voice	Video, voice, data	Analog/FDMA	According to transmission plan approved by Intelsat	C and Ku	Non-Preemptible, Fully Protected Preemptible also available	Long-term lease
	Cross-strapped Video	Video, non-public switched voice	Analog/FDMA	According to transmission plan approved by Intelsat	C and Ku cross-strapped	Preemptible only available	Long-term lease

the 400 transponders on the Intelsat system and 120 on domestic and regional satellite systems (Palapa, Arabsat, Morelos, SBTS and Insat). The Paris based consulting company, Euroconsult, predicted, during 1985, that the proportion of satellite transponders directly owned by developing countries would fall from 10% at the end of 1985 to 6% by 1990, by which time there would be around 2500 transponders available worldwide. The implication is that there will be almost static growth in developing country ownership at a time when the rest of the market would double in size.

The 10% ownership factor is also reflected in the use of the Intelsat system by developing countries. At the end of 1985, some 19 developing countries leased 32 transponders on the system for domestic communications use. Of Intelsat's total revenue, some 7·5% is accounted for by developing countries, either through domestic leases or through use for international traffic. In the latter case the developing countries are almost exclusively linked by Intelsat's thin route pathways. The system employs some 1700 earth-to-earth station pathways. Of these, 850 account for 10% of Intelsat's revenue. Fifty per cent of Intelsat's revenue is accounted for by 150 pathways. The latter are between developed countries. The thin route pathways tend to link developed countries with developing countries or provide direct links between developing countries. Where the link is between a developed and a developing country, it is arguable as to who benefits. If it is assumed that half the benefit of the link is attributable to the developed country, then the weighting of benefits to the Western World increase even further.

The dominance of Western needs in the Intelsat system is also reflected in the organisation's history. A product of US space and foreign policy of the early 1960's, Intelsat was originally seen as a captive export market for the US satellite manufacturing industry as well as a very good method of improving international communications. Since its inception all of the prime contractorships for Intelsat spacecraft have gone to US manufacturers, not only because the USA makes better satellites than any other country but because the USA is the primary user of the Intelsat system. It was only at the instigation of France in the 1960's that Intelsat was not given a global monopoly to the exclusion of regional satellite systems. This would have largely prevented the growth of a European satellite manufacturing capacity. The captive market concept came under pressure in the USA in the early 1980's from a new pressure source.

Until 1984, Hughes Aircraft, who were supplying the Intelsat VI system, were, like other US aerospace manufacturers, vigorous supporters of the Intelsat monopoly. The new pressures came from the telecommunications business in the USA. They saw that their domestic market had been opened up to much more competition and was in the process of being restructured. Similarly, the continued growth of the internationalisation of business and consequential communications demands would create a demand for innovative services not well provided by the Intelsat system. It is clear that some sections of the US telecommunications business were worried that the market would be swallowed up by non-US telecommunications concerns. A central feature of concern covered standards. It has been a major preoccupation of Western industrial policy-makers for electronics and computers that the IBM de facto SNA computer communications standard should not be adopted as a world industrial standard for computers and communications. If it were, it would give the Americans a significant advantage in both computer hardware and telecommunications. For example, 70% of mainframe computers in Europe have been supplied by IBM. As the computing and communications industries are perceived to be converging and are both major growth areas this would have a major impact on both industries in Japan and Europe, neither which yet have a dominant communications standard for computers.

The Europeans in particular caused concern to the Americans during the early 1980's over perceived threats to the Intelsat system. In 1982 the UK announced plans for its Unisat satellite which would have provided international services across the North Atlantic directly aimed at business users. There was also some concern that France's Telecom 1 satellite system was capable of providing transatlantic links between the most populated parts of Europe and the Eastern seaboard of the USA. Both countries are supporters of OSI standards, in direct competition with the de facto IBM SNA standard.

The US response to these changing pressures has been to introduce an open skies policy towards private venture international satellite systems. Its abandonment of a policy of full support for the Intelsat monopoly on international communications is illustrative of the fact that its interests are now no longer convergent with those of the Intelsat organisation. As well the future of the organisation is likely to be determined entirely in the framework of Western rather than developing country interests. Past rhetoric may have suggested that global communications (and the "global village" concept) would lead to greater interdependence between the Developing and Western worlds but such models of international relations appear to fly in the face of events in Washington DC concerning Intelsat. Nevertheless, the events do not fully explain why Intelsat is not capable of meeting the needs of developing countries. The explanation needs a fuller consideration of the Intelsat technology, the needs of developing countries and an overall framework of international relations and economic development.

Dependency, Communications and Intelsat: During the 1970's a body of international relations thinking developed a model known as dependency theory. There have been numerous subsequent criticisms of the approach, which largely originated from studies of the relationships between the USA and South America. It has a Marxist element in its conceptualism. Nevertheless, it is still used extensively in the West and appears to present a more realistic description of reality than the "realism" theories which went out of fashion in the late 1960's or the global village and interdependence theories.

It essentially leads to the conclusion that the type of advanced technologies embodied in satellite communications have been developed in advanced economies and societies for the benefit of and the integration with those societies. Their transfer to developing countries is wholly inappropriate to the needs of those countries and may be damaging to them. The World Association for Christian Communications (WACC) has expressed a similar "economic imperialism" viewpoint. It argues that most new technologies are rigidly controlled by foreign pressures operating through multinational corporations.

The new technologies are volatile. They evolve at such a rate that local cultures rarely have a chance to adapt or integrate them into the local environment. The new technologies have a "shock" effect on third world societies such that they are either thrown into social convulsions or are rapidly assimulated into the metropolitan cultures of the Western world. At its most extreme, the argument concludes that the social elites in the West subordinate elites in developing countries to the extent that they look to the West for "reference points" as well as goods and services. The communications infrastructure may reflect this in that the capital of a developing country often has far better international connections through the Intelsat system than it does with its interior. Ultimately the conclusion is that Intelsat is a product of US imperialism. Even if developing countries master its use to their benefit, the USA (and other Western countries) will move onto other forms of technology to reinforce the "imperialist" structure.

Like many such viewpoints, it fails to suggest a solution to the problem (except, perhaps, for the world to embrace one of the many forms of Marxism). Nevertheless, it may help explain why some countries have been so concerned to develop their own domestic satellite systems or share regional satellite systems outside the framework of Intelsat. Europe, India and the Arab League all began considering satellite systems soon after Intelsat was established in 1964. For Europe (and to a lesser extent, India), industrial policy requirements were involved and there is no doubt that they were concerned about the dominance of US technology in the satellite field.

The Maitland Commission and the needs of Developing Countries: The Maitland Commission's report (ITU 1985) took a more pragmatic view of the communications requirements of the developing world and tended to follow the "interdependency" approach to the international political structure. It concluded unequivocably that improvements in the telecommunications infrastructure of developing countries are of benefit to both them and the Western world. It also recognised major differences between individual developing countries (the lack of such recognition may be the biggest shortcoming of dependency theory). "It has to be recognised that the problems in this field (telecommunications) cannot be generalised. The situation of each developing country, or region at least, is unique. It follows that particular rather than global solutions have to be applied...." Maitland also believed that the interests of western countries trying to export communications technology could be reconciled with the internal social needs of developing countries: "...an effective telecommunications system not only is inherently profitable but also generates wealth... technological advances have widened the options for operators and have reduced the cost of equipment... developing countries represent a fast growing and potentially the largest market for telecommunications equipment and expertise."

Maitland saw telecommunications as an integral part of the development process stating that the commission recognised "the fundamental importance of communications infrastructure as an essential element in economic and social development". It recommended that less developed countries must place a higher priority in investing in telecommunications and that annual investment should rise from $8 billion to $12 billion. Unfortunately, the Maitland Report leaves one with the distinct feeling that it has not explained how this increase will come about in the current international order. All that Maitland suggests is that new sources of finance be arranged.

The Report of the Second United Nations Conference on the Exploration and Peaceful Uses of Outer Space (UN 1982) drew similar conclusions: "communications play a key role in development to the extent that they transfer information, data and ideas. They are the key infrastructure elements for social and economic development." The report is more explicit than Maitland on the role of satellite communications, stating that satellite technology can accelerate national development because it provides a way of leap-frogging over obsolete technologies and getting away from percolating and trickle-down models of development for which developing countries do not have the time. It could effectively deal with the problems of illiteracy, isolation and lack of information afflicting the development process. Nevertheless, the rhetoric concerning the tremendous untapped potential for satellite communications has largely been followed up by more rhetoric rather than orders for satellites. Indeed the World Bank has notably failed to provide funding for any form of telecommunications infrastructure except for the odd case like Thailand.

Specific Benefits of Satellite Communications: In analysing the benefits of satellite communications, there are two distinct approaches which, conveniently, are reflected in two types of telecommunications. One approach looks at the direct economic value of a service in terms of, for example, profits. This may be appropriate for such services as telephony or telex. A second approach looks at the wider impacts and is typically related to the educational use of communications. UNESCO has described the latter type of service as having an informational as distinct from an organisational value. In the case of satellites the type of service involved may be educational programming or national TV and radio being transmitted through the satellite.

However, in practice where satellites have been applied in the Developing world their use has largely not been of an informational value. Instead they have been used for telephony services and distribution of national TV and radio channels to conventional terrestrial transmitters. Put another way, they have provided a cost-effective alternative to conventional land lines or microwave transmitters but have not been used to develop services which otherwise would not exist. Their use is associated and interdependent with conventional communications technologies. Part of the problem is that developing countries have had to purchase off the shelf satellites designed to meet the needs of the dominant US market. There they are used almost exclusively for services whose benefits can only reasonably be measured in terms of commercial terms. The only attempt so far to develop a satellite optimised for the needs of a developing country has been India's Insat system. This is an expensive option which also has the additional risk of using an untested design. The Indians had difficulty in the project management of design and construction of the sophisticated Insat which led to the loss of the first satellite after launch.

Yet there have been many proponents of the idea that the primary value of satellites to third world countries lies in their informational services. As far back as 1968, the Agency for International Development stated that satellites would "be able to reform and modernise education systems more rapidly than would otherwise be possible, give more people access to education and training, place the best teachers within easy reach of large audiences, help to integrate large sections of the population into the social, economic and cultural life of the nation or region, and contribute to international understanding". The earliest practicle demonstrations of such services was the Peacesat project which used NASA's ATS–1 spacecraft. This involved university level education in the Pacific Basin under a programme called Pan Pacific Education and Communications Experiments by Satellite. It was never developed into an operational system.

A much more ambitious project was the Satellite Instructional Television Experiment (SITE) which used NASA's ATS–6 satellite. Television programming was beamed to 2400 villages in India which were outside the very limited reach of the largely urban based terrestrial TV transmitters. The satellite programming was largely development oriented and covered such topics as agricultural training, family planning and adult and childrens' education. The experiment lasted for 12 months and eventually led to the purchase of the Insat satellites. However it was not followed up on an operational basis by other countries. The Peacesat and SITE experiments were, however, followed up by 23 further such experiments for telecommunications to rural areas via satellite of which only one survives. This is that for Alaska, hardly a Developing country.

The US Agency for International Development began in 1979 to look at the issue and to date its conclusion is that there may be an explosion of interest in less developed nations for such services but that a step-by-step evolutionary approach is needed to develop them. It has also concluded that once a reliable communications system is in place, new users and uses will rapidly emerge. For example, in a demonstration project in the Caribbean region, the University of the West Indies is "continually getting requests for use of satellite capacity from institutions that have never used telecommunications before. These institutions are involved in areas such as agricultural training, health services and family planning."

Barriers to Entry for Third World Satellite Communications: The Agency for International Development has provided a significant insight into why developing countries have been unable to take advantage of satellite technologies. It suggests that telecommunications and economic development cultures in developing countries have little interaction with each other. Both are wary of incorporating the other's considerations into their traditional decision making and economic calculations. Telecoms ministries fear the potential economic drain of specialised services to other governmental agencies which might not want to pay. A successful pilot project could spell disaster because it could create a political demand for expansion which the PTT could not meet. Social development ministries, on the other hand, often worry that telecommunications costs will run out of control and that they may become dependent on a technology which they do not control.

For satellite communications, the problem is writ large. Joe Pelton of Intelsat quoted (in *Satellite World*, August 1985) the cost of a dedicated satellite system to be around \$235–365 m and ground terminals for the system to be \$11–31 m. These are essentially up front costs. The money has to be financed before the service starts operation. It also requires a degree of long term planning which the evidence of the last decade suggests that no Western or third world government can guarantee to stick to. Moreover, the commitment to investing in a satellite system leaves a government with little flexibility. A partially completed system has virtually no resale value. The US satellite may offer to take a partially completed satellite off the customer's hands for 10% of the expenditure incurred but will certainly not take it off the customer's hands at cost. Once funds have been committed in terms of progress payments, it would pay the customer to pay for final completion of the system. The problem is illustrated by Mexico's Morelos system which was ordered before the country's economic problems became apparent in 1983. The Mexicans had to store the second satellite in-orbit because it found that it could not afford the foreign exchange to purchase the ground equipment for the system.

The Maitland Commission recognised that there were major problems in financing telecommunications infrastructure but appears to have been silent on analysing the risks between indivisable and divisable high cost exercises. Put another way, most countries tend to operate, politically, on a day-to-day basis with, at best, a time horizon extending to a maximum of two years. To ask a poor country to commit to spending \$300 m or more on a satellite project with no opportunity to either revise its expenditure or pull out of the exercise appears to be politically untenable.

To do so is asking for a level of political competence that is largely absent from the developed world, let alone poor countries in the hands of regimes that are both dictatorships and corrupt. Many third world one party states have served their citizens poorly. The lack of a political infrastructure to reflect the inherent contradictions and conflicts of interest that exist in any society means that the political establishment distorts the allocation of resources towards its own preservation. Typically, a developing country one party state will divert resources towards its close-at-hand urban population and the similarly threatening military. In practice it means that the cost of foodstuffs to urban purchasers will be low, the exchange rate grossly overvalued (so that imports of luxury consumer goods and military hardware are cheap) and as a result the rural hinterland, producing and exporting agricultural products, becomes impoverished and reduces output. The overall economic performance of the country then deteriorates and it suffers from a shortage of foreign currency. Such countries have no inherent political reason to divert investment towards their rural hinterlands.

Fortunately, there are exceptions to this model in the developing world. India's success with its Insat system may be related to its democracy even though there is an element of corruption and inefficiency within its political institutions. Dependent on the political support of an overwhelmingly rural population, the Indian government has to pay attention to the interests of that group. Consequentially agricultural prices in India have been sufficient to encourage the sector to increase output and allow rural incomes to double over the last twenty-five years. It also has a major advantage when it comes to rural use of satellites. For political reasons it pays the central government to invest in the rural communications infrastructure because it is dependent on rural political support. But it also allows the rural population to support the investment involved without total dependence on central government. For example, it provides a rural community with an increasing real income to purchase, operate and maintain a TVRO terminal for reception of educational programming as well as pay for the on-the-spot teachers and classrooms needed to make full use of the service.

But cost remains a major deterrent to adoption of satellite communications by developing countries. The problem of the indivisibility of cost of the space segment has already been mentioned with its consequential risks and inflexibility. However, it has been estimated that for every \$1 invested in the space segment another \$2 is required in the ground segment. Although the ground segment has a degree of divisibility, it still represents a major cost barrier for developing countries. From a foreign exchange viewpoint, ground equipment is capable of being manufactured locally but certain components, such as LNA's, need to be imported. The large 15–30 metre terminals are also almost entirely imported from the USA, Europe or Japan. The experience of developing countries is that if ground equipment is manufactured locally, there is a significant cost penalty over imported terminals. Such countries as Brazil, India and Indonesia which have built up their own domestic ground equipment manufacturing industries have largely been unable to get unit costs down to compete with Japanese, US or European systems. The technical barriers to entry are related to the observation that satellites have been developed in the West for use by the West and Developing countries have been forced to take off-the-shelf hardware. Until recently, the Western satellite industry has tended to concentrate on keeping down the cost of the space segment in an overall communications system. There is an engineering trade-off between the cost of the space segment and the cost of the ground segment.

Because the emphasis in the West has been on providing trunk capacity for telecommunications carriers, it has paid them to build satellites which use large 15–30 metre earth terminals connected by terrestrial means to the end user premises. Developing countries often lack the need for trunk capacity and the local terrestrial communications infrastructure to link the large terminals with end user premises. Their need is, instead, to provide thin route communications with ground equipment geared up to low volumes of traffic and much less dependent on the terrestrial network. The engineering trade off therefore places greater emphasis on the cost of the ground segment than on the cost of the space segment. The position is closer to recent developments in the US domestic satellite communications field which places emphasis on small terminals located on the customer premises and providing a possible complete by-pass of the terrestrial network where needed. Such systems are inherently geared up to thin route traffic.

Moreover, the development of medium power satellites for these services allows the development of quasi-DBS services using TVRO dishes as small as 90 cm diameter. This would be highly suitable for developing countries, allowing them to provide educational services on a telecommunications satellite without resort to untested and specialised DBS systems. On the other hand, it indicates the unsuitability of the Intelsat system for provision of domestic communications. The latter remains a system primarily intended for trunk route international telephony traffic using 15–30 metre dishes linked to local terrestrial networks.

The Benefits of Intelsat to Developing Countries: It would be quite wrong to suggest that Intelsat cannot meet at least some of the needs of the developing world. On the contrary, there is no dispute that its international services are important and that, under certain circumstances, its capacity can provide a useful and flexible entry into domestic satellite communications. Yet, there remain nagging doubts about many of the claims that Intelsat makes about how beneficial it is to developing countries and how far it can meet the telecommunications needs of those countries.

Cross Subsidies to Developing Countries: It has been argued that the Intelsat system transfers telecommunications resources from the developed to the developing world by cross-subsidies inherent in its pricing structure. The problem can be traced back to the earliest days of the organisations when it was created as a product of US foreign policy. At that time the USA had a virtual monopoly on satellite communications technology outside the USSR. There was an obvious and growing demand for international telephony capacity and a secondary market for television programming distribution. The USA turned this to its advantage by creating the international organisation with a world wide monopoly and with the requirement that it provide global and non-discriminatory inter-connectivity. Implicit in this was an assumption that satellite communications were a natural monopoly, as were, it was then believed, national telecommunications networks.

It is an inevitable feature of satellite technology that the cost power telephony circuit is proportional to the number of circuits uplinked from a ground station. Thus heavy routes could provide a cost per circuit lower than on a thin route. But because Intelsat is required to provide both global interconnectivity (that is, any signatory who wants to use the system can as long as they provide the earth terminals in their own country) and to charge on a non-discriminatory basis (that is, the charge per circuit on a thin route is the same as a charge on a heavy route), there is an inevitable cross subsidy from heavy to thin routes. From this argument it has been assumed that there is therefore a cross subsidy from the Developed world, using the heavy routes, to the developing world using thin routes only. This argument has been disputed. In 1983 Orion Satellite Corporation, proposing its private international satellite system, submitted evidence to both the US government and Intelsat parties that no firm conclusion could be made as to whether those cross subsidies existed in practice.

According to Orion "public statements by those involved in the pricing of Intelsat thin route services in the late 1970's indicate that, at least at that time, all space segment prices were cost based." In a 1983 study of the subsidy issue prepared by US Telecommunications consultants Dale N. Hatfield and Associates, it was concluded that, based on publicly available information from Intelsat and Comsat, the claims Intelsat has made regarding inter- and intra-regional subsidies cannot be substantiated. (It also concluded that, to the extent that inter-regional subsidies exist, they appear to flow from the Pacific to the Indian and Atlantic Ocean regions.) Intelsat has claimed that the increases in charges to third world countries resulting from the activities of two of the new private venture satellite systems (Orion and ISI), would be in the order of one-third. The implication, if correct, is that there are substantial cross subsidies.

The Intelsat Organisation has introduced a wide range of new services and charges since 1983 date and has come under considerable criticism that it is using the new offerings to cross-subsidise to prevent new international satellite operators from entering the market. Panamsat has been particularly vocal in this field, arguing that modifications to the Intelsat V F13 and F15 satellites are not being charged to the services to which the modifications are addressed. Similarly PanAmSat has argued that the offer of outright sale of transponders on the Intelsat spacecraft is a method of dumping surplus capacity on the market.

Headquarters of Intelsat, International Telecommunications Satellite Organisation, 3400 International Drive, N.W. Washington, D.C. 20008

There are good arguments that Intelsat should not cross-subsidise. By doing so it effectively becomes a vehicle of foreign aid between the West and the Developing world and consequentially threatens its own position politically. If the West wants to provide aid, so the argument goes, it should do so through formal channels designed for the process rather than through a back-door method which places it where it may not be best spent. Moreover, if it is correct that cross-subsidies to developing countries exist, then there is an implicit argument that it is uneconomic to provide Intelsat services to some countries, at least on that current scale of operations. It means that Intelsat is under pressure to invest in those services that provide the cross-subsidies, i.e. services for the Western World, rather than those addressing the needs of developing countries. Although Intelsat argues that it is a non-profit making organisation, it is required to make a return of 14% on its members investment in the organisation. It therefore has to make some substantial profits somewhere from its system to cross-subsidise.

Intelsat Technology and the Needs of Developing Countries: Intelsat's suitability for developing countries' telecommunications needs is closely linked to the specific technical performance of its satellites. Intelsat suffers from being an early entrant into a market. The technologies of the 1960's and early 1970's meant that it utilised C-band frequencies in conjunction with large beams and low power travelling wave tube amplifiers. This meant that user PTT's had to invest heavily in large 30 metre ground stations. The 1970's and 1980's saw the introduction of new developments in satellite technology which were superior for new satellite systems. These included Ku-band frequency use, shaped and spot beams and higher power travelling wave tube amplifiers. Intelsat has been able to incorporate much of this in its space segment but also has to continue in providing services geared up to the substantial early investment by PTT's in ground equipment.

The technical developments in satellite platforms and payloads since 1975 have not only reduced the overall cost of a satellite circuit but, to a much greater extent, have reduced the overall cost per circuit because of much greater reductions in the costs and performance requirements of ground equipment. It also means that satellite technology has become much more cost effective in meeting the needs of thin route communications. It may well be the case that, if Intelsat is cross-subsidising thin route traffic, it is doing so because the technology it is using on such services is more expensive than need be.

In the early 1980's the European Space Agency and Interim Eutelsat suggested that Intelsat, as a single global system, may have reached its saturation point. ESA suggested that the latest generation of Intelsat spacecraft tended to show that the maximum economy of scale has probably been reached. It added that, in the future, it may not be the best approach to continue procuring successive series of identical spacecraft to satisfy all needs and that it may be more economic to start diversifying the type of satellite needed for various regions and types of services.

Intelsat Services suitable for Third World use: The pressures on the Intelsat organisation in the first half of the 1980's, which largely resulted from plans for competitive private international satellite systems and a slowing down in the rate of growth of international communications, has led to a substantial increase in the number and terms of services on offer by the organisation, although, so far, there has been no announcement of plans to offer radically different satellites.

For 10 years the organisation has been offering its members transponder leases for domestic satellite communications. Until recently these lease terms were restrictive, being preemptible and only available for fixed terms. Brazil used them to develop its domestic satellite communications system but argued that they did not allow the country to plan the services properly. It opted for its own domestic satellites, the SBTS spacecraft, as a consequence. Nevertheless, both the old and the new terms allow developing countries to enter the satellite communications scene without the need to resort to an inflexible and costly dedicated space segment. They also allow the gradual build-up of a satellite ground infrastructure before the commitment is made to go to a dedicated domestic system. In addition to Brazil, India and Mexico have taken this route. Intelsat also argues that the domestic leases can provide emergency back up in the event of failure of a domestic satellite.

From 1984, when it was already apparent that Intelsat was likely to have a lot of surplus capacity, it began to offer both a wide variety of new lease terms and, from 1986, outright sale of transponders for domestic use. The new terms included non-preemptible conditions and a range of lease periods. They have proved to be mildly popular amongst developing countries. Gabon, for example, has used them to develop an internal system for TV and telephony traffic.

The organisation has also been developing specialised services aimed at or useful to developing countries. Again, the experience so far has shown limited results. The Vista service was introduced in 1983 and was specifically aimed at developing countries requiring links of between two and four voice circuits for rural and remote communities. Its effectiveness depends on the availability of small, low cost earth stations of simple design and low power consumption. But even in bulk quantities the terminals cost between $40,000 and $50,000 each. In contrast a two way VSAT terminal for use with a US domestic satellite may cost around $20,000. The first customer for Vista did not appear until mid-1985.

The Intelsat Intelnet service uses much cheaper terminals which may eventually drop below $1000 a unit. However, the service is one-way only from one or more hub earth stations and is intended primarily for data distribution. There is some doubt as to the value of such services in developing countries.
A more long-term project in the Intelsat Organisation is "SHARE", developed in conjunction with the International Institute for communications in London. This is a "loss-leader" offering from Intelsat in which spare capacity is offered free of charge to a member country to conduct demonstrations in the fields of distance-education and health care for rural and remote areas. The participating countries have to pay for their ground equipment as well as the programming going through the Intelsat system. The SHARE demonstrations are all short term, running for periods of several months.

Future generation Intelsat spacecraft: At the time of writing, it appears that the Intelsat organisation may abandon the concept of purchasing a series of standardised and identical satellites. Ex-Director General Richard Colino, prior to his dismissal from the organisation in December 1986, said that the organisation was aiming for dedicated satellites designed for different geographical areas. "They may be bigger or smaller than the Intelsat VI series. We'll probably see some satellites designed for application primarily in the Atlantic Ocean region, for example, which will have an emphasis on video." Decisions to purchase such specialised satellites were not expected until 1988 but subsequently it was revealed that Intelsat had taken an option on the SBS–6 satellite under construction and owned by IBM. It is not clear if the Board of Governors or members would agree to this purchase at a time when the organisation has massive over-capacity. It is due for launch in 1989 and has a Ku-band payload based on a Hughes HS–393 platform. Intelsat has not revealed which geographical area the satellite would serve, although outside observers have suggested the Pacific Ocean Region where demand is growing fastest. Intelsat's option on SBS–6 was subsequently dropped.

SUMMARY OF INTELSAT IV–VI LAUNCHES

Intelsat IV–F1	22nd May 1975	Atlas-Centaur	50° West	
Intelsat IV–F2		Atlas-Centaur	Retired	
Intelsat IV–F3	26th June 1971	Atlas-Centaur	Retired	
Intelsat IV–F4	19th December 1971	Atlas-Centaur	Retired	
Intelsat IV–F5	22nd January 1972	Atlas-Centaur	Retired	
Intelsat IV–F6	13th June 1972	Atlas-Centaur	Retired	
Intelsat IV–F7	24th August 1973	Atlas-Centaur	Retired	
Intelsat IV–F8	21st November 1974	Atlas-Centaur	50° West	
Intelsat IVA–F1	25th September 1975	Atlas-Centaur	31° West	
Intelsat IVA–F2	29th January 1976	Atlas-Centaur	2° East	
Intelsat IVA–F3	6th January 1978	Atlas-Centaur	179° East	
Intelsat IVA–F4	26th May 1977	Atlas-Centaur	21·5° West	
Intelsat IVA–F5	1978	Atlas-Centaur	Not available	Launch failure
Intelsat IVA–F6	31st March 1978	Atlas-Centaur	174° East	
Intelsat V–F1	23rd May 1981	Atlas-Centaur	57° East	
Intelsat V–F2	6th December 1980	Atlas-Centaur	1° West	
Intelsat V–F3	16th December 1981	Atlas-Centaur	53° West	
Intelsat V–F4	5th March 1982	Atlas-Centaur	34·5° West	
Intelsat V–F5 MCS	28th September 1982	Atlas-Centaur	63° East	Carries Inmarsat payload
Intelsat V–F6 MCS	19th May 1983	Atlas-Centaur	18·5° West	Carries Inmarsat payload
Intelsat V–F7 MCS	19th October 1983	Ariane 1 V7	60° East	Carries Inmarsat payload
Intelsat V–F8 MCS	5th March 1984	Ariane 1 V8	53° West	Carries Inmarsat payload
Intelsat V–F9 MCS	9th June 1984	Atlas-Centaur	Not available	Launch failure (insured for $102 m)
Intelsat VA–F10	22nd March 1985	Atlas-Centaur	24·5° West	
Intelsat VA–F11	29th June 1985	Atlas-Centaur F64	27·5° West	
Intelsat VA–F12	28th September 1985	Ariane 2	43·5° East	
Intelsat VB–F13	August 1987	Ariane 2 V23		
Intelsat VA–F14	29th May 1986	Ariane 2 V18	Not available	Launch failure
Intelsat VB–F15	March 1988	Ariane 2 V28		
Intelsat VI–F1	January 1989	Ariane 4 V34		
Intelsat VI–F2	June 1989	Ariane 4 V38		
Intelsat VI–F3	November 1989	Ariane 4 V42		
Intelsat VI–F4	To be announced	Ariane 4 (reservation)		
Intelsat VI–F5	To be announced	Ariane 4 (reservation)		

INTELSAT TRANSPONDERS SOLD TO MEMBERS

DELIVERY DATE	COUNTRY	NUMBER OF TRANSPONDERS	TRANSPONDER CAPACITY	FREQUENCY	SATELLITE
30th January 1986	West Germany	2	72 MHz	Ku-band	60° East
30th January 1986	West Germany	1	241 MHz	Ku-band	60° East
1st February 1986	Israel	2	72 MHz	Ku-band	1° West
1st February 1986	Israel	1	241 MHz	Ku-band	1° West
14th February 1986	Norway	2	72 MHz	Ku-band	1° West
14th February 1986	Norway	1	241 MHz	Ku-band	1° West
30th June 1986	China	1	72 MHz	C-band	66° East
10th August 1986	Turkey	1	72 MHz	Ku-band	66° East
1st October 1986	Gabon	1	72 MHz	C-band	1° West
1st October 1986	Italy	1	72 MHz	C-band	1° West
1st November 1986	Iran	1	72 MHz	Ku-band	63° East
1st November 1986	West Germany	2	72 MHz	Ku-band	60° East
1st November 1986	West Germany	1	241 MHz	Ku-band	60° East
15th December 1986	Niger	1	72 MHz	Ku-band	1° West
15th December 1986	Norway/Gabon	1	n/a	Ku-band	1° West
31st December 1986	China	1	72 MHz	C-band	66° East
1st February 1987	Italy	2	n/a	Ku-band	18·5° West
1st February 1987	Italy	1	n/a	Ku-band	34·5 West
1st April 1987	Turkey	1	241 MHz	Ku-band	66° East
1st July 1987	Iran	1	72 MHz	Ku-band	63° East
1st August 1987	Iran	3	n/a	Ku-band	66° East
1987	Portugal	1	72 MHz	Ku-band	53° West
Total number of transponders sold		29			

Source: Via Intelsat, Volume 1, Number 2, Winter 1986, published by Intelsat and Intelsat press office, December 1986.

NORTH ATLANTIC DEMAND vs. SUPPLY

(WITHOUT TASI AND OTHER CIRCUIT MULTIPLIER TECHNIQUES)

FOR THE YEAR 1995

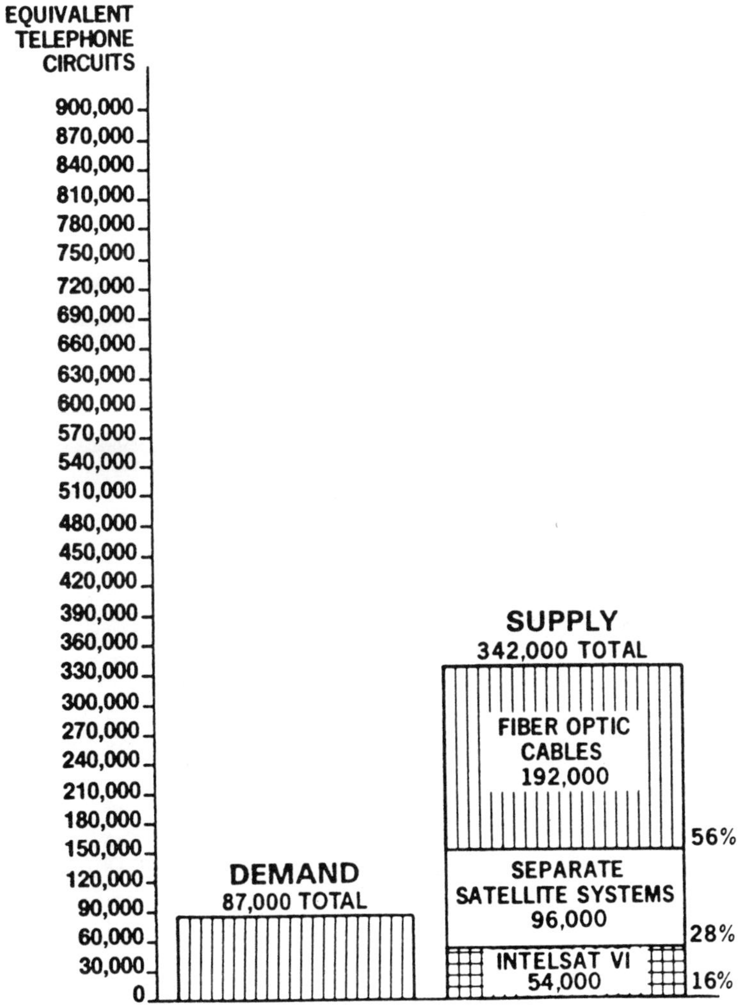

FCC AUTHORIZED FIBER OPTIC CABLES

TAT-8	12,000[2]
TEL-OPTIK 1	12,000
TEL-OPTIK 2	12,000
SLC-TAV 1	144,000[3]
TAT-9 (NO FILING YET) BUT ASSUMING AT LEAST TAT-8's CAPACITY	12,000
TOTAL	192,000

SEPARATE SATELLITE SYSTEMS NORTH ATLANTIC

(PENDING APPLICATIONS)

ORION, CYGNUS, ISI, RCA[4], AND FINANSAT PROPOSE 128 TRANSPONDERS, WITH TOTAL CAPACITY OF 96,000 CIRCUITS

INTELSAT VI SERIES

3 SATELLITES UNDER CONSTRUCTION, WITH TOTAL CAPACITY OF 120,000 CIRCUITS 45% OF WHICH WILL BE USED FOR NORTH ATLANTIC 54,000

SUPPLY OVER DEMAND 4:1

1. NTIA'S ESTIMATE. THE U.S. NORTH ATLANTIC CONSULTATIVE WORKING GROUP'S FIGURE IS 81,888 (ORLANDO, FLA. JAN. 1985).
2. BASIC DESIGNED CAPACITY. WITH TASI, ETC., IT WILL RUN BETWEEN 30,000 AND 40,000.
3. PER ITS FILING, SLCC'S CABLE WILL HAVE 12 TIMES CAPACITY OF TAT-8. HENCE 144,000.
4. SUBSEQUENT TO THE PREPARATION OF THESE MATERIALS, RCA WITHDREW THEIR APPLICATION EFFECTIVE NOVEMBER, 1985.

NORTH ATLANTIC CUSTOMIZED SERVICES

(WITH TASI AND OTHER CIRCUIT MULTIPLIER TECHNIQUES)

FOR THE YEAR 1995

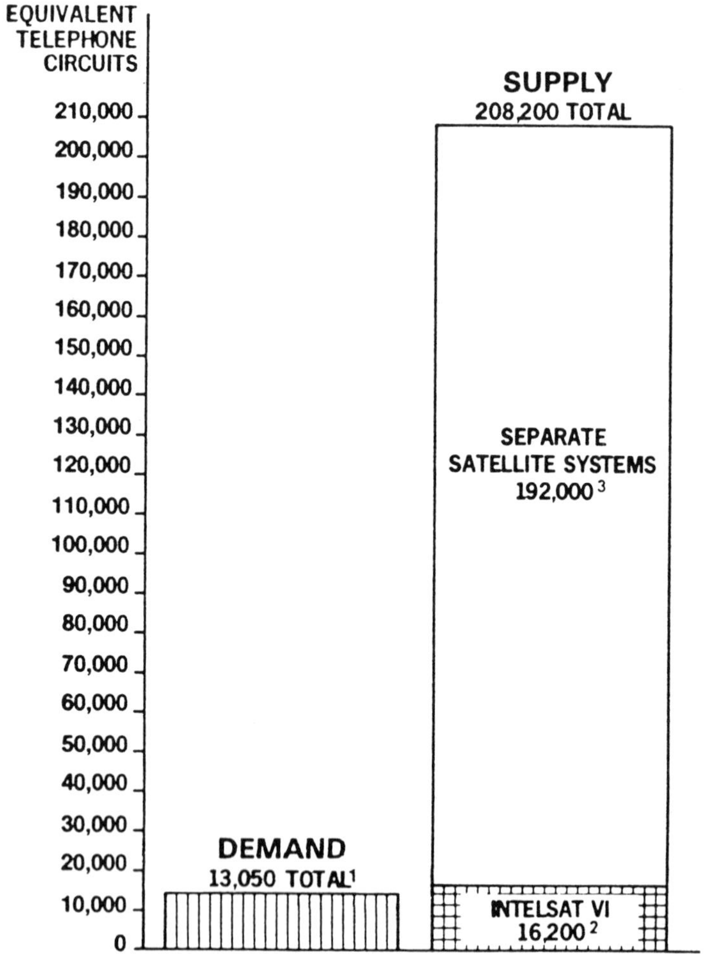

1. BASED ON SIG WHITE PAPER ASSUMPTION THAT 15% OF COMMUNICATIONS TRAFFIC IS "CUSTOMIZED SERVICES." TOTAL NORTH ATLANTIC DEMAND IS ESTIMATED AT 87,000 CIRCUITS BY 1995. 15% IS 13,050 CIRCUITS.

2. 45% OF INTELSAT VI CAPACITY (54,000 CIRCUITS) WILL BE USED FOR NORTH ATLANTIC TRAFFIC WITH TASI, ETC., 15% OF 108,000 = 16,200 CIRCUITS.

3. THE SEPARATE SATELLITE SYSTEMS PROPOSED FOR THE NORTH ATLANTIC HAVE 128 TRANSPONDERS PROVIDING 96,000 CIRCUITS. TASI AND OTHER CIRCUIT MULTIPLIER TECHNIQUES WILL ROUGHLY DOUBLE THIS CAPACITY TO 192,000 CIRCUITS. ALL OF THIS CAPACITY IS PLANNED FOR "CUSTOMIZED SERVICES."

4. THIS EXCESSIVE IMBALANCE IS WITHOUT TAKING INTO ACCOUNT THE CAPACITY OF THE FIBER OPTIC CABLE OPERATORS WHO WILL BE COMPETING FOR THE SAME "CUSTOMIZED SERVICES" BUSINESS. SUBMARINE LIGHTWAVE CABLE, FOR EXAMPLE, IS DESIGNING ITS FACILITY ESPECIALLY FOR VIDEO CUSTOMERS.

SUPPLY OVER DEMAND 16:1[4]

(Source: Intelsat, 1985)

TECHNICAL SUMMARY

	Intelsat I (Early Bird)	*Intelsat II*	*Intelsat III*	*Intelsat IV*	*Intelsat IVA*
PLATFORM	Intelsat I	Intelsat II	Intelsat III	Intelsat IV	Intelsat IVA
PRIME CONTRACTOR	Hughes Aircraft Company	Hughes Aircraft Company	TRW	Hughes Aircraft Company	Hughes Aircraft Company
OTHER MAJOR CONTRACTORS					
STABILISATION	Spin	Spin	Spin (despun antenna)	Spin	Spin
LAUNCH MASS				1418 kg	1516 kg
MASS IN ORBIT	38·6 kg (BOL)	86·4 kg (BOL)	152 kg (BOL)	732 kg (BOL)	863 kg (BOL)
POWER SUPPLY, END OF LIFE	40 Watts	75 Watts	134 Watts	460 Watts	600 Watts
DESIGN LIFE	18 months	3 years	5 years	7 years	7 years
OPERATIONAL LIFE (ACTUAL)	3·5 years	3–5 years	5–7 years	7–10 years	7–10 years
NO. OF SATELLITES IN SERIES	2	5	8	8	6
LAUNCH VEHICLE	Thor Delta	Thor Delata	Thor Delta	Atlas-Centaur	Atlas-Centaur
FIRST LAUNCH	1965	1966	1968	1971	1975
ORBITAL POSITIONS	See launch summary	See launch summary	See launch summary	See launch summary	See launch summary
STATION KEEPING ACCURACY					
ECLIPSE PROTECTION					
SPACECRAFT DIMENSIONS	0·59 m long, 0·72 m diameter	0·67 m long, 1·42 m diameter	1·04 m long, 1·42 m diameter	2·81 m long, (5·26 m extended) 2·38 m diameter	2·81 m long, (5·26 m extended) 2·38 diameter
ACTIVE TRANSPONDERS	2 (C-band)	5 (C-band)	8 (C-band)	12 (C-band)	20 (C-band)
SPARE TRANSPONDERS					
TWTA POWER OUTPUT					
UPLINK FREQUENCIES				C-band only	C-band only
DOWNLINK FREQUENCIES				C-band only	C-band only
TRANSPONDER BANDWIDTH	25 MHz			36 MHz	36 MHz
EIRP, MINIMUM, BEAM EDGE					
BEAM COVERAGE C-BAND	Toroidal Northern only	Toroidal almost full earth	Despun earth cover	2 global downlink beams 2 steerable spot beams	2 global downlink beams 2 steerable spot beams
BEAM COVERAGE Ku-BAND	Not available	Not available	Not available	Not available	Not available
BEAM COVERAGE L-BAND	Not available	Not available	Not available	Not available	Not available
POLARISATION					
RECEIVE COVERAGE (G/T)					
SINGLE CARRIER SATURATION FLUX DENSITY					

TECHNICAL SUMMARY

	Intelsat V	*Intelsat VA*	*Intelsat VB*	*Intelsat VI*
PLATFORM	Intelsat V	Intelsat V	Intelsat V	Intelsat VI
PRIME CONTRACTOR	Ford Aerospace	Ford Aerospace	Ford Aerospace	Hughes Aircraft Company
OTHER MAJOR CONTRACTORS	Aerospatiale, Marconi, MBB, Mitsubishi Electric, Selenia Spazio, Thomson-CSF	Aerospatiale, Marconi, MBB, Mitsubishi Electric, Selenia Spazio, Thomson-CSF	Aerospatiale, Marconi, MBB, Mitsubishi Electric, Selenia Spazio, Thomson-CSF	Spar Aerospace, Thomson-CSF, MBB, British Aerospace, Selenia Spazio, NEC, Comdev, AEG-Telefunken
STABILISATION	3-axis	3-axis	3-axis	Spin
LAUNCH MASS	1870 kg	2016 kg	2141 kg	
MASS IN ORBIT	1072 kg (BOL)	1160 kg (BOL)	1160 kg (BOL)	2004 kg
POWER SUPPLY, END OF LIFE	1288 Watts	1270 Watts	1270 Watts	2200 Watts
DESIGN LIFE	9 years	9 years	9 years	10 years
OPERATIONAL LIFE (ACTUAL)	10 years	Not available	Not available	Not available
NO. OF SATELLITES IN SERIES	9	3	2	5
LAUNCH VEHICLE	Atlas-Centaur & Ariane	Atlas-Centaur	Atlas-Centaur	Ariane
FIRST LAUNCH	1980	1985	1986	1988
ORBITAL POSITIONS	See launch summary	See launch summary	See launch summary	See launch summary
STATION KEEPING ACCURACY				
ECLIPSE PROTECTION				
SPACECRAFT DIMENSIONS	6·44 m long, (15·85 m extended)	6·44 m high at launch 15·85 m long deployed	6·44 m high at launch 15·85 m long deployed	11·8 m high deployed 3·6 m diameter
ACTIVE TRANSPONDERS	21 (C-band), 6 (Ku-band) 2 (L-band MCS only)	32 (C-band), 6 (Ku-band)	32 (C-band), 6 (Ku-band)	36 (C-band), 10 (Ku-band)
SPARE TRANSPONDERS	12 (C-band), 4 (Ku-band)	12 (C-band), 4 (Ku-band)	12 (C-band), 4 (Ku-band)	12 (C-band), 4 (Ku-band)
TWTA POWER OUTPUT	4·5 & 8·5 Watts (C-band) 10 Watts (Ku-band)	4·5 & 8·5 Watts (C-band) 10 Watts (Ku-band)	4·5 & 8·5 Watts (C-band) 10 Watts (Ku-band)	5·5–16 Watts (C-band) 8·5–10 Watts (Ku-band)
UPLINK FREQUENCIES	C-band and Ku-band			
DOWNLINK FREQUENCIES	C-band and Ku-band 77 MHz			
TRANSPONDER BANDWIDTH	C-band: Spot, West Hemi, Zone 72 MHz, East Hemi, global 36 MHz. Ku-band: 77 MHz	As per Intelsat V	As per Intelsat V	
EIRP, MINIMUM, BEAM EDGE	C-band: Spot 35·5, West & Zone 29, East 26, Global 23. Ku: West 44·4, East 41·1	As per Intelsat V	As per Intelsat V	
BEAM COVERAGE C-BAND	2 global downlink beams 2 hemispheric beams 2 zone beams	2 hemispheric beams 2 steerable spot beams 2 zone beams	2 hemispheric beams 2 steerable spot beams 2 zone beams	2 hemispheric beams 4 zone beams
BEAM COVERAGE Ku-BAND	2 steerable spot beams	2 steerable spot beams	2 steerable spot beams	2 steerable spot beams
BEAM COVERAGE L-BAND	MCS earth coverage	Not available	Not available	Not available
POLARISATION		dual	dual	dual
RECEIVE COVERAGE (G/T)	C-Band: Spot & Global −16, West & East −9, Zone −6. Ku: West spot 3·3, East 0	As per Intelsat V	As per Intelsat V	
SINGLE CARRIER SATURATION FLUX DENSITY				

SUMMARY AND ANALYSIS OF INTELSAT MEMBERSHIP AND USE

Country	*Signatory*	*Investment Share*
Africa		
Algeria	Government	000·362102%
Angola	Empresa Publica de Telecomunicacoes (EPTEL)	000·124739%
Burkina Faso	Office des Postes et Telecommunications	000·050000%
Cameroon	Intelcam	000·242399%
Central African Republic	Government	000·050000%
Chad	TIT	000·050000%
Congo	Government	000·050000%
Egypt	Government	000·690000%
Ethiopia	Telecommunications Service of the Government	000·089209%
Gabon	TIG	000·050000%
Ghana	Ministry of Transport & Communications	000·050000%
Ivory Coast	Government	000·264921%
Kenya	Kenya Posts & Telecommunications Corporation	000·374390%
Libya	Government	000·149600%
Madagascar	STIMAD	000·050000%
Malawi	Department of Posts & Telecommunications	000·050000%
Mali	TIM	000·086374%
Mauritania	Government	000·050000%
Morocco	Government	000·257424%
Niger	Government	000·050000%
Nigeria	Nigerian External Telecommunications	001·230844%
Senegal	Government	000·050000%
Somalia	Ministry of Posts & Telecommunications	000·050000%
South Africa	Department of Posts & Telecommunications	001·271510%
Sudan	Government	000·288474%
Tanzania	Tanzanian Posts & Telecommunications Corporation	000·098839%
Tunisia	PTT Administration	000·050000%
Uganda	Ministry of Power, Posts & Telecommunications	000·050000%
Zaire	Office National des Postes & Telecommunications	000·286292%
Zambia	Government	000·098744%
Middle East & Asia		
Bangladesh	Telegraph & Telephone Board	000·104614%
India	Overseas Communications Services, Ministry of Communications	001·060000%
Indonesia	Government	000·498731%
Iran	Telecommunications Company of Iran	000·937963%
Iraq	Government	000·405501%
Israel	The Israel Telecommunication Corporation	000·557462%
Jordan	Government	000·298781%
Kuwait	Ministry of Communications	000·986464%
Lebanon	Government	000·086501%
Malaysia	Telecommunications Department	000·682881%
Oman	Sultanate of Oman	000·509837%
Pakistan	Government	000·393000%
Quatar	Government	000·324020%
Saudi Arabia	Government	003·106999%
Singapore	Telecommunications Authority of Singapore	001·400000%
Sri Lanka	Government	000·050000%
Syria	Government	000·122415%
Thailand	Government	000·424225%
United Arab Emirates	Ministry of Communications	001·736693%
Vietnam	Direction General des Postes & Telecommunications	000·050000%
Yemen Arab Republic	Government	000·219888%
Pacific Basin		
Australia	Overseas Telecommunications Administration	003·159546%
China	Ministry of Poste & Telecommunications	000·496820%
Fiji	FINTEL	000·050000%
Japan	Kokusai Denshin Enwa Company (KDD)	003·330332%
Korea	Korean Telecommunications Authority	000·720852%
New Zealand	Postmaster General	000·402584%
Papua New Guinea	Posts & Telecommunications Corporation	000·050000%
Philippines	Philcomsat	000·644842%

SUMMARY AND ANALYSIS OF INTELSAT MEMBERSHIP AND USE—continued

Country	*Signatory*	*Investment Share*
Latin & South America		
Argentina	Empresa Nacional de Telecomunicaciones	001·152640%
Barbados	Barbados External Telecommunications	000·050000%
Bolivia	Empresa Nacional de Telecomunicaciones	000·123670%
Brazil	Empresa Brasileira de Telecomunicacoes (EmBraTel)	003·041419%
Chile	Empresa Nacional de Telecomunicaciones	000·578405%
Colombia	Empresa Nacional de Telecomunicaciones	001·100000%
Costa Rica	Instituto Costarricense de Electricidad	000·050000%
Dominican Republic	Compania Dominicana de Telefonos	000·050000%
Ecuador	Instituto Ecuatoriano de Telecomunicaciones	000·350831%
El Salvador	Administracion Nacional de Telecomunicaciones	000·050000%
Guatemala	Empresa Guatemalteca de Telecomunicaciones	000·050000%
Haiti	Telecommunications D'Haiti	000·168004%
Honduras	Empresa Hondurena de Telecomunicaciones	000·050000%
Jamaica	Jamica International Telecommunications	000·382145%
Nicaragua	Compania Nicaraguense de Telecomunicaciones por Satelite	000·050000%
Panama	Intercontinental de Comunicaciones por Satelite	000·050000%
Paraguay	Administracion Nacional de Telecomunicaciones	000·116067%
Peru	Empresa Nacional de Telecomunicaciones	000·748993%
Trinidad & Tobago	Trinidad & Tobago External Telecommunications Company	000·050000%
Uruguay	Administracion Nacional de Telecomunicaciones	000·050000%
North America		
Canada	Teleglobe Canada	002·318574%
Mexico	Government	001·818184%
USA	Communications Satellite Corporation	022·462815%
Europe		
Austria	Government	000·300000%
Belgium	Regie des Telegraphes et des Telephones	000·745004%
Cyprus	Cyprus Telecommunications Authority	000·153000%
Denmark	General Direcktoratet for Post-OG Telegrafvaesenet	000·384163%
Finland	General Direction of Posts & Telecommunications	000·092766%
France	Government	005·642236%
Germany (West)	Federal Ministry for Posts & Telecommunications	004·098193%
Greece	Hellenic Telecommunications Organisation	000·813135%
Iceland	Government	000·180000%
Ireland	Irish Telecommunications Board (Bord Telecom)	000·133182%
Italy	Telespazio	002·148906%
Liechenstein	Government	000·050000%
Luxembourg	Government	000·050000%
Monaco	Government	000·050000%
Netherlands	Government	001·128008%
Portugal	Companhia Portuguesa Radio Marconi	000·682498%
Spain	Compania Telefonica Nacional de Espana	001·962657%
Sweden	Swedish Telecommunications Administration	000·664878%
Switzerland	Direction Generale de l'Enterprise des PTT Suisses	001·234140%
Turkey	Government	000·227404%
United Kingdom	British Telecom	013·068645%
The Vatican	Government	000·050000%
Yugoslavia	Community of Yugoslav Posts, Telegraphs & Telephones	

(Source: Intelsat Annual Report 1984–85)

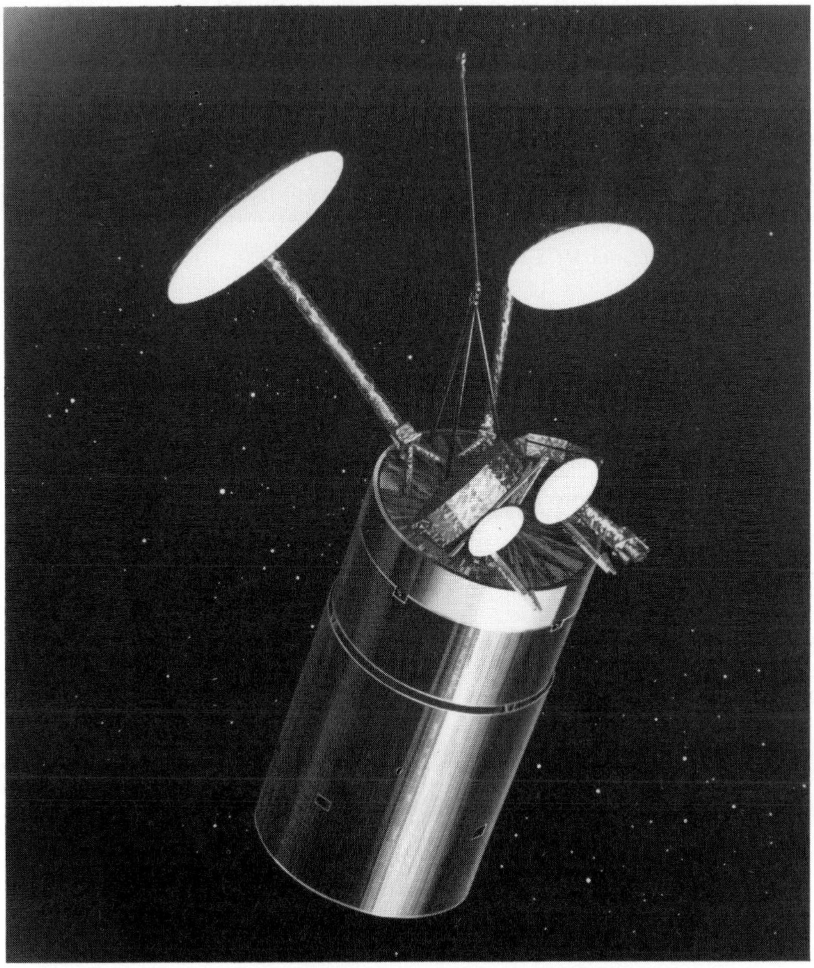

Intelsat V spacecraft with MCS payload

TECHNICAL DETAILS

	INTELSAT V	*INTELSAT VA*	*INTELSAT VI*
PRIME CONTRACTOR	Ford Aerospace	Ford Aerospace	Hughes
STABILSATION	3-axis	3-axis	Spin
LAUNCH MASS	1946 kg	1946 kg	3675 kg (Ariane), 4157 kg (STS)
MASS IN ORBIT	950 kg	950 kg	2243 kg
POWER SUPPLY (EOL)	1354 Watt	1354 Watt	2300 Watt
DESIGN LIFE	7 years	7 years	7 years
ATLANTIC CAPACITY (½-circuits)	25,000	30,000	80,000
NO. OF TRANSPONDERS	21 C-band; 6 Ku-band	32 C-band; 6 Ku-band	38 C-band; 10 Ku-band

COMMERCIAL DETAILS

OPERATOR	International Satellite Incorporated
ADDRESS	c/o TRT Telecommunications, 1331 Pennsylvania Avenue, NW Washington, DC 20004, USA
TELEPHONE NO.	(202) 296-6892
TELEX NO.	197791 trtut
SENIOR STAFF	David H. Lubestzky, President Melvin Barmat, Vice President W. L. Fishman, Regulatory Counsel Garry Lippman, Marketing Director
PURPOSE OF PROJECT	Provision of Ku-band capacity for the USA and Western Europe for international services
PROJECT STATUS	Provisional FCC authorisation given in July 1985. Construction has yet to commence
PROJECT VALUE	1983 filing with the FCC showed total cost of the three satellite systems to be $230 m, including launch and insurance and TT&C
LAUNCH DATES	Not before 1989. No launch has yet been booked

SATELLITE USERS

ISI has claimed that it anticipates that the bulk of the traffic on its system will consist of video programme distribution, teleconferencing and high speed data. It says that its system will primarily operate with small aperture earth stations on customer premises. According to its 1983 submission to the FCC, the company's intent is "to apply the inherent advantages of customer premises Ku-band earth stations to provide direct single-hop, urban location services from the Pacific Coast of the USA to the Adriatic Coast of Italy as well as to US and European points between." It proposes to offer more than half of its capacity for sale to users but some capacity would be made available to the public on a common carrier basis.

PROJECT OVERVIEW

Introduction: The ISI project is a private venture satellite system intended to provide capacity over the North Atlantic. It was the second such US venture to announce such plans and is most notable because of the ownership of the largest group of shares in the company. TRT Communications holds 43% of the voting stock in the company. TRT is a record carrier company which originated in the early part of this century and now has ambitions to become a major force in international communications.

This is not the only satellite project it is considering. It is a partner with the Papua New Guinea telecoms administration in a proposed regional satellite system, Pacstar. TRT is owned by United Brands, a large US conglomerate in the food business and other areas, which has assets of over $1000 m. UB's communications interests include cellular mobile radio, telephone resale, paging systems and applications.

Other partners in the venture are Satellite Syndicated Systems (15%) which provides common carrier distribution of video signals across the USA and the Caribbean, and Kansas City Southern Industries (14%), a diversified conglomerate with interests in transport, data processing, insurance and telecommunications. It held a DBS permit, has been active in obtaining cellular radio licences and operates an extensive private microwave network.

Background to the ISI proposal: ISI submitted its filing with the Federal Communications Commission in August 1983 and was granted conditional authority in July 1985 but no construction work has begun on the satellite. It proposed that its system would be a "North Atlantic Regional system" providing Ku-band only communications to CONUS USA and most of Western Europe. Each satellite would have two beams, one for the entire CONUS USA, the other for Western Europe. The system would allow single hop from any point in the USA to any point in Western Europe, a feature not yet available on the Intelsat system. It placed considerable emphasis on the use of low cost customer premises based earth stations which would also cut out dependence on terrestrial links. The company also proposes that telephony traffic from such discount services as SPRINT, Mercury (UK) and Execunet use the system. In that sense it is significantly different from Orion. It also differs in that it is engineering dominated (Orion was founded by lawyers and has been in the public eye primarily because of its arguments with Intelsat).

Services proposed by ISI: 1983 submission to the FCC stated that the company envisaged that the system would be used primarily for video and audio programme distribution, other video services, and high speed data. The company proposed to sell over half its capacity to users but that some part of the system would be made available to the public on a tariffed common carrier basis. The latter would allow occasional or low volume users to make use of the system. The single hop capability was said to considerably reduce the overall communications costs of customers who would otherwise rely on double hops or extensive use of landlines. The use of Ku-band capacity meant that there was likely to be no interference from terrestrial microwave communications in the USA. Thus customers could more easily locate terminals in urban US areas than would be the case with C-band communications. It also envisaged significant growth in private company telecoms networks which the new system would have encouraged.

Satellite system overview: The company is proposing the use of two in-orbit spacecraft, at 56 and 58° West and one ground spare satellite. Each of the in-orbit satellites would provide simultaneous coverage, in separate and independent beams, of the 48 contiguous US states and most of Western Europe. There will be 16 operational transponders in each beam which will be connected through a variation of services and connectivities. It was envisaged that, although many of the earth terminals would be owned by the users of the system, ISI would provide earth stations at major US cities. For Western Europe it suggested that earth stations would be owned by PTT's rather than by ISI. The spacecraft envisaged would have been in the Delta/half-Ariane class.

TECHNICAL SUMMARY

PLATFORM	not decided
PRIME CONTRACTOR	not decided
OTHER MAJOR CONTRACTORS	not decided
STABILISATION	not decided
LAUNCH MASS	not decided
MASS IN ORBIT	not decided
POWER SUPPLY, END OF LIFE	not decided
DESIGN LIFE	10 years
ORBITAL POSITIONS	USA–SAT 1, 56° West USA–SAT 2, 58° West USA–SAT 3, ground spare
STATION KEEPING ACCURACY	+/−0·05 d (E/W & N/S)
ECLIPSE PROTECTION	Full
SPACECRAFT DIMENSIONS	not decided
SPAN, TIP-TO-TIP	not decided

NO. OF ACTIVE TRANSPONDERS	34
NO. OF SPARE TRANSPONDERS	not known
TWTA POWER OUTPUT	20 W (West), 10 W (East beam)
FREQUENCY RANGE:	
UPLINK	14·000–14·500 GHz
DOWNLINK	11·700–12·200 GHz (West) 10·700–10·950 GHz (East) 12·500–12·750 GHz (East)
TRANSPONDER BANDWIDTH	54 MHz
EIRP MINIMUM, BEAM EDGE	43 dBW (West beam, EOC) 44 dBW (East beam, EOC)
CHANNEL POLARISATION	Orthogonal Linear
RECEIVE COVERAGE (G/T)	+2·0 dBK (West beam, EOC) +5·0 dBK (East beam, EOC)
SINGLE CARRIER SATURATION SATURATION FLUX DENSITY	−82 dBW/m^2 (West beam) −85 dBW/m^2 (East beam)

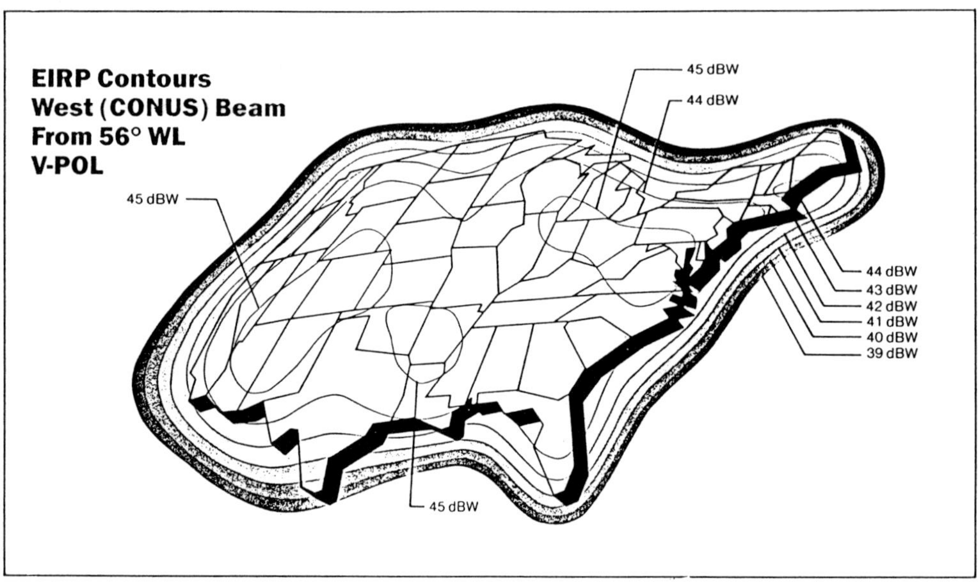

EIRP contours for both the United States and Europe

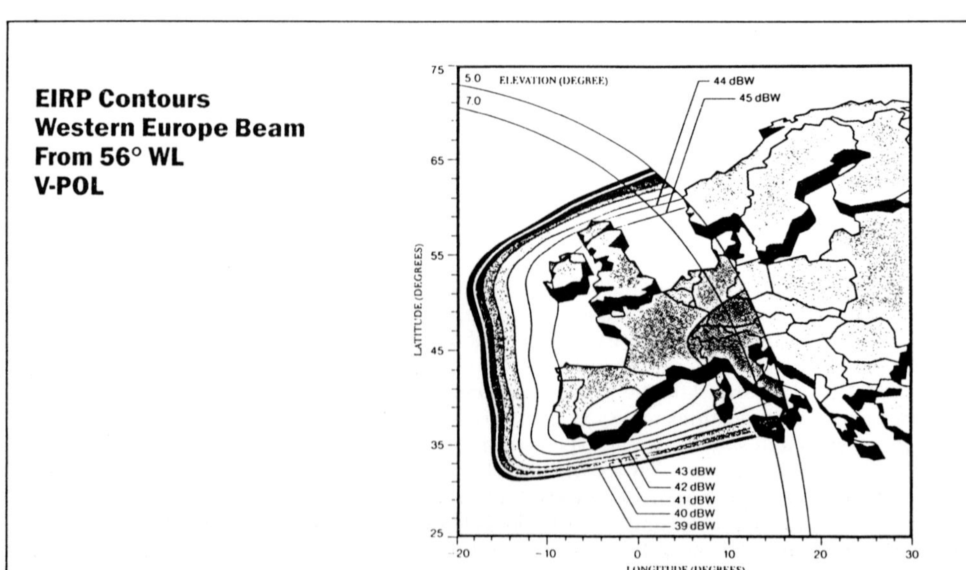

COMMERCIAL DETAILS

OPERATOR	(Expected) Telespazio on behalf of the Agenzia Spaziale Italiana
ADDRESS	00159 Roma, Via Alberto Bergamini 50, Italy
TELEPHONE NO.	49874
TELEX NO.	610654/611596
FAX NO.	4987254
SENIOR STAFF	Raffaele Minicucci, Managing Director, Telespazio Cesare Begingn, General Manager, Telespazio Giovani Di Sorte and Aldlo Calcagn, Press & Public Relations, Telespazio
FINANCIAL BACKING	Funded by the National Research Council as part of the National Space plan
PURPOSE OF PROJECT	Pre-operational domestic telecoms satellite system for integration with the terrestrial network
PROJECT STATUS	Satellite under construction but subject to delays from financing problems
PROJECT VALUE	Italian Lira 391 billion ($250 m) through until 1988
LAUNCH DATES	1990/1991 by Ariane (A ground spare is also under construction)

SATELLITE USERS

The Italsat spacecraft is a pre-operational satellite intended to pave the way for a telecommunications satellite system which is integrated with the terrestrial telecommunications network in Italy.

Telespazio, owned by the state owned IRI/STET group has a concession from the Ministry of Posts and Telecommunications for the exclusive right to install and operate satellite communications systems in Italy. However, it is not clear at this stage who will actually be responsible for either operating Italsat or using the capacity.

The overview of the project indicates that the multibeam capacity will be primarily aimed at the PTT's terrestrial network but it is by no means certain that the PTT will want to use the satellite for a significant proportion of its traffic. Similarly the satellite is also aimed at providing digital capacity for specialised business services but it is expected that this will only be used on an experimental basis unless more satellites are procured to turn it into an operational system. The satellite is also capable of being used for distribution of satellite TV channels although current legislation in Italy effectively limits its use for such purposes to the public service broadcaster RAI. Future legislation may enable programming to be distributed by satellite to Italy's 800 or so "independent" local TV stations, 80% of which in practice are controlled by Silvio Berlusconi.

It is expected that some three or four earth stations will initially be used in conjunction with the satellite.

PROJECT OVERVIEW

Introduction: Many observers of the European telecommunications scene have concluded that European PTT's are attempting to introduce ISDN as quickly as possible as a method of securing the continuation of their monopolies. This policy has not been without major criticism; for example CIT Research in the UK has warned that ISDN could be made obsolete by developments in cable TV network switching.

However, the Italsat programme is directly related to the development of Italy's ISDN network. Although a pre-operational satellite, it is aimed at providing a digital satellite network superimposed on and integrated with the Italian terrestrial network. Unlike Tele-X, for example, it is seen as a highly network integrated satellite system for use by both public telephony traffic and the introduction of new business services ahead of ISDN. Nevertheless, Italsat is an experimental satellite and much of the interest in it lies in its use of new technologies as much as the underlying telecommunications philosophy.

The project provides the Italian aerospace industry with a follow-on R&D project following on from the SIRIO programme in the 1970's and concurrent with participation in ESA's Olympus programme. According to PR for the project "The Italsat programme's objective is the realisation of a telecommunications satellite of an exceedingly advanced conception, working with totally digital telecommunications technologies and that will allow experiments in a pre-operative form the operative solutions of the 1990's." Italsat's origins lie in the Italian National Space Plan of 1979 as a technology (as distinct from scientific) programme, with launch around 1985 or 1986. Phase A configuration studies of the satellite were carried out by Telespazio in 1981–82 under contract to the Italian National Council for Research (CNR). Subsequently phase B studies were completed and construction of the satellite begun. Management of the project remains in the hands of CNR. However, the project has suffered from delays, largely caused by financing problems and it may not now meet its (now) 1988 launch date. The Italian constitution effectively prevents allocation of funds more than three years in advance.

Principal features of the satellite are its use of Ka-band (30/20 GHz) frequencies, use of a six beam antenna with six 110 MHz transponders, and on-board switching of signals, three 40 MHz transparent transponders to provide an elliptical beam over Italy and a propagation payload in the 40/50 GHz bands. The operational follow-on to Italsat has yet to be defined and a timetable to be set. However, the Italians are considering a telecommunications payload for the SARIT DBS proposals. The pre-operational mission of Italsat has been defined as follows:

1) to be representative enough of the operational system given the schedule, financial and technical constraints of the pre-operational satellite,

2) to be capable of demonstrating in orbit at least some of the advanced technologies required for the operational system,

3) to be capable of demonstrating, by field trials, the validity of advanced telecommunications services and their acceptability to users,

4) to conduct new experiments in millimetric bands of high interest to the scientific and telecommunications community.

Multibeam Payload: This can provide capacity for 11,000 telephone circuits. Its configuration reflects studies by the Italian telecommunications industry for a satellite system which would integrate with the terrestrial telephone network. The terrestrial network's hierarchy consists of 21 "compartment centres" at the highest level which in effect represent the major cities, and, below this level, 231 district centres which generally connect with each other through the compartment centres. The satellite capacity of Italsat and its operational follow-on is intended to connect the district level directly.

Such a system would require over 200 earth stations in an operational form and require an equivalent of 40,000 telephone circuits, although it is envisaged that the minimum number of ground systems would be 50 in the operational system. Ground stations would be placed directly in large cities, preferably on the roof of corresponding exchange buildings, and use dishes in the range 3·4–5·0 metres and 50–200 Watt HPA's. The high traffic demands expected in the operational system meant that there were likely to be problems in obtaining sufficient frequency spectrum – hence one reason for choosing 30/20 GHz bands. It was also decided to keep earth stations as small as possible. Given that the Ka-band (30/20 GHz) is subject to high atmospheric attenuation under certain conditions, these factors lead to the multibeam approach. The multibeam payload utilises Satellite Switched Time Division Multiple Access techniques (SS–TDMA). The uplink signals received by the satellite are demodulated to baseband, regenerated, switched and remodulated to form TDM continuous streams on downlinks.

The bit rate chosen for the multibeam payload is 147·456 mbit/s. Low traffic stations can access the satellite at lower rates (24·576 mbit/s) and baseband switching allows bit-rate conversion to allow thinner routes to be merged with the main routes. On-board regeneration allows on-board generation of a Reference Unique Word (RUW) which allows alignment of the 32 ms long TDMA frames and thus avoids the need for reference stations and related complex protocals required by currently operational satellite systems using TDMA techniques. The 6×6 baseband switching matrix controller can accommodate TDMA bursts of fixed lengths and can cope with special configurations required for multipoint video conferencing. The multibeam payload is intended for both telephony and special services, such as video conferencing, computer file transfer and high speed facsimile. Telephony traffic will be served on demand on a call by call basis. The special services will be served on an advance reservation basis. Terrestrial ISDN equipment is not expected to be able to handle special services in a switched mode for the foreseeable future.

The multibeam payload uses 6 regenerative transponders with 20 Watt TWTA's. There are two steerable antennas for this payload with each generating three spot beams of 0·435° diameter coverage.

The Italsat spacecraft: The platform used for the Italsat satellite is essentially a design derived from experience of the OTS/ECS programme but is largely a new configuration using some off the shelf units, modified or otherwise. Prior to selecting the configuration a trade-off analysis was carried out amongst various spin and three-axis stabilised options and a medium sized (Ariane/Sylda or STS/PAM-D) three axis stabilised satellite chosen. Apart from the payloads, major features of the satellite are:

1) The adoption of a Unified Propulsion System (UPS). This was chosen to allow the design to be adapted for future telecoms missions without major satellite and propulsion subsystem design changes. The design is also being evaluated for the SARIT programme which envisages a DBS payload.

2) Attitude control utilises two momentum wheels with axes which form a 45° angle. This provides pointing accuracy of +/−0·15° North/South and East/West and +/−0·24° around the satellite–earth axis (yaw). The attitude control system also uses eight redundant thrusters as part of the UPS.

3) Batteries can provide eclipse protection up to 100 Watts but the two solar wings, each of four panels, provides up to 1450 Watts at end of life.

4) S-Band TT&C system is compatible with ESA and NASA tracking network.

Ground stations: The Italsat project relies on experimental work undertaken in the SIRIO programme on rain attenuation in the 12 and 18 GHz bands. The results indicate that attenuation would be significantly affected from location to location using 20/30 GHz bands. This requires substantial link margins to achieve satisfactory requirements. Quality of signal would range between 4 and 12 dB at 20 GHz and from 8 to 20 dB at 30 GHz. As a result, three types of earth stations are envisaged: 1) Standard A uses a 5 metre dish and its performance in terms of EIRP is 82 dBW with a G/T ratio (dBK^{-1}) of 30·5. 2) Standard B, also using a 5 metre dish, has an EIRP of 76 dBW and a G/T of 30·5 dBK^{-1}. 3) Standard C with 3·5 metre dish has an EIRP of 73 dBW and a G/T of 27·5 dBK^{-1}.

Some three to four stations are expected to be available for initial use with the Italsat system and experimental traffic.

A follow-on to the Italsat project is under study but many details have to be finalised. They may include capacity for DBS services as well as telecommunications for the PTT.

TECHNICAL SUMMARY

PLATFORM	Italsat/ECS
PRIME CONTRACTOR	Selenis Spazio
OTHER MAJOR CONTRACTORS	Aeritalia, Fiar, Matra, Aerospatiale, BAe, AEG/Telefunken, MBB, Ford Aerospace, Marquardt, GTE (Italy), LemDev, Melco
STABILISATION	3-axis
LAUNCH MASS	1650 kg
MASS IN ORBIT	890 kg
POWER SUPPLY, END OF LIFE	1450 Watts
DESIGN LIFE	10 years
ORBITAL POSITIONS	Italsat 1, 13° East Italsat 2, ground spare
TT&C FACILITIES	S-band station located at Fucino
STATION KEEPING ACCURACY	+/−0·15° (E/W N/S)
ECLIPSE PROTECTION	1000 Watts
MULTIPLE ACCESS – MODULATION TECHNIQUES	Spot beams, SS/TDMA Q.PSK. 147·5 Mbit/s (on-board regeneration). National beam, TDMA Q.PSK 24·5 Mbits/s with possible frequency hopping. FM for TV traffic.
NO. OF ACTIVE TRANSPONDERS	6 of 120 MHz (multibeam), 3 of 36 MHz (global)
NO. OF SPARE TRANSPONDERS	4 of 120 MHz (multibeam), 2 of 36 MHz (global)
TWTA POWER OUTPUT	20 Watts (all)
FREQUENCY RANGE:	
UPLINK	27·700–29·500 GHz 29·500–30·000 GHz
DOWNLINK	18·500–20·500 GHz 19·700–20·200 GHz
TRANSPONDER BANDWIDTH	6 of 120 MHz 3 of 36 MHz
EIRP MINIMUM, BEAM EDGE	Spot Beam, 55·2 dBW National Beam, 42·2 dBW
CHANNEL POLARISATION	Linear
RECEIVE COVERAGE (G/T)	Spot Beam, 16·1 dB/K National Beam, 5·5 dB/K
PROPAGATION PAYLOAD	1. 40 GHz circular polarised downlink with PM Modulation by 505 MHz sinusoidal tone 2. 50 GHz linear downlink (H/V switchable), unmodulated

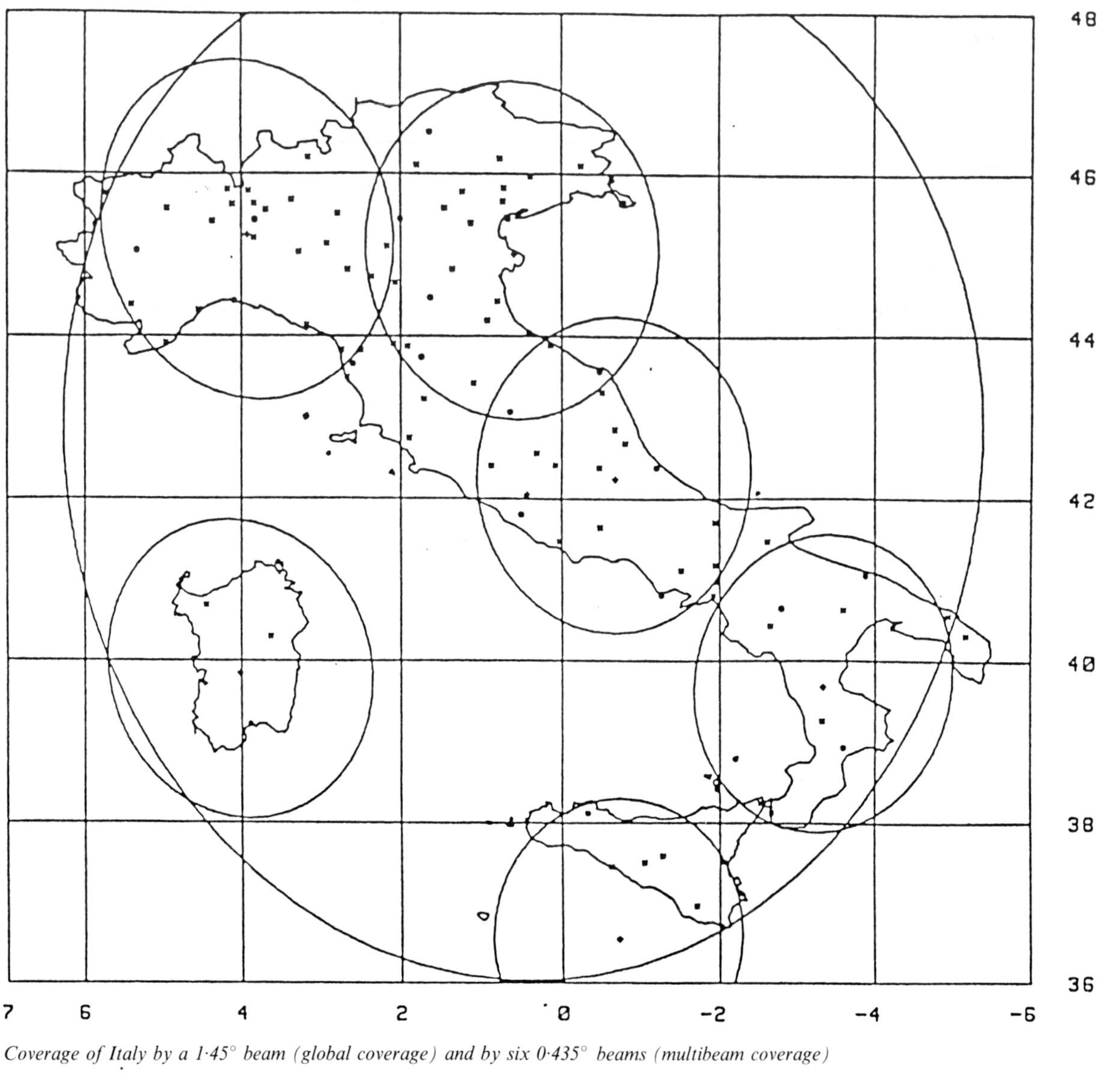

Coverage of Italy by a 1·45° beam (global coverage) and by six 0·435° beams (multibeam coverage)

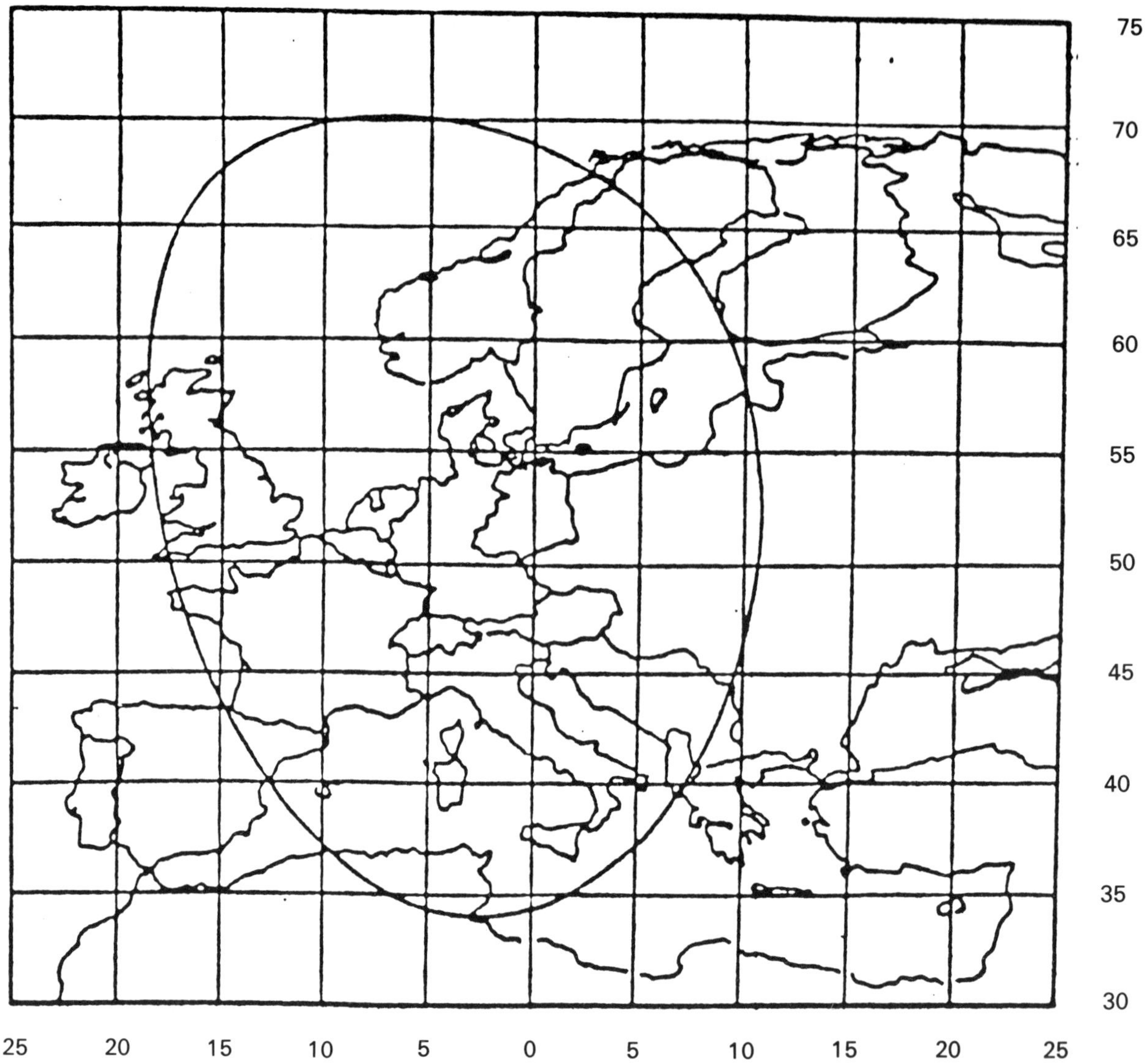

European coverage of the Italsat 40 GHz and 50 GHz propagation experiments

JAPAN COMMUNICATIONS SATELLITE INCORPORATED JCSAT

COMMERCIAL DETAILS

OPERATOR: JCSAT (Japan Communications Satellite Company)

ADDRESS

TELEPHONE NO.

TELEX NO.

FAX NO.

SENIOR STAFF

PURPOSE OF PROJECT

PROJECT STATUS

PROJECT VALUE: Total cost of the project is roughly $300 m, including two satellites, TT&C and launch costs

LAUNCH DATES

SATELLITE USERS

PROJECT OVERVIEW

Japan Communications Satellite is a joint venture between Hughes Communications (30%), C. Hoh (40%) and Mitsui (30%) to provide a Ku-band domestic communications satellite system for Japan under a class I telecoms licence. Hughes are providing two HS-393 satellites, ground control stations and technical support, whilst the other two partners are handling marketing and finance. The origins of the project lie in the capacity and cost shortcomings of the CS–2 and CS–3 domestic satellite systems, the move towards competition in provision of telecoms in Japan and pressure from the US government to open up the Japanese domestic satellite communications market to US industry. Japan communications satellite faces competition from Space Communications Corporation which has a similar licence and is procuring two satellites from Ford Aerospace in the USA.

Both systems required allocation by the Ministry of Posts and Telecommunications of part of the Ku-band frequency, which is heavily used in Japan for terrestrial microwave links. Japan's own domestic satellite systems, CS–2 and CS–3 primarily use the Ka-band. The American partners in the two consortia were both wary of building Ka-band satellites for Japanese operations because of the unproven technology and consequential high costs and higher risks over Ku-band and the known problem of rain attenuation in the Ka-band. Moreover, adoption of Ku-band enabled them to construct the satellites more quickly, an important factor as both Hughes and Ford have financial investments in the operating companies. An application for a third, similar, system is still pending. The third is Satellite Japan Incorporated in which RCA Astro Electronics has an interest.

TECHNICAL SUMMARY

PLATFORM: HS-393
PRIME CONTRACTOR: Hughes Aircraft Company*
OTHER MAJOR CONTRACTORS
STABILISATION
LAUNCH MASS
MASS IN ORBIT
POWER SUPPLY, END OF LIFE
DESIGN LIFE
ORBITAL POSITIONS
STATION KEEPING ACCURACY
ECLIPSE PROTECTION
PROPULSION SYSTEM

SPACECRAFT DIMENSIONS
SPAN, TIP-TO-TIP
NO. OF ACTIVE TRANSPONDERS: 32
NO. OF SPARE TRANSPONDERS
TWTA POWER OUTPUT
FREQUENCY RANGE:
UPLINK
DOWNLINK
TRANSPONDER BANDWIDTH
EIRP MINIMUM, BEAM EDGE
CHANNEL POLARISATION

*Through Hughes Communications Incorporated

COMMERCIAL DETAILS

OPERATOR	Hughes Communications Services Incorporated
ADDRESS	Worldway Postal Center, PO Box 92424, Los Angeles, California 90009, USA
TELEPHONE NO.	(213) 607-4000
TELEX NO.	
FAX NO.	
SENIOR STAFF	John E. Koehler, President and CEO Steven Petrucci, Executive VP Robert Waldron, Senior VP Jack Hayden, Chief Financial Officer Bruce Leaderman, General Counsel Cindi S. Whalen, Director, Public Affairs
PURPOSE OF PROJECT	Provision of military communications capacity for lease to the US Department of Defense
PROJECT STATUS	System now fully operational
PROJECT VALUE	See text for financial arrangements. Each satellite insured for \$85m. A two day delay in the launch of the first satellite left Hughes with a large interest bill
LAUNCH DATES	LEASAT 4–2, 30th August 1984, STS LEASAT 4–1, LEASAT 4–3, 27th October 1985, (see text) LEASAT 4–4, 30th August 1985, STS 51-I LEASAT 4–5, to be launched, STS, 1988

SATELLITE USERS

Although owned and operated by Hughes Communications, the Leasat Satellite System is used entirely by the US Navy, Army, Air Force and Marines. From Hughes' point of view the satellite system is a commercial venture but its user base is entirely military. The satellites provide global coverage. Primary user amongst the US armed forces is the Navy for its VHF satellite communications system which links its ships and P–3C aircraft with shore sites. The Leasat system is a replacement of the FleetSatCom system which operates in the same frequency range. The FleetSatCom system in turn replaced the Gapsat/Gapfiller system. Those two satellite systems were both owned and operated by the US Navy. The decision to lease the Leasat system followed on from lease of capacity on Comsat's Marisat system.

The US Navy uses its VHF Satcom system for six communications subsystems:

1) Fleet Satellite Broadcast Subsystem
2) Common User Digital Information Exchange Subsystem/Naval Modular Automated Communications System (CUDIXS/NAVMACS)
3) Submarine Satellite Information Exchange Subsystem (SSIXS)
4) Officer in Tactical Command Information Exchange Subsystem (OTCIXS)
5) Secure Voice Subsystem
6) Tactical Data Information Exchange Subsystem (TADIXS)

PROJECT OVERVIEW

The Leasat system, known as Syncom IV by the US Navy, is a military communications satellite system leased by the US Department of Defense but owned and operated by Hughes Communications Services Incorporated as a commercial venture. The latter company is a subsidiary of Hughes Communications Incorporated dedicated to operating the Leasat system. In 1978, the (US) Naval Electronics Systems Command contracted with Hughes Communications Services to provide five years of leased service at \$83·75 m per each satellite in orbit. The contract was subsequently modified to include an option to extend the period beyond five years and for the Department of Defense to purchase the satellites after seven years. Under the contract, Hughes Communications Services is responsible for the development, procurement and installation of the Leasat system. This included design and construction of the spacecraft as well as the procurement of launch services on Shuttle.

Hughes Communications Services is also fully responsible for the maintenance and operation of the satellites when in orbit. Naturally, as a commercial venture, Hughes Communications insured the satellite system on the international market. The story of the launch of the Leasat 4–3 satellite by Discovery in April 1985 was one of the more spectacular events in the space world. Following two flawless Shuttle deployments and satellite activations of Leasat 1 and 2 in August and November 1984, Leasat 3 failed to activate itself after successful deployment from Discovery on 13th April 1985 (STS mission 51–D). A rendezvous and "flyswatting" salvage manoeuvre on the same mission on 17th April 1985 failed to activate the satellite. Hughes then put in a claim to underwriters for \$85 m – which was subsequently paid. However NASA and Hughes had already figured out a method of turning the satellite on. This required a subsequent Shuttle rendevous with the errant satellite and the construction of special hardware, not least to permit ground command of the satellite. The Leasat activation sequence is normally performed by an automatic timer on-board the satellite, including turn-on of the telemetry. The special hardware was intended to by-pass this.

Between April and August 1985 the satellite drifted in a low-earth orbit about 230 miles above the surface of the earth, experiencing temperatures well below the design and test limits of the liquid and solid propellant systems, electronic units, batteries and all other components. Neither NASA nor Hughes were particularly confident about recovery of the satellite, with Hughes quoting only a 50:50 chance. The rescue mission took place on Shuttle mission 51–I, which also launched Leasat 4–4. The special hardware was attached to the satellite, enabling the satellite to be turned on and controlled from the ground. It was then oriented, under ground control, to a thermally favourable attitude to warm its solid perigee motor to a temperature acceptable for firing. It was thought that the satellite's propellants were partially or largely frozen and fuel lines may have ruptured. Some two months of thawing out took place and its solid perigee motor was fired on 27th October 1985. The satellite reached its final orbit one week later.

After a further month of verification, it was handed over to the US Navy on 30th November 1985. Under the arrangements with the underwriters a "significant" amount of the $85 m insurance loss will be repaid to the underwriters during the lifetime of the satellite. However, the Leasat 4–4 satellite launched on the same mission, was lost in orbit, leading to an insurance claim for $85 m.

The Leasat spacecraft was designed specifically for Shuttle launches, using a widebody design to make full use of the width of the cargo bay, but being short in length. Shuttle launch charges take into account the length of the cargo bay occupied by the satellite. The length was kept down by employment of an integral propulsion system and folding antenna. The basic structure of the satellite is shown on the right.

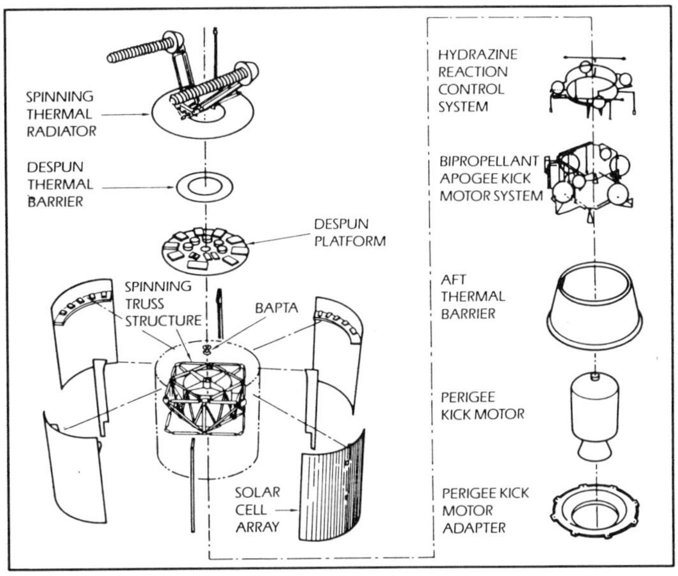

The two large helix antenna atop the satellite are for VHF frequencies, providing receive and transmit capabilities in the 240–400 MHz band. The communications payload handles four types of transmissions: fleet broadcasts (25 kHz), wideband (500 kHz), relay (25 kHz) and narrowband (5 kHz). The fleet broadcast signal is received on the earth coverage receive SHF horn antenna. It is then downconverted by the SHF receiver to the IF required by the fleet broadcast processor. After processing, the resulting signal is routed to the bypass/limiter and upconverted where it is filtered, limited and upconverted to VHF for transmission through the channel 1 35 Watt transmitter. In the bypass mode, the SHF receiver output is routed directly to the bypass/limiter and upconverter. The SHF beacon and transmitter is included to transmit processor status information through the earth coverage SHF horn antenna.

Signals for the remaining three functional groups are received through the common receive VHF helix antenna and preamplifier. A hybrid divider then splits the preamplifier output into the three signals that feed the wideband, relay and narrowband receivers. Each signal is then downconverted to IF in its receiver. The individual channels are split by selective channel filters and separately limited in the respective IF units. The wideband channel and each of the six relay channels are separately upconverted to VHF for transmission through the channel 2 53 Watt transmitters and channel 3 to 8 35 Watt transmitters, respectively. The five narrowband channels are combined after limiting and upconverting to VHF for transmission as a group through the channel 9 to 13 20 Watt transmitter. All VHF transmitter outputs are combined in a nine channel output multiplexer to form one output to the VHF transmit helix antenna.

The Leasat ground segment includes an Operations Control Centre in El Segundo, California, and operated by Hughes Communications. There are four satellite control sites i.e. Guam, Hawaii, Norfolk and Stockton (California), and two moveable ground stations in Guam and Norfolk (Virginia). These provide back-up support to the satellite control sites. The latter acquire telemetry and range data from and transmit commands to individual satellites.

The Operations Control Centre houses computer hardware/software and communications equipment that interfaces with the satellite control sites. The Navy VHF satellite communications system provides communications links between designated mobile units and shore sites. The coverage is worldwide between latitudes 70° North and 70° South. Although any part of the satellite system may be operated as a separate entity the system provides connections for message traffic and voice communications to Department of Defense long haul communications networks. The whole system uses a composite of six information exchange subsystems (see section on satellite users) which each address a specific naval communications requirement.

TECHNICAL SUMMARY

PLATFORM	HS–381
PRIME CONTRACTOR	Hughes Aircraft Company
OTHER MAJOR CONTRACTORS	
STABILISATION	Spin
LAUNCH MASS	6895 kg
MASS IN ORBIT	1388 kg
POWER SUPPLY, END OF LIFE	1200 Watt
DESIGN LIFE	7 years
ORBITAL POSITIONS	Leasat 4–2, 75° East, Leasat 4–1, 176·5° East, Leasat 4–3, 100° West, Leasat 4–5, 23° West
STATION KEEPING ACCURACY	+/−0·5° (E/W & N/S)
ECLIPSE PROTECTION	Full
PROPULSION SYSTEM	Internal transfer stage
SPACECRAFT DIMENSIONS	4·29 m height, stowed, 6·17 m deployed, diameter 4·26 m
SPAN, TIP-TO-TIP	
CONTROL CENTRE	El Segundo, California

NO. OF ACTIVE TRANSPONDERS	12 at VHF, 1 at SHF
NO. OF SPARE TRANSPONDERS	
TWTA POWER OUTPUT	53 Watt, 35 Watt, 20 Watt
FREQUENCY RANGE:	
UPLINK	240–400 MHz, 7975–8025 MHz
DOWNLINK	240–400 MHz, 7250–7500 MHz
TRANSPONDER BANDWIDTH	see below
EIRP MINIMUM, BEAM EDGE	see below

Channel Type	*No.*	*Bandwidth kHz*	*EIRP,* dBW*	*G/T,* dB/K*
Relay	6	25	26	−18
Wideband	1	500	28	−18
Narrowband	5	5	16·5	−18
Fleet Broadcast	1	onboard processing	26	−20

*Specified minimum values over the coverage areas for the full service period

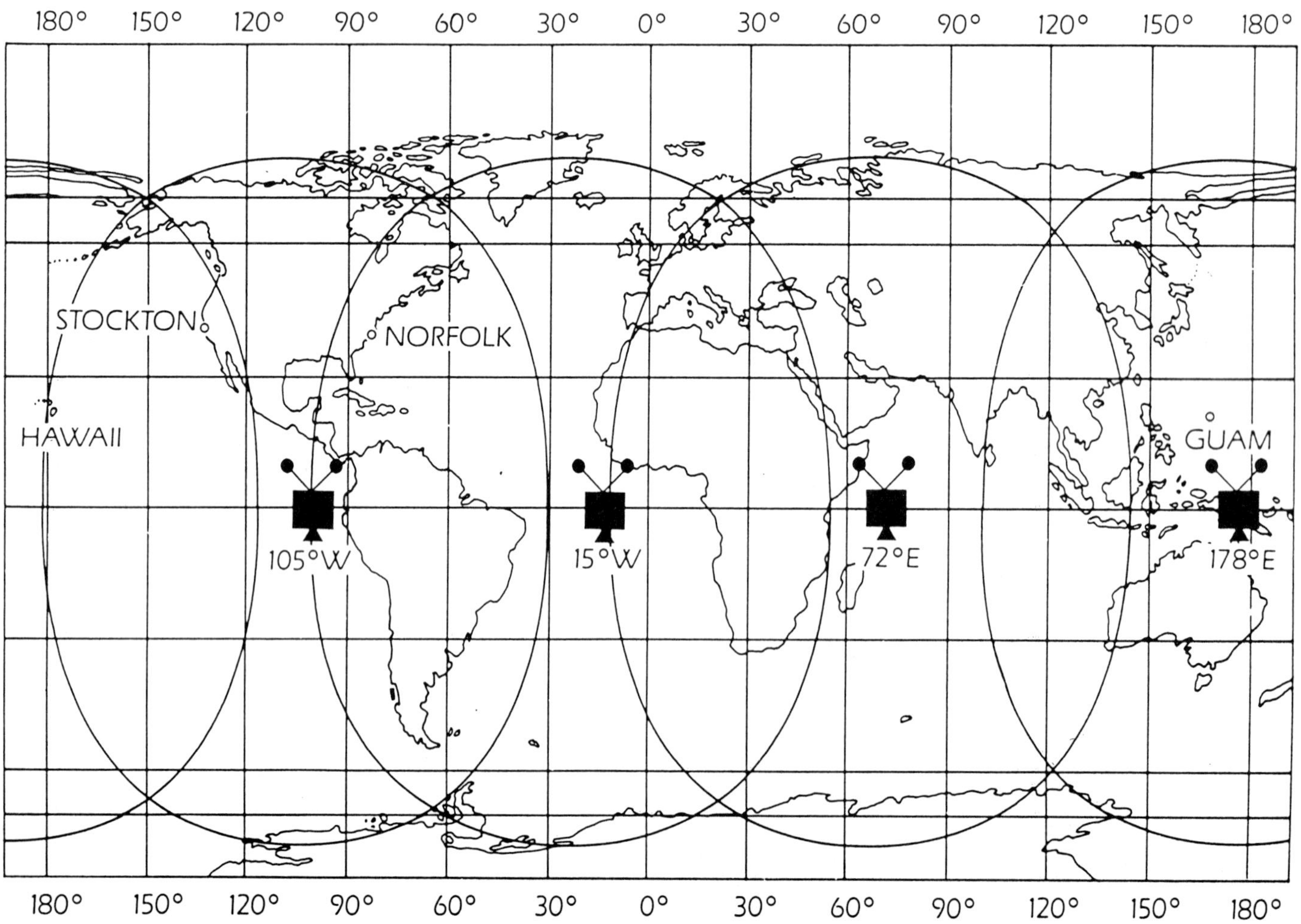

VHF coverage of the four satellite Leasat system. It provides global coverage between latitudes 70° North and 70° South

Artist's impression of the launch of a Leasat (Syncom IV) satellite from the cargo bay of a Shuttle orbiter. Later on the antennae will be deployed from their stowed positions. This satellite system is of interest in that although it is used entirely for military communications, the satellites are owned, operated, launched and insured by a commercial organisation, Hughes Communications. The Leasat spacecraft were built for launch, and can only be launched, by Shuttle. Each Leasat spacecraft fills the full width of the orbiter bay but less than half the length. The Leasat satellites each have their own internal perigee and apogee motors, which not only make for a compact spacecraft but eliminate the need to use separate perigee stages, such as PAM

COMMERCIAL DETAILS

OPERATOR	Locstar
ADDRESS	c/o Centre National d'Etudes Spatiale (CNES), 2 Place Maurice Quentin, F-75001 Paris, France
TELEPHONE NO.	45 08 75 00
TELEX NO.	214.674
FAX NO.	
SENIOR STAFF	Daniel Hernandez, Head of Locstar office, CNES Peter Watkins, Geostar Director of Development Europe
PURPOSE OF PROJECT	Provision of Radio Determination Satellite Services for ITU Region 1 (Europe, Africa and the Middle East) but initially to Europe only
PROJECT STATUS	Project at the development investigation stage. The Locstar operating company is not expected to be formed until the Spring of 1987
PROJECT VALUE	Not available
LAUNCH DATES	Initial Locstar space segment is expected to consist of piggyback payloads carried on a conventional telecommunications satellite to be launched in 1990. It will provide European coverage. A fully dedicated space segment is planned for launch in 1994–1995. This is expected to provide coverage for Europe, Africa and the Middle East

SATELLITE USERS

The Locstar system is initially intended to provide Radio Determination Satellite Services to users in Europe and its off-shore waters up to a distance of around 200 kilometres. these services will consist of positioning, navigation and text message services.

Initial market studies on European demand for Locstar services suggest that the system should be able to capture a portion of the market corresponding to between 300,000 and 400,000 operational transceivers. The capacity of the Locstar 1 system is estimated to be able to handle about 500,000 subscribers. The number of users that can be accommodated depends on the average length of message exchanged. The 500,000 capacity estimate also assumes a traffic rate of one transaction (location/message service) per hour.

CNES claims that the regional nature of the initial Locstar services coupled to its limitations in serving polar and equatorial regions (polar zones are not visible to geostationary spacecraft, while latitude determination accuracy is considerably reduced in the equatorial zone) suggests that demand will come from terrestrial vehicles such as trains, lorries and motor cars. The French state owned railway company, SNCF, has taken a strong interest in Locstar for supplementary signalling purposes. A secondary market would be for marine based mobiles in coastal waters and inland waterway vehicles.

CNES envisages that later, following the establishment of compatible regional systems, scheduled airliner and long haul shipping operators would use the system. In the meanwhile it believes that there may be a demand from general aviation business.

PROJECT OVERVIEW

Introduction: The French national agency, CNES, is planning a radio determination satellite system called Locstar for the International Telecommunications Union's Region 1 (Europe, Africa and the Middle East). The system is being planned in conjunction with the Geostar Corporation and is intended to be compatible with the Geostar RDSS system for North America and part of a global RDSS system, with further satellite operators in the Far East. The Locstar project is still at the development investigation stage and plans for the project remain uncertain at the time of writing. It is intended that the Locstar operating company will be funded from a wide variety of private and public organisations in Europe but that France (and CNES in particular) will have a significant role in the project. Initial operations under the name Locstar 1 will use piggyback payloads carried on conventional communications satellites before a fully dedicated satellite system is launched from 1994/1995. Full three dimensional positioning and navigation services will require payloads on three separate satellites. Given the need for in-orbit back-up capacity, at least four satellites will be needed to carry Locstar payloads. CNES has talked with a number of operators of communications satellites in exploring the possibility of piggyback rides including Société Européenne des Satellites, Eutelsat and Arabsat. At the time of writing no final agreement has been announced.

The Locstar system is intended to provide position location to an accuracy of between 10 metres and 100 metres and an alphanumeric two-way message service between mobiles and their home base. The Locstar 1 system, planned to become operational from 1989 will cover Europe. A fully dedicated system, Locstar 2, is planned to become operational from 1994/1995 and cover the whole of ITU region 1.

Background to the Locstar system: The Geostar Corporation began looking at Europe in 1984 as a possible market for Radio Determination Satellite Services. As a start up venture, it did not plan to operate a space segment for Europe even though the company had ambitions for a global RDSS system. Instead, it decided to find partners in Europe to operate an RDSS system. It located a marketing manager in Europe, Peter Watkins, and signed a memorandum of understanding with CNES in February 1986. This called for the establishment of the Locstar 1 system "at the earliest possible time" but was, in effect, a study agreement. CNES then formed a Locstar office as a precursor to the establishment of a commercial operating company funded by European private and public sector outfits. The operational Locstar company had not been formed at the time of writing. CNES has initiated its own technical studies of the concept and will be sharing test information, technical studies and marketing data with Geostar. CNES is also currently responsible for finding satellite operators willing to allow Locstar packages to be carried piggyback on their satellites.

Geostar's strategy towards forming an international RDSS system is illustrated not just by the agreement with CNES but also by agreements with organisations elsewhere in the world. In December 1986 it signed an agreement with the Indian Space Research Organisation, ISRO, whereby the latter will provide a RDSS package on a domestic Insat spacecraft to provide coverage of India and the Indian Ocean region. An agreement has also been signed with Aussat covering a feasibility study of an RDSS service using the Aussat system for Australasia and part of the Pacific Ocean region. Geostar is also in negotiation with the Chinese for a similar agreement and this is expected to be signed in the first half of 1987. In Canada, it has signed an agreement with a company called AGRA to market Geostar RDSS services which use the US space segment capacity planned by Geostar.

Like Locstar, it is relying on piggyback payloads before it launches its own dedicated satellite system. However, unlike Locstar, its services are due to begin with a one way service, Link 1, using payloads aboard GSTAR 3 and Spacenet F3R. A two way package for Geostar prime services for North America is to be carried on the GSTAR 4 satellite due for launch by Ariane in May 1989. The Geostar payload on GSTAR 2 failed completely soon after the satellite was launched in March 1986. Failure has been put down to a faulty capacitor which was one of a batch that had given problems. A spillage of hydrazine immediately prior to launch may also have been a contributory factor.

System description: The Locstar 1 system currently envisaged draws upon the technical work done by Geostar, such as patents and test results, and will also draw upon the commercial experience from initial US services. The Locstar RDSS system is intended to perform three functions:

1. Radio Location (or positioning) where a central facility, known as "Locstar Central" or "Central" will calculate the position of a mobile. Typically, this may be position information on a lorry to be forwarded to the fleet operator.

2. Radio Navigation where the mobile vehicle is told of its position.

3. Message service (two way) where simple text only messages are transmitted to and from the mobile. Current thinking on Locstar envisages that any such message will be limited to about 100 alphanumeric channels.

The overal Locstar system follows that of Geostar and consists of three distinct segments:

1. The space segment comprising (initially) of packages carried piggyback on geostationary communications satellites. The receive part of the payload relays signals transmitted by mobiles to the "Central" facility, and a "Transmit" package relays signals transmitted by the Central to mobiles (time base and message dissemination).

2. The user segment consisting of Locstar transceivers (terminals), whether mounted on mobiles or fixed points.

3. The Locstar Central, or central ground control segment, comprising the transmit/receive ground connection stations, satellite ground control centre, position computation and interconnecting processing centre and data distribution centre interconnecting with the terrestrial or other communications networks and systems.

The Locstar operating principle for positioning determination of mobiles is based on the measurement of mobile-to-satellite range by either two or three satellites equipped with Locstar payloads. Position computations result in two symmetrical or "mirror" solutions; one for the northern hemisphere and one in the southern hemisphere. According to CNES, the position ambiguity is readily resolved for all latitudes above a few degrees. The number of range measurements required depends on the number of parameters required to be measured. The system user has the choice of either two coordinates, latitude and longitude, or, in addition to these, altitude as well. If the mobile's altitude is known, then two range measurements (one by each of two satellites) are required. Location accuracy is to about 10 metres for an altitude error of less than five metres. If the mobile's altitude is not known, then three measurements (one by each of three satellites) provide three coordinates to a location error of about 100 metres. In some cases the altitude is supplied by the mobile itself, for example with an aircraft altimeter. A vehicle fleet operator may also use a digital terrain model (DTM) to give altitude.

In operation of the positioning service, the Central broadcasts a general timing signal with messages and interrogations to mobiles superimposed on it. The Transmit package on the satellite relays the timing signal to the mobile. Interrogated mobiles, or mobiles wishing to transmit, first synchronise with timing signal, then transmit their messages. Mobiles responses are relayed by either two ro three "receive" satellites to Central. the time lapses between the interrogation of a mobile and the receiption of its response are used to determine the mobile's position. Once the Central has determined a mobile's position, this information can be transmitted with any other message either to the mobile itself or to its home base. Each mobile is allocated a unique identification code. Through use of spread spectrum techniques (with random access, pseudonoise, or PN, code) a large number of low power transceivers can access each satellite at the same time without interference. A complete transaction (interrogation/response/position determination/message and position transmission) lasts less than a second.

The service zone of the Locstar 1 system will cover all of Europe and corresponds to the footprint of a single beam. However, the dedicated Locstar satellites planned for the mid-1990s will use spacecraft with multiple beams.

A critical factor for the success of Locstar and Geostar will be the availability of frequencies. The WARC '87 Mobile Conference will take place in September 1987 and cover worldwide RDSS frequency allocations. Proposed RDSS frequencies are currently partially allocated to other services.

CNES, in conjunction with Geostar, is pushing for frequencies that are common to both Geostar and Locstar. These are as follows:

2·4835 – 2·500 GHz for satellite to mobile links.

1·610 – 1·6265 GHz for mobile to satellite links.

5·117 – 5·183 GHz for satellite to central base links.

6·525 – 6/5415 GHz for central base to satellite links.

To ensure common compatibility, the Locstar and Geostar systems have to share a common frequency allocation plan for the three ITU regions. According to CNES, the proposed allocation plan appears, at present, to be the best one, despite certain coordination problems in Europe.

(See section on Geostar for further details).

Competitive analysis of the Locstar system: Papers presented by European Space Agency staff during 1986 claimed that the majority of demand for mobile communications in Europe will come from voice services, leaving the Locstar system with only specialised markets to address. Nevertheless, voice markets for mobile satellite communications remain an unproven market opportunity and the Geostar/Locstar concept has been viewed on a highly favourable basis by some sectors of the satellite industry in Europe for some years. Geostar claims that its system is complementary to the current and planned Inmarsat mobile communications system rather than competitive. Nevertheless, Inmarsat is looking at providing RDSS services and, unlike Locstar/Geostar, has the space segment capacity to provide such a service on a global basis rather rapidly and has the advantage of the potential for voice and "substantive" text and data communications as distinct from the maximum of 100 character messages envisaged by Locstar. Similarly, US Mobile Satellite Communications proposals envisage addressing the thin route voice communications market for point-to-point communications between non-mobile sites. On the other hand, there is clear evidence that the information technology industries are increasingly recognising that the most substantive and useful information transmitted by a communications system can be accounted for by only a small proportion of actual information available for a given message.

Line of sight requirements for communications with geostationary satellites, particularly in the northerly latitudes where most Western countries are located, naturally place a limitation on the usefulness of mobile communications to land based mobile users. It would, for example, be very easy for a lorry driver to "hide" his lorry behind a building, tree or hill to evade an inquisitive communication through a mobile satellite communications system from the fleet operator owning his lorry. Likewise, if the lorry driver wanted to communicate through a geostationary satellite, he could easily be thwarted by line of sight problems. Given these problems such mobile communications systems as cellular radio remain a major competitor to any system based on satellites. In real terms, RDSS communications services are not cheap. A transceiver is expected to cost several thousand dollars in addition to monthly subscription costs per terminal.

On the other hand, it has been estimated the aeronautical voice services through the Inmarsat system will require the organisation to charge between $30 and $60 per minute for the space segment (compared with Standard A maritime call charges of just under $5 per minute.

For navigation and positioning services, the Locstar/Geostar concept also faces competition from the US NAVSTAR system and, potentially, the European Space Agency's NAVSAT concept. Both are intended to provide truly global communications through use of polar orbiting satellites. Locstar/Geostar are unable to provide communications with polar latitudes and therefore have limitations for use by airliner operators. It has been argued that the Locstar/Geostar system is unsuitable for any navigation use where safety is involved. This is because the commercial aspects of the system require that users' credit ratings be checked on using the system and that the user can be billed. If the "Central" fails, the user could be left without navigation facilities. This would be highly undesirable, for example, to an aircraft in flight.

French thinking on the military value of Locstar is unknown. France already has a mobile military communications system in the form of the military capacity on Telecom 1.

Nevertheless, the original Geostar concept has entrepreneurial origins and, like Locstar, is taking a (relatively) low risk approach to development of its services. The piggyback payload concept avoids the need for expensive dedicated satellites until such time as there are sufficient users to justify them. The arrangements to set up regional RDSS services will allow both a global system to develop and spread the risks (and R&D costs) amongst many organisations. It also allows quicker introduction of services.

Competition to the Locstar concept: Inmarsat has expressed strong interest in providing RDSS services and already has the global space segment capacity to do so. It can already offer RDSS services through the Inmarsat Standard A system, but for those users seeking a low cost mobile communications service, this is unsuitable. The Inmarsat Standard C service offers a limited RDSS service and much lower costs than Standard A. However, both are aimed at the maritime markets and require interfacing with the Coast Earth Stations operated by PTT's as part of the Inmarsat system. Geostar argues that its system is complementary to the Inmarsat system in that Locstar/Geostar services are aimed at land masses. Inmarsat is expected to introduce operational aeronautical communications from 1988. However, these will be much more sophisticated than the Locstar type services in that they will offer voice and data services. Inmarsat is not yet committed to providing RDSS services and potential end user costs have yet to be published. It is not certain that Inmarsat members will agree to the organisation providing them.

Other European developments in mobile communications: The Locstar system lies outside the framework of the European Space Agency which, surprisingly, has remained very quiet about it. Yet the Agency has several programmes in hand covering mobile communications services by satellite and has undertaken a number of market studies for such services. The latter show what the agency describes as "great unsatisfied demand for European-wide mobile communications services of many different types." The studies indicate that plans of European PTTs will not meet the demand. Also indicated is a strong preference for voice services despite the much higher prices envisaged in comparison with RDSS services. Demand for voice services appears to be very price insensitive even if a pan-European cellular system is available. The proportion of subscribers who want access to both cellular and satellite services is about 5% of the total. It is suggested that there is a need for both a private mobile communications system and one for access to the public network. ESA believes that the lack of agreement between European countries on cellular radio standards means that a pan-European cellular radio service is highly unlikely before the next century.

ESA's Prosat project covers electronic mail services for mobile communications users and is using spare capacity on the Inmarsat system for test and demonstration activities. Prosat is a step towards the Agency developing advanced mobile communications systems for land based vehicles, small boats and aircraft. Out of the Prosat programme, a new system, called Prodat is now under development. Demonstrations are expected to take place from the end of 1987, mostly involving road transport companies and airline operators flying across the Atlantic.

ESA argues that within current cellular radio concepts, satellites have a major potential role. Within cellular radio systems, no call destined for a mobile terminal can be initiated without first locating the mobile. This is necessary to establish a transmission path between the caller and the mobile. Without a satellite, this function must necessarily involve continuous tracking of the movements of the mobile communications users when they are moving over large areas and between cellular networks. For mobiles that are far away from their base, this would require a large amount of signalling traffic despite the likelihood that the average numbers of calls with a mobile per day would be low. A Mobile Satellite Service, including RDSS services, could provide the position requirements of cellular radio. The Agency has also called for studies of demand for mobile communications in Europe which could not be met by cellular radio. Such a study would determine the economic trade-offs to establish which proportion of Europe should be covered by cellular radio and which by mobile communications satellite services.

ESA is now defining a demonstration satellite payload for mobile communications which might include a foldable array antenna to produce four spot beams covering Europe, the North Atlantic and the North East of the USA, together with a global beam for the same area. The demonstration payload would support both land and aeronautical services. It is possible that the payload will also incorporate on-board processing. However, the Agency has yet to commit to its development and it remains one of a number of candidate payloads proposed for launch by the Agency on experimental satellites planned for launch in the early 1990s. Although not strictly an RDSS system, ESA's proposals for a global navigation and positioning system under the name NAVSAT and similar in concept to the US military Global Positioning System (confusingly called NAVSTAR) would provide competitive services to Locstar and Geostar. The ESA NAVSAT system has been under study by the Agency and European industry since 1981. Current thinking does not envisage initial operations until the 1992–96 period and global coverage until the period 2010–15.

The full NAVSAT system, as originally envisaged, used a constellation of 24 satellites in 12 hour circular orbits. There would be three orbital planes each with eight satellites. This ambitious concept was seen by many as uneconomic for civil use. Like the NAVSTAR system, to be fully operational this system would require all satellites to be fully deployed and operational. It was also considered by some commentators to be uneconomic for civilian purposes and a hybrid constellation was conceived in reply. This would have consisted of six payloads flown piggyback on geostationary satellites and 12 small payloads in highly eccentric orbits inclined at 63,4 degrees to the equator. The revised concept also allowed regional deployment of the satellites to meet areas of greatest demand with gradual expansion to provide a global communications system. At the time of writing, the NAVSAT system remains a paper concept and appears unlikely to be developed. Although it conceptually offers the advantage of positioning communications in polar regions (useful to airline operators), it has few other advantages over the Locstar/Geostar concept, duplicating the US NAVSTAR system and the Soviet military Glonas system.

The West German R&D organisation DFVLR has also funded a study of a civil navigation system called GRANAS in which a space segment of 20 satellites is envisaged. Unlike NAVSAT, it includes a data communications capability. A study contract has now been placed by ESA to integrate the best features of GRANAS and NAVSAT concepts.

COMMERCIAL DETAILS

OPERATOR	Mailstar AB
ADDRESS	c/o Swedish Space Corpporation, ITI 54, Solna, Sweden
TELEPHONE NO.	8-98 02-00
TELEX NO.	17128 Spaceco
PURPOSE OF PROJECT	Provision of very low cost satellite communications for areas without sufficient terrestrial communications and for Swedish multi-nationals
PROJECT STATUS	Marketing and project design and definition studies underway
PROJECT VALUE	Initial two satellite system is envisaged to cost between \$30 m and \$45 m including launch, insurance and the Kiruna control station
LAUNCH DATES	Mailstar 1 (possibly) April 1989 on Ariane V36 Mailstar 2 (possibly) 1990 on Ariane

SATELLITE USERS

The Mailstar concept is currently subject to only study work, although Sweden has a cooperative agreement with China which may possibly lead to the latter using the concept.

Mailstar is aimed at users, such as business, in developing countries or remote locations where current telex or telefax facilities are overloaded or non-existent. It is essentially an office to office communications facility. The remote terminals are intended to be cheap, relatively simple to use and maintain and consist of a steerable antenna, a simple transmit/receive system and a personal computer.

Suggested users for the system are the Saab group and the Swedish government. At the time of writing, the backers of Mailstar were suggesting that Swedish multinationals could be significant users of the system.

PROJECT OVERVIEW

In August 1986 Sweden formed a commercial company to develop its Mailstar proposals for cheap satellites with limited text/data transmission capabilities. Mailstar is a concept which originated in India to provide very basic communications to areas without any telecommunications infrastructure and avoiding the high cost of conventional satellite communications. The Swedish Mailstar system envisages use by Swedish multinational corporations to provide links between the parent companies in Sweden and their subsidiaries abroad as well as for third world areas with unsatisfactory terrestrial communications.

The company, Mailstar AB, is initially concentrating on further market studies for the Mailstar services and systems definition work. The project was originally made public in 1983 and was, until the late Summer of 1986, handled by the Swedish Space Corporation. It is currently envisaged that the overall cost of an initial two satellite system will be around SKr200–300 m (US\$30–45 m). Current investors in the new company are the Swedish PTT, Televerket, (60%), Saab Space (17%), L. M. Ericsson Radio Systems (17%) and the Swedish Space Corporation (6%). Possible first launch is envisaged in 1989 on Ariane V36 as a piggyback payload to Spot 2. A first launch using the Long March 2 vehicle appears to have been abandoned as being "uneuropean". Second launch could be in 1990–91 with ESA's ERS–1 satellite on an Ariane 4 vehicle.

The Mailstar concept is seen by some observers as a project designed to meet the limited capabilities of the Swedish satellite manufacturing industry rather than as an attempt to provide a competitive plug for a market gap. The industry looks unlikely to get any work from a follow on to the pre-operational TELE–X spacecraft for which no serious use has been found (so far). The Mailstar satellite uses the experience gained in the construction of the small (500 kg) Viking scientific spacecraft. It is currently envisaged that the Mailstar satellites will each weigh around 90 kg. Viking was constructed by Saab Space who have already accepted an order to start up detailed design of Mailstar. The company says that it has also taken a shareholding in Mailstar AB to "acquire know-how" in the design of small spacecraft. It may also use the operational system for its own internal communications requirements. Other potential users that have been suggested are the Swedish government to provide communications with its embassies.

The satellite project initially envisages two in-orbit spacecraft operating in low earth polar orbit. Users would each have a small earth station connected to at least a micro computer and used for both transmitting and receiving text and data but not for voice traffic. In the two satellite system, a satellite would pass over the user terminal between six and 14 times per day. In each pass the terminal would have a maximum of 10 minutes to transmit or receive information. The period would be less if other nearby terminals wanted to use the satellite during the 10 minute period. Uplinked information would be stored on the satellite in a 80–100 mbit semiconductor memory weighing around 12 kg.

Somewhat suprisingly, Saab Space say that the memory could only store the equivalent of 2500 A4 pages of text. The Swedish Space Corporation has suggested that a typical daily traffic volume for a remote terminal would be the equivalent of 200 pages of A4 text. When the satellite passes over another terminal to which the uplinked information is addressed electronically, it is downlinked. The maximum period between uplinking through a remote terminal and downlinking through a remote terminal is several hours. However, the system is intended to include a base station in Kiruna in Northern Sweden, which, because of its Northerly latitude, would give access to the satellite "almost once in every orbit". Information stored on-board the satellite could be downlinked through Kiruna for onward transmission through the Swedish (and International) terrestrial network. By this route it is claimed that information once uplinked to the satellite, can be at its final destination within a maximum of one hour. The Kiruna base can also be used in the reverse of this route.

Only limited details of the satellite design have been released. It uses simple gravity gradient stabilisation and has a total power consumption of around 60 Watts. The communications payload operates at around 1·6 GHz and RF power output is 4 Watts. Operating life of each satellite is set for five years. Polar orbit is between 800 and 1500 km. The satellite could be launched by a Scout class launch vehicle or, with an adapter, as a piggyback payload on Ariane.

Early in 1986 the Swedes announced that they had reached an arrangement with the Chinese for possible launch of the satellites atop the Long March 2 (CZ–2) vehicle. The plan was part of a broader cooperative agreement between the two countries covering space activities. However, for the first two satellites at least, the plan for the CZ–2 launches appears to have been abandoned. Because the satellites are so small, the project is largely unaffected by the current crisis in the launch vehicle market. There are a number of opportunities to launch the spacecraft piggyback with other Ariane payloads and it appears that the Swedes bowed to the reality of the politics of the European space scene in opting for Ariane.

Although the Mailstar project has its critics, it represents a novel method of providing low cost and possibly very reliable communications at a time when traditional satellite communications are proving to be neither low cost nor reliable. Despite much talk over the last twenty five years of the value of satellite communications in building up the telecommunications infrastructure of underdeveloped or developing countries, the cost of domestic or regional conventional satellite systems has proved beyond the reach of many such countries. It is still not clear to a lot of observers why these countries should opt for sophisticated satellite systems when, often, large areas of their interior territories lack even the most rudimentary communications. The lack of a telephone or telegram facility often contrasts with the sophisticated connections to the Intelsat system which tends to reinforce the dominance of an urban elite at the expense of a rural hinterland. Purchase of expensive conventional regional satellite systems would increase this dominance and divert resources away from the rural hinterlands, whether those resources be in basic communications or in other pressing sectors of the economy. Similarly the use of dedicated domestic satellite systems might also reinforce the imbalance and actually reduce the resources available to the rural areas.

It is difficult to see what the value is to such areas of a satellite capable of carrying several TV channels or expensive SCPC telephony when there are insufficient resources to purchase and maintain ground terminals, where electricity supply is often intermittent or non-existent and there are no other communications facilities available. For example, when the Aussat system was being planned, one of the basic features that telecoms planners found was that a simple telephone service was rated by people in the outback much higher in priority than the availability of TV. Although not a developing country example, it illustrates that such priorities do exist. It must also be remembered that satellite communications technology was developed by Western countries in response to their own needs which are very different to those of the Third World. Where satellites have been purchased by Third World countries, existing off the shelf hardware has had to be taken. This even applies to the Intelsat system which is constructed, and primarily used, by developed nations.

Developing countries have made little use of the Intelsat system for their domestic telecommunications requirements, primarily because the system is not designed to meet those needs. Intelsat domestic leases are now very cheap and involve no capital outlay by the purchaser for the space segment. But ground equipment remains costly and requires trained technicians to install and maintain. Given this environment the Mailstar system may prove to be of significant use in the Third World, with simple terminals costing no more than, say, \$3000 per unit and with transmission costs per A4 document equivalent of no more than \$3. This contrasts with Intelsat's Vista ground terminals aimed at thin route traffic in developing countries. Such terminals, purchased in bulk, cost around \$40,000 to \$50,000 per unit. The much cheaper Intelsat Intelnet service (yet to be introduced) is one-way only.

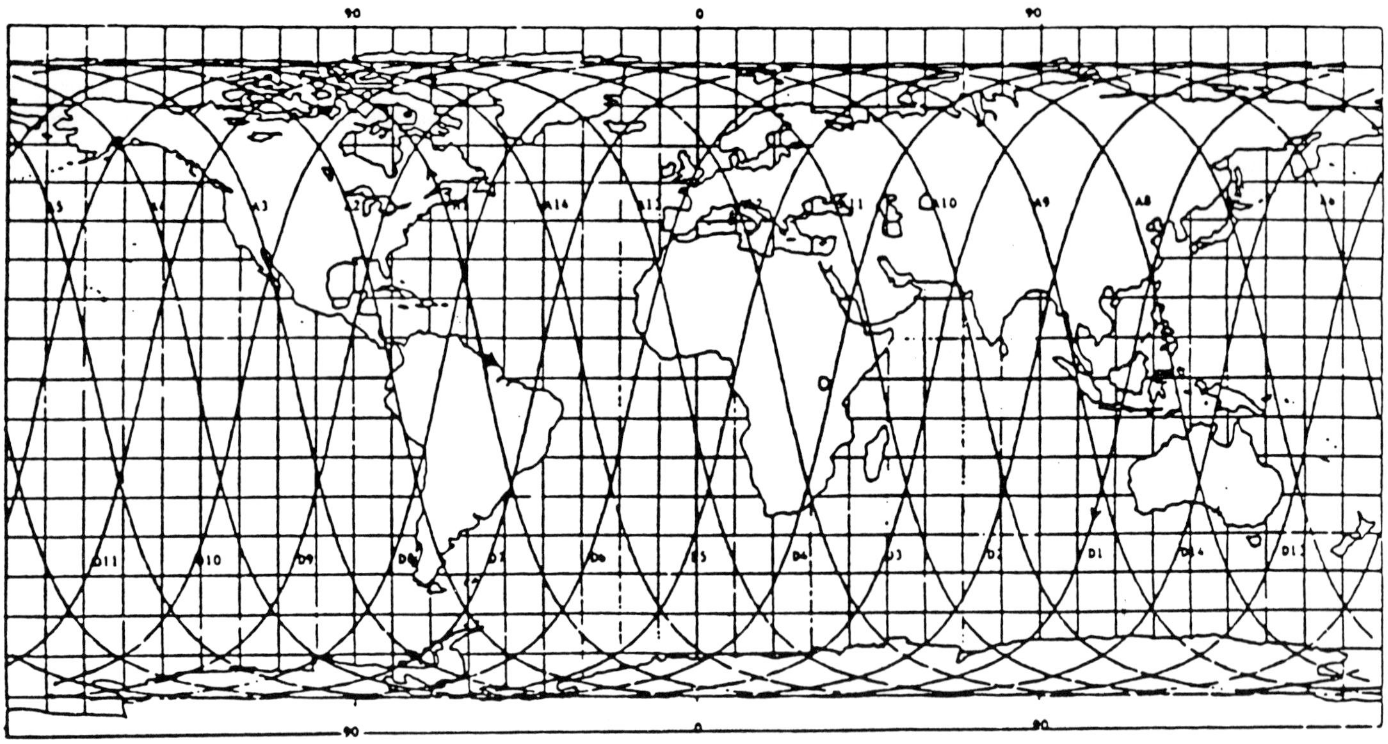

The figure shows an example of how the satellite passes the world during 24 hours. The satellite that performs 14 orbits every 24 hours can be seen approximately 2400 km on each side of the trace

TECHNICAL SUMMARY

PLATFORM	New
PRIME CONTRACTOR	Saab Space
OTHER MAJOR CONTRACTORS	not available
STABILISATION	Gravity-gradient
LAUNCH MASS	90 kg
MASS IN ORBIT	90 kg
POWER SUPPLY, END OF LIFE	50 Watts
DESIGN LIFE	5 years
ORBITAL POSITIONS	Not applicable; low earth polar orbit used, between 800 and 1500 km

STATION KEEPING ACCURACY	not applicable
NO. OF ACTIVE TRANSPONDERS	1
NO. OF SPARE TRANSPONDERS	not available
TWTA POWER OUTPUT	4 Watts
FREQUENCY RANGE:	
UPLINK	} 1·6 GHz
DOWNLINK	} 1·6 GHz

COMMERCIAL DETAILS

OPERATOR	Martin Marietta Communications Systems Incorporated
ADDRESS	6801 Rockledge Drive, Bethesda, MD 20817, USA
TELEPHONE NO.	(301) 897-6511
SENIOR STAFF	
FINANCIAL BACKING	Martin Marietta Communications Systems Incorporated is a wholly owned subsidiary of the Martin Marietta Corporation
PURPOSE OF PROJECT	Providing a "full range" of telecoms services using Ku-band capacity. US domestic service
PROJECT STATUS	The Company received FCC authority for the system in July 1985 but construction is delayed
PROJECT VALUE	The Company plans to spend $404 m in construction and pre-operational costs on the system
LAUNCH DATES	Late 1988 (proposed)

SATELLITE USERS

Martin Marietta says that the potential customers for this system include "a variety of users" among them other common carriers, communications service companies, business enterprises and government agencies.

PROJECT OVERVIEW

Martin Marietta Communications Systems is a wholly owned subsidiary of Martin Marietta Corporation and has received FCC approval for a Ku-band domestic communications satellite system to provide "a full range" of telecommunications services. The Company applied for Federal Communications Commission authority in November 1983 and received approval in July 1985. Two in-orbit satellites will each carry eight 54 MHz transponders and 10 250–500 MHz spot beams. A third of the capacity will be offered for lease or sale, the other two-thirds will be offered as a common carrier. Proposed Martin Marietta domsats would provide up to four and a half times the capacity of "today's typical operating systems" and would be used for point-to-point voice, data, facsimile, teleconferencing, video and other services. The eight 40 Watt, 54 MHz channels on each satellite would provide a contiguous United States beam. The 10 spot beams would be located for use over major population centres. Martin Marietta requested a six-month extension from the Federal Communications Commission of its July 1986 deadline to begin construction of the system. The Commission usually requires holders of construction authority to commence building their satellites one year after the authorisation is granted. However, Martin Marietta requested the extension primarily because of uncertainties over the future of Shuttle launches. Some observers, however, are suggesting that the request reflected broader uncertainties about the demand for satellite capacity.

TECHNICAL SUMMARY

	Nationwide (CONUS) beam	Spot beams
PLATFORM		
PRIME CONTRACTOR		
OTHER MAJOR CONTRACTORS		
STABILISATION		
LAUNCH MASS		
MASS IN ORBIT		
POWER SUPPLY, END OF LIFE		
DESIGN LIFE	10 years	
ORBITAL POSITIONS	73° West, 75° West	
STATION KEEPING ACCURACY	+/−0·05°	
ECLIPSE PROTECTION	Full	
FREQUENCY RANGE:		
UPLINK	14·0–14·25 GHz	14·25–14·5 GHz
DOWNLINK	11·7–11·95 GHz	11·95–12·2 GHz
NO. OF DOWNLINK CHANNELS	8	1 (250 MHz) or 2 (500 MHz)
TRANSPONDER BANDWIDTH	54 MHz	235 MHz
POLARISATION	Dual Linear	Linear Dual or Single
TWTA POWER OUTPUT	40 Watt	Several*
EIRP MINIMUM, BEAM EDGE	46·5 dBW	52·5 dBW (typical)*
RECEIVE G/T	+1·0 dB/°K	+1·0 dB/°K
SATURATION FLUX DENSITY	−85 dBW/m^2	not applicable†

*Note: The transmit power in each spot beam will be selected for local rain and traffic conditions. The individually selected linear power level into the antenna is expected to range from 2·0–10·0 Watts and the resultant linear EIRP from 49·0–56·0 dBW.

†The output power of the spot beam must be operated in a linear mode and will be prevented from being operated at saturation by spacecraft circuitry.

COMMERCIAL DETAILS

OPERATOR	Secretaria de Comunicaciones y Transportes
ADDRESS	Unidad Proyectos Especiales, Torre Central de Telecomunicaciones, Colona Narvarté, Avenidá Lazaro Cardenas 567, Piso 11 a la sur Mexico, DF03028 Mexico
TELEPHONE NO.	519-45-67
TELEX NO.	170902 upe me
PURPOSE OF PROJECT	Provision of C- and Ku-band domestic satellite communications for Mexico and some transborder communications
PROJECT STATUS	Satellite system already launched and fully operational
PROJECT VALUE	Procurement of two satellites plus control facilities cost \$92 m. Two PAM–D motors cost \$11·2. Launch costs for both was \$24 m. Total project value excluding insurance was \$127·5 m
LAUNCH DATES	Morelos A, 17th June 1985, STS 51G Morelos B, 27th November 1985, STS 61B

SATELLITE USERS

Morelos B is in storage orbit until it is activated some time in 1988.

Morelos A therefore currently provides all the used capacity on the system.

Principal user of the system is the National Telecommunications Company TelMex. Its digital trunk telephony system employs the six 72 MHz wideband C-band transponders using FDMA/FM techniques. It encompasses some of the 32 MHz C-band capacity for thinner route analogue telephony. Some of the Ku-band capacity is also used for its rural telephony.

Data and private line networks also make use of the Ku-band capacity. Petroleos Mexicanos (PEMEX) uses this for its SCPC private network which links oil platforms and provides links between Ciudad del Carmen and Mexico City.

TV traffic is either carried on the 36 MHz C-band transponders or the Ku-band transponders. Users are listed in the project overview.

Like Brazil, with its SBTS system, Mexico is looking to use Morelos for "social applications" including educational television, health and family planning and agricultural education.

PROJECT OVERVIEW

The Mexican Morelos satellite system has many similarities with the Brazilian SBTS system. Both were procured to replace and expand upon leased Intelsat capacity rather than for development of a satellite network from scratch. Both systems became operational in the same year each using HS–376 platforms and both are essentially C-band systems, although Morelos has a small Ku-band capacity. Both have been purchased by countries with a major foreign debt problem.

Prior to the launch of the Morelos system, Mexico had already developed an extensive domestic satellite communications system using capacity leased from Intelsat. It began this service on 1st May 1980 and eventually leased four spot beam transponders on Intelsat IV F1 at 53° West. Primary use was for distribution of TV channels although PEMEX used a half-transponder for SCPC traffic. Mexico also lease a transponder on Galaxy 1 for transborder distribution of live programming to the Spanish International Network. The Intelsat domestic system used 174 earth stations although nearly all of these were television receive only terminals. However, once the Morelos system became operational and most of this traffic transferred to the system, Mexico's shortage of foreign currency meant that expansion had to be curtailed. There was not enough hard currency to pay for the ground stations. One consequence is that Morelos B has been placed in in-orbit storage. Brazil got round the same problem by manufacturing the earth stations domestically. It was intended that some 10,000 earth stations would be installed for the Morelos system in the decade up to 1995. Mexico has also had to delay introduction of the Morelos A Ku-band services because it could not afford the ground segment.

Upon the Morelos A satellite entering into service, Mexico let its leases with Intelsat expire. In any case, the satellite it was using had already exceeded its design life and was time expired. The Intelsat users transferred to Morelos. The first four TV channels on Morelos A were as follows:

XEW–TV, out of Mexico City using transponder 12 (C-band). This is essentially a "middle class" conservative channel providing news coverage of national and international affairs. It is the most established of Mexico's TV channels.

XHITM–TV, also out of Mexico City uses transponder 14 (C-band). This is an educational network aimed at the rural population.

XHDF–TV out of Mexico City, uses C-band transponder 22. It is operated by the Ministry of Government Affairs, although it is financed by advertising. It is a general entertainment channel with some educational and cultural programming thrown in.

XETV, out of Tijuana/San Diego uses transponder 24 (C-band). XETV picks up the local terrestrial CBS affiliate TV station in San Diego and uplinks it to Morelos to distribute to cable networks in Mexico. A bilateral arrangement between Mexico and the USA allows Mexico to distribute US network programmes on Mexican cable networks.

Lack of space segment capacity on Morelos, because Morelos B is in in-orbit storage, has delayed other programming services going up on the spacecraft. Amongst TV channels eventually expected to use the channel are XHGC–TV (Channel 5), XHTM–TV (Channel 9) and XHIPN (Channel 11 – the Mexican equivalent of the BBC in terms of cultural content). Other possible channels are Mexico City Cable Channel 10, Television de la Republica de Mexicana and Mexican Educational Television.

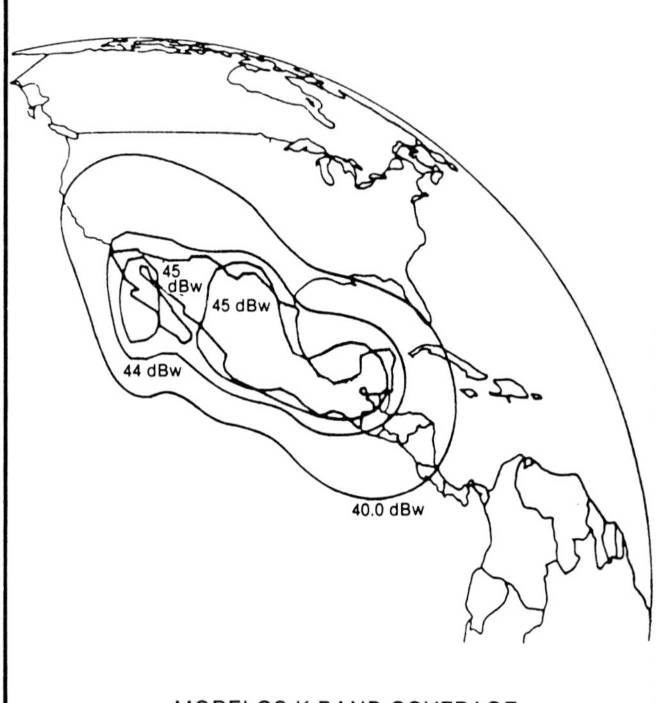

MORELOS K BAND COVERAGE
(108 MHz, 20 W)

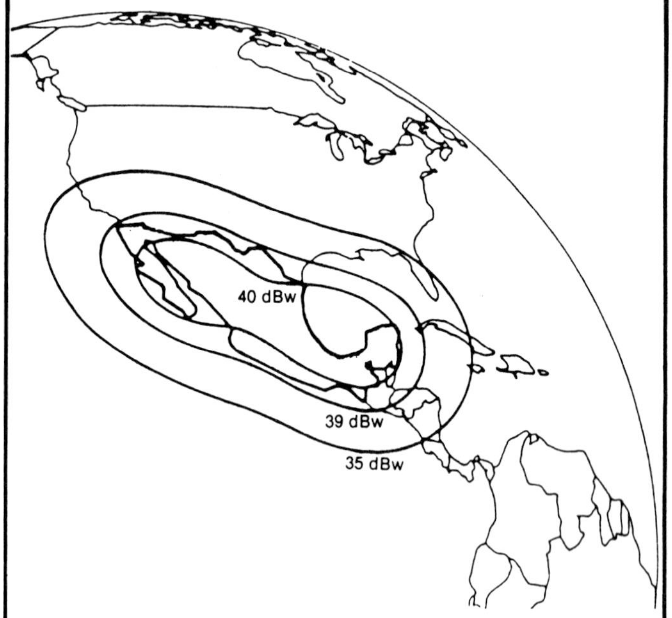

MORELOS C BAND COVERAGE, WIDEBAND CHANNELS (72 MHz, 10.5 W)

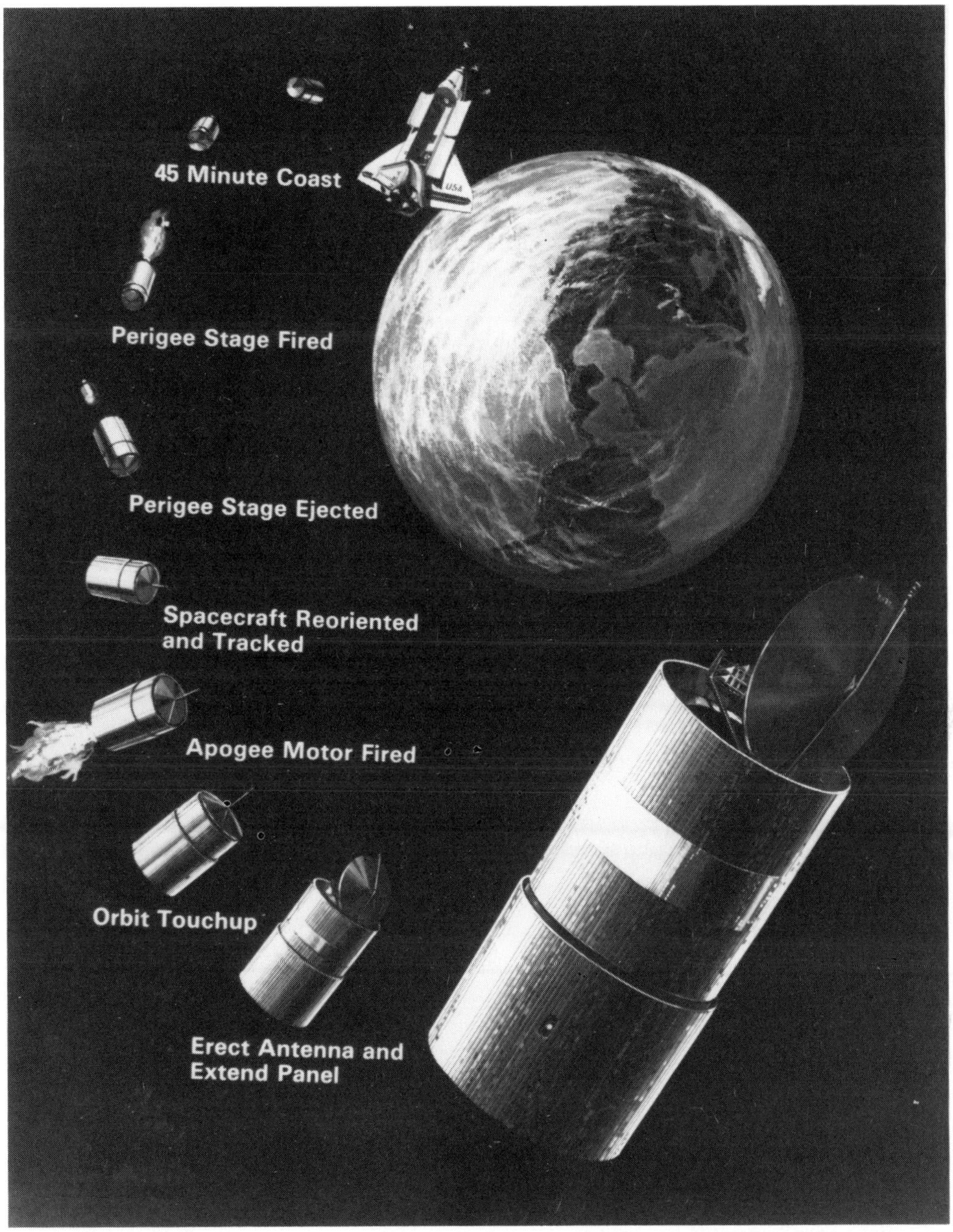

Deployment sequence of the Morelos satellites on launch by Shuttle. Both satellites were launched in 1985 by Shuttle

TECHNICAL SUMMARY

PLATFORM	HS–376
PRIME CONTRACTOR	Hughes Aircraft Company
OTHER MAJOR CONTRACTORS	
STABILISATION	Spin stabilised
LAUNCH MASS	1140 kg
MASS IN ORBIT (BOL)	666 kg
POWER SUPPLY, END OF LIFE	800 Watts
DESIGN LIFE	9 years
ORBITAL POSITIONS	Morelos A, 113·5° West Morelos B, 116·5° West
STATION KEEPING ACCURACY	
ECLIPSE PROTECTION	
PROPULSION SYSTEM	
SPACECRAFT DIMENSIONS	2·16 m diameter, 7·09 m height, (deployed)

SPAN, TIP-TO-TIP	
TT&C STATION	Iztapalapa (Mexico City)
NO. OF ACTIVE TRANSPONDERS	22 (18 at C-band, 4 at Ku)
NO. OF SPARE TRANSPONDERS	6 (4 at C-band, 2 at Ku)
TWTA POWER OUTPUT	7 Watts, 10·5 Watts & 20 Watts
FREQUENCY RANGE:	
UPLINK	
DOWNLINK	
TRANSPONDER BANDWIDTH	12 at 36 MHz (7 Watt C-band) 6 at 72 MHz (10·5 Watt, C-band) 4 at 108 MHz, (20 Watt, Ku-band)
EIRP MINIMUM, BEAM EDGE	
CHANNEL POLARISATION	

COMMERCIAL DETAILS

OPERATOR	Telesat Canada (see project overview)
ADDRESS	333 River Road, Ottawa, Ontario K1L 8B9, Canada
TELEPHONE NO.	(613) 746-5920
TELEX NO.	053 4184 telesatott
FAX NO.	
SENIOR STAFF	David Sward, Manager, MSAT programme and Manager Business Network Planning Debbie Look, Senior Market Analyst
FINANCIAL BACKING	Satellite to be financed, owned and operated by Telesat Canada
PURPOSE OF PROJECT	Provision of mobile communications capacity in the UHF band and thin route capacity in L-band for Canada in conjunction with an identical system for the USA
PROJECT STATUS	Awaiting institutional arrangements and spectrum allocation with the USA. Dependent on decisions from the Federal Communications Commission on US mobile satellite communications
LAUNCH DATES	Early 1990's; launch vehicle not yet chosen

SATELLITE USERS

The majority of MSAT users are expected to be organisations involved in transportation, mining, forestry, services and government; this is to be expected since much of the activities within these industries takes place in rural and remote areas of Canada. Telesat says that some applications include:

Forest fire fighting communications.
Remote area medical emergency response.
Communication to and from remote construction or mineral exploration camps.
Data transmission to and from remote oil well sites.
Communications on inter-city passenger transportation.
Inter-city trucking.
Law enforcement voice and data communications.
Data acquisition and control platforms.
Wide area paging.

Telesat says that it forecasts some 80,000 users by the year 2001. Some organisations have already enlisted for early use of MSAT.

MSAT is intended to operate in both the UHF and L-band frequencies and will provide compatibility with terrestrial, aeronautical and maritime mobile satellite services.

However, the final system configuration will depend on which US MSS applicant is chosen as a partner with Telesat. The FCC has decided to allocate only 4mHz of UHF spectrum, leaving, in effect, only the L-band for MSS services. Canada may defer use of the UHF band to future MSS satellites.

PROJECT OVERVIEW

Telesat Canada, intending to operate the MSAT mobile satellite communications system for Canada, is proposing a single satellite in the system for Canada and one for the USA. Each satellite would act as back up for the other in the case of in-orbit failure. The proposed "joint system" operated by separate operators in each country not only requires institutional agreements with the US operator, which has yet to be chosen from 12 applicants, but also use of the same frequencies.

Studies on the MSAT project began in Canada in the late 1970's by the Department of Communications in conjunction with industry. The Canadian government has designated Telesat Canada to be the owner and operator of the Canadian MSAT systems. Telesat says it has been working closely with several of the 12 US mobile satellite applicants towards providing the service on a North American basis. The objective is to reduce capital and development costs of the project. Studies of the economics of MSAT suggest that the first generation system would be only marginally profitable, although a second generation system, taking advantage of a market place where penetration had already built up, would generate substantial profits.

The first generation Canadian MSAT system would be capable of serving 35 UHF users and 25,000 L-band users. The UHF band would be used for true mobile applications. The L-band would be "better suited" to fixed and transportable applications that would allow the use of higher gain antennas. MSAT is aimed at providing services for mobile radio and telephone users outside metropolitan centres and organisations not currently using mobile communications because services are unavailable or uneconomic. Telesat appears somewhat unsure about the extent of this market. Telesat proposes MSAT services to be premium price services. It plans to work in cooperation with end user service providers and revenues would be divided between service providers and Telesat Canada. Service providers would retail mobile satellite services to end users. This would involve marketing, sales and customer services, installation and maintenance of mobile terminals and base stations.

A major problem in providing any first generation satellite service is the rate at which the market is penetrated. Telesat studies indicate a relatively rapid growth of users up until 1996 when growth falls off. It forecasts about 80,000 users by the year 2001. However, some US applicants for mobile satellite communications are highly sceptical about the reliability of such forecasts. This first generation satellite is capable of accommodating 60,000 users assuming an allowable peak busy hour blocking factor of 15%. Telesat Canada assumes in its baseline forecast that a mobile terminal will cost $4500, access charge will be $50 per month and airtime usage fee of $1·50 per minute (all figures in Canadian $). This gives a total revenue potential in the first seven years of operation of $650 m. It is looking to capture 5–7% of the total mobile market in Canada.

TECHNICAL SUMMARY

PLATFORM	not available
PRIME CONTRACTOR	not available
OTHER MAJOR CONTRACTORS	not available
STABILISATION	not available
LAUNCH MASS	not available*
MASS IN ORBIT	not available
POWER SUPPLY, END OF LIFE	not available
DESIGN LIFE	not available
ORBITAL POSITIONS	not available
STATION KEEPING ACCURACY	not available
ECLIPSE PROTECTION	not available
PROPULSION SYSTEM	not available
SPACECRAFT DIMENSIONS	not available

SPAN, TIP-TO-TIP	not available
DOWNLINK FREQUENCY:	
SATELLITE TO MOBILE, UHF	821–825 MHz
SATELLITE TO MOBILE, L-BAND	1·544–1·559 GHz
MOBILE TO SATELLITE, UHF	866–870 MHz
MOBILE TO SATELLITE, L-BAND	1·6455–1·6605 GHz
UPLINK FREQUENCY:	
BASE TO SATELLITE	14 GHz
SATELLITE TO BASE	11 GHz
DOWNLINK EIRP	26·5 dBW EOC (UHF) 28·5 dBW EOC (L-band)
MOBILE ANTENNA GAIN	8–12 dBi
VOICE CHANNEL BANDWIDTH	5 kHz

*Satellite is expected to be in the 2500 kg class.

Left: Alternative coverage patterns proposed for MSAT. Future of UHF proposals remains uncertain.

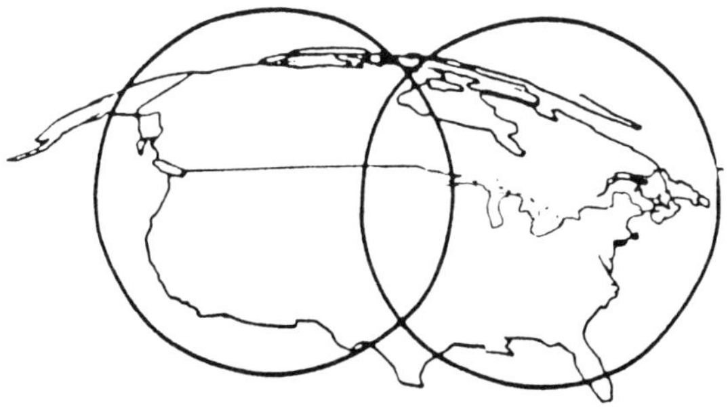

2 - BEAM UHF COVERAGE

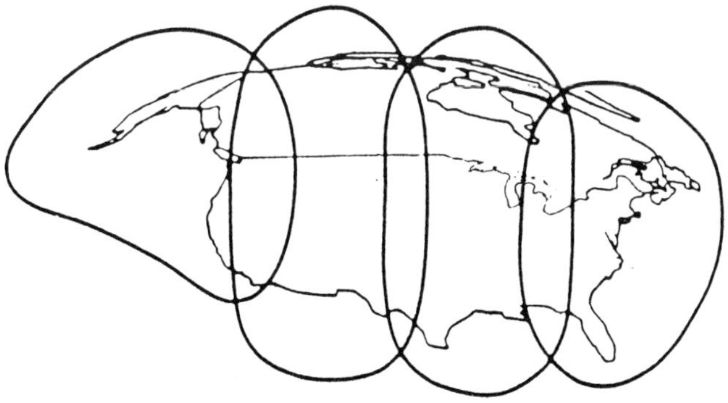

4 - BEAM L-BAND COVERAGE

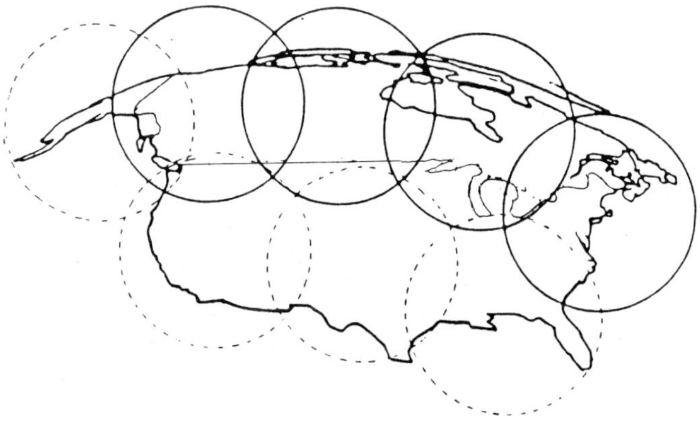

8 BEAM L-BAND COVERAGE

COMMERCIAL DETAILS

OPERATOR	It is expected that the Federal Communications Commission will grant construction, launch and operating authority for only one US Mobile satellite service operator for the time being. Some twelve companies have applied to the FCC for authorisation. It is not clear at the time of writing whether the FCC will give authorisation to one of these or a consortium of several or whether the authorised system will operate in conjunction with Telesat Canada's MSAT programme
PURPOSE OF PROJECT	Various applicants are proposing mobile communications capacity offering a variety of services aimed at differing definitions of the market. Some are offering thin-route FSS type services and Radio Determination Satellite Services (RDSS)
PROJECT STATUS	Twelve applications and allocation of frequencies are currently under consideration by the FCC which is also pushing various applicants to form a consortium of applicants. FCC go-ahead is expected in 1987
LAUNCH DATES	The first satellite is not expected to be launched until at least 1990. "Interim" services using FSS micro earth stations may provide an earlier start, although, so far no applicants appear to have shown an interest

SATELLITE USERS

Currently some 11% of the land area and 7% of the population of the USA is not served by basic telephone services. MSS proposals envisage this as a market opportunity for thin-route traffic of an FSS nature. The market may well be served by Ku-band FSS VSAT systems. Proposals are also aimed at pure mobile voice communications, including mobile telephony (full duplex) and mobile radio (half duplex). Other services for both mobile and fixed include data communications (up to 2400 bit/s) and messaging (packets of information). Three of the twelve applicants also propose radio determination satellite services (MCCA, McCaw and Omninet). All the proposed services face strong potential competition from alternative technologies.

Services on offer are directed at the the following markets:
Trucking; interstate, intrastate, courier and package delivery
Railroads
Aeronautical, aircarrier, general aviation and air traffic control
Resource allocation
Marine; coastal and inland waterways and harbour
Public utilities
Public safety; police, fire, rescue and emergency medical
Forestry
Construction
Agriculture
Motor cars; rental and private fleets
Governmental.

The markets aimed at are characterised by low communications density and wide-area coverage. Addressable market estimates range from 300,000 to 3 m user terminals in use by 1990.

PROJECT OVERVIEW

The Mobile Satellite Services market in the USA is complex. It is perceived as consisting of four main elements: 1) pure mobile communications for voice and data; 2) thin-route communications in rural areas to and from fixed terminals and as an alternative to terrestrial services; 3) aeronautical mobile communications and 4) Radio Determination Satellite Services (RDSS) providing locational information and text only information communications. The extent of these markets is uncertain and they are all subject to competition from other existing and new communications services. Moreover the final allocation of frequencies for MSS has yet to be finalised. The FCC believes it desirable that the MSS service be implemented jointly with Canada's MSAT proposals. Canada proposes use of both the UHF band at 800 MHz and L-band at 1·5 GHz for mobile satellite services. US industrial interests have been attacking the prospects for allocation of 800 MHz for MSS and the FCC has subsequently allocated only 4 MHz of the UHF band to MSS. This leaves Canada's system potentially incompatible with that of the USA.

L-band frequencies to be allocated for FSS services make use of frequencies which were originally allocated for aeronautical satellite services and aeronautical emergencies but which have been fallow for 25 years. The market for MSS systems also appears to be changing rapidly. With increasing competition the rural market may not exist in five years. If the FCC gives authorisation in 1987, it will take at least a year to draw up system specifications and issue an RFP and a further three years before the first satellite becomes operational. Meanwhile it is becoming apparent that digital/alpha-numeric services are taking an increasing lead over voice traffic demand. RDSS positioning/navigation systems including ancillary paging and alpha-numeric messaging addresses much of the MSS market and FCC authorisation for an RDSS system is expected early in 1986. Geostar is the early entrant into the market.

US MOBILE SATELLITE SERVICE APPLICANTS: SUMMARISED DETAILS

Operator	Beams per satellite		Channel bandwidth (kHz)	Channels per satellite		Normalised to provide "truly mobile" services*			Satellite G/T at Edge dBi/K		Satellite EIRP at Edge dBW/Channel		No. of satellites in total system (ground spare)	Orbit location (all West)
	UHF	L-band		UHF	L-band	Normalised channel capacity	Spectrum usage factor	Capital cost per channel per year ($)	UHF	L-band	UHF	L-band		
i) High Capacity Systems														
Global Land Mobile	6	6	15	653	653	2428	1·5	21,700	+11	+11	55	50		94° & 129°
MCCA American Satellite	10	—	30	640	—	1600	2·0	20,700	+11·1	—	35·8	—		96°
MCCAW Space Tech.	20	2	5	1700	90	1543	1·3	24,500	+7·3	−0·3	27·7	24·7		101°
North America Mobile Sat.	4	—	4–5	800	—	984	1·2	41,600	+5·5	—	30 & 33	—		129°
Omninet Corporation	12	17	5	2208	3840	6048	2·2	10,800	+4	+10	26·2	29·3	4 (1)	89° & 119°
Satellite Mobile Telephone	12	40	5	1200	4200	10460	3·7	8,400	+5	+11	29	35		75° & 120°
ii) Low Capacity Systems														
Globesat Express	1	—	5	790	—	490	0·13	316,500	−25	—	7	—	50	50 satellites in LEO
Hughes Communications	1	1	5	400	732	606	0·31	51,100	−1	0	25·3	14 & 24		135°
Mobile Satellite Corporation	1	3	4, 5, 15	800	4725	940	0·26	57,900	−0·8 −9·2	+4·5 −0·4	20·3	22·2	3	85° & 125°
Mobile Satellite Service	1	1	4, 15, 30	667	1000	248	0·05	171,600	−5	0	25·6	24·6		85° & 125°
Skylink Corporation	2	8	5	400	800	609	0·31	68,000	−0·5	+3·7 28·6	22·6 30·3	26·4	1	136°
Transit Communications	—	4–5	5–10	—	736	412	0·82	106,300	—	+4·2	—	24·7		80° & 113°

*For comparative purposes, the 12 applications have been "normalised" to deliver "truly mobile" communications services. That is to say, the required link budget to support 5 kHz channels over the full amount of spectrum with a 5 dBi omni-directional user mobile antenna has been imposed on all of the applications. This analysis results in a normalised channel capacity for each system and a spectrum usage factor which reflects frequency re-use. Similarly, the system capital costs have been normalised to reflect an equivalent cost per channel per year.

(Diagram Courtesy of Richard L. Anglin Jnr., Omninet Corporation, USA)

Note: Subsequent to these plans being submitted, the Federal Communications Commission decided to allocate only L-band frequencies to US mobile satellite services, with a nominal 4 MHz in the UHF band to placate Canada.

US position also differs from that of Canada in that, in general, the Americans do not see MSS as a premium price service yet the proposals still envisage an investment requirement of $400–600 m. Moreover, the exercise requires considerable coordination between the satellite operator, the network for marketing, selling and maintaining the ground segment of the services and also the manufacturers of individual receivers. Forecasts of take up of the service generally estimate that one-third of users will take up the service immediately it becomes operational. It is not clear at this stage if a consortium of applicants could be created with the necessary experience to deliver services required by the users and there are commercial objections to sub-contracting out work on major services that are integral to the system. Mobile satellites require market aggregation of a variety of services – no one market can support it.

The history of decision taking which has led to the current position for MSS services in the USA dates back to the 1960's with discussions between Canada and the USA. WARC '79 allocated frequencies for MSS. In 1982 NASA petitioned the FCC to establish a US domestic MSS system using the UHF spectrum in the 800 MHz range. The FCC issued a Notice of Proposed Rulemaking on 28th January 1985 to establish a Mobile Satellite Service and licence a domestic service provider. The Notice also allowed ancillary fixed services. On 30th April 1985, 12 companies filed applications and over the Summer of 1985 the FCC was considering petitions for denials and rejection of applications. The FCC is legally required to allocate frequencies and issue the service rules one year after the issue of Notice of Proposed Rulemaking. However the FCC has ways of getting round this problem, primarily by making a preliminary and perfunctory order. The issue is so complex that the FCC is believed to require a considerable amount of time before it can finalise its position.

WARC '79 accepted an allocation in the 806–890 MHz band in Region 2 (the Americas). The Canadian Department of Communications and the FCC proposed an initial allocation of 8 MHz in the 821–825 MHz and 866–870 MHz and invited comments on 28 MHz of the possible L-band (1·5–1·6 GHz) band. NASA had initially requested 20 MHz of UHF for MSS. Many of the twelve MSS applicants argue that 8 MHz of frequency is insufficient. Existing terrestrial cellular radio system operators have also requested retention of adjacent 800 MHz spectrum for further expansion. The public safety community has argued that UHF be preserved for their future needs. Of the L-band, 1·53–1·544 GHz and 1·6265–1·6455 MHz is allocated to Maritime Mobile Satellite services and is used by Inmarsat. The only unused spectrum at L-band (1·545–1·559 GHz) and (1·6455–1·6605 GHz) is allocated to aeronautical mobile satellite service but has yet to be used for this service.

Canada and the US are expected to propose to WARC '87 that this band be allocated to MSS. An L-band only MSS simplifies the design of the system and offers operational advantages if it is assumed that 20 MHz of frequency is required for MSS. The FCC has decided on minimal frequency allocation in the UHF band for MSS services frequency allocation. The US Federal Aviation Administration (FAA) has said that the 28 MHz of L-band allocated for aeronautical MSS services may be insufficient for aviation requirements by the year 2010. Inmarsat is now looking at providing aeronautical MSS services and will use a slice of L-band Aeronautical MSS frequency for this.

The bulk of the twelve applicants have been described as outfits with "big plans, little money".

COMMERCIAL DETAILS

OPERATOR National Christian Network Incorporated

ADDRESS 1150 West King Street, Cocoa, Florida 32922, USA

TELEPHONE NO. (305) 632-1000

SENIOR STAFF Ray Kassis, President

FINANCIAL BACKING

PURPOSE OF PROJECT Provision of DBS capacity for contiguous USA

PROJECT VALUE $700 m including TT&C facilities, launch and insurance costs, costs of first year of operations and programming facilities

LAUNCH DATES NCN–1, 1988, STS
NCN–2, 1989, STS

SATELLITE USERS

See Project Overview.

PROJECT OVERVIEW

National Christian Network was a second-round application and has conditional approval to begin construction of its spacecraft. The Company intends to use a scaled up version of the RCA Astro-Electronic Satcom satellite each with six 230 Watt channels. It has requested two orbital slots from FCC at 101° West and 148° West but the FCC has already allocated all 32 channels available at 101° West to Hughes Communications. National Christian Network envisages that programming on the satellite will be advertisement supported although it will also offer one or two pay channels.

TECHNICAL SUMMARY

PLATFORM	Advanced Satcom
PRIME CONTRACTOR	RCA Astro-Electronics
STABILISATION	3-axis
LAUNCH MASS	not available
MASS IN ORBIT	not available
POWER SUPPLY, END OF LIFE	not available
DESIGN LIFE	not available
ORBITAL POSITIONS	101° West and 148° West requested
STATION KEEPING ACCURACY	not available
ECLIPSE PROTECTION	not available
PROPULSION SYSTEM	not available
SPACECRAFT DIMENSIONS	not available
SPAN, TIP-TO-TIP	not available
NO. OF ACTIVE TRANSPONDERS	6
NO. OF SPARE TRANSPONDERS	not available
TWTA POWER OUTPUT	230 Watts
TRANSPONDER BANDWIDTH	27 MHz
EIRP MINIMUM, BEAM EDGE	54 dBW
CHANNEL POLARISATION	Circular

COMMERCIAL DETAILS

OPERATOR	Nordiska Telesatellit Aktiebolaget
ADDRESS	Marievik 1A, S-117 43 Stockholm, Sweden
TELEPHONE NO.	(8) 763-0240
SENIOR STAFF	Karl Eriksson, President John R. Veastad, Vice President Lars Anderson
FINANCIAL BACKING	Notelsat is currently owned on a 50:50 basis by the Swedish and Norwegian PTT. It is not clear at this stage what the financial structure of Notelsat will be when TELE–X becomes operational and further satellites are launched
PURPOSE OF PROJECT	Follow on to the TELE–X programme to provide DBS and Data/Video capacity to Nordic countries on a fully operational basis
PROJECT STATUS	Plan under consideration
PROJECT VALUE	not available
LAUNCH DATES	Late 1980's with launch by Ariane or STS

SATELLITE USERS

Notelsat has not been specific about the intended customers for the TELE–X or Notelsat–1 satellites but, like TELE–X, Notelsat is intended to provide data/video and DBS capacity. A combination of TELE–X and Notelsat–1 can provide seven DBS channels. When TELE–X is phased out and replaced with a second Notelsat spacecraft, eight channels can be provided. The Nordic countries were allocated eight DBS channels under WARC '77. Seven TV channels are currently produced by public service broadcasters in Norway, Sweden, Denmark, Finland and Iceland. Due to low population densities, distances and in Norway in particular, mountainous terrain, terrestrial systems for extra channels are expensive in per-household terms. There has not yet been any agreement to use TELE–X or follow on Notelsat spacecraft by either existing Nordic broadcasters or new enterprises interested in DBS TV. Moreover, only two countries have a state-holding in Notelsat – Sweden and Norway.

According to Notelsat, terrestrial telecoms networks in Scandinavia for data communications are rapidly expanding through investment in fibre optic cables. Data/video communications via satellite "will only be a complement to the terrestrial network". Notelsat again is somewhat vague about who will use these services.

PROJECT OVERVIEW

The Notelsat–1 and 2 hybrid DBS/business communications satellites are under consideration as follow-on spacecraft for the pre-operational TELE–X satellite which will initially be operated by the Swedish Space Corporation before handing over to the Notelsat company in 1988. The Notelsat–1 and 2 satellites described here are no more than "possible specifications" for an operational system. Notelsat–1 would operate in conjunction with TELE–X and provide preemptible capacity in the event of failure of either satellite. Notelsat–2 is envisaged as a replacement for TELE–X when it is phased out of operation.

The specifications of the operational Notelsat spacecraft are expected to reflect technical developments in ground equipment resulting in a requirement for a smaller spacecraft platform but with twice the TV channel capacity and twice the data/video capacity. Notelsat says it is looking at three European platforms for the Notelsat system – DFS, Eurostar and Spacebus 100. Following the decision to use C–MAC modulation on Nordic TV services and general developments in home receiver technology it is possible to reduce satellite power output by 4 dBW compared with TELE–X but retaining the original antenna size and picture quality envisaged with TELE–X. Consequently TWTA power output can be reduced to 100 Watts or less and the number of operational TV channels increased from two to four. For data/video traffic, the introduction of forward error correction and adaptive output power in the earth stations the performance has been improved by 3–4 dBW. The 2·5 m earth station in TELE–X has a G/T = 22 dB/K. Developments in antenna design could increase G/T by 3 dBW without any increase in cost. The CCIR working groups on video conferencing are working to introduce as standard a bitrate of 384 kbit/s in 1988. Current standard is 2 mbit/s which was used in formulating payload requirements for TELE–X.

This development reflects new compression techniques. The two operational 100 Watt transponders on Notelsat–1 dedicated to video/data traffic will be capable of carrying 560 × 383 kbit/s per channel, despite a reduction in TWTA power output from 230 Watts in TELE–X. TELE–X is capable of 300 × 384 kbit/s channel capacity. Total bandwidth available on each Notelsat satellite for data/video is 460 MHz corresponding to 1120 channels of 384 kbit/s.

TECHNICAL SUMMARY

COMPARISON OF NOTELSAT REQUIREMENTS WITH TELE–X

	TELE–X	*NOTELSAT–1*
TV MISSION:		
NO. OF TV CHANNELS	2	4
EIRP, SCANDINAVIA	60 dBW	56 dBW
EIRP, ICELAND	not available	50 dBW
TWTA POWER OUTPUT	230 Watts	100 Watts (or less)
MODULATION SYSTEM	C–MAC	C–MAC
DATA/VIDEO MISSION:		
NO. OF TRANSPONDERS	2	2
BANDWIDTH	40 + 86 MHz	115 + 115 MHz
NO. OF 384 kbit/s CHANNELS	300	560
TWTA POWER OUTPUT	230 Watts	100 Watts
OVERALL SATELLITE CHARACTERISTICS:		
PLATFORM	DFS, Eurostar or Spacebus 100	Spacebus 300
TWTA REDUNDANCY	3:2	2:1
LIFETIME	7 years	10 years plus
SOLAR ARRAY POWER	3200 Watts	2200 Watts
MASS AT LAUNCH	2130 kg	less than 1500 kg
LAUNCH VEHICLE	Ariane	Ariane or STS

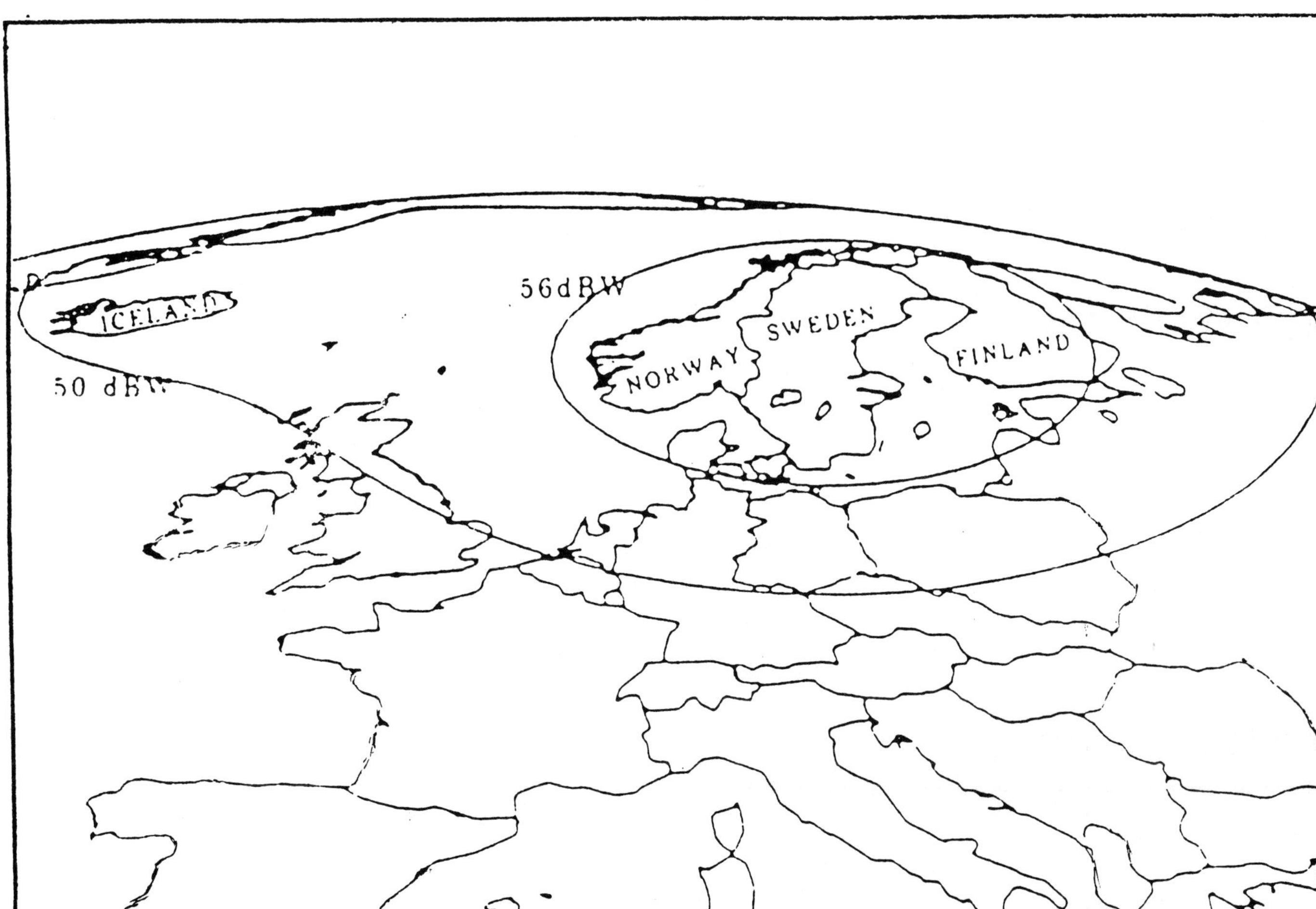

COMMERCIAL DETAILS

OPERATOR	European Space Agency
ADDRESS	8-10 rue Mario Nikis, 75738 Paris, Cedex 15, France
TELEPHONE NO.	14 273 7654
TELEX NO.	202 746 F
FAX NO.	14 273 7560
SENIOR STAFF	Reimur Lust, Director General Giorgio Salvatori, Director, Telecoms Programme George van Reeth, Director, Administration
PURPOSE OF PROJECT	Experimental satellite intended to provide ability in construction of large satellite platforms and to test new payloads for future satellite designs
PROJECT STATUS	Delayed. The satellite is not expected to be launched until October 1988
PROJECT VALUE	462 m Accounting Units at 1981 price levels including launch, insurance and operational costs
LAUNCH DATES	Due for launch on Ariane 3 V32, in October 1988 or Ariane 3 V30 in June 1988

SATELLITE USERS

It is intended that the satellite will be made available to organisations outside ESA for experimental demonstration, pilot and quasi-operational services. Such users would be limited to those under the jurisdiction of ESA member states. During 1985 the Agency considerably stepped up its efforts to find such users but it is understood that the response has been disappointing. Launch of the satellite is still some two years away, and users may not materialise until after launch.

However, agreements have already been signed for the use of the DBS payload. One of the two high power transponders has been allocated to the Italian public service television broadcaster, Radiotelevisione Italiana (RAI). The latter has two years free use of the transponder and will pay only for the third year if Italy goes ahead with its own domestic DBS satellite system. The other transponder has been allocated to the European Broadcasting Union and was intended for use by Europa TV, transferred from Eutelsat I–F1. Europa TV would have free use of the transponder for three years. The DBS payload will also be available to the experimental user. However, Europa TV folded in November 1986 and RAI is understood to be having doubts about its plans for use of Olympus.

PROJECT OVERVIEW

The Olympus satellite is the European Space Agency's second major R&D communications satellite project, following on from its OTS programme. Originally known as L–SAT (an acronym for Large–SATellite), the origins of the Olympus satellite date back to the 1970's, when industry decided that the satellite market would require a larger platform than ECS. In parallel, a general study of the future prospects for European telecoms satellites was undertaken by the European Space Agency at its ESTEC facilities. This study, completed early in 1979, coupled with the work done by industry, led to an ESA Council decision in late 1979 to proceed with a Definition Phase Study for this large R&D project. This was to develop and demonstrate a large, multipurpose platform and new communications hardware on board the spacecraft. However, the project did not gain universal support amongst ESA member states. France and Germany opted out of the project in favour of a bilateral R&D programme to develop both a large platform and a DBS payload.

The outcome of this was the TDF–1/TV–SAT programme. France and Germany thought that Olympus was too ambitious and that a larger platform was needed to meet market demand before Olympus was tried and tested. In the event, TDF–1/TV–SAT have been delayed by a combination of over-runs in construction of the spacecraft and the non-availability of the Ariane launch vehicle. It is likely that TV–SAT will be operating about a year before Olympus, which has similarly suffered from construction delays, with the latter being launched in mid-1988. Olympus will then be delivered into orbit some three years after the date envisaged in 1979. So far no orders have materialised for satellites using platforms derived from the Olympus design. In 1979 ESA calculated that the Western World market for satellites in the Olympus class would total 150 spacecraft in the decade after the demonstration satellite was launched. It further calculated that Europe would sell between 20 and 30 of such satellites. British Aerospace, prime contractor for the first Olympus satellite, similarly calculated that the "open market" would require 150 large satellites by the year 2000 of which Olympus would capture around 40.

However, despite the current lack of firm orders, the platform in a much modified form, is likely to be used for the Canadian Radarsat earth observation project. The platform has been variously associated with plans for UK DBS, the Intelsat VII series and the Italian Sarit project. To be price competitive, a "production-line" for the platform needs to be built up. The

Artist's impression of the Olympus 1 spacecraft in orbit (courtesy of British Aerospace)

current prospects for this appear limited. It is possible that ESA will order a second Olympus satellite as part of its R&D programme but the organisation has not announced the intended payload.

Organisation of the Olympus Project: Like most ESA projects, the industrial contracts for Olympus were parcelled out to ESA member states in rough proportion to their contribution to the project. Nevertheless, within that framework, there was some competition between individual contractors. British Aerospace was selected as Prime Contractor at the outset of the definition phase in 1980, following a "competitive" assessment by ESA. Similarly a second tier of major subcontractors was selected. Finally there is a tier of equipment level contractors selected by tender action. The industrial structure of the Olympus programme is detailed below.

When the contract for the construction of the Olympus satellite was signed in 1981, the overall project budget was 462 m Accounting Units (roughly equivalent to $462 m) at 1981 prices. This covered the cost of three development model spacecraft and procurement of two flight models, launch and launch insurance, operations and payload tests and demonstrations, studies, test facilities and ESA programme management. The second flight model has yet to be ordered. Inflation, delays and possibly much increased insurance costs, are expected to push the final bill up to at least $550 m, excluding the second flight model. The cost of the project is met by the eight ESA members participating in the programme. Principal contributors are the US, with 34·3% of the total budget, Italy with 32·8%, Holland with 11·8% and Canada with 9%. Belgium has a 3·7% stake, Spain 2·6%, Denmark 1·3% and Austria 0·75%.

Design of the Spacecraft: There is little about the basic Olympus platform design which is novel, except for its size. Launch mass of the satellite is 2450 kg and power supply 3300 Watts. The basic Olympus design can be used to produce a satellite with a launch mass of up to 3700 kg with a power supply of 7700 Watts. One of the principal problems that faced British Aerospace in marketing the Olympus platform was the lack of a suitable transfer stage for Shuttle launches. Under contract to Scott Science and Technology of California, BAe has designed a Satellite Transfer Vehicle. However, as commercial satellites have now been banned from Shuttle the transfer stage has little impact on Olympus platform sales prospects and appears to have been abandoned. Of greater technical interest is the Olympus payload. This involves steerable spot beams, use of Ka-band frequencies and TDMA satellite switching. The satellite has four basic payloads:

1) a DBS payload; 2) a Ku-band specialised services payload for advanced communications between small earth stations; 3) a Ka-band communications payload for point-to-point and multipoint teleconference and other experimental applications; 4) a 12/20/30 GHz propagation package to complement and verify propagation statistics in the higher frequency range.

Direct Broadcast Satellite Payload: This consists of two 230 Watt travelling wave tube amplifiers to provide a pre-

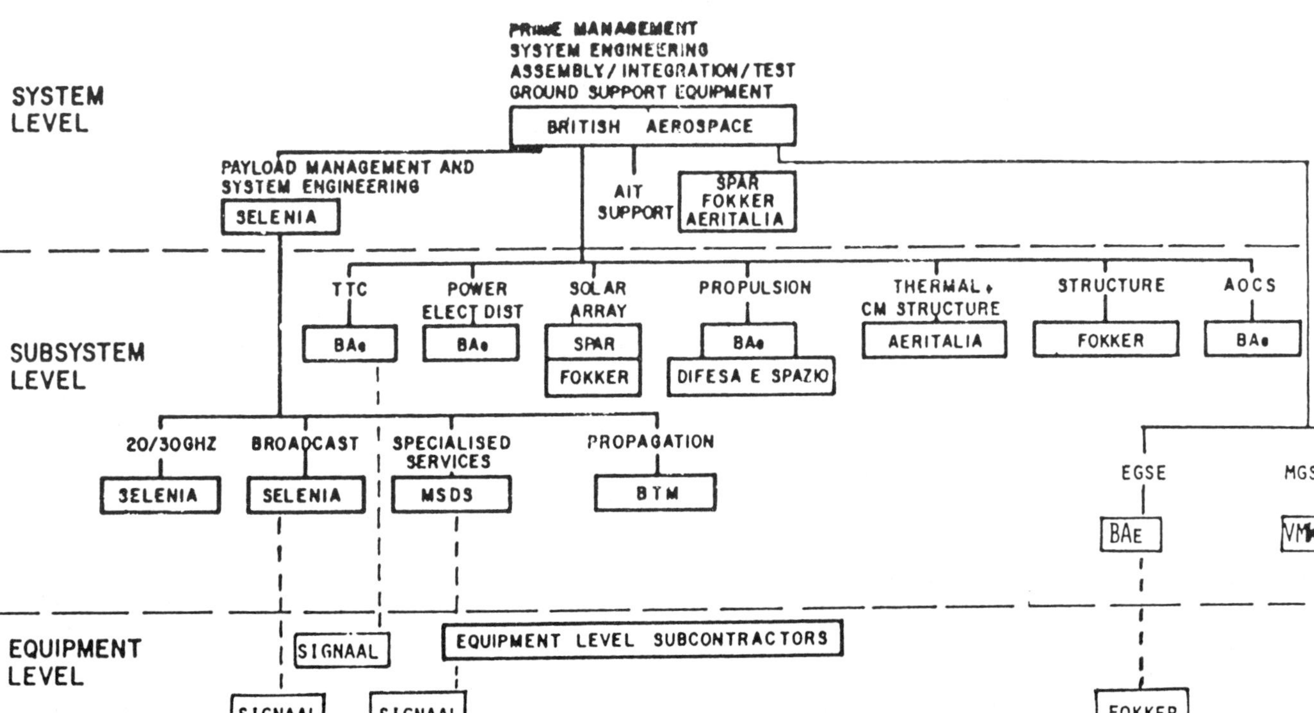

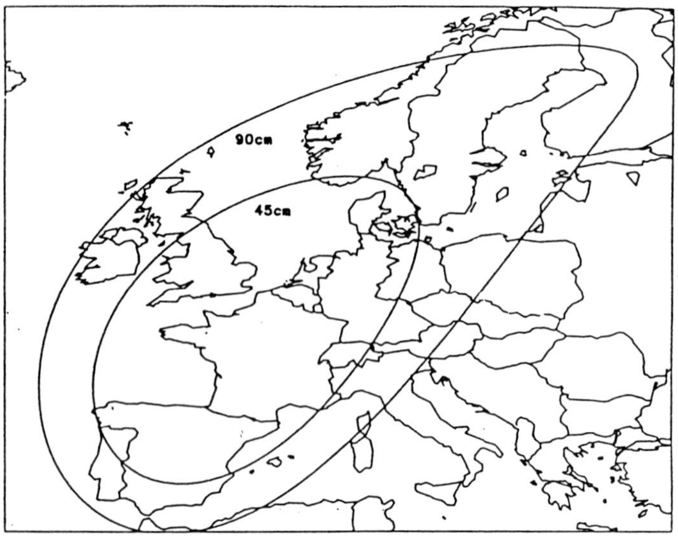

Typical example of "fair quality" reception from the Olympus European DBS beam. Example is in clear sky and provides a C/N ratio of 9 dBW. The European beam is steerable to provide a different pattern of coverage in Europe

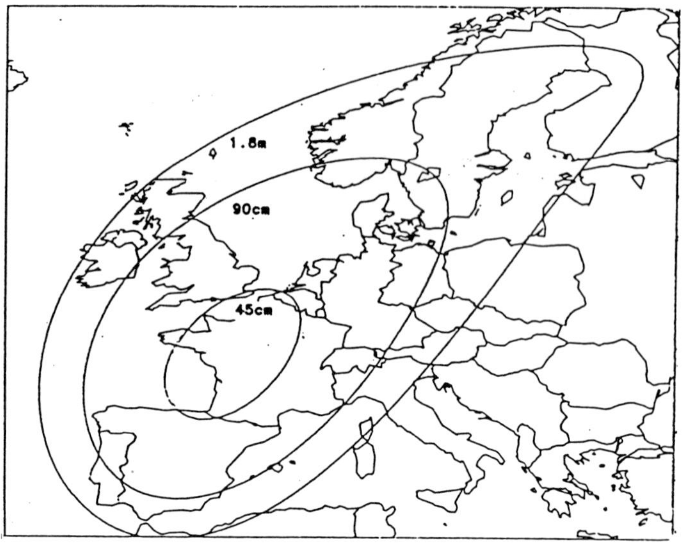

Footprint shows same beam pointed in the same direction and the size of dish needed for "excellent quality" reception in clear sky. C/N ratio is 14 dBW. This would provide the quality of reception needed by, say, cable TV networks

operational channel for Italy and an experimental channel with a steerable transmit beam to provide coverage of all parts of Europe. The Italian beam is nominally fixed but can also be repointed. Both beams employ a moderate fall-off of the downlink footprint shape beyond the basic 3 dBW beam width. This enables both beams to be used to transmit to larger diameter earth stations well outside their coverage area for DBS services. This is illustrated through the following footprint maps. The European beam uses frequency allocated to Austria under WARC '77 – the left-hand circular polarised channel number 20. The Italian DBS channel also uses an Italian WARC '77 frequency allocation (but with right-hand circular polarisation) channels 24 and 28. The Italian frequency can be shifted between channels 24 and 28. The European beam, if necessary, can be switched to right-hand circular polarisation. The European beam downlinking unavoidably conflicts with the country-by-country frequency allocations established at WARC '77, but given that Europe has largely, so far, failed to establish National DBS TV, this is likely to have limited impact.

The Olympus satellite is not intended as a fully operational spacecraft. However, its DBS payload looks like being used for close-to-operational services. Public Service television broadcaster Radiotelevione Italiana (RAI) will be using the Italian beam to extend its experience in satellite television including possible use for High Definition TV. Italy has a plan called Sarit for its own domestic DBS operations, taking over from Olympus. RAI currently transmits its first national TV channel through Eutelsat I–F1. This is due to transfer to Eutelsat I–F2 before transferring to the Italian transponder on Olympus. More important is the use of the Olympus European transponder by the European Broadcasting Union. Five members of the EBU, Ireland, Portugal, West Germany, Holland and Italy own a public service broadcasting type channel, Europa TV. This was to be transferred to Olympus once the latter is operational, but folded in 1986.

The DBS payload on Olympus will also be made available to other users outside the transmit hours used by RAI and Europa TV. It is envisaged that these users will use Olympus to demonstrate a variety of services to peoples' homes and point-to-multipoint business and professional services. Dutch media company VNU has submitted a proposal to ESA to demonstrate a "home information service."

The European channel can be pointed at any European country, but full European coverage can only be obtained by combining the Italian and European beams.

Specialised Services Payload: This payload is notable because it employs multibeam coverage patterns with separate receivers and transmitters for each beam coupled with satellite-switched time-division multiple access techniques (SSTDMA). The latter overcomes a major problem with multibeam payloads in that it cuts down on the number of transmit chains on-board the satellite. SSTDMA allows part-time interconnection of each pair of receive beams and transmit beams, according to the expected traffic. Each beam therefore only requires one transmit chain. It also requires a switch matrix on board the satellite which is able to set up a pre-selected sequence of different connections between the receivers and the transmitters. The multibeam payload allows considerable frequency re-use, sufficient, in the Olympus case, to avoid frequency re-use by polarisation means. Nevertheless, there are likely to be possible interference problems with the Intelsat system in both up and downlinks used by the Olympus specialised services payload.

The specialised services payload consists of five spot beams and four 30 Watt transponders. There are four receive chains and four transmit chains interconnected by a SSTDMA switch matrix, receiving and transmitting over a single multi-beam antenna which serves the five spot beams. The antenna is fully steerable, enabling all five spot beams to be moved en bloc. Individual spot beams are not steerable. The system is intended to demonstrate business services with small ground terminals (3–3·5 metres). For SSTDMA demonstrations on the satellite, up to four repeater chains can be used. Each would operate over a different spot beam and all work with an 18 MHz nominal bandwidth. Through use of spatial frequency re-use two repeater chains could operate on one frequency and two operate on another. For other demonstrations two repeater chains can be switched to wideband operation (36 MHz). EIRP at the edge of the beam is 44·3 dBW with a G/T ratio of 1·1 dB/K. Earth terminals for use with the SSTDMA mode, using 24·576 mbit/s QPSK signals are intended to use 3–3·5 metre diameter dishes, FET receivers with 290 K noise temperatures and transmitters with 200 Watt power output.

Channels	Italian Beam: Channel 24 "European" Beam: Channel 20 or 28
Bandwidth	27 MHz
Uplink Frequency	Channel 20: 17691·90 MHz Channel 24: 17768·62 MHz Channel 28: 17845·34 MHz
Uplink Polarisation	RHCP (Right-Hand (Clockwise) Circular Polarisation)
Uplink Coverage	European (TVB3) Beam: 3·3° × 2·0°
Downlink Frequency	Channel 20: 12091·90 MHz Channel 24: 12168·62 MHz Channel 28: 12245·34 MHz
Downlink Polarisation	Italian Beam: RHCP "European" Beam: RHCP or LHCP
Downlink Coverage	Italian (TVB1) Beam: 1·0° × 2·4° (3 dB) stability: better than ±0·1° steerability: European "European" (TVB2) Beam: 1·5° diam. (3 dB) stability: better than ±0·2° steerability: European
G/T	0 dB/K at edge of coverage
Peak EIRP	Italian Beam: 62·4 dBW "European" Beam: 62·7 dBW

Summary of Olympus DBS performance characteristics. (courtesty ESA)

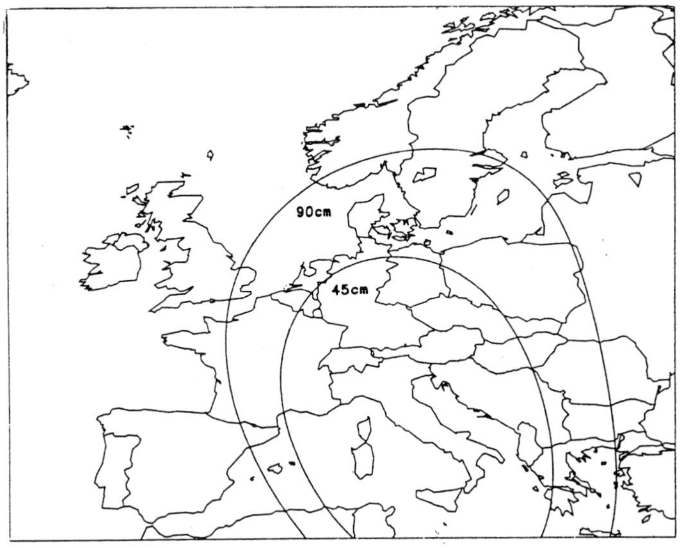

"Fair quality" reception from the Italian channel on Olympus. Reception with a C/N ratio of 9 dBW with various receive antenna diameters is shown in clear sky conditions. The illumination of Italy requires a highly elliptical (1° × 2·4°) downlink beam. This in turn requires a high level of pointing accuracy, using radio-frequency tracking of an earth-station beacon. Pointing accuracy is +/−0·1°. The antenna used by the Italian downlinking beam can be repointed. The European channel, in contrast to the Italian beam, employs a circular 1·6° beam. "Excellent quality" reception, with a C/N ratio of 14 dBW would require a 1·8 metre receive antenna where a 90 cm antenna is shown. Similarly, "excellent quality" reception would need a 90 cm antenna where a 45 cm is shown in the footprint. "Excellent quality" reception would be available with a 45 cm antenna in most of Italy except the far North and far South-West

Amongst the demonstrations proposed for the specialised services payload is a multi-point video teleconference which would use up to four carriers at 8·448 mbit/s, two per repeater channel. The proposed experiment uses a power splitter after the low-noise receiver and a multiplexer after the TWTA transmitter to combine two channels over one spot beam. This will allow a single videoconferencing session held between four locations. The 36 MHz wideband channel on the payload is also intended to demonstrate carriage of a broadcast-quality FM television signal which can be switched between beams or steered on command.

Ka-band Payload: Inclusion of a demonstration Ka-band payload on Olympus reflected concern about congestion in the Ku-band frequency spectrum and the consequential need to gain experience in the higher Ka-band frequencies. Whether, in the long term, there will be overcrowding in the satellite communications assigned Ku-band frequencies remains a matter for some speculation. The lack of interest in DBS may open up a major slice of Ku-band frequency for FSS services. In the longer term, technological developments in spatial frequency re-use through rapidly steerable multi spot beams and in on-board switching, may result in a high degree of reluctance to invest in Ka-band technology. The technology developments involved are best demonstrated by NASA's ACTS satellite rather than the much less ambitious Olympus programme. Nevertheless, the interest in the Olympus Ka-band payload has been widespread. The payload design was supported by an ad hoc group representing governments, PTT's, research institutions and industry.

The Ka-band payload uses two transmit/receive spot beam antennas. Each is independently steerable and each uses a one metre diameter reflector, the largest which could be accommodated on the Olympus satellite. Each spot beam provides coverage of only a limited area in Europe, say, for example, England and Wales, but the steerability allows the beams to cover any selected area of the same size visible from the satellite. EIRP in the centre of each beam, if pointing at Europe, is 53·7 dBW and 52·1 dBW at beam edge. The frequency plan of the Ka-band payload uses two 40 MHz channels. Expected demonstrations of this involve videoteleconferencing with digital compressed video signals at 2 mbit/s or 8 mbit/s, Tele-Education with point-to-multipoint analog (FM) transmission of a TV signal within a limited region, low-speed data transmission for business and videophone trials. Also available is a 700 MHz wideband facility. This involves the input from one antenna being fed directly to the receiver. The 40 MHz bandwidth channel filters would be short-circuited to allow use of the full 700 MHz repeater bandwidth. The repeater gain would be reduced by about 12 dBW to avoid noise loading. As far as it is known, there have been no serious requests to use the 700 MHz wideband option.

The 40 MHz narrow band experiments can be operated with a 3 metre earth terminal with a 100 Watt transmitter. However, this would probably result in unacceptable rain attenuation for operational services. ESA suggests that at the edge of the spot beam, a 5 metre antenna with autotrack or a higher-power transmitter, say 500 Watts, will be needed to offset the lower gain from the satellite. The Ka-band facility has also been considered for an intersatellite link between Olympus and the Eureca orbital platform.

Beacons: In support of radio propagation research, the satellite includes s 20/30 GHz experiment operating at 19·8 and 29·66 GHz. Some 40 organisations are involved in an experimental group to use the beacons. There is also a 12·5 GHz beacon transmitting an unmodulated carrier in order to provide a tracking facility for Olympus earth stations. This is also available for propagation measurements.

COMPARISON OF THE ESA OLYMPUS SATELLITE AND GROWTH POTENTIAL OF THE DESIGN

PARAMETER	*ESA OLYMPUS SATELLITE*	*GROWTH POTENTIAL*
LAUNCH MASS	2420 kg	3700 kg
PAYLOAD	334 kg	600 kg
PLATFORM	895 kg	not available
FUEL	1168 kg	not available
MARGIN	23 kg	not available
POWER SUPPLY, END OF LIFE		
PAYLOAD	2281 Watts	not available
PLATFORM	863 Watts	not available
MARGIN	531 Watts	not available
ARRAY CAPABILITY	3675 Watts	7700 Watts
ECLIPSE CAPABILITY	1285 Watts	4000 Watts
BODY DIMENSION	2·1 × 1·75 × 3·5 m	same as ESA Olympus satellite
ARRAY SPAN	26·3 m	up to 48 metres
DESIGN LIFE	5 years	10 years plus
PAYLOAD CONTENT	8 redundant transponders 3 beacons 9 antennas	up to 12 230 Watts DBS channels up to 50 FSS channels High felexibility for multiple antenna configuration
LAUNCH VEHICLE	Ariane 3	Ariane 3, Ariane 4, Ariane 2, Jarvis, Atlas G/Centaur. STS unlikely until well into 1990's H–II

TECHNICAL SUMMARY

PLATFORM	Olympus
PRIME CONTRACTOR	British Aerospace
OTHER MAJOR CONTRACTORS	Selenia Spazio, Spar, Fokker, Aeritalia, Marconi Space Systems, BTMC
STABILISATION	3-axis
LAUNCH MASS	2420 kg
MASS IN ORBIT	1244 kg (dry mass)
POWER SUPPLY, END OF LIFE	3675 Watts
DESIGN LIFE	5 years
ORBITAL POSITION	19° West
STATION KEEPING ACCURACY	+/−0·07° (N/S & E/W)
ECLIPSE PROTECTION	Partial, 1100 Watts available
PROPULSION SYSTEM	Combined bi-propellant
SPACECRAFT DIMENSIONS	2·1 × 1·75 × 3·5 m
SPAN, TIP-TO-TIP	26·3 m
TT&C CENTRE	Fucino, Italy
DBS PAYLOAD:	
NO. OF ACTIVE TRANSPONDERS	2
NO. OF SPARE TRANSPONDERS	0
TWTA POWER OUTPUT	230 Watts
FREQUENCY RANGE:	
UPLINK	17·6919–17·84534 GHz
DOWNLINK	12·0919–12·24534 GHz
TRANSPONDER BANDWIDTH	27 MHz
EIRP MINIMUM, BEAM EDGE	62·4 dBW
CHANNEL POLARISATION	RHCP (channel 24), RH or LHCP
KA-BAND PAYLOAD:	
NO. OF ACTIVE TRANSPONDERS	3
NO. OF SPARE TRANSPONDERS	1
TWTA POWER OUTPUT	30 Watts
FREQUENCY RANGE:	
UPLINK	28·072255–28·622255 GHz
DOWNLINK	18·925–19·475 GHz
TRANSPONDER BANDWIDTH	40 MHz (two transponders), 700 MHz
EIRP MINIMUM, BEAM EDGE	52·1 dBW (in Europe)
CHANNEL POLARISATION	Y (downlink)
SPECIALISED SERVICES PAYLOAD:	
NO. OF ACTIVE TRANSPONDERS	4
NO. OF SPARE TRANSPONDERS	1
TWTA POWER OUTPUT	30 Watts
FREQUENCY RANGE:	
UPLINK	14·081–14·169 GHz 13·157–13·249 GHz
DOWNLINK	12·507–12·598 GHz
TRANSPONDER BANDWIDTH	18 MHz and 36 MHz
EIRP MINIMUM, BEAM EDGE	44·3 dBW

COMMERCIAL DETAILS

OPERATOR	Orion Satellite Corporation
ADDRESS	Suite 200, 2000 L Street, NW Washington, DC 20036, USA
TELEPHONE NO.	(202) 862-3430
TELEX NO.	4298726
SENIOR STAFF	Chris Vizas, President Steven C. Halstead, Chairman Ronald A. Fried, Director, Goverment and Regulatory Affairs Gustave M. Hauser, Director Greggory B. Mendenhal, Director John V. Saeman, Director
PURPOSE OF PROJECT	Provision of Ku-band capacity for North Atlantic route traffic. Capacity would be leased by Orion to heavy users
PROJECT STATUS	Still no more than a plan, although the company has conditional FCC authority
PROJECT VALUE	not available
LAUNCH DATES	Prior to the loss of Challenger in January 1986, the Shuttle manifest showed the following: Orion A, January 1989, STS Orion B, April 1989, STS Orion C, July 1989, STS Orion D, October 1989, STS

SATELLITE USERS

Orion's proposals envisage outright sale of capacity to private organisations and public sector bodies for their internal communications needs. The purchasers will need authority from the FCC and/or from their PTT administrations to resell the capacity to outside parties. Orion will not be offering telecommunications services directly. Purchasers of capacity will use their own dedicated ground equipment to access the spacecraft.

PROJECT OVERVIEW

Introduction: Orion Satellite was the first of a series of US proposals for private sector international satellite communications capacity to compete with the Intelsat system although the Orion backers have consistently claimed that it is complementary to Intelsat. The company dates back to 1982 and was founded by a couple of lawyers, Chris Vizas and Tom McKnight (McKnight left the company in late 1985). Like all the other similar proposals from such companies as ISI, Cygnus, Finansat, PanAmSat and so forth, Orion envisaged selling capacity to other carriers or heavy users of communications capacity, rather than acting as a carrier itself. Thus Orion would be limited in its activities to operating and selling satellite capacity. The Orion system would be a private facility for private users and the transponders would be owned by the users to meet their own needs for private, user owned (or non-public dedicated) facilities.

Neither Orion nor those who owned and used the satellite transponders could offer services to any one else without authorisation from the FCC. Orion claimed that its private facilities were similar to such facilities within the conventional telecommunications infrastructure and would not divert traffic from the public network. In contrast, the Intelsat network was intended to offer public telecommunications services. The argument is not easy to follow, particularly as it is being presented from a legal standpoint, but there is little doubt that Orion would have impacted on Intelsat's traffic and its plans for future communications services. There is also no doubt that the prospect of competition from such ventures as Orion led to Intelsat introducing a whole range of new services and terms and conditions during the mid-1980's.

The company was granted conditional authority to construct, launch and operate its system on 6th September 1985 by the Federal Communications Commission. This had followed Presidential support for the private venture concept announced in November 1984. Nevertheless, despite Orion's early entrance into the field and its initial high public profile, its future appears to be in grave doubt. Most observers believe that there is no money to be made in the North Atlantic market and that the days of start up satellite companies has now passed as the risks have become too great. The growth in international communications has slowed down and the Intelsat organisation is suffering from gross overcapacity as a consequence. Moreover fibre optic cable may become a formidable competitor on the trunk North Atlantic route. Orion's proposals now appear remarkably dated. In late 1985, Tom McKnight said that alternative systems to Intelsat were still viable but that "the quest for financing and customers must enter a new phase".

The Orion Satellite System: At the time of writing, the prospects for the Orion system appear bleak. The company has yet to raise the finance for the space segment and order a satellite (details of the proposed payload capacity are given in the technical summary). The company initially envisaged launch of two satellites for the North Atlantic market. It has not considered servicing other regions. Capacity would be sold to large corporations, governmental agencies and other institutions whose offices and other facilities are located in Europe and North America. The owners would use the space segment in conjunction with dedicated ground stations to carry internal communications. Orion has yet to select a contractor for its space segment although at one time it had reservations for two launches on Shuttle in 1989. Footprints would cover the whole of Western Europe and the Eastern seaboard of the USA.

Orion and Intelsat: Orion's arguments with the position of Intelsat make for some interesting reading as to the status and activities of the latter organisation. In a letter sent to all parties to the Intelsat agreement in September 1983, Orion outlined its case. The latter is summarised below:

1) Orion claimed that its "non-public dedicated facilities" could increase the amount of traffic on the Intelsat system, although it was not specific as to how this would be achieved. The company also claimed that Intelsat had no plans to transfer ownership of a portion of its system to the control and operation of dedicated end users and therefore Orion was not competing with the organisation.

2) The Orion proposals would not affect Intelsat's role as a single global system. Orion was not intending to offer public telecommunications services which was the primary objective of the Intelsat organisation. Neither was Orion intending to enter the public telecoms market, such as message toll telephony, telex, data and conventional TV transmission. The European Space Agency and Eutelsat had previously pointed out that Intelsat may have already reached saturation point as a global system. The Intelsat V series had, it was argued, reached maximum economy of scale and in the future it may not be the best approach to continue procuring successive series of identical spacecraft to satisfy all needs. Alternatively, it may be more economic to start diversifying the type of satellites needed for various regions and utilisations. To some extent this has already happened with the growth of regional and domestic satellite systems.

3) Orion denied that its system would divert revenue away from Intelsat because Orion was offering private facilities. Intelsat had argued that taking revenue away from the North Atlantic would cause rates to rise for thin route users. Orion suggested that it would be offering services that did not exist at the time. Intertwined with the revenue diversion argument is the issue as to whether cross subsidisation existed in the Intelsat system. Orion suggested that the evidence pointed towards all Intelsat space segment prices being cost based. In an analysis of the Intelsat cross subsidy issue prepared in 1983 by telecoms consultants Dale N. Hatfield and Associates, it was concluded that Intelsat claims of inter and intra regional cross subsidisation could not be substantiated. But it did suggest that where evidence did exist it pointed out to inter regional subsidies from the Pacific Ocean region to the Indian Ocean and Atlantic Ocean regions.

4) The company claimed that its proposals met with both the letter and the spirit of the Intelsat agreement. This has of course been subject to a lot of questioning and clarification of the rules but the interpretation of the argument has finally been dependent on broader US policy changes rather than a strict interpretation of the legal framework of Intelsat.

TECHNICAL SUMMARY

PLATFORM	not decided
PRIME CONTRACTOR	not decided
OTHER MAJOR CONTRACTORS	not decided
STABILISATION	3-axis or spin
LAUNCH MASS	not decided
MASS IN ORBIT	not decided
POWER SUPPLY, END OF LIFE	not decided
DESIGN LIFE	7·5–10 years
ORBITAL POSITIONS	37·5° West, 47° West and 50° West
STATION KEEPING ACCURACY	+/−0·1° (N/S & E/W)
ECLIPSE PROTECTION	Full
PROPULSION SYSTEM	
SPACECRAFT DIMENSIONS	not decided
SPAN, TIP-TO-TIP	not decided
NO. OF ACTIVE TRANSPONDERS	22
NO. OF SPARE TRANSPONDERS	not available
TWTA POWER OUTPUT	20 Watts
FREQUENCY RANGE:	Ku-band
UPLINK	14·0–14·5 GHz
DOWNLINK	10·45–10·7 GHz 10·95–11·2 GHz
TRANSPONDER BANDWIDTH	36 MHz
EIRP MINIMUM, BEAM EDGE	44·9–47·4 dBW (East) 43·9–46·4 dBW (West)
CHANNEL POLARISATION	Horizontal and vertical
RECEIVE COVERAGE (G/T)	+1·9 dBi/K (East) +0·9 dBi/K (West)
SINGLE CARRIER SATURATION FLUX DENSITY	Adjustable between −80 and −92 dBW/m² (East & West)

COMMERCIAL DETAILS

OPERATOR	Pacific Satellite Incorporated
ADDRESS	c/o TRT Telecommunications, 1331 Pennsylvania Avenue, NW, Washington DC 20004, USA
TELEPHONE NO.	(010) 1 202 879 2200
TELEX NO.	197600
SENIOR STAFF	Kim Degnan, Manager, Financial Planning and Service Planning David Lubetzky Donald Jansky Father Danny Coyle, PTC
PURPOSE OF PROJECT	Regional satellite system for the Pacific Ocean region with special provision for islands within the area, including thin route telephone and TV distribution
PROJECT STATUS	In the planning stage
PROJECT VALUE	not available
LAUNCH DATES	Intended "before the end of the decade" by either Ariane or Shuttle

SATELLITE USERS

Through use of "high-power" C and Ku-band spot beams, the system aims to provide capacity for thin-route telephony, data and television traffic for the domestic and inter-country needs of the Pacific Island nations and links with the Western USA and Eastern Asia. It is assumed that the most likely users would be island PTT's and the international telecoms operators wanting access to these island nations. The system allows customers to lease or purchase various increments of capacity which will be non-preemptible. Customers will own and operate their own earth stations.

C-band capacity performance will enable reception of TV channels with 2–3 metre diameter TVRO terminals. According to TRT, a domestic TV service could be established on a full transponder lease rate of US$1·25 m or outright purchase for $5·0 m.

Publicity from TRT indicates that the system could be used for 1) wide area video and facsimile distribution and communications (TV, videoconferencing, education, publishing) 2) data communications 3) SCPC domestic and regional telephony, audio and data services.

PROJECT OVERVIEW

The origins of this novel satellite system lie in the ambitions of the Papua New Guinea Post and Telecommunications Corporation and deregulation of telecommunications in the USA which enabled TRT Telecommunications to develop plans to become a worldwide force in telex, packet-switched data and private-line voice communications. TRT started out at the turn of the century as the Tropical Radio and Telegraph Company, a subsidiary of the United Fruit Company whose economic and political dominance in some countries created the term "Banana Republic".

Both United Fruit and TRT are now subsidiaries of United Brands, a large US coglomerate, which has let TRT loose on the world markets. TRT describes itself as a provider of US domestic and international data and voice communications. It is a partner in the trans Pacific fibre optic cable TPC-3, currently being developed, an Intelsat IBS satellite earth station at the New York Teleport and a 43% shareholder in ISI which aims to provide international services in both the Atlantic and Pacific basins. The story goes that Papua New Guinea asked TRT Telecommunications to finance and operate the Pacstar satellite system on its behalf. The Posts and Telecommunications Corporation informed the International Frequency Registration Board early in 1985 that it wanted two orbital slots for the system. Papua New Guinea has a population of around three million; although widely dispersed over 463,000 square kilometres it was obvious that Papua New Guinea was looking at much wider market opportunities than its own domestic requirements. This lead to some speculation as to what those ambitions were.

Subsequently, TRT has been more specific. The project is now described as a joint venture between the Post and Telecommunications Corporation and Pacific Satellite Incorporated. The latter is a US corporation in which TRT Communications is the principal shareholder. PSI was formed to serve as a vehicle for the participation of other countries and organisations in the development of a Pacific satellite system. It appears that the C-band capacity on the hybrid satellites is aimed at providing capacity for the island nations in the Pacific Basin; particularly as some suffer from very heavy rainfall. The Ku-band capacity is aimed at "more densely populated areas" where there may be problems of co-ordination with terrestrial uses of this frequency range, and at the Pacific rim nations.

However, illustrative EIRP contour maps issued by TRT show a heavy concentration of C-band over Eastern China, Japan, Korea and Indo-China, amongst the most heavily populated and/or industrialised areas of the world, as well as the Pacific island nations. Ku-band capacity is available on the West coast of the USA. Typical transmit EIRP in C-band beams is between 34 and 38 dBW; Ku-band is 43·5 dBW. There is also a cross-strapped C/Ku-band transponder arrangement on the spacecraft to "interconnect small terminals in locations along the Pacific Rim and within the Pacific Basin". TRT claims that the planned Pacstar system is compatible with the existing international satellites operated by Intelsat but that its regional and national services "complement" the international services provided by Intelsat.

An essential feature of the system is that it will require small aperture earth stations between 1·8 metres and 4·5 metres in diameter. Early in 1986, a case study of use of Pacstar for the Solomon Islands was produced by TRT showing the requirements for a system interconnecting six sites. The hub of the system was the Islands' capital Honiara, with five remote sites on other islands. This gave the results shown below.

The figures are presented purely as an illustrative example and TRT provides no comparison with terrestrial alternatives. Moreover, the figures cover satellite services only, with no provision for buildings, power equipment, switching, etc. There is no provision for redundancy either.

One point of interest is that the Pacstar satellites would allow use of 2·4 metre diameter earth stations for outlying areas but 3·7 metre would provide a more optimal cost position. This would be because the savings in the space segment would offset the extra ground station costs. Conversely, stations above 3·7 metres "would not realise a significantly greater savings in space segment costs". TRT points out that the cost of the earth stations could be considerably reduced through cooperative purchases of earth stations in larger quantities than the six quoted. The $118,334 transponder lease cost is also reduced to $97,464 per annum by use of a five year lease contract, although as any good economist or accountant knows, such a saving is largely, if not totally off-set by the increased risk/commitment required by a long-term lease.

Location	*No. of subscribers per island*	*No. of circuits from Capital*	*Earth Station diameter*	*Earth Station costs, US$ k*	*Fraction of transponder capacity required*	*Transponder capacity cost per year US$*
Honiara Hub	HUB	77	4·5 metres	$478,150	·0513	$ 64,167
Gizo	160	18	3·7 metres	$ 94,200	·0150	$ 18,750
Kira Kira	35	7	3·7 metres	$ 44,700	·0058	$ 7,292
Auki	140	17	3·7 metres	$ 89,700	·0142	$ 17,708
Buala	30	7	3·7 metres	$ 44,700	·0058	$ 7,292
Lata	10	3	3·7 metres	$ 26,700	·0025	$ 3,125
TOTAL				$778,150	·0946	$118,334

TECHNICAL SUMMARY

PLATFORM	not available
PRIME CONTRACTOR	not available
OTHER MAJOR CONTRACTORS	not available
STABILISATION	not available
LAUNCH MASS	not available
MASS IN ORBIT	not available
POWER SUPPLY, END OF LIFE	not available
DESIGN LIFE	10 years
ORBITAL POSITIONS	167° East, 175° West
STATION KEEPING ACCURACY	+/−0·05°
ECLIPSE PROTECTION	Full service throughout life
PROPULSION SYSTEM	not available
SPACECRAFT DIMENSIONS	not available
SPAN, TIP-TO-TIP	not available
NO. OF ACTIVE TRANSPONDERS	8 Ku- and 17 C-band
FREQUENCY RANGE:	
UPLINK	6·495–7·075 GHz 14·0–14·5 GHz
DOWNLINK	3·615–4·195 GHz 11·7–12·2 GHz 12·5–12·75 GHz
TRANSPONDER BANDWIDTH	C-band, 28, 38 and 60 MHz Ku-band, 50 MHz
EIRP MINIMUM, BEAM EDGE	C-band, 34–38 dBW Ku-band, 43·5 dBW
CHANNEL POLARISATION	Orthogonal Linear

COMMERCIAL DETAILS

OPERATOR	Perum Umum Telekkomunikasie (PERUMTEL)
ADDRESS	Jalan Cisanggarumg 2, Bandung, Indonesia
TELEPHONE NO.	+62 22 59 100
TELEX NO.	288220 diruttel bd
SENIOR STAFF	Boedi Santoso, Deputy Director General for Telecommunications
PURPOSE OF PROJECT	Nominally to provide domestic communications for Indonesia but also used for regional communications
PROJECT STATUS	System operational but short of capacity because of non-availability of launch vehicles
PROJECT VALUE	The Palapa B1 & B2 satellites cost $74·5 m in total, excluding launch cost of $18 m and insurance but including PAM-D motors
LAUNCH DATES	Palapa B1, 19th June 1983, STS-7 + PAM–D Palapa B2, 6th September 1984, STS 41-B + PAM–D (see Project Overview) Palapa B3, to be decided

SATELLITE USERS

The Palapa system is nominally an Indonesian domestic C-band satellite system intended to operate the country's PTT traffic, notably telephony and television, radio and governmental traffic. Prior to purchase of the original Palapa A system the country's terrestrial system was insufficient to provide quality links between the major urban centres and links between the 5000 islands which constitute the country.

However, the system has been used on a regional basis with capacity being leased by the Philippines for internal telecommunications use and by Thailand and Malaysia for distribution of national TV channels. The Palapa B system provides coverage of all of the ASEAN (Association of South East Asian Nations). However, due to problems in launching the series, the Indonesians appear to have insufficient capacity to take full advantage of the commercial possibilities of regional services.

SUMMARY OF SATELLITE USERS

Palapa B1
Transponder 08, TBS 7 TV Bangkok
Transponder 10, RTM 1 (WARNA 1) TV Kuala Lumpur
Transponder 12, Army Channel 5, TV Bangkok
Transponder 15, TV1–TV Indonesia

PROJECT OVERVIEW

Introduction: The Palapa B system is Indonesia's second generation domestic satellite communications system. Unlike the first generation system, it is geared up to providing regional as well as domestic services. It has also been one of the more successful attempts at third world satellite communications but, like so many current projects, has been affected by the crisis in the launch vehicle market. Because of Indonesia's heavy dependence on the system, this has become a political issue in the country. The problem has arisen because the second spacecraft in the series, Palapa B2, was lost on launch in February 1984. Although subsequently recovered, the Indonesians decided to purchase a replacement rather than purchase the satellite back from the underwriters, there is currently no method of launching the replacement satellite.

The entire Palapa A system is now at the end of its design lifetime and the country has not invested in alternative terrestrial communications systems. It has placed political pressure on the USA to launch the replacement Palapa B3 spacecraft either on Delta or on the earliest Shuttle launch but has now opted for an Ariane launch instead. It has recently tried to repurchase the Palapa B2 satellite but the spacecraft has been sold to California based Sattel instead. Perumtel is also currently considering the requirements of its third generation satellite system. Satellite communications have long been criticised in Indonesia as an expensive and over engineered solution to the country's communications problems. The criticisms date back to the 1970's when, after some delay, the Palapa A system was ordered. Nevertheless, the two Palapa systems have enabled the country to become a regional communications carrier with the consequential impact on foreign earnings.

Background to the Palapa System: Indonesia was the first country outside of the USA and the Soviet Union to have a domestic satellite communications system. The two Palapa A satellites were launched in 1976 and 1977 and followed some extensive negotiations between the country and US satellite manufacturers. The Palapa A satellites were procured from Hughes Aircraft Company and used the HS-333 platform with a payload of 12 transponders. The Indonesians used the system to become a regional force in the satellite communications market place. The Palapa A2 satellite, which was nominally the in orbit spare, was used by the Philippines, Malaysia and Thailand. Annual lease charges per transponder were a modest $693,000. The Philippines leased one and a half transponders and Thailand and Malaysia one each. The follow on Palapa B system provides a footprint covering the entire ASEAN region and can hardly be considered a domestic satellite system.

The Palapa A system was seen by many observers as a model for use of satellites to improve the telecommunications infrastructure of the developing world. It permitted direct dialling between major Indonesian cities as well as the opportunity to upgrade the country's TV network. Nevertheless during the 1970's and early to mid-1980's, Indonesia was earning a considerable sum from petroleum exports, whilst other developing countries were paying the high prices involved. The Palapa system not only required a considerable amount of foreign debt but also a major investment in ground stations. For the original system some forty major earth terminals were required.

It is not widely known that in 1982 proposals were made

in the UK to transfer the Palapa A satellites for use in Africa. At that time it was envisaged that they would be surplus to requirements once the Palapa B satellites were launched. They would have provided a cheap introduction to African regional satellite communications. It was envisaged that the exercise would have cost around $15 m and have had a three year operational lifetime. In the end the proposal was rejected by the British government who were expected to put up the funds.

The Palapa B2 story: Like so many of the world's satellite systems, Palapa has become a victim of the crisis in the launch vehicle market. The Palapa B2 satellite was launched along with the similar Westar VI satellite on Shuttle mission 41-B. This took place on 6th February 1984. The perigee kick motor failed on both satellites, apparently because of a flaw in the composite material used in the nozzle cone of the motor. Perumtel made an insurance claim in full for the satellite from the insurance underwriters. The latter decided to pay NASA for the recovery of both spacecraft and under salvage terms took ownership of them. This legal position was subject to a lot of discussions with both Perumtel and Western Union (the latter owned the Westar satellite prior to launch). The principal underwriters involved in the recovery arrangements were Merrett Syndicates (UK) and International Technology Underwriters (INTEC, USA).

Both satellites were recovered on Shuttle mission 51-A in November 1984, NASA having been paid $4·8 million for each satellite recovered. There was also an agreement in the NASA contract that the satellites had to be relaunched by Shuttle. Once recovered the satellites were returned to the manufacturer, Hughes Aircraft Company where they were refurbished and stored until the underwriters were able to find customers for them. Perumtel decided that it did not want to repurchase Palapa B2 and ordered a virtually identical B3 spacecraft from Hughes instead. However, it was reported in 1986 that it was looking again at repurchasing B2. However, the satellite was sold to a new California based company Sattel who say that they are aiming at providing C-band services for the Pacific Ocean region.

The underwriters took two years to sell the satellite during which period there were negotiations with a number of other potential operators. Thailand considered purchase for domestic communications including use by its military but eventually rejected the idea. A lone satellite, somewhat customised for the original customer would have required modifications and the order of a back up spare. The price being offered by the underwriters was, initially, not particularly cheap at around $30 m. Moreover, the satellite had originally been ordered in 1980 and, by the mid-1980's, was somewhat dated in concept. A US new venture company, Teresat discussed purchase of both Palapa B2 and Westar VI to provide US domestic capacity and went as far as booking launch reservations on the Chinese Long March 3 vehicle. It was widely reported, however, that the company did not have the finance to purchase the spacecraft and Palapa B2 was sold to Sattel instead. Westar VI has not, at the time of writing, been sold.

The events have left Perumtel with a major problem indicative of the previously under-rated risks in the satellite business. At the time of writing there is no firm date as to when Palapa B3 will be launched. It is unlikely before 1989, some five years after the Palapa B2 satellite was launched. The Indonesians have had discussions with NASA and the US State Department about possible launch on one of the remaining Delta vehicles and, under the new arrangements for Shuttle, it is possible that it will have priority on that vehicle.

Nevertheless, Indonesia has invested heavily in the ground segment of its system and is in no position to abandon it because of the delays and high risks in the space segment. The country has a major financial problem caused by the collapse in the price of oil and abandonment would leave large areas of the country without communications and without the resources to invest in alternative means. For this reason, the Palapa system has become a major internal political issue in Indonesia. The country has made the US political system aware of the problem, hence State Department involvement. It is also a very bad piece of publicity for the satellite industry in its attempts to market new systems to other countries.

TECHNICAL SUMMARY

PLATFORM	HS-396
PRIME CONTRACTOR	Hughes Aircraft Company
STABILISATION	Spin
LAUNCH MASS	1200 kg
MASS IN ORBIT	635 kg
POWER SUPPLY, END OF LIFE	830 Watts
DESIGN LIFE	8 years
ORBITAL POSITIONS	Palapa B1, 108° East Palapa B2, 118° East (intended) Palapa B3, 118° East 113° East also allocated
STATION KEEPING ACCURACY	+/−0·05° (E/W & N/S)
ECLIPSE PROTECTION	Full

SPACECRAFT DIMENSIONS	2·1 m diameter, 2·7 m height (6·6 m after deployment)
NO. OF ACTIVE TRANSPONDERS	24
NO. OF SPARE TRANSPONDERS	6
TWTA POWER OUTPUT	10 Watts
FREQUENCY RANGE:	
UPLINK	
DOWNLINK	3·720–4·160 GHz
TRANSPONDER BANDWIDTH	36 MHz
EIRP MINIMUM, BEAM EDGE	34 dBW (Indonesia), 32 dBW (ASEAN)
CHANNEL POLARISATION	
RECEIVE COVERAGE (G/T)	−5 dBW/K (Indonesia only)

PAN AM PACIFIC SATELLITE CORPORATION PANAMPACSAT

COMMERCIAL DETAILS

OPERATOR	Pan Am Pacific Satellite Corporation
ADDRESS	200 Park Avenue, Suite 901, New York, NY 10166, USA
TELEPHONE NO.	(212) 880-1897
SENIOR STAFF	Michael Johnson C. C. Wong
PURPOSE OF PROJECT	To provide capacity for the Pacific Ocean region PTT's
PROJECT STATUS	Satellite purchased and due to be launched in 1988
PROJECT VALUE	The satellite was purchased from underwriters for \$20 m. Owners will also have to pay for launch and insurance (say \$50 m) and further satellite capacity
LAUNCH DATES	Not available at time of writing. Company plans to have the ex-Westar satellite operational from mid- to late-1988

SATELLITE USERS

The company has announced that it intends to offer the capacity to Pacific Ocean region PTT's who want C-band capacity for domestic use. As at September 1986, two countries were understood to have a strong interest in using the system.

PROJECT OVERVIEW

Pan Am Pacific Satellite Corporation intends to provide a Pacific Ocean regional satellite system and in September 1986 announced that it had bought the Westar VI satellite from insurance underwriters to inaugurate its services from 1988. Apparently, it has been studying the opportunity for some two years prior to purchase. It plans to "manage" the space segment which will be operated by "various" PTT's. The system is not intended to provide services with Japan. The company plans to have the satellite refurbished by its original manufacturer, Hughes Aircraft Company. This will take between 12 and 14 months. The company is understood to be looking at a second satellite for this system.

Pan Am Pacific Satellite Corporation has no relationship with PanAmSat. It is a joint venture between Pan Am Commercial Services, a subsidiary of the Pan Am Corporation, and Johnson Geneva, telecommunications consulting firm specialising in third world communications.

The Westar VI satellite was originally ordered by Western Union for US domestic C-band services but its launch failed because of problems with the PAM–D motor. It was recovered by Shuttle, along with Palapa B2, in November 1982 after it had passed into the hands of the insurance underwriters who also paid \$11 m for its recovery. Western Union purchased a new satellite, Westar VI–S, to replace it. The recovered satellite was subject to a number of potential bids during 1986, including one from a company called Terasat (which also unsuccessfully bid for Palapa B2 before it was sold to Sattel).

TECHNICAL SUMMARY

PLATFORM	HS–376
PRIME CONTRACTOR	Hughes Aircraft Company
STABILISATION	Spin
MASS IN ORBIT	585 kg
POWER SUPPLY, END OF LIFE	700 Watts
DESIGN LIFE	10 years
ORBITAL POSITIONS	not available
STATION KEEPING ACCURACY	$+/-0{\cdot}05°$ (N/S & E/W)
ECLIPSE PROTECTION	Full
NO. OF ACTIVE TRANSPONDERS	24
NO. OF SPARE TRANSPONDERS	None
TWTA POWER OUTPUT	10 Watts
FREQUENCY RANGE:	
UPLINK	5·925–6·405 GHz
DOWNLINK	3·700–4·180 GHz
TRANSPONDER BANDWIDTH	36 MHz
EIRP MINIMUM, BEAM EDGE	Not available until details of modifications are released. Probably at least 33 dBW
CHANNEL POLARISATION	Linear, horizontal & vertical
RECEIVE COVERAGE (G/T)	Not available until details of modifications are released

COMMERCIAL DETAILS

OPERATOR	RCA American Communications
ADDRESS	400 College Road East, Princeton, New Jersey 08540, USA
TELEPHONE NO.	(609) 734-4200
TELEX NO.	710 480 4286
SENIOR STAFF	Dr James J. Tietjen, President E. F. Murphy, Chief Executive Officer Jack F. Underwood, VP, Commercial Communications Harold W. Rice, VP Video and Audio Services R. E. Smylie, VP, Business & Government Network J. J. Christopher, Technical Operations Director Walter W. Pioli, Director, Commercial Sales
PURPOSE OF PROJECT	Provision of C-band capacity for US cable television channels and radio. The system was also used for other RCA Americom services now on Advanced Satcom
PROJECT STATUS	First two satellites (Satcom I and II) now retired. Satcom III lost on launch. Satcom IIIR and IV fully operational
PROJECT VALUE	Full system, including satellites, launch, insurance and control facilities estimated at \$200 m before insurance claims and financing charges
LAUNCH DATES	Satcom I, 12th December 1975, Delta 3924, F118 Satcom II, 26th March 1976, Delta 3924 Satcom III, 6th December 1979, Delta 3924 Satcom IIIR, 19th November 1981, Delta 3910, F158 + PAM–D Satcom IV, 15th December 1982, Delta 3910, F159 + PAM–D

SATELLITE USERS

Of the original five RCA Satcom spacecraft, only two now remain in operation, Satcom IIIR and Satcom IV. These are also known as Cable Net–1 and Cable Net–2 respectively, reflecting their use for distribution of Cable TV channels to cable networks. The first RCA Americom satellite, Satcom I, was the first satellite to provide capacity for distribution of cable TV channels, before Satcom IIIR became the primary cable TV satellite (see Project Overview). Satcom IIIR now carries 24 cable TV channels and Satcom IV carries 22. In sub-carrier mode, the current Satcom and Advanced Satcom series also carry 25 radio channels. These are split as follows: Satcom IR, one; Satcom IIIR, 20; Satcom IV, four. These channels are for Turner Broadcasting, Satellite Radio Network, Genesis Cable Storytime, United Video, Satellite Syndicated Systems, ESPN, Studioline Cable Stereo, Georgia Radio, Family Radio Network and the National Broadcast Museum.

The Satcom II satellite was used to develop RCA Americom's other satellite carried services which are described below. However, as this satellite is now retired, the services are now carried on the Advanced Satcom series described in the next section.

Southern Pacific Communications Corporation used Satcom II for development of its Sprint Service and as such was the first US specialised common carrier to use a large amount of satellite capacity.

PROJECT OVERVIEW

The RCA Americom Satcom series became the second US domestic satellite system to become operational. It's Satcom I spacecraft was launched some 19 months after Western Union's Westar I satellite. RCA Americom Communications Incorporated, commonly known as RCA Americom, was formed in 1976 as a wholly owned subsidiary of RCA Corporation to own and operate the parent company's domestic communications satellite system. The decision by RCA to go into operating a domestic satellite system dates back to the Federal Communications Commission's decision in 1982 to open up the domestic communications satellite market. Its "Domsat" (Domestic Communications Satellite Facilities report and order) decision paved the way for RCA, Comsat and Western Union but needed further liberalisation to attract GTE (GStar), SPCC (Spacenet), SBS, American Satellite Corporation, Hughes (Galaxy), Ford (Fordstar), SBS and AT&T (Telstar) into the field of operating satellites.

The early entrance of RCA has placed it in a strong competitive position which, so far, has only been matched by Hughes, and perhaps GTE with Galaxy and Spacenet. RCA Americom was able to take early advantage for the demand for satellite capacity for distribution of TV channels and programming. The development of cable RV was critically dependent on three factors. Firstly the FCC's Domsat decision allowed customer premises earth stations which were owned by the cable operators. Prior to the decision all earth stations were owned by common carriers. Secondly, the ground equipment industry applied the Gallium arsenide Field Effect Transistor (GAS–FET) to TVRO's. The GAS–FET is a solid state high frequency microwave device which provides low noise figures and which can be produced, in volume, very cheaply. The first TVRO systems, sold to Comsat in the later 1960's, cost \$250,000 a piece. The GAS–FET, and increasing volume production, brought the price figure down to the level which cable operators could begin to purchase TVRO's – by the mid-1980's the minimum price of a consumer TVRO had fallen to a retail price below $1000. The third factor was that US cable systems were able to develop in the 1970's because there was limited frequency allocated to off-air broadcasters (most areas in the USA had only three TV channels available) and because the Video Casette Recorder had yet to become a commercial alternative.

RCA Americon was the original operator of the "cable hot bird" concept where a satellite was dedicated to providing capacity for cable television channels and programming. RCA sub-leased capacity for its customers as far back as 1973 when it leased two transponders on Canada's Anik A2 C-band satellite. In 1975 Home Box Office leased two transponders from RCA and went on the air with two earth stations, and the satellite TV industry began to grow.

Ted Turner was the other visionary who saw the value of satellite distribution to build up the penetration of his cable TV programming. He joined HBO on Satcom I in 1986 using it to distribute his WTBS channel. Initially other cable TV programmers were reluctant to join WTBS and HBO in satellite distribution but by 1979 Satcom I was a "hot cable bird" and RCA Americom were looking to make Satcom III, due for launch in that year, a "hot cable bird". However, Satcom III was lost on launch, resulting in a $77 m insurance claim for loss of the satellite and initial revenue. RCA then leased capacity on Comstar D2 as interim capacity before the replacement Satcom IIIR satellite became operational. This was launched on 19th November 1981. However, such was the demand for capacity for cable TV programming that the Satcom IV satellite was dedicated to that traffic. The Satcom IIIR and Satcom IV were modified versions of the original Satcom I–III spacecraft, none of which are now operational. The Satcom IIIR/IV satellites have 18 5 Watt operational tubes and six 8·5 Watt tubes plus four spare tubes. Satcom I–III had 24 5·5 Watt tubes with no back-up spares.

Despite the 1972 Domsat decision by the Federal Communications Commission, RCA Americom faced major regulatory problems in developing its cable TV service. It took until 1979 for RCA Americom to make operating profits, having to offer the early cable TV users of Satcom reduced rates until such time as the users had built up an adequate user base. In 1979, the FCC insisted that the transponder pricing, should be kept low but subject to a maximum rate, and offered on a first come first served basis. As demand for capacity on the Satcom system soared instant millionaires appeared. They purchased capacity on Satcom at a fixed tariff and resold the capacity on to cable TV channels at a large profit. RCA Americom was faced

Satcom I, the first of RCA Americom's domestic satellites, prior to launch in 1975. It was retired on 4th June 1984 and replaced by Satcom IR

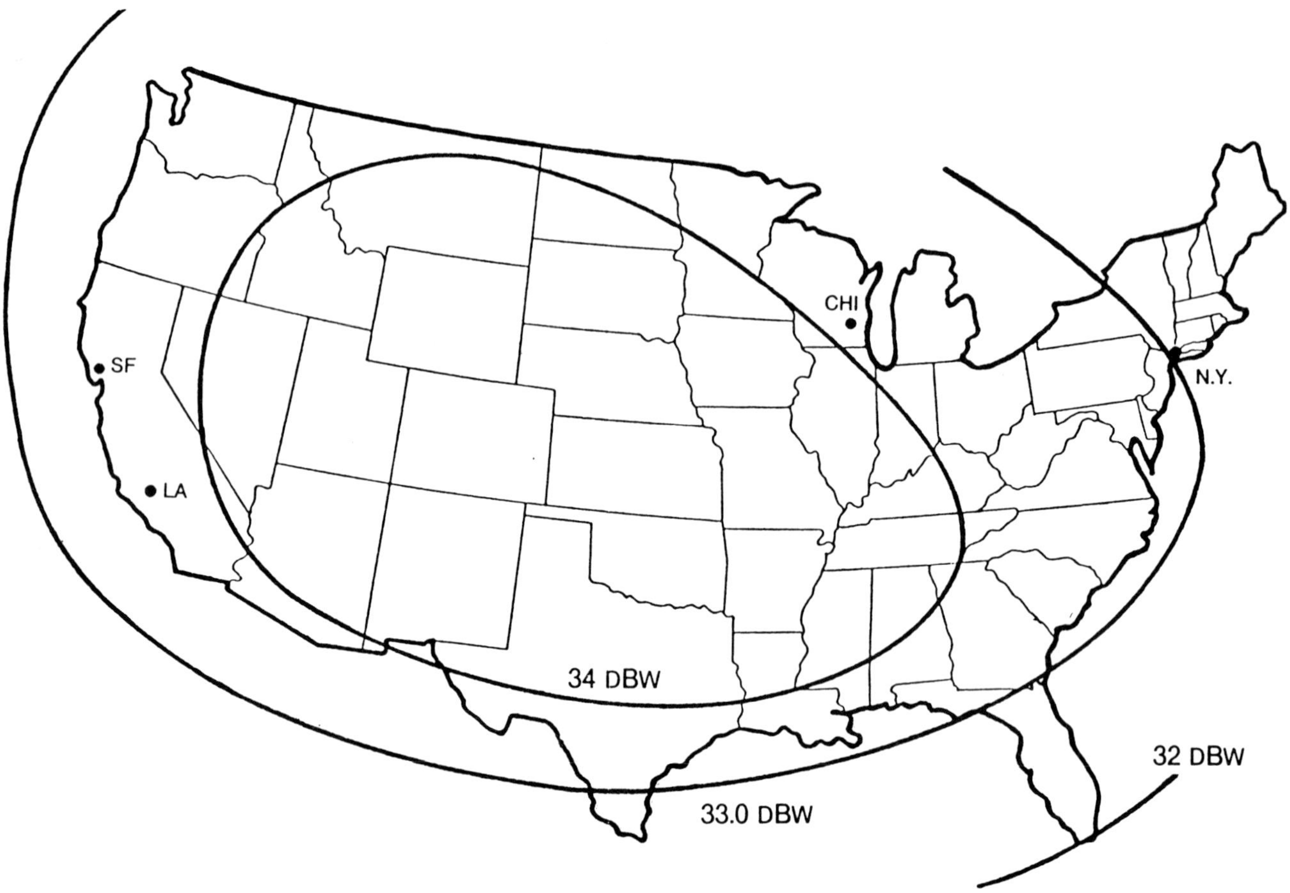

Satcom IIIR EIRP pattern at 130° West. Transponders 3, 7, 11, 15, 19 and 23 which use 8·5 Watt TWTA's have a dBW figure 2 dB higher than the figures quoted

with the position that when demand was slack in its early days, it could price on a demand basis, at below "cost" but had to abandon demand base pricing when demand increased, thus effectively being unable to offset "losses" against "profits". Given that the satellite operating business was and is a commercially risky exercise, RCA Americom was being regulated as an established, low risk monopoly telecommunications operator. In March 1982, the FCC abandoned its policy in favour of a fixed rate demand-based tariff. The Commission later approved the sale of transponders under demand-based pricing.

The use of the Satcom series of satellites has not been limited to cable TV traffic. The other major use of the Satcom I and II satellites has proved to be the development of and support of Alascom, a subsidiary of Pacific Power and Light, which provides basic telecommunications infrastructure and services for Alaska. Alascom used capacity on Satcom II to develop telecommunications capacity before purchasing the Satcom V satellite from RCA Americom. RCA Americom has a division called Alascom Relations which now provides Alascom with telemetry, tracking control and service protection for satellite communications within the state of Alaska.

RCA was the first of the US manufacturers of commercial satellites to enter into the business of operating satellites, creating a captive customer for RCA Astro-Electronics and facilitating the building up of a production line for its commercial satellites. The Satcom design faced considerable competition from the Hughes–376 design during the late 1970's and early 1980's but the 3-axis design proved capable of generic development, which, along with developments in payload design, allowed the basic design to evolve into the Advanced Satcom (RCA–3000), RCA–4000 and RCA–5000 designs. Coupled with the captive RCA Americom requirements, RCA Astro-Electronics was roughly level pegging with Hughes Aircraft Company by 1986 in the supply of commercial communications satellites.

The original Satcom spacecrafts were large satellites by the standards of 1976, requiring an upgrading of the Delta launch vehicle to accommodate the higher weight. They were the first commercial satellites to employ frequency re-use by alternating polarisation between vertical and horizontal. This doubled the capacity of the satellite to 24 transponders from the 12 that previously represented the limit that could be put into orbit by a Delta-class launch vehicle. On each Satcom spacecraft, four groups of six transponders are served by cross-polarised antennas having overlapping grided reflectors. One antenna handles vertically polarised transponders, the other horizontally polarised transponders. Each 36 MHz transponder can carry the equivalent of 1500 two-way telephone calls or 64 mbit/s of data or one high quality vide/audio channel. The transponder capacity has been substantially improved over the lifetime of the Satcom series by improvements in ground equipment.

RCA Americom operates Satcom V as a joint licensee with Alascom. Satcom V, which is called Aurora by Alascom is part of the Advanced Satcom series, Satcom V, IR, IIR and VI, which incidentally, are also known as Satcom F, G, H and I. The original Satcom series, Satcom I, II, III, IIIR and IV of the original series remain in service. RCA Americom's other services are now carried on the Advanced Satcom series and the company's new Ku-band spacecraft.

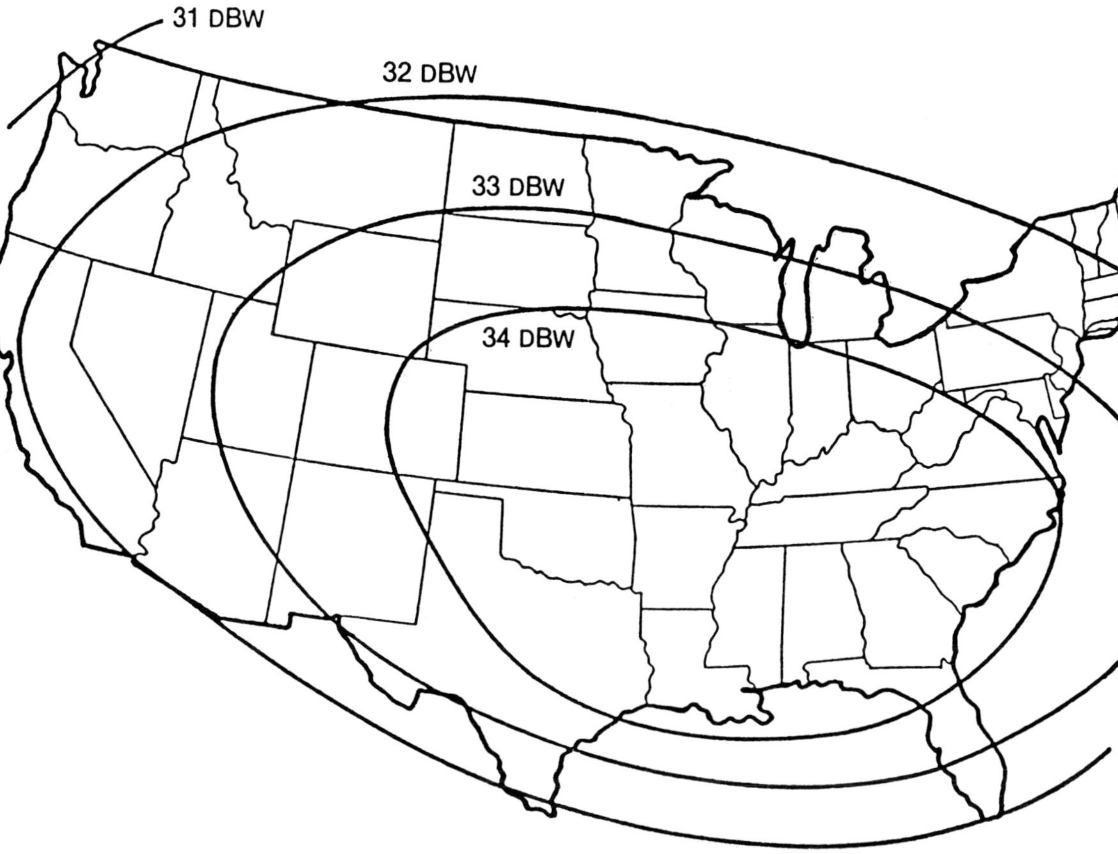

Satcom IV EIRP pattern at 81° West. Transponders 3, 7, 11, 15 and 23 are 2 dB higher than the figures quoted

Satcom IIR, replacing Satcom II, is now used to carry commercial and government voice, data and video traffic which had been built up on Satcom II. The latter was also used to build up Southern Pacific Communications Corporation Sprint long distance digital telephone/data service. This now uses capacity on GTE's Spacenet system since SPCC was taken over by GTE. RCA Americom's Commercial Communications Services group provides private leased channels to businesses for voice, data and facsimile applications. It uses a series of earth stations located near to major US business centres, notably New York, Chicago, San Francisco, Los Angeles, Houston, Atlanta, Miami, Seattle and Honolulu. These stations can also transmit and receive TV and radio traffic.

The Commercial Communications Services group also provides other carriers with lines to connect the operating and switching centres of the carriers. The group also carried a series of "specialised" services using dedicated earth stations at customer sites. The major growth period for the latter is likely to take place over the next decade, the original Satcom series has largely missed this market. However it was used for a "56 Plus Service" which provides a 56 kbit/s digital service for voice, data and facsimile. The Private Leased Channel Service can be used for data at 9·6 kbit/s.

RCA Americom's dedicated earth station facilities for business and government are now under the responsibility of its Business and Government Network Services group. For government requirements it installs, operates and maintains dedicated earth station facilities providing communications services to government agencies. These include NASA, the General Services Administration (GSA), Department of Defense, NOAA, Voice of America and the American Forces Radio and Television Service (AFRTS). Services include voice grade to high speed data traffic and video. The Business Network segment of this group began installing systems in 1985. RCA provides a 50 mbit/s data link in support of the Shuttle programming.

TECHNICAL SUMMARY

PLATFORM	Satcom
PRIME CONTRACTOR	RCA Astro-Electronics
STABILISATION	3-axis
MASS IN ORBIT	462 kg (I–III), 581 kg (rest)
POWER SUPPLY, END OF LIFE	550 Watts (I–III), 740 Watts (rest)
DESIGN LIFE	8 years
ORBITAL POSITIONS	Satcom IIIR, 130° West Satcom 4, 81° West Satcom I and II retired
LAUNCH VEHICLE	Delta and PAM–D
SPACECRAFT DIMENSIONS	5·3 × 4·1 × 4·1 m
SPAN, TIP-TO-TIP	11 metres (Satcom V)
NO. OF ACTIVE TRANSPONDERS	24
NO. OF SPARE TRANSPONDERS	4 (Satcom IIIR/IV only)
TWTA POWER OUTPUT	5·5 Watts*
FREQUENCY RANGE:	
UPLINK	5·928–6·425 GHz
DOWNLINK	3·700–4·200 GHz
TRANSPONDER BANDWIDTH	34 MHz
EIRP MINIMUM, BEAM EDGE	33 dBW, 26 dBW
CHANNEL POLARISATION	Linear, horizontal & vertical
RECEIVE COVERAGE (G/T)	−5 & −10 dB/K
SINGLE CARRIER SATURATION FLUX DENSITY	−80 & −75 dBW/sq. m.

*Satcom IIIR and Satcom IV carry six 8·5 Watt TWTA's.

ALLOCATION AND USE OF TRANSPONDERS BY BROADCASTERS ON THE RCA AMERICAN SATELLITES

SATCOM IIIR (C-BAND AT 130 DEGREES WEST) ALSO KNOWN AS CABLE NET–1		
NUMBER	*TRANSPONDER USE*	*USER(S)*
1	TV	Nickleodeon East (TV)
2	TV & Radio	The Learning Channel (TV), Open Public Events Network (TV), Home Shopping Network (TV), Satellite Radio Network (Radio)
3	TV & Radio	Trinity Broadcasting Network (TV), Genesis Cable Storytime (Radio)
4	TV & Radio	Financial News Network/Score (TV), Satellite Jazz Network (Radio)
5	TV	Viewers Choice (TV – STMC)
6	TV & Radio	Tempo TV and 7 Radio Channels from Tempo
7	Radio	ESPN Information Network (Radio)
8	TV & Radio	CBN Cable Network (TV) and three Radio Channels from Tempo
9	TV	USA Network – West (TV)
10	TV	Showtime West (TV)
11	TV	MTV (TV)
12		
13	TV	HBO West (TV)
14		
15	TV	Video Hits One (TV – MTV)
16	TV & Radio	Home Theatre Network (TV), Home Shopping Network (TV?), Commodity Corporation (Radio)
17	TV	Lifetime (TV)
18		Cable Value Network (TV), Eternal World TV Network (TV), Reuter Monitoring Service (TV)
19	TV	The Weather Channel (TV)
20	TV	Black Entertainment Network (TV)
21		
22	TV	Home Shopping Network (TV?)
23	TV	Cinemax West (TV)
24	TV	Arts & Entertainment (TV)

Viacom Networks Group signed an agreement in late 1986 with AT & T taking 21 transponders on the Telestar system. It is therefore expected that many of the channels on SATCOM IIIR will Transfer to Telstar including Showtime.

SATCOM IV (C-BAND AT 81 DEGREES WEST) ALSO KNOWN AS CABLE NET–2		
NUMBER	*TRANSPONDER USE*	*USER(S)*
1	TV	Home Shopping Network (TV)
2	TV	Bravo (TV), The Heartbeat Network (TV), National Jewish TV (TV)
3	TV	Rock Christian Network (TV)
4	TV	Nickleodeon West (TV)
5	TV	HSE Affiliate Feeds (TV)
6	TV & Radio	Biznet (TV), Madison Square Gardens Net (TV), Family Radio Network (East & West) (Radio)
7	TV	Liberty Broadcasting Network (TV)
8	TV	C–Span 11 (TV)
9	TV	Sportavision (TV)
10	TV	American Movie Classics (TV)
11	TV	Home Sports Entertainment (TV)
12	TV	Shop at Home (TV)
13	TV	New England Sports Network (TV)
14	TV	Television Shoppers' Showcase (TV)
15	TV	American Music Entertainment (TV)
16		
17		
18	TV	Hit Video USA (TV)
19	TV	WPIX New York (TV)
20	TV	Prime Ticket (TV)
21	TV	The Nostalgia Channel (TV)
22	TV	Home Team Sports (TV)
23	TV	Sports Channel NE (TV), The Silent Network (TV)
24	TV	The Playboy Channel (TV), Private Ticket (TV), CDN 2 (TV)

SATCOM K1 (KU-BAND AT 85 DEGREES WEST)		
NUMBER	TRANSPONDER USE	USER(S)
1		
2		
3	TV	HBO West (TV)
4		
5		
6		
7		
8		
9		
10		
11		
12		
13	TV	Cinemax West (TV)
14		
15	TV	Festival (HBO) (TV)
16		

All three video programming channels on Satcom K1 are owned by HBO. The company is using the satellite on a "temporary basis" until such time when it is possible to transfer the channels to Satcom K3 which still has to be launched. Primetime, which uses Satcom IIR, has an option for up to three transponders on Satcom K1.

SATCOM K2 (Ku-BAND AT 81 DEGREES WEST)		
NUMBER	TRANSPONDER USE	USER(S)
1	TV	NBC Pacific (TV)
2		
3		
4		
5	TV	NBC Central (TV)
6	TV	CONUS Communications (SNG TV)
7		
8	TV	CONUS Communications (SNG TV)
9	TV	NBC Affiliate Feed (TV)
10		
11	TV	NBC Affiliate Feed (TV)
12		
13	TV	NBC Affiliate Feed (TV)
14		
15	TV	NBC Affiliate Feed (TV)
16		

All 16 transponders on this satellite have been allocated, AT&T has two transponders, Hubbard Broadcasting has four (of which two are used by CONUS Communications, two are used by RCA Americom for Syndication System programming) and eight are allocated to NBC. It is normal practice for two channels to use each transponder on this satellite. The satellite is used primarily for network feeds rather than cable TV distribution. RCA Americom also offers SCPC services for radio broadcasters on Satcom K2.

SATCOM F1R (ADVANCED SATCOM) (C-BAND AT 139 DEGREES WEST)		
NUMBER	TRANSPONDER USE	USER(S)
1	TV	Occasional Video
2	TV	Netlink-KWGN (Networking)
3	Radio	Digital Radio
4	TV	Netcom (O/V)
5	TV	Occasional Video
6	TV	Netlink-KRMA/PBS (Networking)
7	TV	Prime Ticket Network
8	TV	NBC East (TV)
9	TV	Occasional Video
11	TV	WTN
12	TV	FNN/Netcom, Netlink-WNYW/Fox (Networking)
13	TV	Maryland Horse Racing
15	TV	Occasional Video
19	Radio	NBC, CBS Digital Radio Channel Networking
20	TV	Netlink-KCNC/NBC (Networking)
22	TV	Hi-Net/Netcom, Netlink-KMGH/CBS (Networking)
23	Radio	ABC Digital Radio Networking (Full transponder)
24	TV	RAI Uno (TV), Netlink-KUSA/ABC (Networking)

Satcom IR is designated as in-orbit spare but carries occasional TV traffic and network radio channels.

SATCOM F2R (ADVANCED SATCOM) (C-BAND AT 72 DEGREES WEST)		
NUMBER	TRANSPONDER USE	USER(S)
2	TV	Occasional Video
3	TV	Primetime-WABC/ABC*
11	TV	Prime time-WBBM/CBS*
13	TV	NASA Contract Channel (TV)
22	TV & Radio	American Radio & TV Service (TV)
23	TV	Primetime-WXIA/NBC*

*Primetime channels due to begin transmitting in the first quarter of 1987.

Satcom IIR is primarily used to carry commercial and government voice data and video services.

SATCOM V (ADVANCED SATCOM) (C-BAND AT 43 DEGREES WEST) ALSO KNOWN AS AURORA		
NUMBER	TRANSPONDER USE	USER(S)
24	TV	Learn Alaska & Alaska Satellite TV Project

Details are correct as at September 1986. However, use of these satellites by TV and radio channel providers is subject to frequent changes.

COMMERCIAL DETAILS

OPERATOR	RCA American Communications
ADDRESS	400 College Road East, Princeton, New Jersey 08540, USA
TELEPHONE NO.	(609) 734-4200
TELEX NO.	710 480 4286
SENIOR STAFF	Dr James J. Tietjen, President E. F. Murphy, Chief Executive Officer Jack F. Underwood, VP, Commercial Communications Harold W. Rice, VP, Video and Audio Services Donald O. Reinert, Director, CATV Services Walter W. Pioli, Director, Commercial Sales John H. Williamson, Director, Public Affairs
PURPOSE OF PROJECT	Provision of replacement and further C-band capacity for RCA Americom's US domestic services and for Alascom Incorporated
PROJECT STATUS	System fully operational but launch of the Satcom VI spacecraft delayed by non-availability of launch vehicles
PROJECT VALUE	Approximately $75 m per satellite including construction and launch costs
LAUNCH DATES	Satcom V, 27th October 1982, Delta 3924 Satcom IR, 11th April 1983, Delta 3924 Satcom IIR, 8th September 1983, Delta 3924 Satcom VI, 1987–88

SATELLITE USERS

The Advanced Satcom series are used primarily for RCA Americom's non-video broadcasting services, essentially voice telephony and data. These have been transferred from the original Satcom series and subsequently expanded. More recent video traffic captured by RCA Americom is using its Ku-band satellites rather than the C-band Advanced Satcom spacecraft.

Satcom V is owned by Alascom and used primarily by that organisation, although RCA Americom leases four transponders on the satellite. Satcom IR is designated as an in-orbit spare but carries occasional TV programming and network radio. Satcom IIR is used to carry commercial and government voice, data and video services. This includes the Armed Forces Radio and Television Service.

RCA Americom considered modification of Satcom VI to provide transatlantic capacity in C-band. It requested FCC authorisation to modify the satellite to provide six transponders for the service which would link the USA with Europe and Africa. However RCA Americom eventually withdrew its FCC application. The Satcom VI satellite is not likely to be launched until 1988.

The main users of the RCA C-band capacity are detailed in the section on the Satcom series.

PROJECT OVERVIEW

The Advanced Satcom series consists of Satcom V, IR, IIR and VI which are also known as Satcom F, G, H and I respectively. Satcom V is owned by Alascom Incorporated where it is known as Aurora. It was sold to Alascom in 1982 but is operated by RCA Americom on behalf of Alascom. RCA Americom leases four transponders on the satellite.

Satcom IR and IIR replace the time-expired Satcom I and II satellites. Although having the same number of operational transponders, the Advanced Satcom design provides up to a 50% increase in voice/data capacity. Traffic capacity improvement has come through use of 8·5 Watt solid state power amplifiers and a shaped-beam antenna design. The new antenna design on Advanced Satcom increased gain over the coverage areas. Shaping the antenna beam pattern is accomplished by a large-aperture antenna reflector combined with multiple feed horns whose size, location, power division and relative phasing are the design variables to achieve the desired contour.

The Advanced Satcom spacecraft were the first operational satellites to use all solid state power amplifiers rather than Travelling Wave Tube Amplifiers. The Satcom Ku-band system returned to using Travelling Wave Tube Amplifiers but, at the time the Ku-band birds were ordered, state of the art technology had not produced commercially reliable 45–50 Watt SSPA's. The 8·5 Watt SSPA's were developed at RCA's David Sarnoff Research Laboratories which subsequently developed 45—50 Watt SSPA's. RCA argues that its SSPA's offer greater reliability over TWTA's because they are solid-state devices. They also offer increased traffic handling capacity due to improved linearity that results from lower inter-modulation between carriers on the same transponders. As with the Satcom IIIR and IV spacecraft, the Advanced Satcom spacecraft have a spare transponder for each of the four groups of six channels.

MAJOR SUB-CONTRACTORS ON ADVANCED SATCOM

Major Sub-contractors	*Responsibility*
COM-DEV Limited Cambridge, Ontario, Canada	Transponder Input and Output Multiplexers and Filters
Lockheed Space Systems Division Sunnyvale, California	Earth Sensor Assembly and Horizon Sensor Assembly
Thiokol Corporation Elkton, Maryland	Apogee Kick Motor
Frequency Sources, Incorporated Wacecom Division Northridge, California	Antenna Power Dividers and Combiners
Rocket Research Redmond, Washington	Reaction Engine Assembly
Composite Optics, Incorporated	Antenna Reflectors
Hercules	Antenna Tower Structure
Parsons of California Stockton, California	Structure and Solar Panels
Adcote Corporation Waltham, Massachusetts	Sun Sensor Assembly
Northrop Corporation Norwood, Massachusetts	Rate Measuring Assembly

TECHNICAL SUMMARY

PLATFORM	Advanced Satcom
PRIME CONTRACTOR	RCA Astro-Electronics
OTHER MAJOR CONTRACTORS	see above text
STABILISATION	3-axis
LAUNCH MASS	1080 kg
MASS IN ORBIT	598 kg
POWER SUPPLY, END OF LIFE	1100 Watts
DESIGN LIFE	10 years
ORBITAL POSITIONS	Satcom V, 143° West Satcom IR, 139° West Satcom IIR, 72° West Satcom VI, 67° West Satcom VII, 62° West
ECLIPSE PROTECTION	Full

LAUNCH VEHICLE	Delta 3924
SPACECRAFT DIMENSIONS	5·3 × 4·1 × 4·1 m
SPAN, TIP-TO-TIP	11 metres
NO. OF ACTIVE TRANSPONDERS	24
NO. OF SPARE TRANSPONDERS	4
SSPA POWER OUTPUT	8·5 Watts
FREQUENCY RANGE:	
UPLINK	5·925–6·425 GHz
DOWNLINK	3·700–4·200 GHz
TRANSPONDER BANDWIDTH	36 MHz
EIRP MINIMUM, BEAM EDGE	38 dBW, 34 dBW, 28 dBW
CHANNEL POLARISATION	Linear, horizontal & vertical
RECEIVE COVERAGE (G/T)	−5 dB/K, −10 dB/K

COMMERCIAL DETAILS

OPERATOR	RCA American Communications Incorporated (RCA Americom)
ADDRESS	400 College Road East, Princeton, New Jersey 08540, USA
TELEPHONE NO.	(609) 734-4000
TELEX NO.	710 480 4286
SENIOR STAFF	E. F. Murphy, Chief Executive Officer Dr James J. Tietjen, President Harold W. Rice, VP, Video and Audio Services John Williamson, Public Affairs Director John F. Clark J. J. Christopher, Technical Operations Director
FINANCIAL BACKING	To be financed by RCA Americom which is now owned by General Electronic
PURPOSE OF PROJECT	Provision of DBS capacity for continental USA
PROJECT STATUS	Being planned. Company has conditional FCC construction authority. However project appears to be a plan to be used in the unlikely event of DBS being successful
PROJECT VALUE	Current plans envisage capital expenditure of \$377 m for construction, launch and insurance of satellites and TT&C facilities
LAUNCH DATES	DBS–1, late 1989 DBS–2, 1991

SATELLITE USERS

RCA Americom envisages offering capacity on this system as a common-carrier but as DBS opportunities appear limited it is unlikely that this (and most other) DBS systems will ever be constructed. To some extent, the role of this DBS system has been taken over by the Satcom Ku-band satellite system. That system is capable of distributing broadcast quality network TV to network affiliates. Many industry commentators believe that the three major networks, NBC, CBS and ABC (and CBS in particular) would have taken DBS capacity for this role had DBS developed. Such a role would have required a satellite with considerably more channel capacity than, say, the STC DBS birds and the RCA DBS bird would have fitted the bill. The Satcom Ku birds are capable of providing a quasi-DBS service although, so far, there have been no takers. It is noted with interest that RCA originally envisaged a six transponder (230 Watt) design but increased the number of transponders to 16 at 100 Watts, moving towards the FSS concept for a DBS satellite. Meanwhile, the considerable delays in establishing a US DBS system has resulted in its underlying market shrinking as cable and VCR penetration increases and as the terrestrial broadcasting market begins to restructure itself (partly with the assistance of Satcom Ku type satellites) with the possibility of a fourth terrestrial network or a break-up of the conventional network system.

PROJECT OVERVIEW

RCA Americom has conditional approval from the Federal Communications Commission to begin construction of its current proposals for a two satellite system to become operational in the late 1980's. It has until September 1986 to meet the Commission's due diligence requirements. It has requested that the two satellites, each with 16 channels, be located at 110° West at which Comsat/Satellite Television Corporation already has six channels assigned and at which there are requests from other potential DBS operators. If RCA Americom decides to go ahead with its DBS project it may not receive all 32 channels at 110° West.

Current plans at RCA Americom envisages 100 Watt transponders giving an EIRP at beam edge of 51 dBW. Beam coverage would illuminate either the Eastern or Western half of the country and also Puerto Rico and the Virgin Islands. It proposes to launch the first satellite in the fourth quarter of 1989. The second spacecraft would be launched within two years of the first launch.

However, RCA's proposals for this system have (by satellite world standards) a fairly long history and reflect both the doubts about the financial viability of DBS services and technical developments in and commercial thinking about ground receivers. RCA Americom originally proposal to operate its own DBS system for contiguous USA services back in 1982. It received a construction permit from the Federal Communications Commission in 1982 but not launch and operating authority. These proposals envisaged four in-orbit satellites plus one ground spare and TT&C facilities. Each satellite would have six channels using 230 Watt transponders. Two channels would be capable of being switched to carry broadband high definition TV. The full system was designed to permit remote uplinking from any time zone, allowing real-time programming. RCA envisaged that most uplink facilities would be customer owned, although uplink facilities would be provided by RCA at the customer's request. RCA would act as a common carrier.

However, these proposals were expensive. The total

investment required by RCA Americom was \$760 m and initial investment was \$400 m. Initial services would be provided by one in-orbit operational satellite only, providing coverage of the Eastern zone. A possibility of a quasi-DBS service using Fixed Satellite Service systems was also considered as an interim measure. On the DBS system RCA Americom estimated that the annual lease rate for a single channel on all four of the in-orbit satellites (thereby providing full US coverage) was \$75 m.

By 1984, the DBS scene in the USA saw the first of its big shake-outs it was becoming clear that there was neither an unlimited consumer demand for television and that DBS TV would require lifting of a large number of viewers from terrestrial and cable TV. The big potential DBS operators, CBS, Western Union and RCA Americom withdrew their construction permits from the FCC. Later, Comsat's Satellite Television Corporation in effect withdrew from the market. RCA Americom notified the Federal Communications Commission in July 1984 that it wanted to reduce its proposed transponder power from 230 Watts to 100 Watts and that it only wanted two instead of four in-orbit operational satellites, each with 12 transponders (now 16). This change in plans forced RCA Americom to withdraw from the first round of DBS appointees. These revised plans reduced RCA Americom's capital investment requirements for DBS from \$760 m to \$377 m.

TECHNICAL SUMMARY

PLATFORM	RCA–6000B
PRIME CONTRACTOR	RCA Astro-Electronics
OTHER MAJOR CONTRACTORS	not available
STABILISATION	3-axis
LAUNCH MASS	2629 kg
MASS IN ORBIT (EOL)	1087 kg
POWER SUPPLY, END OF LIFE	5075 Watts
DESIGN LIFE	10 years
ORBITAL POSITIONS	RCA 6B1, 110° West RCA 6B2, 110° West
STATION KEEPING ACCURACY	0·1° (E/W & N/S)
ECLIPSE PROTECTION	Partial
LAUNCH VEHICLE	STS + SCOTS or Ariane IV
NO. OF ACTIVE TRANSPONDERS	16
NO. OF SPARE TRANSPONDERS	not available
TWTA POWER OUTPUT	100 Watts
FREQUENCY RANGE:	
UPLINK	17·3–17·8 GHz
DOWNLINK	12·2–12·7 GHz
TRANSPONDER BANDWIDTH	24 MHz
EIRP MINIMUM, BEAM EDGE	51 dBW

Note that the technical details provided here are the latest presented by RCA Americom. The construction permit granted by the FCC in 1982 covered a four in-orbit satellite system, each with six operational 230 Watt transponders, a satellite bus weighting 1096 kg at launch and providing a power supply of 3660 Watts. Satellite EIRP was 59 dBW. Design life was 7 years.

RCA Direct Broadcast Satellite

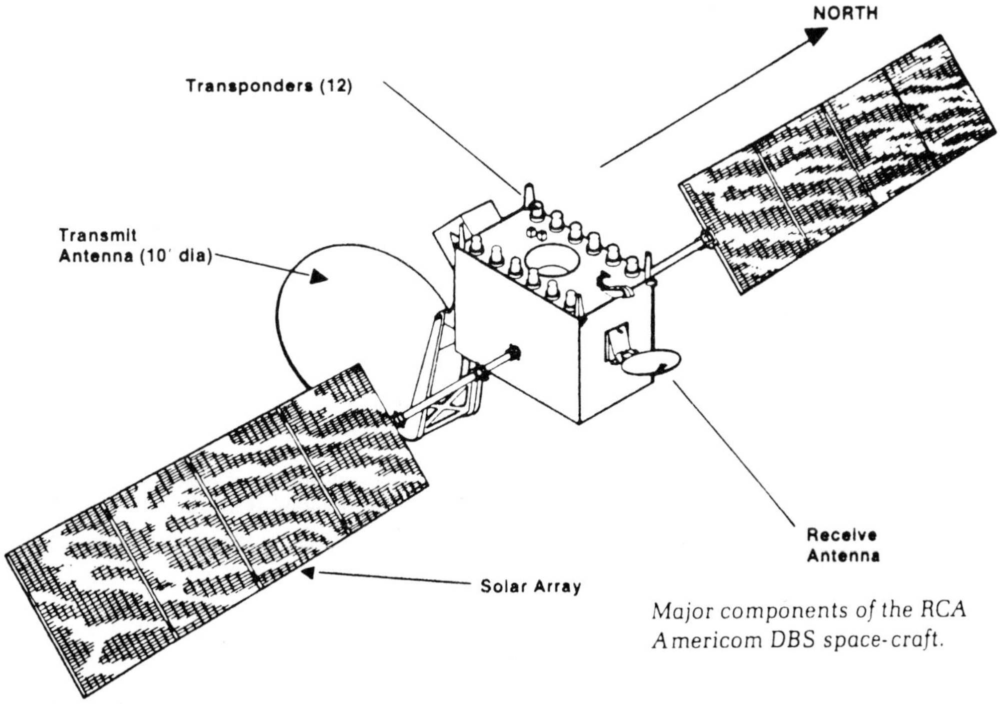

Major components of the RCA Americom DBS space-craft.

COMMERCIAL DETAILS

OPERATOR	RCA American Communications
ADDRESS	400 College Road East, Princeton, New Jersey 08540, USA
TELEPHONE NO.	(609) 734-4200
TELEX NO.	710 480 4286
SENIOR STAFF	Dr James J. Tietjen, President E. F. Murphy, Chief Executive Officer Jack F. Underwood, VP, Commercial Communications Harold W. Rice, VP, Video & Audio Services R. E. Smylie, VP, Business & Government Network J. J. Christopher, Technical Operations Director Walter W. Poli, Director, Commercial Sales
PURPOSE OF PROJECT	Provision of medium power Ku-band capacity for domestic US use, primarily by the broadcasting industry but also for VSAT type data and voice traffic
PROJECT STATUS	The first two satellites are in-orbit and operational but launch of a third satellite is delayed
PROJECT VALUE	RCA Americom claims that its investment in Satcom K1–4 will total $306 m
LAUNCH DATES	Satcom K2, 26th November 1985, STS 61–B* Satcom K1, 19th December 1985, STS 61–C* Satcom K2, 1989, Shuttle Satcom K4, ground spare *with PAM D–II

SATELLITE USERS

Capacity on Satcom K2 is split between the following:

NBC: Eight channels for use by NBC and its affiliate TV stations (206 in total) for distribution of NBC network TV, sports, news and current events
RCA Americom: Two transponders retained for piecemeal lease to television syndicators for distribution of their programming
AT&T: Two transponders for data transmission to small dish terminals
CONUS Communications: One transponder for exchange and distribution of news
Hubbard Broadcasting Incorporated: Three transponders for provision of a national distribution service for Hubbard and other broadcasters and TV stations.

RCA and HBO have a deal to purchase K3 and re-sell capacity on it. RCA Americom will retain two transponders and HBO four.

PROJECT OVERVIEW

Some people in the US television industry believe that the Satcom Ku-band satellites will have an impact on US terrestrial TV comparable to the impact that Satcom I had on the cable television scene. The combination of the use of Ku-band and high power transponders (45 Watts for Satcom K1 and K2) requires smaller transmit/receive dishes. These (by virtue of their small size) can be located on the premises of broadcasters. Typically broadcast quality reception will use a dish from 3·5 to 6·1 metres diameter. With conventional C-band satellites, the low transponder power output (a maximum of 8·5 Watts on Satcom/Advanced Satcom) means that the receive dish may have to be located outside the city where the broadcaster is operating. It would be linked to the broadcaster by a terrestrial microwave link which, particularly during thunder storms, would result in interference.

Moreover, prior to the introduction of the Satcom Ku-band system only about two-thirds of US commercial TV stations had receive C-band capacity. This meant that the national TV networks, ABC, NBC and CBS, had to maintain a terrestrial system to distribute their networks to local television stations, even if they were using satellites capable of national distribution. Moreover, distribution of network TV was fragmented amongst several satellites and satellite operators. This either required the local TV station to invest in several receive terminals or re-orient a dish to different satellites at different times, perhaps missing some programming as a result.

Two other factors place the Satcom Ku-band in a position to alter the structure of US terrestrial TV. Firstly, by the mid-1980's the position of having (only) three dominant national networks, with 596 out of 765 commercial broadcast stations affiliated to them (the three networks owned only 15 stations between them because of FCC restrictions on the ownership of TV stations) had become increasingly anachronistic. The structure of US broadcasting had been fossilised by decisions of the FCC in 1945 and 1952 which, in effect, carved up the frequency bands in favour of NBC and CBS, leaving ABC to struggle for two decades as a third runner and squeezing out any newcomer. There were no provisions for regional networks either.

One result of the carve up was the growth of cable television in the 1970's which began to give the consumer considerable choice. The networks began to suffer from a significant decline in audience viewing, threatening their advertising revenue base. The limitations of and FCC carve up of terrestrial frequencies prevented the networks providing new channels and services. Moreover, their plans for involvement in DBS came to nothing when it became apparent how great the risk is in that business. But a second critical factor which resulted in RCA's decision to go ahead with the Satcom Ku concept was that RCA owned NBC.

To attract broadcasting organisations to the Satcom Ku system, RCA offered to all television stations in the USA a free receive earth station dedicated to reception from the K2 satellite. The NBC affiliate stations had to buy their own but NBC took eight transponders on Satcom K2 to entirely replace its terrestrial distribution system. The NBC transponders are not only used to distribute its regular network but also sports, news and special events. RCA Americom retained two transponders "for piecemeal lease to television syndicators for distribution of their programming." This capacity is complementary to MBC's capacity. AT&T took two transponders for data distribution. The remaining four transponders went to Hubbard Broadcasting, an ambitous operator of five TV stations which is still looking to develop its own DBS satellite system. The four transponders gives it instant access to nearly all US commercial TV stations, the first non network company to gain such access, NBC, of course, being the only network to have such access. Hubbard and NBC now each have access to three times as many commercial TV stations as CBS and ABC. Of the four Hubbard transponders, one is owned by CONUS Communications, which is a subsidiary of Hubbard. CONUS Communications is involved in Satellite News Gathering and exchange of live news and daily package news feeds.

Having put this package of users together on Satcom K2 and restricting the use of the "free" earth terminals to reception of K2 transmissions, RCA Americom has created a "dominant" position for the users, putting them in a position to bypass the traditional organisational arrangements in network broadcasting. Under traditional network arrangements, a network television company pays its affiliate stations to broadcast its programmes. The networks derive revenue from its sale of advertising time. The local affiliate could intersperse its own programming between the networks programming. It would derive revenue from selling its own advertising airtime and typically gain about three times the revenue from broadcasting network TV.

However, it would also have to pay for the programming which may be in-house programming or bought in syndicated re-runs. However, unlike the networks, the local TV station does not have the easy capability of distributing its own programming to other TV stations, particularly on a national basis. Moreover, any independent company wishing to distribute, say live sports programming, to a large number of TV stations is at a disadvantage compared with the networks which could automatically distribute to a large number as part of its network service. Using Satcom K2 the independents can now distribute to more stations than any one network organisation could serve prior to K2. With a larger potential audience, the independent could afford to, say pay more for coverage of live sports events, or produce more lavish drama, or pay stations more for transmitting the programming.

Some US commentators believe that the K2 type approach will result in the development of a fourth national TV network, pointing to Mr Rupert Murdoch's current attempts to create one. However, others dispute this, saying that the K2 type system may rapidly destroy the current programming structure leaving the network companies to produce programming and arrange distribution, leaving advertising sales to new structures. The

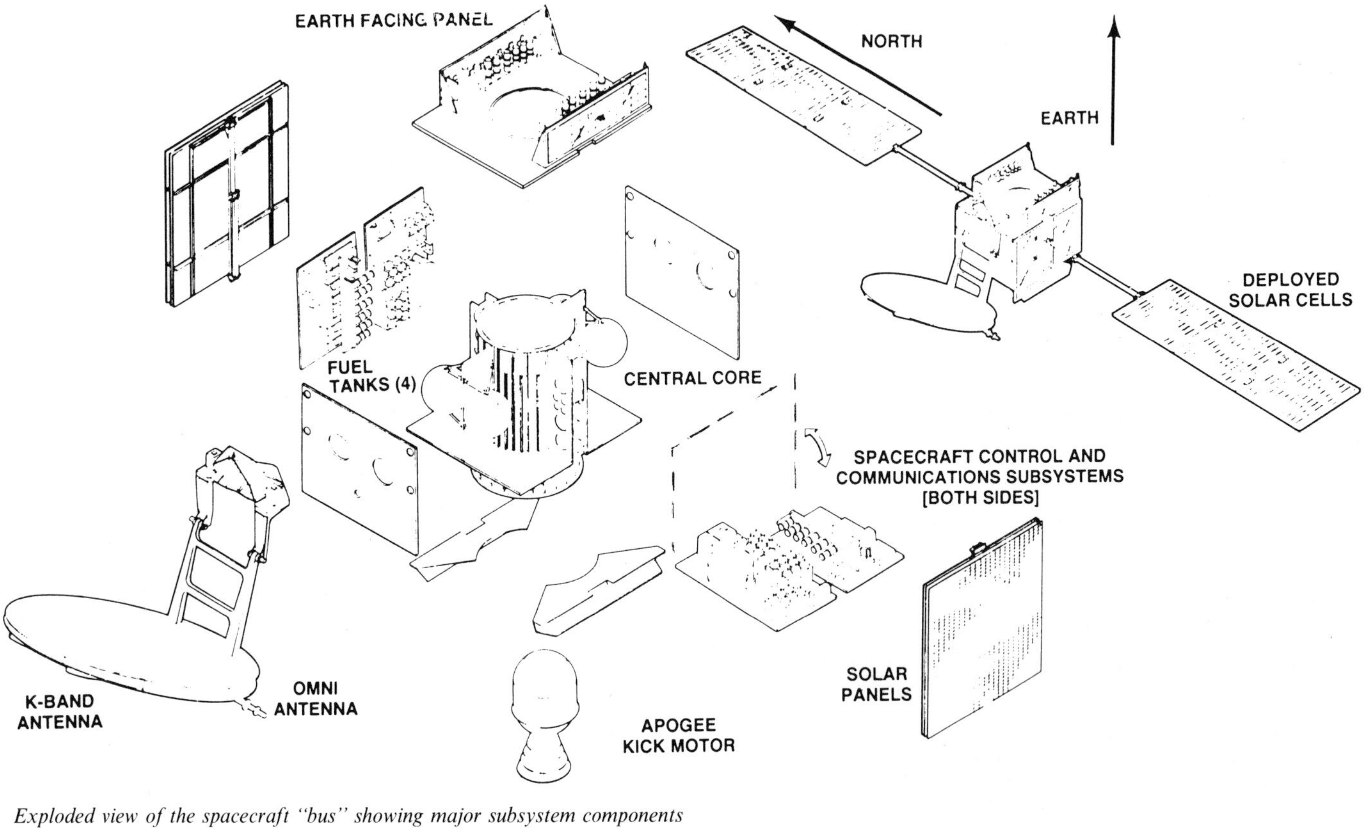

Exploded view of the spacecraft "bus" showing major subsystem components

complete destruction of networks is probably wishful thinking; the network companies are powerful enough to adjust to, and to some extent, control the changing structure of the industry.

The Satcom K2 concept has a particular impact on television news programming arrangements. The use of satellite news gathering equipment allows film crews to transmit back to their station immediately (and at the same time distribute footage nationally). With traditional arrangements, a station would have to arrange for its news film to be brought back from a local event by physical means, either road or for special events, a helicopter. This, in effect, meant that it could only cover news within a 40-mile radius before facing logistical problems, and even within 40 miles, live coverage was expensive and, for sudden events, impracticable. Moreover, the local station would have difficulty in distributing either live or recorded news nationally without going through network organisations.

TECHNICAL SUMMARY

PLATFORM	RCA 4000
PRIME CONTRACTOR	RCA Astro-Electronics
STABILISATION	3-axis
MASS IN ORBIT (BOL)	664 kg
POWER SUPPLY, (BOL)	1395 Watts
DESIGN LIFE	10 years
ORBITAL POSITIONS	Satcom K1, 67° West* Satcom K2, 81° West Satcom K3, 85° West
NO. OF ACTIVE TRANSPONDERS	16
NO. OF SPARE TRANSPONDERS	4
TWTA POWER OUTPUT	45 Watts
TRANSPONDER BANDWIDTH	54 MHz
EIRP MINIMUM, BEAM EDGE	43 dBW, all 50 states, 47 dBW, spot beams
CHANNEL POLARISATION	

*Satcom K1 is initially located at 85° West

COMMERCIAL DETAILS

OPERATOR	RCA American Communications
ADDRESS	400 College Road East, Princeton, New Jersey 08540, USA
TELEPHONE NO.	(609) 734-4200
TELEX NO.	710 480 4286
SENIOR STAFF	Dr James J. Tietjen, President E. F. Murphy, Chief Executive Officer Jack F. Underwood, VP, Commercial Communications Harold W. Rice, VP, Video and Audio Services R. E. Smylie, VP, Business and Government Network J. J. Christopher, Technical Operations Director Walter W. Pioli, Director, Commercial Sales
PURPOSE OF PROJECT	Provision of Ku- and C-band capacity for US domestic operations providing replacement and extra capacity for the RCA Americom system
PROJECT STATUS	FCC authorisation was granted in July 1985 but construction has yet to begin
PROJECT VALUE	In its 1983 submission to the FCC, RCA Americom said that it would invest $439 m in this system, including launch and insurance costs
LAUNCH DATES	The 1983 submission gives launch dates as June 1989, June 1990 and September 1992. Last Shuttle launch manifest prior to 1986, showed June 1990 launch of RCA 3001

SATELLITE USERS

In its 1983 submission to the FCC, RCA Americom said that this satellite system would be used to provide new capacity for both existing markets and for emerging new telecommunications markets. The company forecasted that of the markets it serves, a substantial potential for growth would be in commercial communications services. Much of this would stem from the surge in office automation and the technological and regulatory developments related to DTS/DMS, spread spectrum techniques for data networking, digital teleconferencing and other telecommunications services. These also encompass a long-distance private-line service, satellite links to offshore and remote area drilling and exploration rigs, private telecommunications networks and wideband digital data networks. In TV distribution RCA indicated eight markets in addition to cable television as areas of growth. These include DBS TV, SMATV, Lower-Power TV, Subscription TV, Multipoint Distribution Service, Broadcast TV, International TV and Occasional TV. In the government services area, RCA Americom would utilise new capacity to provide expanded services to a variety of its customer groups within the Federal government. These include NASA, NOAA, the Department of Defense and the Government Services Administration.

At the time that RCA Americom made the submission to the FCC it was not as apparent as it is today that the VSAT market would develop into a major market opportunity for the period up to 1995.

PROJECT OVERVIEW

In November 1983 RCA Americom revealed some details of its thinking on its space segment requirements for the 1990's. Its 1983 filing with the FCC was for a plan to launch and operate three new hybrid C/Ku-band communications satellites, or at the company's option, three collocated C-band and Ku-band spacecraft in the 1989–92 time frame. The system would provide replacement capacity for the Advanced Satcom and Satcom K1–4 satellites and meet projected "substantial increases in demand for satellite communications capacity in the markets" (served by RCA Americom). RCA Americom requested orbital slots at 61°, 63° and 65° West, indicating that these locations had been assigned by the FCC for both C-band and Ku-band services. The spacecraft would provide service to the 48 contiguous United States plus the Caribbean basis, specifically Puerto Rico and the US Virgin Islands.

The Ku-band portion of the hybrid satellite concept would use 16 operational transponders with a power output of 50 Watts and a bandwidth of 54 MHz. This would provide a signal strength of 46 to 52 dBW in the coverage area. The satellite Ku-band antennas would have an in-orbit switching capability for full CONUS and the Caribbean basin or half-CONUS with the Eastern beam also providing service to the Caribbean basin. Reliability would be enhanced through an 11:8 redundancy provided through three backup transponders serving each bank of eight operating transponders. The C-band portion of the communications payload would be provided through 24, 10 Watt solid state power amplifiers operating at a bandwidth of 36 MHz and delivering a signal strength of 36 dBW across CONUS. Each bank of six operating transponders would have a backup to give a 7:6 redundancy for reliability. The satellites would be designed for launch by Shuttle. This requirement is presumably being abandoned.

TECHNICAL SUMMARY

PLATFORM	RCA 5000
PRIME CONTRACTOR	RCA Astro-Electronics
STABILISATION	3-axis
LAUNCH MASS	
MASS IN ORBIT (BOL)	1017 kg
POWER SUPPLY, (BOL)	2440 Watts
DESIGN LIFE	12–13 years
ORBITAL POSITIONS	Not available
ECLIPSE PROTECTION	Full
NO. OF ACTIVE TRANSPONDERS	16 at 50 Watts, 24 at 10 Watts
NO. OF SPARE TRANSPONDERS	6 at 50 Watts, 2 at 10 Watts
TWTA POWER OUTPUT	50 Watts (Ku), 10 Watts (C)*
TRANSPONDER BANDWIDTH	54 MHz (Ku), 36 MHz (C)
EIRP MINIMUM, BEAM EDGE	46–52 dBW (Ku)

*C-band transponders will use solid state power amplifiers

COMMERCIAL DETAILS

OPERATOR	Satellite Business Systems (SBS)
ADDRESS	8283 Greensboro Drive, McLean, Virginia 22102, USA
TELEPHONE NO.	(703) 442-5500
SENIOR STAFF	Marvin L. Mann, President Robert J. Perry, VP, Finance and Administration Richard T. Liebhaber, Senior VP, Engineering Eldon Blust, International Services Betty Byrne, SBS Videolink Marketing Tedd Wilson, Marketing Manager, Space Resources Larry Weekly, Director, Public Affairs
PURPOSE OF PROJECT	Provision of Ku-band US domestic satellite capacity for digital transmission for business requirements and specialised common carrier services
PROJECT STATUS	The system is fully operational, but the company has chalked up considerable losses since first satellite launched in 1980
PROJECT VALUE	It is estimated that IBM, Comsat and Aetna sunk \$1000 m into SBS–1 to 5, the SBS ground segment and development and operational expenses
LAUNCH DATES	SBS–1, 15th November 1980, Delta F153 + PAM–D SBS–2, 27th September 1981, Delta F156 + PAM–D SBS–3, 11th November 1982, STS–5 + PAM–D SBS–4, 30th August 1984, STS–41D + PAM–D SBS–5, not before 1988, Ariane

SATELLITE USERS

Most of the capacity on the SBS system is used for provision of SBS's own telecommunications services, which are detailed in the text. As originally envisaged, the SBS system was intended to provide capacity for high speed digital data transmission for business. However, the market was very disappointing and SBS rapidly moved into digital voice services. Ku-band transponders are available for sale or lease for television traffic. NCB through Comsat, took transponders 3, 5, 6 and 9 on SBS–3 for distribution of its network programming and affiliate feeds but has now transferred to RCA Satcom K2. CONUS Communications leased transponder 7 on SBS–3 but likewise has moved to Satcom K2. Private Satellite Network uses transponder 8 on SBS–3 for TV traffic.

Now that SBS is owned by MCI, it is expected that the company's increasing traffic will transfer to SBS satellites. IBM also retains an interest in SBS (see Project Overview) which may provide a future user base.

SBS also uses the Intelsat system to provide services for its customers who may or may not also be using SBS's space segment.

PROJECT OVERVIEW

There has been much talk and analysis about Satellite Business System's failure to meet its original objectives and its dire financial performance during the first half of the 1980's. However, its failure to meet those objectives is not particularly suprising, nor does its future need to be written off. Satellite Business Systems was formed as a partnership between IBM, Comsat and Aetna Life to provide digital data transmission capacity at a time when not only was the demand for such capacity increasing (as it still does) but also when the computer industry saw an increasing convergence between computer facilities and communications facilities. IBM, with a market share of the computer hardware industry in excess of all the rest of the world's computer manufacturers combined, had a strategic interest in communications, not least to protect its de facto computer connection industry standard, SNA. The partnership also gave Comsat further access to the US domestic satellite scene.

A number of companies in the US attempted to introduce digital transmission services during the 1970's but of these early attempts, only SBS now survives. For example Datran foundered in 1976 after an early start in the market. Xerox Corporation sank some \$30 m into its XTEN system which would have used satellite broadband links between major cities. However Xerox abandoned the project in 1981 before the system ever became operational. SBS's early performance was equally as disappointing. Formed in 1977, its first year of operation was in 1981, when its revenue was only \$5 m. The following year was as equally disappointing, with revenue of \$39 m. It began to creep up to \$142 m in 1983, reaching \$290 m in 1984, when Comsat dropped out of the partnership (adding \$2 to Comsat's share price). Comsat sold its shareholding in December 1984 to IBM and Aetna Life for an almost peppercorn fee of \$50 m less its share of 1983/4 losses.

These figures were not revealed but it is estimated that SBS lost at least \$140 m in 1983 and \$100 m in 1984. SBS as a going concern was probably worth nothing in 1984, but Comsat had its own financial problems, including financing the STC spacecraft. In 1985, IBM traded in its then 60% share of SBS. It exchanged the share with specialised common carrier MCI in exchange for a 20% shareholding in MCI. At the same time IBM assumed responsibility for \$400 m of SBS's debt. Finally IBM then bought out Aetna Life's 40% share in SBS. The complex arrangement resulted in IBM owning SBS–4 to 6 and MCI owning SBS–1 to 3, but leaves SBS in a position to make a profit. Comsat, Aetna Life and IBM, under the original partnership arrangement, each (eventually) provided \$237 m in equity.

The failure of SBS to achieve its original objectives and the need to write-off such large sums of money has been put down to conceptual errors in the early stages of the project. SBS opted for Time Divisional Multiple Access as the transmission mode. Although this allowed efficient use of the transponder capacity on-board the spacecraft ground stations were hideously expensive, particularly as volume production did not build up rapidly.

Moreover the ISDN concept behind SBS was, and is, engineering rather than market led. Despite opting for Ku-band space capacity and a then relatively high transponder power output, the ground stations for high data rates were large (in excess of 5·5 metres). The National Network service also required a considerable amount of dependence on local telephone operators' networks to connect earth stations and SBS's collocated switching centres with customer premises. This meant that part of SBS's services were under the physical control of other organisations not least, prior to its breakup, AT&T. SBS experimented with local by-pass with tests to distribute data on Manhattan and San Francisco cable networks but was unable to take the idea up. At the end of 1984, SBS was leasing 39,000 telephone lines from AT&T and other telephone companies for origination and completion of long distance telephone calls.

When it became apparent early in SBS operations that its Communications Network Services and Skyline Network Service would not provide the customer base to support its huge capital investment (at the end of 1982, its three partners had invested $591 m in equity and it had further debt funding of $253 m) it rapidly changed course to address two other markets. It introduced a Skyline service, basically providing long-distance telephone services for business and homes and decided to lease transponders. However, it missed the market for cable TV traffic, partly because its initial strategy was muddled, offering only preemptible capacity (later dropped in favour of non-preemptible arrangements) and charging a high annual price (up to $3·2 m per year). Nor was the concept of a "hot cable bird" developed. In any event, RCA, with Satcom IIIR and Hughes with Galaxy I, stitched this market up with C-band capacity.

This left SBS with its Skyline services, in competition with GTE Sprint and MCI as well as the conventional terrestrial network, to fill its expensive satellite capacity which was not designed for such a service. The service also saw a structural shake-out of the market in 1985/86 which involved SBS's takeover by MCI. But by that time, it was evident that its space segment was not the most suitable method of carriage of its Skyline service. SBS had begun to invest in fibre-optic capacity for long distance links. Prior to its takeover by MCI, SBS had become the number three company, after MCI and Sprint, in the long distance alternative telephone service business.

SBS Services: The list of services offered by SBS clearly shows that the company is (now) primarily in the telecommunications rather than the satellite operating business. Unlike such satellite operators as Hughes Communications and RCA Americom, it is clear that most of SBS's revenue derives from customers who are indifferent as to whether their traffic is transmitted through a satellite or other means and are wholly indifferent as to which satellite the traffic is routed through.

Skyline Services Division: The primary Skyline Service was introduced in 1983 in competition with (the then Southern Pacific Communications Corporations) Sprint Service and MCI. The service is marketed within five separate concepts, Skyline MTS (Message Toll Service), Skyline WATS (Wide Area Telephone Service), Skyline FX (Foreign Exchange – nothing to do with currency – US "FX" telephone services provide a single link from the customer's premises to another city (say) in which the customer has a "local" telephone number for incoming and outgoing telephone calls), Skyline Toll-Free and SBS Hostline (for Hotels). These are basically the same voice telephony services but with different charging methods, and, generally, a higher quality of signal than provided by POTS (plain Old Telephone Systems). They are similar to services provided by MCI and GTE Sprint who, by and large, are not primarily dependent on satellite capacity. The skyline service has proved to be SBS's primary business activity. The SBS–4 satellite, launched in August 1984, was described by SBS as its "Skyline" satellite because half of its 10 transponders were dedicated to the service. The then President of SBS, Stephen B. Schwarz, said, at the time that "Satellites are only one element of SBS's advanced nationwide communications system. But ... without them we could not have established our system as quickly or made it as flexible for meeting changing customer needs." In marketing the Skyline Service, SBS has also emphasised its IBM connection which, it says, "customers recognise as the foundation of SBS's permanence and commitment to, service excellence."

In plain English this means that despite bad publicity about its profitability SBS is claiming that it is reputable because it is backed by Big Blue (IBM). Apart from price and quality of signal, Skyline can also offer business users features not available on POTS. For example the Skyline WATS service offers a T1 data service capable of 1544 mbs capacity lines (equal to 24 telephone lines) at a lower price than basic WATS services. The Skyline Long Distance Service, its basic long distance telephone call service for both domestic and business services, has suffered from some arrangements in the break up of AT&T.

The new Bell Operating Companies (BOC) have tended to route all long distance calls through AT&T unless the customer has stipulated otherwise. According to SBS, "this passive approach has resulted in large numbers of (potential) customers being defaulted to AT&T". The Skyline service was originally conceptualised as an interstate service. But despite the existence of a national communications regulatory body, the Federal Communications Commission, regulation is also in the hands of individual states, as well as the Federal Courts. Since 1984, SBS has had some success in obtaining state authorisations for long distance services on an intrastate basis. Such authorisation is pursued by SBS on a state-by-state basis. The Skyline services are essentially available in cities and large towns in contiguous USA, Southern Canada, Puerto Rico and the US Virgin Islands. It is expected that by the end of 1986 some 80% of US telephone users will have "equal access" to the basic Skyline services for long distance telephone calls.

National Accounts Division: This area of SBS's activity encompasses its core activity as envisged when the company was set up in 1977 plus the obvious extensions of such activities. The four areas are: 1) Communications; Network Service, providing private networks for large business, using earth stations located on customer premises; 2) The Skyline Network Service, providing "virtually" private networks for large businesses but using SBS earth stations shared with other users. Both Communications Network Service and Skyline Network Service are capable of providing integrated voice, data and video traffic, although it is still not clear that such integration is desirable from a user operational viewpoint. Within the Skyline Network Service, SBS offers two specialised services, Data Service and Videolink. For data services, SBS claims it has a bit error rate of 10^{-7} in this area. The figure is not exceptional by Ku-band satellite transmission standards but is better than envisaged before the SBS satellite system became operational in 1981. Also included within the Skyline Network Service is SBS's Videolink service linking public videolink conference rooms. The service is also available for links to dedicated customer premises using the Communications Network Service. It is separate from SBS's broadcast video transmission service which is for point to multipoint services rather than point-to-point; 3) The SBS Interlink service is for high speed international communications services, and is essentially the international link to all SBS services through operating agreements with PTT's outside the USA. These include the UK, Belgium, Canada, France, Japan, Mexico, Switzerland, Hong Kong, Italy and Holland. SBS has

direct access, through its own earth stations, to the Intelsat system for these services. However, during 1985, it considered modification of its own domestic satellites to provide coverage for Europe for the Interlink services. Included in the Interlink services are motion videoconferencing, alternative voice/data 56 kbps and 1544 Mbps channels on a full-period, part-time or occasional use basis. The Interlink service is available from virtually all US cities with SBS ground stations. For SBS's Skyline services, the company has an agreement with Telecom Canada for routing of Long distance, WATS and Skyline Network Services to Canada; 4) SBS Dataline is a digital termination service for business data communications. Used to overcome capacity bottlenecks in the distribution of high-speed data in large cities. It is aimed at Communications Network Service users who need local distribution. It offers 56 kbit/s digital transmission. Digital Terminating Technology (DTS) is used for intercity transmission and SBS satellites for long haul.

Space Resources Division: Not, so far, one of the more spectacularly successful SBS services, this offers full or part time capacity on a full transponder. This service was introduced when SBS's core business failed to materialise in 1980/1. Most noticeable amongst the users of this service has been NBC, who has used SBS capacity for distribution of its network TV. NBC now uses RCA Americom's Satcom K2 satellite instead. Mr Rupert Murdoch, still looking to form a fourth US TV network, leased capacity on SBS in 1983 through a company called United States Satellite Broadcasting Company. This was intended to provide an interim quasi-DBS service with six TV channels, three on each satellite. When Murdoch signed the agreement with SBS he said that either the project will make money or "there will be red ink" all over the place. The project was quickly abandoned, leaving Murdoch with a healthy loss on his transponder leases. There have been no similar takers of this SBS service. Also within the Space Resources Division is the SBS Broadcast Video service aimed at the videoconferencing market. SBS was a pioneer in videoconferencing, having worked on the service prior to its space segment becoming operational in 1981. The Space Resources Division is also responsible for the broadcast video service. This transmits a business's point-to-many point programming used, for example, in employee training or product announcements. Service requires only a small dish (4–6 feet) to receive the programming which is transmitted in an analogue format. Most of SBS's videoconferencing, mostly part of its Communications Network Service or Videolink Service uses a digital transmission format.

SBS Real Estate Communications Company: Although originally envisaged as a service for large businesses, SBS has attempted to enter the market for telecommunications services for small businesses through providing communications packages for individual office buildings with several tenants. SBS Real Estate Communications Corporation (known as RealCom) is a subsidiary of SBS. The services provided by RealCom are shared between the various companies occupying an office block. RealCom provides a central PBX exchange in the building together with telephones and cabling to connect each building's tenants to local and long-distance telephone companies. It also contracts with the individual tenants in the office block to provide turnkey telecommunications services, equipment and features to meet the tenant's particular requirements. This service is essentially aimed at new or refurbished office blocks. The individual tenants by routing their telecommunications traffic through a central shared switchboard are able to take advantage of bulk discounts through aggregation of long-distance lines. Equipment for the RealCom installations is provided by Ameritech Communications Incorporated, a subsidiary of Ameritech, one of the seven regional holding companies resulting from the divestiture of AT&T. Ameritech also installs, maintains and administers the equipment.

Prior to 1984 SBS and IBM were prohibited by the FCC from selling or promoting each others's products and services and from making joint marketing presentations. This prohibition, dating from the formation of SBS in 1977, was dropped in 1984 and an IBM–SBS joint marketing venture was set up.

TECHNICAL SUMMARY

PLATFORM	HS–376
PRIME CONTRACTOR	Hughes Aircraft Company
STABILISATION	Spin
LAUNCH MASS	1060 Kg
POWER SUPPLY, END OF LIFE	900 Watts (SBS–5, 1150 Watts)
DESIGN LIFE	7 years (SBS–1/2) 8 years (SBS–3/4)
ORBITAL POSITIONS	SBS–1, 99° West; SBS–2, 97° West; SBS–3, 95° West; SBS–4, 91° West; SBS–5, 122° West
PROPULSION SYSTEM	Delta 3920, STS, Ariane 3/4
SPACECRAFT DIMENSIONS	2·6 m diameter, 6·6 m high, deployed, 2·82 m high, stowed
CONTROL CENTRE	McLean, Virginia
TT&C STATIONS	Castle Rock, Colorado Clarksburg, Maryland
NO. OF ACTIVE TRANSPONDERS	10
NO. OF SPARE TRANSPONDERS	6
TWTA POWER OUTPUT	20 Watts
FREQUENCY RANGE:	
UPLINK	14·025–14·466 GHz
DOWNLINK	11·725–12·166 GHz
TRANSPONDER BANDWIDTH	43 MHz
EIRP MINIMUM, BEAM EDGE	44 dBW
CHANNEL POLARISATION	Linear horizontal (SBS–5 dual)
RECEIVE COVERAGE (G/T)	+/−2 dB/K
SINGLE CARRIER SATURATION FLUX DENSITY	−76 to −82 dBW/sq. m.
PRIMARY TRANSMISSION MODE	TDMA

Note: SBS–5 is basically similar to SBS–1–4 but includes four 110 MHz wideband 20 Watt transponders, in addition to the 10 43 MHz transponders. Three of the wideband and two of the narrowband transponders have the capability of operating with parallel TWTA's. This allows transponder power output for the five to increase to 40 Watts. It has a total of 24 transponders. Spot beam coverage is also provided for Alaska and Hawaii. It has a design life of 10 years. Unlike SBS–1–4, SBS–5 uses both horizontal and vertical polarisation.

COMMERCIAL DETAILS

OPERATOR	Satellite Business Systems (SBS)
ADDRESS	8283 Greensboro Drive, McLean, Virginia 22102, USA
TELEPHONE NO.	(703) 442-5500
SENIOR STAFF	Marvin L. Mann, President Robert J. Perry, VP, Finance and Administration Ray H. Fentriss, Senior VP, Marketing Betty Byrne, SBS Videolink Marketing Tedd Wilson, Marketing Manager, Space Resources Larry Weekly, Director, Public Affairs
FINANCIAL BACKING	Satellite Business Systems is now a wholly owned subsidiary of International Business Machines
PURPOSE OF PROJECT	Second generation satellite system providing Ku-band domestic capacity for digital transmission services for US business and specialised common carrier services
PROJECT STATUS	SBS–6 ordered and due for launch in 1988–89. Remaining satellites still to be ordered
PROJECT VALUE	Contract price with Hughes for SBS–6 is \$60 m, excluding launch and insurance costs
LAUNCH DATES	SBS–6, March 1989 by Ariane 4, V35 SBS–7, 1989 (earliest date) SBS–8, 1989 (earliest date) SBS–9, 1989 (earliest date)

SATELLITE USERS

SBS–6 is intended to replace SBS–1, SBS–8 will replace SBS–2. SBS–9 will replace SBS–3. SBS–7 will provide new, rather than replacement capacity. However, the future of all four satellites remains uncertain.

PROJECT OVERVIEW

SBS–6 is to use the new Hughes HS–393 platform; previous SBS satellites have used the HS–376 platform. SBS–6 was originally designated a ground spare, but is now to be launched in 1987 or 1988. It will replace the ageing SBS–1 satellite.

Orders have not yet been placed for SBS 7–9. However SBS plans SBS–7 to be a new satellite. SBS–8 will replace SBS–2 and SBS–9 will replace SBS–3.

Locations of the satellites will be SBS–6 at 99° West, SBS–7 at 93° West, SBS–8 at 97° West, SBS–9 at 95° West.

SBS–6 is owned by IBM which has created a subsidiary, the Satellite Transponder Leasing Corporation, to lease or sell capacity on SBS IV–VI. Intelsat took up an option to purchase SBS–VI in 1986 for transatlantic use but subsequently decided not to renew the option.

TECHNICAL SUMMARY

PLATFORM	HS–393 (SBS–6)
PRIME CONTRACTOR	Hughes Aircraft Company
STABILISATION	Spin
ORBITAL POSITIONS	SBS–6, 99° W; SBS–7, 93° W; SBS–8, 97° W; SBS–9, 95° W (only SBS–6 authorised)
PROPULSION SYSTEM	Bi-propellant
NO. OF ACTIVE TRANSPONDERS	19

COMMERCIAL DETAILS

OPERATOR	Empresa Brasileira de Telecomunicacoes (EMBRATEL)
ADDRESS	Av. Presidente Vargas 1012, CP 2586, 20071 Rio de Janeiro, Brazil
TELEPHONE NO.	(021) 2168182
TELEX NO.	2121185 ebtl br
SENIOR STAFF	Benjamin Himelgryn, Director, Domestic Satellite Departments Ovidio Barrandas, Head of Domestic Operations
PURPOSE OF PROJECT	Provision of domestic C-band capacity for telephony, TV distribution, military and high speed data links and to replace leased Intelsat capacity
PROJECT STATUS	System fully operational
PROJECT VALUE	Two satellites plus TT&C stations cost $131·1 m. Ariane space launch charge per satellite was $29 m. SBTS–1 insurance cost $55 m and SBTS–2 insurance cost $45 m
LAUNCH DATES	SBTS–1, 8th February 1985, Ariane 3, V12, SBTS–2, 28th March 1986, Ariane 3, V17

SATELLITE USERS

Embratel, established in 1965, is a division of the Brazilian PTT, Telebras, and is responsible for international communications and long distance domestic communications. The SBTS system is primarily used for long distance domestic telephone. Between 14 and 19 transponders are used for this traffic.

Second major use of the system is for relay of television programmes. Earliest users of the capacity for this service are the two TV networks, Globo and Bandeirantes, transferring from the leased Intelsat capacity. These channels can also be picked up with 3 metre TVRO terminals. There is no cable TV in Brazil so there are no cable TV channels on the satellite.

Two transponders are allocated through the government to provide communications capacity to the armed forces.

Capacity is also available for private communications systems. Known users are the two banks, Itau and Bradesco and the state owned Petrobras which uses the system for links to offshore platforms. TDMA is being studied for use with the system. FDMA and SCPC are currently used.

The national air traffic control agency is looking to use the system for centre-to-centre communications.

PROJECT OVERVIEW

The origins of Brazil's two satellite domestic communications system date back over a decade. In the mid-1970's Brazil developed a plan to have its own satellite system. The country is an obvious place to have a domestic satellite system with a large geographical area, generally high growth rates and a relatively advanced commercial sector. However, in 1977 Embratel cancelled its invitations to bid for the SBTS system. Instead it opted to lease capacity from Intelsat to provide domestic communications capacity. This was somewhat surprising as Embratel had begun preliminary studies for the system as far back as 1968 and subsequently justified the final order for SBTS, placed in July 1982, on the grounds that capacity leased from Intelsat could not provide sufficient flexibility for improvements to Brazil's internal communication.

Yet Brazil's economy had run into major problems between 1977 and 1982, with the country having difficulty in financing external debt in excess of $90 billion. The SBTS project was almost wholly dependent on foreign financing. Nevertheless, the Brazilians were able to obtain offset trading opportunities with Canada and France over a period of eight years. Loans were also raised from Citibank and Exim Bank in the USA to finance the project. Although Embratel begun leasing Intelsat transponders in 1978, leasing up to a total of seven, it apparently regarded this as a stop gap action and go-ahead for the purchase of SBTS was given in 1981. The use of Intelsat transponders required large, and expensive earth stations and in the early 1980's only preemptible transponder leases were available from Intelsat. The Intelsat earth stations were 10 metres in diameter, whereas the equivalent for SBTS were six metres.

Procurement of SBTS: The procurement process for SBTS provided a scoop for Spar Aerospace which won its first export order for a satellite system – apart from sales to international organisations the Spar export order has been the only order for a satellite system placed outside the USA apart from Arabsat. However, Spar was offering a Hughes platform, the HS–376, which was already proven in-orbit and Hughes itself received contracts from Spar for 32% of the total order value which included TT&C facilities. Aerospatiale, with Ford, offered an Arabsat design to Embratel in 1981. Back in 1976, Matra, as part of the Mesh consortium, offered a design.

The total value of the final contract with Spar was for $131·1 m. This included provision of the ground control centre and TTC equipment by SED Systems Incorporated and staff training. The control centre is located at Guaratiba, near Rio de Janeiro.

Design of Satellites: The two satellites use a Hughes HS–376 platform. There are 24 operational C-band transponders, each of 36 MHz bandwidth and a TWTA power output of 9·6 Watts. The basic design is close to that of the Anik D satellites, also manufactured by Spar Aerospace. There is only one downlink beam. The peculiar shape of Brazilian territory has required assembly of a set of feed arrays for optimisation at the scanned border areas. This subsystem employs six feed arrays in vertical polarisation and 9 in horizontal polarisation. Reception on the satellite is achieved by two overlapping parabolic antennas, each capable of receiving signals in one polarisation. In line with general trends in the early 1980's, frequency reuse is achieved by using horizontal and vertical polarisation. Overall, the SBTS satellites are very conventional.

Commercial aspects of the system: The system is primarily used to carry long distance telephony between local and regional terrestrial telephone networks. As with most satellite systems, it has required a considerable investment in the ground segment.

Embratel already has a considerable investment in long distance terrestrial microwave links, particularly in the most developed sections of the country. However, such systems are expensive, particularly in the relatively flat and inaccessible Amazonian jungle. That area has required considerable investment in communications as it is rapidly opened up. Not surprisingly, Embratel has used SBTS considerably to improve communications in that area. As at Spring 1986, 13 of Ebratels 21 communications earth stations for SBTS were located in the Amazonian area.

Three national networks, out of a total of six, use SBTS. These are Rede Globo, said to be the fifth largest television network in the world, SBT-TV and Banderrantes network. Globo has 25 "TV-SAT" earth terminals (receive only) for distribution to major cities. Bandeirantes has 21 similar terminals. The Embratel receive/transmit terminals are also used to distribute television. The national terrestrial microwave grid, on which construction started in 1967, is also used to relay programmes to transmitters in the more distant settlements. A fourth TV channel, Machele TV was due to start using the satellite in 1986. Rede Globo has also been looking at taking a second transponder, either on SBTS–1 or SBTS–2.

Use of the satellite for "social applications" is under study. These follow on from experiments undertaken on NASA's ATS–6 satellite and include tele-education and tele-medicine programmes. The latter could include student training course and remote diagnosis. The EBT-TV educational TV service has already begun transmissions on SBTS–1, transponder 24. Programming material using the prime satellite, SBTS–1, is allocated transponders as follows: Rede Globo, transponder 1; SBT-TV, transponder 2; Bandeirantes TV, transponder 5; Machete TV, transponder 6; Embratel for occasional video news and sports feeds, transponder 15 and EBT-TV (educational), transponder 24.

Reception of these channels is available through a "Videosat" TVRO service using 3 metre dishes. This is aimed at blocks of flats, hotels and isolated households. All services are in Portuguese and PAL transmission standards are used. The government allowed private ownership of individual TVRO's in 1984, but in 1985, such systems were still retailing for around $7000.

The Brazilian government has given approval to Embratel to lease surplus capacity on SBTS to neighbouring countries. So far there have been no takers but Peru and Paraguay have both expressed interest. Lease rates quoted in 1985, were said to be close to those of Intelsat, although the latter has subsequently increased the range of its lease terms and now offers outright sales of transponders. The Embratel leases would only be for the domestic needs of neighbouring countries, rather than for international communications.

Above: launch of SBTS–1 on Ariane V12, 8th February 1985

The introduction of the SBTS system has given the Brazilian ground equipment manufacturing industry a further opportunity to develop its capabilities. The Brazilians relied heavily on domestic suppliers for the ground segment of its leased Intelsat capacity, although the manufacturers had to import some more complex systems such as LNA's and transmitters. Telebras has a research centre, CPqD, which works on the development of equipment for satellite operations before they are handed over to domestic manufactures. The earth station market receives considerable protection from foreign competitors.

TECHNICAL SUMMARY

PLATFORM	HS–376
PRIME CONTRACTOR	Spar Aerospace, Canada
OTHER MAJOR CONTRACTORS	Hughes Aircraft Company, SED
STABILISATION	Spin
LAUNCH MASS	1140 kg
MASS IN ORBIT, (BOL)	671 kg
POWER SUPPLY, (EOL)	799 Watts
DESIGN LIFE	8–10 years
ORBITAL POSITIONS	SBTS–1, 65° West SBTS–2, 70° West
SPACECRAFT DIMENSIONS	2·16 m diameter, 7·09 m height (deployed)
NO. OF ACTIVE TRANSPONDERS	24
NO. OF SPARE TRANSPONDERS	6
TWTA POWER OUTPUT	9·6 Watts
FREQUENCY RANGE:	
UPLINK	5·925–6·425 GHz
DOWNLINK	3·700–4·200 GHz
TRANSPONDER BANDWIDTH	36 MHz
EIRP MINIMUM, BEAM EDGE	32–34 dBW
CHANNEL POLARISATION	Frequency re-use through horizontal and vertical polarisation

COMMERCIAL DETAILS

OPERATOR	Satellite Development Trust
ADDRESS	11 East 71st Street, New York, NY 10021, USA
TELEPHONE NO.	(212) 517-8686
SENIOR STAFF	Nick Cowan, Chief Executive Officer
PURPOSE OF PROJECT	Provision of DBS TV capacity to continental USA and programming four of 16 channels
PROJECT STATUS	Has conditional FCC construction authorisation
PROJECT VALUE	not available
LAUNCH DATES	SDT–E1, third quarter 1988 SDT–E2, fourth quarter 1988 SDT–W1, third quarter 1989 SDT–W2, first quarter 1991

SATELLITE USERS

Satellite Development Trust's plans envisage offering a 16-channel TV service in the bottom half of contiguous USA. Of the 16 channels, 12 would be leased or sold to programmers and SDT would programme four itself.

Programming plan is as follows:

Channel 1	SDT Feature Films and Special Events
Channel 2	SDT Special Interest Films
Channel 3	SDT Music Video Channel
Channel 4	SDT Speciality Music Channel
Channel 5	News Channel
Channel 6	Sports Channel
Channel 7	Superstation
Channel 8	General Entertainment, Advertisment supported
Channel 9	General Entertainment, Advertisment supported
Channel 10	General Entertainment, Advertisment supported
Channel 11	Speciality Channel
Channel 12	Foreign Minorities Channel
Channel 13	Public Service Entertainment Channel
Channel 14	Public Service Health Channel
Channel 15	Weather Channel
Channel 16	Educational Channel

PROJECT OVERVIEW

Satellite Development Trust had until December 1985 to meet the Federal Communications Commission's requirements for "due diligence" on which its conditional approval to begin construction of its DBS systems was granted. The company plans a four satellite system, each satellite with eight operational channels and with two spacecraft located at 110° West and two at 148° West. The first two satellites launched, SDT–E1 and E2, would provide coverage to the Eastern half of the USA. The second two satellites, launched two years later, would provide coverage to the Western half of the USA. Sixteen channels would be available in each half of the USA. In an amendment filed with the Federal Communications Commission in early 1985, the company suggested that it would use a Hughes HS–393 design of satellite for the system.

TECHNICAL SUMMARY

PLATFORM	HS–393
PRIME CONTRACTOR	Hughes Aircraft Company
STABILISATION	Spin-stabilised
ORBITAL POSITIONS	SDT–E1 & E2, 110° West SDT–W1 & W2, 148° West
NO. OF CHANNELS PER SATELLITE	8
TWTA POWER OUTPUT	100 Watts
FREQUENCY RANGE:	
UPLINK	17·3–17·8 GHz
DOWNLINK	12·2–12·7 GHz
TRANSPONDER BANDWIDTH	24 MHz
EIRP MINIMUM, BEAM EDGE	50 dBW

COMMERCIAL DETAILS

OPERATOR	Pan American Satellite Corporation (PanAmSat)
ADDRESS	Suite 100, 2 Greenwich Plaza, Greenwich, Conneticut 06830, USA
TELEPHONE NO.	(203) 622-1324
TELEX NO.	
FAX NO.	(203) 622-9163
SENIOR STAFF	Reynold V. Anselmo, Chairman Fred Landman, President Luli Saralegui
PURPOSE OF PROJECT	Provision of C-band capacity for international and domestic Latin American services and Ku-band capacity across the North Atlantic
PROJECT STATUS	Satellite constructed and awaiting launch but the company still has coordination problems with Intelsat and requires landing rights in Europe (see Project Overview)
PROJECT VALUE	PanAmSat has said that the first satellite, including launch, TT&C facilities, legal and other expenses will cost around $110–120 million. The full project will require expenditure of around $155 million. It is not clear if this is an estimate of project expenditure or a financing estimate for the company. Launch cost of the first satellite on Ariane 4 (401) was $8 million
LAUNCH DATES	Panamsat 1, July 1987 by Ariane 4, V21 (demonstration launch)

SATELLITE USERS

At the time of writing, the only firm customer for the first PanAmSat spacecraft is Peru which has taken five transponders on the system. Italian public service broadcaster RAI is also discussing the possibility of taking capacity. Other names that have been linked to the project are Venezuela and (nominally) Cygnus Satellite Corporation (see Project Overview).

The system will offer three kinds of service.

1. Latin American C-band regional service covering Central and South America and the Caribbean and Miami. This will use the "Latin American" beam on the system.

2. Domestic C-band services for Central and South America and the Caribbean but excluding Brazil and Mexico which have their own domestic satellite systems.

3. Transatlantic Ku-band services for the USA and Western Europe. It is possible that the Ku-band capacity could be used for intra-European traffic.

Transponders will be either leased or sold on a condominium basis to television companies, governments, communications companies and large businesses with internal communications needs.

US regulations prevent the system from being used to carry international telecommunications traffic from the public switched networks.

PROJECT OVERVIEW

Introduction: Pan American Satellite Corporation was established in 1984 with the intention of providing "sub-regional Western hemisphere" satellite capacity for Latin America, the Caribbean region and links between the two and the USA. It also envisaged services across the North Atlantic to serve the Iberian peninsula. Subsequently it modified its plans to serve all of Western Europe. It was one of a spate of projects attempting to complete with Intelsat in the light of changing US telecommunications policies towards Intelsat's monopoly. At the time of writing a number of observers believe that it will be the only one of that round of attempts to prove successful. The company is essentially a venture capital project and is therefore unusual amongst such US proposals in that it has proceeded so far. It has relied heavily on a cheap, partially constructed satellite originally intended for the American Satellite Corporation and a very cheap launch on the Ariane 4 demonstration flight vehicle not really intended for launch of commercial satellites. The project has had to face considerable opposition from a hostile Intelsat, although the latter organisation has considerably modified its attitude. To become a reasonable risk to users it will also have to purchase a further satellite.

The first PanAmSat spacecraft is called Simon Bolivar after the Latin American liberator who gained the continent independence from Iberian control. The spacecraft is due for launch in July 1987 but the company still faces several hurdles before it can offer all of the capacity on it to customers. These problems include further economic coordination with Intelsat and obtaining agreements from European telecommunications administrators, many of whom may prove not to be favourably disposed towards the competitive threat to their shares in the Intelsat organisation.

Current position of PanAmSat: Nominally PanAmSat requires approval of the extraordinary assembly of parties of Intelsat on economic coordination under Article XIV (d) of the Intelsat Articles of Association before it can proceed with its system. Article XIV (d) coordination applies only to the non-domestic services on PanAmSat, Article XIV (c) applying to the domestic services. The latter case appears to require only "rubber stamping" by Intelsat. All 24 transponders on the first satellite have received technical clearance from Intelsat.

However, PanAmSat's position with Intelsat on Article XIV (d) is not entirely clear, with two conflicting camps arguing the significance of past agreements with Intelsat.

In December 1986, the Intelsat Board of Governors recommended that economic coordination should take place of five C-band transponders approved by the Assembly of Parties in March 1987. The five C-band transponders are intended primarily for international TV services between Peru and the USA and, according to Intelsat would result in a diversion of 0·9% of revenue away from the Intelsat system. The Board of Governor's recommendation followed on from recommendations from the Intelsat executive that it accept economic coordination of all 24 transponders. The Board of Governors rejected that in favour of five only whilst agreeing that all 24 transponders were technically compatible with the Intelsat system. The recommendations by the Board are being interpreted in two different ways.

1) PanAmSat is suggesting that the December 1986 recommendation means that the Extraordinary Meeting of the assembly of Parties will approve economic coordination of the five transponders and that approval of coordination of other C-band and Ku-band transponders will be straightforward. It is arguing that it has the Intelsat executive favouring coordination along with the US State Department and Comsat, the US signatory to Intelsat. According to PanAmSat's attorney, Henry Goldberg, "... Intelsat would have to scramble quite a bit" to find that new services (over and above the five C-band transponders) would cause economic harm since the executive have already told the Board of Governors that neither the 18 C-band nor the six Ku-band would do so. The domestic use "requires little more than registration with Intelsat ... the coordination of five C-band transponders under XIV (d) was a "de facto" coordination of other transponders under the same clause, particularly as all 24 C- and Ku-band transponders had been subject to technical coordination", a PanAmSat spokesperson told the author in December 1986. The fact that PanAmSat would not cause economic harm had already "been determined by the Director General's office".

2) Nevertheless, the failure of the Board of Governors in the December meeting to recommend economic coordination of all the satellite's capacity not only reflects hostility to the potential competition to Intelsat from some Intelsat members and political positioning by other members but also a degree of uncertainty about future economic coordination. The executive had recommended that coordination be for 10 years. France objected to the latter point on the grounds that all recent coordination approvals had been for five years. It appears that France has not been favourably disposed towards PanAmSat and that it is concerned about the five year approval once given to Eutelsat. France is supplying the Eutelsat II satellite system and may also be concerned about the potential impact of PanAmSat on the Eutelsat organisation. It is understood that France may again object to the PanAmSat coordination at the March 1987 meeting.

PanAmSat requires Intelsat approval on both economic and technical aspects before it can launch its satellite. Its approval from the Federal Communications Commission is conditional on meeting the Intelsat procedures. PanAmSat's satellite is currently due for launch in July 1987, only some four months after the Extraordinary Assembly of Parties meeting. It is understood that if the Assembly approves the five C-band transponders then launch can go ahead but only those five could be operated until further approval is forthcoming.

PanAmSat and Intelsat: These proposals, along with those from other private venture satellite projects caused much concern at Intelsat, which was suffering from gross overcapacity following a major decline in the growth rate of international communications and the prospect of competition from fibre optics. Intelsat's position is detailed elsewhere in this book and it is sufficient to say that the organisation followed up by offering very cheap prices for outright sale or lease of capacity to those very countries that PanAmSat is trying to market in. Needless to say PanAmSat has been unhappy about these moves and there has been an endless process of claim and counter claim amongst the interested parties.

PanAmSat's position is that Intelsat has been in the process of predatory pricing, charging below cost for capacity that it should not even be launching, given that it has planned too much capacity. PanAmSat has made claims that Intelsat has delayed receipt of lease payments to make its charges more attractive, and has modified the design of its Intelsat V F13 and F15 spacecraft intended to compete with private veture international satellite systems. It has also objected to Intelsat proposals to use bulk capacity for a regional satellite service for the Caribbean basin. It argues that this Caribnet service would be priced at lower rates than the Intelsat international services. At the time of writing talks were still being held between Comsat (the US signatory to Intelsat), Intelsat, Peru, the (US) National Telecommunications and Information Agency, the FCC and PanAmSat to try to sort out this acrimonious state of affairs. This was compounded in June 1986 when Intelsat's executive turned down a PanAmSat request for an extraordinary meeting of the Assembly of Parties to decide whether PanAmSat's plans would be detrimental to Intelsat on technical or economic grounds. By the end of 1986, the acrimony appeared to have sorted itself out.

Background to the project: PanAmSat describes itself as a partnership between Rene Anselmo, its Chairman, and Fred Landman, its President. It received conditional approval from the Federal Communications Commission on 23rd July 1985 and further approval in September of that year to acquire a partially constructed satellite originally destined for the American Satellite company but which was surplus to that company's requirements. PanAmSat had originally intended to procure a Hughes HS–393 satellite with 12 Ku-band transponders to serve North Atlantic transatlantic national traffic and 24 C-band transponders for American domestic use. This proposal would have cost $198 million in full. It would have also taken longer than purchase of a second hand satellite and given other organisations a chance to enter a marketplace already with a lot of surplus capacity on offer from Intelsat. PanAmSat looked at the Westar IV satellite, rescued by Shuttle in late 1984 and owned by insurance underwriters. This was rejected and a partially constructed hybrid RCA 3000 bird purchased instead.

PanAmSat originally planned that the Ku-band capacity on the system be allocated to transatlantic traffic to the Iberian peninsula but subsequently notified the FCC that it intended to offer it to other private venture international systems, notably ISI, RCA Americom, Cygnus or Orion.

PanAmSat's services: The C-band capacity on the satellite can provide four beams to South America, excluding Brazil. These are a South spot beam covering the Argentine and Chile, a West spot beam covering Peru and Ecuador and a North spot beam covering Colombia and Venezuela. A hemispheric beam combines all three footprints. Thirteen C-band transponders are allocated to the spot beams and five more can be switched between the spot beams and the hemispheric beam. It is intended that a US international gateway be provided in Miami for US-Latin America links.

Early in 1986 PanAmSat announced that it had reached an agreement with Cygnus Satellite Corporation whereby the two agreed to combine the use of the 45 degrees West orbital slot then allocated to Cygnus. Cygnus was a start up venture planning a Ku-band satellite system to serve the North Atlantic market, including the Caribbean and Central America but not South America. It has applied for FCC authorisation in March 1984 and subsequently received provisional authorisation.

PanAmSat had intended to use the 57 degrees West slot allocated to it by the FCC. Under the new arrangements the PanAmSat 1 satellite will use the 45 degree West slot allocated to Cygnus to provide coverage to all of Western Europe (not just the Iberian penensula as envisaged in the original PanAmSat FCC filing). Cygnus also has a slot at 43 degrees West allocated to it. The move did not prevent Cygnus from eventually operating its own system and it was announced that Cygnus could lease capacity from PanAmSat as a precursor to operating its own satellites.

However, in November 1986 PanAmSat requested authority from the Federal Communications Commission to purchase Cygnus whose assets were reported to be negligible. The purchase was apparently necessary for PanAmSat to use the six Ku-band transponders on its Simon Bolivar spacecraft. The Commission is involved in the sale because it must determine that Cygnus is not selling its licence at a profit. The licence is seen by outside observers as Cygnus' only significant asset although some $221,000 had been paid by the company in legal fees and $105,000 in consulting fees. Purchase price offered by PanAmSat was $350,000. PanAmSat was expecting FCC authorisation for the purchase of Cygnus to be received early in 1987.

Ku-band transatlantic services: The six Ku-band transponders on the first PanAmSat spacecraft can all be either uplinked in Europe or the USA. Three are downlinked on a US beam providing a minimum signal strength of 43 dBW for most of Continental USA. Three can be downlinked to Europe providing a minimum EIRP of 45 dBW for France, the Low Countries, Germany and Northern Italy. EIRP performances for the UK is between 41 and 44 dBW. The C-band capacity is not intended for North Atlantic traffic.

The company is looking for a European partner in order to obtain landing rights on the continent. It was talking with several European PTT's and major telecommunications administrations. It has not revealed who it has been talking to. Until the end of 1986, the company's efforts appear to have been concentrated on its C-band services and Intelsat coordination, although efforts in Europe were not absent. It is known that the company has had discussions with the Department of Trade and Industry in the UK although these were described by one source as "somewhat half-hearted".

PanAmSat envisages broadcasters as the major users of the Ku-band capacity, although its publicity material suggests their use for voice and data applications, including VSAT systems, as well. It confirmed that it was still talking with Italian public broadcaster RAI, which, currently distributes its RAI Uno channel in the USA and Canada using a transatlantic Intelsat hop. This is in addition to European distribution of the channel on Eutelsat I F1. RAI uses the RCS Satcom C-band system for distribution within the USA. PanAmSat has said that it is talking with other European broadcasters looking to distribute in the USA and US broadcasters looking to distribute in Europe.

The Ku-band transponders on the first satellite, located at 45 degrees West, are being offered either for sale or for long term lease. Company says that the lease or sale prices are proprietary information but that they will be competitive with Intelsat Ku-band charges and would avoid the need for cross strapping, as used currently by CNN for its European service. In 1984, US trade press carried advertisements from PanAmSat offering outright sale of both C-band and Ku-band transponders for $6,950,000 each. The deal would have covered non-preemptive terms and the buyers of the first 24 transponders would have become limited partners in PanAmSat. These prices presumably have changed because both market conditions and the nature of the services offered have changed. Baseline Intelsat rates for sale of C-band and Ku-band transponders are presented here for an indication of the current state of the market. Actual Intelsat prices differ from these in that they are charged pro-rata depending on the life left in the satellite carrying them. The PanAmSat sale charges are based on a design life of 11·5 years, although RCA guarantees cover only for the first ten years of life.

BASELINE INTELSAT TRANSPONDER SALE PRICES		
TRANSPONDER TYPE	*BAND WIDTH*	*PRICE*
C-band global	36 MHz	$4,321,000
C-band hemi zone	36 MHz	$2,602,000
C-band hemi zone	72 MHz	$3,472,000
C-band spot beam	36 MHz	$4,489,000
Ku-band spot	72 MHz	$5,700,000
Ku-band spot	241 MHz	$5,700,000

It is possible that telecommunications administrations allowing PanAmSat landing rights may arrange for higher charges to the end user.

The Intelsat transponders offered for sale are not available for North Atlantic traffic or for intra European traffic. Spot beam Ku-band transponders on Intelsat provide only limited European coverage.

C-band transponder services: The first PanAmSat spacecraft will offer three spot beams and one regional beam in C-band to serve Latin America and the Caribbean. The three spot beams, North, Central and South, will offer focused clusters of power to groups of countries. The beam contours have been shaped so that the maximum power is provided within the national boundaries of each country within the cluster. Within each beam satellite power range will be from 37 dBW to 40 dBW, depending on the transponder bandwidth and location in relation to the beam centre. The spot beams are intended for development of domestic and subregional telephone, data and television networks. The North beam will serve Central America, Columbia, Venezuela and the Caribbean. The Central beam will cover Ecuador, Peru and Bolivia. The South beam will cover Argentina, Chile, Paraguay and Uruguay. Neither Mexico nor Brazil are served by the beams as each of these two countries has a domestic satellite system of its own.

The Latin American beam covers all Central and South America and the Caribbean bar Mexico. It allows up and downlinks in Southern Florida. The beam is intended to allow television, data and voice services "amongst all of the Spanish speaking peoples of the Western Hemisphere". EIRP signal strengths are between 33 and 39·9 dBW which PanAmSat claims is equal to the power of a domestic satellite in the USA and far in excess of anything now available from any system providing services to all of Latin America.

The second PanAmSat spacecraft: As at the end of 1986, PanAmSat had not ordered a second satellite. Indeed, it still had to issue an RFP but had talked with several manufacturers in preparation of issue of such a document. The company appears to be reluctant to discuss when it intends to order a second spacecraft. Past comments indicate that it will order a new satellite either of the RCA 4000 or Hughes HS–393 design. In its 1984 filing with the Federal Communications Commission, PanAmSat envisaged its first satellite as a Hughes HS–393 platform carrying 24 C-band and 12 Ku-band transponders.

PANAMSAT KU–BAND COVERAGE

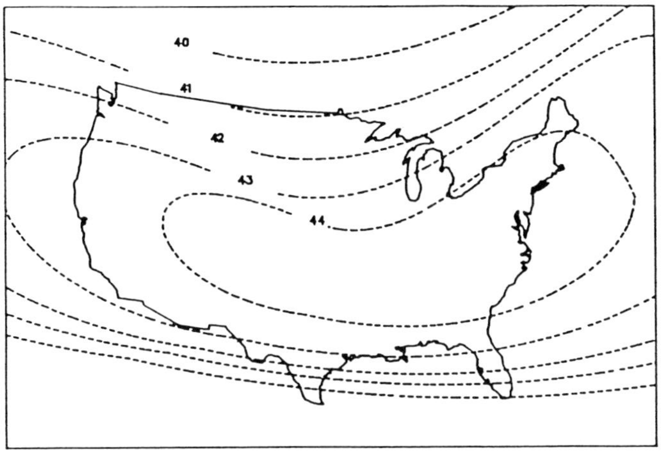

CONUS BEAM

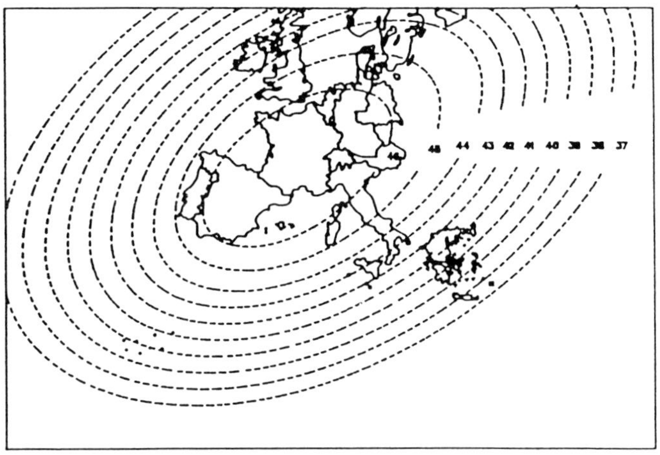

EUROPEAN BEAM

TECHNICAL SUMMARY

(PANAMSAT 1 ONLY)

PLATFORM	RCA–3000
PRIME CONTRACTOR	RCA Astro-Electronics
OTHER MAJOR CONTRACTORS	Hughes (TWTA's), Morton Thiokol (AKM), Parsons (structure), ComDev, ASEC, Heath Tecna, Exxon, Adcole, Lockheed, Eagle
STABILISATION	3-axis
LAUNCH MASS	1273 Kg
MASS IN ORBIT	692 Kg
POWER SUPPLY, END OF LIFE	1300 Watts
DESIGN LIFE	11·5 years
ORBITAL POSITIONS	PanAmSat 1, 45° W, originally 57° W requested PanAmSat 2, 43° W
STATION KEEPING ACCURACY	+/−0·05 d (E/W & N/S)
ECLIPSE PROTECTION	Full

SPACECRAFT DIMENSIONS	1·29 m × 1·59 m × 3·15 m
SPAN, TIP-TO-TIP	15·8 m
NO. OF ACTIVE TRANSPONDERS	24 (18 at C and 8 at Ku)
NO. OF SPARE TRANSPONDERS	4 at C-band, 1 at Ku-band
TWTA POWER OUTPUT	16·2 W (C & Ku), 8·5 W (C, SSPA)
FREQUENCY RANGE:	
UPLINK	14·000–14·500 GHz 5·925–6·425 GHz
DOWNLINK	11·500–12·000 GHz 3·700–4·200 GHz
TRANSPONDER BANDWIDTH	72 MHz (6 at C & 6 at Ku-band), 36 MHz (12 at C-band)
EIRP MINIMUM, BEAM EDGE	45 dBw (Europe, Ku-band), 43 dBW (USA, Ku-band), 37–40 dBW (Spot, C-band), 33–35·9 dBW (Latin America C-band)
CHANNEL POLARISATION	Linear, Horizontal, Vertical

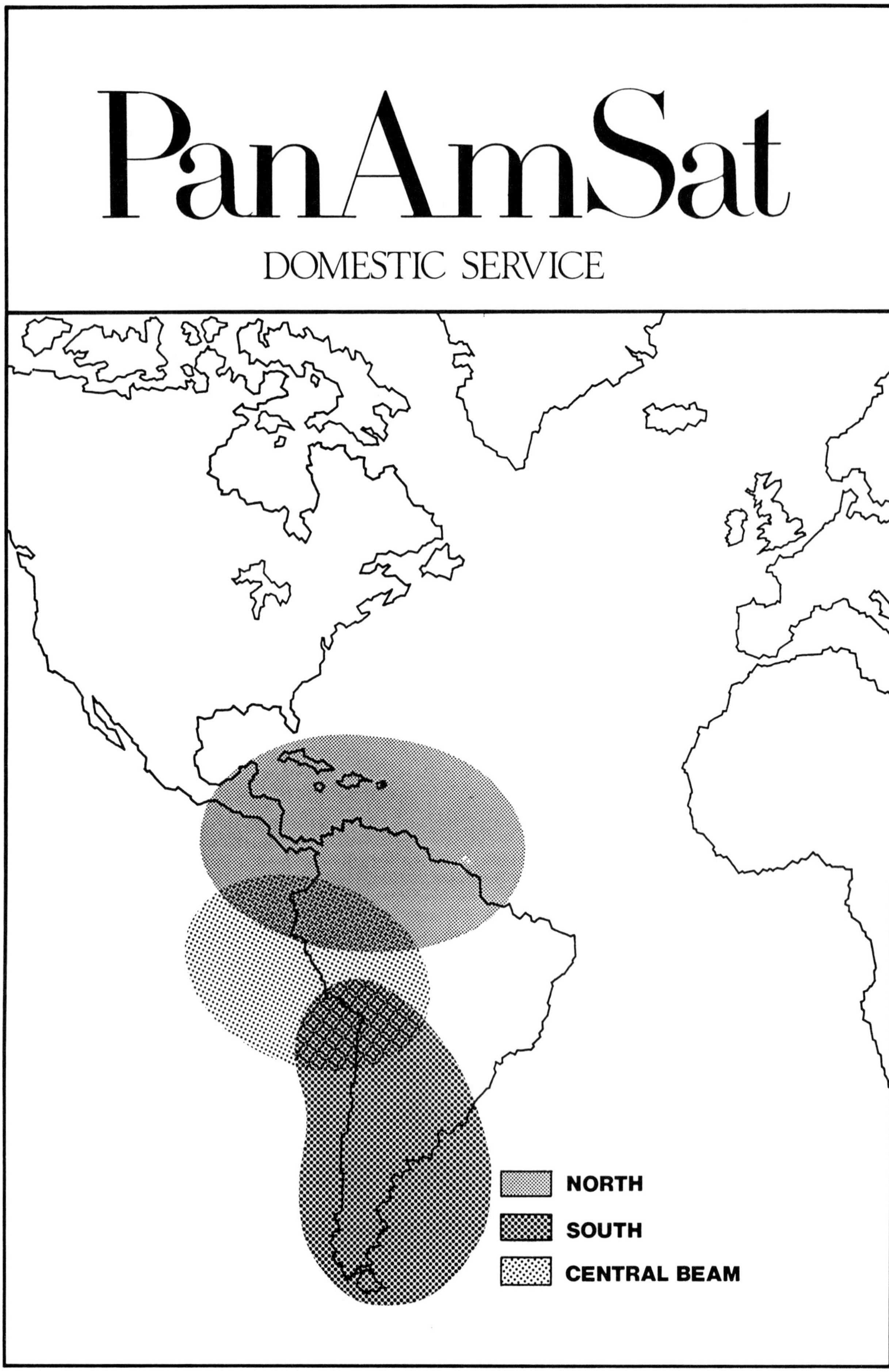
PanAmSat
DOMESTIC SERVICE
NORTH
SOUTH
CENTRAL BEAM

PanAmSat

INTERNATIONAL SERVICE

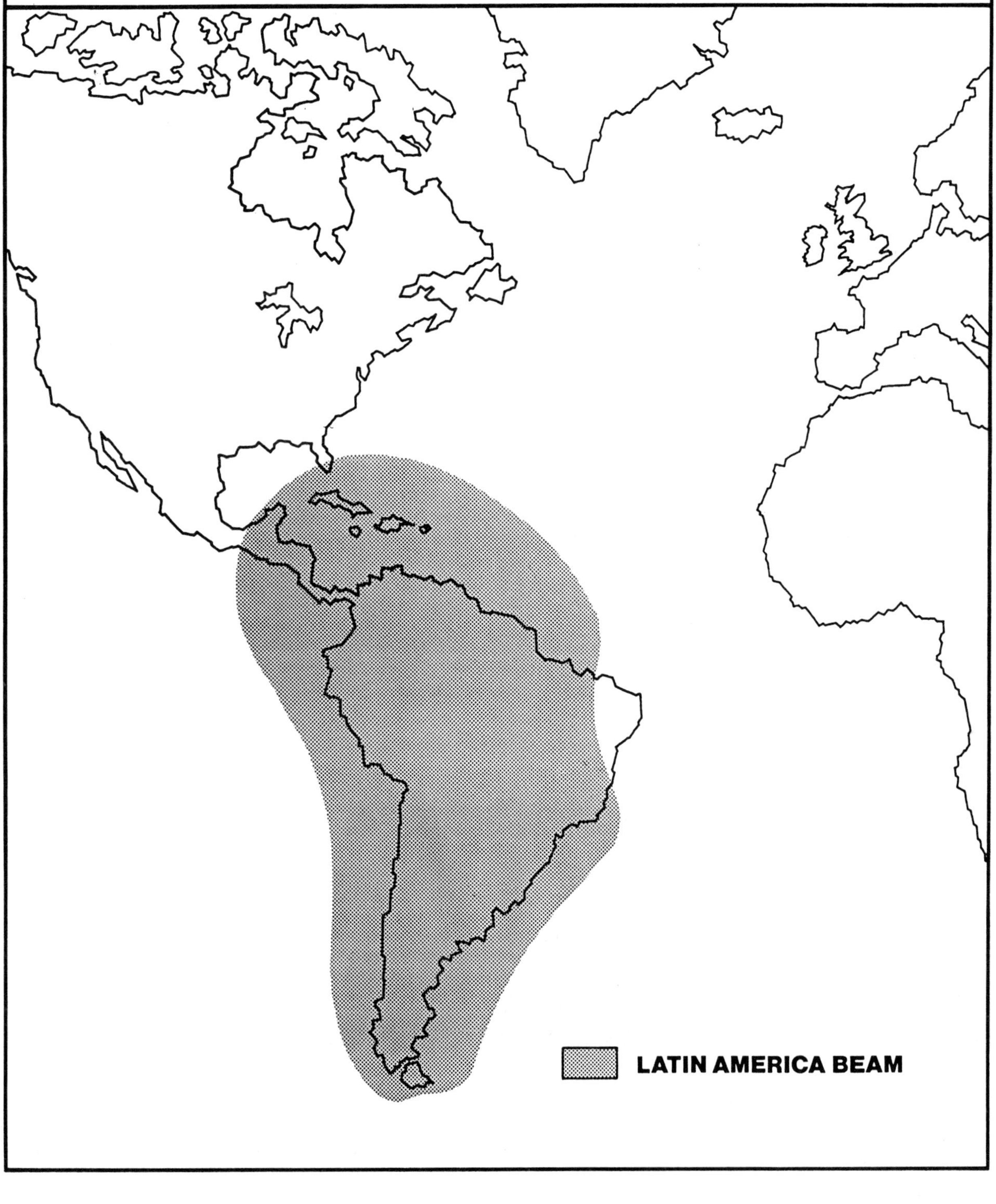

SATELLITE JAPAN CORPORATION

SJC

COMMERCIAL DETAILS

OPERATOR	Satellite Japan Corporation (joint venture between RCA and Sony)
PURPOSE OF PROJECT	Domestic satellite communications
PROJECT STATUS	Proposal requiring a licence
PROJECT VALUE	The total project, including two satellites, was expected to cost \$300 m
LAUNCH DATES	Were scheduled for launch in November 1987 and February 1988

SATELLITE USERS

PROJECT OVERVIEW

Satellite Japan Corporation is a joint venture between RCA and Sony to establish a Japanese domestic communications satellite system using satellites procured from RCA Astro-Electronics. Also in the consortium is Nissho-Iwai Trading Company and Marubeni Trading Company. The Ministry of Posts and Telecommunications has deferred action on the consortium's application for construction and operation of the system on the grounds that it could lead to overcapacity in Japanese domestic communications, either satellite or terrestrially based.

TECHNICAL SUMMARY

PLATFORM	RCA 4000
PRIME CONTRACTOR	RCA Astro-Electronics
STABILISATION	3-axis
MASS IN ORBIT	1·5 tonnes
NO. OF ACTIVE TRANSPONDERS	36 (16 at Ku and 20 at Ka)

COMMERCIAL DETAILS

OPERATOR	GTE Spacenet Corporation
ADDRESS	1700 Old Meadow Road, McLean, Virginia 22102, USA
TELEPHONE NO.	(703) 790-7700
SENIOR STAFF	Dr C. J. (Jerry) Waylan, President Daniel French, Controller Michael Jeye, Director of Marketing, Space Services Susan Kalla, Account Manager Susyn Conway, Business Promotion Manager Ivan Riley, VP, Marketing Leslie Taylor, Director, Regulatory Affairs
PURPOSE OF PROJECT	Provision of C and Ku-band capacity for US domestic services
PROJECT STATUS	System now operational but with one launch still to go
PROJECT VALUE	Original four satellite system, including ground spare, earth segment, launch, insurance TT&C and control cost in excess of \$300 m. Original contract for three satellites was for \$100 m
LAUNCH DATES	Spacenet F1, 22nd May 1984, Ariane 1, V9 Spacenet F2, 9th November 1984, Ariane 3 Spacenet F3, 12th September 1985, Ariane 3, V15 Spacenet F4, 1990, by Ariane Spacenet F3R, by Ariane V24 or V25, September or November 1987

SATELLITE USERS

The Spacenet 1 satellite was initially intended to carry mainly video traffic. In C-band it can cover all 50 US states and the Caribbean. At Ku-band it can only cover 48 CONUS states. Spacenet 2 can only cover 48 CONUS states in C-band. As at February 1986, the following TV channels were being transmitted in C-band through Spacenet 1.

1) Hospital Satellite Network on transponder 11
2) American Christian TV on transponder 15
3) Baptist Telecommunications on transponder 21
4) Vanderbilt Medical TV on transponder 21

Spacenet 2 was, and is, intended for business traffic and is primarily used for the GTE Sprint Service. The Sprint network is essentially separate from GTE's Private Channel Service Network which tends to use GSTAR. The Sprint Service also uses Spacenet 1 and is intended to use Spacenet 3R.

Lease charges for Spacenet transponders, on an annual protected basis, are believed to be \$2·1 m for a 36 MHz C-band transponder, \$3·84 m for a 72 MHz C-band transponder and \$4·2 m for a 72 MHz Ku-band transponder. The 72 MHz transponders were intended to each carry two TV channels.

Spacenet F3R is expected to carry a Geostar two way communications payload.

PROJECT OVERVIEW

The GTE Spacenet system is one of two US domestic communications satellite systems operated by GTE, a major US telecommunications equipment manufacturer and operator. The other satellite system is GSTAR. Unlike GSTAR which originated as an in-house GTE project, Spacenet originated with the Southern Pacific Railway. This had a subsidiary called Southern Pacific Communications Corporation which operated a highly successful long distance public telephone network called Sprint which utilised the Southern Pacific Railway's right of way. SPCC became a major user of other companies' satellites in producing its Sprint service and therefore decided to purchase its own satellite system which it called Spacenet.

However financing the satellite system proved difficult for the Southern Pacific Railway, which was also looking to purchase other US railway companies in the wake of deregulation of railway transportation. SPCC look for partners to finance the Spacenet system and amongst the organisations it talked to were General Electric, American Express, Federal Express and Cox Broadcasting. It obtained a "cheap" launch service from Arianespace as one of the first US users of Ariane. In the event, Southern Pacific Railway sold SPCC outright to GTE in 1983. The Spacenet system is now operated by GTE Spacenet, which, along with GTE Sprint (the major user of Spacenet), GTE Telenet and GTE TeleMessager is part of GTE Communications Service Group. The GSTAR satellite system is under the wing of GTE Satellite Corporation but the management of the two satellite systems are being combined under GTE Spacenet.

Largest user of the Spacenet system is GTE's Sprint service which started in 1978 with 977 customers. This resulted from a decision by the Federal Courts in 1977 which followed the pressures of another specialised common carrier, MCI. Prior to 1977, MCI had only been offering its customers private line services and had been losing a lot of money. In 1977 it was prohibited by the FCC from offering a new switched voice service called Execunet. The Courts over-ruled that decision and confirmed MCI's right to use the local distribution networks of the established telecoms operators.

The Courts confirmed that the local telephone companies had to provide interconnections between specialised carriers' (such as MCI and SPCC) switch terminals and customers' telephones. SPCC quickly established itself as the second largest specialised (common) carrier after MCI. Like MCI, prior to 1977 the majority of its business involved private line services for large commercial, industrial and governmental organisations. Since 1978 it has been marketing switched services. It was the first of the specialised common carriers to employ satellite capacity, starting with the RCA Americom system. By 1981 some 5% of Southern Pacific's communications voice traffic was routed through RCA Americom. SPCC's gross revenue increased from

$95 m in 1979 to $153 m in 1980, increased its market share from 0·17% to 0·24%. The specialised common carrier business was also becoming profitable since, in 1977, the law courts allowed them to use local distribution facilities owned by such companies as Bell.

On 4th December 1980 the FCC authorised construction, launch and operating of 25 domestic satellites, at the time there were only eight such satellites in-orbit. The authorisation included the systems for SPCC, GSTAR, AT&T's Telstar 3 and further RCA Satcom satellites. In 1982, the US Federal Court made a decision which furthered the development of the Sprint service. It approved AT&T's divestiture of its local operating companies. The divested local units were transformed into regional companies that were required to treat Sprint and other long-distance services as equals when granting access to local networks. That equal access became effective in 1984, when Sprint's customer base reached 1·5 m.

By 1986 the long distance market was in the process of structural change with MCI and Sprint emerging as dominant forces alongside AT&T. MCI took control of the first three SBS satellites in 1985 at a time when SBS was becoming a commercial proposition, having fallen flat on its face with its initial SBS market concept. The Sprint/Spacenet project is thus illustrative of the dependence of satellite communications on the changing telecommunications regulatory environment as well as technical developments.

For Sprint, which is also heavily dependent on terrestrial microwave facilities, GTE has integrated Spacenet capacity into the system. Sprint uses both C- and Ku-band frequencies divided between several subnetworks on three satellites. Wideband transponders are used to carry heavy route traffic and narrowband transponders for lighter route traffic. Earth stations provide connectivity to cities that are "on" and "off" the Sprint backbone network. On-net earth stations are located at major Sprint network nodes and carry high to medium volume long-haul traffic. Off-net earth stations carry traffic to gateways on the terrestrial network. Earth station equipment is shared between networks which access the same satellite.

Technicians assembling the Spacenet 2 satellite at RCA Astro-Electronics. This satellite, now located at 69° West is used primarily for Sprint services

Artist's impression of the GTE Spacenet design, using the RCA 3000 (Advanced Satcom) platform and a mixed C-/K-band payload

The initial Sprint Digital Satellite Network used 11 earth stations at Atlanta, Chicago, Denver, Honolulu, Houston, LosAngeles, New York City, Orlando, San Francisco, Seattle and Washington DC. The satellite network is entirely digital using Time Division Multiple Access (TDMA) and ADPCM/DSI speech compression. Spare segment capacity is used more efficiently with TDMA rather than Frequency Division Multiple Access, although ground equipment tends to be much more expensive. The Sprint satellite network also uses Digital Speech Interpolation to improve spare segment utilisation. Sprint was also the first satellite user to adopt Adaptive Differential Pulse Code Modulation (ADPCM) which GTE claims achieves a tripling of channel capacity on the satellite.

The full Spacenet space segment should have been operational by 1985. However, the Spacenet 3 satellite was lost on launch on 12th September 1985 when the third stage of the Ariane V15 vehicle failed. This lead to an insurance claim of $107 m from GTE. A launch of Spacenet 3R is now expected in 1987, atop Ariane. Although originally envisaged as a three satellite system, with two in-orbit. GTE later decided it wanted three satellites in-orbit plus a ground spare. Because Satcom 3 was lost on Ariane, Arianespace is giving GTE priority in its 1987 launch schedule.

Each of the Spacenet satellites has 24 transponders. Twelve are 36 MHz C-band (narrowband) with horizontally polarised downlinks. Six transponders are at 72 MHz C-band (wideband) with vertically polarised downlinks. There are six 72 MHz Ku-band transponders with horizontally polarised downlinks. The narrowband C-band transponders use 8·5 Watt solid-state power amplifiers. The C- and Ku-band wideband transponders use 16 Watt TWTA's The SSPA's are manufactured by RCA, the TWTA's by Hughes. The satellites use the RCA 3000 bus and are based on the Advanced Satcom design. The Spacenet 3 satellite differed from Spacenet 1 and 2 in that its Ku-band transponders were modified to provide East and West coast regional coverage beams.

SUB-CONTRACTORS ON THE SPACENET SATELLITES

Communications	
SSPA's C- and Ku-band Receivers/ Ku-band Channel Amplifiers	RCA Corporation, Princeton, New Jersey 08540
TWTA's	HEDD
Multiplexers	ComDev, 155 Sheldon Drive, Cambridge, Ontario, Canada NIR 7H6
Tower (C-band)	Heath Tecna, 19819 84th Avenue South, Kent, Washington 98031
Tower (K-band)	Exxon Corporation, US 276 and Old Laurens Road, Fountain Inn, S Carolina 29644
Reflectors	Composite Optics, 9623 Distribution Avenue, San Diego, California 92121
Attitude Control	
Sun Sensor	Adcole Corporation, 330 Bear Hill Road, Waltham, MA 02154
Earth/Horizon Sensors	Lockheed Space Systems Division, PO Box 504, Sunnyvale, CA 94088
Rate Measurement Assembly	Northrop Corporation, 100 Morse Street, Norwood, MA 02062
Attitude Control Electronics	RCA Corporation, Princeton, New Jersey 08540
Electrical Power	
Battery Cells	Eagle Picher, PO Box 47, Joplin, MO 64801
Solar Cells	ASEC, 15251 East DonJulian Road, City of Industry, CA 91746
Solar Array Substrate	Heath Tecna, 19819 84th Avenue South, Kent, Washington 98031
Power Supply Electronics	RCA Corporation, Princeton, New Jersey 08540
Propulsion	
Thrusters	Rocket Research, Yorke Center, Redmond, Washington 98052
Tanks	PSI, 2017 Cambield Avenue, Los Angeles, California 90004
Valves	Pyronetics/Hydraulic Research
AKM	Morton-Thiokol, PO Box 241, Elkton, Maryland 21921
Structure	Parsons of California, 3437 Airport Way, PO Box 6189, Stockton, California 95206

TECHNICAL SUMMARY

PLATFORM	Advanced Satcom
PRIME CONTRACTOR	RCA Astro-Electronics
OTHER MAJOR CONTRACTORS	See above
STABILISATION	3-axis
LAUNCH MASS	1195 kg
MASS IN ORBIT, (BOL)	705 kg
POWER SUPPLY, (EOL)	1200 Watts
DESIGN LIFE	10 years
ORBITAL POSITIONS	Spacenet 1, 120° West Spacenet 2, 69° West Spacenet 3R, 87° West*
STATION KEEPING ACCURACY	+/−0·05° (E/W & N/S)
ECLIPSE PROTECTION	100%
LAUNCH VEHICLE	Delta 3920, STS, Ariane 3 or 4
SPACECRAFT DIMENSIONS	64 × 52 × 39 inches
SPAN, TIP-TO-TIP	14·5 metres
CONTROL CENTRE	McLean, Virginia
TT&C CENTRES	Oxford, Connecticut Grand Junction, Colorada
NO. OF ACTIVE TRANSPONDERS	24 (12 C-band, 12 Ku-band, 6 of C-band are narrowband)
NO. OF SPARE TRANSPONDERS	4
SSPA POWER OUTPUT	8·5 Watts (for C-band narrow)
TWTA POWER OUTPUT	16 Watts (C- and Ku-band)
FREQUENCY RANGE:	
UPLINK	5·925–6·425 GHz (narrow) 5·945–6·425 GHz (wide) 14·0–14·5 GHz
DOWNLINK	3·7–4·2 GHz (narrow) 3·72–4·2 GHz (wide) 11·7–12·2 GHz
TRANSPONDER BANDWIDTH	36 MHz (C, narrowband) 72 MHz (others)
EIRP MINIMUM, BEAM EDGE	34·4 to 37·3; 30·7; 26·4 dBW (C-band, narrow) 36·3 to 38·5; 33·2; 26·4 dBW (C-band, wide) 39·7 to 44·8 dBW (Ku-band)
CHANNEL POLARISATION	Linear, horizontal and vertical
RECEIVE COVERAGE (G/T)	−0·9 to +4·2; −6·0; −4·5 dB/K (C-band) −0·2 dB/K
SINGLE CARRIER SATURATION FLUX DENSITY	−94 to −86·6 dBW/sq. m.

*Spacenet 4 is due for launch in 1987–88 for location at 136° West

COMMERCIAL DETAILS

OPERATOR	National Exchange Incorporated
ADDRESS	1726 San Vincente Boulevard, Los Angeles, California, CA 90049, USA
TELEPHONE NO.	(213) 820-5454
SENIOR STAFF	Dr Clay Whitehead, President George S. Kush, Vice-President for Development
PURPOSE OF PROJECT	Provision of C-band and Ku-band capacity for US domestic services for the USA. Company has also proposed a quasi-DBS system
PROJECT STATUS	Remains a paper plan and currently lacks any form of FCC authorisation
PROJECT VALUE	Not available. The following satellites have been proposed: Spotnet K1 Spotnet K2 Spotnet K3 Spotnet K4 Spotnet K5 (ground spare) Spotnet C1 Spotnet C2 Spotnet C3 (ground spare) Spotnet DBS1 Spotnet DBS2 Spotnet DBS3 (ground spare)
LAUNCH DATES	The original FSS filing with the FCC envisaged launch of its first two Ku-band spacecraft in 1987. It also planned to launch its DBS spacecraft in 1988 and 1990

SATELLITE USERS

National Exchange has said that its FSS system would be used for voice, electronic mail, teletext, videoconferencing and other wideband information to and from customer-premises earth terminals. It anticipates that its primary customers would be large business users.

PROJECT OVERVIEW

Introduction: National Exchange is a "new venture" company set up by Clay Whitehead which has proposed three satellite systems for domestic US services; one in C-band, one in Ku-band and one for DBS. The generic name Spotnet applies to all three and derives from the fact that each satellite would use high power spot beams. In the four satellite Ku-band system, 96 spot beams would be available. The company has proposed separate C- and Ku-band systems because it did not believe that cross-strapped services were needed except under very specialised conditions. National Exchange (NEX) has a fairly colourful history amongst US potential satellite operators. It was formed by ex-Hughes Communications Galaxy President Clay Whitehead and in 1983 submitted filings to the Federal Communications Commission for a four satellite Ku-band domestic satellite system.

This system involved two satellites at 101° West and two satellites at 75° West plus a ground spare. It also then applied to the FCC for a C-band satellite system. This involved one satellite at 101° West and one at 75° West plus a ground spare. It then followed this up with requests for a two satellite DBS system, with one satellite at 101° West and one at 75° West. As an entrepreneurial start up company, it is not surprising that these plans have not proceeded far. The company lost out in the July 1985 licensing awards by the FCC and later sold out to the railway company, Burlington Northern. It essentially lost out because it had financial difficulties and subsequent to the Burlington Northern aquisition, sought to amend its FCC applications to reflect its "sounder" financial position.

In the past, NEX has argued that it intends to market the C-band capacity for high density point-to-point traffic, the Ku-band capacity for low density point-to-point and the DBS capacity for point-to-multipoint traffic. Services would most likely be offered on a common carrier basis. The company continues to argue that its spot beam concept could succeed where other systems fail and that it is now eager to start construction once it receives FCC authorisation.

TECHNICAL SUMMARY

PLATFORM	not available
PRIME CONTRACTOR	not available
OTHER MAJOR CONTRACTORS	not available
STABILISATION	not available
LAUNCH MASS	not available
MASS IN ORBIT	not available
POWER SUPPLY, END OF LIFE	not available
DESIGN LIFE	9 years
ORBITAL POSITIONS	Spotnet K, two slots at 101° West and two at 75° West requested from the FCC. Spotnet C, one slot at 101° West and one at 75° West requested from the FCC. Spotnet DBS, two at 101° West
STATION KEEPING ACCURACY	not available
ECLIPSE PROTECTION	not available
SPACECRAFT DIMENSIONS	not available
SPAN, TIP-TO-TIP	not available
NO. OF ACTIVE TRANSPONDERS	Spotnet K, 24; Spotnet C, 24; Spotnet DBS, 16
NO. OF SPARE TRANSPONDERS	
TWTA POWER OUTPUT	Spotnet K, 9 Watts; Spotnet C, 9 Watts; Spotnet DBS, 50 Watts
FREQUENCY RANGE: UPLINK	14·0–14·5 GHz (Spotnet K) 5·925–6·425 GHz (Spotnet C)
DOWNLINK	17·7–12·2 GHz (Spotnet K) 3·7–4·2 GHz (Spotnet C)
TRANSPONDER BANDWIDTH	Spotnet K, 54 MHz; Spotnet C, 54 MHz
EIRP MINIMUM, BEAM EDGE	not available
CHANNEL POLARISATION	not available
RECEIVE COVERAGE (G/T)	not available
SINGLE CARRIER SATURATION FLUX DENSITY	not available

COMMERCIAL DETAILS

OPERATOR	National Exchange Incorporated
ADDRESS	1505 Planning Research Drive, McLean, Virginia 22102, USA
TELEPHONE NO.	(703) 883-8833
SENIOR STAFF	Dr Clay Whitehead, President George Kush, VP, Development
FINANCIAL BACKING	Not known for certain, but apparently financing still has to be arranged
PURPOSE OF PROJECT	Provision of DBS capacity for US services
PROJECT STATUS	A plan. National Exchange needs a substantial amount of finance for its satellite systems
PROJECT VALUE	Not known
LAUNCH DATES	Spotnet DBS–1, 1988 Spotnet DBS–2, 1989

SATELLITE USERS

National Exchange's satellite plans envisage both C-band and Ku-band Fixed Satellite Service systems and a (Ku-band) DBS system to offer "a myriad of integrated services through a variety of tailored communications networks". C-band capacity would be aimed at high density point-to-point traffic. Ku-band (FSS) capacity would be aimed at low density point-to-point traffic. DBS capacity is intended for point-to-multipoint traffic.

Company envisages a variety of services on its DBS capacity – video entertainment direct to private homes, videoconferences, instructional video to business, digital information transmissions to homes and offices.

Each Spotnet DBS transponder will be able to transmit either a single colour video signal, 16 digital signals at a bit rate of 1·544 mbit/s or up to 400 digital signals at a bit rate of 64 kbit/s.

PROJECT OVERVIEW

National Exchange is an entrepreneurial start up with grand plans but little money to provide Fixed Satellite Service and Direct Broadcast Satellite Service capacity for domestic US operations. Established by Clay Whitehead, formerly President of Hughes Communications, it was a second-round applicant to operate a DBS system and received conditional approval in December 1984 from the Federal Communications Commission to begin constructing a DBS system for launch in the late 1980's. It was required to meet FCC due diligence requirements by December 1985 before the FCC would assign orbital slots and channels. It had requested the 101° West slot, already fully assigned to Hughes Communications. It planned to also locate two Ku-band spacecraft and one C-band spacecraft at 101° West. This appears to have created problems for the company as it intended to offer 16 DBS channels each on two spacecraft co-located at the same orbital location. However, at the time of writing the only orbital slots with 32 channels not requested or assigned are 175° West and 157° West which are not suitable for providing full Continental US coverage. If National Exchange satisfies FCC "due diligence" requirements it is expected that it will be awarded 16 DBS channels at each of two orbital slots, one for coverage of the Eastern half of the USA, the other for the Western half.

Company has established a "preliminary" design for its DBS spacecraft, based on RCA Astro-Electronics' Advanced Satcom design but again, at the time of writing, no order has been placed. Payload of 16 100 Watt transponders would be capable of transmitting by means of East or West half-CONUS beams.

TECHNICAL SUMMARY

PLATFORM	Advanced Satcom*
PRIME CONTRACTOR	RCA Astro-Electronics*
STABILISATION	3-axis*
DESIGN LIFE	9 years*
ORBITAL POSITIONS	101° West requested
NO. OF ACTIVE TRANSPONDERS	16
TWTA POWER OUTPUT	100 Watts
EIRP MINIMUM, BEAM EDGE	51 dBW

*Based on tentative selection of an RCA Astro-Electronics design.

COMMERCIAL DETAILS

OPERATOR	Satellite Television Corporation (see Project Overview)
ADDRESS	c/o Comsat, 955 L'Enfant Plaza SW, Washington DC 20024, USA
TELEPHONE NO.	(202) 863-7250
TELEX NO.	89-2668
SENIOR STAFF	Bob Kince
FINANCIAL BACKING	STC is a wholly owned subsidiary of the Communications General Corporation who continues to fund the space segment hardware of the system
PURPOSE OF PROJECT	Originally envisaged as a "Cable in the Sky" DBS project. See overview for current position
PROJECT STATUS	Construction of satellites has been completed and Comsat is looking for a customer or use for the satellites
PROJECT VALUE	Two satellites and a ground spare were ordered under a contract valued at $113 m
LAUNCH DATES	STC–1, June 1986, STS STC–2, December 1986, STS

SATELLITE USERS

Comsat has been looking to either sell or lease the satellites to US or non-US operators. Amongst opportunities it has looked at are the UK, China, where Comsat assisted in preparation of the now defunct CBSC project and Dominion Video Satellite and other US DBS applicants.

The original intended use of the satellites was to provide "Cable in the Sky" DBS pay-TV services to continental USA. Six channels, three on each satellite, would have provided coast-to-coast services from an orbital position at 110° West. The satellites' power would have been focused on two of the most popular areas of the USA – a triangular area bounded by Boston, Atlanta and Georgia, and a strip down the West Coast including San Francisco and Los Angeles. This contained 58% of US households and reception with a 60 cm dish was envisaged (EIRP level was 54 dBW). Most of the rest of the USA (32% of households) could have received the channels with a 75 cm dish. The remaining 10% would have required a 90 cm dish. The latter were located in a strip along the Canadian Border and in the rainiest areas of southern Texas and Florida.

PROJECT OVERVIEW

Seen by some industry observers as the "technological dinosaurs" of the satellite communications world and as monuments to the folly of high power DBS TV, Comsat appears to have major problems in trying to get rid of the two satellites built for its aborted DBS activities. Some observers believe that the two satellites may never be used.

Satellite Television Corporation (STC) was formed by the Communications General Corporation (Comsat) to develop and operate what was expected to be the first operational DBS system for the USA. It signed a contract in October 1982 with RCA Astro-Electronics for supply of two in-orbit satellites and a ground spare (although the ground spare will not now be constructed). It had been the first DBS US company to submit for Federal Communications Commission authority when it filed for an experimental DBS system in 1980. It was therefore seen as the "father" of US DBS services. But by 1984, it was apparent that many companies who had been looking at DBS TV were getting cold feet. Amongst these was CBS Incorporated who had been studying DBS as a vehicle for high-definition TV. STC in late 1983 began courting CBS Incorporated as a potential backer for its DBS project which was then seen to require some $400 m in start-up costs. But in June 1984 CBS Incorporated announced that it was not going into DBS TV. STC then found potential partners in the Prudential Insurance Company and the two principal owners of the press agency, UPI, Messrs Douglas Ruhe and William Geissler. Prudential had invested $50 m in a quasi-DBS operation, USCI, which used the Anik C2 satellite to provide five channels to mid-west and north-east USA. This required 1·2 metre antennas with some consequently hefty installation costs.

Rupert Murdoch had also considered a similar quasi-DBS system under the name Skyband and using Ku-band capacity on the SBS system. However, Murdoch quickly pulled out. USCI also proved to be a dead end, never having more than 10,000 customers. The Prudential perhaps saw STC as a way of recouping its costly investment in USCI which folded in April 1985. STC had been looking at using SBS–4 as part of an interim quasi-DBS service before the 1986 launch of its high power DBS birds.

By November 1985 Comsat's shareholders were relieved to hear that it was pulling out of DBS TV and cutting its losses. Most of STC's staff were transferred out of the company, leaving a handful to handle the odd remaining bits of business. Comsat took a $24 m pre-tax write off on STC's non-satellite assets. It was left with the problem of what to do with the two spacecraft then under construction. It decided to continue construction and hang on to its FCC construction permits. These then became members of an exclusive group of satellites which have been described as "waifs and strays" of the satellite world – custom built and "new" satellites up for sale on the international market place and of which, at the time of writing, none have been sold. Comsat has looked at a lot of potential customers for the satellites. One caustic European potential DBS operator said it would be difficult to make money out of them even if Comsat gave them away.

TECHNICAL SUMMARY

PLATFORM	RCA–3000
PRIME CONTRACTOR	RCA Astro-Electronics
OTHER MAJOR CONTRACTORS	AEG Telefunken (TWTA's)
STABILISATION	3-axis
LAUNCH MASS	1247 kg
MASS IN ORBIT, (BOL)	670 kg
POWER SUPPLY, END OF LIFE	1700 Watts
DESIGN LIFE	9·7 years
ORBITAL POSITIONS	STC–1, 110° West* STC–2, 110° West (FCC channels 3–8)
STATION KEEPING ACCURACY	+/−0·1° (N/S & E/W)
ECLIPSE PROTECTION	None
PROPULSION SYSTEM	STS plus PAM–D or Ariane 3–4 dual launch
SPAN, TIP-TO-TIP	15 metres
NO. OF ACTIVE TRANSPONDERS	3
TWTA POWER OUTPUT	230 Watts
FREQUENCY RANGE:	
UPLINK	17·3–17·8 GHz
DOWNLINK	12·2–12·7 GHz
TRANSPONDER BANDWIDTH	24 MHz
EIRP MINIMUM, BEAM EDGE	57 dBW
CHANNEL POLARISATION	Circular

*STC originally granted orbital slots at 115° West, 135° West, 155° West and 175° West and slots at 115·05° West and 174·95 West for orbital spares. The 115° West slot would have initially been used for an experimental DBS service for the Eastern time zone.

COMMERCIAL DETAILS

OPERATOR	Space Communications Corporation of Japan
ADDRESS	c/o Mitsubishi Corporation, Building Annex 3–1, Marunochi, 2-Chome, Chiyoda-Ku, Tokyo, Japan
TELEPHONE NO.	03-210-8160
FAX NO.	03-210-8164
SENIOR STAFF	H. Minayawa, President Y. Taniguchi, Executive VP K. Hoh, Managing Director M. Kishimoto
PURPOSE OF PROJECT	
PROJECT STATUS	Satellites under construction. Operator has already been granted a class 1 telecommunications licence
PROJECT VALUE	Contract for the two satellites was worth \$200 m. Launch costs were budgeted at \$100 m
LAUNCH DATES	Superbird (SCS) 1A, November 1988, Ariane 4, V33 Superbird (SCS) 1B, May 1989, Ariane 4, V37

SATELLITE USERS

System is intended to provide service to users in cable television, videoconferencing, telephone backup, newspaper transmission, business applications, banking and other services.

PROJECT OVERVIEW

This is one of three projects to establish domestic satellite systems for Japan using US manufactured satellites. Space Communications Corporation is a joint venture between Ford Aerospace, who are supplying the two satellites, Mitsubishi Trading Company and Mitsubishi Electric Corporation. The other two ventures are Japan Communications Satellite, with Hughes built satellites, and Satellite Japan Corporation, with RCA built spacecraft. Unlike Space Communications Corporation and Japan Communications Satellite, Satellite Japan has not yet received a licence for its system.

The Space Communications Corporation is to use two "Supersat" spacecraft developed by Ford from the Intelsat V design. Payload will operate in the Ku-band section of the frequency spectrum. Japanese built domestic communications satellites have so far relied primarily on the Ka-band, with its associated problem of rain attenuation, primarily, it appears, because the Ku-band spectrum is used heavily for Japanese terrestrial microwave links.

TECHNICAL SUMMARY

PLATFORM	Supersat
PRIME CONTRACTOR	Ford Aerospace & Communications
OTHER MAJOR CONTRACTORS	Mitsubishi Electric
STABILISATION	3-axis
MASS IN ORBIT	1·5 tonnes
POWER SUPPLY, (BOL)	4000 Watts
DESIGN LIFE	10 years
PROPULSION SYSTEM	Bi-propellant
NO. OF ACTIVE TRANSPONDERS	12 Ku-band, 23 Ka-band

COMMERCIAL DETAILS

OPERATOR	To be established. Company is currently being formed by Claude Contamine, head of Telediffusion de France
ADDRESS	Telediffusion de France, 21–27 rue Barbes, F–92120 Montrouge, France
TELEPHONE NO.	(1) 4567 11 15
TELEX NO.	250738 tdf f
SENIOR STAFF	Claude Contamine, Director General, TDF
PURPOSE OF PROJECT	Initially an R&D exercise. Then envisaged as a DBS system for French domestic TV programming; now considered as a European DBS system. Current position is in doubt
PROJECT STATUS	See Project Overview
PROJECT VALUE	Full cost of TDF–2 has been set at FFr1758 m ($266 m) including launch and insurance (see Project Overview). The contract for the development and construction of TDF–1 and TV–SAT 1 was for Dmk520 m ($236 m) at 1982. There were subsequent modifications to the two satellites which cost Dmk50 m ($24 m). TDF–1/TV–SAT costs exclude launch, insurance and TT&C facilities. Full cost of TDF–1 is expected to reach FFr2000 m ($300 m)
LAUNCH DATES	TDF–1, Ariane V24, September 1987 or Ariane V25, November 1987 TDF–2, Ariane V36, April 1989 or Ariane V41, October 1989. Due to problems with the Ariane launch vehicle, TDF–1 launch is likely to slip by at least two months

SATELLITE USERS

At the time of writing, the future of this project remains in doubt and the only potential user with any commitment to the satellite is the public sector cultural channel, La Sept.

The use of the satellite by CLT of Luxembourg has been a constant feature of the project since 1983. CLT may take one channel as a ticket to getting into the French terrestrial TV market, but such a move would probably be on political rather than narrow commercial criteria. It is unlikely that CLT will take a second channel for its German language RTL-Plus channel as this is now due to go on TV–SAT.

It also appears unlikely that either Robert Maxwell or Silvio Berlusconi will have any continuing interests in using the satellite.

Amongst past proposals for channels on the system have been the following:

TV5, currently on Eutelsat I F1
Music Box (UK) with Europe No. 1
La Cinq (French language version)
TV6
RTL-Plus
CLT–RTL French language service
European Satellite Television Broadcasting Company
MirrorVision
Sky Channel
Super Channel
Berlusconi Italian language channel
Berlusconi Italian, English, Spanish and German language channel
Radio France (radio broadcasts only)
Cable News Network

PROJECT OVERVIEW

Introduction: The TDF–1/2 satellite system is a product of WARC '77 and Franco–German industrial policies of the late 1970's and early 1980's. In establishing the project, little attention was paid to the needs of the broadcasting industries. Consequently, France has repeatedly failed to find users for the system. Whilst it may be argued that this reflects the political nature of French broadcasting policies, it appears that the inherent uneconomic nature of this high-power DBS satellite system has placed TDF–1/2 low in the list of priorities of French broadcasting policy.

The full cost of TDF–1/2, including R&D and financing costs is likely to exceed $650 m. The satellites are of an unproven design using high power transponders that have no in-orbit track record. Launch of the first satellite is now expected to be nearly three years late. There are still no finalised plans to use the satellites on a commercial basis. This is now necessary as the French government has refused to finance the completion and launch of TDF–2; funding must now come from private interests. From an engineering viewpoint, TDF–1 remains an interesting project. It is likely to be the first satellite system to use high power 200 Watt plus TWTA's.

There are still many proponents of high-power DBS in Europe who will be looking eagerly at the AEG manufactured TWTA's on the satellite. At the moment AEG is the only source of such tubes since Thomson appears to have continuing problems with its own such tubes. The platform is also large and may form the basis of bids for other European FSS systems. It is marketed under the name Spacebus 300.

Current position of TDF–1/2 Project: At the time of writing, France is in the process of a third attempt to commercialise the TDF DBS system. The attempt was initiated in July 1986 after the government refused to put any more money into TDF–2 and followed pressure from the electronics industry not to abandon DBS TV. The attempt has been organised by Claude Contamine, head of Telediffusion de France. The latter owns and operates all of the French terrestrial transmitters and was expected, in the past, to operate TDF–1/2. The current attempt appears to reflect a governmental policy to wash its hands of DBS TV and allow private interests to bail the exercise out. The TDF–1 satellite is being given away to whoever wants to use it on condition that they pay for the TDF–2 satellite. If there is no taker, it is understood that TDF–1 will be launched but operated only on an experimental basis. It is highly unlikely that the ground equipment industry will gear itself up to volume production of terminals to serve a single satellite system.

The position appears to be that the last socialist government, unwilling to admit that DBS was a major failure, handed the problem over to the current Conservative administration with a remit saying "you sort it out". Although the last government arranged users for the satellite, none put up a financial commitment to using it. The current government cancelled the agreements to use TDF–1/2 (see later). The Conservatives had originally supported the creation of the Franco-German TDF–1/TV–SAT programme and cannot blame the failure on someone else. To save face, it has offered the TDF–1 satellite free of charge for commercial operations but said that it is not prepared to pay for the TDF–2 satellite. Those who argue that DBS TV is a commercial proposition are now placed in the position that they have to prove their point with a satellite which is given away free of charge and with a back-up satellite provided at cost, without any R&D expenditure. It also keeps the government's hands clean in that it cannot be criticised for not supporting French technology; it is still subsidising DBS through the offer of a "free" TDF–1.

There is no doubt that the pressure for continuation of commercialisation of the French DBS programme came from industry involved in the manufacture of TVRO ground equipment. What has been absent is pressure from broadcasters. The current attempts are distinctly supply led as they have mostly been since TDF–1 was originally envisaged in 1977. The European broadcasters have experienced a string of cancellations of agreements to use the satellite and therefore are reluctant to commit again.

The current Conservative administration in France has said that it "has done its duty" towards the project; that is to say it has neither abandoned it nor thrown good money after bad.

The Contamine plan for commercialisation of TDF–1/2 is similar to the 1985 plan for "Television par Satellites" headed up by Jacques Pomonti but which was abandoned in the Autumn of that year. Pomonti has subsequently said in public that TDF–1/TV–SAT are engineering led concepts which are now obsolete; he argues in favour of the Europesat concept under study by Eutelsat.

Contamine's plan involves raising intitial funding of FFr900 m ($137 m) rising to FFr1758 m ($268 m) by 1989. This is in addition to the FFr2000 m ($304 m) that the French government has already sunk into the TDF–1 satellite which is largely being given away to the new operating company. The French government had, by September 1986, paid 45% of the construction costs of TDF–2 and committed a further 25%. Under the Contamine proposals, it is implicit that this will be repaid by the commercial operating company. Contamine says that he wants to get the new company established by Christmas with shareholders from three sources both from France and the rest of Europe:

1) Aerospace and Electronics companies already involved in the space and ground segments of the system. These include Aerospatiale, Alcatel and Philips. The ground receiver industry says that it has already invested FFr400 m ($60 m) in DBS ground equipment production.

2) Broadcasters intended to use the satellites. Contamine says that these include Messrs Rupert Murdoch, Robert Maxwell, Robert Hersant and Silvio Berlusconi, CLT, Hachette and Havas. CLT has said that it is not prepared to invest in the new company. However, it has said that it may still be interested in one or two transponders. CLT is also looking to take a stake in the French terrestrial channel, La Cinq. Access to the French terrestrial TV scene (TF–1 is also up for privatisation) may require a DBS transponder as an "access ticket". At the time of writing, the only certain user of the satellite is the French cultural channel, La Sept, which was originally set up to provide a DBS channel. It is state owned and therefore may be in no position to invest in TDF–1/2. The investment will be over and above the annual lease charges for transponders which are expected to be in excess of the previously quoted FFr100 m. The users of the satellite will require authorisation from the new FCC-like body, the National Commission for Communications and Liberties (CNCL). Thus investment does not guarantee access to the transponders.

3) Financial institutions. these are understood to be wary about investing in TDF–1/2.

In addition to FFr900 m in equity, Contamine is planning to raise FFr800 m in loan finance. Primary use of the funding will be to finance TDF–2; Contamine will have to work quickly as he is already being invoiced for the satellite. It remains to be seen which of the proposed users will want to take capacity on the system; his basic plan is for a French based but European financed company with European programming.

Background to the TDF–1/2 Project: TDF–1 is very much a product of WARC '77 which allocated high power DBS channels on a country-by-country basis. Groundwork for the project was set out with the formation of the Eurosatellite consortium in 1978. This was formed by French and German aerospace and electronics interests (MBB, Thomson, AEG and Aerospatiale), later joined by ETCA of Belgium. The European Space Agency had been proposing its H–SAT design since the early 1970's. This eventually evolved into Olympus but was always intended as a platform for use with DBS payloads.

It is well known in Europe that ESA was very disappointed that France and Germany opted out of Olympus in 1979, choosing instead to develop their own platform and payload through the joint TDF–1/TV–SAT project. The basic reason for this was that they believed that Olympus was too ambitious and would take too long to develop a platform and payload suitable for DBS TV.

The finalised order for the two satellites was not placed with the Munich based Eurosatellite until 1982. Eurosatellite then placed prime contractorships with Aerospatiale for TDF–1 and MBB for TV–SAT. Eurosatellite was intended to be responsible for marketing the platform to other countries and organisations. It did succeed in obtaining an order for the Swedish TELE–X system, but subsequently marketing was taken back by the consortium members; Aerospatiale and MBB are ultimately in competition to supply the design to other customers.

The Eurosatellite platform has been proposed to a number of DBS operators, including the China Broadcasting Corporation, USSBCI in the USA and, more recently, for UK DBS. Eurosatellite, France and Germany have also tried to improve the attractiveness of the design and DBS in general by pushing for other European countries to share DBS satellites located at the 19° West orbital slot, which would become the dominant European DBS slot at which most TVRO dishes would be pointed at. Countries approached with this proposal include the UK, Switzerland and the low countries.

However, there is a contradictory feature in this approach in that further DBS systems would be competitive with the TDF–1/2 and TV–SAT 1/2 systems. France has been fairly clever in specifying the downlink performance of TDF–1/2. The footprints are apparently very much larger than needed to provide coverage of France and overspill into most of the rest of Western Europe (this is not the case with TV–SAT). Hence France has been able to offer its satellites for European, rather than just French domestic services. The only problem with this approach is that the satellite is not eclipse protected, meaning that TV transmissions will have to cease earlier at night (local time) during eclipse periods in the UK than for continental

European transmissions or for a UK DBS satellite located at 31° West.

The biggest problem to have surfaced in the hardware side of the TDF–1 project has been with the Thomson-CSF tubes; in 1984 Thomson asked for a further six to eight months to deliver the hardware. At the end of 1985, use of Thomson tubes was abandoned and three AEG tubes substituted. It is understood that Thomson tubes will be used on TDF–2. TDF–1 and TV–SAT were each originally intended to carry three Thomson and three AEG tubes. The Germans abandoned the Thomson tubes some time prior to France. Whilst Thomson has achieved a reputation for producing unreliable tubes, it must be remembered that these tubes are very advanced, and that overall, there is only limited evidence to suggest that Thomson Ku-band tubes are any less reliable than those of AEG.

The order for the TDF–2 satellite was not formally placed until the Autumn of 1985. By 1986, the French government failed to renew the launch reservation for the satellite on Ariane and although it is still shown on the manifest for launch in 1989, it is not clear if the reservation is still valid. At the time of writing, the satellite is still under construction, but with the bills going to Claude Contamine rather than the French government, there is some sense of urgency to commercialise the project.

CLT and TDF–1/2: It appears that CLT is looking to secure its position with the French terrestrial channel, La Cinq and in Germany before opting to use the Astra satellite, now being modified to serve "specific markets" (i.e. specific users). Media International, a joint venture between News International (owners of Sky Channel) and Groupe Bruxelles Lambert (GBL – it has the largest single shareholding in RTL) in a commissioned report on new media opportunities in Europe, identified that the TV–SAT system offered more opportunity for broadcasters than TDF–1/2. The latter was no longer aimed at a French market, which was being served by new terrestrial channels (La Cinq, TV6 and Canal Plus as well as three established channels, TF–1, Antenne 2 and FR–3). TV–SAT gave access to a German language market which was under-served and presented more opportunities and less risk. CLT is already transmitting its RTL–Plus channel terrestrially to Southern Germany and to cable networks in that country via the Eutelsat system. TV–SAT would allow the already established and successful channel to reach consumers not already served by the two existing methods (a sizeable proportion of the potential market) and presumably allow CLT to abandon use of Eutelsat with some off-setting savings.

CLT's relationship with the TDF–1/2 project dates back to 1983 when Luxembourg abandoned its LuxSat DBS proposal, partly because France "offered" CLT capacity on TDF–1 to do so. Most Frenchmen believe that CLT is a French company; it was originally established by French interests and with the approval of the French government in the 1930's to provide cross-border commercial broadcasting and therefore undermine such broadcasting from more hostile countries. or from none-French commercial interests.

However, by the 1980's, the French influence on CLT had waned, with a subtle shift in control to Belgium. The last socialist government eventually ruled out CLT from both use of TDF–1/2 and the French terrestrial network. Part of the reasoning behind this was that CLT was critical of socialist policies in its news and current affairs reporting; that is to say, in the politically charged environment of French media, CLT was too far from control by the French government.

Despite the 1983 offer to CLT, it was not until the Summer of 1984 that plans began to finalise for use of the TDF–1 satellite. In December 1984, a company called Société de Commercialisation des Satellites Telediffusion (later called Television par Satellites) was announced, and charged with the responsibility of marketing and leasing the transponders on the satellite. Two months previously, an accord had been signed between France and Luxembourg giving CLT two channels on TDF–1 and exclusive rights over French and German language advertising supported programming on the spacecraft. Related to this move was the Luxembourg government's support for a medium power broadcasting satellite system, then under study by Coronet and later taken up by Société Européenne des Satellites. The French move effectively isolated CLT from this project, even though, to this day, three transponders are reserved on the satellite for CLT. France was threatened by Coronet through four different facets of the project:

1) The system envisaged use of US manufactured satellites, which undermined the French space industry's semi-captive European market.

2) There were US broadcasting interests in Coronet (HBO) and the prospect of more. The French were concerned about both the cultural impact of the "Coca Cola" project and the commercial impact on its own broadcasting interests. The system had the capability of beaming at least 16 channels into France. Luxembourg was seen as a "captive" for French broadcasting interests and Coronet was using the base to threaten French interests.

3) Coronet and SES were and are competitive with both TDF–1/2 and the Telecom 1 system. Although the latter was not originally intended to carry TV traffic, the failure to attract business users has left the DGT with surplus capacity. It subsequently began marketing Telecom 1 capacity for use by cable TV channels and commercial radio broadcasters.

4) Coronet/SES is a threat to France's interests in the Eutelsat system, both in terms of its financial investment and the contract for the Eutelsat II satellites and options.

The continued support by the Luxembourg government of the medium power GDL/Astra satellite system appears to have influenced the last socialist government in eventually ruling out CLT use of TDF–1. In November 1985 it was announced that the TDF–1 transponders were allocated to interests associated with the then newly allocated terrestrial French channels, La Cinq and TV6. The arrangement was seen as part of a political fix to ensure that broadcasting remained in the hands of those sympathetic to the French socialist government.
The allocation of the channels were as follows:

1) Two channels allocated to the European Satellite Television Broadcasting Corporation, a consortium of UK newspaper proprietor, publisher and printer and former socialist MP, Robert Maxwell, Italian commercial TV entrepreneur, Silvio Berlusconi, West German media and film entrepreneur, Leo Kirch and the French Seydoux family interests.

2) One channel allocated to La Cinq (i.e. Silvio Berlusconi). It is understood that Berlusconi planned to use the channel for Italian language services to strengthen his Italian network, rather than to transmit the French language La Cinq to areas of France outside the range of terrestrial transmissions or to French speaking parts of Belgium and Switzerland. At one time, Berlusconi also said that he intended to use the channel for a two language service, Italian and English, and possibly in Spanish and German as well.

3) One channel to La Sept. This is the only allocation which survived the French General election in March 1986, which the socialists lost.

It is interesting to note that Maxwell has subsequently taken no interest in using UK DBS, although there have been unsubstantiated reports that he has submitted a letter of intent for one to three channels on the Astra system. The combination of allocations of new French terrestrial channels (La Cinq and TV) to Berlusconi and Seydoux interests and the allocation on TDF–1 was seen as part of a structuring of new European media interests into a North-South divide, with German interests lining up with CLT (through the Bertlesmann Group), GBL and News International (right-wing) against the Southern (left-wing) group from Italy and France. This framework now appears to have broken down, although News International is still associated with GBL and therefore CLT. The right-wing group mostly have their major interests in Northern Europe rather than France, Italy and Spain. The November 1985 allocations of DBS capacity led to CLT–RTL taking legal action against the French government on the grounds that it was in breach of its 1984 agreement giving CLT exclusive rights on TDF–1. However, at the time of writing, CLT is again in a position to use the TDF–1 satellite and the legal issue no longer remains.

The arrangements for the use of TDF–1 by the above users involved some apparent very heavy subsidies to the broadcasters. Maximum annual transponder rental was set at FFr60 m ($9 m at late 1986 exchange rates, $6 m at the then rates). This was about half the rate originally envisaged a year earlier. CLT had argued in 1985 that the lease rate had to be in line with that offered on FSS systems, around $3 m. The November 1985 price lead to criticisms within France that foreign broadcasting interests were being subsidised (again). However, the lease arrangements also involved considerable payments in advance to finance the construction and launch of TDF–2. Moreover, the satellite had been promoted during 1985 as a pan-European multi-language TV opportunity. France no longer had any real requirements for it for French domestic TV. It was creating broadcasting opportunities by opening up the terrestrial frequencies, which it could retain control over.

The socialists eventually offered CLT the fifth channel on the TDF–1/2 system, but CLT replied that it would be suicidal to take a DBS channel without having a share in the French terrestrial TV scene. This was a further indication that CLT has never really been interested in TDF–1 and has only seen it as an entrance ticket into terrestrial TV. When the current French government was elected in March 1986, it is understood that it wanted to cancel the entire TDF–1/2 project, but was quickly influenced by the industrial lobby to at least continue with the TDF–1 satellite and try and find a way to get the system set up on an operational basis. However, the new government refused to put any more money into TDF–2. Two other factors influenced the current government in bailing out TDF–1. These were that Germany was continuing with the TV–SAT project; abandonment of TDF–1 would have caused a diplomatic rift between the two countries. The second issue was over transmission standards.

The Transmission Standards Problem: This has always been a major issue in the TDF–1/2 exercise. The French have opted for the D2–MAC transmission standard for the satellite, which is claimed will provide an evolutionary opportunity towards High Definition TV (HDTV). Japan and the USA have backed an HDTV transmission standard using the Muse transmission technique. Although adoption of the standard on a world-wide basis was rejected at the May 1986 meeting of the International Consultative Committee on Radio Communications (CCIR) after strong objections from Europe.

However, the European electronics industry remains concerned that the US/Japanese backed standard will become a de facto standard, undermining their markets. It would hand an irrecoverable advantage to Japanese industry, already dominating much of the European audiovisual equipment market. French industry argues that abandonment of TDF–1/2 would undermine the MAC transmission standard. The transmission issue also has an impact amongst European competitors. The UK had opted for the C–MAC standard, which has been proposed as a European standard, unsuccessfully, by the European Broadcasting Union. It is possible that European industry will produce a micro-chip making TVRO equipment compatible with both standards. However, French and German industry are already manufacturing TV sets which are compatible with D2–MAC only. This is despite an agreement between European broadcasters and manufacturers in 1985 that a dual standard chip be applied to TV receivers.

Nevertheless, the transmission standards problem ignores another difficulty faced by the proponents of either D2–MAC or C–MAC. This is that both will initially increase the problem with incompatible standards in Europe, where the market is already divided between SECAM and PAL standards. There are many in the DBS industry who believe that the adoption of either C–MAC or D2–MAC for initial transmissions will make DBS TV a wholly unviable commercial proposition. There will neither be a sufficient "population" of compatible TV sets and the consumer would be unwilling to purchase both a new TV set and TVRO system. Moreover, there are many in the broadcasting industry who believe that there is no significant consumer appeal to HDTV, leaving its adoption as a commercial non-starter.

The Costs of TDF–1/2: The TDF–1/2 satellites have proved to be immensely costly, leaving major doubts as to whether such high-power satellites could ever prove to be a commercial proposition. The TDF–1 and TV–SAT spacecraft were ordered together and included a considerable R&D expenditure. The initial contract for the two satellites was for DM520 m ($236 m) at 1980 economic conditions. This was a fixed price contract. Subsequent modifications to the satellites, including extra fuel to increase design life from seven to nine years and increasing the number of operational transponders from three to four, added to the overall contract price.

According to information issued by Claude Contamine, the cost of TDF–2 is as follows:

	FFr (m)	US$ (m)
Construction	647	98
Launch	499	76
Launch Accessories*	130	20
Launch Insurance	285	43
TT&C Facilities, D2–MAC Development	197	30
Total	1758	266

Applying the same launch and insurance costs to TDF–1, gives costs for the latter as follows:

	FFr (m)	US$ (m)
Construction	779	118
Launch	400 (est.)	60
Launch Accessories*	130	20
Launch Insurance	285	43
Total	1594	242

*Launch accessories refers to transport to launch site and integration with vehicle.

The French government claimed in September 1986 that TDF–1 cost FFr2000 m ($303 m). Until the Spring of 1986, some FFr1140 m ($218 m) had been spent on TDF–1, according to one French governmental source. Some FFr1000 m ($152 m) had been provided directly from central public funds. FFr440 m ($67 m) had been provided by Funds de Development Economique et Social, a public sector entity providing funds for new enterprises. It is not clear if this includes interest charges and costs of development of the Thomson TWTA's and other costs not included in the Eurosatellite contract.

Nevertheless, the costs of TDF–1 and 2, from the above analysis, stand at a minimum of $500 m. This compares with the estimated cost of the two satellite Astra system, including launch, insurance and TT&C facilities, of around $190 m. Société Européenne des Satellites has said that the base price annual lease charge for an Astra transponder is $5 m. Applying an extrapolating formula to TDF–1/2, the annual charge of a transponder on the latter would have to be around $32 m to

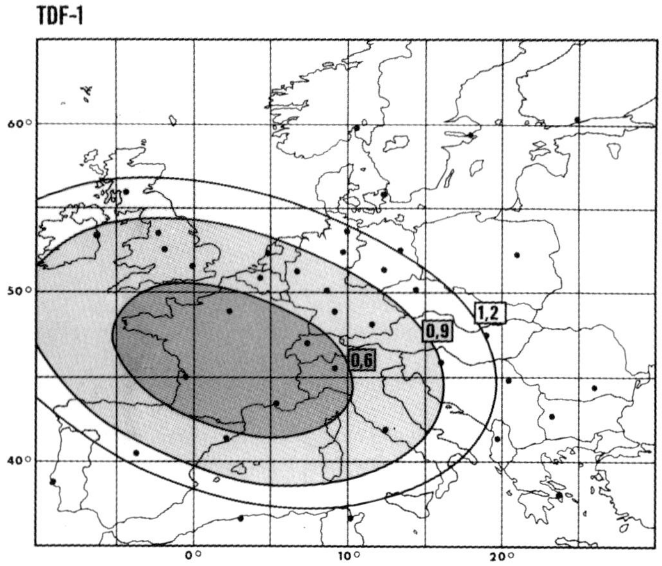

TDF–1 footprint showing expected antenna sizes needed to receive channels. Note that the footprint covers virtually all of Eastern Europe except the Nordic countries and Portugal

recover full costs. (The TDF–1/2 system costs two-and-a-half times as much as a two satellite Astra system and has one-quarter of the number of transponders.) Other sources have suggested that the economic charge for a TDF–1/2 transponder, after write-off of R&D costs (quoted by the French government as \$60 m), is around FFr127 m (\$19 m) per year. During 1985, Jaques Pomonti was asking around FFr150 m (\$23 m) per year but later reduced the figure on the grounds that it was unacceptable to broadcasters.

The French government has said that it will sell the TDF–1 satellite to the company being proposed by Claude Contamine "for a nominal sum". It is not clear how much the final annual lease charge planned by Contamine will be, although the general feeling is that it will be considerably in excess of the FFr55 m (\$8 m) established early in 1986 under the Socialist administration. One source has suggested around \$15 m.

Telecom I, the DGT and TDF–1/2: The French DGT, with responsibility for Telecom I and an interest in DBS through the Eutelsat Europesat proposals and concerned about the Astra project, has been strongly opposed to TDF–1/2 claiming that the satellite is an obsolete concept. It is not surprising that the DGT is so opposed to TDF–1/2. Its opposition can be put down to the following factors:

1) The project is very expensive and could divert resources away from the DGT, including those for the Telecom II system, which is increasingly looking like a broadcasting satellite, as well as Eutelsat in which it has a financial stake.

2) The DGT has a stake in cable TV which is a competitor to TDF–1/2 for TV distribution. Cable also produces traffic for the Telecom I and II systems.

3) In the artificial allocation of frequencies between DBS and FSS services, Telediffusion de France has been allocated responsibility for communications services which the DGT sees as its natural preserve. There remains a widespread belief that the future of satellite TV, whether it involves direct reception by private individuals, TV networking or distribution of cable TV channels, is best served by an FSS satellite system, that is to say, a satellite system under the control of the PTT.

Demand for TDF–1/2 DBS TV: Despite the high costs of the TDF–1/2 project, evidence remains to suggest that there is considerable unsatisfied consumer demand for extra TV in Europe. Cabling outside the Low Countries and odd-ball markets like Switzerland remains low and with limited growth rates. At the time of writing, France has suspended its plan to cable the major urban areas and the UK experience has yet to recover from past "shot in the foot" policies of government. Spain and Italy have no effective cable policies, leaving only West Germany amongst the major European countries with an effective cabling policy. Within France, Italy, Spain and the UK there are some 17·9 m, 17·6 m, 9·4 m and 20·1 m households as potential markets for DBS TV. There is evidence to suggest that, out of these markets, some 7 m households would purchase a DBS TVRO terminal. This is sufficient to provide a base market for broadcasters, although they would probably need more viewers through cable networks to become attractive to advertisers. Some indications show that the most important markets, in order, for TDF–1/2 DBS services are the United Kingdom, France, West Germany, Spain and Italy. Nevertheless, the analysis available in Europe fails to provide arguments that these markets are not better served by an FSS service offered by either Eutelsat or SES where the consumer is offered 16 or more channels and the broadcaster both a larger potential market (covering all of Europe) and the lowest cost per transponder.

A further problem relates to the quality of programming offered on either DBS or new FSS services. The experience of Eutelsat I F1 is that the programming is generally not very attractive (with some notable exceptions; Music Box, Sky and Europa) or overspecialised (RAI). Without any indication as to

TDF–1 spacecraft in Aerospatiale's clean room in Cannes before mechanical tests

the intended programming for TDF–1/2, it remains virtually impossible to predict its attractiveness to consumers.

Conclusions about the TDF–1/2 Project: The most significant conclusion to be derived from the project is that satellite television is about television, not about satellites. The project has failed to place French industry in a position to sell satellites outside its captive markets and there is little chance that it will be able to compete on cost, delivery time or reliability criteria. Much of the recent justification for continuing the project has come from electronic interests looking at the market for consumer equipment, rather than the capital equipment markets. Nevertheless, the real underlying market remains disinterested in the project. This is that of the TV programmers and broadcasters who have to sell the final product to the consumer.

It is a lesson that Europe is a long way from seeing a project through and it is expected that it will lose from its inability to put an alternative in place. By accident, France has taken an obvious step in opening up terrestrial frequencies to provide more TV (and more jobs).

At the time of writing most other European countries are unable to comprehend the relationship between consumer demand and economic activity. They remain protective of existing domestic broadcasters and reluctant to recognise that the prime economic reason behind that protection has disappeared. That reason is the shortage of frequency spectrum for terrestrial broadcasting. Cable and satellite television has broken the link. TDF–1/2 also appears to have proven that there is no future in the kind of high power DBS and national carve ups of orbital slots and frequencies envisaged at WARC '77.

TECHNICAL SUMMARY

	TDF–1	*TDF–2*
PLATFORM	Spacebus 300	Spacebus 300
PRIME CONTRACTOR	Eurosatellite	Eurosatellite
PILOT CONTRACTOR	Aerospatiale	Aerospatial
OTHER MAJOR CONTRACTORS	AEG-Telefunken (TWTA's), ANT (power supply), MBB Alcatel Espace	
STABILISATION	3-axis	3-axis
LAUNCH MASS	2139 kg	
MASS IN ORBIT	1190 kg	
POWER SUPPLY, END OF LIFE	3240 Watts	
DESIGN LIFE	8·9 years (75% probability)	
ORBITAL POSITIONS	19° West	19° West
STATION KEEPING ACCURACY	Better than +/−0·05°	Better than +/−0·05°
ECLIPSE PROTECTION	None	
SPACECRAFT DIMENSIONS	6·5 m height 2·4 m width	
SPAN, TIP-TO-TIP	19·3 m	

	TDF–1	*TDF–2*
NO. OF ACTIVE TRANSPONDERS	4*	5
NO. OF SPARE TRANSPONDERS	2*	
TWTA POWER OUTPUT	230 Watts	230 Watts
FREQUENCY RANGE:		
UPLINK	17·3–18·1 GHz	17·3–18·1 GHz
DOWNLINK	11·7–12·1 GHz	11·7–12·1 GHz
TRANSPONDER BANDWIDTH	27 MHz	27 MHz
EIRP MINIMUM, BEAM EDGE	60 dBW (France)	60 dBW (France)
CHANNEL POLARISATION	R/H circular	R/H circular
RECEIVE COVERAGE (G/T)		
SINGLE CARRIER SATURATION FLUX DENSITY		

*Of the six Travelling Wave Tube Amplifiers intended for this satellite, it was originally envisaged that three would be operational and three spare. The contract with Eurosatellite was subsequently modified to provide four operational transponders. Three of the six TWTA's were supposed to have come from Thomson and three from AEG-Telefunken. Troubles with the Thomson tubes mean that the final configuration of TDF–1 consists only of AEG tubes.

COMMERCIAL DETAILS

OPERATOR	Space Communications Corporation (Spacecom)
ADDRESS	1300 Quince Orchard Boulevard, Gaithersburg, Maryland 20878, USA
TELEPHONE NO.	(301) 258-6800
SENIOR STAFF	William P. Kamachaitis, President and CEO
PURPOSE OF PROJECT	Provision of data relay services for Shuttle missions and other low earth orbit space vehicles and satellites
PROJECT STATUS	Supposed to be operational but plagued by satellite and launch vehicle failures
PROJECT VALUE	NASA agreed to pay Spacecom $250m for all the capabilities of an in-orbit system with a ground spare
LAUNCH DATES	TDRSS–A, 5th April 1983, STS–6 TDRSS–B, 28th January 1986, STS Prior to the loss of TDRSS–B, the remaining satellites were due for launch as follows: TDRSS–C, 16th February 1986, STS 71D TDRSS–D, 22nd July 1986, STS 61M TDRSS–E, 1989, STS TDRSS–F, 1990, STS

SATELLITE USERS

The only customer for this satellite system is NASA. Although it was envisaged as far back as 1976 that Western Union would use the system for US domestic commercial satellite communications, this plan was abandoned. There have also been subsequent plans by Systematics General to offer the surplus capacity on the system for transatlantic communications. The company has, however, withdrawn its request at the FCC for authorisation.

NASA's primary use of the system is for communications with Shuttle and eventually other satellites in low earth orbit. The system is particularly useful for such missions as Spacelab which produce large amounts of data. NASA will also need a considerable amount of data relay satellite capacity to support the International space station from the mid-1990's. As the current TDRSS satellites are all grounded and not required until Shuttle begins operations again from mid-1988, the TDRSS system is likely to find work in that field. In any event, the TDRSS satellites can only be launched by Shuttle.

The system can be used with such low earth orbiting satellites as Landsat and NOAA TIROS–N to provide continuous communications from the spacecraft to the USA.

PROJECT OVERVIEW

Introduction: The Tracking and Data Relay Satellite System (TDRSS) has been one of the most prestigious satellite systems of the first half of the 1980's but has also proved to be one of the most troubled. Its primary role is to provide continuous communications for Shuttle missions as an alternative to a worldwide system of ground stations needed to communicate directly with Shuttle as it orbits the Earth. Unfortunately, the Shuttle system has been grounded for at least two and a half years. The second TDRSS satellite was the primary payload on the last mission that failed on 28th January 1986. The first TDRSS satellite was due to be decommissioned in 1985 having virtually run out of station keeping fuel.

Therefore, the system is now not able to fulfil its primary mission until the next satellite in the series can be launched by Shuttle. This is not likely until at least the second half of 1988, some five years after the first satellite was launched and nearly eight years after the first Shuttle mission. In the meantime there is little need for the system. The TDRSS satellites are integral parts of the STS system, providing transmission capacity for the very high data rates needed for such missions as Spacelab and, eventually, continuous coverage to and from Shuttle no matter where the latter is located over the Earth. The NASA ground facilities can only provide communications links with Shuttle for between 15% and 40% of the time it is in orbit whereas a two satellite TDRSS system can provide coverage of at least 85% of the time. The TDRSS system uses a single ground station at White Sands in New Mexico, with the capability of replacing NASA's world-wide network needed for direct communications with Shuttle.

History of the Project: The origins of TDRSS date back to 1976 when Western Union won a contract from NASA to develop the system. The company formed a subsidiary called Western Union Space Communications to carry out the exercise and the satellite design was also known as Advanced Westar. Western Union intended to use the C-band and some of the Ku-band capacity on the satellites for US domestic services. In 1979, Fairchild Industries and Continental Telephone Corporation each took a 25% share in the company. In the following year, Space Communications Corporation was formed as a partnership between the three companies and took over the project. Western Union pulled out of the partnership in the middle of 1983 but the use of the capacity for US domestic services had already been abandoned. However there have from time to time been plans to use the capacity on TDRSS which is built around the original Advanced Westar requirements of 1977–79, for other commercial needs. Amongst these have been proposals from Systematics General for transatlantic traffic and for Pacific Island use of the C-band transponders. In 1985 Fairchild Industries pulled out of the partnership, essentially because, like Western Union, it was facing financial problems in its main core of business. Spacecom is now wholly owned by Continental Telecom.

The first satellite in the system was launched with the Orbiter Challenger in April 1983. Ironically it was this Orbiter, carrying the second TDRSS satellite in January 1986, which was destroyed on the last Shuttle mission. Immediately after the first TDRSS satellite was deployed from STS–7 there were problems with the two-stage IUS transfer stage needed to place the satellite into geosynchronous orbit. Approximately two thirds of the way through the second stage burn of the Inertial Upper Stage, the rocket failed and the combination of the satellite and IUS began to tumble. The ground engineers were able to save the satellite after about three hours but some 20 seconds before that the IUS batteries died. They were subsequently able to use the satellite's attitude control thrusters to nudge it into its geostationary orbit but used up so much fuel that it had to be retired in September 1986, some eight years before the end of its design life.

The second satellite was, at the time, due to be launched in 1984 but delays in the Shuttle launch manifest put this back until the Spring of 1985. However, a week before the satellite was due for launch (on 7th March 1985), NASA cancelled its launch on the grounds that problems had been identified in its batteries. In the meanwhile, the TDRSS–A satellite was suffering from other problems.

Spacecom's Relationship with NASA: Spacecom owns and operates the TDRSS system using its White Sands ground terminal in New Mexico. The plan has been for NASA to lease three satellites on a full time basis from Spacecom. Three additional satellites were to be available as ground flight ready spares. Communications and tracking signals are relayed through the satellite to and from the White Sands facilities. The latter are connected to NASA ground facilities elsewhere in the USA by NASA's NASCOM communications network which may also use normal commercial communications satellite systems.

Under the three in-orbit satellite plan, one satellite is located over the Pacific, one over the Atlantic and one as an in-orbit spare. The satellites are all launched by Shuttle using a Boeing built two stage IUS system. The system could provide tracking and communications services for up to 40 separate user spacecraft (up to 26 simultaneously) with coverage from 85% to 100% of the user satellite orbit. The intention had been to close down NASA's world wide network of ground tracking stations which, typically can provide communications coverage of low earth orbit satellites for about 15% of an orbit's time. Each station was also limited in the number of satellites it could handle at any one time, typically one.

Under the terms of the agreement between NASA and Spacecom for the three in orbit satellites, NASA owns all of the capacity on each spacecraft including the FSS capacity originally designated and intended for Advanced Westar. For this NASA paid Spacecom $250 m.

Technical features of the TDRSS Spacecraft: When the TDRSS–A satellite was launched in 1983, it was the largest communications satellite to have ever been launched in the West, weighing 2200 kgs in orbit. It was also the first satellite to employ three communications bands (C-, Ku- and S-bands). The satellite

Artists impression of the Spacecom TDRSS satellite in orbit (courtesy TRW)

used a combination of fixed, phased-array and large, moveable antennas. Each satellite has an FSS payload (formerly known as Advanced Westar) and a Tracking and Data Relay payload. The former operates in the Ku (14/12 GHz) and C (6/4 GHz) bands whereas the latter operates in S (2 GHz) and the Ku (15/13 GHz). Each of the two Ku-band Return Service channels on each spacecraft can receive digital information at a rate of 300 mbit/s. The ground facilities at White Sands can handle two such channels simultaneously. Each satellite can also relay signals from up to 20 users at the same time through an array of 30 helix antennas whch make up the Multiple Access receiving equipment.

Signals from each helix antenna are received at the same frequency, frequency-division multiplexed into a single composite signal and transmitted to the ground. The multiple-access system uses 12 of the 30 helix antennas on each satellite to form a single transmit beam in the direction of the desired user. The beam steering is performed under ground computer control. The satellite also incorporates Spacecraft Switched Time Division Multiple Access (SSTDMA) allowing four 250 mbit/s signals to be transmitted simultaneously to four US geographic zones.

TECHNICAL SUMMARY

PLATFORM	TDRSS
PRIME CONTRACTOR	TRW
OTHER MAJOR CONTRACTORS	Harris Corporation
STABILISATION	3-axis
MASS IN ORBIT	2120 kg
POWER SUPPLY, END OF LIFE	1700 Watts
DESIGN LIFE	10 years
ORBITAL POSITIONS	TDRSS–A, 41° West TDRSS–B, 61° West TDRSS–C, 171° West TDRSS–D, not available TDRSS–E, not available TDRSS–F. not available
ECLIPSE PROTECTION	Full
SPACECRAFT DIMENSIONS	17·41 × 12·98 m
NO. OF ANTENNAS	7 (3 steerable)
TELECOMS CHANNELS	2 26 Watt (+2 spare) SSPA's S-band duplex Single Access 1 28 Watt (using a minimum of 8 of 12 3·5 Watt TWTA's) S-band Multiple Access return 2 25 Watt (+4 spare) Ku-band transponders 2 1·5 Watt (+2 spare) Ku-band transponders 12 5·5 Watt C-band transponders (FSS services) 2 1·5 Watt Ku-band transponders (+2 spare) (FSS services) 4 30 Watt Ku-band (+2 spare) transponders (FSS)
EIRP MINIMUM, BEAM EDGE	34 dBW (Multiple Access Service); 46·4 dBW (S-band Single Access Service); 49·4 dBW (Ku-band Single Access Service); 49·9 dBW (Space to Ground Link)

COMMERCIAL DETAILS

OPERATOR	France Cables et Radio (Direction Generale des Telecommunications)
ADDRESS	25 rue des Jeuneurs, F-75002 Paris, France
TELEPHONE NO.	(33-14) 233 51 25
TELEX NO.	214.233 F
SENIOR STAFF	Henri Bustarret, Director of Industrial and International Affairs, DGT J. Serris, Director, Satellite Projects
FINANCIAL BACKERS	French PTT or, more specifically, the Direction Generale des Telecommunication with France Cables and Radio (a subsidiary of the DGT) responsible for managing the project
PURPOSE OF PROJECT	Provision of domestic satellite communications capacity for civil and military use in France and French overseas departments (DOM). Capacity is also leased to Eutelsat and Deutsche Bundespost
PROJECT STATUS	The system is fully operational
PROJECT VALUE	Three satellites cost $264 m plus launch and insurance
LAUNCH DATES	Telecom IA, 4th August 1984, Ariane 3, V10 Telecom IB, 8th May 1985, Ariane 3, V13 Telecom IC, Sept 1987, Ariane 3, V24 or November, 1987, Ariane 2, V25

SATELLITE USERS

Television transmission: Telecom I's Ku-band 36 MHz transponders (six per satellite) are eminently suited to FM-TV transmissions but apart from demonstrations have not yet been used for this traffic. Telecom IA has an actual peak EIRP of 52 dBW, suitable for quasi-DBS services for much of France, Germany and the low countries.

Telecom 1 system is intended for three missions. C-band capacity is used for telecommunications with French overseas departments in the Caribbean. X-band capacity is used for French military communications. Ku-band capacity was intended primarily for business communications using TDMA techniques, with some capacity leased to Eutelsat for European coverage. However, there was little demand for this and it is being leased for three categories of broadcasting traffic – distribution of terrestrial TV channels (La Cinq and TV–6), distribution of cable TV channels and SCPC radio channels. Telecom 1B is nominally the in-orbit spare but is used for this traffic.

PROJECT OVERVIEW

The Telecom I satellite system is Europe's first such domestic project to become operational. The main mission of the satellites, which has determined the baseline specification for the system and which accounts for nearly half of the payload capacity, is data/business traffic. However, this is not a market segment which has been spectacularly successful for satellite communications and the problem has been compounded for Telecom I by its "over-engineering". Business capacity uses TDMA and ground equipment manufacturers have not been able to produce low cost stations for the system. A receive/transmit terminal system appears to cost up to $1 m and unless the spacecraft can offer business substantially lower transmission charges Telecom I may prove to be a commercial liability for the French PTT. Moreover, the ground equipment manufacturers have also had difficulty in supply equipment on time. France Cables et Radio have been looking at making the business capacity available for TV distribution following poor market demand for data traffic. However the French PTT ministry has let broadcasters know that they will not permit this – apparently because it conflicts with policy towards Eutelsat.

There are nine Ku-band transponders on the satellite, each with a power output of 20 watts and a bandwidth of 36 MHz. Six of these nine transponders are active.

Of the six active transponders, five are used for business services. These five have each got a capacity of 32 Kbit/s to 2 mbit/s.

The ground segment was planned to be equipped with "over 200" 3·5 metre fixed antennas.

The sixth Ku-band transponder is allocated to routing of video programming for cable networks, using receive only terminals in the 1·8–2·3 metre range.

The third role of the Telecom 1 system is to replace the Intelsat and Symphonie satellites for telephone and TV traffic between France and its territories in the Caribbean region.

As well, each satellite has six transponders operating in C-band (6/4 GHz) four of these are active. Of the four, two have a bandwidth of 120 MHz and carry up to 1800 telephone circuits.

The other two have a bandwidth of 40 MHz and are used for TV traffic and occasional services. Their footprints cover the Eastern seaboard of the USA, a factor seen as an astute move by the French (and followed up with the IFRB notification for a "videosat" spacecraft at 37·5° west) to take advantage in changes in regulation of North Atlantic traffic.

The four C-band transponders use FDMA and Intelsat B standard stations.

In addition to the business services, the video for cable networks and the replacement of Intelsat/Symphonie satellites, the fourth service is the "syracuse" payload for military and governmental services.

This service is covered by two transponders operating in the 8/7 GHz band, and with a 36 MHz bandwidth. The TWTA power output for each transponder is 20 watts, to provide data, telegraphy and telephony capacity.

This telephony capacity is for transmission between fixed, transportable or mobile terminals on board ship.

TECHNICAL SUMMARY

PLATFORM	ECS
PRIME CONTRACTOR	Matra
OTHER MAJOR CONTRACTORS	Thomson-CSF (comm. payload), BAe (power supply), Aeritalia (structure), ERNO (propulsion), Saab Scania (coders/decoders), Aerospatiale (solar array and thermal subsystems), SEP (apogee motor)
STABILISATION	3-axis
LAUNCH MASS	1185 kg
MASS IN ORBIT, (BOL)	684 kg
POWER SUPPLY, END OF LIFE	1080 Watts
DESIGN LIFE	7 years minimum
ORBITAL POSITIONS	Telecom IA, 8° West Telecom IB, 5° East Telecom IC
STATION KEEPING ACCURACY	East–West ±0·06° North–South ±0·05°
ECLIPSE PROTECTION	Full
SPACECRAFT DIMENSIONS	2·18 m × 2·96 m
SPAN, TIP-TO-TIP	15·98 m
TRANSPONDERS	9 at 14/12 GHz, 6 of which are active 6 at 6/4 GHz of which 4 are active 2 at 8/7 GHz
NO. OF ACTIVE TRANSPONDERS	6 at Ku, 4 at C, 2 at X
NO. OF SPARE TRANSPONDERS	3/2 redundancy of TWTA's
TWTA POWER OUTPUT	14/12 GHz, 20 Watts 6/4 GHz, 8·5 Watts 8/7 GHz, 20 Watts
FREQUENCY RANGE: UPLINK	14–14·25 GHz, 5·925–6·425 GHz, 7·98–8·095 GHz
DOWNLINK	12·5–12·75 GHz, 3·7–4·2 GHz, 7·255–7·455 GHz
TRANSPONDER BANDWIDTH	14/12 GHz, 36 MHz 6/4 GHz, one at 40 MHz 6/4 GHz, one at 40 MHz 8/7 GHz, 36 MHz
EIRP MINIMUM, BEAM EDGE	44 dBW (Ku-band only)
CHANNEL POLARISATION	Ku-band, linear C-band, circular (right hand) X-band, circular (right hand)

ALLOCATION AND USE OF TRANSPONDERS BY BROADCASTERS ON THE TELECOM I SYSTEM

TELECOM IA (8 DEGREES WEST)

NUMBER	*TRANSPONDER USE*	*USER(S)*
C 1	TV	C-band occasional video (France)
Ku 4	Radio	Ku-band radio subcarrier services; Chic FM, Radio Energie
Ku 5	TV	Ku-band occasional video (France)
Ku 6	Radio	Ku-band radio subcarrier services; Rock FM, Radio Nostalgie

TELECOM IB (5 DEGREES EAST)

NUMBER	*TRANSPONDER USE*	*USER(S)*
Ku 1	Radio	Ku-band radio subcarrier services; Rock FM, Radio Monte Carlo, spare
Ku 2	TV & Radio	Ku-band TV; Antenne-2 (Test, France), Auxiliary Radio subcarriers
Ku 3	TV & Radio	Ku-band TV; La Cinq (France), radio subcarriers europe No. 1, RTL
Ku 4	TV	Ku-band TV; TV–6
Ku 5	TV	Ku-band TV; La Cinq (test mode)
Ku 5	TV	Ku-band TV; Worldnet
Ku 5	TV	Ku-band occasional video
Ku 6	TV	Ku-band occasional video

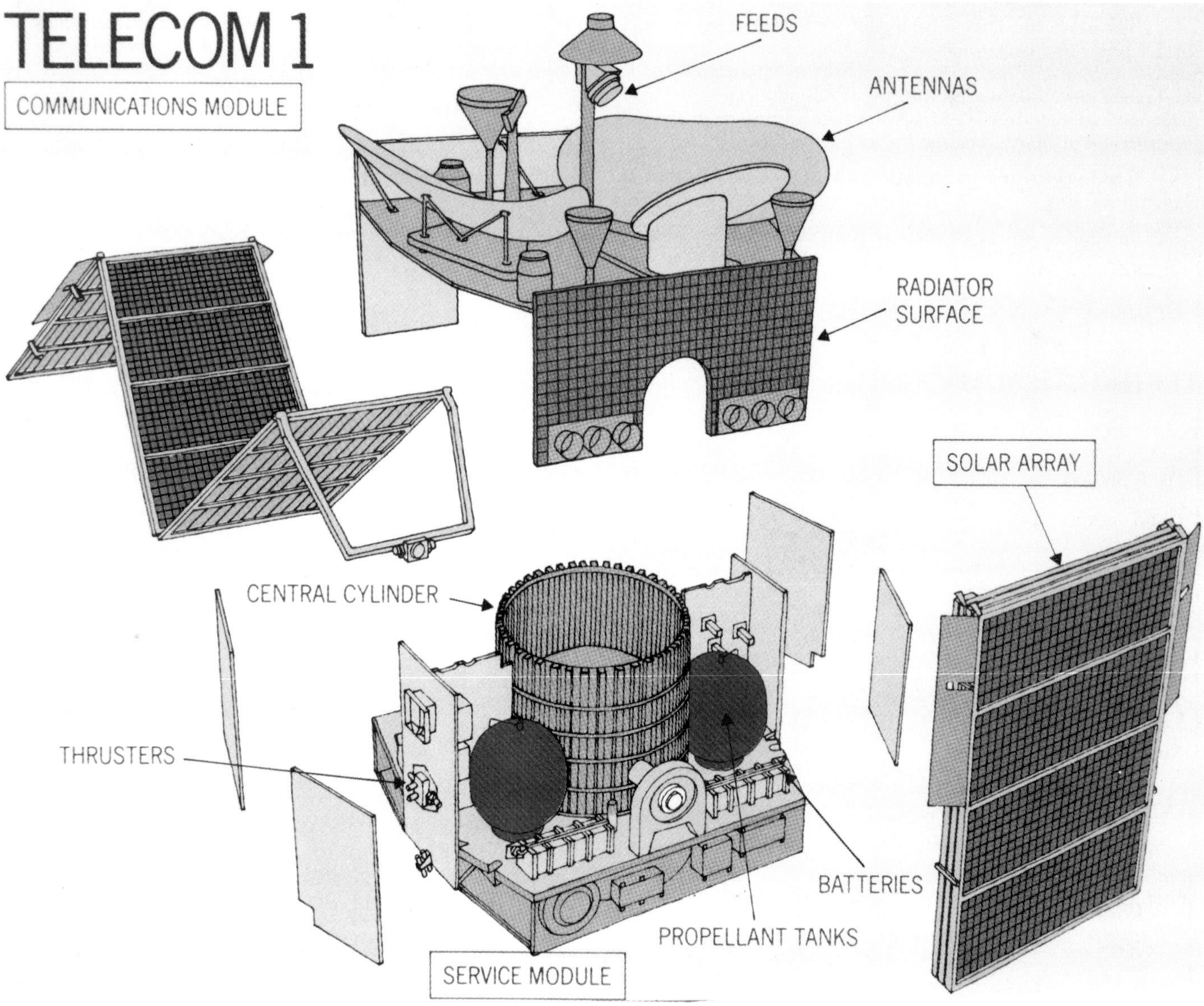
TELECOM 1
COMMUNICATIONS MODULE
FEEDS
ANTENNAS
RADIATOR SURFACE
SOLAR ARRAY
CENTRAL CYLINDER
THRUSTERS
BATTERIES
PROPELLANT TANKS
SERVICE MODULE

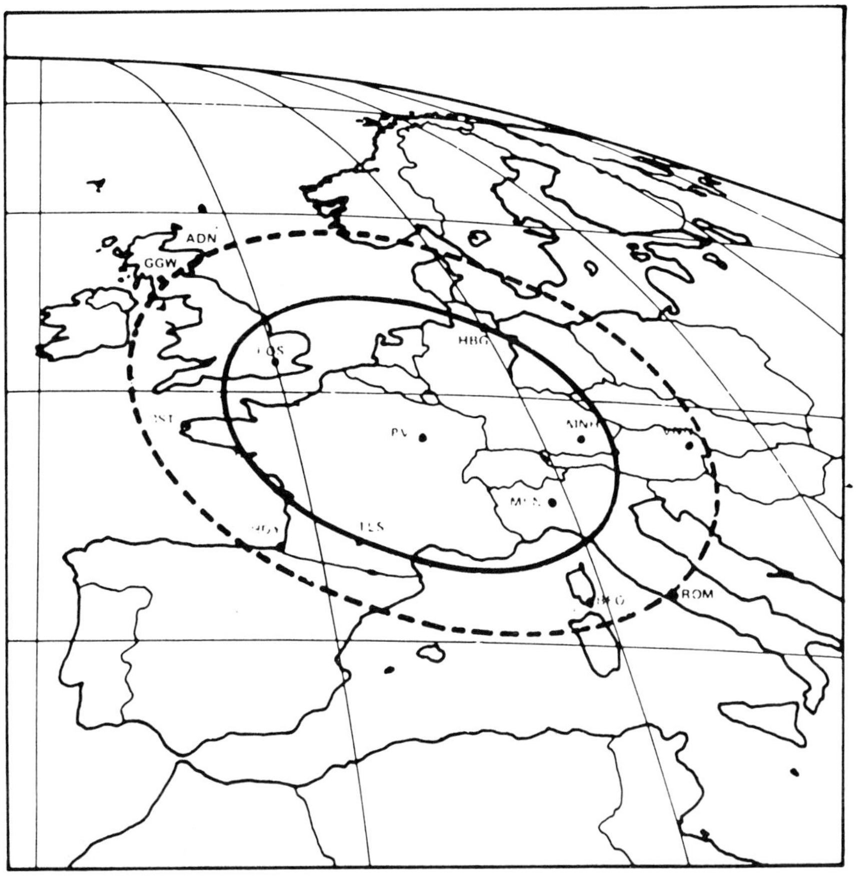

Ku-Band

- **6 transponders**
  - 12/14 GHz
  - 36 MHz bandwidth
- **Transponder capacity**
  - 25 M bits/s
- **Ground station**
  - Antennas Ø = 3.5 m and 4.5 m
  - Time division multiple access
- **Use**
  - Computers connection
  - Videoconference
  - Videotransmission

—— Main coverage area, 47 dBW

---- Extended coverage area, 44 dBW

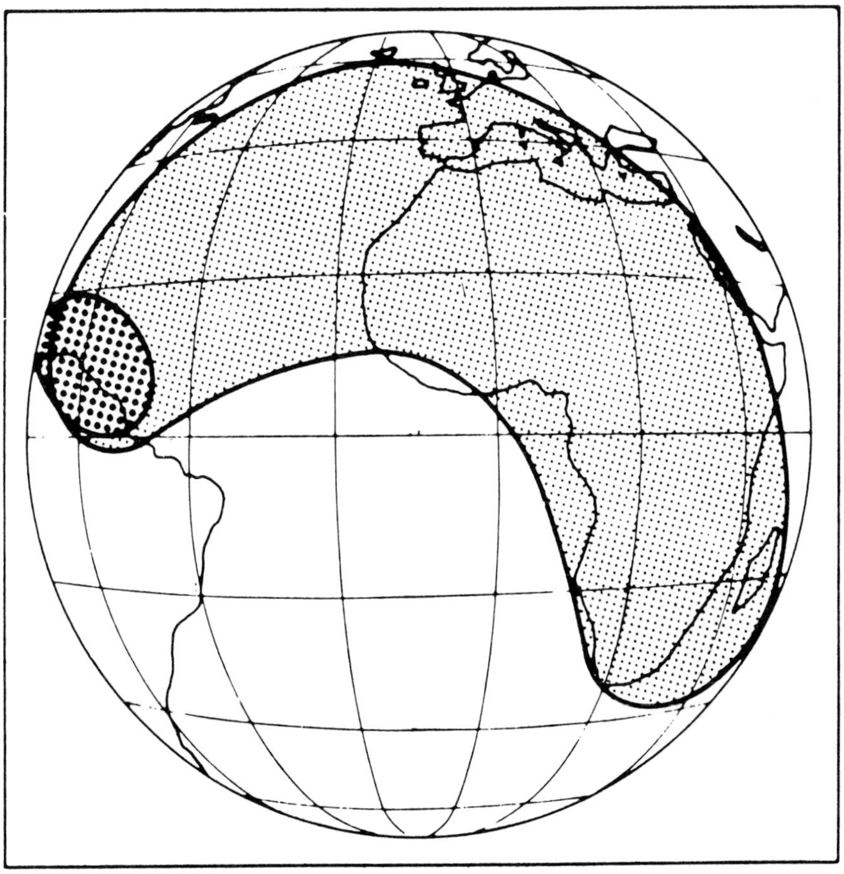

C-Band

- **4 transponders 6/4 GHz**
- **Transponder capacity**
  - 1000 telephony channels
  - 1 TV program
- **Ground stations**
  - Intelsat B standard Ø = 11.8 m
- **Use**
  - Telephony
  - Television
  - Main coverage area, 26 dBW
  - Spot beam, 34.8 dbW

COMMERCIAL DETAILS

OPERATOR	Direction Generale des Telecommunications (France Cables et Radio for marketing)
ADDRESS	Immeuble Perisud, 7 Boulevard Romain Rolland, F-92128 Montrouge Cedex, France
TELEPHONE NO.	+14 564 1994
TELEX NO.	214.233F
SENIOR STAFF	Henri Bustarret, Director, Industrial and International Affairs J. Serris, Director, Satellite Projects
PURPOSE OF PROJECT	Provision of C-, Ku- and X-band communications to follow on from Telecom I
PROJECT STATUS	Still at the planning stage but an order is likely in 1987 with first launch in 1991
PROJECT VALUE	Not available
LAUNCH DATES	First launch in 1991, almost certainly by Ariane 4

SATELLITE USERS

Principal user of this system is likely to be the PTT, using C-band capacity for overseas links. France Cables et Radio is responsible for marketing the system to other users such as broadcasters and radio channels. The design of the Ku-band payload may prove attractive to new French cable TV channels which are expected to develop as the country is cabled over the next few years. The experience of Telecom I suggests that French business is highly unwilling to invest in expensive TDMA equipment for business services on the system.

The Ku-band payload is likely to provide an alternative quasi-DBS opportunity to the troubled TDF–1/2 system which the DGT has consistently opposed.

It is also likely that the system will carry the Syracuse II payload for the French Armed Forces.

However, it is still too early to state who the intended users are to be.

PROJECT OVERVIEW

Introduction: The Telecom II system is currently still being defined and is the fellow space segment to the current French domestic satellite system, Telecom I. The contract for the system was expected to be placed in the first half of 1987 with launch of the first satellite some time in 1991.

Requirements for the Telecom II system: Although Telecom I is primarily a domestic satellite system, there are elements that have close similarities with an international system. The C-band capacity provides coverage of the Caribbean region and the Ku-band capacity provides good coverage of most of the heavily populated parts of Europe both in and outside France. It may reasonably be expected that the DGT in France is looking at the possibilities of using the Telecom II system outside the domestic market. Back in 1983, France filed with the International Frequency Registration Board for frequencies and orbital slots for two satellite systems that have, to this day, not been publicised.

One was for a system known as F–SAT using orbital slots at 11° West and 7° East and capable of providing good European coverage. Frequencies were in Ku-band only and 12 active transponders of 50 Watt power output were intended.

The other was filed for slots at 37·5° West and 32° East for a system known as Videosat. The latter would have been capable of providing coverage of North America, Europe and the Middle-East. Trade sources advised the author immediately prior to going to press that the new Telecom II system contained elements of the thinking behind F–Sat and Videosat.

It also appears likely that the Telecom II satellites will include a Syracuse II payload for use by the French defence forces. The existing Telecom I system similarly includes a Syracuse I payload. The requirements of military forces suggest that global coverage may be appropriate for Syracuse II. The French armed forces continue to be dispersed world wide and the Syracuse II system is also used for mobile communications for the French Navy.

Current position of project: The French DGT was, in the latter half of 1986, treating the evaluation of payload options for the Telecom II satellite system with some degree of urgency. Increasing use of the C-band capacity on the Telecom I system may mean that there is insufficient capacity for the overseas services of the PTT. The C-band capacity is used for communications with French overseas territories (DOM) and each of the current Telecom I satellites carries four such active transponders with two back-ups. (The French overseas departments are too small and remote to justify investment in trunk submarine cables.) The C-band transponders are also capable of providing communications with the Eastern seaboard of the USA but have not (yet) been used for such services. It is not clear if the French intend to expand the C-band services to compete in the new open skies environment initiated by the Americans to the cost of Intelsat.

Mr J. Serris, responsible for satellite projects at the French DGT, said in September 1986 that "if we plan the first Telecom II launch in 1991 as scheduled, the order for the satellite will be signed during the first half of 1987. We are finalising the payload configuration for its approval in January (1987)". The DGT appears to have a major problem in selecting the final payload configuration having to take into account the need to service three separate market segments.

The DGT wants both C-band and Ku-band capacity, the latter to follow on from the Ku-band capacity used on Telecom I for domestic operations. The Ministry of Defence has not decided whether it can go ahead with a separate Syracuse II satellite system to meet its 8/7 GHz communications requirements. If it decides not it will have 8/7 GHz capacity incorporated on the Telecom II satellites.

A separate Syracuse II satellite system is an expensive option. It would require at least two dedicated satellites, one of which would be operational and the other an in orbit spare. This would be in addition to the minimum of two satellites needed for the Telecom II system.

The DGT's requirements for Ku-band capacity for broadcasting and business communications also (euphemistically) take into account the services to be developed and offered by Eutelsat through its Eutelsat II system. This is seen by some observers as a method of looking at how the DGT can muscle in on Eutelsat's patch. The DGT has still not decided how many Ku-band transponders it wants on each satellite. However, Serris has said that Telecom II's "Ku-band repeaters will use 50 Watt TWTA's because they represent the best technology in 1990 with the advent of the Eutelsat II satellites."

Bidders for the Telecom II system: Aerospatiale and Matra have both replied to the RFP's for the Telecom II system. Aerospatiale is proposing use of either a Spacebus 100B platform as used for Eutelsat II or a Spacebus 200 platform. Matra is offering a Eurostar bus developed by Satcom International and currently being used for the Inmarsat II spacecraft. The current provisional payload configuration for Telecom II consist of up to ten C-band transponders, each of 10 or 20 Watt power output, and up to 10 Ku-band transponders (of 50 Watt power output).

The Telecom I satellites each have nine Ku-band transponders. Those on Telecom IA are used for specialised services and for distribution of eight radio channels. Those on Telecom IB are used for distribution, within France of the TV channels La Cinq, TV6 and Antenne 2.

The Telecom IA satellite was launched on the 4th August 1984, having a design life of seven years. It will be life expired by the end of 1991. Telecom IB is expected to last until the end of 1992 at the earliest. Telecom IC, now due for launch in the Autumn of 1987, is expected to last until 1995.

Both the two consortia had, early in 1985, submitted bids for separate Syracuse II and Telecom II systems. But the DGT subsequently requested revised bids for a combined system. The Ministry of Defence has said that it does not have budgetary funding for its own separate system but the possibility had not been ruled out. The military communications payload on the Telecom IA satellite is not functioning properly, leaving the Forces with a shortfall in in-orbit operational capacity. This is the main reason why the Telecom IC satellite is being launched instead of being held as a ground spare.

TECHNICAL SUMMARY

PLATFORM	Eurostar or Spacebus 100B or Spacebus 200
PRIME CONTRACTOR	Matra or Aerospatiale
OTHER MAJOR CONTRACTORS	Satcom International Consortium (with Matra Eurosatellite consortium members with Aerospatiale
STABILISATION	3-axis
LAUNCH MASS	not available
MASS IN ORBIT	not available
POWER SUPPLY, END OF LIFE	not available
DESIGN LIFE	10 years approximately
ORBITAL POSITIONS	Probably 7° East and 11° West allocated for F–SAT. Possible use of 37·5° West, 32° East
STATION KEEPING ACCURACY	not available
ECLIPSE PROTECTION	Full
SPACECRAFT DIMENSIONS	not available
SPAN, TIP-TO-TIP	not available
NO. OF ACTIVE TRANSPONDERS	Up to 20 (10 at C, 10 at Ku)
NO. OF SPARE TRANSPONDERS	not available
TWTA POWER OUTPUT	10–20 Watts (C-band), 50 Watts (Ku-band)
FREQUENCY RANGE:	
UPLINK	not available
DOWNLINK	not available
TRANSPONDER BANDWIDTH	not available
EIRP MINIMUM, BEAM EDGE	not available
CHANNEL POLARISATION	not available
RECEIVE COVERAGE (G/T)	not available
SINGLE CARRIER SATURATION	not available
FLUX DENSITY	not available

COMMERCIAL DETAILS

OPERATOR	(Initially) Swedish Space Corporation (After tests) Nordiska Telesatellit Aktiebolaget (Notelsat)
ADDRESS (SSC)	Swedish Space Corporation, Tritonvagen 27, S-171 54 Solna, Sweden
ADDRESS (Notelsat)	Nordiska Telesatellit Aktiebolaget, Marievik 1A, S-117 43 Stockholm, Sweden
TELEPHONE NO.	(8) 763 0240
TELEX NO.	54 13 906 telsat s
FAX NO.	(8) 744 41 11
SENIOR STAFF	Karl Eriksson, President, Notelsat John R. Veastad, VP, Notelsat Lars Anderson
FINANCIAL BACKING	Funding is provided by Swedish government (64%), Finnish government (10%) and Norwegian government (26%)
PURPOSE OF PROJECT	A pre-operational satellite intended as a precursor to a fully operational system for DBS TV and business traffic
PROJECT STATUS	Satellite under construction and due for launch in 1987
PROJECT VALUE	Total cost of TELE–X project is SKr1250 m ($205 m) at 1982 prices, cost of satellite alone is $100 m
LAUNCH DATES	1987 by Ariane

SATELLITE USERS

The satellite is a pre-operational spacecraft intended initially (for up to two years) for experiments in DBS and video/data business services. It is likely that the satellite would after this phase be handed over to Notelsat to possibly become part of an operational system. Experiments are being coordinated by Notelsat and will be undertaken by DTT's, broadcasting companies and industry in the three countries involved. Given problems with the prospects for DBS TV, the most interesting uses for TELE–X are likely to be in the data/video field. The following types of earth stations are envisaged:

Low Speed	
Data Station	64 kbit/s, data communication, all solid state low cost stations
Extensible Data Station	64 kbit/s, data communication, modular concept
Data & Video Station	2 mbit/s, data communication, four parties video conference, modular concept
Tele Education Station–Teacher	64 kbit/s, sound and 2 mbit/s video, remote education
Tele Education Station–Student	64 kbit/s, sound and 2 mbit/s video (receive only), remote education
Transportable Sound Station	2 mbit/s, outside broadcasting for sound radio
High Speed	
Transportable Video & Data Station	64 kbit/s, 2, 8 and 34 mbit/s, TV conference, outside broadcasting for TV, video distribution, modular design
Fixed Video & Sound Station	2 and 34 mbit/s, outside broadcasting for TV and sound radio, modular design
Video Receive only Station	34 mbit/s, receiver of special TV transmission

PROJECT OVERVIEW

TELE–X is a pre-operational satellite being developed by Sweden and Norway and intended for Direct Broadcast and business services. The sole satellite in the system is likely to be launched in 1987 and could become part of an operational Notelsat system at the end of the decade. The satellite is intended to be initially operated by the Swedish Space Corporation after which it is expected to be handed over to Notelsat for pre-operational activities. The two principal services which TELE–X can carry are Direct Broadcast TV and digital and video communications for business use. Each of these services are served by two active transponders on the spacecraft.

History of the TELE–X Project: Origins of the TELE–X project date back to the mid-1970's when a regional DBS satellite for Scandinavia, called Nordsat, was envisaged. The Nordsat system was a non-starter, partly because Scandinavian countries could not agree amongst themselves on it. This was despite an interest in exchange TV programming between the countries which dated back to the late 1950's. None of the Scandinavian countries involved, Sweden, Norway, Finland, Denmark and Iceland is well served by TV, mostly because the small domestic markets will not support extensive or expensive programming. It was envisaged that the Nordsat DBS system would not require extra programming. It would have essentially been used for exchange of National TV channels. Feasibility studies on the Nordsat system began in 1976 and with this in mind, the Nordic Telecommunications Administration negotiated a "special deal" at WARC '77. This gave each Nordic country five TV channels, of which three would be within national boundaries, and two covering all of the region. Thus there were eight channels available for the region. In addition, Iceland, Greenland and the Faroes were also given five channels.

The Swedish Space Corporation began looking at the

project from the point of view of industrial policy, including how Swedish companies could contribute to development of the system and how it could utilise the technology of Sweden's fellow ESA members. It was at this stage that the same platform and other elements of Eurosatellites' TDF–1/TV–SAT design were chosen for the space segment. The Swedish Space Corporation also developed the idea of a multi-mission payload after discussions with telecommunications organisations. The opportunities for a multi-mission payload reflected the availability of the large Eurosatellite platform.

At the same time, two other events of significance occurred. In 1979 the budgets for space R&D in Sweden were increased threefold, mostly for work on communications satellites.

Secondly Sweden was reviewing the impact of new technology on telecommunications, particularly the introduction of an extensive switched digital terrestrial network and in respect of the business payload, a balance between terrestrial and satellite transmission requirements was fixed. It was decided to opt for satellite capacity capable of carrying high speed data communications.

A third mission was also considered. This was known as Trucksat and was a mobile telex system providing communications between fixed standard telex machines and a large number of small mobile terminals. This interesting payload was eventually dropped as was a propagation experiment at 30/20 GHz.

The Swedes were also uncertain at this stage as to whether DBS TV or data services would develop first.

In 1980, the Swedish Telecommunications Administration and the Swedish Space Corporation began work on the feasibility of the project and gave it the name TELE–X. A definition study was started in the Spring of 1981 and discussions began on making it a joint venture with Scandinavian countries with a follow-on for a Nordic telecommunications system. Norway joined in the project in April 1983, taking a 26% share, and Finland in October 1983 with the remaining 10%. Denmark and Iceland decided not to join in the project. Sweden also tried to get the UK interested in the Trucksat mission which would have used an L-band transponder for coverage of Europe, North Africa and the Middle East but the cost, at $30 m, appeared too high. Phase C/D of the TELE–X project started in late 1982 and the contract for procurement of the spacecraft was signed in August 1983, after the Swedish government made a final go-ahead decision in April 1983.

Design of the Spacecraft: The platform is the same as that used on the Eurosatellite. TDF–1/TV–SAT spacecraft with modifications to the structure and thermal control subsystems which have a different interface to the payload. The satellite carries two operational data/video transponders. These are of a completely new design. In designing TELE–X it was believed that 50–100 Watt Travelling Wave Tube Amplifiers could be used for a follow-on satellite design. Thus the choice of high power, 220 Watt TWTA's, of an existing (Thomson–CSF) design on TELE–X was regarded as an interim solution to achieve the necessary RF output power and linearity performance.

The DBS TV transponders used in TELE–X are identical to those used on TDF–1 (i.e. Thomson–CSF based) but modified for C–MAC transmissions. The most unique feature of the satellite is probably the antenna subsystem which has been adapted for the East Nordic service area and includes a number of new design concepts. The design is based on an astigmatic offset Cassegrain concept. Aerospatiale is prime contractor for the satellite. Saab–Space is designated co-prime contractor and shares technical and administrative management with Aerospatiale. It is responsible for the structure. Ericsson of Sweden has overall responsibility for the payload and design of the antennas, as well as the earth segment.

Data/Video Mission: Given the gloomy outlook for DBS TV throughout Europe and North America, the business communications payload of TELE–X appears the most interesting. It takes a radical approach to the problem. ESA's Pierre Bartholome in reviewing digital satellite networks in Europe (in *Proceedings of the IEEE*, Vol. 72, No. 11, November 1984) describes the thinking as follows. It is a move "in the direction of a truly user-oriented system". The high power TWTA's "enable links to be established between very small stations installed on the premises of the end users and connected directly to the terminal equipment. The goal is that the network should connect at least 1000 subscribers each with their own station. The underlying assumption is that the cost of stations for the Eutelsat/SMS system will be prohibitively high and that the economic threshold in the cost/production run equation will never be reached.

Furthermore, since the Nordic countries are in the process of installing an extensive switched digital terrestrial network operating at 64 kbit/s, emphasis has been placed right from the outset on providing services at higher speeds, i.e. from two to 140 mbit/s." The business service also contrasts with that offered by the French Telecom I system which uses TDMA access techniques and some very expensive ground equipment for users.

TELE–X uses a Single Channel per Carrier/Frequency Division Multiple Access technique with correspondingly low costs for earth stations. Antenna dish diameters are between 1·5 and 2·5 metres depending on location. Two standards of earth terminals are intended reflecting two different bandwidths for the two operational transponders. The 40 MHz transponder is geared up for low bit rates (64 to 2048 kbit/s) and the 86 MHz transponder for higher rates (2 to 140 mbit/s).

The 40 MHz transponder can carry 500 channels at 64 kbit/s or 20 at 2 mbit/s. The 86 MHz transponder can carry 25 channels at 2 mbit/s, six at 8 mbit/s, two at 34 mbit/s or one at 140 mbit/s. High bit rates are aimed at videoconferencing (2 mbit/s upwards) and high-definition TV requiring up to 140 mbit/s. Transportable video/data stations require up to 34 mbit/s. The data/video services full Demand Assignment Multiple Access (DAMA) capability. This means that a channel between two earth stations is set up only on demand from a subscriber. The system has an initial capacity of handling 500 earth stations and three calls/second. The capacity is extendable to 5000 earth stations and 30 calls/second.

The network will be managed by a central station which, through special signals, will control access to the satellite by the subscribers and assign frequencies to them. Cost of the receive/transmit terminals to the user are expected to start at around $30,000. Uses envisaged are outside broadcasting of TV and radio, video conferences, Tele-education, Tele-medicine, Video distribution, Electronic mail, Computer to computer communications, remote newspaper printing.

DBS TV Mission: The TELE–X satellite is a pre-operational spacecraft and it is unlikely at this stage that consumers, broadcasters or receiver manufacturers will commit themselves to the prospects in their current form. Discussions have taken place about who could programme the two DBS channels but at the time of writing it appears no one is willing to commit themselves on a commercial basis. It appears that at least initially, use of this capacity will be limited to experiments using the C–MAC transmission standards conducted by Sweden and Norway.

TECHNICAL SUMMARY

PLATFORM	Eurosatellite/Spacebus 300
PRIME CONTRACTOR	Aerospatiale
OTHER MAJOR CONTRACTORS	Saab Space (TT&C, Structure) MBB (AOCS, Harness, Propulsion) AEG (Power conditioning) Ericsson (Antennas, Tracking receiver) Thomson–CSF (Repeaters)
STABILISATION	3-axis
LAUNCH MASS	2130 kg
MASS IN ORBIT	1200 kg
POWER SUPPLY, END OF LIFE	3300 Watts
DESIGN LIFE	5·3 years
ORBITAL POSITION	5° East
STATION KEEPING ACCURACY	+/−0·3°
INSTALLED CHANNELS	3 TV and 3 Data/Video channels
OPERATING MODE	2 TV and 2 Data/Video channels
TWTA POWER OUTPUT	230 Watts (all)
FREQUENCY RANGE (DBS):	
UPLINK	17·7–18·1 GHz
DOWNLINK	12 GHz
EIRP DATA/VIDEO CHANNELS	30 dBW per 64 kbit/s channel
FREQUENCY RANGE:	
UPLINK, DATA/VIDEO	14·085–14·250 GHz
DOWNLINK, DATA/VIDEO	12·585–12·750 GHz
EIRP, MINIMUM BEAM EDGE	60·6 dBW
TRANSPONDER BANDWIDTH:	
DATA/VIDEO	1 channel at 40 MHz
DATA/VIDEO	1 channel at 86 MHz
DBS TV	2 channels of 27 MHz each
DBS CHANNELS, PER WARC '77	Numbers 26, 32 and 40

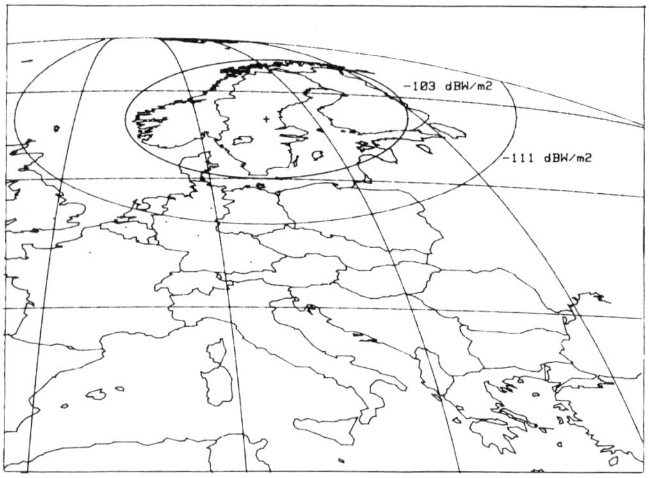

Artists impression of the TELE–X satellite as deployed in orbit (courtesy SSC)

COMMERCIAL DETAILS

OPERATOR	AT&T Long Line Communications Satellite Systems
ADDRESS	Route 202–206, Room 2A101, Bedminster, New Jersey 07921, USA
TELEPHONE NO.	(201) 234-4000
SENIOR STAFF	Karl Savatiel, Director of Satellite Services Sam Willcoxson, Executive VP, Marketing Randall Tobius, Chairman & CEO Robert Kleinwort, President John Harrington, Chief Financial Officer Alfred A. Green, General Counsel J. Walton Cannon, VP, Public Relations
PURPOSE OF PROJECT	Provision of US domestic C-band capacity to replace Comstar capacity
PROJECT STATUS	Now fully operational
PROJECT VALUE	The three satellites are believed to have cost \$113 m, excluding launch and insurance. Control facilities cost \$17 m and in-house support \$7 m (prices quoted to FCC in 1980)
LAUNCH DATES	Telstar 301, 28th July 1983, Delta 3920 + PAM–D, F 171 Telstar 302, 1st August 1984, STS 41D + PAM–D Telstar 303, 19th June 1985, STS 51C + PAM–D

SATELLITE USERS

Principal non-AT&T users of the Telstar 3 system or listed in the Project Overview below. As a major carrier in its own right AT&T uses the system for its long distance communications needs.

PROJECT OVERVIEW

Introduction: The Telstar 3 series is a conventional US domestic satellite system operating in C-band and primarily used for telephony traffic. It is AT&T's first attempt at a fully commercial satellite system, Telstar 1 & 2 being pre-operational experimental satellites. The Telstar I satellite was launched nearly a quarter of a century ago (1962) with the promise of cheap and reliable communications. Today many observers believe that satellites have failed in this mission and will, in the future, be limited to providing point-to-multipoint communications only. Their role in providing conventional point-to-point communications, such as trunk telephony links, will be taken over by fibre optic systems. This suggests that there is no future in Telstar 3 like systems. Moreover, at the present date Telstar 3's C-band capacity is addressing a market which is over-supplied.

Background to the Telstar 3 System: The origins of this system date back to the early 1960's when Bell Systems developed the Telstar 1 satellite. This demonstrated the feasibility of using broad-band microwave repeaters for commercial telecommunications. However, US regulations prevented Bell/AT&T from operating their own domestic satellite systems.

In 1972, the Federal Communications Commission's Domsat decision began to open up the US domestic market for satellite operations. The ruling allowed Western Union, GTE and RCA to develop their own domestic satellite systems but exclude AT&T for a period of three years. This was because AT&T was a dominant monopoly and the FCC decided that time was needed for new entrants to establish themselves before they faced the full force of competition from AT&T. The latter used capacity on Comsat's Comstar system as a consequence. The Commission initially limited AT&T's use of the Comstar system to the provision of Message Toll Services (MTS), Wide Area Telephone Services (WATA) and Autovon services. In 1979 it authorised Bell to use Comstar for its own private line services. On the 4th December 1980, the FCC authorised the launching of twenty-five second generation domestic satellites, including AT&T's Telstar 3 system. This allowed AT&T to wind down the use of Comstar during the 1980's. At one time it had been paying Comsat \$40 m per year for the use of three Comstar spacecraft. The first of the Telstar 3 satellites was launched in July 1983 and by 1986 AT&T had stopped using the ageing Comstar spacecraft. Telstar 302 and 303 were launched in 1984 and 1985 respectively. A fourth satellite in the series, Telstar 304, was not proceeded with.

Telstar 3 Services: The Telstar 3 system is intended to provide voice, high-speed data transmission, radio and TV services within CONUS, Alaska, Hawaii, Puerto Rico and the US Virgin Islands. For this system AT&T used eight 20 metre and one 15·5 metre major earth stations serving principal population areas. Not all of these are owned by the company; GTE G–SAT owns three and Alascom and All-American Cable and Radio one each.

AT&T also uses 10 metre terminals with the system for governmental, private line, network TV and experimental uses. The individual services on the system are marketed under the following categories:

1) Skynet 1·5 Service. This is a high speed digital private line service for voice, data and video traffic. Customers can either use dedicated or AT&T shared earth stations. They can also use multiple data channels or use it as a single conduit for bulk data, facsimile, electronic mail, videoconferencing or other uses at a speed of 1·544 mbit/s.

2) Skynet Audio Service offers transmission of either mono- or stereo-audio programme signals on a point-to-point or point-to-multipoint basis. Such services have had a major impact on the structure of the radio networks in the USA.

3) Skynet Transponder Service offers lease of whole transponder capacity on either a part-time or a full-time basis. Period ranges from one hour to ten years. The service is used for TV and data traffic.

4) Skynet Television Service offers two-way non-simultaneous services to distribute TV programming and to receive feeds from remote locations.

All of these services are marketed under the overall name of Skynet Satellite Services. The company also offers a Ku-band service, Skynet Star Network Service but as Telstar 3 is C-band only, this service uses capacity leased from other satellite operators. (See also Fordstar).

SUMMARY OF SATELLITE USERS

TELSTAR 301
Transponder 02, CBS Regional Programmes
Transponder 03, ABC, Occasional Use
Transponder 04, ABC, Los Angeles–New York feed
Transponder 05, Occasional Video
Transponder 06, ABC, Occasional Use
Transponder 07, CBS, Regional Programmes
Transponder 09, World Satellite Network
Transponder 10, ABC, Regional Programmes
Transponder 11, Occasional Video
Transponder 12, ABC, Regional Programmes
Transponder 13, Hughes TV Network INN
Transponder 14, Independent Network News, Occasional Video
Transponder 15, CBS, Occasional Video
Transponder 16, Occasional Video
Transponder 17, Occasional Video
Transponder 18, Occasional Video
Transponder 19, Occasional Video
Transponder 20, Occasional Video
Transponder 23, World Satellite Network
Transponder 24, Occasional Video

TELSTAR 302
Transponder 01, Occasional Video
Transponder 02, Occasional Video
Transponder 07, Occasional Video
Transponder 08, Occasional Video
Transponder 10, ABC, Pacific Time Feed
Transponder 11, ABC, Occasional Video
Transponder 13, Occasional Video
Transponder 14, Occasional Video
Transponder 15, CBS, Los Angeles–New York feed
Transponder 16, CBS, Pacific Time Feed
Transponder 17, CBS, Toronto, Occasional Video
Transponder 19, CBS, Test Channel
Transponder 20, CBS, Regional Programmes
Transponder 21, Netcom, Occasional Use
Transponder 22, Occasional Video
Transponder 23, Occasional Video
Transponder 24, Occasional Video

TELSTAR 303
Transponder 01, Country Music TV
Transponder 02, Selec-TV
Transponder 08, Occasional Video
Transponder 18, American Extasy FM America
Transponder 21, FNN, Alternative Feed, Sat. Forum
Transponder 22, KTVT, Dallas–Fort Worth

Source: World Satellite Directory, 1986

TECHNICAL SUMMARY

PLATFORM	Hughes HS–376
PRIME CONTRACTOR	Hughes Aircraft Company
STABILISATION	Spin
MASS IN ORBIT	865·3 kg
POWER SUPPLY, END OF LIFE	670 Watts
DESIGN LIFE	10 years
ORBITAL POSITIONS	Telstar 301, 126° West Telstar 302, 97° West Telstar 303, 85° West Telstar 304, satellite option cancelled
ECLIPSE PROTECTION	Full
SPACECRAFT DIMENSIONS	2·1 m diameter, 2·7 m height (6·6 m after deployment)
NO. OF ACTIVE TRANSPONDERS	24
NO. OF SPARE TRANSPONDERS	6
SSPA/TWTA POWER OUTPUT	5·5 Watts*
FREQUENCY RANGE:	
UPLINK	5·925–6·425 GHz
DOWNLINK	3·700–4·200 GHz
TRANSPONDER BANDWIDTH	36 MHz
EIRP MINIMUM, BEAM EDGE	34 dBW (CONUS) 33 dBW (Hawaii etc.)
CHANNEL POLARISATION	Linear, horizontal and vertical
RECEIVE COVERAGE (G/T)	−5 dB/K
SINGLE CARRIER SATURATION FLUX DENSITY	−84·7 dBW/sq. m

*Each satellite has 18 solid state power amplifiers and 12 TWTA's.

TERASAT

COMMERCIAL DETAILS

OPERATOR	Terasat
SENIOR STAFF	Henry Schwartz, President Charles Abrahamson Richard Crawford
PURPOSE OF PROJECT	Provision of US domestic C-band capacity
PROJECT STATUS	At the time of writing, it remains a paper plan
PROJECT VALUE	Company's two satellite system plan will cost \$160–180 m including purchase of two second-hand satellites, launch and insurance
LAUNCH DATES	Terasat 1, November 1987, Long March 3? Terasat 2, February 1988, Long March 3?

SATELLITE USERS

Not available at this stage.

PROJECT OVERVIEW

Introduction: This company announced in the Spring of 1986 that it intended to purchase the Westar VI and Palapa B2 satellites then owned by underwriter Merrett Syndicates. It also announced that it intended to launch the satellites via the Chinese Long March 3 vehicle.

The claims were treated with some scepticism, not least because the company did not appear to have the funds to do so and because the underwriters appeared sceptical about the claims. Terasat said that it intended to use the satellites for domestic US operations.

The Palapa B2 satellite was sold for \$20 m to a California based company, Sattel. It arranged for the satellite to be modified and has said that it intends to resell it. It is not clear if Terasat is a possible purchaser. Westar VI was sold for \$18 m to a company called Pan Am Pacific Satellite Corporation.

Terasat is a satellite leasing company jointly founded by Universal Satellite Corporation and International Capital and Technology Corporation. However, Donaldson, Lufkin and Jenrette Securities Corporation and Telstar Corporation purchased a controlling interest in September 1986. They claimed that they would be instrumental in raising the \$160–180 m needed to finance the purchase, insurance, launch and operation of a satellite system.

As of September 1986, Terasat was saying that it was interested in the purchase of a used RCA satellite, although it is not clear if this means an RCA Americom or an RCA Astro satellite operated by another company (GTE Spacenet, American Satellite and RCA Americom all operate RCA Astro-Electronics built satellites). The company also said that it was planning to launch the first satellite in November 1987 and the second in February 1988. It had a letter of intent for the long term lease "of a significant portion of the transponders on the two satellites".

TECHNICAL SUMMARY

PLATFORM
PRIME CONTRACTOR
OTHER MAJOR CONTRACTORS
STABILISATION
LAUNCH MASS
MASS IN ORBIT
POWER SUPPLY, END OF LIFE
DESIGN LIFE
ORBITAL POSITIONS
STATION KEEPING ACCURACY
ECLIPSE PROTECTION
PROPULSION SYSTEM

SPACECRAFT DIMENSIONS
SPAN, TIP-TO-TIP
NO. OF ACTIVE TRANSPONDERS
NO. OF SPARE TRANSPONDERS
TWTA POWER OUTPUT
FREQUENCY RANGE:
UPLINK
DOWNLINK
TRANSPONDER BANDWIDTH
EIRP MINIMUM, BEAM EDGE
CHANNEL POLARISATION

COMMERCIAL DETAILS

OPERATOR	Deutsche Bundespost
ADDRESS	Telecommunications Central HQ, Am Kavalleriesand 3, 6100 Darmstadt, West Germany
TELEPHONE NO.	(6151) 83 49 00
TELEX NO.	4 19 511 Ftz d
FAX NO.	(6151) 83 49 00
FINANCIAL BACKERS	TV–SAT is financed by the West German Research and Technology Ministry (BMFT), partly through DFVLR
PURPOSE OF SYSTEM	Pre-operational DBS TV capacity, provision of a second satellite and agreement amongst Lander governments are pre-requisites for operational services
PROJECT VALUE	The contract for development of TV–SAT and TDF–1 were signed in 1982 and covered expenditure of DM420 m, in addition to DM100 m spent on preliminary development. At the time of writing, long lead items for a second TV–SAT spacecraft have been ordered, although a contract for the second satellite has yet to be signed
LAUNCH DATES	TV–SAT, July 1986, by Ariane 2 A launch reservation for 1988 for TV–SAT 2 has been made with Arianespace

SATELLITE USERS

In practice allocation of channels on this satellite system is subject to choice by political parties. There are four transponders on the pre-operational channel. The CDU/CSU political parties can allocate "two and one-third transponders", SAT–1 may take one transponder and the Bertelsmann Group another. A third possibility is the ARD Eins Plus channel which was intended to use FSS transmissions to cable networks.

Manfred Lahnstein, Head of Bertelsmann's electronic media division which has a shareholding in RTL–Plus and is interested in Pay–TV has queried the point of renting a DBS channel, not least because of the high cost. He has said that the adoption of D2–MAC has drastically reduced market chances "maybe we should stay for a while on ECS (used for RTL–Plus) and wait for the next generation of Direct Broadcast Satellites which will be much cheaper as systems and will offer five or six times as many channels."

Hanns Eberhard Schleyer, Head of Media Policy in the Rheinland–Pfalz government headquarters has also said that TV–SAT is more of a technical experiment and may prove its value in export rather than domestic use.

The TV–SAT spacecraft's original specifications were for three channels but four operational channels are now being provided.

PROJECT OVERVIEW

TV–SAT is a product of Franco–German industrial policy and, like DBS satellites in general, has been described as technology in search of a market. It is a pre-operational satellite developed in conjunction with the French TDF–1 satellite as an alternative to Franco–German participation in ESA's Olympus satellite (which includes DBS capacity).

One major complaint about TV–SAT was that it was developed without (widespread) discussion with West German broadcasters. The satellite is fully compliant with WARC '77 technical specifications. For broadcasters, DBS still has an uncertain future, it is a new product in a new market and requires considerable investment by broadcasters before a return is seen.

Moreover, it also requires the ground receiver industry and retailers to make available a considerable amount of equipment when the service is introduced. The German situation is made more complex because individual Lander states have a substantial degree of control over broadcasting policy and their agreement is required on services carried on TV–SAT. This has not been forthcoming as yet. However current thinking is that two of the four channels now available on TV–SAT will be made available to the existing public service broadcasters, ARD and ZDF, and the other two to new channels.

One problem with TV–SAT is that its footprint is limited to Germany, Holland, Switzerland and Belgium with limited overspill into France, Italy and Austria. It has virtually no overspill into the UK. In that sense it is not as suitable as TDF–1 for broadcasters aiming at a European market. But the satellite is co-located with TDF–1 at 19° West which is in some sense a "dominant" orbital slot. The Deutsche Bundespost has approached other European countries, notably Switzerland and Austria to develop the combination of TV–SAT and the 19° West slot.

Eurosatellite says that long lead items for about 25% of a second satellite have been ordered, although elements for a third ground spare to the original two TV–SAT/TDF–1 spacecraft were ordered as part of the original programme. It has always been envisaged that any operational follow-on satellite to either TV–SAT or TDF–1 would carry five operational channels. Both TV–SAT and TDF–1 are late; in March Eurosatellite was stating in public that "launches are foreseen in September (1985) for TV–SAT and November 1985 for TDF–1.

TECHNICAL SUMMARY

PLATFORM	Spacebus 300
PRIME CONTRACTOR	Eurosatellite Consortium, headed by MBB
OTHER MAJOR CONTRACTORS	Aerospatiale, Alcatel-Thomson, ANT, AEG-Telefunken
STABILISATION	3-axis
LAUNCH MASS	2065 kg
MASS IN ORBIT	1028 kg
POWER SUPPLY, END OF LIFE	3353 Watts (Equinox)
DESIGN LIFE	7 years with 80% probability
ORBITAL POSITION	19° West
STATION KEEPING ACCURACY	+/−0·1°
ECLIPSE PROTECTION	None

PROPULSION SYSTEM	Bi-propellant	
NO. OF ACTIVE TRANSPONDERS	5, of which 4 are operational	
TWTA POWER OUTPUT	220 Watts (EOL)	
FREQUENCY RANGE:		
UPLINK	17·7–18·1 GHz	
DOWNLINK	11·7–12·0 GHz	
TRANSPONDER BANDWIDTH	27 MHz	
WARC '77 CHANNELS ALLOCATED & MAXIMUM EIRP	Channel No.	EIRP dBW
	2	65·5
	6	65·6
	10	65·6
	14	65·7
	18	65·7
CHANNEL POLARISATION	Right-hand circular	

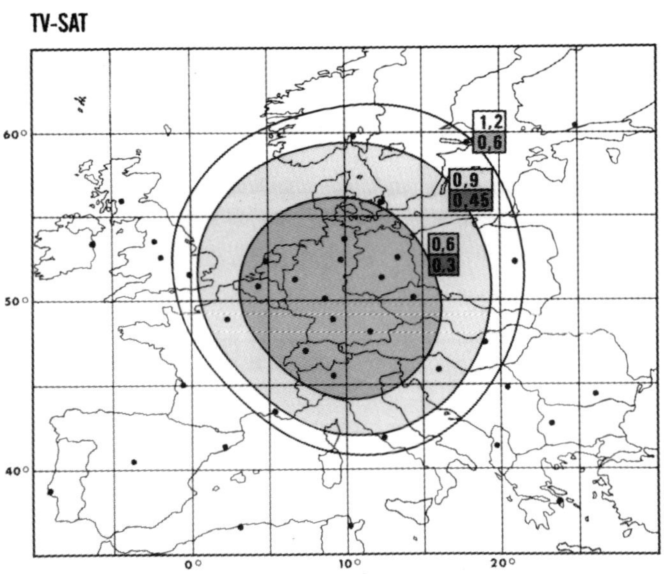

Footprint of TV–SAT spacecraft showing diameters (in metres) of antennas needed to receive TV (larger figures) or radio (smaller figures underneath). It was originally envisaged that one transponder on the satellite would be used to transmit several radio channels

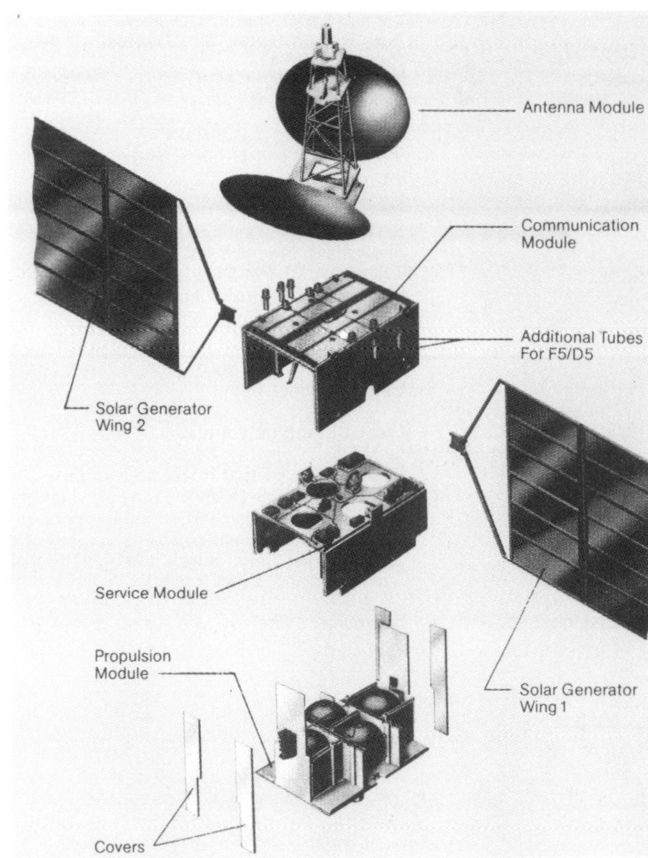

Exploded view of the TV–SAT spacecraft (courtesy of ANT Telecommunications)

COMMERCIAL DETAILS

OPERATOR	A consortium called British Satellite Broadcasting has been awarded the contract to supply three DBS TV channels for the United Kingdom. It was one of five consortia bidding to supply the three channels. The future plans of the losing consortia remain unknown but it is possible that they, or their individual members may join the winning consortium. Alternatively, they may seek to use the Astra, Eutelsat II or Atlantic Satellites spacecraft to provide direct-to-home TV to the UK or Europe. At the time of writing, British Satellite Broadcasting had not chosen a satellite and has several options open to it
WINNING CONSORTIUM	British Satellite Broadcasting c/o Super Channel/Music Box, 19/21 Rathbone Place, London W1P 1DF Telephone: 01-489 1893 Contact: Gillian Pattison or c/o Granada Satellite & Cable, 36 Golden Square, London W1R 4AH Telephone: 01-734 8080 Contact: Andrew Quinn BSB Coordinator
LOSING CONSORTIA	DBS UK c/o Carlton Communications, New Roman House, 10 East Road, London N1 6AJ Telephone: 01-251 1533 Contact: Michael Green Direct Broadcast Ltd (DBL) 9–11 Kensington House, London W8 5NP Telephone: 01-836 4702 or 01-938 2151 Contacts: John Jackson, Anthony Davey SAT UK Broadcasting Shotesham, Norwich NR15 1YD Telephone: 01-994 9462 Contacts: Muir Sutherland, Jimmy Hartley National Broadcasting Services 17 Great Cumberland Place, London W1A 1AG Telephone: 01-262 8040 Contact: James Lee
PROJECT STATUS	Currently still only a plan
PROJECT VALUE	BSB has quoted that it will need to raise 500 million pounds ($720 million) of which its members have pledged 80 million pounds. Total includes costs of setting up channels and construction, launch and operation of satellites and TT&C facilities
LAUNCH DATES	Not before 1990

SATELLITE USERS

Three of the five DBS channels allocated to the UK were given to British Satellite Broadcasting in December 1986 by the Independent Broadcasting Authority. The other two channels will not be allocated until the 1990's. The organisations and consortia which lost in bidding for contracts from the IBA may look to other satellite projects to find capacity for DBS or quasi-DBS services.

In the process of awarding the contracts, the IBA received bids for all three channels from five consortia and bids for two individual channels from two companies. A proposal to provide Teletext Services was also received.

The five consortia were:
SAT UK Broadcasting
British Satellite Broadcasting
Direct Broadcast Ltd
National Broadcasting Service
DBS UK

The individual channel bids were from:
Starstream
44–46 Whitfield Street, London W1P 5RF
Telephone: 01-580 6611
Contact: Richard Wolf

Independent Television News
48 Wells Street, London W1P 4DE
Telephone: 01-637 2424

(ITN subsequently announced that it would be providing programming for BSB).

The teletext proposal came from:
Direct Business Satellite Systems

Amongst the proposals for existing TV channels to use the DBS system were the following:
Sky Channel (DBL proposal)
Superchannel (DBS UK proposal)
Children's Channel (Starstream proposal)
BBC–1 (DBL proposal)
BBC–2 (DBL proposal)
ITN 24-hour news channel (ITN proposal)

Of these, Sky Channel has apparently now abandoned plans to enter the UK DBS scene. ITN's plans for a 24 hour news channel are uncertain. The Children's Channel and Super Channel may use Astra. It is highly unlikely that the BBC will use DBS.

Robert Maxwell, operator of the Premiere film channel and MTV Europe did not attempt to try for a place on UK DBS. It is understood at the time of writing that Maxwell intends to take three channels on Astra, two of which are for his existing satellite TV channels and one for a general entertainment channel. This will allow Maxwell to offer quasi-DBS TV to the UK and Europe from the Spring of 1988, some two years before the earliest start date of UK DBS. Maxwell is also likely to transmit the Home Video film channel by satellite from 1987 but it is not known if he has any intentions to offer this as part of a quasi-DBS package.

PROJECT OVERVIEW

Introduction: The United Kingdom had a third attempt in 1986 to establish a commercial, privately funded DBS system. As of September 1986, five consortia had submitted bids to the Independent Broadcasting Authority (a regulatory agency covering commercial TV and radio) for franchises to provide three DBS channels. The Authority granted a franchise at the end of 1986. Each of the bidders has considered and proposed a variety of space segment options. It appears that the most well thought out of the proposals for the space segment comes from a UK company called Britsat, which has been proposing use of RCA–4000 satellites for the UK since 1983. Also strong contenders are Atlantic Satellites and Comsat, using the two surplus to requirements STC DBS spacecraft. Most of the larger UK broadcasters have decided not to get involved in the latest attempt at DBS and those that have are, in the most part, commercial lightweights.

It is expected that the attempts will, yet again, fall apart. However, those broadcasters that are serious may turn to FSS systems for quasi-DBS operations. Société Européenne des Satellites is offering Astra capacity to UK broadcasters as an alternative to the DBS proposals. It is also possible that some broadcasters go to the French TDF–1/2 satellite system.

Summary of Bids for UK DBS Channels: Five bids were submitted to the Independent Broadcasting Authority on 29th August 1986 for all three channels. Two bids were also received for single channels and one bid for teletext services. Of the bids, that of British Satellite Broadcasting (backed by the Granada group) is considered to be the best, although DBL submitted what appears to be the best thought out plan. The bids are as follows:

1) SAT UK. This company was established in early 1986 as a new venture by Messrs. Jimmy Hartley and Muir Sutherland. It proposed three channels; an advertising supported general entertainment channel, a subscription based film channel ($7·50 per month) and a subscription based family channel ($3 per month). It has the backing of the Bond Corporation from Australia which has interests in DBS TV on the Aussat system. Lonrho and Trillion of the UK are also potential investors. Muir Sutherland is well connected in the international TV programming markets having played a major role in building up Thames Television's programming sales activities before being pased over for promotion to Managing Director of the company. Sutherland also owns a small programming production company, Celtic Films.

The company proposed to use the Britsat system, at least as its preferred option.

2) British Satellite Broadcasting (BSB). Led by the Granada group whose interest in DBS stems from both its ownership of a major UK commercial terrestrial TV station and of high street retail TV rental outlets suitable for both renting TVRO equipment and collecting DBS subscription charges. Other members of the consortium are Anglia TV (another UK commercial TV broadcaster), the Pearson Group (with interests in publishing and existing cable TV operations), Virgin Vision (with interests in SuperChannel/Musicbox) and Amstrad (with interests in electronics distribution and marketing).

It proposes subscription based film channel and advertising supported channels for live news and sport, entertainment, drama and games shows and Disney programming.

At the time of writing, BSB has not stated publicly which satellite option it prefers. It has appointed Telesat Canada as consultant to study and advise on the options available.

3) DBS UK. Backed by Carlton Communications (independent programme producer, also manufacturing TVRO equipment), Columbia Pictures, Dixon's (high street retail chain), London Weekend TV (commercial terrestrial TV station), Hambro's Bank, Robert Flemming (merchant bank) and Saatchi and Saatchi (advertising agency). It was looking for $105 m in equity finance from these partners, $135 m in equity from further investors and the balance in debt or leasing arrangements. It proposed to offer SuperChannel, an entertainment channel aimed at "Yuppies" and children and a news, business and sports channel. All three would be financed from advertising. The aim is to provide programming for market segments underserved by the national terrestrial TV network.

DBS UK did not commit itself to any particular satellite option but is understood to have paid particular attention to the Comsat, STC and Hughes offers.

4) Direct Broadcast Limited (DBL). This company was once closely associated with Britsat, who were formative in its establishment and who have a common shareholder in Ferranti. However, Ferranti and Britsat fell out and DBL appears to be no longer associated with the latter. DBL has not indicated which satellite option it proposed to the IBA. It is known to have studied various options, one of which is for a satellite procured from the Eurosatellite consortium and operated by Marconi Space Systems.

DBL is backed by Ferranti (defence and electronics manufacturer), British and Commonwealth Shipping, Cambridge Electronic Industries, Electronic Rentals, Sears and News International (owners of Sky Channel). It proposed to carry Sky Channel, a 24-hour premium channel of feature films, news and live entertainment and a Showcase family entertainment channel. It has also proposed that the two spare DBS channels allocated to the UK at WARC '77 be used for subscription based transmissions of the two national channels of the British Broadcasting Corporation. This is seen as a step towards implementing the Peacock recommendations to phase out publicly funded TV in the UK during the 1990's.

5) National Broadcasting Service. Seen as the lightweight amongst the five franchise applicants, it is backed by James Lee (formerly Managing Director of Goldcrest Films) and Robert Holmes a Court's Bell group from Australia. It proposed two advertising supported channels, one for news and sport and the other for "best of British TV" and classics (i.e. repeats) and a third, subscription channel for films and educational and arts material.

The company has proposed that it uses a space segment funded by the Bell Group and operated by a company called Bell Satellite Services. It has also said that if it fails to win a DBS franchise, it may consider using the Atlantic Satellites system.

Independent Television News (ITN). This company, which provides the news programming to the UK commercial TV network, submitted for a franchise for one channel to carry a 24-hour news channel loosely based on Ted Turner's Cable News Network channel. It appears that the channel is also being proposed for FSS distribution to cable networks in Europe; it is likely to face considerable competition from CNN Europe.

Starstream. This company is owned by British Telecom, D. C. Thomson Group (publishing) and Thorn-EMI (electronics and music) and requested a single DBS channel to carry the existing Children's Channel. Like ITN, it is expected to join a winning consortium.

Background environment to UK DBS: Continuing interest in DBS TV in the UK can be ascribed to the following factors:

1) Failure of governmental policies to stimulate the growth of cable TV which remains a small and fragmented market. UK broadcasters are having to look to continental cable networks to expand their operations to the point of commercial viability. That is not an easy option because there is limited demand for pan-European advertising and in certain categories of programming, such as films, copyright arrangements limit distribution in other countries.

2) The UK is a major production centre of English language programming. Some broadcasters are looking for DBS to provide English language channels to Europe, where English is the most acceptable common language.

3) Changes in governmental policies towards DBS which have given the broadcasters the freedom to choose the cheapest and/or most reliable space segment available. In practice, this allows them to use US built satellites, either directly procured or through Atlantic Satellites based in Dublin and controlled by Hughes Communications.

4) Lobbying of government by electronics equipment and

FRANCHISE BIDDER	DIRECT BROADCAST LTD (DBL)	BRITISH SATELLITE BROAD-CASTING GRP (BSB)	DBS UK	NATIONAL BROAD-CASTING SERVICE	SAT UK BROAD-CASTING LTD	INDEPENDENT TELEVISION NEWS	STARSTREAM
No. of channels bid	3 (plus 2 BBC channels proposed)	3	3	3	3	1	1
No. of channels supported by							
1) advertisements	2 (Sky, Showcase)	2 (Now, Galaxy)	3	2	1	1	1
2) subscriptions	1 (Premium)	1 (Screen)	0	1	2	0	0
No. of existing channels proposed	1 (Sky Channel)	0	1 (Super Channel)	0	0	1	1
No. of new channels proposed	2 (see notes below)	3	2	3	3	0	0
Geographic market	UK but potential for CATV/SMATV in Europe	UK	UK	Europe	Europe	UK	UK (?)
Programming Contents							
Channel A	PREMIUM: general entertainment incl. feature films, live spectaculars, news and information	SCREEN (subscription) and DISNEY (advertisement): feature films plus Disney programming	News, sport, business plus College of the Air at weekends	CHANNEL Y (evenings), BRITISH BEST (daytime), CLASSIC VIDEO (overnight): Channel Y aimed at under 35's. British best from cable TV programming	ENTERTAIN-MENT CHANNEL (advertisement sponsored). 10 hours per day of drama serials, comedy and music	24 hour news channel along the lines of Cable News Network and primarily aimed at UK audiences. Will draw upon resources of ITN and Worldwide Television News but will have a separate news room and news team. Features will include C-SPAN coverage of the US Senate and Congress, similar coverage of the UK Parliament and possible coverage of the European Parliament	Starstream is a consortium of D. C. Thomson, Thorn-EMI and British Telecom proposing that the existing cable TV channel, the Children's Channel, be carried on DBS TV
Channel B	SHOWCASE: home centred entertainment for children and families. Part of SuperChannel may be used	NOW: 24 hour live events, covering sport, concerts and news	Segmented entertainment aimed at different groups at different times of the day	NEWS: consisting of NBS News and Sportsworld, World News and US News	FILM CHANNEL: (subscription). Daytime children's programmes and evening feature films		
Channel C	SKY CHANNEL: existing TV channel on Eutelsat I–F1. General entertainment. Possible modification to cater for sole UK market	GALAXY: general entertainment such as soap operas and series/serials	SuperChannel (if the Independent Broadcasting Authority agrees – if not a selection of programming from three other cable TV channels)	PAY SERVICE: consisting of Film Firsts (evenings/weekends), Video Arts (weekday daytime) and Educational Video (overnight)	THE FAMILY CHANNEL (subscription). News, sport, women's programming, educational and family viewing		
Teletext	On all three	Yes, but not finalised	Yes, but not finalised	None so far proposed	Not known	Not proposed	Not proposed
Transmission standard	MAC	C–MAC	C–MAC	C–MAC	C–MAC	C–MAC	C–MAC
Subscription rate	£1·00 per week plus 10p per hour for evening schedules of Sky and Showcase	£10 per month	Not available	Not available	Film Channel at £5 per month Family Channel at £2 per month	Not available	Not available
NOTES	DBL also proposes carriage of BBC 1 and BBC 2 on its satellite						

Source: New Media Markets, Cable and Satellite Express, Interspace

component manufacturers and high street TV equipment rental organisations interested in the commercial prospects of the TVRO ground segment. Related to this has been past lobbying by Britsat aiming to allow procurement of a foreign built space segment.

5) The preponderence of the public service broadcasting ethos amongst the existing four national terrestrial TV channels. Some broadcasters believe that there remains considerable unmet demand for popular entertainment led programming.

Offsetting these factors are a number of features of the existing UK broadcasting scene, namely:

1) TV advertising in the UK is well developed and well supplied by the existing two commercial TV channels. In the medium term, the advertising industry cannot support much extra TV. The current applicants for DBS franchises nearly all propose a mixture of subscription and advertising funded programming.

2) Despite claims that there is unmet demand for more popular entertainment led programming, there is evidence to suggest that the UK consumer is well satisfied with the existing terrestrial TV.

3) Most of the major broadcasters are disinterested in DBS. The BBC, by far the largest broadcaster, has not become involved in this round. The absence of most of the broadcasting industry may leave DBS TV with a supply problem for its programming.

4) UK DBS TV, if it does go ahead, is unlikely to be available before 1990. It may therefore be too late into the new media markets. By that time, it is likely that SES will have an operational Astra system and it is possible that the French TDF-1/2 DBS system will be beaming other channels into the UK.

5) Although cable TV has not developed to the extent seen in the low countries and the USA, the UK broadcasting market may have bypassed the need for cable by serving the customer through the VCR (mostly by time shifting abilities of the VCR).

6) Governmental policies towards broadcasting are incremental and often contradictory. The record since 1980 has frequently involved the government shooting itself in the foot. This creates a degree of uncertainty in the industry.

7) The UK broadcasting scene remains heavily regulated by a number of different organisations. The BBC is effectively regulated by a combination of Parliament and its Board of Governors. The commercial terrestrial TV companies and the potential DBS channels are regulated by the Independent Broadcasting Authority. Cable TV channels are unregulated but the cable network operators are regulated by the Cable Authority. This leads to contradictions. For example, News International, as a foreign owned organisation, is not allowed by the IBS to own a terrestrial or DBS channel but can own a cable TV channel. Past experience with the IBA appears to have made some potential satellite TV broadcasters wary of entering the New Media scene.

History of DBS in the UK: The origins of interest in DBS TV in the UK date back to 1977 when the World Administrative Radio Conference (WARC '77) allocated Ku-band frequencies on a country-by-country basis for such services. This stimulated interest amongst commercial organisations in the UK, particularly satellite manufacturing companies and what is now British Telecom.

By 1979 Industry had got its lobbying act organised and was beginning to have some impact on governmental thinking. A new conservative government was elected in 1979 and was quickly faced with a rapidly deteriorating economic position. It had no coherent policy towards new media and communications but soon began to realise that the "sunrise industries" involved could create a substantial number of jobs. It embarked upon a series of short reports on communications on which incremental policy decisions were taken. One of these was a Home Office (equivalent of the Department of the Interior) report called Direct Broadcasting by Satellite which was published in 1981. The report was heavily influenced by lobbying from such companies as British Aerospace and who were concerned that unless the UK ordered a satellite system from the company, it would be unable to build up its expertise to compete with other satellite manufacturers.

Yet the very fact that the report was issued by the Home Office indicated some gross shortcomings in British communications policy thinking. The Home Office has responsibility for broadcasting in the UK but has only obtained it because previous governments could not find a better home for it. The Home Office has no overall responsibility for communications and no responsibility for industrial policy. Much of the current conservative government's success in communications has come through the Department of Trade and Industry. The Home Office backed the idea that the initial DBS channels in the UK should be allocated to the British Broadcasting Corporation, which is the major public service broadcaster for the UK. Its rival, the commercial sector regulatory agency, the Independent Broadcasting Authority was, at the time tied up in the creation of a fourth national television channel, Channel Four, and did not see the significance of the then developments in DBS TV. British Telecom also climbed on the bandwagon in that it joined with British Aerospace and GEC to form a consortium to operate the space segment.

The outcome was a company called United Satellites which proposed a hybrid telecommunications/DBS satellite system capable of providing two high power DBS channels to the UK and low power Fixed Satellite Service (FSS) transponders for transatlantic communications. The apparent willingness of British Telecom to break the Intelsat monopoly on such FSS capacity was a major factor in subsequent US decisions in favour of open skies policies in that field. However, the BBC failed to put together a workable plan for DBS despite having signed heads of agreement with United Satellites for supply of the satellites. In 1984 the government bailed out DBS TV by allowing commercial interests to join the BBC in a new consortium.

The IBA advertised for new franchise members in May 1984 and by the end of the summer of that year a new consortium of 21 companies, headed up by Andrew Quinn of Granada, began to put together a financially oriented business plan for DBS.

The government continued to insist that the consortium procure the space segment from UK sources which meant that United Satellites continued in existence. Much to the consternation of the "21 consortium" United Satellites considerably bumped up the price of the space segment, from US$18 m per annum per transponder to US$26 m. By this time British Telecom had abandoned plans to have a FSS payload aboard the satellite system and entire satellites were dedicated to DBS payloads. The government was offering three DBS channels to the consortium. It was already apparent in the winter of 1985 that the new consortium was having major problems in getting together a workable plan and in June 1985 it collapsed. The consortium never had time to put together a proper plan for programming the three DBS channels. Quickly afterwards United Satellites closed up shop.

Yet, despite this second failure there remained concern that if the UK did not get its satellite broadcasting act together there would be widespread ramifications within the television hardware industry. This particularly applied to the next generation of transmission standards to follow on from PAL and the introduction of high definition TV (HDTV). A government enquiry headed by Sir Anthony Part had in 1983 recommended that C–MAC transmission standards be adopted for DBS TV and the UK subsequently tried hard to get European countries to adopt it. Part had made his recommendations on technical grounds rather than broader industrial policy needs.

Related to these concerns were Japanese efforts in both HDTV and DBS TV. As soon as the "21 consortium" collapsed the government began to look to see if any interest remained in DBS TV. To some people's surprise it did. But because of past failures the government was forced to concede on several issues. In February 1986, for example, the Home Secretary announced that prospective operators were free to buy foreign satellites. The government also said that it intended to introduce legislation to allow franchises up to fifteen years in length. It also awarded the IBA a major role in organising DBS TV and at the beginning of April 1986 the IBA advertised for franchise applicants.

The failure of the "21 consortium" is indicative of some muddle and disjointed thinking by UK policy makers which does not auger well for the future. However, the apparent shortcomings have led to the current revival in DBS TV as the players have got smarter in establishing the ground rules. The failure of the "21 consortium" can be put down to the following factors:

1) Governmental insistence that the satellites be procured from the UK. The consortium was said to be "puzzled" by continuing such insistence, even when it was apparent that the DBS project was in jeopardy. However, the government had interests of its own to look after. It had invested heavily in satellite R&D through the European Space Agency, being the primary source of finance for the ECS and Olympus programmes and could argue that British Aerospace was the largest satellite manufacturer in the world outside of the USA. Had the satellite order gone to the USA or Europe, it would have been seen that the British built satellites were uncompetitive and that past governmental expenditure was wasted.

The United Satellites spacecraft were approximately twice the price, per transponder year, than was available from US manufacturers. Moreover, delivery time for the first satellite was 42 months against 28 months from the USA. There was also a strong element of R&D expenditure in the Eurostar DBS design. The potential users questioned the reliability of a new design which had yet to be tested in orbit.

2) Nevertheless, it appears that whatever satellites had been offered for satellites, the "21 consortium" could not have achieved a viable plan. The government was only offering a ten year franchise period and at least a thirteen year period was needed to obtain a reasonable rate of return. The government had granted concessions to those commercial terrestrial TV companies involved in the consortium to the effect that their terrestrial franchises would be extended. Nevertheless, the consortium was dominated by the BBC and included other interests and therefore the concessions had a marginal impact on the rate of return of the overall project.

3) The consortium was unwieldy, having 21 members and with the BBC holding a 50% share. The BBC was not allowed to cross-subsidise or finance its DBS activities from its licence fee revenue. It therefore had to take a no risk approach to the project. Yet, in contradiction to this, it had to finance its involvement from debt finance. In retrospect, it appears odd that the BBC was given a half share in the consortium when it had failed to make a go of a full share of DBS TV.

The commercial TV companies in the consortium were also competing with the BBC in the national terrestrial scene. Much of the BBC's financial problems relate to the fact that it is steadily losing market share to the ITV companies. The DBS proposals, had they worked, would have placed the BBC in a stronger position over its commercial rivals.

4) The "final straw" that broke the back of the consortium was the BBC's financial problems. In June 1985, the government announced that it was setting up an enquiry into the finances of the Corporation (the Peacock Committee). It was expected that the Committee would recommend that the BBC carry advertising to supplement its licence fee revenue (in the event it did not and that helped in the current revival of DBS interest). Nevertheless, the possibility alarmed the commercial broadcasters who saw both their terrestrial operations being undermined and the DBS operations.

A major factor in UK DBS TV appeared in 1983 when a small, new venture company, Britsat, was formed by ex-civil servant and telecommunications consultant, John Anderson and US engineer Lloyd Ludwig. It was established to exploit the inherent weaknesses in governmental thinking on UK DBS and, in particular, the lack of an adequate UK satellite design.

Anderson had looked at various options to provide FSS services for the UK (and, indeed, had at one time looked at providing low cost satellite services to Africa using Indonesia's Palapa A spacecraft). He had found that there was no market opportunity on the North Atlantic FSS market. He then turned to DBS and soon began arguing in favour of an RCA based high power design. Britsat has consistently argued that the best price/technical solution for the UK DBS space segment is the use of RCA Astro-Electronic satellites. At the time of writing, the Britsat proposals for UK DBS remain the most thought out and well prepared space segment proposals available.

After the "21 consortium" collapsed and was disbanded Anderson continued to use his political connections to influence the government to change the rules of the game. In this way he was very effective and, in the late summer of 1985, the government asked the Independent Broadcasting Authority to find out what interest remained in DBS TV. The Authority found sufficient interest for the government to announce that it would take a major role in reviving DBS. In the spring of 1986 the Authority advertised for bids for franchises for three channels for a period of fifteen years. It was also explicit in allowing procurement of satellites from any source. The final date for submission of franchise bids was the end of August 1986 and, at the time of writing, these bids are being assessed by the IBA.

Summary of Past Proposals for UK DBS: The original plan for UK DBS envisaged two TV channels provided by the British Broadcasting Authority using satellites operated by a company called United Satellites PLC (now defunct). The latter was a consortium of British Aerospace, GEC and British Telecom and would have procured the satellites from British Aerospace with Marconi (a GEC subsidiary) as principal co-contractor. The satellite system was known as Unisat and would have consisted of three spacecraft; one as an in-orbit operational spacecraft located at 31° West, an in-orbit spare and a ground spare. Each satellite would have carried two operational 200 Watt DBS transponders and six operational Ku-band FSS transponders. The latter were variously envisaged for use for transatlantic business communications, cable TV traffic or lease to Intelsat.

This plan was concocted together in 1981–82 and was announced by the government in 1982. The partners had considered using an Olympus platform capable of carrying five high power DBS channels. The government had funded the development of Olympus 1 in 1981 as a European Space Agency project but the United Satellite members felt that it would take too long to develop an operational satellite based on the design. They also considered use of the ECS platform with separate satellites for the FSS telecommunications payload and separate satellites for DBS. This was rejected on the grounds of cost. But at the time, Intelsat had not "woken up" to the need to provide transatlantic business communications.

The eventual choice of satellite was the Eurostar platform,

an enhanced ECS design. Unisat would have been the launch order for the platform, which required considerable expenditure on development. The two satellite manufacturers, BAe and Marconi appeared including all such costs in the price charged to United Satellites and this eventually created major problems in the project. However, the hybrid DBS/FSS design was in itself complex requiring some ingenuity in such areas as thermal control. It also eventually transpired that the delivery period from commencement of construction was 42 months. US satellite manufacturers can currently deliver within 28 months. The BBC was told by the government that it could not procure the space segment outside of the UK so had no choice in going to other manufacturers.

The BBC eventually found that it could neither take the risks of DBS nor put together a sound commercial plan for it. It had only been given a 10 year franchise period. Its plan collapsed in early 1984 and the government bailed out DBS by allowing a new consortium to form including the BBC (with 50%) and the commercial TV companies and other interests with the remainder. The IBA now took role in DBS, having smarted from losing out to the BBC in the first attempt. The BBC had signed heads of agreement with United Satellites in March 1983 for supply of the satellites and construction had begun. This had been suspended late in 1983 when it became apparent that the BBC could not go it alone in DBS TV. No official figures have ever been made public as to how much was spent on the original Unisat spacecraft and the matter is still subject to legal action between the BBC and BAe/Marconi. Trade sources have suggested that the figure was around US$100 m.

Satellite Proposals for UK DBS: The proposals for the space segment of UK DBS represent a mixed bag of alternatives from most of the major satellite suppliers. They fall into two groups, those based on 115–120 Watt TWTA's and those based on the conventional high power (200 Watt plus) TWTA's. The manufacturers have each put in proposals for the two. The options are as follows:

115–200 Watts. These meet the Independent Broadcasting Authority's requirement of a minimum EIRP at selected UK locations of 59 dBW and each satellite provides three operational channels.

1) Aerospatiale with a variant of the Spacebus 100 (Arabsat) platform, also being used for Eutelsat II.

2) British Aerospace in conjunction with its Satcom International partner, Matra, with the Eurostar platform.

3) RCA Astro-Electronics with the RCA 3000 (Satcom) platform.

4) Hughes Aircraft Company with the HS–376 platform.

5) Atlantic Satellites/Hughes Communications with a medium power payload sharing a FSS payload on a Hughes HS–393 platform.

Absent from this list is GEC/Marconi, despite having received RFP's from the broadcasting consortia. The company also declined to respond to requests for payload supply from British Aerospace and RCA Astro-Electronics.

200–240 Watts. These provide either three or five channels with various options for European coverage.

1) Comsat/STC. An RCA 3000 platform, each satellite having three channels. The footprint would be modified according to the final customer requirements.

2) Eurosatellite Consortium (Aerospatiale/MBB). A variation on the TDF–1/TV–SAT design. This proposal is able to offer five full power channels.

3) Hughes Aircraft Company with an HS–393 platform (this proposal is separate from the Atlantic Satellites' proposal). This offers a maximum of three full power channels (240 Watts). Hughes are not offering the HS–394 platform which has been designed to meet DBS requirements.

4) Britsat with an RCA 4000 platform carrying five high power channels.

5) British Aerospace with a high power payload on a Eurostar platform. It is not clear how many high power transponders this could carry.

Société Européenne des Satellites is also proposing the use of the Astra system to the consortia, to the Independent Broadcasting Authority and to other UK broadcasters for quasi-DBS operations.

The analysis of each offer raises some doubts as to the suitability of some of the platforms for DBS operations. From a technical and operational viewpoint, three come out as clear leaders, but, as the requirements of the final winning consortium are unknown, we can only speculate as to which is most likely to be taken up. The three that appear to be front runners on these criteria are Atlantic Satellites, Britsat and Comsat/STC. The Astra proposals fall outside the scope of this analysis but, like Eutelsat II, it has its own distinct advantages. Moreover, the Astra design is now subject to modifications which target it at selected markets in Western Europe. It is somewhat surprising that so little attention has been paid to the Astra option. Proposals based on Hughes platforms. Hughes have been approached by the DBS consortia directly and through Atlantic Satellites. It appears that the HS–376 platform is only marginally capable of delivering three channels of 115–120 Watts to end of life. It is a telecommunications satellite design with limited capabilities for power supply inherent in the spin stabilised concept. It is likely that it would have a shorter design life than the RCA 3000 and 4000 platforms adapted for DBS payloads.

All of the proposals for 115–120 Watt channels suffer from the problem that tubes of this power output are not in general production and have not been space-qualified. Hughes have conducted development work on 100 Watt TWTA's. The 100 Watt Thomson/General Electric tubes used on the Japanese BS–2 DBS satellites have not proved to be reliable.

The Hughes HS–393 proposal has some similar disadvantages to its HS–376 offering. The HS–393, like the HS–376, is a telecommunications satellite design, derived from the Intelsat VI platform. It is power supply limited because of its spin stabilised concept. Just as the HS–376 may be too small for the 115–120 Watt three channel offering, the HS–393 may be too small for the 200–240 Watt five channel proposals. It would be limited to three 240 Watt operational transponders.

The Atlantic Satellites option has not, so far, been well received by the consortia. It too uses the HS–393 platform but will only have its available capacity dedicated to 120 Watt DBS TV. The proposal has a cost advantage in that the operator is able to spread overall risks and costs amongst a variety of FSS services as well as DBS TV. It is estimated that with 2 m DBS subscribers in the UK, Atlantic Satellites could provide the space segment at $1·50 per subscriber per month. Like the HS–376 offering, this proposal is dependent on untried DBS TWTA's.

The HS–394 platform appears to be a non-starter. It was offered to Hubbard Broadcasting in 1984 for the USSBCI six channel DBS system, but the customer rejected it on the grounds of very high cost and technical risks involved in the new design.

Spacebus proposals. It is not known what prices are being offered for either the Spacebus 100 or TV–SAT/TDF–1 platforms. Indications from other sales of the two designs are that the price is up to twice that available for comparable US manufactured satellites. For example, the RCA Astra 1 satellite cost SES \$45 m when ordered in 1985. The Eutelsat II satellites, in payload performance requirements almost identical to Astra and ordered some six months later, each cost \$71 m (assuming that 1AU = \$1, at todays rate 0·73AU = \$1, giving a price of nearly \$100 m per unit). Some of the consortia are also understood to have questioned Aerospatiale's ability to deliver a Spacebus 100 based design on time and with reliability comparable with US built satellites. This is particularly the case now that Ford Aerospace are no longer involved with the platform – Ford contributed 70% of the Arabsat system. Nevertheless, Aerospatiale are understood to be very active in pushing the platform to the UK consortia and, like British Aerospace, have recognised the cost problem.

Eurosatellite originally claimed that it was only interested in providing the UK space segment if it was part of a shared system using the 19° West orbital slot along with France and Germany. The Independent Broadcasting Authority has stipulated use of the 31° West slot and, subsequently, both the West German and French DBS plans have been subject to further uncertainties. Neither look like succeeding at the present time.

British Aerospace/Matra proposals. BAe has put a considerable amount of effort into its bids for the UK DBS segment and it is understood that the price, in real terms, is considerably below that offered by the now defunct United Satellites consortium. The BAe offers have been seen as politically attractive by some broadcasting consortia concerned that the Independent Broadcasting Authority would be unwilling to give franchises to those proposing US manufactured satellites. BAe has considerable lobbying influence in political circles as well as a very efficient and effective public relations machinery.

Nevertheless, the consortia are questioning the bids from BAe on the following grounds:

1) The reliability of the Eurostar platform, which has yet to fly, and, because it is still in its early days of development, the ability to meet its intended weight budget. The launch customer for the platform is Inmarsat and there are indications that a weight problem has arisen on this. The higher the actual weight against planned, the less station keeping fuel can be carried. This in turn reduces the operational life of the satellite. The projected weight of the Eurostar 3 channel DBS satellite is greater than the three channel 240 Watt satellite from RCA.

Some observers also believe that the development of the Eurostar DBS system would seriously affect the delivery time involved. Both the platform and many sub-systems require original development. Similarly, BAe is proposing that it take responsibility for the DBS payload and antennas. With Unisat,

British Aerospace proposal for the current round of DBS franchise applications in the UK. Platform is Eurostar with three 100 Watt operational TWTA's

this had been the responsibility of Marconi. BAe's experience in this field is limited.

2) BAe is proposing TWTA's of 105 Watt power output. There is some indication that this is not sufficient to meet the minimum EIRP requirements laid down by the Independent Broadcasting Authority.

3) Also questioned is BAe's current resources to handle the exercise. It already has a large amount of work on hand. The company has, perhaps, been too successful in obtaining orders for satellites, having won the Inmarsat II contract and (still to be officially announced) the NATO IV contract. Further orders from Inmarsat are also likely. These are on top of the ECS and Skynet IV systems which still have some time to run. There is a shortage of quality technical and engineering staff in the UK which further affects the position.

RCA Astro-Electronics. The RCA 3000 proposal suffers from the problem of using unproven 115–120 Watt TWTA's. However, it appears to be the most cost effective and reliable option amongst 115–120 Watt options with the exception of Atlantic Satellites.

The RCA 4000 proposal is that put forward by Britsat, and uses the AEG-Telefunken TWTA's developed for TV-SAT/TDF-1. These are the only tubes on offer for the high power options. They have yet to be flight tested but represent the most proven technology around. The Britsat proposals are more expensive than some of the other proposals in that the satellites provide five channels and the Independent Broadcasting Authority is only allowing three channels. The company estimates that the overall annual cost of transmission service is 7% higher than with a three channel 240 Watt RCA built satellite. It would be lower per channel if the full five channels were used.

Britsat envisages changeable beams on its satellites which will either provide full European coverage with a minimum EIRO performance of 52 dBW over all of Europe or primary coverage of the UK, Northern France and the Low Countries with a 58 dBW strength and for the UK with 61 dBW. It is recommending a three satellite system, each satellite with five transponders. The company envisages allowing other TV channels onto the system besides the three authorised by the IBA. The others may be continental channels, creating a dominant UK orbital slot at 31° West. The company also considers that the capacity will be used for VSAT type business traffic, arguing that high power transponders are the way forward for such traffic.

The Comsat/STC satellites also use a proven RCA 3000 platform with AEG-Telefunken TWTA's. Nevertheless, the consortia have generally become less enthusiastic about using these second-hand satellites. They are no longer available with

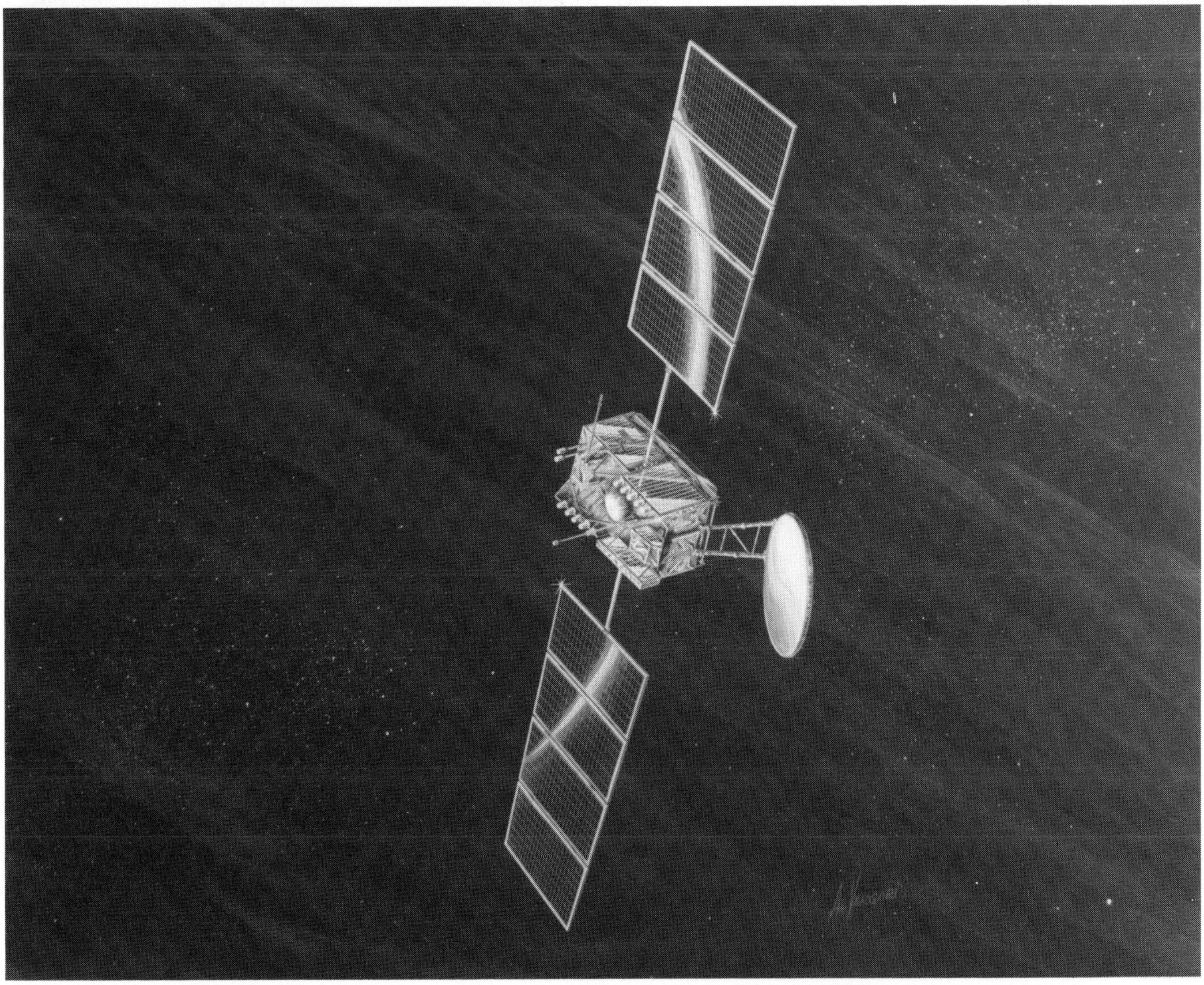

In contrast is the RCA 4000 platform being proposed by Britsat. This is equipped with five 230 Watt transponders (operational) and five spares

cheap and early launch because Shuttle has been grounded and have, in comparison with new satellites, a limited design life (seven years).

Moreover, for a full three satellite system, a third spacecraft would be required. Comsat are offering the satellites at a "distress" price (with easy payment terms) which may be objected to by the Independent Broadcasting Authority. Those objections may also delay DBS. It is also possible that the satellites may be sold elsewhere before a winning UK consortium is able to put in a final bid. On a strict cost comparison basis, the Comsat spacecraft may be uncompetitive as they require modifications which the satellite manufacturer may be unwilling to provide performance guarantees for (and invalidate the performance guarantees for the main satellite) and because of their limited design life.

Two interrelated factors have further impact on all of the proposals. These relate to whether the broadcasters should aim for a European market or limit themselves to the UK and the reliability of TWTA's. Given that there are no reliable 100–120 Watt TWTA's available at the moment, the implication is that the broadcasters should aim for 200 Watt plus thereby being able to provide European coverage.

Nevertheless, there remain good economic reasons for aiming at a European market. The reasons are summarised as follows:

1) The Astra, Eutelsat II and Atlantic Satellites systems will provide the capacity to enter the European markets if UK DBS does not. All three are capable of providing quasi-DBS services.

2) Limiting to a UK market not only restricts the opportunities for UK DBS broadcasters but makes it easier for non-UK broadcasters to compete in the UK market by allowing them to build up their European audiences and programme quality before entering the UK. Under certain circumstances, non-UK channels could be received on TVRO's initially purchased by the consumer for reception of UK DBS only. In that case, it would be easy for any such broadcaster to enter the UK market.

3) By limiting to a UK market, the broadcasters may not be able to provide the variety which some see as the key long-term feature of the new media industry. A three channel system may be the right way to go initially but sooner or later the consumer will be looking for further variety.

4) Serving Europe requires only a marginal increase in the cost of the space segment, but with a several fold increase in the total market size.

5) The broadcasting market is developing along international lines and by restricting to a UK only market, the broadcasters would forever be condemned to being small timers against up and coming broadcasters like Rupert Murdoch, Ted Turner and CLT. Their national operations and small markets would prevent them gaining control of quality programming.

6) It would mean that the UK would be dependent on decisions on transmission standards and broadcasting techniques adopted by more dominant non-UK broadcasters and countries at the expense of the UK equipment suppliers.

7) A European outlook would justify more transponders and allow the development of new broadcasting and business services via satellite.

Notes on the Technical Requirements for UK DBS: WARC '77 allocated five channels to the UK at 31° West. These channels are numbers 4, 8, 12, 16 and 20 (Ireland has numbers 2, 6, 10, 14 and 18). All five downlink in the 11·700–12·100 GHz range and use right-hand circular polarisation.

The Independent Broadcasting Authority has several requirements of its own that broadcasters are expected to meet in their selection of the space segment. These are as follows:

1) The TT&C facilities must be located in the UK. To meet this, Atlantic Satellites is proposing that the TT&C facility for one of its two satellites be located in the UK. However, some broadcasters are considering using the European Space Agency's facilities at Redu in Belgium. Nominally, the requirement may rule out the Astra satellite which has TT&C facilities in Luxembourg.

2) The IBA is initially offering only three of the five channels allocated under WARC '77. The downlink channels are numbers 4, 8 and 12. The use of the 31° West orbital slot is also stipulated, which again weighs against Astra and possible shared use of the 19° West slot with TDF–1/2 and TV–SAT.

3) The IBA requires use of the C–MAC transmission standard. If it continues with this requirement, there may be major problems for the project. Europe has failed to agree on a common transmission standard and therefore C–MAC will be incompatible with continental standards. UK DBS may therefore be blocked out of the European market if continental viewers purchase equipment geared up to current PAL or SECAM standards or another variant of MAC. C–MAC may also make TVRO's more expensive because patent protection blocks out cheap Far Eastern imports of suitable TV receivers and TVRO's.

Moreover, European TVRO manufacturers and industry have yet to gear up production of C–MAC compatible equipment. This may mean that most consumers have incompatible TV receivers and there will be insufficient numbers of TVRO's available on launch of UK DBS.

4) The IBA is stipulating a minimum EIRP performance of 59 dBW for UK coverage. Société Européenne des Satellites says that it is modifying the first Astra satellite to provide improved EIRP performance for "targeted" markets. This may allow UK DBS without the cost of a dedicated system.

FOOTPRINTS PROPOSED BY BRITSAT FOR UK DBS

Primary coverage of the UK and North West Europe
Inner Contour EIRP = 61 dBW
Outer Contour EIRP = 58 dBW

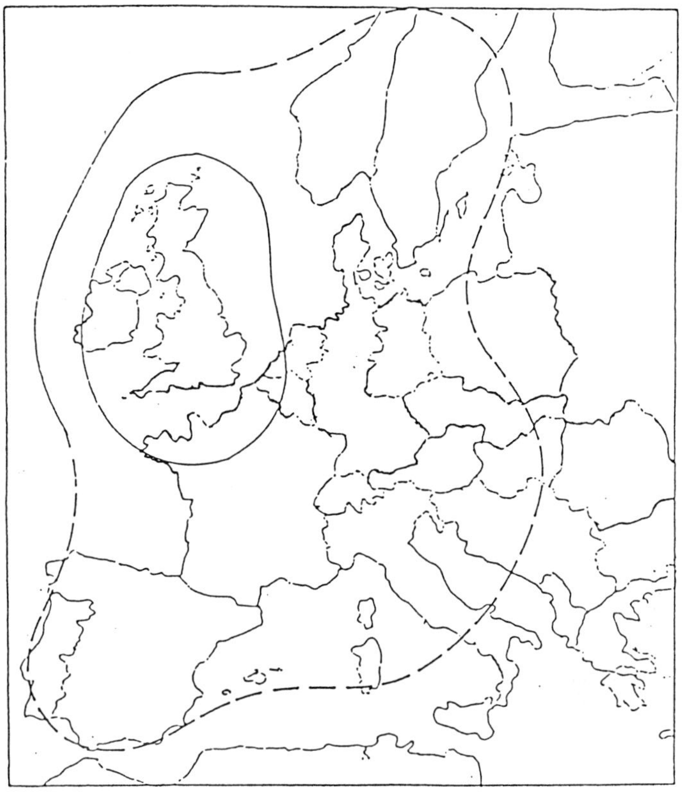

Extended European Beam
Inner Contour EIRP = 60 dBW
Outer Contour EIRP = 52 dBW

TECHNICAL SUMMARY

PROPOSAL	Full European high power DBS coverage	Full European high power DBS coverage	Full European high power DBS coverage*	Full European high power DBS coverage	European FSS coverage and UK medium power DBS
PLATFORM	RCA 4000	RCA 3000/STC	Spacebus 300 (TV–SAT)	HS–393	HS–393
PROPOSED BY	Britsat	Comsat	Eurosatellite	Hughes Aircraft Co.	Atlantic Satellites
NO. OF SATELLITES PROPOSED	3	2	1–2	At least 2	2
PRIME CONTRACTOR	RCA Astro-Electronics	RCA Astro-Electronics	Eurosatellite consortium	Hughes Aircraft Co.	Hughes Aircraft Co.
OTHER MAJOR CONTRACTORS	AEG Telefunken for TWTA's. Possible involvement by Marconi Space Systems	AEG Telefunken for TWTA's	Marconi Space Systems, Aerospatiale, MBB	AEG Telefunken for TWTA's	Possible BAe involvement
STABILISATION	3-axis	3-axis	3-axis	Spin	Spin
LAUNCH MASS	2381 kg	1247 kg	2450 kg		
MASS IN ORBIT	750 kg (empty)	670 kg			
POWER SUPPLY, END OF LIFE	3400 Watts	1700 Watts	3240 Watts	2700 Watts	2700 Watts
DESIGN LIFE	12·8 years	9·7 years	13 years	not available	not available
ORBITAL POSITIONS	31° West	31° West	19° or 31° West	31° West	31° West
STATION KEEPING ACCURACY	+/−0·05° (E/W & N/S)	+/−0·01° (E/W & N/S)	0·05° (N/S & E/W)		
ECLIPSE PROTECTION	partial, 2 channels per spacecraft	none	not available	not available	not available
SPACECRAFT DIMENSIONS	3·49 × 1·94 × 1·6 m		6·35 m high		
SPAN, TIP-TO-TIP	19 m		19 m		
NO. OF ACTIVE TRANSPONDERS	5	3	5	3 max.	5 DBS and 24 FSS
NO. OF SPARE TRANSPONDERS	5	3	not available	not available	not available
TWTA POWER OUTPUT	240 Watts	240 Watts	240 Watts	240 Watts	115 Watts (DBS) 16 Watts (FSS)
FREQUENCY RANGE:					
UPLINK		17·3–17·8 GHz			
DOWNLINK		12·2–12·7 GHz			
TRANSPONDER BANDWIDTH		24 MHz			
EIRP MINIMUM, BEAM EDGE	UK beam 61 dBW, European beam 52 dBW		62 dBW	62 dBW	59 dBW (UK DBS) 47 dBW (European FSS)

PROPOSAL	UK medium power DBS coverage	UK medium power DBS only	UK medium power DBS only	UK medium power DBS only
PLATFORM	HS–376	Spacebus 100B	Eurostar	RCA 3000
PROPOSED BY	Hughes Aircraft Company	Aerospatiale	British Aerospace	RCA Astro-Electronics
NO. OF SATELLITES PROPOSED	At least 2	At least 2	At least 2	At least 2
PRIME CONTRACTOR	Hughes Aircraft Company	Aerospatiale	British Aerospace	RCA Astro-Electronics
OTHER MAJOR CONTRACTORS		Eurosatellite members. Possible involvement by Marconi Space Systems	not available	not available
STABILISATION	Spin	3-axis	3-axis	3-axis
LAUNCH MASS		1650 kg (est.)		
MASS IN ORBIT		985 kg (est.)		
POWER SUPPLY, END OF LIFE		3000 Watts max. (est.)		
DESIGN LIFE	not available	13 years	not available	13 years (est.)
ORBITAL POSITIONS	31° West	31° West	31° West	31° West
STATION KEEPING ACCURACY	0·05° (N/S & E/W)	+/−0·05° (E/W & N/S)		0·05° (N/S & E/W)
ECLIPSE PROTECTION	not available	not available	not available	not available
SPACECRAFT DIMENSIONS	6·35 m high	2·5 m high		
SPAN, TIP-TO-TIP	19 m	22·4 m		
NO. OF ACTIVE TRANSPONDERS	3	5	3	5
NO. OF SPARE TRANSPONDERS	not available	not available	not available	not available
TWTA POWER OUTPUT	115–120 Watts	115–120 Watts	105 or 115–120 Watts	115–120 Watts
FREQUENCY RANGE:				
UPLINK				
DOWNLINK				
TRANSPONDER BANDWIDTH				
EIRP MINIMUM, BEAM EDGE	59 dBW (UK DBS)	59 dBW	59 dBW	59 dBW
CHANNEL POLARISATION				
RECEIVE COVERAGE (G/T)	Not worse than 55 dBW			
SINGLE CARRIER SATURATION FLUX DENSITY	Not more than −80 dBW square metre			

*Eurosatellite proposed performance shown here is based on the assumption that the satellites would be as close as possible in design to TV–SAT and TDF–1.

COMMERCIAL DETAILS

OPERATOR	United States Satellite Broadcasting Company Incorporated (USSBCI)
ADDRESS	3415 University Avenue, St. Paul, Minnesota 55114, USA
TELEPHONE NO.	(612) 645-4500
TELEX NO.	910-563-3598
FAX NO.	
SENIOR STAFF	Stanley S. Hubbard, President Stanley E. Hubbard II, VP and Director of Special Projects Bernie Weiss, Senior Project Manager Gerald D. Deeney, Treasurer and Chief Financial Officer Robert Hubbard, Special Projects Paul R. Heinerscheid, Director of Technical Operations Robert F. Fountain, Executive VP and Chief Operating Officer Carlos W. Luis
FINANCIAL BACKING	The project is currently backed by Hubbard Broadcasting but is looking for finance for the satellite project. At the time of writing it appears that USSBCI has "some" but "not all" of the financing
PURPOSE OF PROJECT	Provision of DBS capacity for the USA
PROJECT STATUS	Has full FCC authorisation for the system and has met due diligence requirements but still requires further funding to complete project
PROJECT VALUE	Contract for two satellites is worth $160 m
LAUNCH DATES	USSBC–1, July 1988, STS USSBC–2, December 1988, STS

SATELLITE USERS

USSBCI planning envisages that both pay-TV and advertisement supported channels will be carried on its satellites. The type of programmes envisaged are "in-depth" news coverage, sports and "much-needed" children's programming.

PROJECT OVERVIEW

If any of the original first round DBS applicants in the USA are likely to survive, it is likely that United Satellites Broadcasting Company Incorporated will be one of them. It is one of the most interesting of the proposals for DBS capacity. The company is owned by the Hubbard family who also control Hubbard Broadcasting which operates five TV and five radio stations and CONUS Communications, a satellite news gathering service. Hubbard Broadcasting is also developing another satellite TV service, K–2, which uses RCA Americom's Satcom K2 Fixed Satellite Service spacecraft. This provides instantaneous national interconnection of (potentially all 900) US TV stations. This is essentially a programme distribution service with vague overtones of a fourth network.

The implications are that USSBCI and its backers are tackling the DBS exercise from a programming point of view based on considerable experience and in that has an edge over all of the other US (and European) companies planning DBS TV capacity. Nevertheless, the USSBCI proposals face many of the same problems facing other US DBS operators. Most US households already have access to, if not actually connected to, cable TV as well as Terrestrial Transmitters. C-band TVRO installations are mopping up part of the demand for satellite TV and VCR penetration continues. By the time DBS TV becomes available, the market will be even better served by alternative distribution means than it currently is.

Moreover, DBS TV economics require rapid penetration to achieve profits which in turn require manufacturers to commit to large-scale production of receivers and organise distribution networks.

USSBCI is unusual amongst current companies looking to provide DBS TV capacity in that it has consistently stuck to and defended the use of high-power TWTA's (200 Watts plus) to provide beam edge EIRP levels of around 57 dBW, as originally envisaged by the first round of DBS applicants. USSBCI argues that DBS will only succeed if the size and expense of the home receive equipment are kept to a minimum but the quality of transmitted signal remains high. According to USSBCI, this requires an EIRP minimum of 54 to 55 dBW.

The USSBCI satellites provide EIRP levels from 54·5 dBW to 56·5 dBW in the Eastern or Western CONUS coverage areas or 57 dBW to 58 dBW within the quarter CONUS beams on the satellite at 110° West which is intended to serve the Eastern part of the USA. All channels on the spacecraft at 110° West are individually switchable between half and quarter CONUS beams. The high-power, 240 Watt TWTA's required for this level of performance are manufactured by AEG-Telefunken in West Germany.

USSBCI has also taken a strong position on technical standards for DBS TV, arguing that the industry needs technical standards to make all DBS reception equipment compatible with all operating systems authorised at any orbital location assigned to the USA. For example it argued that Comsat's STC could never have made money with its "Cable System in the Sky" DBS project because, amongst other things, it envisaged non-compatible exclusive equipment. STC strongly attacked the need for technical standards. In June 1984 a public advisory committee to the FCC completed its work on high-power DBS standards which gave seven recommendations for mandatory standards, fourteen recommendations for voluntary standards and twelve suggested industry guidelines.

In response to this the Federal Communications Commission released a Notice of Proposed Rulemaking in February 1985 which proposed to adopt the recommendations. The results of this Notice will not become clear until 1986. However, in response, the Direct Broadcasting Satellite Association, which supports voluntary DBS transmission and receiver standards, conducted comparative tests of one Frequency Division Multiplexing System (General Instrument's Starlock NTSC) and one Time Division Multiplexing System (Scientific Atlanta's B–MAC) at the NASA Lewis Research Centre. It was expected to make recommendations to the Commission late in 1985, on preferred standard video and audio modulation and encryption formats.

TECHNICAL SUMMARY

PLATFORM	RCA-5000
PRIME CONTRACTOR	RCA Astro-Electronics
STABILISATION	3-axis
DESIGN LIFE	10 years
ORBITAL POSITIONS	110° West, 148° West
ECLIPSE PROTECTION	Full
NO. OF ACTIVE TRANSPONDERS	8 (operational)
NO. OF SPARE TRANSPONDERS	Not available
TWTA POWER OUTPUT	240 Watts (EOL)
FREQUENCY RANGE:	
UPLINK	17·3–17·8 GHz
DOWNLINK	12·2–12·7 GHz
TRANSPONDER BANDWIDTH	24 MHz
EIRP MINIMUM, BEAM EDGE:	
HALF CONUS COVERAGE	54·5 dBW
QUARTER CONUS COVERAGE	57·0 dBW
CHANNEL POLARISATION	Circular

COMMERCIAL DETAILS

OPERATOR	Western Union Westar Satellite Services
ADDRESS	1 Lake Street, Upper Saddle River, New Jersey, MJ 07458, USA
TELEPHONE NO.	(201) 825-5000
TELEX NO.	642499
SENIOR STAFF	Robert Leventhal, Chairman & CEO W. R. Pope, VP & COO Fred M. Knip, AVP, Transmission Systems Nancy E. Salvati, Product Manager, Westar Transmission Services J. T. Regan, VP, Commission Systems J. J. Albert, AVP, Westar Satellite Systems Planning W. G. Bradley, AVP, Westar Satellite Services J. K. Livingstone, Senior Director, Partial Transponder Services M. M. Abrahamson, Senior Product Manager, Video Services Tim McCudy, Senior Product Manager, TI Channel Services Warren Bechtel, Corporation Commissioner Don Virgilit, SNR Product Manager, Sky Data Services May Abramsom, Senior Product Manager, Westar Video Services
PURPOSE OF PROJECT	Provision of US domestic C-band capacity
PROJECT STATUS	The system is now at the end of its design life
PROJECT VALUE	The Westar I satellite cost $100 m
LAUNCH DATES	Westar I, April 1974, Delta F101 Westar II, October 1974, Delta Westar III, August 1979, Delta

SATELLITE USERS

See Westar IV–VIS for details of users of the Western Union system.

Western Union is a common carrier. The Westar I–III satellites constituted the first US domestic satellite system. Westar I was retired in 1983, Westar II in 1984. The latter was due to be replaced by Westar VI (now VIS). Westar III remains in use for voice, video, audio and data traffic.

Westar IV and V, launched in 1982, are used primarily for broadcasting and cable traffic, but also carry voice and data traffic.

PROJECT OVERVIEW

The Westar I–III System, a brief description: The first generation of Western Union satellites, this system is at the end of its design life with the first two satellites already retired and the third due for retirement in 1986–87. The now small C-band satellites are a product of the Federal Communications Commissions 1972 Domsat decision which opened up the US domestic satellite communications market of competitive satellite operators. The first two of this series of satellites were launched in 1974 and retired in 1983 and 1984. The third was launched in 1979, some years after the first two and hence the reason why it is still operational.

TECHNICAL SUMMARY

PLATFORM	HS–333
PRIME CONTRACTOR	Hughes Aircraft Company
STABILISATION	Spin
MASS IN ORBIT	249 kg
POWER SUPPLY, END OF LIFE	262 Watts
DESIGN LIFE	7 years
ORBITAL POSITIONS	Westar I, retired Westar II, retired Westar III, 91° West
NO. OF ACTIVE TRANSPONDERS	12
TWTA POWER OUTPUT	5 Watts
FREQUENCY RANGE:	
UPLINK	5·925–6·425 GHz
DOWNLINK	3·700–4·200 GHz
TRANSPONDER BANDWIDTH	36 MHz
EIRP MINIMUM, BEAM EDGE	35 dBW (CONUS), 32 dBW (Alaska), 24·5 dBW (Hawaii), 23 dBW (Puerto Rico)
CHANNEL POLARISATION	
RECEIVE COVERAGE (G/T)	−7·4 dB/K (CONUS), −14 dB/K (Alaska), −14·4 dB/K (Hawaii)
SINGLE CARRIER SATURATION FLUX DENSITY	−80 dBW/sq. m. (CONUS) −72 dBW/sq. m. (Alaska/Hawaii)

COMMERCIAL DETAILS

OPERATOR	Western Union Westar Satellite Services
ADDRESS	1 Lake Street, Upper Saddle River, New Jersey, MJ 07458, USA
TELEPHONE NO.	(201) 825-5000
TELEX NO.	642499
SENIOR STAFF	Robert Leventhal, Chairman & CEO W. R. Pope, VP & COO Fred M. Knip, AVP, Transmission Systems Nancy E. Salvati, Product Manager, Westar Transmission Services J. T. Ragan, VP, Commission Systems J. J. Albert, AVP, Westar Satellite Systems Planning W. G. Bradley, AVP, Westar Satellite Services J. K. Livingstone, Senior Director, Partial Transponder Services M. M. Abrahamson, Senior Product Manager, Video Services Tim McCudy, Senior Product Manager, TI Channel Services Warren Bechtel, Corporation Commissioner Don Virgilit, SNR Product Manager, Sky Data Services May Abramsom, Senior Product Manager, Westar Video Services
PURPOSE OF PROJECT	Provision of US domestic C-band capacity
PROJECT STATUS	The system is operational but due to launch failures and lack of new launch capacity the third satellite is delayed until 1989. Western Union also has financial problems
PROJECT VALUE	Western Union has stated that the approximate cost of Westar VI was \$30 m plus \$10 m launch, \$5 m insurance and \$6–8 m for PAM–D. Insurance cover was for \$105 m
LAUNCH DATES	Westar IV, 25th February 1982, Delta 3910 F160 + PAM–D Westar V, 8th June 1982, Delta 3910 F162 + PAM–D Westar VI, 3rd February 1984, STS 41B + PAM–D (see project overview) Westar VIS, launch date not available Westar VI relaunch to be decided

SATELLITE USERS

The principal users of the Westar satellite system are summarised in the project overview. The system is used primarily for video traffic in the US domestic market.

PROJECT OVERVIEW

Introduction: The financially troubled Western Union is now relying for the rest of the decade on the Westar IV and V satellites as its space segment capacity. It appears that the company is hard pressed to finance its next generation satellite system and is sitting on the Westar VIS spacecraft as there is no capacity to launch it until 1989 at the earliest. The Westar VI satellite is currently owned by insurance underwriters, having been recovered by Shuttle after its PAM–D motor failed on the launch attempt of 3rd February 1984.

The satellite has been up for sale for two years but there has been no taker so far. The current Westar satellites are Hughes HS–393 C-band spacecraft used entirely for US domestic services. Western Union was the first US domestic satellite operator, having launched the now retired Westar I satellite in 1974. It has had plans to launch a Ku-band segment towards the end of the decade and has conditional FCC authorisation for the new system.

Background notes on Western Union's involvement in Satellites: Western Union's primary business has been record messages (better known in Europe as telegrams). Its attempts to diversify out of this dated communications technology have not offset the underlying weaknesses in the company's financial performance and, at the time of writing, it continues to make substantial losses.

Some of its plans for involvement in satellites have long fallen by the wayside. Until the early 1980's it had a 50% shareholding in the TDRSS project which was also known as

Advanced Westar. For this it formed, in 1976, the Spacecom partnership with Continental Telecom and Fairchild Industries to develop the TDRSS system for use by NASA with Shuttle. In 1983, following substantial losses in its main business activities, Western Union pulled out of the partnership. Western Union had intended to operate one TDRSS design of spacecraft on its own account, sharing a second satellite with NASA as a back up.

It was also an early DBS applicant, having received FCC authorisation in 1982 but in July 1984 it notified the commission of its intention not to begin construction of the space segment. At that time other applicants were already beginning to have doubts about the commercial prospects of DBS. Western Union was also associated with the American Satellite Company which now operates the ASC spacecraft. American Satellite had a 20% share in the Westar space segment, initially preferring to use Western Union capacity than take the risk of developing its own satellite system.

Services on the Westar System: The system is used primarily for broadcasting and cable TV distribution although both of the in-orbit spacecraft also carry voice and data traffic. Principal satellite services offered by Western Union on the Westar system are as follows:

1) Video and Audio. The users of this service are listed in the section called Summary of Satellite Users. It accounts for approximately half of the capacity on the system.

2) News Media Services. A remote printing service. Users include Dow Jones for facsimile transmission of the Wall Street Journal to East Coast, West Coast and Midwest printing plants, Time Incorporated, US News and World Report, Associated Press and UPI. Dow Jones leases two transponders on Westar V for this service.

3) Corporate Networks. Offered in conjunction with the company's terrestrial communications facilities. This means that it can offer bypass of the local telecom network by using its own rights of way and microwave links to connect the nearest earth station with the customers' premises. Customers include Citicorp, IDB and DCA.

In terms of capacity the company can offer full and partial transponder terms, radio programme services, video channel services, uplink and downlink facilities and, in addition, international services and consulting. The full transponder service offers either lease or outright sale of full transponders. A similar partial transponder service is available on lease or purchase terms. The Radio Program Service offers one way, two way or multipoint channels. Video channel services are provided on a monthly, weekly or daily basis. The company also deals in transponder time for other owners and lessees.

The international services include Eurolink, a joint venture with Mercury Communications for voice and data traffic between New York and London. It also has a shareholding in Brightstar, along with Visnews. This provides two way transatlantic transmission of video programming.

As from the Spring of 1986 Western Union has been offering TDMA services on the Westar system.

The Westar VI/VIS Saga: The Westar VI story is one of the more interesting events in the satellite world. The satellite was launched by Shuttle in February 1984 but the PAM–D transfer stage mis-fired due to a fault in the carbon fibre cone of the stage. The satellite was nudged back into an orbit capable of allowing recovery of the satellite on a subsequent Shuttle mission. The satellite was fully insured and Western Union made a claim in full, receiving $105 m from the underwriters. As a consequence, the underwriters took over ownership of the spacecraft and paid NASA to recover it, along with the Palapa B2 satellite which failed on the same mission for the same reason. The satellites were brought back to earth in September 1984 and were refurbished and stored by Hughes Aircraft Company.

The underwriters put the satellites up for sale but Western Union decided that it would purchase a new, and almost identical, spacecraft (Westar VIS) from Hughes. The baseline price being offered by the underwriters was around $30 m for each of the recovered satellites. At the time of writing only the Palapa B2 spacecraft has been sold. Western Union may be regretting not having purchased Westar VI instead of ordering Westar VIS for there is now no launch vehicle capacity available to launch it before 1989.

In the meanwhile an expensive spacecraft, which had been due for launch in mid-1986 is sitting idle, and the company publicly admits that it is short of capacity for the next few years. Amongst the potential purchasers of the Westar VI satellite have been a new US start up venture, Teresat who have a reservation to launch it on the Chinese Long March 3 vehicle but, apparently, no money to purchase it. Teresat claims that it intends to use it for US domestic communications.

The Westar VIS satellite includes modifications to give enhanced coverage of the Caribbean basin, Central America and the northern part of South America. Six transponders are allocated to this. The Caribbean service has had to be co-ordinated with Intelsat. The satellite also has greater downlink power than Westar VI and has TWTA redundancy and improved filter characteristics.

SUMMARY OF SATELLITE USERS, VIDEO TRAFFIC

WESTAR IV (99° West)
Transponder 06, World Communication and World of Faith
Transponder 08, Occasional Video
Transponder 09, Occasional Video, KSCI, CTNA
Transponder 10, World of Faith Satellite Network
Transponder 11, World Communications and Occasional Video
Transponder 14, Occasional Video
Transponder 15, Public Broadcast Service, Schedule A (Eastern)
Transponder 16, CNN contract channel and Turner Broadcasting
Transponder 17, Public Broadcast Service, Schedule B (Central)
Transponder 19, International TV Network
Transponder 20, CNTA and World Communications
Transponder 21, Public Broadcast Service, Schedule C (Mountain)
Transponder 22, Occasional Video
Transponder 23, Public Broadcast Service, Schedule D (Pacific)

WESTAR V (122·5° West)
Transponder 01, New York Racing Authority
Transponder 02, University Network
Transponder 03, Group W Satellite Communications
Transponder 05, JSIO/Brighstar
Transponder 07, CBS contract channel
Transponder 08, PASS Detroit
Transponder 11, Group W Satellite Communications
Transponder 14, Occasional Video
Transponder 15, TVSC Pittsburg and Penn Racing
Transponder 16, Caribbean Superstation
Transponder 18, TVSC and Group W Satellite Communications
Transponder 20, Synsat Hollywood and Santa Anna Racing
Transponder 21, FNN and Satellite Forum
Transponder 22, Taft Broadcasting and Meadows Racing
Transponder 24, Taft Broacasting and Fantasy Channel

(Channel 18 is also used for Sheridan Broadcasting Network radio, and the Muzak radio channel)

TECHNICAL SUMMARY

PLATFORM	HS-376
PRIME CONTRACTOR	Hughes Aircraft Company
STABILISATION	Spin
MASS IN ORBIT	585 kg
POWER SUPPLY, END OF LIFE	700 Watts
DESIGN LIFE	10 years
ORBITAL POSITIONS	Westar IV, 99° West Westar V, 122·5° West Westar VI, to be decided Westar VIS, 79° West
STATION KEEPING ACCURACY	+/−0·05 d (N/W & E/W)
ECLIPSE PROTECTION	Full
NO. OF ACTIVE TRANSPONDERS	24
NO. OF SPARE TRANSPONDERS	None
TWTA POWER OUTPUT	7·5 Watts (10 Watts for VI/VIS)
FREQUENCY RANGE:	
UPLINK	5·925–6·405 GHz
DOWNLINK	3·700–4·180 GHz
TRANSPONDER BANDWIDTH	36 MHz
EIRP MINIMUM, BEAM EDGE	CONUS 37 dBW (IV/V), 38 dBW (VIS): Alaska 31 dBW (IV/V), 32 dBW (VI): Hawaii 29 dBW (IV/V), 33 dBW (VI): Puerto Rico 27 dBW (31·5 dBW VI)
CHANNEL POLARISATION	Linear, horizontal and vertical
RECEIVE COVERAGE (G/T)	IV/V: −13·5; −3·5; −4·5; −4 dBW VI; −4; −4; −6; −6 dBW for CONUS, Alaska, Hawaii and Puerto Rico
SINGLE CARRIER SATURATION FLUX DENSITY	IV/V: −84; −75; −76; −75·5 dBW/sq m.VI: −75 to −82 dBW/Sq.m for CONUS, Alaska, Hawaii and Puerto Rico respectively

COMMERCIAL DETAILS

OPERATOR	Western Union Westar Satellite Services
ADDRESS	1 Lake Street, Upper Saddle River, New Jersey, NJ 07458, USA
TELEPHONE NO.	(201) 825-5000
TELEX NO.	642499
SENIOR STAFF	Robert Leventhal, Chairman & CEO W. R. Pope, VP & COO Fred M. Knip, AVP, Transmission Systems Nancy E. Salvati, Product Manager, Westar Transmission Services J. T. Ragan, VP, Commission Systems J. J. Albert, AVP, Westar Satellite Systems Planning W. G. Bradley, AVP, Westar Satellite Services J. K. Livingstone, Senior Director, Partial Transponder Services M. M. Abrahamson, Senior Product Manager, Video Services Tim McCudy, Senior Product Manager, TI Channel Services Warren Bechtel, Corporation Commissioner Don Virgilit, SNR Product Manager, Sky Data Services Mary Abramsom, Senior Product Manager, Westar Video Services
PURPOSE OF PROJECT	Provision of US domestic C-band capacity
PROJECT STATUS	The company has FCC authorisation for two satellites in this series of plans which take the company until the year 2000
PROJECT VALUE	Estimated at $1900 m
LAUNCH DATES	Not available

SATELLITE USERS

The Ku-band satellites in the proposals are intended for VSAT type services as well as more conventional satellite services. The C-band satellites are intended as replacements for time expired Western Union satellites. See Westar IV–VIS for further details of Western Union services.

PROJECT OVERVIEW

Summary: Financially troubled Western Union has filed with the FCC over the last three years a number of plans for its satellite systems which take the company until the end of the century. Part of the reasoning behind the filings with the FCC has been to obtain orbital slots at a time (now past) when there was considerable concern as to whether enough orbital slots would be available for either C-band or Ku-band US domestic satellite systems. More cynically, some observers have commented that the company made the filings early to avoid a considerable increase in FCC charges for processing applications. The filings are for both C-band and Ku-band systems for launch up until the year 2000. The first filing was for a Ku-band system, originally known as Westar IX–XI but now known as Westar A–C.

The company was granted conditional FCC authorisation in 1985 to construct, launch and operate two satellites in this series. The overall design of the satellites is close to that of the RCA Satcom K spacecraft, providing 16 medium power Ku-band transponders each with 54 MHz bandwidth and providing an edge of CONUS EIRP of 41 dBW. The FCC has granted orbital slots at 73° and 132° West for these satellites. The company has also filed for three replacements for these satellites for launch in 1998 to 1999. They are known as Westar D–F. Its plans for C-band satellites cover two systems. Westar VII–XIII are for launch between 1986 and 1994 and each satellite would carry 24 36 MHz transponders available for lease or sale to broadcasters, private networks, message services, occasional video and transborder services. The final series is Westar XIV–XVII for launch between 1995 and 2000. These are again C-band birds.

TECHNICAL SUMMARY

PLATFORM	
PRIME CONTRACTOR	Ford Aerospace
STABILISATION	3-axis
LAUNCH MASS	2580 kg
MASS IN ORBIT	1400 kg
POWER SUPPLY, (BOL)	4800 Watts
DESIGN LIFE	10 years
ORBITAL POSITIONS	101° West 148° West
STATION KEEPING ACCURACY	+/−0·1° E/W
ECLIPSE PROTECTION	2 half-CONUS service area channels minimum
NO. OF CHANNELS PER SATELLITE	6 service areas, 12 spot beam channels. Revised to 16 operational per service area
NO. OF ACTIVE TRANSPONDERS	16
TWTA POWER OUTPUT	6 at 200 Watts } Revised to 16 12 at 20 Watts } at 100 Watts
TRANSPONDER BANDWIDTH	24 MHz
EIRP MINIMUM, BEAM EDGE	54 dBW Service area 54–58 dBW Spot
CHANNEL POLARISATION	Circular
RECEIVE COVERAGE (G/T)	0 dB/K Service area 10 dB/K Spot Beam
SINGLE CARRIER SATURATION FLUX DENSITY	-75 to $-106\,dBW/m^2$
REDUNDANCY	9:6
SUBCHANNELS	Multiple high fidelity (15 kHZ) audio channels, 56 kb/s data, 1·544 Mb/s data

COMMERCIAL DETAILS

OPERATOR	Satellite and International Affairs Office, Telecommunications Company of Iran
ADDRESS	Dr Shariati Avenue, Tehran 16314, Iran
TELEPHONE NO.	+98 21 81 11
TELEX NO.	212799 tci ir
SENIOR STAFF	Ahmad Reza Sherafat Gholam Ali Zoghi
PURPOSE OF PROJECT	Provision of domestic (or regional?) Ku-band capacity
PROJECT STATUS	RFP due to be issued in the Autumn of 1986
PROJECT VALUE	Not available
LAUNCH DATES	Not before 1989–90

SATELLITE USERS

Not available at this stage. The country has two (now) state owned TV channels. TVI was nationalised in 1969 and operates on 525 line NTSC standards. NIT, founded as a public service broadcast channel in 1966, uses 625 PAL standards.

PROJECT OVERVIEW

Iran was planning to issue a Request for Proposal (RFP) in the Autumn of 1986 for a four satellite Ku-band domestic system. It is being issued by the Telecommunications Company of Iran's Satellite and International Telecommunications Affairs Office. The project dates back to pre-revolution days and has been in the pipeline again recently. Advanced notification of the current frequency and orbital slot requirements have been published by the ITU. Orbital slots planned are at 26°, 34°, 41° and 47° East. The satellite system is known as Zohreh.

Past indications are that the country has been aiming for a 1988 start to operations of the system. However, it appears that the Iranians are concerned about their domestic ability to handle the programme and say that they are contacting satellite and ground equipment manufacturers for help in assessing state of the art technology. Moscow News has announced that it is having discussions with the Iranians about supplying both the satellites and the launch vehicle services. It is understood that the Iranians are looking for a medium power FSS system with 40 Watt transponders.

Iran signed a letter of intent in 1986 with China's Great Wall Industry Corporation for use of the Long March vehicle to launch its proposed satellites.

Glavcosmos, now responsible for all cooperative and commercial space ventures between the Soviet Union and the rest of the world, has said that it is discussing with Iran the lease of a C-band Girozont satellite, possibly one that is already in orbit.

TECHNICAL SUMMARY

PLATFORM	not available
PRIME CONTRACTOR	not available
OTHER MAJOR CONTRACTORS	not available
STABILISATION	not available
LAUNCH MASS	not available
MASS IN ORBIT	not available
POWER SUPPLY, END OF LIFE	not available
DESIGN LIFE	not available
ORBITAL POSITIONS	26° East, 34° East, 41° East, 47° East
STATION KEEPING ACCURACY	not available
ECLIPSE PROTECTION	not available
PROPULSION SYSTEM	not available
SPACECRAFT DIMENSIONS	not available
SPAN, TIP-TO-TIP	not available
NO. OF ACTIVE TRANSPONDERS	not available
NO. OF SPARE TRANSPONDERS	not available
TWTA POWER OUTPUT	not available
FREQUENCY RANGE:	Ku-band
UPLINK	not available
DOWNLINK	not available
TRANSPONDER BANDWIDTH	not available
EIRP MINIMUM, BEAM EDGE	not available
CHANNEL POLARISATION	not available

SECTION III

LAUNCH VEHICLES

CURRENT OPERATIONAL STATE OF VEHICLE	Grounded until February 1987 with a fundamental design flaw in the third stage ignition process
LAUNCH VEHICLE OPERATOR	Arianespace
ADDRESS	Boulevard de l'Europe, BP 177, F–91006, Every Cedex, France
TELEPHONE NO.	33 (01) 60.87.60.00
TELEX NO.	692 392 aresp f
SENIOR STAFF	Frederic d'Allest, Chairman & CEO Charles Bigot, President Roland Deschampes, Secretary Klaus Iserland, Deputy DG, International Affairs Jean-Claude Biget, VP Marketing Xavier Picard, VP, External Relations Jacques Boloh, VP, Planning Claude Quievre, Director, Engineering Hubert Palmieri, VP, Operations Douglas Heydon, Washington Office Patrice Larcher, Marketing
PRIME CONTRACTOR FOR LAUNCH VEHICLE	CNES
OTHER MAJOR CONTRACTORS	Aerospatiale (industrial architect), Air Liquide (third stage tank), BAe (Spelda), Contraves (nose fairing), Matra (vehicle equipment bay), MBB–ERNO (second stage), SEP (propulsion system, first, second and third stage), SNIA BPD (solid propellant boosters and separation rockets)
PAYLOAD CAPABILITIES	Ariane 1: 1825 kg to GTO Ariane 2: 2175 kg to GTO Ariane 3: 2580 kg to GTO or two payloads of 1195 kg to GTO Ariane 4 AR40: 1900 kg to GTO Ariane 4 AR42P: 2600 kg to GTO Ariane 4 AR44P: 3000 kg to GTO Ariane 4 AR42L: 3200 kg to GTO Ariane 4 AR44LP: 3700 kg to GTO Ariane 4 AR44L: 4200 kg to GTO
PAYLOAD LAUNCH COSTS	Due to a shortage of capacity in the satellite launching market, launch charges have increased considerably in 1986. The costs quoted below may be an underestimate. Basic cost of an Ariane launch of a PAM–D class satellite is $30–36 m on a dual launch. The Ariane 4 price is based on a linear weight relationship. Charge for a 2500 kg payload on Ariane 4 is $48 m. These are current charges and exclude S3R or any other insurance costs
FIRST LAUNCH OF VEHICLE	24th December 1979
FIRST COMMERCIAL PAYLOAD LAUNCH	20th December 1981
NEXT AVAILABLE PAYLOAD OPPORTUNITY	Spring/Summer 1990
MAXIMUM LAUNCH RATE	8 vehicles per year
LAUNCH SITE(S)	Centre Spatial Guyanais, Kourou, French Guyana with two launch pads, ELA–1 and ELA–2

Introduction: The Ariane launch vehicle is Europe's only launcher and is a conventional three stage expendable vehicle. It was nominally developed by the European Space Agency but 60% of the funds, and 60% of the work were provided by France. The vehicle has had a poor reputation for reliability and its launch manifest has slipped frequently. At the time of writing, the vehicle is grounded until February 1987 following two failures in the ignition sequence of the third stage. Nevertheless, there is a considerable order book for launches on the vehicle until 1990. It is due to be phased out in the mid-1990's in favour of an entirely new vehicle, Ariane 5.

Background to the Ariane Vehicle: Despite being plagued by technical problems, the Ariane vehicle has proved to be one of the major commercial successes of the European aerospace industry. It has achieved commercial viability by exploiting the market gap caused by problems with the Shuttle programme and the run down of the US Expendable Launch Vehicle programmes which Shuttle was supposed to replace. It is currently the only Western launch vehicle in production designed and intended to carry commercial communications satellites.

The vehicle is in a commercial position which was not envisaged when it was originally conceived in the early 1970's. At that time the primary interests of the Europeans was to achieve independence from US launch services. Ariane was primarily intended to launch two or three European payloads per annum, primarily ESA scientific or development payloads. It became apparent in the late 1970's that the Shuttle programme was already in difficulties and that its launch rate would be considerably lower and its launch costs much higher than envisaged. (For example, in 1975, it was planned that Shuttle would be launched over 500 times by 1986 – so far it has managed 25 launches.) It also became apparent in the late 1970's that the demand for launch services was growing rapidly, with conventional FSS satellites proving to be commercially viable for domestic communications, with the Intelsat organisation growing rapidly and the prospect of a new major industry, DBS TV, creating even more demand.

To take advantage of this, Ariane production rates were stepped up to the current level of eight vehicles per year and the marketing of the vehicle put on a commercial basis through the formation of Arianespace. This increase in production rate and the move to a commercial type of operation is seen by some industry observers as the principal reason why the vehicle has had such a poor reliability record. The production of the third stage motor by SEP had not fully moved from a development, one-off type production, to serial production in quantity. This became apparent when the French government arranged for the company to be taken over by SNECMA, the French aero-engine manufacturer. When Ariane V18 failed on 29th May 1986, it became public that SEP was still struggling to turn out its third stage machinery and did not have a spare unit in hand. In other words the equipment was still coming off the production line late and straight onto the Ariane vehicle.

The delays in launch subsequent to the loss of Ariane V18 have given some breathing space and it appears that SEP was close to solving its problems when V128 went down.

The Ariane vehicle is technically unambitious and has limited scope for stretching beyond the Ariane 4 version. Europe's technical ability in launch vehicle systems has fallen considerably from its position in the late 1950's and early 1960's when the UK was still trying for total independence from the USA in its nuclear deterrent field, including ICBM capacity. It failed in this objective but looked to the satellite launch vehicle scene to apply the otherwise wasted R&D expenditure. Out of this grew the Europa 1 and 2 launch vehicles developed by an organisation known as ELDO (European Launcher Development Organisation). The Europa project was plagued with technical problems, particully due to the form that the ELDO organisation took. Each stage of the vehicle was the development responsibility of a different country and there were problems in overall management of the project.

In 1973, ELDO collapsed, leaving Europe without any prospect of an independent launch vehicle capability. The French were particularly concerned about this. The USA was the only alternative source of launch vehicle capacity and was unlikely to launch European satellites which threatened their own technological dominance or markets. The French devised a scheme where overall control of development of a new launch vehicle would be in the hands of one country (France needless to say), overcoming the organisational problems of ELDO. The British government had had enough of launchers and was not willing to participate in any significant way. France let it be known that it would go ahead with the project if no other country was willing to participate.

In the event others joined them under the framework of the European Space Agency, formally established in 1975. Participation in the Ariane programme was as follows:

West Germany	20·12%
Belgium	5·00%
Denmark	0·50%
Spain	2·00%
France	63·87%
Netherlands	2·00%
Italy	1·74%
United Kingdom	2·47%
Sweden	1·10%
Switzerland	1·20%

The first Ariane vehicle flew in 1981. The first four flights were demonstration flights handled entirely by ESA and the R&D establishment behind the vehicle. From flight five, Arianespace began to take over control of the commercial side of the programme although ESA still provides the R&D costs needed to upgrade the vehicle through its various marques and CNES provides the launch facilities in French Guiana. Arianespace took over full control from Ariane V9 on 22nd May 1984.

Launch Record of Ariane: First launch of Ariane took place in 1981 and the nineteenth launch is due in February 1987. There have been four launch failures (F2, F5, F15 and F18) and a problem with the upper stage on F6 which damaged a small AMSAT German satellite when, after separation, the stage caught up with the satellite. All of the failed launches involved commercial payloads. The failures of F15 and F18 were directly attributed to problems with the ignition sequence of the third stage. This is currently subject to redesign and testing. The third stage motor is the most technically advanced element of the vehicle.

Versions of the Ariane Vehicle: There are four different marques of Ariane, divided into two families, Ariane 1–3 and Ariane 4 40–44L. So far, all launches of Ariane have used Ariane 1–3. These vehicles are now being phased out and replaced by Ariane 4. The first launch of the latter is V22 in mid-1987 and from 1989 it will be the only vehicle available. Ariane 4 comes in six different versions, which vary according to which combination of strap on boosters are used:

Ariane 40 is the basic version and has a payload performance equivalent to Ariane 2 at 1900 kg to GTO. It has no strap on boosters.

Ariane 42P is the equivalent of Ariane 3, carrying 2600 kg to GTO. It uses two solid strap on boosters.

Ariane 44P (four solid strap on boosters), 42L (two liquid strap-ons), 44LP (two liquid and two solid strap-ons) and 44L (four liquid boosters) have a payload capability of 3000, 3200, 3700 and 4299 kg to GTO respectively.

The Ariane 5 vehicle has little in common with Ariane 1–4 and is an entirely new vehicle. It is expected to be man rated to carry the Hermes manned spaceplane to LEO (18,000 kg in weight).

Ariane 3 is identical to Ariane 2 except that the former has strap on solid boosters.

Charges for use of the Ariane Vehicle: Arianespace is the marketing agent for the Ariane vehicle with responsibility for finance, production, marketing and launching of the vehicle. The company is owned by contractors for the Ariane vehicle and European banks. The first Ariane 4 vehicle is the responsibility of ESA. Transpace Carriers Incorporated of the USA, aiming to commercialise the Delta launch vehicle, and having lost a launch contract to Ariane, petitioned that Arianespace was practising predatory pricing in offering lower prices to US than captive European customers. In the absence of details of the economic cost of each Ariane vehicle and the known presence of substantial subsidies for Shuttle launches, the claim was complex.

Arianespace admitted its pricing policy is flexible. Actual prices charged can vary to within 10% according to negotiations and reflect currency fluctuations and specific launch requirements. Arianespace was eventually cleared of predatory pricing, primarily because Shuttle was so heavily subsidised by the US taxpayer.

In February 1986, Arianespace announced a 20% increase in its launch charges. This increase was a direct result of the grounding of Shuttle of the loss of Challenger on 28th January 1986, although Arianespace claimed that it reflected currency changes between the US$ and the Franc. After the loss of Shuttle, Arianespace had virtually a captive market.

Arianespace's pricing policy is somewhat difficult to follow. Ariane 1–3 launch costs are based on a straight cost per satellite. These costs were, in 1985, roughly \$25–30 m per satellite of the PAM–D class on a dual launch. As there are no longer any available launch slots Ariane 1–3 and the vehicle is now being phased out, the increased 1986 charges no longer apply to new customers. SES is paying \$37 m for its Ariane 3 launch but this includes insurance cover. Moreover, it is a European customer.

Arianespace has now adopted a linear element in pricing, such that for an Ariane 4 dual launch, price starts at FFr260 m (\$26 m) for a 1140 kg satellite to FFr480 m (\$48 m) for a 2500 kg spacecraft. Dedicated Ariane 4 launch charges are based on a single cost per launch. However, because there are several marques of this vehicle, each suitable for different payload weights, the price varies on a stepped scale. The basic version AR40 now costs FFr480 m (\$48 m) and AR44L FFr600 m (\$60 m). Ariane prices are also affected by changes in foreign exchange rates and subject to variations in individual customer requirements. The highest launch charge to date is that for the Intelsat Vi F3 satellite finalised in September 1986. This was for a 1989 dedicated Ariane 4 launch and cost Intelsat \$84·1 m, including S3R launch insurance coverage. The latter is optional. The new Intelsat charge compares with \$40 m for each of the Intelsat VA dedicated Ariane II launches.

The terms for launching satellite on Ariane use a best effort clause. This means that Arianespace is neither obliged to offer a relaunch (unless under the S3R arrangements) or reimburse the satellite owner if the launch vehicle fails. Moreover, Arianespace is not obliged to pay any compensation if the launch manifest slips. It can also bump customers off their designated launch under certain conditions. The European Space Agency has priority in launches if it books more than three years in advance. Under three years its priority is subject to modifications.

A launch reservation on Ariane costs \$100,000. Progress payments are required once the firm order is signed.

Arianespace claims that the use of the Kourou launch site gives a payload performance to GTO 12% better than if the vehicle was launched from the Kennedy Space Centre.

Future Prospects for Ariane: Arianespace is currently riding high with its largest order book to date. It has no available payload opportunity until the Spring or Summer of 1990. None of the potential US operators of Expendable Launch Vehicles is likely to be able to take a confirmed booking for a launch until

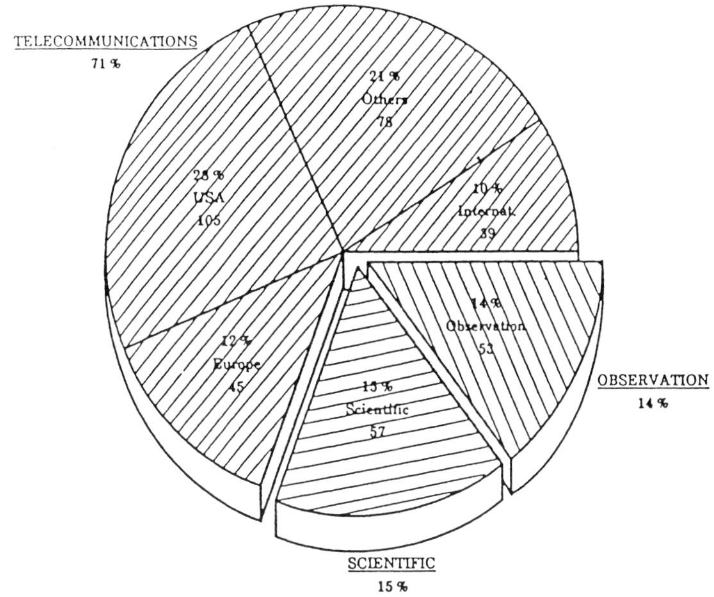

Telecommunications 267				Observation 53	Scientific 57	Total 377
Europe	USA	Others	Internat.			
45	105	78	39	53	57	377
12%	28%	21%	10%	14%	15%	100%
71%				14%	15%	100%

Distribution of the commercial market in number of satellites

Sector	Regions	Satellites	%	Arianespace Part Max.	Min.
Telecommunications	—	88	40	36	31
	Europe	13	80	11	10
	USA	35	25	9	7
	Others	29	38	11	10
	International	11	40	5	4
Observation		21	40	9	7
	Europe	8	80	7	6
	USA	7	3	—	—
	Others	6		2	1
Scientific		4		2	2
Total		113	40	47	40
Number of satellites per year				10	8

the early Spring of 1987 and is also unlikely to be able to launch a commercial satellite until mid-1989. For some years after that there is likely to be a shortage of capacity. Moreover, any newly reopened US ELV production line may have teething troubles and make satellite operators uneasy about being early users. This indicates that Arianespace is likely to be taking a lot of orders for the next few years. The company says that it has no intention of increasing its production and launch rate above eight vehicles per year. On a cost basis, an established Ariane vehicle is likely to remain competitive with the commercialised US ELV industry.

The first modern launch vehicle (apart from Shuttle) will be the Japanese H–II vehicle, due to fly in 1992. This places the most significant threat to Arianespace in the mid-1990's. The Ariane 5 vehicle, using a much more dated open cycle engine technology than the H–II vehicle, will be available to Arianespace from the mid-1990's offering, it is planned, a reduction of 20% in payload costs to GTO. Nevertheless, Ariane 5 is not optimised for launch of satellites. Like Shuttle, it is aimed at serving a variety of space transportation needs including manned spaceflight and large payloads for the International Space Station.

It may place Arianespace in the position of operating a launcher with problems created by its design for other missions. It is unlikely that Shuttle will return to the commercial satellite launch vehicle market. There is likely to be some limited competition from the Chinese Long March 2, 3 and 4 vehicles and possibly from Soviet launchers (Proton).

Arianespace is likely to retain its partial hold over its domestic market. Under terms of membership of the European Space Agency, European countries are obliged to give preference to Ariane in launching their satellites.

According to a paper presented in late May 1986 by Arianespace's Marketing Manager, Patrice Larcher, it was estimated by the company that until the year 2000 there would be 380 satellites available for launch on Ariane. Of these, nearly three quarters would be telecommunications satellites. The market is summarised on page 319.

THE ARIANE LAUNCH VEHICLES SERIES

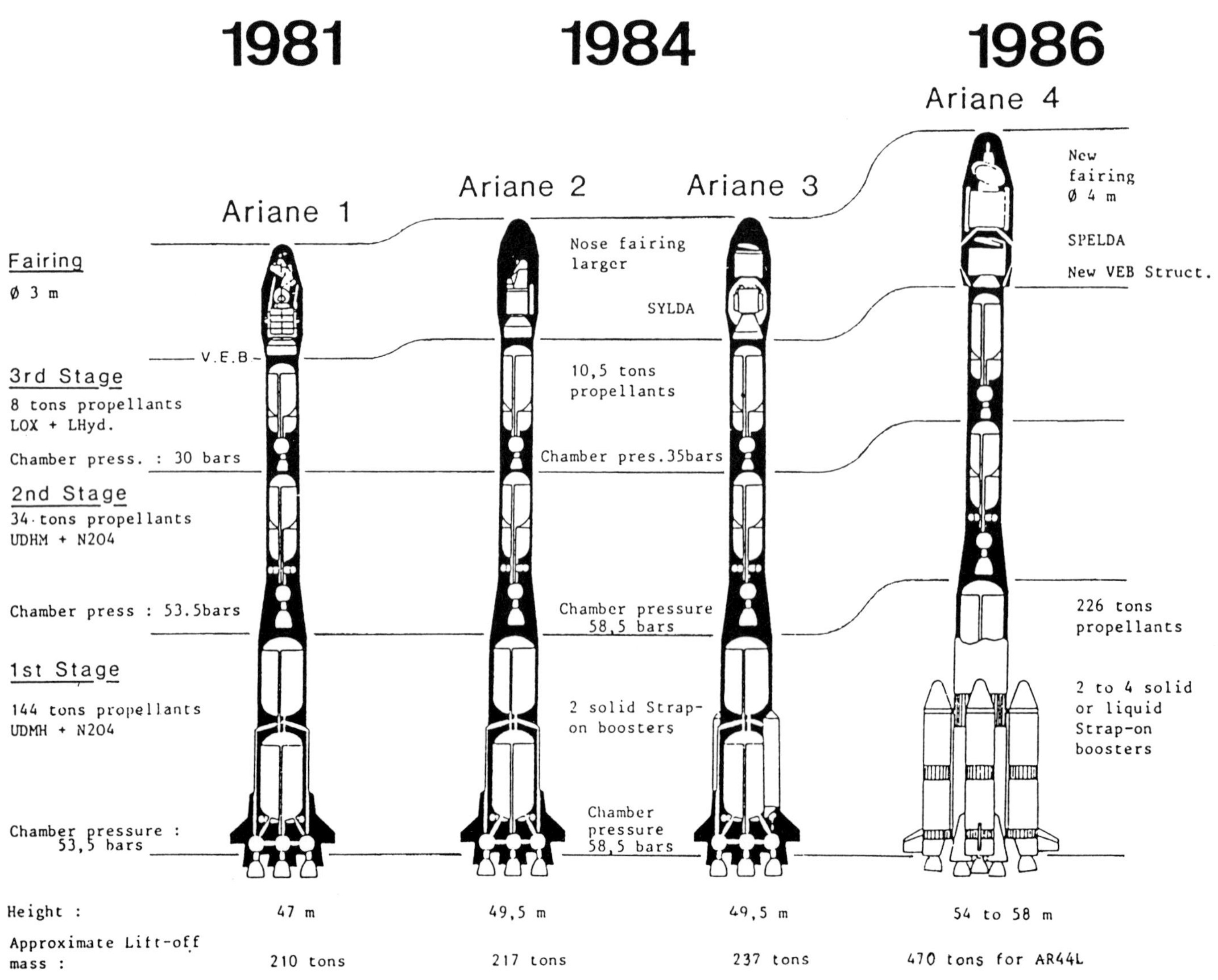

THE DIFFERENT VERSIONS OF ARIANE 4

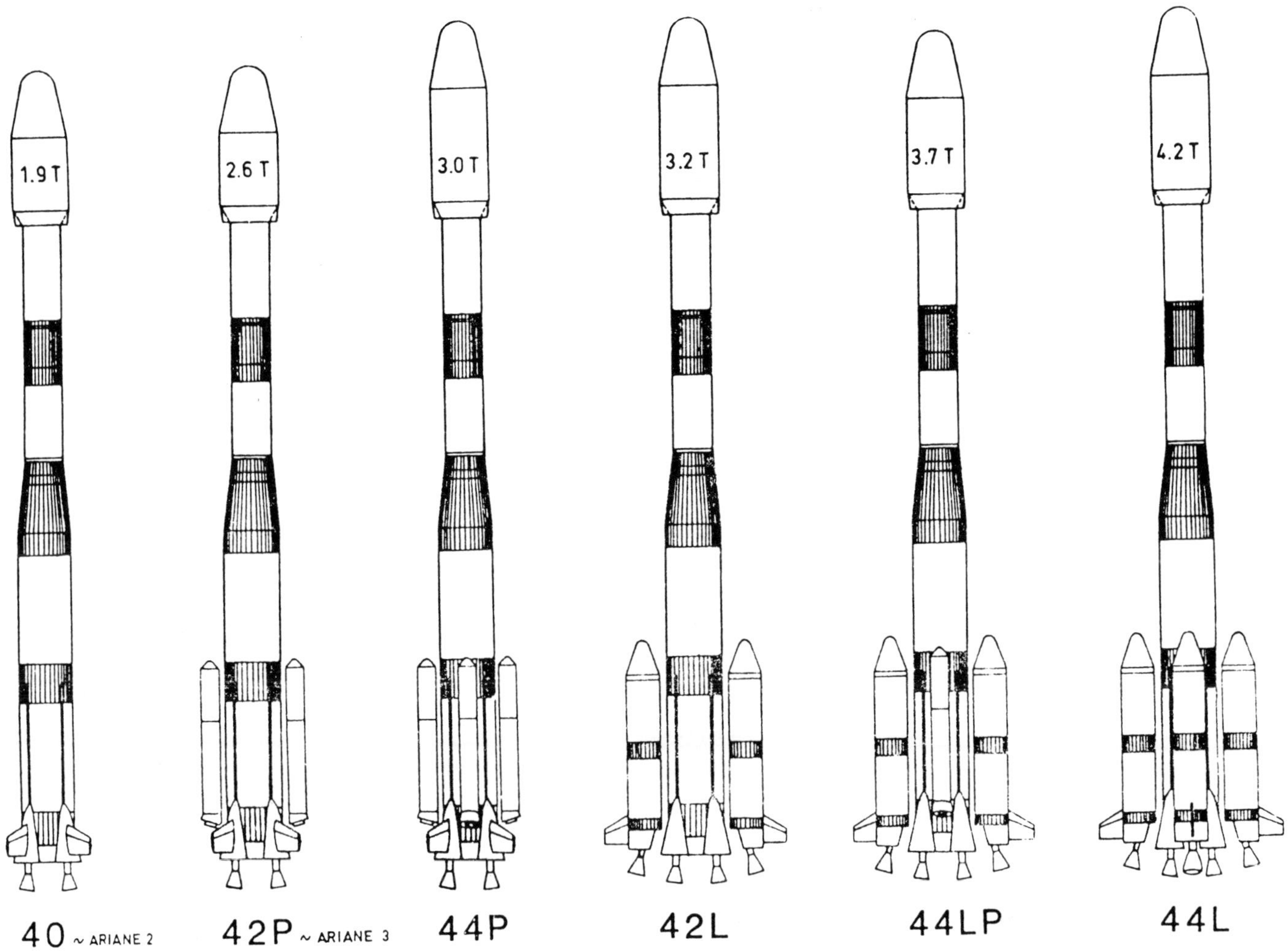

Launch	Date	Payload
Ariane 1 L01	24th December 1979	Technological Ariane Capsule CAT
Ariane 1 L02	23rd May 1980	Cat, Firewheel and Oscar 9 (launch failure)
Ariane 1 L03	19th June 1981	CAT, Meteosat 2 and Apple
Ariane 1 L04	20th December 1981	CAT and MARECS A
Ariane 1 L5	9th September 1982	MARECS B and SIRIO 2 (launch failure)
Ariane 1 L6	16th June 1983	ECS–1 and Amsat Phase IIIB (partial launch failure)
Ariane 1 L7	18th October 1983	Intelsat V F7
Ariane 1 L8	5th March 1984	Intelsat V F8
Ariane 1 V9	22nd May 1984	Spacenet 1
Ariane 3 V10	4th August 1984	Telecom 1A and ECS–2
Ariane 3 V11	9th November 1984	Spacenet 2 and MARECS B2
Ariane 3 V12	8th February 1985	Brasilsat S1 and Arabsat F1
Ariane 3 V13	8th May 1985	GStar 1 and Telecom 1B
Ariane 1 V14	2nd July 1985	Giotto
Ariane 3 V15	12th September 1985	Spacenet 3 and ECS–3 (launch failure)
Ariane 1 V16	22nd February 1986	Spot 1
Ariane 3 V17	31st March 1986	Brasilsat S2 and GStar 2
Ariane 2 V18	30th May 1986	Intelsat V F14 (launch failure)
Ariane 3 V19	February 1987	ECS–4 and Aussat (or GStar III)
Ariane 2 V20	April 1987	TV–SAT
Ariane 4 V21	June 1987	Meteosat P2 and Amsat and Panamsat (Apex 401, Flight 401)
Ariane 3 V22	July 1987	GStar III (or Auusat K3) and SBS 5
Ariane 2 V23	August 1987	Intelsat V F13
Ariane 3 V24	September 1987	(Telecom 1C and Spacenet F3R) or TDF–1
Ariane 2 V25	November 1987	TDF–1 or (Telecom 1C and Spacenet F3R)
Ariane 3 V26	January 1988	ECS–5 and Insat IC
Ariane 4 V27	February 1988	Astra 1 and MOP 1
Ariane 2 V28	March 1988	Intelsat V F15
Ariane 4 V29	May 1988	Skynet IVB and TELE–X or JC SAT–1
Ariane 2 or 3 V30	June 1988	TELE–X or Olympus
Ariane 4 V31	July 1988	DFS–Copernicus 1 and JC SAT–1 or Hipparcos
Ariane 2 or 3 V32	October 1988	Olympus or Spot 2
Ariane 4 V33	November 1988	Superbird 1A and Hipparcos or Inmarsat II F1
Ariane 4 V34	Janauary 1989	Intelsat VI F1
Ariane 4 V35	March 1989	SBS–6 and Inmarsat II F1 or DFS–Copernicus 2
Ariane 4 V36	April 1989	(DFS–Copernicus 2 and TDF–2) or Spot 2
Ariane 4 V37	May 1989	Superbird 1B and Geostar or Inmarsat II F2
Ariane 4 V38	June 1989	Intelsat VI F2
Ariane 4 V39	July 1989	Eutelsat IIA and MOP 2
Ariane 4 V40	September 1989	RCA K3 and Inmarsat II F2 or Skynet IVC
Ariane 4 V41	October 1989	TDF–2 or TV–SAT 2 and Skynet IVC or ASC–2
Ariane 4 V42	November 1989	Intelsat VI F3
Ariane 4 V43	Janauary 1990	Eutelsat IIB and TV–SAT 2 or ASC–2

SALES PROSPECTS LIST FOR ARIANESPACE

Likely Launch Date	Satellite	Customer
1990	Italsat	CNR
1990–91	Intelsat VI F4	Intelsat
1990–92	Intelsat VI F5	Intelsat
1990–92	Eutelsat II F3	Eutelsat
1990–91	Telecom IIA	DGT France
1990–92	Telecom IIB	DGT France
1988–89	Mailstar 1	Mailstar AB Sweden
1989–90	Mailstar 2	Mailstar AB Sweden
1991–92	Spot 3	Spot Image
1992–93	Spot 4	Spot Image
1990–91	Astra 2	Société Européenne des Satellites
1990–91	Italsat	Italian National Space Agency
1990–91	JC–SAT 2	Japan Satellite Communications
1990–91	Inmarsat II F3	Inmarsat
1989	PacSat	TRT/Papua New Guinea

ARIANE LAUNCH RECORD

FLIGHT No.	L01	L02	L03	L04	L5	L6	L7	L8	V9
LAUNCH PAD	ELA 1	ELA 1	ELA 1	ELA 1	ELA 1	ELA 1	ELA 1	ELA 1	ELA 1
LAUNCH VEHICLE	AR1	AR1	AR1	AR1	AR1	AR1	AR1	AR1	AR1
LAUNCH DATE (U.T.)	24-12-79	23-05-80	19-06-81	20-12-81	10-09-82	16-06-83	19-10-83	05-03-84	23-05-84
NAME (MISSION)	CAT/BALLAST (TECHNOLOGICAL)	FIREWHEEL (SCIENTIFIC) · OSCAR 9 (RADIO AMATEURS) · CAT (TECHNOLOGICAL)	METEOSAT (METEOROLOGICAL) · APPLE (COMM. & TECHNO) · CAT (TECHNOLOGICAL)	MARECS-A (MARITIME COMM.) · VID (TECHNOLOGICAL) · CAT & THEBEE (TECHNO & SCIENTIFIC)	MARECS-B (MARITIME COMM.) · SIRIO-2 (COMM. & TIME SYNCHRO)	ECS-1 (COMMUNICATIONS) · AMSAT Ph III C (RADIO AMATEURS)	INTELSAT V F7 (COMMUNICATIONS)	INTELSAT V F8 (COMMUNICATIONS)	SPACENET 1 (COMMUNICATIONS)
OVERALL MASS (kg)	1646	1649	1678	1690	1617	1301	1861	1861	1737

FLIGHT No.	V10	V11	V12	V13	V14	V15	V16	V17	V18	V19
LAUNCH PAD	ELA 1	ELA 1	ELA 1	ELA 1	ELA 1	ELA 1	ELA 1	ELA 2	ELA 1	ELA 1
LAUNCH VEHICLE	AR3	AR3	AR3	AR3	AR1	AR3	AR1	AR3	AR2	AR3
LAUNCH DATE (U.T.)	04-08-84	10-11-84	08-02-85	08-05-85	02-07-85	12-09-85	22-02-86	28-03-86	31-05-86	
NAME (MISSION)	ECS-2 (COMMUNICATIONS) · TELECOM 1A (COMMUNICATIONS)	SPACENET 2 (COMMUNICATIONS) · MARECS-B2 (MARITIME COMM.)	ARABSAT-F1 (COMMUNICATIONS) · BRASILSAT-F1 (COMMUNICATIONS)	G-STAR 1 (COMMUNICATIONS) · TELECOM 1B (COMMUNICATIONS)	GIOTTO (SCIENTIFIC)	SPACENET 3 (COMMUNICATIONS) · ECS-3 (COMMUNICATIONS)	SPOT 1 (EARTH OBSERVATION) · VIKING (SCIENTIFIC)	G-STAR II (COMMUNICATIONS) · BRASILSAT-F2 (COMMUNICATIONS)	INTELSAT V F14 (COMMUNICATIONS)	SPACENET 3 (COMMUNICATIONS) · ECS-8 (COMMUNICATIONS)
OVERALL MASS (kg)	2876	2446	2997	2848.4	1068.5	2566.1	2477	2831.1	2004.3	

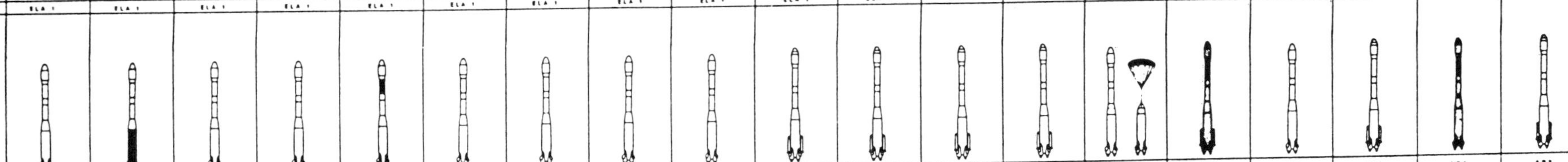

PAYLOAD CONFIGURATION

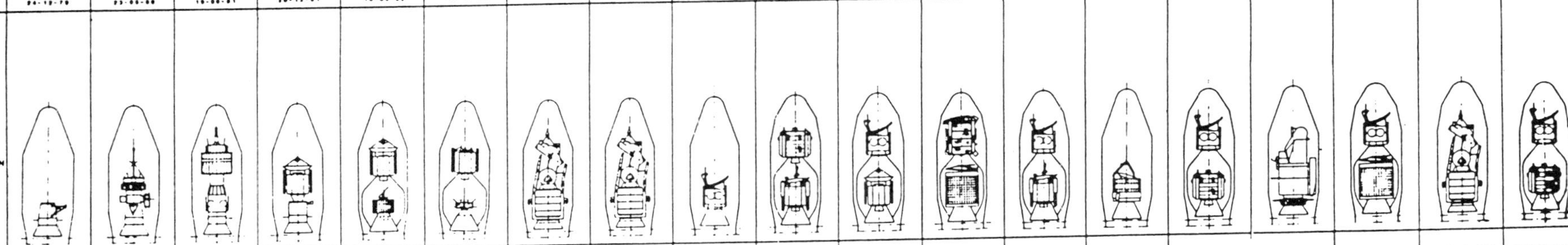

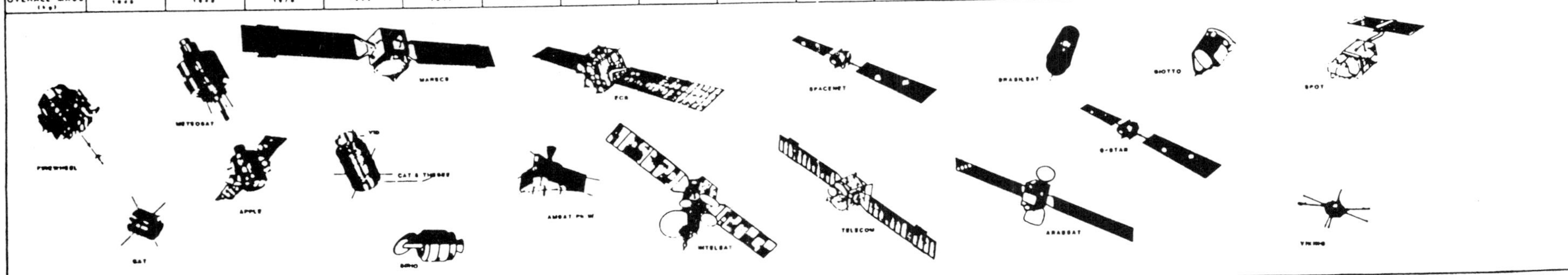

CURRENT OPERATIONAL STATE OF VEHICLE	Operational, but only two vehicles left and production line closed (see text)
LAUNCH VEHICLE OPERATOR	Currently NASA but possibly General Dynamics and the USAF. General Dynamics (proposed), Space Systems Division
ADDRESS	PO Box 85357, MZ-16-9505, San Diego, California 92138, USA
TELEPHONE NO.	(619) 573-8000
TELEX NO.	
FAX NO.	
SENIOR STAFF	General Dynamics: Lee R. Scherer, Director, Commercial Space Roger Lynch, Director of Commercial Space William F. Rector, VP & Program Director, Space J. Isabel, Corporate Manager, News and Information
PRIME CONTRACTOR FOR LAUNCH VEHICLE	General Dynamics
PAYLOAD CAPABILITIES	Atlas G/Centaur: 2360 kg to GTO Atlas K: 1280–1460 kg to GTO Atlas H: 1220 kgs to LEO Atlas–Centaur MELV version: 2400 kg to GTO and 5500 kg to LEO
PAYLOAD LAUNCH COSTS	The final Intelsat launches by Atlas–Centaur cost $61 m each. General Dynamics has not announced a price for commercialised launches. It is expected that dual launches will be available on Atlas–Centaur
NEXT AVAILABLE PAYLOAD OPPORTUNITY	For Atlas–Centaur, not before 1989
LAUNCH SITE(S)	Kennedy Space Center (complex 36B) for Atlas–Centaur, Vandenberg for Atlas H and E

Current Position of the Atlas–Centaur Vehicle: As at September 1986, there were two Atlas–Centaur launch vehicles in final assembly and the production line has been closed. The two vehicles are intended for the launch of two Fleet Satellite Communications (FleetSatCom) spacecraft for the US Navy. Though the production line is now closed tooling and equipment remain in existence. According to some industry observers, the Atlas–Centaur vehicle and Atlas variants are the most likely of US ELV's to be commercialised. General Dynamics has been trying since 1984 to turn the vehicle into a commercial operation.

Background to the Atlas–Centaur Vehicle: Atlas has been one of the workhorses of the US space industry since the early 1960's. It began as an ICBM and first flew in 1957. However, General Dynamics first launched the MX–774 ICBM in 1951, from which Atlas ICBM was derived. It was man-rated in the 1960's when it was used in the Mercury missions. The vehicle was due to have been phased out by 1981 in favour of Shuttle launches, but production was extended when it became apparent that Shuttle would be late. The final version of the vehicle to appear was the Atlas G/Centaur.

General Dynamics' interest in commercialising the vehicle dates back to 1983 when President Reagan announced that the country's space policy would "fully endorse and will facilitate commercial operations of expendable launch vehicles by the US private sector... it is the US governments intent to establish a full cost recovery for commercial and foreign STS (Shuttle) operations" starting in fiscal year 1989. At that time forecasts indicated that the total number of launches of commercial satellites between 1983 and 1986 would equal the total of all previous such launches. It was still not clear that there would be minimal demand for DBS commercial launches.

General Dynamics took to the bait and set up an organisation called CommSpace to plan commercialisation of the Atlas–Centaur vehicle. Space Services Incorporated of Houston (who are still involved with launch vehicles through the Conestoga project) also expressed an interest at the time. For the following three years General Dynamics were active in trying to sell the vehicle. They needed to launch at least two vehicles per year to make it worthwhile to continue production and operate it on a commercial basis. It was not until the loss of Challenger in January 1986 that the company's proposals began to be taken very seriously by commercial satellite operators. But by that time the Atlas–Centaur production line had closed down and the remaining vehicles were those in stock for NASA.

In July 1986, it was announced by the USAF that the organisation wished to procure twelve Medium Expendable Launch Vehicles (MELV's) and that a study contract had been placed with General Dynamics for a variation on the Atlas–Centaur design. The move was seen as part of a governmental strategy to place the ELV launch vehicle market on a sound footing by giving an initial captive market to a commercial ELV operator. General Dynamics followed up the move with an announcement that it had been approached by a number of commercial satellite operators to launch at least 21 satellites with the vehicle.

The placement of the contract for the MELV system is not expected until February 1987 and, at the time of writing, General Dynamics had made no public announcement that it intends to go ahead with commercialisation whether or not it wins the MELV contract. It is, however, apparent that the Atlas–Centaur vehicle is a favoured choice by the USAF.

General Dynamics Proposals for a Commercialised Atlas–Centaur: The company has been offering two versions for launch of commercial satellites. The first is a slighly modified version of the current Atlas–Centaur and the second, aimed at the requirements of smaller commercial satellites is a new variant known as Atlas K. This would use a variety of solid perigee kick motors instead of the large Centaur stage. Both the improved Atlas–Centaur and the Atlas K vehicles would use new options for fairings including a four metre fairing developed by Contraves of Switzerland for the Ariane 4 vehicle. For the MELV requirements for the USAF, General Dynamics is proposing a version of the Atlas–Centaur launch vehicle with minimal modifications. It would be able to launch about 5500 kg to low earth orbit and 2400 kg to GSO.

Current and Proposed Versions of the Atlas Vehicle: Since 1983 a number of versions of the Atlas vehicle have been proposed for commercial operations. In practice, most of these have centred around the Atlas G/Centaur version which was introduced in 1984.

1) Atlas G/Centaur. Atlas G is 81 inches longer than its immediate predecessor, SLV–3D, and develops 438,000 lbs (1950 kN) of thrust on lift-off. The narrow bodied Centaur stage is a liquid hydrogen/oxygen fuelled upper stage developing 33,000 lbs (147 kN) of thrust at altitude. It can deliver 2360 kg to GSO and with minor modifications, this can be increased to 2495 kg. Two launch complexes at Cape Canaveral, 36A and 36B, can be used to launch the vehicle although only 36B is now being used for it.

COMMERCIAL LAUNCH VEHICLE FAMILY

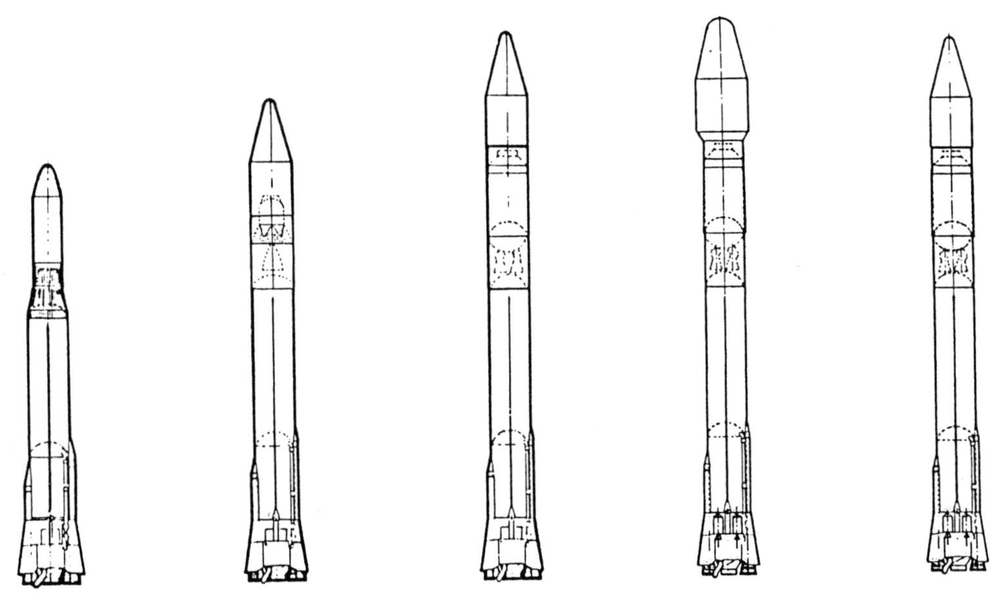

Launch vehicle	Atlas H	Atlas J/SSPS	Atlas G/Centaur	Performance Improved A/C	
Performance Status	3,000 lb LEO In production	3,000 lb+ GTO Proposed	5,200 lb GTO In production	5,600 lb GTO Proposed	5,800 lb GTO Proposed

LEO = Low earth orbit
GTO = Geosynchronous transfer orbit

Variations on the Centaur upper stage are used with the Titan vehicle. Two wide bodied versions, Centaur G and G Prime, were intended for use with Shuttle but were abandoned in 1986 for safety reasons. The G Prime version is to be used with the USAF's Titan IV vehicle.

2) Atlas K. The Atlas–Centaur vehicle has been used primarily for Intelsat launches. The Intelsat V satellites are large, and to address the needs of smaller commercial satellites, General Dynamics have made two proposals. The first is a dual launch adaptor for the large Atlas–Centaur. This appears to have been abandoned in favour of the second, the Atlas K vehicle. This is capable of carrying 1280 to 1460 kg to GSO and uses a variety of solid fuelled upper stages.

3) Atlas H. This is only available for launches to low earth orbit and is used by the USAF. General Dynamics are proposing its use for commercial customers with such payload requirements. It has a payload capability of 2000 kg. It is intended that the three metre fairing currently used on Atlas–Centaur be used on Atlas–H.

Potential Customers for a Commercialised Atlas Vehicle: As at August 1986, General Dynamics was claiming that it had received fourteen expressions of interest for the use of Atlas–Centaur. It had received no firm customer commitments. Of the satellites that were likely to be removed from the Shuttle manifest, there were about 18 which would provide suitable customers. It is understood that General Dynamics would not commercialise the vehicle if launch rates were less than a minimum of two vehicles per year. NASA projections now indicate that there will be 10–15 commercial satellites in the free world market requiring launch per year by the end of the decade.

Reliability of the Atlas Vehicle: As at the end of 1985, there had been 486 space launches of the Atlas vehicle since 1957 of which 206 were of the version that flies with Centaur. There had been 65 flights of Atlas–Centaur and seven flights of Centaur on Titan. The operational record of all versions of Atlas was 96% and of current D–1 version of Centaur, 97%.

ATLAS–CENTAUR LAUNCH MANIFEST (as at September 1986)

Launch	*Date*	*Payload*
Atlas–Centaur	8th November 1986	FleetSatCom (US Navy)
Atlas–Centaur	1987	FleetSatCom (US Navy)
Atlas–Centaur	1987	FleetSatCom (US Navy)

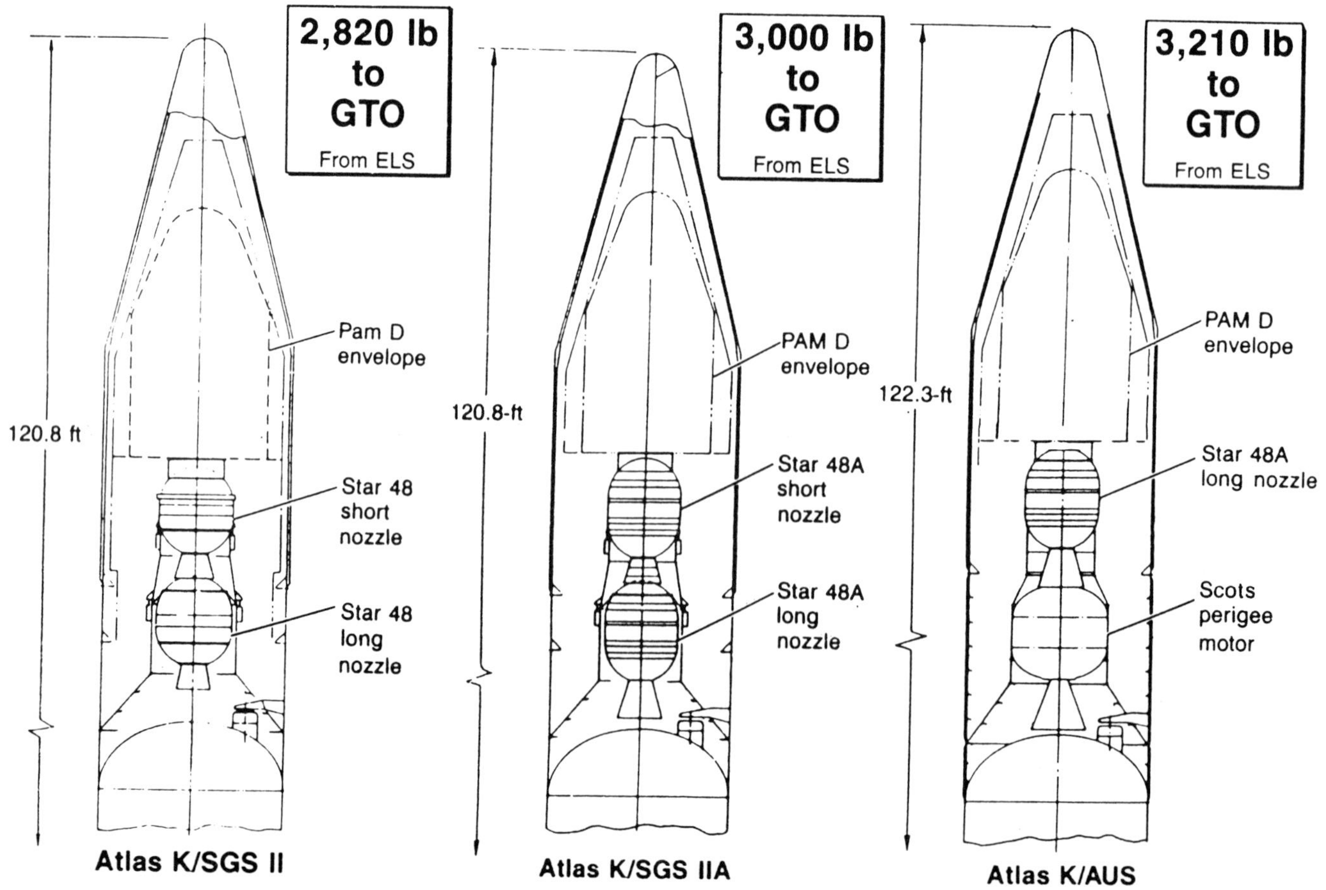

Atlas K upper-stage configurations

Atlas/Centaur is compatible with a variety of payload configurations. Fairing options now available include the standard 10 ft (3 metre) fairing and a 10·5 ft diameter (3·2 metre) fairing. The 13·1 ft (4 metre) payload fairing option will be available for 1988 launches.

A tandem adapter in development for the Atlas–G/Centaur will permit simultaneous launch of two 2400 lb (1100 kg) payloads to geosynchronous transfer orbit.

PAYLOAD OPTIONS COMMERCIALISED ATLAS CENTAUR

Single spacecraft

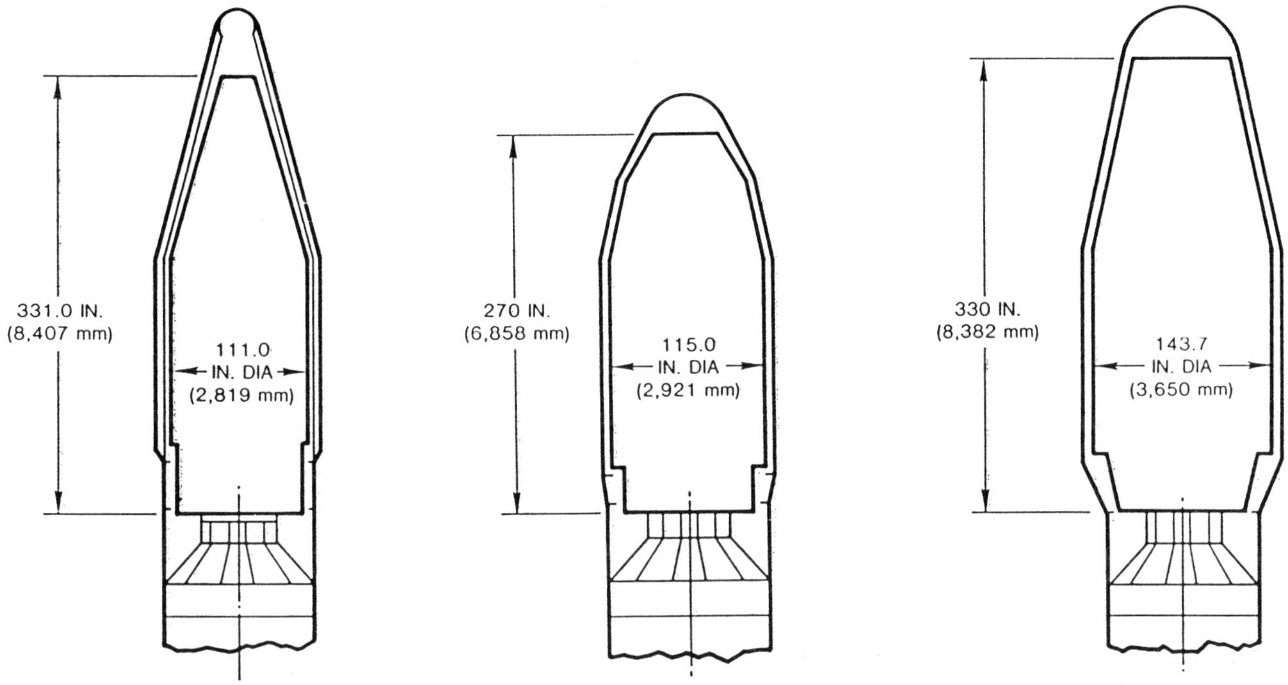

Atlas G	10 ft fairing (3·0 m)	10·5 ft fairing (3·2 m)	13·1 ft fairing (4·0 m)

Dual spacecraft

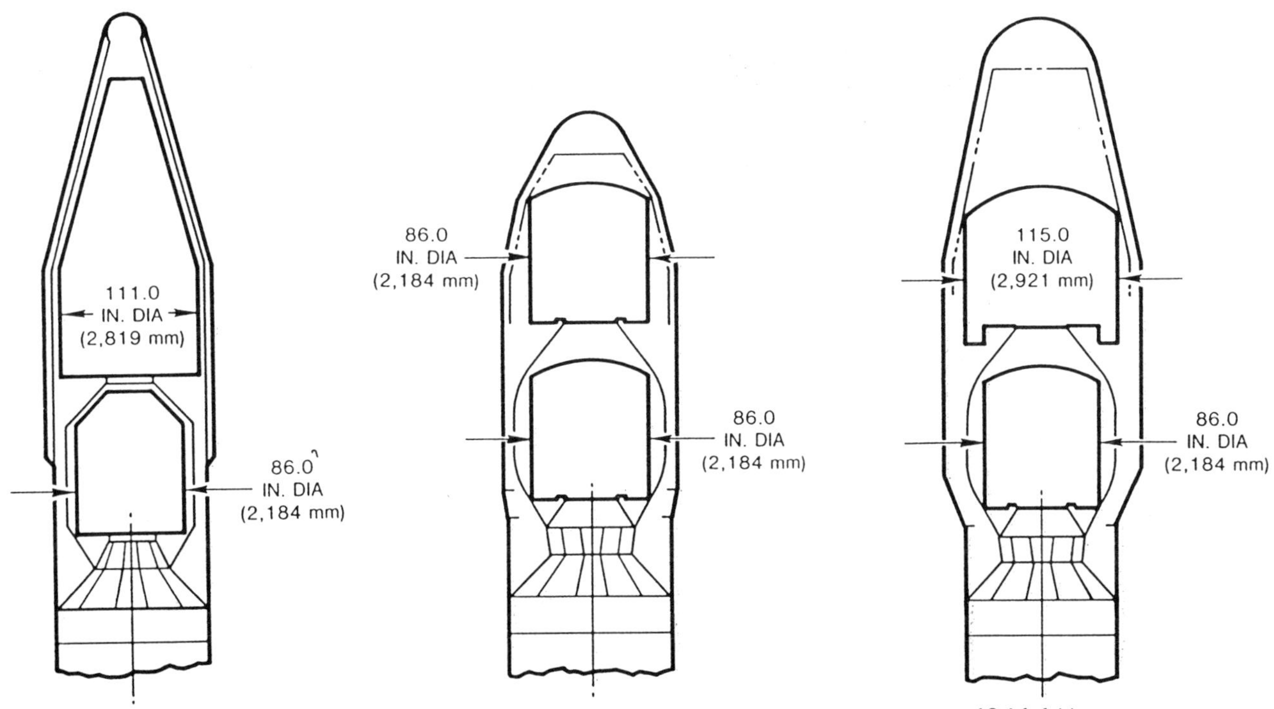

Atlas G	10 ft fairing (3·0 m)	10·5 ft fairing (3·2 m)	13·1 ft fairing (4·0 m)

Some of the features described are under development and not immediately available

CURRENT OPERATIONAL STATE OF VEHICLE	Operational but production line now closed down. There are two plans to revive it (see text)
LAUNCH VEHICLE OPERATOR	NASA, but with an agreement with Transpace Carriers Incorporated (see text)
ADDRESS	Transpace Carriers Incorporated, 6411 Ivy Lane, Suite 500, Greenbelt, Maryland MD 20770, USA
TELEPHONE NO.	(301) 982-7800
SENIOR STAFF	*Transpace Carriers* David W. Grimes, Chairman Antonio L. Savoca, President and Chief Executive Officer Dennis Ahearn Herbert F. Balzuweit, VP, Finance
PRIME CONTRACTOR FOR LAUNCH VEHICLE	McDonnell Douglas Astronautics
PAYLOAD CAPABILITIES	Delta 3914: 937 kg to GTO Delta 3924: 1429 kg to GTO Delta 3910/PAM: 1111 kg to GTO Delta 3920/PAM: 1497 kg to GTO Delta MELV: 1600 kg to GTO *TCI Commercial Deltas:* Delta 3914A: 1045 kg to GTO Delta 3924A: 1273 kg to GTO Delta 3925A: 1409 kg to GTO Delta 3925B: 1375 kg to GTO
PAYLOAD LAUNCH COSTS	Up to early 1986, TCI were quoting a commercial Delta launch cost "in the low \$30 m". It is believed that, as at September 1986, the figure had increased to around \$40 m. For the launch of Palapa B2 in May 1987, NASA is charging \$54 m
FIRST COMMERCIAL PAYLOAD LAUNCH	Telstar 1, 1963
NEXT AVAILABLE PAYLOAD OPPORTUNITY	Not available (see text)
LAUNCH SITE(S)	Eastern Space and Missile Center, Florida (possibly using two launch pads) for launches to GTO. Western Space and Missile Center, California for other launches

Introduction: The Delta Expendable Launch Vehicle (ELV) has been a workhorse of the US space scene since 1960. It was due to have been phased out in favour of Shuttle launches but, following the 1984 policy decision to commercialise US ELV's, there have been plans by a start-up venture, Transpace Carriers Incorporated, to have the production line restarted. McDonnell Douglas are working on proposals for a further stretch of the design to meet the USAF's MELV requirements and for possible commercial use. No decision to use the vehicle is expected from the USAF before the Spring of 1987. It is unlikely that McDonnell Douglas will commercialise the vehicle without the MELV contract. At the time of writing, there are sections of about nine Delta vehicles in stock and it would therefore take no more than 18 months from restarted production to first launch.

As at September 1986, there were two remaining launches of Delta, both operated by NASA. One was an SDI launch; the other is for the Indonesian Palapa B3 satellite. NASA has said that it will purchase ELV's "wherever it can" for specialised payloads and it must therefore be assumed that it may operate the vehicle for some time to come.

Background of the Delta Launch Vehicle: The origin of Delta is as an ICBM. The USAF initiated development of the

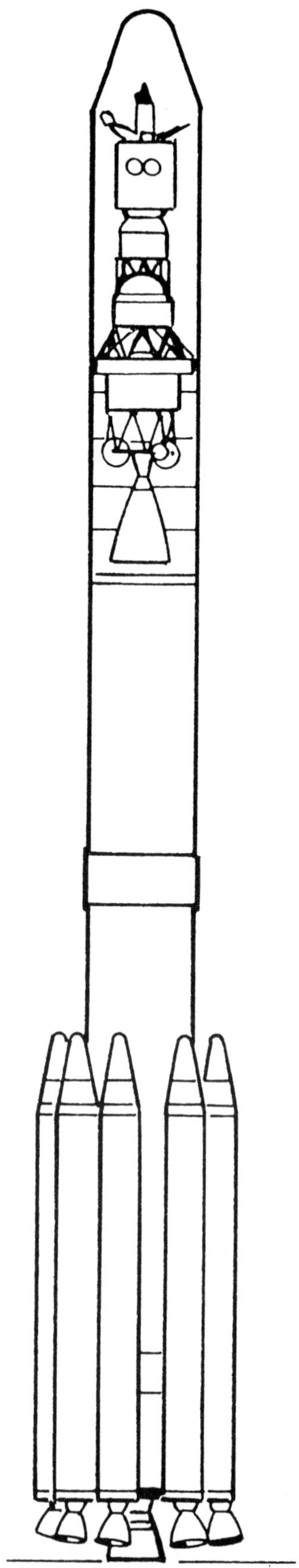

DELTA 3920 W/AEROJET SECOND STAGE AND PAM

Thor Intermediate Range Ballistic Missile in 1955. In 1959 NASA awarded a contract to McDonnell Douglas to develop 12 launch vehicles as an interim space launch vehicle. Thor was modified for use as the first stage. A Vanguard second stage was placed atop this and a Vanguard solid rocket motor was used for the third stage. The vehicle first flew on 13th May 1960 carrying the Echo 1 passive satellite. This launch was not a success. The second, and successful, launch was on 15th November 1960. The vehicle was progressively stretched and uprated over the following twenty years, culminating in the Delta 3910 and 3920 versions. A summary of the versions is provided at the end of this section.

Stretching of the vehicle beyond its current capacity has been considered for some years. For example, McDonnell Douglas considered a 4920 version in the early 1980's which would have provided a payload capacity of 3000 lbs (1360 kg) to GTO. The USAF MELV requirements involve a payload capacity of 3500 lbs (1600 kg) to GTO.

The Delta vehicle has been used as the basis of the Japanese launch vehicle industry, being produced under licence in Japan and progressively upgraded through the N–I, N–II and H–I vehicle. The latter is the test bed for the H–II vehicle which, from 1992, will place Japan in the international launch vehicle market. At the moment, the licence agreements between the USA and Japan prevent the use of the Delta derived vehicles for launch of other than Japanese satellites. In any case their payload capacity is too small to launch most commercial satellites available since the late 1970's.

Final forms of the Delta vehicle: The Delta vehicle now being phased out comes in four different versions. All are three stage vehicles with a liquid fuelled first stage augmented by nine solid propellant strap-on boosters. These are Delta 3910, 3914, 3920 and 3924. The 3910 and 3920 use the Payload Assist Module (PAM–D) as the third stage while the 3914 and 3924 versions use a Morton Thiokol TE–364–4 solid propellant motor. The Delta 3920 and 3924 versions are the final versions of the vehicle developed from 1979 when it had become apparent that Shuttle would be considerably late. A larger payload capacity was required to meet the requirements of such satellites as Landsat D.

Delta 3910 and 3914: Both vehicles use Castor IV strap-on boosters and a second stage TRW–201 engine.

Delta 3920 and 3924: These have a higher payload capacity than the 3910 and 3914 vehicles. A different second stage using the Aerojet liquid rocket company Improved Transtage Injector Programme (ITIP) AJ10–118K adds about 159 kg to the payload to GTO performance. The two versions also use a different firing sequence for the nine solid rocket boosters.

MELV proposals: McDonnell Douglas are proposing only minor modifications to the 3910/3920 design to meet the USAF's need to launch Navstar satellites. However they are proposing a version of the SGS–2 upper stage using a single Star 48 motor as third stage. The Delta modifications include lengthening of the first stage fuel tanks by twelve feet, upgrading the propellant in

Nightime launch of Delta carrying the IRAS scientific satellite

the solid rocket motors and enlarging the main engine nozzle expansion ratio from 8:1 to 12:1 for improved specific impulse. The payload fairing will also be increased to 9·3 feet.

Transpace Carriers Incorporated: This company was founded in 1982 by David Grimes, previously Delta Project Manager. The company has aimed to operate Delta on a commercial basis. Delta was, in 1982, being phased out by NASA in favour of Shuttle launches. It was apparent to many in the industry that Shuttle was a problem vehicle and that overdependence on one launch system could create too much risk for the satellite operators.

On 16th May 1983 President Reagan issued a directive on the Commercialisation of Expendable Launch Vehicles. This set the stage for a number of attempts to put the aging ELV's then being phased out on an operational basis. Two proposals were to finally merge. These were TCI's plans for Delta and general Dynamics' plans for Atlas-Centaur. In September 1983, NASA issued an RFP for transfer of the vehicle to commercial operations. There was little serious response to this. Shuttle launch charges were priced too low to allow private sector companies to take the risks of commercialised launch operations.

In March 1984, TCI and NASA signed an agreement on the terms and conditions under which the transition to commercialised operations would begin. The final agreement, which gave TCI exclusive marketing rights for the vehicle, included clauses which required TCI to sign firm launch contracts before the vehicle was handed over. This the company has never been able to do. It has also had a financial problem in that it cannot raise the finance needed for operational activities until it has both found customers and had the vehicle handed over by NASA.

Prior to the loss of Shuttle on 28th January 1986, TCI was offering Delta launches at 10% above comparable Shuttle launches. The price was in the "low \$30 m".

In May 1984 TCI petitioned the United States Trade Representative (a branch of the White House) claiming that Arianespace was practising predatory pricing in obtaining US customers for Ariane. The claim was based on the differential price charge between Arianespace's US and its European customers. It was claimed that the difference was up to 33% in favour of US customers. Arianespace's reaction was to point to the subsidies then being given to users of Shuttle. The petition was rejected, basically because the White House agreed with Arianespace. Nevertheless, the exercise was indicative of the position that TCI was in. The White House found that commercial conditions did not (in 1985) exist in the launch vehicle industry because space transportation services were developed as governmental monopolies by various countries, unconstrained by the requirements of the commercial market place. The White House, partially as a consequence and partially because Shuttle was proving to be vastly more expensive to operate than originally envisaged, established a new pricing policy for Shuttle launches. This was a step in the right direction for TCI.

When Shuttle blew up on the 28th January 1986, TCI was still without the three firm customers it needed to commercialise Delta. The event was to change the whole US launch scene although the developments over the following eight months proceeded more slowly than some observers expected. Firstly, the potential commercial launch vehicle operators rapidly dusted down old plans for commercial ELV operations. It appears that the USAF also tried to "hi-jack" Delta away from TCI.

Nevertheless, the White House took time to develop a new policy towards launch vehicles, taking into account the demands of both the commercial satellite operators and the US military. This policy was finalised in the summer of 1986 when it was announced that Shuttle would be rapidly wound down as a launch vehicle for commercial satellites and that the USAF would procure a series of ten ELV's of a design suitable for commercial operations as well. This latter proposal is known by the acronym, MELV. TCI's past proposals for upgrading the Delta launch vehicle provided a payload capacity whch was insufficient to meet the USAF's requirements for MELV.

Transpace Carriers position looks fairly weak. As at September 1986, it had only two letters of intent from potential commercial customers. One of these was from American Satellite Corporation, who are looking for an earlier launch than Transpace can offer. Not only will it have to finance its own working capital, but also, if Delta does not win the MELV contract, have to finance a stretch of the vehicle to accommodate the RCA–4000/5000 and Hughes HS–393 class of satellite. In its current form, the Delta vehicle is capable of launching no more than 1270 kg to GTO. Current indications are that McDonnell Douglas are highly unlikely to finance a stretch out of their own resources if it fails to win the MELV contract (which is to launch Navstar satellites). Moreover Delta is not considered by some US trade sources to be a likely winner of the MELV contract; Atlas Centaur appears to be the clear favourite.

Current position of the Delta Launch Vehicle: At the time of writing (September 1986), the Delta vehicle remains operational. There are two remaining Delta launch vehicles in stock (Delta F179 and F182). F179 is due to carry the Geos H weather satellite on 20th November 1986 and F182 is due to carry the Indonesian Palapa B3 satellite in May 1987. NASA has also ordered three further Delta vehicles from McDonnell Douglas, numbers 181, 183 and 184. These are to be assembled from long-lead items and components already in stock. They are to be delivered to NASA between the early Autumn of 1987 and the early Autumn of 1988. Over and above these three, NASA now says that McDonnell Douglas has long lead items in stock for another eight Delta vehicles. NASA says that it intends to launch 29 commercial payloads that have been bumped off Shuttle. This statement presumably rules out a commercial launch vehicle operator and NASA will have to find ELV capacity of its own.

DELTA PAYLOAD MANIFEST

Launch	Date	Payload
Delta 180	5th September 1986	Strategic Defense Initiative Organisation Payload
Delta 179	20th November 1986	Geos Weather Satellite
Delta 181	1987	Strategic Defense Initiative Organisation Payload
Delta 182	May 1987	Palapa B3
Delta 183	1987/1988	Strategic Defense Initiative Organisation Payload
Delta 184	1988	Strategic Defense Initiative Organisation Payload
Delta 185	?	NASA customer? (long lead items only procured)
Delta 186	?	NASA customer? (long lead items only procured)
Delta 187	?	NASA customer? (long lead items only procured)
Delta 188	?	NASA customer? (long lead items only procured)
Delta 189	?	NASA customer? (long lead items only procured)
Delta 190	?	NASA customer? (long lead items only procured)
Delta 191	?	NASA customer? (long lead items only procured)
Delta 192	?	NASA customer? (long lead items only procured)

GROWTH HISTORY OF DELTA

Year	*Fairing*	*First Stage*	*Second Stage*	*Third Stage*	*Delta Model Designation*
1960		Modified Thor	Vanguard propulsion system	Vanguard X–248	Delta
1962		Uprated MB–3 engine			A
1962			Lengthened tankage Higher energy oxidiser Transistorised guidance electronics		
1963	Bulbous fairing			Scout developed X–258 replaced X–248	C
1964		Augmented thrust using Thor developed strap-on solid rocket motors			D
1965	Larger fairing from USAF Agena stage		Larger diameter propellant tanks from the Able-Star stage. Restart capability	USAF developed FW–4 replaced X–258	E
1968				Surveyor retro-motor modified TE–364–3 replaced FW–4	J
1968		Lengthened LOX and RP–1 tanks. Increased diameter of RP–1 tanks			L, M, N
1970		Added three solid rockets to total six			M6
1972		Introduced UBT to accommodate nine solid rockets	Titan transtage engine. Strap-down inertial guidance		903
1972	Increased diameter and length	Increased length of tanks. Isogrid structure	Added interstage and suspended stage within	Increased motor length TE–364–4	1914
1973		Replaced MB–3 engine with more powerful RS–27 engine			2914
1975		Increased size of solid rockets. C–IV replaced C–II			3914

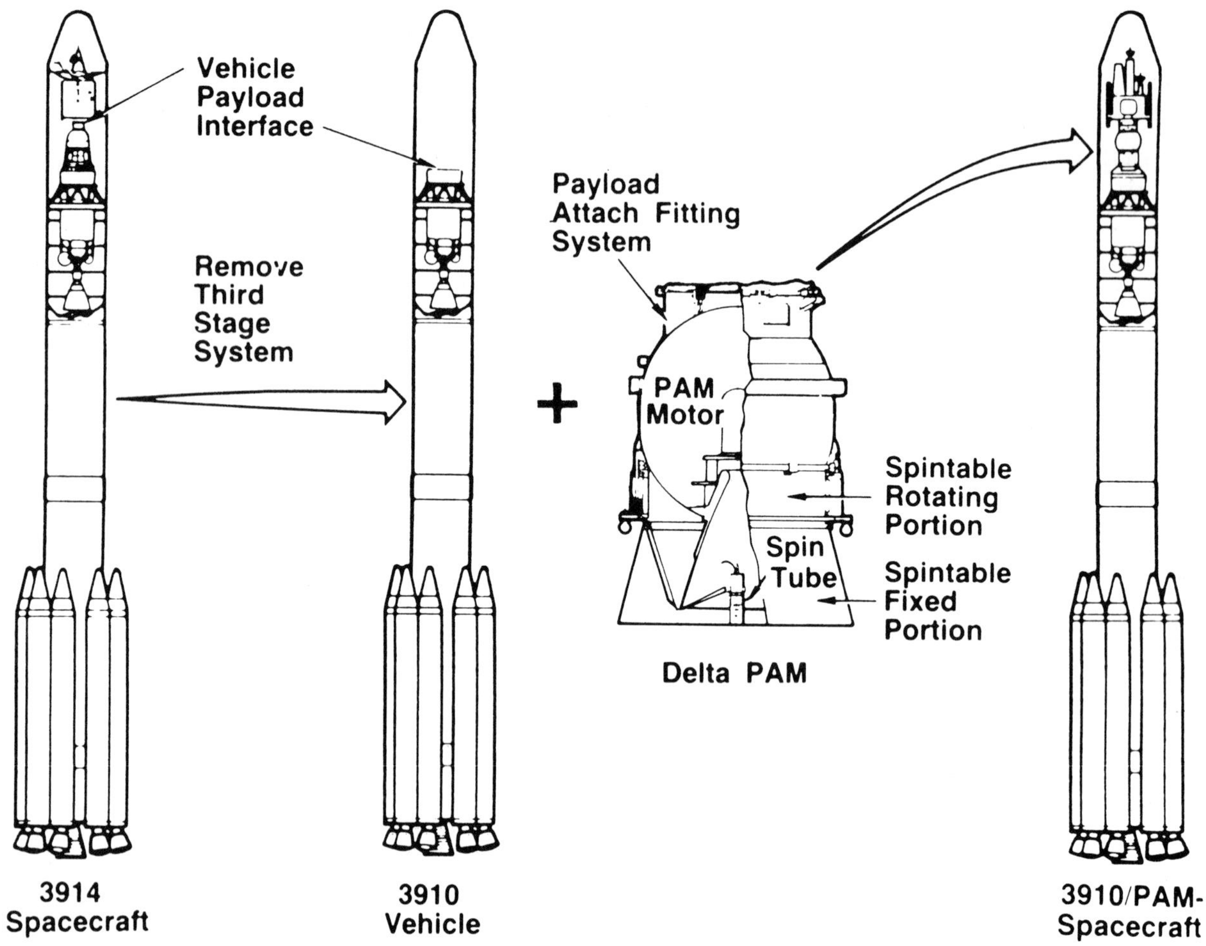

STANDARD DELTA AND 3910/PAM RELATIONSHIPS

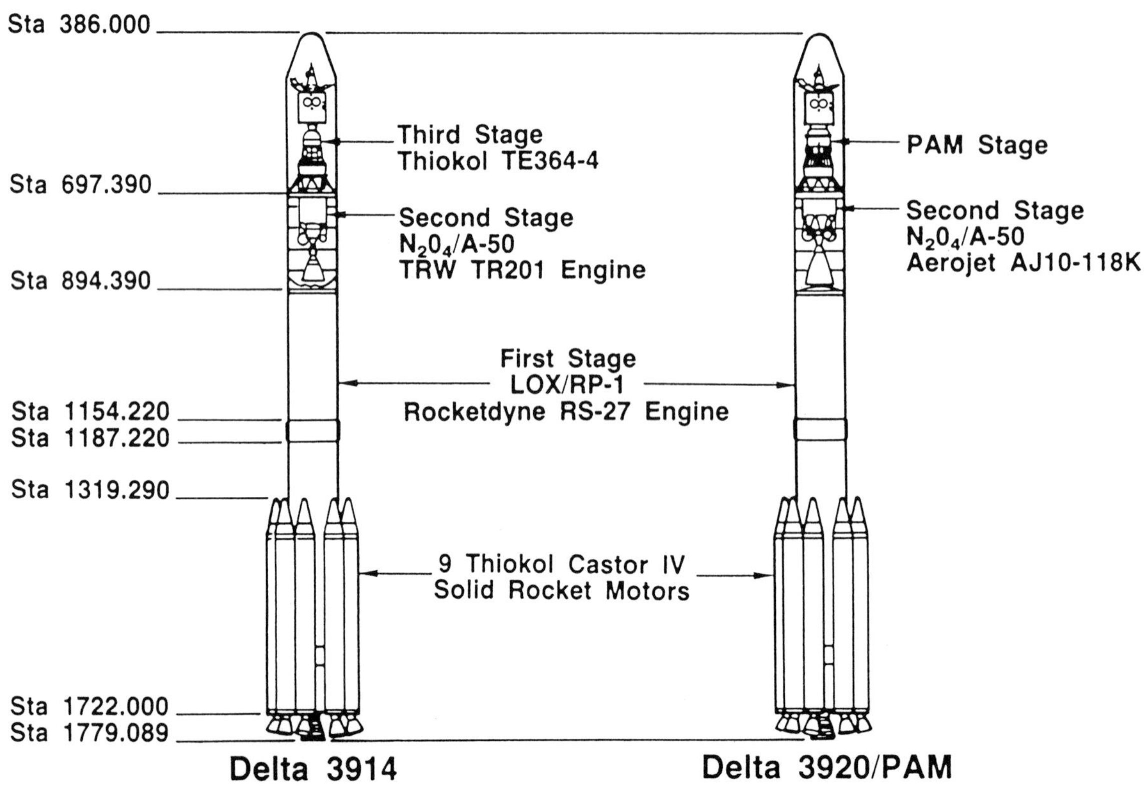

DELTA LAUNCH VEHICLES COMPARISON

■ **Third Stage**	
• Motor	Thiokol TE-364-4 (Solid Propellant)
• Control	Spin-Up
• Fairing	8-Ft (2.44 M) Diameter
■ **Second Stage**	
• Airframe	8 Ft (2.44 M) Diameter
• Tankage	4.5 Ft (1.39 M) Diameter
• Engine	TRW TR-201 9,825 Lb Thrust
• Propellants	N_2O_4/Aerozine 50
• Control	Gimbal/Cold Gas
• Guidance	Inertial
■ **First Stage**	
• Airframe	8 Ft (2.44 M) Diameter
• Engine	Rocketdyne RS-27 207,000 Lb Thrust
• Propellants	LOX/RP-1
• Control	Gimbal/Verniers
• Guidance	Inertial
• Solid Augmentation	9 Thiokol Castor IV Motors (85,000 Lb Thrust Each)

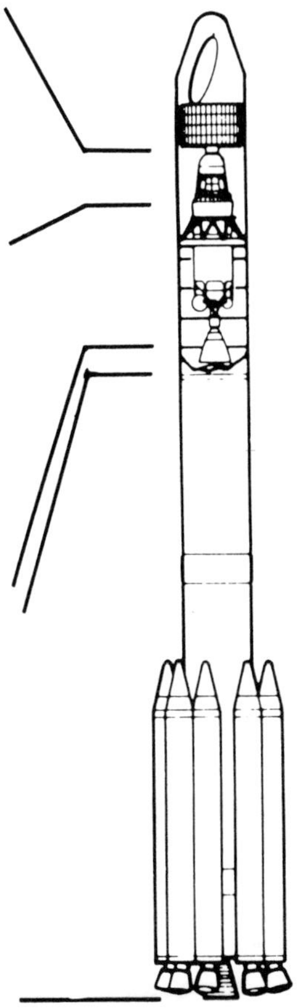

DELTA 3914 VEHICLE PROPULSION AND G&C SYSTEMS

CURRENT OPERATIONAL STATE OF VEHICLE	Operational, but with only one flight
LAUNCH VEHICLE OPERATOR	National Space Development Agency of Japan (NASDA)
ADDRESS	World Trade Center Building, 4–1 Hamamatsu-cho 2-chome, Minato-ku, Tokyo 105, Japan
TELEPHONE NO.	81-3-435-6111
TELEX NO.	J28424
FAX NO.	81-3-433-0796
SENIOR STAFF	Hiroyuki Osawa, President
PRIME CONTRACTOR FOR LAUNCH VEHICLE	
OTHER MAJOR CONTRACTORS	
PAYLOAD CAPABILITIES	550 kg to GSO
PAYLOAD LAUNCH COSTS	not available
FIRST LAUNCH OF VEHICLE	13th August 1986
FIRST COMMERCIAL PAYLOAD LAUNCH	
NEXT AVAILABLE PAYLOAD OPPORTUNITY	None (see text)
MAXIMUM LAUNCH RATE	
LAUNCH SITE(S)	Tanegashima Space Center, Japan

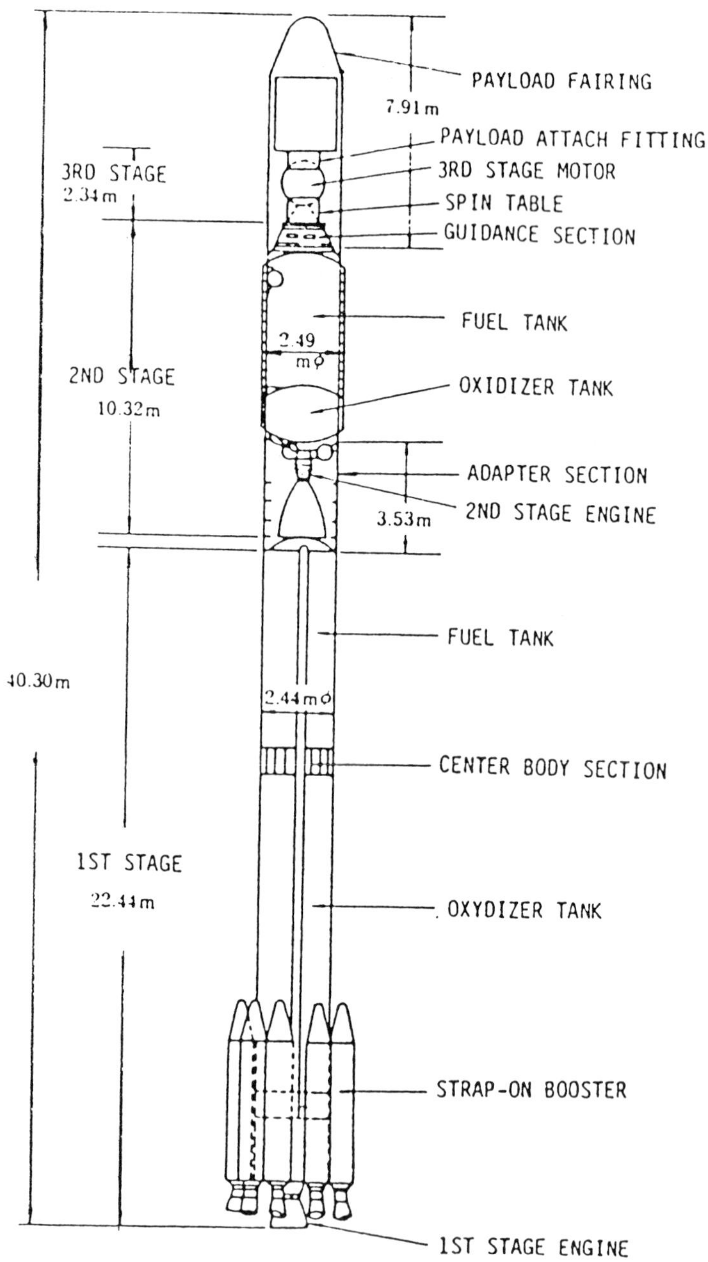

H-I LAUNCH VEHICLE

Introduction: The Japanese H–I vehicle is the current operational launch vehicle of that country. It is available as either a two or three stage system. H–I is, strictly speaking, not a commercial launch vehicle in that it can only launch Japanese satellites (because the first stage is built under licence with clauses to prevent its use for non-Japanese satellites) and because its payload capability is only 550 kg to GSO, well below the 2000 kg plus needed for commercial vehicles in the West. Nevertheless, the vehicle is important in that it is a stepping stone to the development of the advanced H–II vehicle due for launch in 1992. The Japanese have spent $1000 m in developing H–I and are taking the launch vehicle market very seriously.

Background: The H–I vehicle is the result of ten years effort to reduce Japanese dependence on US hardware. The original N–I vehicle, which first flew in 1975, used the US Delta 2914 concept to acquire a launch capability for applications satellites as quickly as possible. It was upgraded to the N–II version capable of launching a 350 kg payload to GSO. The H–I vehicle uses the first stage and strap on boosters of N–II but with the new LE–5 liquid oxygen/hydrogen second stage, a new solid fuelled third stage and a new inertial guidance system. H–I also uses the N–II launch facilities. It is now the main Japanese launch vehicle until H–II becomes available in the early 1990's. N–II has now been phased out.

Technical Summary of the H–I Vehicle: The H–I vehicle uses the same first stage as the N–II vehicle but a new second stage developed in Japan and known as the LE–5. The first stage is basically the Delta 2914 first stage with an MB–3 motor built under licence from McDonnell Douglas and Morton Thiokol by Nissan. The vehicle is being used as a test bed for the LE–5 stage which will form the second stage of the H–II vehicle from 1992. The Japanese are also planning an advanced version of LE–5 for H–II. The LE–5 stage, although using open cycle engine technology, is also useful in the development of the H–II first stage which uses closed cycle technology. It was the first cryogenic propulsion system developed in Japan and is, generally, rather conservative. Unlike the Ariane third stage, it has a restart capability. The possibility of use of the LE–5 stage with the US Delta (as an advanced upper stage) and the European Ariane (as a new third stage) vehicles has also been considered.

LE–5 stage cost some $108 m to develop over 10 years. The H–I vehicle also employs a new solid propellant spin stabilised third stage for use with geosynchronous mission. The payload fairing is the same as used on N–II, having a 2·4 m diameter.

H–I LAUNCH MANIFEST

Launch	*Date*	*Payload*
H–I F1	13th August 1986	Geodetsic Laser Reflector and Fuji (amateur satellite)
H–I F2	August–September 1987	Engineering Test Satellite V (ETS–V)
H–I F3	January–February 1988	CS–3A
H–I F4	August–September 1988	CS–3B
H–I F5	January–February 1989	Marine Observation Satellite 1B (MOS–1B)
H–I F6	August–September 1989	GMS–4
H–I F7	August–September 1990	BS–3A
H–I F8	January–February 1991	ERS–1
H–I F9	August–September 1991	BS–3B

CURRENT OPERATIONAL STATE OF VEHICLE	Under development, with first launch in 1992 and first commercial launch in 1993–94
LAUNCH VEHICLE OPERATOR	National Space Development Agency of Japan (NASDA)
ADDRESS	World Trade Center Building, 4–1 Hamamatsu-cho 2-chome, Minato-ku, Tokyo 105, Japan
TELEPHONE NO.	81-2-435-6111
TELEX NO.	J28424
FAX NO.	81-3-433-0796
SENIOR STAFF	Hiroyuki Osawa, Secretary General
OTHER MAJOR CONTRACTORS	Mitsubishi Heavy Industries Ishikawajima-Harima Heavy Industries Nissan Motor Company
PAYLOAD CAPABILITIES	Initially this vehicle is set to launch 2000 kg to GSO. Growth capability, through addition of further solid rocket boosters is to beyond 4000 kg. Initial design will also be able to launch payloads of 9000 kg to LEO and 4400 kg to sun-synchronous orbit. There will be between two and four launches per year
PAYLOAD LAUNCH COSTS	Not available at this stage. Vehicle is designed for dual payload launches
FIRST LAUNCH OF VEHICLE	1992
FIRST COMMERCIAL PAYLOAD LAUNCH	1993–94
NEXT AVAILABLE PAYLOAD OPPORTUNITY	1993–94
LAUNCH SITE(S)	Tanegashima Space Center, Japan

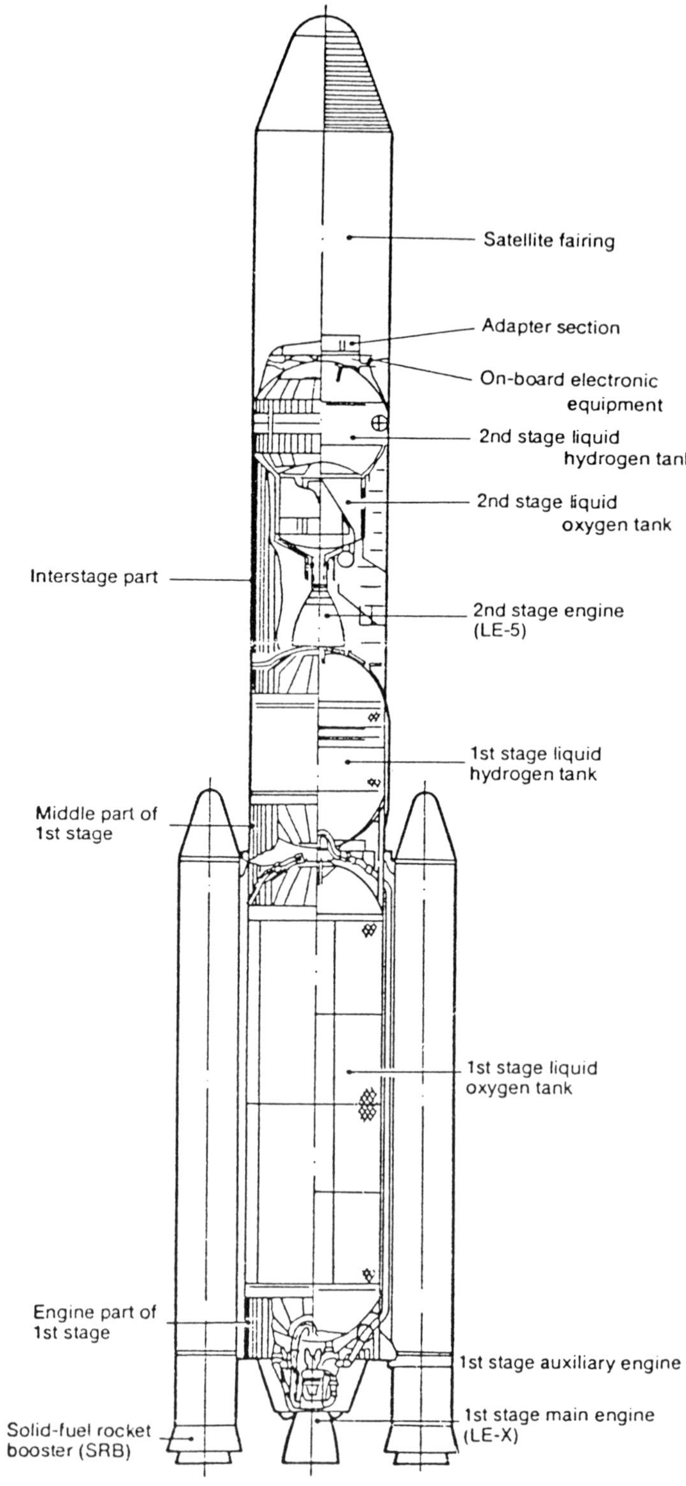

H-II Rocket

Introduction: Development of this advanced expendable launch vehicle began in 1986. The vehicle represents a significant development of Western ELV technology as well as being Japan's entrance ticket into being a major force in space activity. As a commercial launch vehicle it is likely to provide a very competitive launch vehicle price as well as be available to non-Japanese satellite operators. It is also envisaged that the vehicle will be subject to considerable development after it is initially proven on an operational basis. There are plans to upgrade it to man-rated status to launch a Japanese built spaceplane to service the country's participation in the International Space Station.

Background to the H–II Vehicle: The decision to develop an advanced heavy launch vehicle dates back to February 1984 when the Japanese Space Activities Commission revised its 1978 Space Development Policy and opted for the new vehicle instead of an upgraded version of H–I. Japan's efforts in space activities have been modest to date, relying heavily on US technology and a protected domestic market for its overpriced and small satellites. Its launchers, the N–I, N–II and H–I have been based on the US Delta 2914, have not had the payload capability to launch most commercial communications satellites and have not been available to non-Japanese customers because of licensing arrangements between Japan and the US manufacturers of Delta. They have been used to launch Japanese domestic built satellites; their communications satellites have therefore been small and uncompetitive with European and US built spacecraft.

Nevertheless, the Japanese have progressively upgraded their launch vehicles and reduced their dependence on US technology. The latest launch vehicle to become operational, H–I, is seen as a test-bed for H–II, using the advanced LE–5 stage which will become the second stage of H–II. The latter will use closed cycle technology for its first stage and will therefore be the only such launch vehicle (apart from Shuttle) available during the 1990's and will give the Japanese a considerable technological lead over Europe in particular. Studies which have led to the H–II design date back to 1979 when test firings of the LE–5 engine were carried out.

Unlike the LE–5 motor, the first stage LE–7 motor is of the closed cycle type but, like LE–5, it is entirely new. The Japanese appear to be very proud of having the opportunity to develop the LE–7. They have already approached McDonnell Douglas about the use of the LE–5 for an upgraded commercialised Delta third stage and apparently have also looked at the possibility of using it as an Ariane fourth stage. India has also looked at using it. Whether similar export opportunities arise for the LE–7 remains an open question but there is some clear indication that the Japanese are looking at long term industrial policies for its space engineering.

Development cost of the H–II vehicle is Yen200 b ($900 m). This is considerably lower than the cost of developing Ariane but the second stage of H–II has already been developed. The Japanese are considering using the H–II vehicle to launch a small Shuttle by the end of the 1990's as well as for launch of a 2200 kg platform in February 1993. The latter will be on the third launch of the vehicle.

Reliability of the H–II Vehicle: The Japanese claim that the realiability of the H–II vehicle has been set at 0·91 at a 60% level of confidence. The LE–5 second stage has a restart capability.

Technical Summary of H–II: H–II is a two-stage vehicle with two solid rocket boosters. Mass at lift-off is 250 tons. Overall length is 48 metres and diameter is 4 metres. The LE–7 first-stage motor uses liquid oxygen/liquid hydrogen fuel. The two solid rocket boosters also provide attitude control. The satellites are placed in a transfer orbit in two-stages using a re-ignition function on the LE–5 motor. The satellites are housed in a 4 metre diameter fairing.

LEADING DIMENSIONS OF H–II

Item	*First-stage*	*SRB*	*Second-stage*
Propellant	LOX/LH2	Solid	LOX/LH2
Propellant Weight (ton)	80	50 × 2	13·5
Average Thrust (ton)	95 (sea level)	135 × 2 (sea level)	10·5 (vacuum)
Burn Time (sec)	300	90	550
Specific Impulse (sec)	447 (vacuum)	270 (vacuum)	445 (vacuum)
Weight (ton)	90	60 × 2	16
Guidance	Strapped-down intertial guidance system		
Overall Length (m)	46		
Body Diameter (m)	4		
Overall Weight (ton)	235		
Fairing Diameter (m)	4/3·65 (outer/inner)		
Fairing Length (m)	12/10 (outside/inside)		

Note: All figures are approximate. Second-stage engine has re-start capability.

Payload Fairing

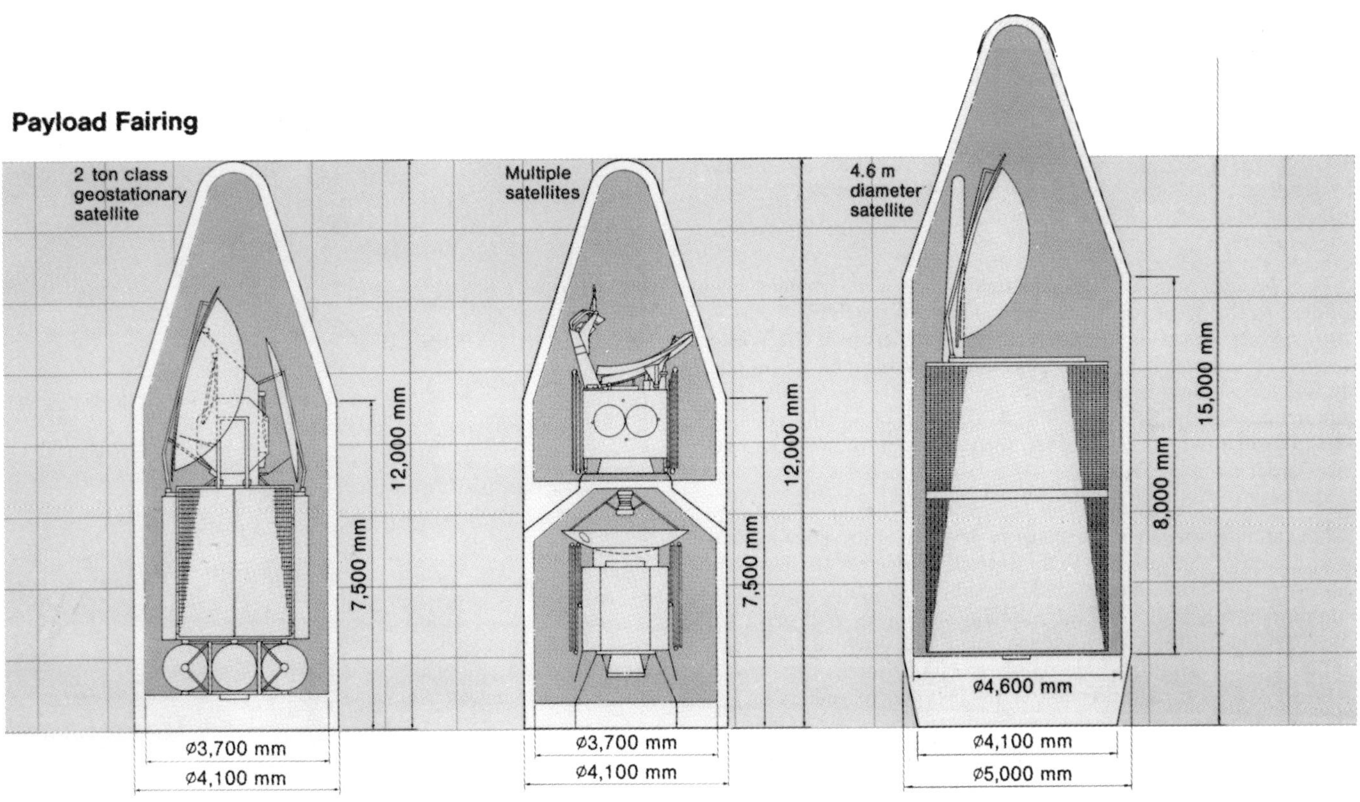

LONG MARCH 2/3

CURRENT OPERATIONAL STATE OF VEHICLE	Long March 2 and 3 versions now operational and Long March 4 version being considered
LAUNCH VEHICLE OPERATOR	Ministry of Astronautics
ADDRESS	PO Box 848, Beijing, China China Great Wall Industry Corporation PO Box 847, Beijing, China
TELEPHONE NO.	895040, 896655 (MoA)
TELEX NO.	20026 MOAFA CN (MoA) 22651 CGWIC CN (CGWIC)
FAX NO.	not available
SENIOR STAFF	Xie Guang Xuan, Chief Engineer Ministry of Astronautics Zhang Jiqing, Director General of Foreign Affairs Bureau, Ministry of Astronautics Fan Shi-he, Vice Chief Engineer, Ministry of Astronautics Chen Shouchin, Chief Engineer, Science Research & Production, Ministry of Astronautics U. Ke Li, Vice President, CGWIC
PRIME CONTRACTOR FOR LAUNCH VEHICLE	Beihing Wanyuan Industry Corporation Ministry of Astronautics Shanghai Astronautical Administration (first and second-stage engines)
PAYLOAD CAPABILITIES	Long March 1: 300 kg to LEO Long March 1C: 600 kg to LEO Long March 2: 1800 kg to LEO Long March 3: 1400 kg to GTO Long March 4: 2500 kg to GTO
PAYLOAD LAUNCH COSTS	15% below those charged by Arianespace with competitive insurance organised through the Chinese Peoples Insurance Company
FIRST LAUNCH OF VEHICLE	Long March 1, 24th April 1970 Long March 2, 15th November 1974 Long March 3, 29th January 1984
NEXT AVAILABLE PAYLOAD OPPORTUNITY	Late 1987
LAUNCH SITE(S)	CZ–3, Xichang launch site (Sichuan province) CZ–1 and CZ–2, Jiuquan (Gansu province)

Introduction: The Long March launch vehicle is now being offered to the West for launch of commercial satellites. It has attracted considerable interest because of the crisis in the Western launch vehicle market. It is a conventional vehicle based on ICBM technology dating back to the 1950's and originally obtained from the Soviet Union. At the time of writing and despite some two years of effort, the Chinese have not obtained a firm order from the West.

Background to the Long March: Until the early 1980's, China shunned foreign assistance in developing its space industry and for more than twenty years it patched together its domestic industry. Its "home made" rocket engines use stainless steel and aluminium substitutes for the titanium and alloys used in Western launch vehicles. Its space policy was oriented towards military observation and, despite large labour forces (the Ministry of Astronautics employs 100,000 people, mostly in missiles and other purely defence hardware), was parsimoniously funded. The space effort dated back to 1959 when the Chinese began modifying Soviet supplied SS–3 and SS–4 medium range ballistic missiles. There also appears to have been a mis-match between China's nominal production and launch capabilities and the actual rate it launched spacecraft from 1970. The reason is not fully understood.

The Long March 3 (CZ–3) vehicle

Customers for the Long March: The Chinese have been very active in trying to obtain Western customers for the vehicle as well as transfer its use to elsewhere in the World. It has, for example, claimed that it has reached an agreement with Hughes Aircraft Company in the USA to operate the vehicle on a joint basis from an island in the Pacific. Hughes have denied that there has been any such agreement. The Chinese have also talked with most of the existing and potential satellite operators in the Western World. It has also attempted to find marketing outlets in the West and has had talks with Transpace Carriers in the USA.

Nevertheless, satellite operators are wary of using the vehicle, partly because of concern about the launch facilities and partly because it is so new onto the market. The Long March 2 launch site currently requires that the spacecraft be mated with the launch vehicle in the open air.

So far the Chinese have received three letters of intent from Western launch vehicle operators. These are from Teresat, who have been trying to purchase the Palapa B2 and Westar VI satellites for US domestic operations. One of the satellites has been subsequently sold to a California based company, Sattel. Western Union also has a letter of intent for Westar VIS and is looking at the launch facilities and the legal limitations on export of technology to China. It is not expected to sign a final agreement before the end of 1986.

Sweden has a cooperative agreement with China covering space technology and in 1986 announced that it was considering launch of a Mailstar satellite by Long March 2. Apparently it backed down under pressure from the European Space Agency and will launch on Ariane instead. China also has a cooperative agreement with Brazil which may lead to the launch of two or more remote sensing satellites on Long March 2 in the early 1990's.

History of Long March: The first launch of Long March was in 1970 when the CZ–1 version successfully launched a 173 kg satellite into LEO. The launcher had been under development for ten years and is a three-stage vehicle. The first two stages are liquid fuelled and the third is solid fuelled. No further CZ–1 launches have taken place although it is understood that the version is still available in a modified form known as Long March 1C with a liquid fuelled third-stage.

The first CA–2 version flew in 1974. This is a two-stage vehicle for LEO payloads and is liquid fuelled. The first flight was a failure, put down to a "broken wire" in the gyro electronics. The vehicle was apparently destroyed some 200 metres above the ground and destroyed a building on the launch site. There have been seven successful launches of Long March 2; only the first vehicle failed. China also developed a variation of the CZ–2, known as FB–1. It was similar to CZ–2 except that it employed different sub-systems. It was abandoned in 1981 because it displayed no better performance than CZ–2.

Feasibility studies on the Long March 3 version began in 1975 and development work in 1977. It first flew in 1984 and as at September 1986, there have been two subsequent flights. The 1984 test flight was only a partial success because of problems with the third stage. Its payload, an experimental telecoms satellite, was placed into an elliptical orbit rather than geostationary orbit when the third-stage failed to ignite on the second burn during transfer orbit. The problem was put down to a fault in the prepellant turbo pump. The modifications were minor and the second flight took place 70 days later. The third stage of the Long March 3 vehicle is liquid oxygen/hydrogen fuelled.

The first available commercial launch is at the end of 1987 or early 1988; one domestic launch is expected before then. It claims to have a production and launch capability of up to 4–6 Long March 3 and 8–10 Long March 2 vehicles per year, although the current rate is about two to three per year for both.

Proposed Developments of Long March: Like most launch vehicle operators, the Chinese have various plans to upgrade their launch vehicles. There are proposals to add both solid and liquid boosters to the Long March 3 and a new third-stage. This would not be available until 1990. Maximum payload capability to GTO is being quoted as 2500 kg. The most advanced version is Long March 4. This would have a payload capability of 4000 kg to GTO.

Cost of Using Long March: The Chinese have quoted a charging policy giving rates at 15% below those of Arianespace. It is also offering insurance rates below those available in the West through its Chinese People's Insurance Company.

A $50,000 deposit is required for a launch reservation. The Chinese take a memorandum of understanding to be a launch reservation. Reservations need to be made two or more years in advance of launch date.

USABLE VOLUME INSIDE FAIRING

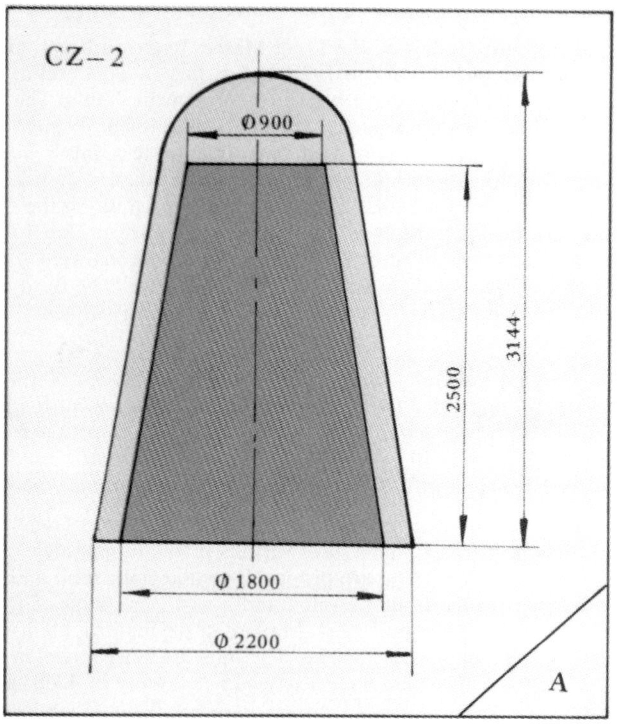

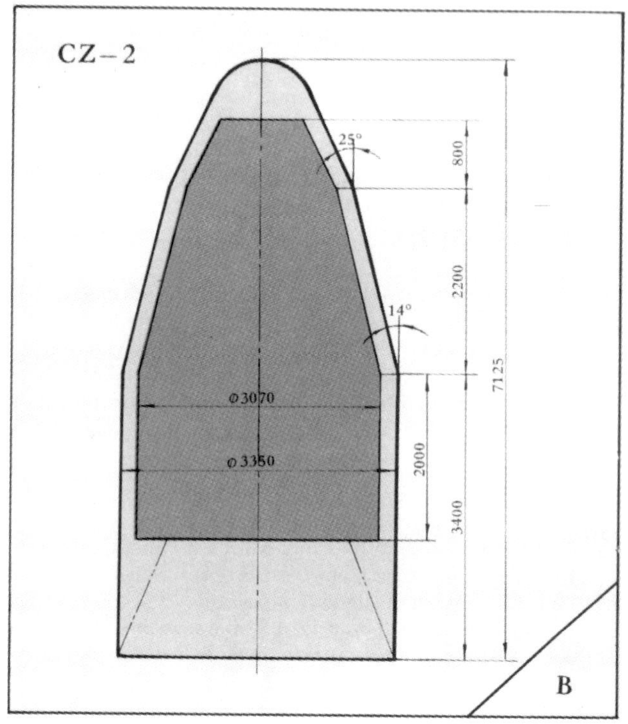

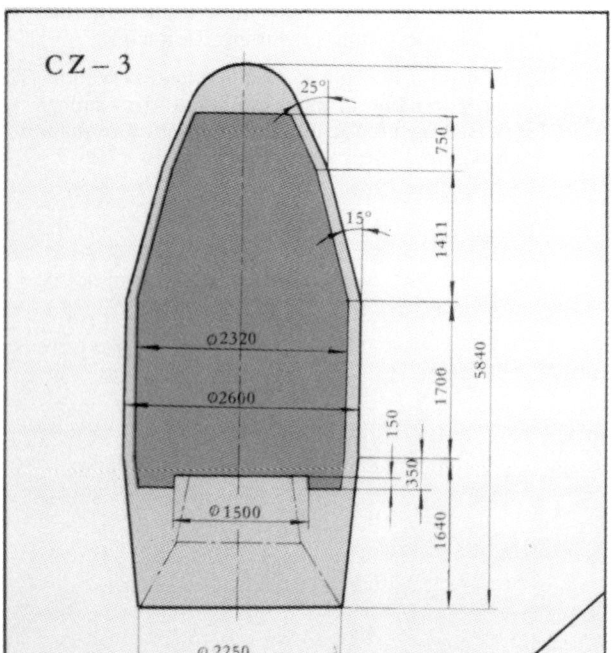

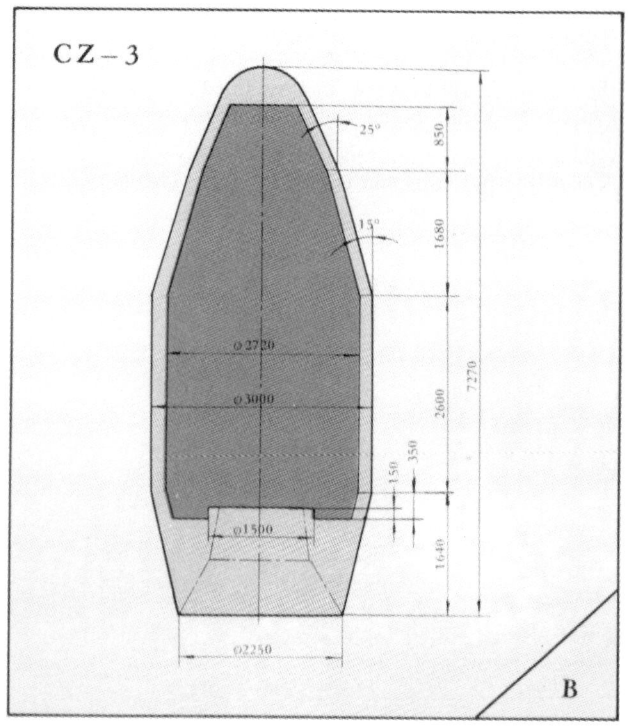

MECHANICAL INTERFACE BETWEEN PAYLOAD AND LAUNCH VEHICLE COULD BE DESIGNED ACCORDING TO USER'S REQUIREMENT

LONG MARCH 2/3—(continued)

CHARACTERISTICS OF THE VEHICLE

	CZ–1 (Long March 1)	*CZ–1C (Long March 1C)*	*CX–2 (Long March 2)*	*CZ–3 (Long March 3)*
General configuration	3-stage	3-stage	2-stage	3-stage
Overall length	29·45 m	33 m	31·65 m	43·25 m
Lift-off mass	81·6 tons	88 tons	191 tons	202 tons
Diameter (1st-stage)	2·25 m	2·25 m	3·35 m	3·35 m
Operational	in 1970	in 1975?	in 1974	in 1984
Payload dimensions:				
Diameter	1 m?	1·56 m	3·17 m	2·23 m
Length	2 m?	2·40 m	2·2 m/3 m	3·1 m/3·8 m
Orbital inclination	70°	50–70°	42–63°	31·1°
Performances:				
400 km circular	300 kg	600 kg	2 tons	3·8 tons
200 km/1500 km	—	500 kg	1/8 tons	4·3 tons
Geostationary transfer	—	—	—	1·4 tons

LAUNCH RECORD OF LONG MARCH

Launch	*Date*	*Payload*
Long March 1 F1	24th April 1970	Red East Satellite (SKW–1)
Long March 1 F2	3rd March 1971	Red East Satellite (SKW–2)
Long March 2 F3	15th November 1974	(launch failure)
Long March 2 F4	26th November 1975	
Long March 2 F5	7th December 1976	
Long March 2 F6	26th January 1978	
Long March 2 F7	9th September 1982	
Long March 2 F8	19th August 1983	
Long March 3 F9	29th January 1984	Test Flight (partial failure)
Long March 3 F10	8th April 1984	Experimental Telecommunications Satellite
Long March 2 F11	12th September 1984	
Long March 2 F12	18th September 1985	
Long March 3 F13	1st February 1986	Experimental Telecoms & Broadcasting Satellite

CURRENT OPERATIONAL STATE OF VEHICLE	Established and operational
LAUNCH VEHICLE OPERATOR	Glavkosmos USSR (see text)
ADDRESS	9 Krasnoproletarskaya Ul, 103030 Moscow, USSR Foreign marketing is handled by Licensintorg
TELEPHONE NO.	250-22-20
SENIOR STAFF	Dmitri Poletayev, Head of Glavkosmos Yuriy Lavkovskiy, VP, Morsviasputnik Leonid Dubonossov, Director, Marketing & Finance, Morsviasputnik
PAYLOAD CAPABILITIES	2200 kg to GSO
PAYLOAD LAUNCH COSTS	The Soviets are still apparently offering a dedicated launch on this vehicle for $24 m. As the vehicle is capable of carrying up to three satellites this suggests a minimum price of $8 m. There is also a re-launch clause where the Soviets will give a free launch in case of failure of the first launcher. The vehicle does not require that the satellites carry an AKM, giving further cost savings
FIRST LAUNCH OF VEHICLE	1965
NEXT AVAILABLE PAYLOAD OPPORTUNITY	not known
LAUNCH SITE(S)	Baikonur Cosmodrome

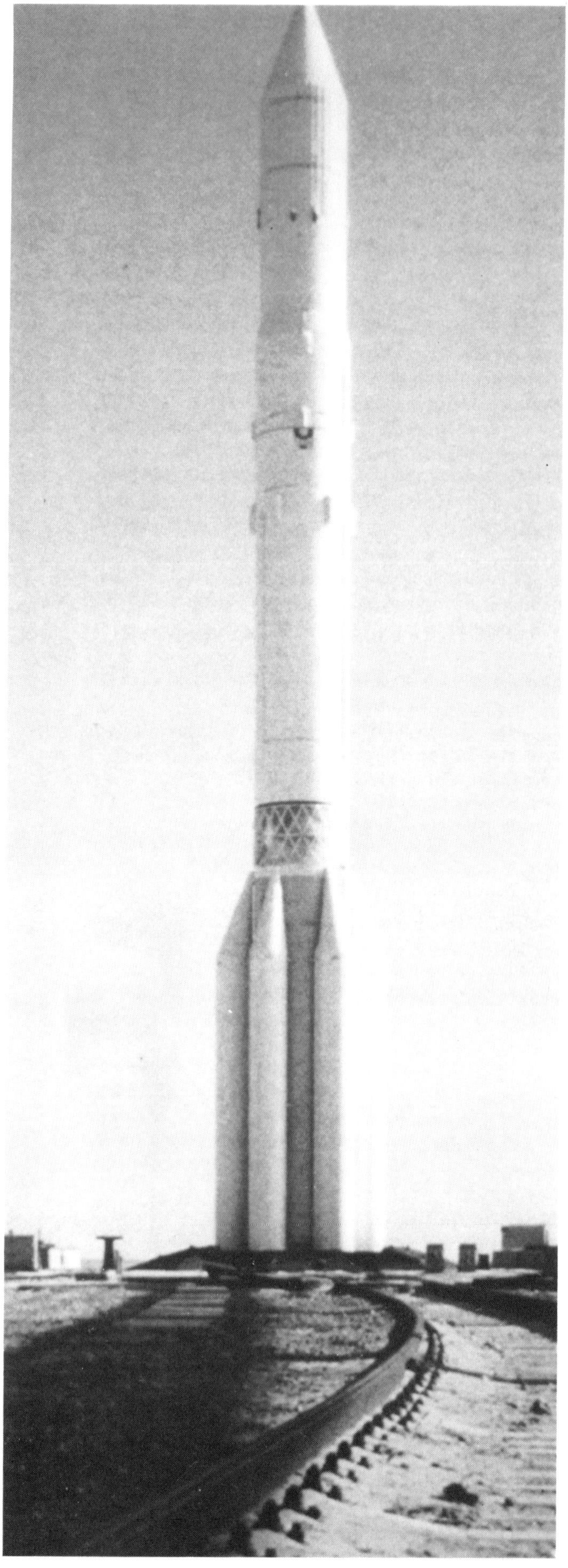

Introduction: This is a technically and possibly commercially underestimated launch vehicle which is being offered to the West to launch commercial satellites. Despite press reports early in 1985 that the Soviets were no longer offering it to Western countries, at the time of writing it was still being "marketed" to Inmarsat and is a possible contender to launch Insat II for India. The Inmarsat council was due to discuss use of the launch vehicle in November 1986. General comments received from satellite operators and manufacturers in the West by the author is that launch by Proton is a political non-starter as the US government would not allow satellites containing US systems of technology into the Soviet Union. Nevertheless, the use of Proton has been considered by a number of satellite operators in the West following the 1986 launch vehicle crisis and it is possible that the Soviets may have some success. The Soviets have given assurances about the technical transfer problem but they have been received with some scepticism. They have also set up a marketing organisation, Glavkosmos, to handle marketing of capacity on the vehicle and have continued to step up efforts during 1986. International sales are still handled by Licensintorg, a Soviet overseas trade body.

Potential Customers for Proton: Apart from Inmarsat, of which the Soviet Union is a member, a number of other satellite operators have been mentioned as possible users of the vehicle. These include Intelsat (of which the Soviet Union is not a member) and Eutelsat. It is difficult to see how the latter could launch on Proton as its members are mostly members of ESA and are therefore obliged by terms of their ESA membership, to use Ariane. There have been some discussions over the last two years between Intelsat and the Soviet Union about the latter joining the organisation. Intelsat has not denied that it is looking to launch Intelsat VI–F4 or F5 with Proton but this still appears to be an unrealistic proposition as the USA is still the largest member of the organisation. The USA has confirmed in September 1986 that the prohibition on bringing US made satellite components into the USSR remains. The Soviets have also offered ESA use of the vehicle to launch the Ulysses interplanetary probe in 1989 or 1990.

Background to the Proton Launch Vehicle: Proton is an old vehicle dating back to the mid-1960's and it is possible that it is no longer in production. However, the propulsion technology used is advanced by the standards of Ariane 5, Ariane 1–4, Atlas–Centaur and Delta. It uses closed-cycle engines. The Soviets are introducing a new series of launch vehicles which will supersede Proton. It appears that the Soviets have a number of Proton vehicles in store and available for use in the future.

Marketing of Proton: Prior to the establishment of Glavkosmos in 1985, the Soviets were weak in their marketing policy to the West. However they have been moving up the learning curve and recent attempts have been more professional and open. Criticisms of past marketing efforts are as follows:

1) A "need-to-know" limitation on supply of technical information. Sources suggested considerable friction between Soviet officials working through their Embassy in London and their Western counterparts, particularly in respect of technical information. Until late 1984, photographs of Proton were not publicly available in the West. When the Soviets are willing to supply information they are often slow in doing so.

2) Bad presentation of the launch vehicle details has compounded this problem. Past presentations, for example, to the IAF Congress, have been verbose and badly translated leading to a blinkered view in the West of the launch vehicle. Literature produced by Licensintorg (responsible for marketing the vehicle before Glavkosmos) has been described as disjointed.

3) The learning curve problem. The Soviets have limited experience in marketing anything and have had to start from scratch in developing their expertise in marketing a "new" product in a competitive market new to them. Many details will still have to be hammered out before a finalised deal could be struck for launch of a Western satellite by Proton.

The attempts to sell the vehicle to Inmarsat is significant in that it presents a solution to the technology transfer problem and also calls into question the constitution of Inmarsat if, as the cheapest tender, Proton is not properly considered. During the late 1970's the Soviets, in a little understood and largely forgotten gesture, indicated that their launch vehicles could be available for ESA launches of the Marecs satellites. (The Soviets also made an offer to ESA in 1986 to launch scientific satellites.) By the Spring of 1983, the Soviets had become highly active in marketing Proton directly to Inmarsat and have remained insistent as to its usefulness to the organisation.

Cost of Using Proton: It is understood that the Soviets are still offering to Inmarsat and other satellite operators a launch cost some \$10 m to \$15 m below costs offered elsewhere for launch of a single satellite. The 1983 bid by the Soviets, denominated in Swiss Francs, quoted a price of \$24 m and this appears not to have changed despite a slide in the value of the US\$ against the Swiss Franc. This covers the cost of a dedicated launch to Geostationary Orbit, but cannot directly be compared with an Ariane launch price. It is in fact considerably cheaper. Using a yardstick of \$80 m for an Intelsat VI launch by Ariane, Proton offers a number of savings:

1) Proton can place a satellite directly into Geostationary Orbit. This eliminates the need for an apogee boost motor and a transfer orbit ground operation. The lack or need for an apogee burn is estimated by the Soviets to add three to four years to the life of the Inmarsat II spacecraft, if launched by Proton.

2) Dedicated single launch cost on Proton can be reduced substantially by opting for multiple payloads. Dual launch is estimated at \$12 m and triple launch at \$8 m. Soviets are offering shared launches with their own military and "commercial" satellites.

3) There are also unquantifiable potentially large savings on insurance. Proton launch costs include an element of insurance. The Soviets offer, in the event of launch failure, a guaranteed re-flight. It is not clear at this stage if the re-flight is offered free of charge or at 50% of original cost or whether a third launch is guaranteed. In comparing with Ariane prices, Arianespace has been considering a form of re-flight guarantee through its S3R insurance arrangements. This is included in the Intelsat price of \$84 m at a nominal rate of 11%.

4) Real insurance costs, either through self-insurance or purchase of cover, may also be reduced by the fact that the Soviets can offer re-flight quickly after initial launch failure. The precise time is not clear although some sources suggest ten days as a minimum – the Soviets appear to have several launch pads available for Proton at the Baykonoor Cosmodrome. A quick flight reduces the need for consequential loss of revenue insurance.

5) Finally, evidence suggest that Proton is a reliable vehicle, clearly comparable with Delta and other US ELV's. There were 13 successful Proton launches during 1984 and Proton has been around for nearly 20 years. To offset against the vehicle's reliability is the relatively harsh launch environment to which its payload is subject.

On the other hand, there are a number of additional costs in using Proton:

1) Accuracy of injection of the satellite into Geostationary Orbit is not believed to be high and would therefore require modifications to the satellite to carry extra fuel for accurate positioning or result in a potentially shorter lifetime.

2) Because the top (escape) stage of Proton is 3-axis stabilised one side of its payload will be subject to intense heat, the opposite to intense cold, during the launch into Geostationary Orbit. This may require modification to the thermal systems of satellites using Proton or development of a spun platform for use with the top stage.

History of the Proton Vehicle: Proton has been an enigma for over 20 years. Until recently there have been no block disclosures of its size and capabilities, only glimpses and snippet releases of data. There have been a host of published articles purporting to give details of the vehicle, most of which have been fanciful. Recent revelations not only reflect improved marketing of Proton but also acknowledge the imminent introduction of a complete new family of launch vehicles with more advanced technology. Indeed the introduction into service of Proton itself facilitated release of details of the "A" series of launchers, including exhibition of an example at the 1967 Paris Air Show.

Released Data: After Proton launches of Vega 1 and Venus 2 on 15th and 21st December 1984, the Russians released comprehensive video material of the vehicle, its control room and first 50 seconds of flight. The Soviets made more details public in 1985 and 1986. Fully comprehensive details have yet to be released and there are a number of techical details that still remain unknown or subject to speculation. But several factors have emerged. The vehicle is tandem staged with an unusual first stage and with fuel and oxidiser stored concentrically. It has six fuel tanks each with its own engine and clustered around a larger diameter central oxidiser tank. The engines employ a pre-burner cycle as no turbine exhausts are visible. The evident oxidiser rich start also is consistent with the statement that the pre-burner is oxidiser rich. Film of launch in 1984 shows that the vehicle lifts off from the pad almost immediately after engine ignition, unlike the Soyuz launcher which takes some time for thrust built-up before lift-off. Also, the pad does not have the counter-weighted hold down arms used on the Soyuz pad, suggesting more conventional hold down clamps.

Proton Engines: The six RD253 engines which power the first stage of the Proton vehicle were, at the time of their introduction, the most advanced in the world. They were developed between 1961 and 1965. Propellants are similar to those used on Ariane, UDMH and N204 and the engine operates at the maximum chamber pressure attainable with these propellants. The RD253 is of the closed cycle or pre-burner design. It is still the most sophisticated storeable propellant engine and without a Western counterpart. The thrust to weight ratio is very high (130 quoted) and the specific impulse must also be high from the known characteristics (chamber pressure and area ratio). At the same time of their introduction the RD253 engines would have been second only to the Saturn V–F1 engines in thrust, but considerably superior in performance.

The high launch accelerations used on Soviet launches suggest that their engines have relatively high thrust/weight ratios. The Soviets have stated that development of these engines and similar ones gave considerable problems with combustion instability. Proton has never been man-rated; presumably the noise/vibration environment of the vehicle precludes this.

The Soviet Union paid a high price and accepted considerable risk in developing these engines and deserve to reap the rewards. New LOX/LH2 engines for the new generation of Soviet launch vehicles, once again are likely to be the most advanced high-thrust engines in use.

Reliability of the Proton Vehicle: The launch of the MIR space station in February 196? was the 120th successful launch of the vehicle. The Soviets claim that there have been 97 launches since 1970 (as at September 1986) of which seven have failed. The last 35 launches have been without loss. It is understood that the Soviets suffered from a number of losses prior to 1970, particularly in the early launches in 1965 and 1966.

HISTORY OF PROTON LAUNCH VEHICLE

1961	RD253 engine development begins – first ever closed cycle design produced in volume.
1965	Initial version of vehicle, SL–9(D) intoduced, comprising six strap-on boosters, the core and a third and fourth-stage. The fourth-stage is placed in a low earth parking orbit. The fourth-stage places the payload into an escape trajectory or into GEO. First launch, Cosmos 146 failed to leave parking orbit. It may have been a re-entry vehicle or part of a Soviet manned lunar mission that went wrong. The SL-12 version of Proton is the version currently being offered to the West.
1968	SL–13(D–1) version introduced. Essentially an SL–12 minus the escape stage. Used to place the Salyut and MIR space station (up to 19T) and associated star modules into LEO.
1968–70	Proton vehicle used for ZOND 5–8 missions around the moon.
1971	First Salyut space station launch (at 17·5T, now 19T).
1971–73	Venera 9–16 Venus probes launched.
1984	General release in the West (from US sources) of details of completely new range of Soviet launch vehicles including Shuttle craft and heavy ELV. Release by Soviets of film of Vega launches – first pictures of Proton clarity. Further details were made public in 1985 and 1986.
1985	Latest attempt to market Proton to Inmarsat. Announcement by USA of new initiative in launch vehicles including second generation Shuttle.
1986	Crisis in the Western launch vehicle market as Shuttle is grounded for two years and Ariane is found to have a generic design fault. Casual interest in use of the Proton vehicle increases.

CURRENT OPERATIONAL STATE OF VEHICLE	Grounded until February 1988. Operations curtailed after that
LAUNCH VEHICLE OPERATOR	National Aeronautics and Space Administration (NASA)
ADDRESS	600 Independence Avenue SW, Washington DC 20546, USA
TELEPHONE NO.	(202) 755-8370
TELEX NO.	8953 nasa wsh
SENIOR STAFF	James Fletcher, Administrator Chester Lee, Director, STS Customer Services
PRIME CONTRACTOR FOR LAUNCH VEHICLE	Rockwell International
OTHER MAJOR CONTRACTORS	Rocketdyne (main engines), United Technologies (booster), Morton Thiokol (booster engines), McDonnell Douglas (booster structure), Martin Marietta (external tank)
PAYLOAD CAPABILITIES	Total payload capacity to LEO is 29,500 kg. In practice this means a maximum of four commercial communications satellites. Payload bay dimensions are 18·28 metres long and 4·57 metres wide. Each Orbiter can carry a maximum of seven crew and return a payload of 14.515 kg to Earth
PAYLOAD LAUNCH COSTS	See text
DATE OF FIRST LAUNCH	14th April 1981 (STS–1)
FIRST COMMERCIAL PAYLOAD LAUNCH	11th November 1982 (STS–5)
NEXT AVAILABLE PAYLOAD OPPORTUNITY	1993 or beyond (see text)
LAUNCH SITE(S)	Kennedy Space Center Vandenberg (mothballed)

Introduction: Shuttle was developed to revolutionise the launch market, providing cheap, frequent and reliable services from the late 1970's. Its failure to do so has probably put the US space satellite and space programme back ten years and has had ripple effects throughout the satellite and space industries worldwide. Shuttle has now been withdrawn from the market for launching commercial satellites, although some 14 to 17 such payloads already booked on the vehicle are expected to fly between February 1988 and the end of 1992. The loss of Challenger on 28th January 1986 followed several years of disappointing results in the programme. The major impacts of those and the grounding of the vehicle for two years are:

1) Launch costs have increased considerably (threefold, including insurance, since 1982).

2) The future of both NASA and the US space programmes are in doubt.

3) New commercial opportunities in space are being curtailed.

4) Production lines of existing US expendable launch vehicles are either likely to be opened up or expanded.

5) The commercial viability of satellite communications has been undermined but not to the extent that some observers believe.

6) It places further pressure on Europe and Japan to achieve autonomy from the USA in space.

At the time of writing, the full implications of the Shuttle problems have yet to be fully understood. It appears that the White House in Washington DC and the US political system have been slow to respond to the problems involved.

Background to Shuttle: There is no doubt that Shuttle has employed some of the most advanced engineering and technology in any concept developed in the Western world. Nevertheless, despite the many glowing attributes that may be given to it, the system has manifestly failed to provide the performance that its development was originally justified on. So great is the difference between what was originally expected and what it has actually delivered that the entire policy and decision taking process behind US space policy have been rendered obsolete.

When Challenger was lost on the 28th January 1986, it was the 25th flight of the system. When the project was given go-ahead, the Shuttle was expected to have flown some 500 times. The difference between planned and actual performance is so great as to lead to the conclusion that the system has not worked at all. There has been considerable coverage of the technical and managerial problems which resulted in the loss of Challenger. Much less has been publicised about the broader aspects of the project. An understanding of the broader policy and decision making processes follows.

Current Position of Shuttle Availability: As at September 1986, Shuttle has been ruled out as a commercial satellite launch

option and a committee is considering which commercial payloads booked prior to its grounding should be launched by it in the future. Recommencement of launches is now set for February 1988, with early priorities including launch of a TDRSS spacecraft. Some 14 to 17 commercial payloads are expected to remain on the launch manifest, although not all of these will be communications satellites. NASA says that it will launch the remaining 25 to 28 commercial payloads by other means, although it has not disclosed which launch vehicles will be used. It is expected that Delta will take some of the payloads. Some observers believe that most of the payloads will defect to privately operated launchers.

Those commercial payloads that are expected to be selected for Shuttle launch are those that have significant foreign policy or national security implications or are designed so that they can only be launched with Shuttle.

Current thinking is that NASA is expected to fly Shuttle about 60 times between 1988 and 1992.

There are a number of payloads which are expected to be carried for certain on Shuttle during this period. These include two TDRSS satellites which are, in effect, part of the Space Transportation System anyway. The Space Telescope is also likely to fly early because of the cost of storage on the ground and concern about losing the scientific team behind the project. Four Spacelab missions are expected before 1990. These are Astro 1, Spacelab Sciences Mission, Earth Observation Mission and the International Microgravity Mission. The West German Spacelab D2 and Japanese D2 missions are not expected to fly before 1992.

Impact of the Shuttle Problems on the Satellite and Space Markets: The problems with the Shuttle programme appear to have set the US space programme back about a decade and has had a significant impact on the communications satellite market at a time when it is increasingly vulnerable to competition from fibre optics. As these are two separate issues, the impact on the space programme will be dealt with here. The introduction and the section on Intelsat cover the fibre optics issue.

The loss of Challenger on 28th January 1986 lost NASA much credibility and came after it was apparent that the vehicle was unable to provide cheap and reliable launch services. Although proponents of Shuttle argued that the insured losses on satellites launched via Shuttle were not caused by failures of the vehicle, the losses were closely related to the US space programme. This is particularly the case with the failures of upper stages experienced with the TDRSS–A, Palapa B2 and Westar VI satellites; the IUS and PAM–D stages were designed for use with Shuttle.

Similarly, the problems with the Leasat spacecraft appear to be related to their Shuttle unique design. The design was intended to be launched by Shuttle in an electrically dead mode, the electronics being turned on when the satellite was ejected from the payload bay. Some engineers believe that had the electronics been live the problem with Leasat 3 would have been known before it was ejected from the payload bay. Thus it is concluded here that the performance of Shuttle commercial satellite launches has not provided the level of reliability needed by commercial satellite operators and this was apparent before the vehicle was grounded.

Secondly, Shuttle has not provided the cheap launch services expected by the satellite industry. This has been primarily caused by the inability of NASA to launch at anything like the rate originally envisaged. It was apparent that prior to the loss of Challenger that the organisation was struggling very hard to handle even the then modest launch rate. This meant that the substantial fixed costs of Shuttle launches were spread over only a handful of launches, putting up the cost of each launch to in excess of \$250 m. NASA said that prior to the loss it was covering (only) the incremental cost of each launch through its charges. The discrepancy between NASA's charges for launch via Shuttle and the actual launch costs led to the White House administration taking corrective action from 1983. It is unlikely that even had the vehicle not been gounded that the new pricing policy could have ever resulted in there not being substantial launch subsidies.

Thirdly, the continuous history of slippages in both the development of the system and in its subsequent launch manifest, placed at times intolerable pressures on the vehicles' customers. They were (and are) faced with uncertainty as to when their spacecraft are launched, and, as a consequence have had to build in contingency plans, which, if realistic, make the operation of commercial satellites unviable. The dependence of the US (NASA and UK) military on the vehicle has also impacted on defence issues. In 1984, NASA lost some of the goodwill of the USAF when it opted to Purchase the Titan IV vehicle because of concern that Shuttle could not deliver the goods. The Titan IV vehicles are an expensive alternative to Shuttle at \$250 m a piece.

Similarly, the delays in the programme have further increased Europe's reluctance to have any dependence on the USA. The Spacelab programme was intended to be a major joint programme with the USA. Well before Shuttle was gounded, the Europeans were concerned that much of their effort and expenditure on the programme had been wasted as the number of

Spacelab missions were continuously wound down or delayed. The delays also have significant impact on the scientific community which, when there is a delay in launch of one of their payloads, has to fund the team using the payload until such time as it can begin to use the information from the operational payload.

The overall impact has been a considerable increase in the costs of being in the space business, for example, the cost of launch of a PAM–D satellite has increased from \$9 m in the early days of the Shuttle programme to \$54 m in 1987. In addition, insurance costs have increased from 10% to 30% of the total value of the satellite and launch costs. A quick comparison of the increases is as follows:

	1982 \$ m	1987 \$ m
Cost of PAM–D class satellite, 1982*	40	40
Launch charge†	9	54
Insurance	4	29
Interest payments from launch delay‡	4	16
Total	53	139

*Including transfer stage.
†1982 Shuttle charge, 1987 Delta charge.
‡Estimated two year delay on value of satellite at 20% per annum.

These costs ignore other costs of delay in any satellite programme, such as the need to use alternative communications systems, financial loss of revenue and satellite storage charges. They do, however, indicate a considerably negative shift in the economics of satellite communications to the extent of a threefold increase. Most of the increase can be put down to the problems with Shuttle.

Changing Political Climate and Shuttle: The problems with the Space Transportation System appear to have made the White House Administration very wary of funding the US space programme. It has produced a half hearted plan to finance the replacement Orbiter to Challenger. The plan gives the current or future Administration a lot of freedom to change its mind. Under the plan, only \$272 m has been initially allocated to the replacement and that only for Fiscal Year 1987. There is also a vague promise to use money for the replacement saved by the two-year grounding of the system. There appear to be three critical issues underlying the thinking in the White House:

1) The broader framework of the future US space programme. This appears not to have full political commitment. There are two major programmes involved; the development of advanced space and transatmospheric vehicles and the International Space Station. Both require expenditure in excess of \$5 b and both are comparable in size to the Shuttle development programme. The advanced launcher concept is also directly competitive with Shuttle funding. Although there is a strong public lobby in the USA supporting the US space programme, there are two reasons why the high level of public expenditure may be unrealistic:

A) The expenditure requires long term political commitment. The political environment in the Western world makes it difficult to commit to major projects beyond two years. The reason lies in the uncertainty of the economic environment. During the 1960's, politicians were protected from this uncertainty by economic growth rates which allowed them to practise welfare economics; that is to say they could expand expenditure without having to cut back in other areas (and thus adversely affect someone's economic welfare). For various reasons their ability to do so has declined in the 1970's and 1980's. However, politicians are loath to break the welfare principles and, at the same time, want to be seen to be doing things. To avoid having to withdraw allocated funding from a project or programme when economies in public spending are required, the simple way out is to commit only to short-term programmes.

B) Lack of political will. The US space programme on the 1960's was both a matter of national prestige and a response to a Soviet challenge. There was also a commercial spin-off in the form of communications satellites. In today's environment, another (third) major space programme may not have the same political charisma. It may be that the Americans are tired of the formula and now look elsewhere for prestige. Related to this is the SDI programme which has placed technology policies squarely in the military camp and outside of the civilian scientific sector.

2) Also open to question is the future level of demand for space transportation. That demand comes from six fields:

A) Commercial satellites. Expected demand has declined considerably over the last four years as growth rates in international telecommunications have levelled off and fibre optic is taking the market for public switched services. The DBS market has (so far) failed to materialise. The Remote Sensing market also remains small. The problems in the launch industry have also contributed to declining interest in satellites.

B) Strategic Defense Initiative Organisation payloads. There remain nagging doubts about the scientific and technical viability of this programme. Some observers believe it may collapse by the end of 1987, leaving Shuttle with even less future demand. SDIO is seen as a significant future user of the vehicle.

C) US Department of Defense payloads. The USAF is already investing heavily in the Titan vehicle (notably Titan IV but also refurbishing Titan II vehicles) and has issued plans for its Medium Expendable Launch Vehicle (MELV) for use from 1989. These launch vehicles will be available when the Shuttle is just returning to "normal" operations.

D) Private venture space commercialisation payloads, such as Spacelab and other manned modules, space platforms (SPAS, Eureka) and support services, and foreign users such as Intospace in West Germany. Many of the commercial organisations are either unlikely to survive the grounding of Shuttle or are likely to wind down their plans. Potential users of the modules and platforms are unlikely to make their plans clear until such time as Shuttle proves itself to be reliable enough to commit investment. In any event the future of such activities as Materials Processing in Space (MPS) looks as uncertain as ever.

E) The International Space Station. This project is already suffering from delays and looks uncertain to reach a 1995 operational basis; in 1984, a 1992 operational date was planned. Moreover, the international partners (Europe and Japan) are looking for autonomy in the exercise, including their own space transportation capabilities. There is therefore both a less and later need for Shuttle to support this. It has been a consistent argument of NASA that a fourth orbiter is required to support the Space Station.

F) Scientific and planatary spacecraft and missions. NASA's own scientific programme is under threat from the Shuttle funding problems; funding may have to be diverted to Shuttle leaving no future in house programmes. Fifty science missions were due to have flown on Shuttle before the end of 1992. Fifteen Spacelab missions have now been dropped in this period, leaving only four to fly by 1990. The Europeans are now very reluctant to enter into joint ventures with Americans following disappointments with Spacelab and the ISPM mission and now prefer projects with complete autonomy, using the Ariane vehicle instead.

There appears to be an almost implicit assumption in the vague plans for the replacement orbiter that the Administration is covering its bets. If the demand for Shuttle services prove to require an extra orbiter, then construction can continue; if it is not, then construction can be terminated without much loss of face or sunk costs. Also involved is implicit insurance against the loss of another orbiter.

3) Concern about NASA's ability to handle the US space

programme has been widely voiced by US politicians since the loss of Challenger. NASA is also in competition for funding of space activities with the Department of Defense, the latter of which appears to have been attempting, over the last three years, to reduce its dependence on NASA. NASA is essentially an R&D organisation rather than a service organisation to others. There is little doubt that many Americans feel that it has failed with Shuttle; the reasons may ultimately undermine the organisation and lead to its collapse.

The organisation's expenditure shrank considerably after the end of the Apollo missions. It had virtually been given an open cheque book by the US government in 1962 to get men on the moon and rapidly grew without having to face the political pressures of obtaining funding faced by other governmental organisations. It is argued here that it mishandled the politics of obtaining the funding for Shuttle as a result of that rapid and unconstrained growth. To survive, it needed a major project to follow on from Apollo, but political circumstances had changed between 1962 and the early 1970's.

After failure in Vietnam, the USA was uncertain about itself and had already proved its technological need. Its attention was diverted elsewhere. In order to obtain the funding for Shuttle, NASA tried to keep everyone happy. It agreed to go ahead with a Shuttle design which was a compromise between conflicting interests and needs. It tried to be an orbiting laboratory, a launch vehicle for large military payloads and a launch vehicle for small commercial payloads. It was perhaps done on the cheap to keep the politicians happy. What was really needed was a much smaller manned reusable launch vehicle carrying two or three people and having a limited in-orbit time. For longer durations a separate space station (Skylab) was needed. Large military payloads would have been left to an expensive ELV system. NASA compromised and promised cheap and reliable launch service to all. In the event, it has failed to deliver, leaving unhappy customers and some indifferent politicians.

NASA's organisational culture may help explain why it compromised itself. Flush with the undoubted successes of the 1960's and its reputation for scientific and technical capability (which it still retains), it may have neglected its own management process. The organisation developed a large bureaucracy in its years of expansion during the 1960's. As any student or organisational behaviour knows, bureaucracies tend to have lives of their own, serving their internal structures rather than either the production side of their organisation's activities or the outside needs to which all organisations ultimately depend upon. Moreover, the organisation has a science and engineering dominated culture rather than a more managerially dominated culture.

There has also been a lot of criticism of the White House's handling of the US space programme since January 1986. It is surprising that announcements were still being made about the future of ELV's and the future payloads of Shuttle as late as September 1986, some eight months after the disaster. Much of what has been said has left both the satellite and launch industries confused. Many believed that the future of the programme should and could have been decided more rapidly and decisively, particularly as the White House had, hitherto, been a keen supporter of space activities. Its efforts to provide commercial alternatives to Shuttle date back three years and therefore it is unlikely that the Administration was not aware of the commercial as well as the long term implications of the loss of Shuttle.

Shuttle Payload Capabilities: Shuttle has a nominal payload capability to low earth orbit of 30,000 kg. For commercial satellites this needs to be treated with some caution. It can carry a maximum of four PAM–D class satellites with their associated transfer stages. However, it is now highly unlikely that insurance underwriters would be willing to take the risks of four possible losses on one launch and as a maximum of 17 commercial payloads are expected to be launched on the vehicle, it is probably not likely to occur anyway.

Shuttle launches require use of a transfer stage to lift the geostationary satellite from the 180 mile high low earth orbit to the elliptical transfer orbit prior to firing of the apogee burn. The transfer stage may or may not combine the perigee and the apogee burn functions. For those that do not, a separate apogee motor is required. The problems with Shuttle have led to several transfer stage plans being abandoned and the Centaur stage being banned from the system on safety grounds. At the current time, the following stages are available:

Stage	*Status*	*Capability to geosynchronous transfer orbit (GTO)*
PAM–D	Operational	1250 kg
PAM–DII	Operational	1890 kg
PAM–A	Operational	1950 kg
IUS	Operational	2240 kg*
DTS	Under development	9070 kg
AMS	Under development	2000 kg
TOS	Under development	5900 kg
TOS + AMS	Under development	2950 kg†

*The Inertial Upper Stage is a two stage vehicle. Capability quoted here is to GEO.

†Payload on GEO with stage, propellant, payload and ASE limited to 7000 kg.

Shuttle Launch Charges: As there are to be no more new customers for Shuttle, its pricing policy is somewhat of academic interest. The final form of pricing policy was based on a minimum of $71 m for the whole cargo bay, with the capacity being auctioned off with this as the reserve price. It was generally felt in the industry that this policy would not have allowed commercial ELV operations to become established.

Prior to the introduction of the policy, the policy had been to charge the customers the incremental launch costs and fully allocated costs but not development costs. Moreover, these costs were based on average costs over a number of years. The early launch costs were expected to be higher than later costs as the frequencies of launches were fewer in the early period. But launch rates failed to improve significantly and the whole exercise was run at well below cost. It is estimated that the typical Shuttle mission incremental and fully allocated costs were still running at

Commercial payloads booked on Shuttle prior to January 1986 and still awaiting launch:

Payload	*Customer*
ASC–2	American Satellite Corporation
GStar 4	GTE Spacenet
Palapa B2	Unsold, but now with Sattel
Westar VI	Unsold
Intelsat VI F3	Intelsat, now on Ariane 4
Intelsat VI F4	Intelsat
Intelsat VI F5	Intelsat
Palapa B3	Perumtel, now on Delta
RCA Satcom	Five satellites for RCA Americom, one now on Ariane
Hughes	Ten satellites
SBS–6	Satellite Business Systems
Skynet IVA	UK MOD
Skynet IVB	UK MOD, now on Ariane
Syncom IV–5	Hughes Communications
Westar IVS	Western Union
STC 1	Comsat
STC 2	Comsat
Imarsat	Two satellites

\$250 m in 1985. As late as 1984, the charge for a fully dedicated payload bay was \$41 m. Some indication of the impact on launch charges resulting from the Shuttle problems is that a dedicated Titan IV launch is expected to cost \$250 m.

The payload charges to customers also took into consideration both the weight of the payload and the length of the cargo bay it occupied.

At the time of writing it has not been decided which of these will fly on Shuttle. About 15 can be accommodated by the end of 1992, with more slots possible later. All of the commercial satellites will receive low priority on the manifest. Those that are allowed to fly on Shuttle will reflect three requirements: 1) that they are Shuttle unique; 2) or that there are foreign policy implications; 3) or that they involve US national security.

Launch	*Date*	*Payload*
Shuttle STS–1	14th April 1981	Test flight
Shuttle STS–2	12th November 1981	Test flight
Shuttle STS–3	22nd March 1982	Test flight
Shuttle STS–4	27th June 1982	Test flight
Shuttle STS–5 (31–A)	11th November 1982	(Columbia) SBS–3 + PAM–D and Telesat E + PAM–D
Shuttle STS–6 (31–B)	4th April 1983	(Challenger) TDRSS–A + IUS*
Shuttle STS–7 (31–C)	18th June 1983	(Challenger) Telesat–F + PAM–D, Palapa B1 + PAM–D, SPAS, OSTA
Shuttle STS–8 (31–D)	30th August 1983	(Challenger) Insat 1B + PAM–D, PDRS, OIM
Shuttle STS–9 (41–A)	28th November 1983	(Columbia) Spacelab 1
Shuttle STS–11 (41–B)	3rd February 1984	(Challenger) Palapa B2*, Westar VI*, SPAS–01A
Shuttle STS–13 (41–C)	6th April 1984	(Challenger) LDEF–1, SMM Repair
Shuttle STS–14 (41–D)	30th August 1984	(Discovery) SBS–4 + PAM–D, Telesat 3C + PAM–D, Syncom IV2, OAST
Shuttle STS–17 (41–G)	5th October 1984	(Challenger) OSTA–3, ERBS, LFC/ORS
Shuttle STS–19 (51–A)	8th November 1984	(Discovery) Telesat–H + PAM–D, Syncom IV–1, HS–376 retrieval
Shuttle STS–20 (51–C)	24th January 1985	(Discovery) DOD payload (classified)
Shuttle STS–23 (51–D)	12th April 1985	(Discovery) Telesat–I + PAM–D, Syncom IV–3†
Shuttle STS–24 (51–B)	29th April 1985	(Challenger) Spacelab 3
Shuttle STS–25 (51–G)	17th June 1985	(Discovery) Morelos–A + PAM–D, Arabsat–1B + PAM–D, Telstar 3D + PAM–D, Spartan–1
Shuttle STS–26 (51–F)	29th July 1985	(Challenger) Spacelab 2
Shuttle STS–27 (51–I)	27th August 1985	(Discovery) Aussat–1 + PAM–D, ASC–1 + PAM–D, Syncom IV–4 †
Shuttle STS–28 (51–J)	3rd October 1985	(Atlantis) DOD payload (classified)
Shuttle STS–30 (61–A)	30th October 1985	(Challenger) Spacelab D1
Shuttle STS–31 (61–B)	21st November 1985	(Atlantis) Morelos–B + PAM–D, Satcom K1 + PAM–DII, Aussat–2 + PAM–D, ease/access
Shuttle STS–32 (61–C)	12th January 1986	(Columbia) Satcom K2 + PAM–D, Syncom IV–5, MSL–2
Shuttle STS–33 (51–L)	28th January 1986	(Challenger)‡ TBRSS–B + IUS, Spartan-Halley
Shuttle Flight 26	18th February 1988	(Discovery) TDRSS–C + IUS
Shuttle Flight 27	26th May 1988	(Atlantic) DOD (classified)
Shuttle Flight 28	28th July 1988	(Columbia) DOD (classified)
Shuttle Flight 29	22nd September 1988	(Discovery) TDRSS– + IUS
Shuttle Flight 30	17th November 1988	(Atlantis) HST
Shuttle Flight 31	19th January 1989	(Columbia) ASTRO–1
Shuttle Flight 32	2nd March 1989	(Discovery) DOD (classified)
Shuttle Flight 33	25th April 1989	(Atlantis) Magellan + IUS
Shuttle Flight 34	2nd June 1989	(Discovery) DOD Spacelab Long Module
Shuttle Flight 35	21st June 1989	(Columbia) Navstar–1 + PAM–D2, Navstar–2 + PAM–D2, MSL–3
Shuttle Flight 36	20th July 1989	(Atlantis) DOD (classified)
Shuttle Flight 37	1st September 1989	(Discovery) DOD (classified)
Shuttle Flight 38	21st September 1989	(Columbia) Navstar–3 + PAM–D2, Navstar–4 + PAM–D2, MSL–4
Shuttle Flight 39	1st November 1989	(Atlantis) Planetary Opportunity
Shuttle Flight 40	7th December 1989	(Discovery) SLS–1
Shuttle Flight 41	18th January 1990	(Columbia) GRO
Shuttle Flight 42	15th February 1990	(Atlantis) DOD (classified)
Shuttle Flight 43	20th April 1990	(Discovery) IML–1
Shuttle Flight 44	4th May 1990	(Columbia) Navstar–5 + PAM–D2, Pathfinder, EOS–1, Share
Shuttle Flight 45	31st May 1990	(Atlantis) DOD (classified)
Shuttle Flight 46	12th July 1990	(Discovery) DOD (classified)
Shuttle Flight 47	26th July 1990	(Columbia) Navstar–6 + PAM–D2, Skynet IV + PAM–D2, MSL–5
Shuttle Flight 48	31st August 1990	(Atlantis) DOD (classified)
Shuttle Flight 49	5th October 1990	(Discovery) Planetary Opportunity
Shuttle Flight 50	25th October 1990	(Columbia) Navstar–7 + PAM–D2, Insat 1D + PAM–D, TSS–1
Shuttle Flight 51	15th November 1990	(Atlantis) Leasat IV–5, LDEF Retrieval
Shuttle Flight 52	17th January 1991	(Discovery) Atlas–1, COFS–1
Shuttle Flight 53	14th February 1991	(Columbia) Navstar–8 + PAM–D2, Navstar–9 + PAM–D2, MSL–6, SSBUV–1
Shuttle Flight 54	1st March 1991	DOD (classified)
Shuttle Flight 55	4th April 1991	(Atlantis) Navstar–10 + PAM–D2, Skynet IV + PAM–D, Eureca

Details of launches from 1988 onwards were supplied by NASA in October 1986.
*Upper stage failure, †Satellite failure on launch, ‡Launch vehicle failure.

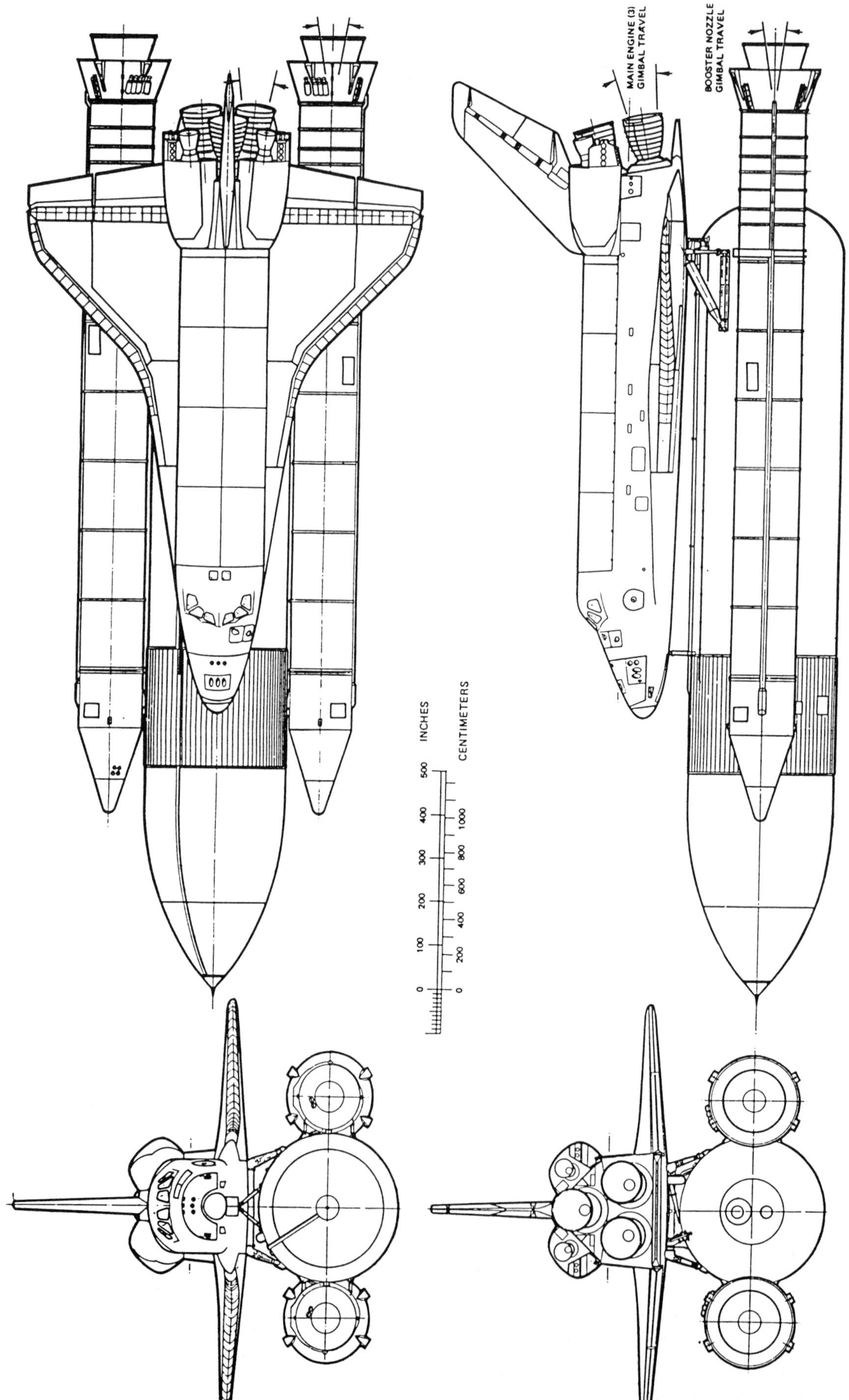
MAIN ENGINE (3) GIMBAL TRAVEL
BOOSTER NOZZLE GIMBAL TRAVEL
INCHES
0 100 200 300 400 500
CENTIMETERS
0 200 400 600 800 1000

CURRENT OPERATIONAL STATE OF VEHICLE	Currently in use for military payloads but proposed for commercial use from 1979
LAUNCH VEHICLE OPERATOR	(proposed) Martin Marietta
ADDRESS	Martin Marietta Denver Aerospace, PO Box 179, Denver, Colorado 80201, USA
TELEPHONE NO.	(303) 977–3000
TELEX NO.	910-935-0718
SENIOR STAFF	Richard Brackeen, VP Business Development Thomas Mayfiedl, Director of Business Development International Robert J. Polutchko, VP Space Electrical Systems John Boyd, Director, Public Relations
PRIME CONTRACTOR FOR LAUNCH VEHICLE	Martin Marietta Denver Aerospace
OTHER MAJOR CONTRACTORS	United Technologies, Chemical Systems Division (solid rocket motors), Aerojet Technical Systems Company (liquid propellant engines), General Motors Delco systems (inertial guidance components), McDonnell Douglas Astronautics (payload fairing)
PAYLOAD CAPABILITIES	See text for further details. Titan III can launch 1905 kg to geosynchronous orbit from Cape Canaveral or 14,470 kg to Low Earth Orbit. Payload to geosynchronous orbit is 6000 kg
PAYLOAD LAUNCH COSTS	Not available at time of writing. Fully dedicated 1989 launch is not expected to cost less than $70 m excluding insurance and payload preparation and integration
FIRST LAUNCH OF VEHICLE	29th July 1986
FIRST COMMERCIAL PAYLOAD LAUNCH	Martin Marietta is offering launches of commercial satellites from 1989
NEXT AVAILABLE PAYLOAD OPPORTUNITY	Early 1989
LAUNCH SITE(S)	Martin Marietta has formally requested to the USAF to use a launch complex at Canaveral Air Force Station for commercial Titan III use

TITAN III SUMMARY

Solid rocket boosters	
Length:	90·4 feet
Diameter:	10·2 feet
Motor Thrust:	1·4 m pounds per motor
Weight:	552,000 pounds per motor
Propellants:	Solid
Contractor:	United Technologies
Core stage	
Length:	78·6 feet
Diameter:	10 feet
Engine Thrust:	529,000 pounds
Propellants:	Liquid*
Stage Contractor:	Martin Marietta
Second stage	
Length:	29·5 feet
Diameter:	10 feet
Engine Thrust:	106,000 pounds
Propellants:	Liquid*
Stage Contractor:	Martin Marietta
Payload fairing	
Diameter:	13·1 feet
Overall Length:	Up to 55 feet

Titan III can deploy a single satellite or two smaller spacecraft to low, transfer, or geosynchronous Earth orbits. It can accommodate all anticipated commercial requirements with minimal impact for spacecraft already designed to fly on the Space Shuttle.

The Titan III can deliver up to 31,900 pounds (14,470 kg) into low-Earth orbit when launched from Cape Canaveral in Florida. Using a Transtage, it can place 4200 pounds (1905 kg) into geosynchronous orbit.

In anticipation of commencing commercial launch operations, Martin Marietta has made a formal request to the Air Force to use a launch complex at Canaveral Air Force Station, Florida, for this purpose.

Introduction: In July 1986, following the January 1986 failure of Shuttle and considerable lobbying from both the launch vehicle manufacturers and US satellite manufacturing and operating interests, President Reagan announced that launch of commercial satellites would be phased out from Shuttle. At the same time the USAF issued study contracts to four companies for a Medium Expendable Launch Vehicle (MELV) which, in effect, will be the "launch" customer for a commercialised ELV. The study contracts run until February 1987 after which the USAF will select one of the four for launch of its Navstar satellites. One of the four launch vehicles involved is the Titan vehicle. Despite not having yet been awarded the contract from the USAF, Martin Marietta began active marketing of Titan to commercial customers in the Summer of 1986, with an emphasis on its advantages over Ariane. Trade sources indicated to the author in early September 1986 that the Titan vehicle was not the front runner for the MELV contract and that it ranked second to Atlas-Centaur.

However, European satellite operators who have been

approached by Martin Marietta appear keen on the vehicle. It is not clear that the market could support more than one US commercial ELV, which will be in addition to Ariane and, from 1992, the Japanese H–II. Martin Marietta has said that it intends to proceed with Titan III as a commercial launch vehicle whether or not it wins the MELV contract from the USAF. The company says that it began holding discussions with potential commercial users of Titan III in March 1986.

Like all of the current US ELV's, Titan originated as an ICBM. The vehicle is the only such US vehicle with a production line that is currently open. Titan is a family of vehicles of which the Titan III version is being proposed for launch of commercial customers.

Background to the Titan vehicle: The Titan I vehicle dates back to 1956 when it was produced as an ICBM. If ever anyone wanted some evidence that the loss of Shuttle in January 1986 was a major set back for US space technology, the revival of a 30 year old launch vehicle (which, even in 1956, was nothing special) presents the case. The Titan I was developed into the Titan II ICBM which in turn was developed into a man-rated space launch vehicle for the Gemini programme of the early 1960's. Titan II launched 10 two man Earth orbiting missions over 19 months in 1965 and 1966. Titan III began service in 1964, although it did not have its first operational launch until 29th July 1966. For twenty years Titan has been the USAF's principal launch vehicle. Martin Marietta is currently working on several versions including the Titan III.

Titan II: This is currently intended for use for military payloads to low earth orbit. There is no public plan to use the version for commercial payloads. Martin Marietta is currently under contract to the USAF to refurbish decommissioned Titan II ICBM's for use as space launch vehicles. The contract covers eight vehicles and options for five more. The vehicles are being taken from an inventory of 56 decommissioned Titan II's. First launch is due in 1988 from Vandenburg. The refurbished vehicles can launch 2000 kg to low Earth polar orbit.

Titan III: This is currently used by the USAF for Department of Defense payloads. From the 29th July 1966 Titan has (at September 1986) flown 134 times with 129 successful launches. This is a 96·3% success rate. Launches have been either for NASA or (mostly) the USAF. It was used by NASA to launch the Viking spacecraft to Mars in 1975 and the Voyager deep space probe in 1977. The current versions of Titan III used by the USAF are Titan 34B (without solid rocket boosters and used for polar and other high inclination orbits) and Titan 34D (for geosynchronous as well as low and high inclination low-Earth orbits).

As a commercial vehicle, the company says that it will be available either for a dedicated launch or for a dual launch. It is looking at a new fairing greater than 4 metres diameter. Several transfer stage options are also available. These include PAM–D, PAM–DII, Transtage, TOS/AMS and IUS.

Titan IV: Also known as the Complimentary Expendable Launch Vehicle (CELV) or Titan 34D7, this is an upgraded version of Titan III with larger solid rocket boosters and a larger payload fairing. Its core is essentially the same as Titan III. It is being developed for the USAF to launch Shuttle class payloads. The company was, in 1985, awarded a contract for 10 vehicles for launch from 1988. The US congress has approved procurement of an additional 13 vehicles. Titan IV was procured because of continued delays in the Shuttle programme during the early 1980's and USAF concern that it was overdependent on the vehicle. Subsequent events have indicated that the USAF's decision proved wise.

Production rates of the Titan family: Martin Marietta claims that Titan manufacturing and launch facilities have excess capacity. It says it would have no difficulty in producing 20 vehicles per year (compared with Ariane's maximum of eight). There are two Titan launch complexes in Florida and two in California that provide a launch rate capability of 20 vehicles per year.

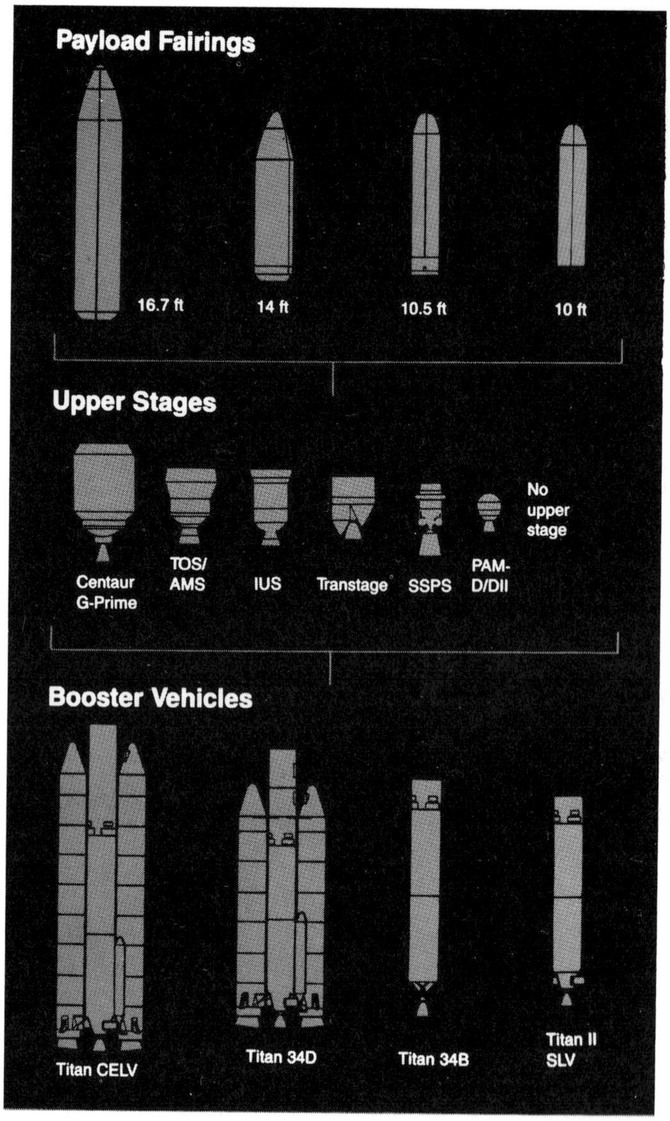

Potential users of the commercialised Titan III: Martin Marietta announced in August 1986 that Federal Express had signed a launch reservation with them for launch of an Expresstar satellite. According to a press release from Martin Marietta, Fedex will deposit $100,000 to establish a launch manifesting priority for one Expresstar satellite in 1989 and would take an option to fly a second satellite on Titan III in 1990. The agreement takes the form of a memorandum of understanding and is preliminary to negotiation of a full launch services contract between the two companies. It is the first signed launch reservation agreement since it was announced on the 19th August 1986 that Martin Marietta is offering Titan III for commercial satellite launches beginning 1989. The company has not released full details of the expected market for Titan III.

Nevertheless, in August 1986 it said that it had received enquiries about the launch of 21 satellites. Amongst the operators involved were Federal Express, Intelsat, Telesat Canada, Britsat and Société Européenne des Satellites. Eutelsat is also understood to be looking at using the vehicle. The Titan MELV vehicle is intended to launch Navstar global positioning satellites from

TITAN II

Low Earth Polar Orbit
Performance—4,200 lb

Maximum Payload
Envelope
9.3-ft dia x 25-ft long

TITAN III

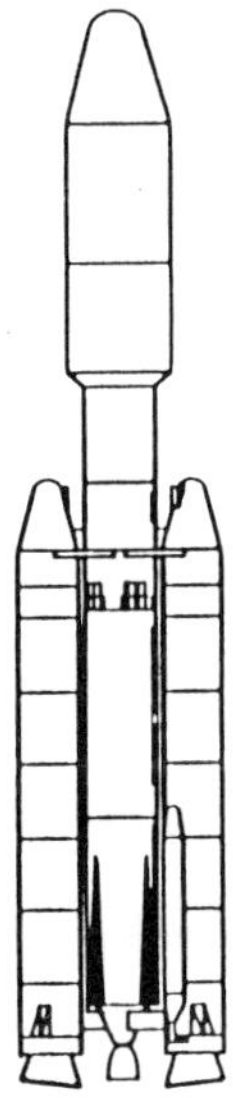

Low Earth Orbit
Performance—31,900 lb

Maximum Payload
Envelope
13-ft dia x 47-ft long

TITAN IV

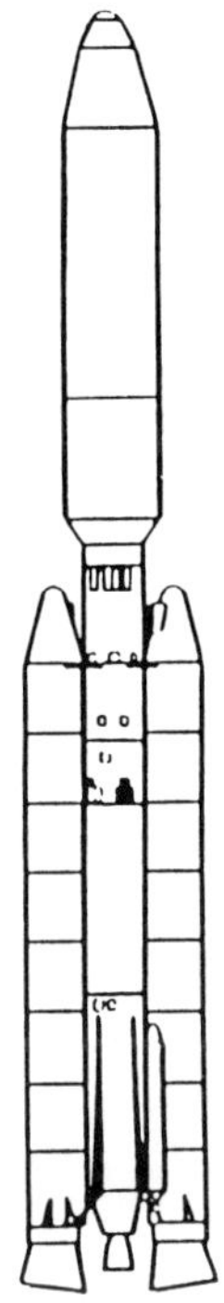

Low Earth Orbit
Performance—39,000 lb

Geosynchronous Orbit
Performance—10,000 lb

Low Earth Polar Orbit
Performance—32,000 lb

Maximum Payload
Envelope
15-ft dia x 67-ft long

Ariane Vehicles

Background

First Operational Launch:
1982
Total Operational Launches:
14
Total Successful: 11
Reliability: 78.6%

Vehicle Performance

Ariane 3—Operational Now
5,700 lb to 5° Geotransfer
Orbit

Ariane 4—Initial Launch 1987
9,500 lb to 5° Geotransfer
Orbit

Titan III Vehicles

First Operational Launch: 1966
Total Operational Launches: 134
Total Successful: 129
Reliability: 96.3%

Titan 34D—Operational Now
12,500 lb to 26° Geotransfer
Orbit (with Transtage Upper
Stage)
31,900 lb to Shuttle-Type Park
Orbit

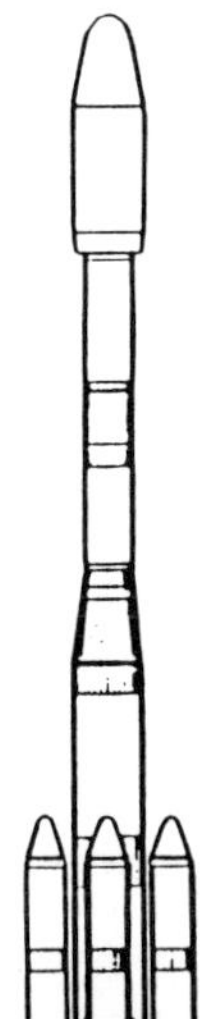

ARIANE 44L

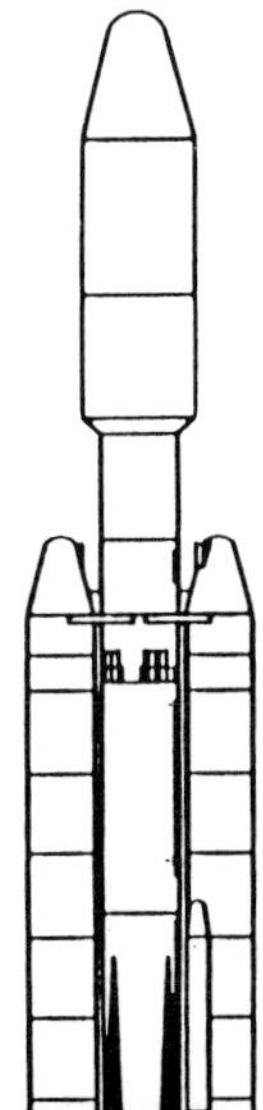

TITAN III

1989. The company was active at the Farnborough Air Show in September 1986 in promoting the vehicle. Titan III was presented as a direct competitor to Ariane on which there are no payload opportunities before the Spring of 1990. Something like one half of Martin Marietta's large stand at Farnborough was given over to the vehicle.

Pricing of a commercialised Titan III: The company has not released details of the intended launch charges for the vehicle It is initially expected to be expensive as there is a shortage of available launch capacity. Presumably its finalised pricing policy will depend on the amount of competition from other US companies (if any) who enter the ELV market. According to a company spokesman, the launch charges "will be competitive with Ariane ... we have a base price from which we will work out a price (for the customer). There are a variety of things we can offer from a complete turnkey service including payload preparation and integration and launch insurance to a basic minimum launch service. We may offer some kind of relaunch service in the event of launch failure". Martin Marietta has received proposals from three insurance consortia to arrange insurance cover for commercial customers on the vehicle. Details have not been released but the company has said that it would be able to compete with Arianespace's captive insurance arrangements (S3R).

UPPER STAGES FOR USE WITH SHUTTLE

Stage Name *Actual status*	*Prime contractor* *First flight*	*Propulsion type* *Manufacturer*	*Thrust* *Burning time*	*Length/diameter* *GTO/GEO payload*	*Remarks about development and commercialisation*
PAM–D Operational	McDonnell Douglas Astronautics 1982	Solid Star 48 Thiokol	66·3 kN 85 s	2·4 m/1·3 m 1250 kg/635 kg	11 uses with Delta, 16 uses with Space Shuttle, with two failures, other 28 missions scheduled
PAM–A Operationally available	McDonnell Douglas Astronautics Not yet used	Solid Minuteman 3 Thiokol	157 kN 60 s	2·3 m/1·3 m 1950 kg/998 kg	Ready for operations in 1982, presently in storage for Atlas–Centaur class spacecraft
IUS/two-stage Operational	Boeing Aerospace + US Air Force 1983	Solid SRM–1/Solid SRM–2 CSD	200/81·2 kN 153/104 s	5 m/2·9 m 4·5 t/2·3 t	Used with the Space Shuttle and with the Titan 34D vehicle
Centaur G Cancelled	General Dynamics 1986	LO_2/LH_2 RL–10A–3–3B Pratt & Whitney	133/64·7 kN 436 s	5·9 m/4·3 m Not available/4540 kg	Developed from the Atlas and Titan upper stage for military payload
Centaur G Cancelled	General Dynamics 1986	LO_2/LH_2 RL–10A–3–3A Pratt & Whitney	147/74·4 kN 609 s	8·9 m/4·3 m Not available/up to 8 t?	To be used for the launch of the interplanetary Galileo and Ulysses probes in solar orbit
TOS Development	Martin Marietta + Orbital Sciences 1986?	Solid SRM–1 CSD	200 kN 150 s	3·3 m/3·4 m 5·9 t/3·1 t	Commercial venture of Orbital Sciences Corporation looking for customers
AMS Development	Martin Marietta + Orbital Sciences 1988?	N_2O_4/MMH RS–41 Rocketdyne	11·8 kN 1050 s	1·7 m/3 m 1·6 t/862 kg	Commercial venture of Orbital Sciences Corporation looking for customers
IRIS Development	Aeritalia 1988	Solid HTPB BPD Difesa Spazio	Not available Not available	2·3 m/1·5 m 900 kg/454 kg	Italian programme for a small upper stage to launch light payload
PAM–DII Development	McDonnell Douglas Astronautics 1985	Solid ISTP Thiokol	78·3 kN 121 s	2 m/1·6 m 1·9 t/967 kg	Uprated PAM–D stage, to be used for the first time in November 1985 to launch RCA Ku-band 1
SCOTS Development	RCA Astro 1987	Solid 63–E Thiokol	156 kN 125 s	2·8 m/1·8 m 2·5 t/1130 kg	Commercial stage developed to launch the RCA DBS-type spacecraft
STV Study	British Aerospace 1988	N_2O_4/MMH engine Not available	Not available Not available	2 m/4·4 m up to 3 t/1·5 t	Commercial venture of Scott Science & Technology, pending on financial support
LPM Study	Aerojet TechSystems Company 1988?	N_2O_4/MMH Transtar Aerojet TechSystems Company	16·7 kN 1660 s	1·5 m/4·1 m 4·4 t/1·5 t	Commercial venture looking for partnership (with Ford Aerospace, to develop the HPPM?
UPS Suspended	Snia-BPD Unknown				Product a result of ESA concern about availability of an upper stage for Olympus class satellites

Abbreviations
AMS = Apogee and Manoeuvre Stage
HPPM = High Performance Propulsion Module
IRIS = Italian Research Interim Stage
IUS = Inertial Upper Stage (proposed by Boeing for an Orbital Transfer Vehicle)
PAM = Payload Assist Module
SCOTS = Shuttle Compatible Orbit Transfer Stage
STV = Satellite Transfer Vehicle
LPM = Liquid Propulsion Stage
TOS = Transfer Orbit Stage

SECTION IV

SUPPORTING INFORMATION

SUMMARY OF PLATFORMS FOR COMMERCIAL SATELLITES

Platform	*RCA–3000*	*RCA–4000*	*RCA–5000*	*NATO III*	*Supersat*	*ECS*	*Eurostar*	*Olympus*	*Spacebus 100*	*Spacebus 100B*
PRIME CONTRACTOR	RCA Astro-Electronics	RCA Astro-Electronics	RCA Astro-Electronics	Ford Aerospace	Ford Aerospace	BAe or Matra (see notes)	Satcom International (see notes)	Mesh consortium (see notes)	Eurosatellite consortium (see notes)	Eurosatellite consortium (see notes)
ADDRESS	PO Box 800 Princeton NJ 08540 USA	PO Box 800 Princeton NJ 08540 USA	PO Box 800 Princeton NJ 08540 USA	3939 Fabian Way Palo Alto CA 94303 USA	3939 Fabian Way Palo Alto CA 94303 USA					
TELEPHONE NO.	609-426-3400	609-426-3400	609-426-3400	415-852-6980	415-852-6980					
TELEX NO.				34-8481	34-8481					
PRINCIPAL OFFICERS	A. Manna, Dir Mrktg L. Erberle, Mgr, Coms	A. Manna, Dir Mrktg L. Erberle, Mgr, Coms	A. Manna, Dir Mrktg L. Erberle, Mgr, Coms	H. Hockeimer, President R. Kueber, VP C. Stiles, WDL Mrkt R. Rusch, Mgr, Space	H. Hockeimer, President R. Kueber, VP C. Stiles, WDL Mrkt R. Rusch, Mgr, Space					
LAUNCH MASS						865–1250 kg	Up to 2000 kg	Up to 3700 kg	Shuttle 1270–1460 kg Ariane 1550 kg	Shuttle 1720 kg Ariane 1850 kg
MASS IN ORBIT	580–670 kg (3000 lbs)	(4000–5500 lbs)	(5000 lbs plus)	670–1100 kg 350–550 kg	950–1528 kg	610–680 kg		Up to 2500 kg Not available		
BUSS DRY MASS									460 kg	560 kg
PROPELLANT MASS								Up to 1200 kg	743 kg	942 kg
STABILISATION	3-axis	3-axis	3-axis	Spin	3-axis	3-axis		3-axis	3-axis	3-axis
POWER SUPPLY	740–2260 Watts			400–800 Watts	1600–8000 Watts	1100 Watts	Up to 5000 Watts	3500–8000 Watts		
DESIGN LIFE	10 years	10–13 years		5–7 years	10 years	7 years	10 years	10 years	7 years	10 years
APOGEE MOTOR	solid propellant				liquid propellant	solid propellant	bi-propellant	bi-propellant		
TRANSFER STAGE	PAM–D PAM–DII	SCOTS		Not applicable	HPPM (cancel.) PAM–DII	PAM–D		STV, UPS, IUS	PAM–D PAM–DII	PAM–DII
BODY DIMENSIONS				3·1 m dia.					2·1 × 1·6 × 1·4 m	2·1 × 1·8 × 1·4 m
LAUNCH VEHICLE COMPATIBILITY	Ariane, Shuttle, Delta	Ariane 3–4, Shuttle	Ariane, Shuttle	CS–2 (2), CS–3 (2)	Ariane, Shuttle, Titan	Ariane 1–4, Shuttle, Delta	Ariane 1–4, Shuttle	Ariane 3–4, Shuttle	Ariane 1–4, Shuttle	Ariane 2–4, Shuttle
FIRM ORDERS & OPTIONS	ASC (2) Anik B (1) PanAmSat (1) RCA Americom (4) Spacenet (4) STC (2) GStar (4) BS–3 (2)	SES (1) RCA Americom (4)		NATO III (4)	Fordstar (3) Space Com. Corp. (2) Telecom 1 (3)	Skynet IV (3) MARECS (3) ECS (5) DFS–Copernikus (3) OTS (2)	Inmarsat II (9)	ESA–Olympus (1)	Arabsat (3)	Eutelsat II (3)
POTENTIAL CUSTOMERS	Helvesat (2)	Britsat (3) SES (1) PanAmSat (2)	Telesat Canada (3) Celestar (3) Expresstar (3) RCA DBS (2) USSBCI			No more orders are likely as the design is obsolete	UK DBS (3) Telecom II (3)	ESA–Olympus (1) Sarit (Italy) (3) Radarsat (1)	Satcom (Colombia) CBSC (China)	Eutelsat II (6) Telecom II

SUMMARY OF PLATFORMS FOR COMMERCIAL SATELLITES—(continued)

Platform	*Spacebus 200*	*Spacebus 300*	*Spacebus 300B*	*Spacebus 400*	*HS–376*	*HS–381*	*HS–393*	*HS–394*	*HS–399*
PRIME CONTRACTOR	Eurosatellite consortium (see notes)	Eurosatellite consortium (see notes)	Eurosatellite consortium (see notes)	Eurosatellite consortium (see notes)	Hughes Aircraft Company Space & Communications Group	Hughes Aircraft Company Space & Communications Group	Hughes Aircraft Company Space & Communications Group	Hughes Aircraft Company Space & Communications Group	Hughes Aircraft Company Space & Communications Group
ADDRESS					PO Box 92919, Los Angeles, CA 90009, USA	PO Box 92919, Los Angeles, CA 90009, USA	PO Box 92919, Los Angeles, CA 90009, USA	PO Box 92919, Los Angeles, CA 90009, USA	PO Box 92919, Los Angeles, CA 90009, USA
TELEPHONE NO.					(213) 648-0942	(213) 648-0942	(213) 648-0942	(213) 648-0942	(213) 648-0942
TELEX NO.									
PRINCIPAL OFFICERS					F. Judge, Mrkt Dev Manager	F. Judge, Mrkt Dev Manager	F. Judge, Mrkt Dev Manager	F. Judge, Mkt Dev Manager	F. Judge, Mrkt Dev Manager
LAUNCH MASS	Shuttle 2050 kg Ariane 1850–2200 kg	Shuttle n/a Ariane 2400 kg	Shuttle 3300 kg Ariane 2950 kg	Shuttle 5000–5300 kg Ariane 5000–5300 kg	7300 kg	7711 kg			
MASS IN ORBIT					540–661 kg	1388 kg	1090 kg		320 kg
BUS DRY MASS	600–690 kg	750–870 kg	1000–1140 kg	1100–1250 kg	1050 kg				
PROPELLANT MASS	1220 kg	1180 kg	1550 kg	2800 kg	250 kg				
STABILISATION	3-axis	3-axis	3-axis	3-axis	Spin	Spin	Spin	Spin/body	Spin
POWER SUPPLY	1–2000 Watts	2–6000 Watts	2–6000 Watts		9000–1118 Watts	1200 Watts	2250 Watts	4000 Watts	300 Watts
DESIGN LIFE	10 years	7–9 years	10–12 years	10–12 years	7–10 years	10 years	10 years	7 years	10 years
APOGEE MOTOR					solid propellant	liquid propellant	liquid propellant	liquid propellant	liquid propellant
TRANSFER STAGE	HPPM	Not applicable	TOS–S	TOS	PAM–D	solid fuel, integral	solid fuel	solid fuel	single bipropellant for perigee & apogee
BODY DIMENSIONS	2·4 × 1·8 × 1·6 m	2·4 × 2·3 × 1·6 m	3 × 2·3 × 1·6 m	3·2 × 2·9 × 1·7 m	21′8″ h × 7′1″ m dia.	6·17 h × 4·26 m dia.	10 m h, 3·6 m dia.	3·6 m dia.	2·8 m h, 0·9 m dia.
LAUNCH VEHICLE COMPATIBILITY	Ariane 4, Shuttle	Ariane 3–4,	Ariane 44P, Shuttle	Ariane 4, Shuttle	Ariane, Shuttle, Delta	Shuttle only	Ariane 4, Shuttle	Shuttle	Shuttle, Ariane
FIRM ORDERS & OPTIONS		TV–SAT (1) TDF–1/2 (2) TELE–X (1)			Aussat (4) SBS (5) Anik C (3) Anik D (3) Westar (5) Palapa B (3) SBTS (2) Morelos (2) Galaxy (4) Telstar (3)	Leasat (5)	SBS (1) JCSAT (2)	None so far	None so far
POTENTIAL CUSTOMERS	ESA–DRS	US DBS (2) TV–SAT 2 (1)			None now likely as design is now obsolescent		Atlantic Satellites (2) PanAmSat (2) Galaxy (3) SBS (3)	Hughes Communications Galaxy (2)	

Notes: Although the Summary of Platforms indicates a straightforward relationship between the design of a satellite and the prime contractor, the relationship is not as straightforward as initially appears. Only in the USA do commercial satellite manufacturers have the wherewithall and market demand to offer their own designs to customers. Other satellite manufacturers either work as part of a consortium, as in Europe, where more than one member may offer the design as prime contractor, or, as in Canada, offer the platform of a US manufacturer. For example, the Spacebus platform, is being offered with either Aerospatiale and MBB–Erno as prime contractor, depending on who the prospective customer is. MBB–Erno and Aerospatiale are both members of the Eurosatellite consortium. In Canada, Spar is acting as prime contractor to Telesat for the Anik D satellites, using a Hughes HS–376 platform, and for the Anik E system using an RC 5000 platform.

The basic reasons behind these arrangements are relatively straightforward. The communications industries in Europe and Canada are major employers and have been a significant factor in industrial policy. European and Canadian governments have therefore been keen to build up an indigenous capability in satellite manufacturing and either directly or through the European Space Agency have funded a substantial effort into R&D and helped build up the industry through procurement of satellites from domestic manufacturers. There are also military reasons for building up capabilities independent of the USA.

However, the European market is fragmented and, in total, considerably smaller than the US market. The small geographic size of Europe compared with North America and the dominance of monopoly PTT organisations with an interest in protecting their investment in terrestrial communications means that, in a broad sense, the underlying demand for satellites is smaller in Europe than North America. Moreover, individual governments in Europe like to see domestic orders for satellites placed with domestic manufacturers. The natural tendency is therefore towards small production lines and high unit costs. One ex-Hughes senior manager has calculated that European satellites cost around 70–80% more than their US equivalents although Telesat Canada calculated a figure of around $12\frac{1}{2}$%. The small demand for satellites in Europe and the heavy expenditure on R&D required to keep abreast of technology has also favoured the consortium approach in Europe where R&D expenditure is also split between governments. The approach is being institutionalised for other European R&D and pre-competitive research efforts in advanced technology, notably the European Commission's RACE and ESPRIT programmes and the broader Eureka programme which is Europe's reply to President Reagan's Star Wars Initiative. These three programmes and the experience of the satellite manufacturing industry are indicative of Europe's weak position in competing with the USA on advanced technology.

In contrast, Canada has built up its domestic satellite manufacturing industry by gaining expertise through sub-contracting to US firms before reaching the lofty peaks of prime contractorship in the Anik D and Anik E series and the SBTS system. This has meant that Canada has not had to reinvent the wheel in developing satellite technologies and can cope with limited domestic demand for satellites.

The Japanese approach, which, so far, has probably been less successful than Europe, combines both the Canadian and European methodologies. Its domestic CS–2/3 and BS–2/3 satellites have relied heavily on US technology but are largely not economic propositions because of high costs and limited capacity. As an interim solution and partly as a result of broader trading pressures it is now importing US satellites from Hughes and Ford and, possibly RCA, to provide domestic communications capacity. Japanese industry is not involved in the manufacturing of the satellites. However it is maintaining a substantial R&D programme to eventually compete with US industry. This involves the Super–CS satellite, the H–II launch vehicle, concentration on Ka-band technology and integration with broader strategies on dominating the world market for television hardware including development of HDTV capabilities.

The weakness of Japan's current satellite manufacturing industry and the current need to import US-built satellites is paralleled in Europe where the high costs, long delivery times and perceived unproven reliability of European manufactured satellites has led to three potential commercial satellite operations, SES, Britsat and Atlantic satellites, to chose US built satellites.

The dominance of the US satellite manufacturing industry in the world markets is explained by: 1) Intelsat, the USA was the force behind setting up this organisation to provide an international market for its satellites. The Intelsat orders for satellites have always resulted in prime contractorship going to US companies with only modest sub-contract orders being placed outside the USA. The orders have enabled US manufacturers to build up economies of scale and obtain further expertise through Intelsat funded R&D. 2) An early start in satellite communications coupled with substantial government funding of R&D through NASA and the Department of Defense. 3) Substantial domestic demand for communications satellites created by a liberal communications market place, geography and the requirements of defence. Consequently, the US satellite manufacturing industry has not had to depend on consortia.

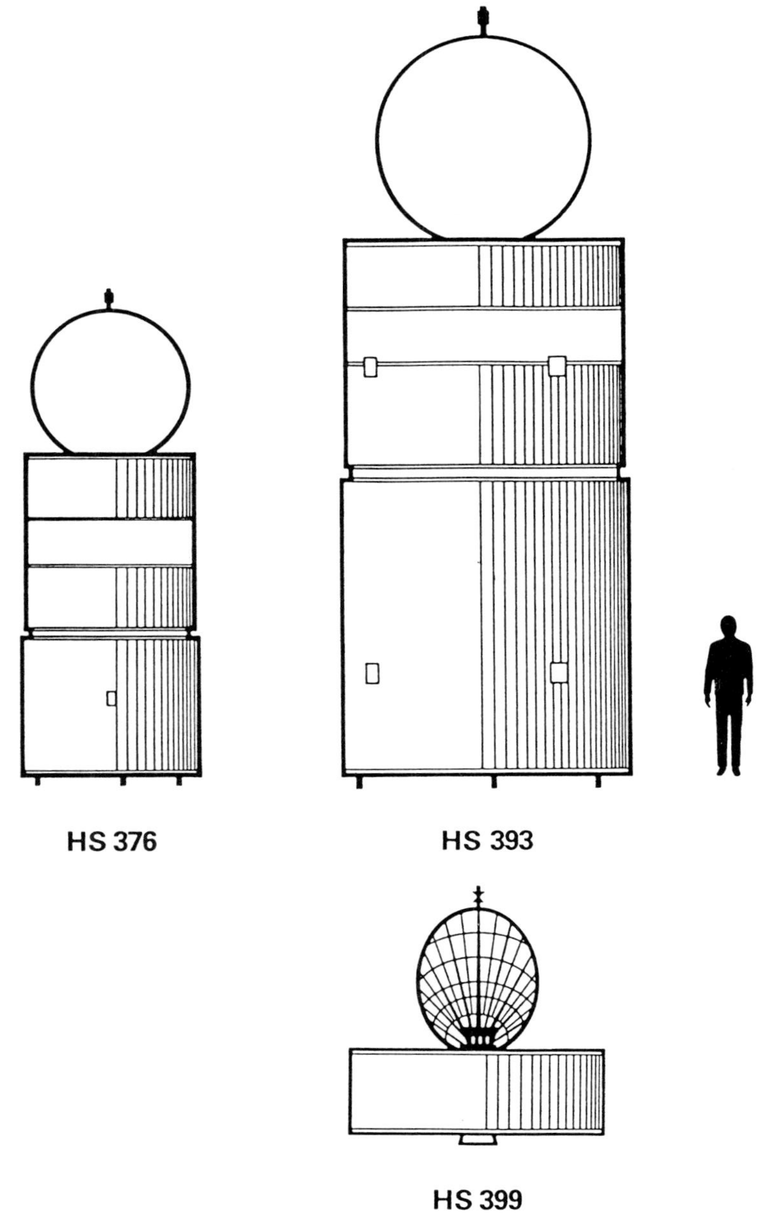

Hughes Aircraft Company's HS–376, HS–393 and HS–399 buses, drawn to comparative scale

LIST OF ACRONYMS

Ack	Acknowledgement (positive)
ACS	Attitude Control System
ACTS	Advanced Communications Technology Satellite
ADC	Analogue-to-Digital Coding
ADCCP	Advanced Data Communications Control
ADM	Adaptive Delta Modulation
ADPCM	Adaptive Differential Pulse Code Modulation
AFC	Automatic Frequency Control
Afsat	African regional satellite system (under study)
AKM	Apogee Kick Motor
Alascom	Alaskan Communications Satellite Corporation
AM	Amplitude Modulation
AMS	African Mediterranean System
ANSI	American National Standards Institute
AOTS	Advanced Orbital Test Satellite
APCM	Adaptive Pulse Code Modulation
Arabsat	Arab Satellite Corporation, also name of its satellite space segment
ARQ	Automatic Retransmission on Request
ASC	American Satellite Company
ASCII	American Standard Code for Information Interchange
ASSA	Austria Solar and Space Agency
AT&T	American Telephone and Telegraph Corporation
Aussat	Australian Satellite Corporation, also name of its satellite space segment
BAe	British Aerospace
BAeDG	British Aerospace Dynamics Group
BBC	British Broadcasting Corporation
BCC	Block Check Character
BER	Bit Error Rate
BIT	BInary digiT
BNSC	British National Space Centre
B-MAC	*see* MAC
BMFT	Bundesministerium fur Forschung & Technologie (Germany) (Federal Ministry for Research and Technology)
BOL	Beginning Of Life
BPSK	Binary Phase-Shift Keying
BRU	Broadcasting Research Unit
BSC	Binary Synchronous Communication
BSS	Broadcast Satellite Service (or Systems)
BT	British Telecom (Telephone Company)
BTI	British Telecom International
CATV	Community Antenna Television
CCIR	Comite Consultatif International de Radio (International Radio Consultive Committee)
CCITT	Comite Consultatif International de Téléphone et de Télégraph (International Telephone and Telegraph Consultative Committee)
CDC	Control and Delay Channel
CDMA	Code Division Multiple Access
CEC	Commission of the European Communities
CEPT	Conference Européene de Postes et Telegraphs (European Conference of PTT's)
CISI	Consortium of the Italian Space Industry
C/N	Carrier to Noise Ratio
C-MAC	*see* MAC
CNES	Centre National d'Etudes Spatiales (French National Space Agency)
CNN	Cable News Network
CNR	National Research Council (Italy)
CNS	Compagnie Nazionale Satellite per Telecomuniazioni
Comsat	Communications Satellite Corporation
CONUS	Contiguous, or Continental USA – i.e. the USA excluding Hawaii and Alaska
CoSpaR	Committee of Space Research
CRTC	Canadian Radio-television and Telecommunications Commission
DAMA	Demand Assignment Multiple Access
dB	Decibel
dBi	Decibel Isotropic
DBP	Deutsche Bundespost
DBS	Direct Broadcast Satellite
dBW	Decibel Watt
DCTE	Data Circuit Terminating Equipment
DES	Data Encryption Standard
DFVLR	Deutsche Forschungs-und Versuchsanstalt fur Luft-und Raumfahrt (German equivalent of NASA)
DGT	Direction Generale de Telecommunications (French PTT)
DM	Delta Modulation
Domsat	Domestic (Communications) Satellite
DPCM	Differential Pulse Code Modulation
DRS	Data Relay Satellite
DSCS	Defense Service Communications Satellite
DSI	Digital Speech Interpolation
DTE	Data Terminal Equipment
DTI	Department of Trade and Industry (UK)
DTS	Digital Termination Service
EBU	European Broadcasting Union
ECS	European Communications Satellite
EIRP	Equivalent Isotropic Radiated Power
ELDO	European Launcher Development Organisation
ELV	Expendable Launch Vehicle
ENT	Equivalent Noise Temperature
EOL	End Of Life
E-PAL	Extended-PAL
ESOC	European Space Operations Centre
ESA	European Space Agency
ESRO	European Space Research Organisation
ESTEC	European Space Research and Technology Centre
Eutelsat	European Telecommunications Satellite Organisation
EVA	Extra Vehicle Activity
FACC	Ford Aerospace and Communications Corporation
FAX	Facsimile
FCC	Federal Communications Commission
FDM	Frequency Division Multiplex
FDMA	Frequency Division (or Domain) Multiple Access
FEC	Forward Error Control
FET	Field Effect Transistor
FM	Frequency Modulation
FSK	Frequency Shift Key
FSS	Fixed Satellite Services
G	Gain
GaAsFET	Gallium Arsenide Field Effect Transistor
GDL	Grand Duchy du Luxembourg – old name for Astra satellite system
GEO	Geostationary Operating Orbit, or Geostationary Earth Orbit
GHz	Gigahertz (1000 MHz)
GPS	Global Positioning System
GSO	Geostationary Orbit
G/T	Gain-to-Temperature Ratio
GTE	General Telephone Electronics Corporation
GTO	Geostationary Transfer Orbit
GWIC	Great Wall Industry Corporation (China)

LIST OF ACRONYMS—(continued)

H–I/H–II	Japanese Launch Vehicles
HACBSS	Homestead and Community Broadcasting Satellite Service
HBO	Home Box Office
HDLC	High-Level Data Link Control
HDTV	High Definition TV
HPA	High Power Amplifier
HPPM	High Performance Propulsion Module
Hz	Hertz
IAF	International Astronautics Federation
IBA	Independent Broadcasting Authority (UK)
IEEE	Institute of Electrical and Electronics Engineers (USA)
IF	Intermediate Frequency
IFRB	International Frequency Registration Board
IIC	International Institute of Communications
Inmarsat	International Maritime Satellite Communications Organisation
Insat	Indian Satellite
Intelsat	International Telecommunications Satellite Organisation
Intersputnik	International Space Telecoms Organisation
ISDN	Integrated Services Digital Network
ISRO	Indian Space Research Organisation
ISO	International Organisation for Standardisation
ITN	Independent Television News
ITU	International Telecommunications Union
IUS	Inertial Upper Stage
KHz	Kilohertz (1·000 Hz)
Lapan	Indonesian National Institute for Aeronautics & Space
LEO	Low Earth Orbit
LNA	Low Noise Amplifier
LNB	Low Noise Block Converter
MAC	Multiplexed Analogue Component
Marecs	Maritime European Communications Satellite
MATV	Master Antenna Television
MCPC	Multiple Channel Per Carrier
MHz	Megahertz (1·000·000 Hz)
MITI	Ministry for International Trade and Industry
MOA	Ministry of Astronautics (China)
MOD	Ministry of Defence (UK)
Modem	Modulator/Demodulator
MSK	Medium Shift Key
M–Sat	Mobile Satellite
MSS	Mobile Satellite Service
MSS	Marconi Space Systems
NAk	Negative Acknowledgement
NASA	National Aeronautics and Space Administration
NASDA	National Space Development Agency (Japan)
NATO	North Atlantic Treaty Organisation
NBP	No Baseband Processing
NIC	Nearly Instantaneous Companding (Coding System)
NIVR	Netherlands Agency for Aerospace programmes
NORAD	North American Aerospace Defence Command
NPR	Noise Power Ratio
NTIA	National Telecommunications & Information Agency (USA)
NTNF	Norwegian Council for Scientific & Industrial Research
NTSC	National Television System Committee (USA Colour Television Transmission System)
NTU	Network Terminating Unit
OBDH	On-Board Data Handling
OOK	On/Off Keying
OSI	Open Standards Interconnection
OTS	Orbital Test Satellite
OTV	Orbital Transfer Vehicle
PAL	Phase Alternation System (Colour TV System used in Europe)
PAM	Payload (or Perigee) Assist Module
PAM	Pulse-Amplitude Modulation
PCM	Pulse Code Modulation
PSK	Phase Shift Keying
PSS	Packet Switched Service
PSI	Pacific Satellite Incorporated
PSTN	Public Switched Telephone Network
PTT	Postes, Téléphone & Télégraph
QPSK	Quaternary Phase Shift Keying
RAE	Royal Aircraft Establishment (UK)
RARC	Regional Administrative Radio Conference
RCA	Radio Corporation of America
RCS	Reaction Control System
RDSS	Radio Determination Satellite Service
RF	Radio Frequency
RFP	Request for Proposal
RTE	Radio Telefis Eireann
RTG	Radioisotope Thermo-electric Generator
S3R	Societé de Reassurance des Risques Relatifs aux Applications Spatiales
Sarsat	Search and Rescue Satellite
Satcom	Satellite Communications
SBAC	Society of British Aerospace Companies
SBS	Satellite Business Systems
SBTS	Sistema Brasileiro de Telecommunicacoes por Satellite
SCPC	Single Carrier per Channel
SCPT	Single Carrier per Transponder
SDI	Strategic Defence Initiative
SDLC	Synchronous Data Link Control
SECAM	Sequentiel Couleur a Memoire
SES	Société Européenne des Satellites
SHF	Super High Frequency
SLV	Satellite Launch Vehicle
SMATV	Satellite Master Antenna Television
SNA	Systems Network Architecture
SNG	Satellite News Gathering
SNIAS	Société Nationale Industrielle Aerospatiale
S/N	Signal to Noise Ratio
SPADE	Single Channel Per Carrier PCM Multiple Access Demand Assignment Equipment
SSB	Single Sideband
SSC	Swedish Space Corporation
SSMA	Spread Spectrum Multiple Access
SSME	Space Shuttle Main Engine
SSPA	Solid State Power Amplifier
SS/TDMA	Satellite Switched Time Division Multiple Access
SSUS	Solid Spinning Upper Stage
STC	Satellite Television Corporation
STS	Space Transportation System or Space Transport Service
STV	Satellite Transfer Vehicle
TASI	Time Assigned Speech Interpolation
TDF	Telediffusion de France
TDM	Time Division Multiplexing
TDMA	Time Division Multiple Access
TDRS	Tracking and Data Relay Satellite
Telesat	Telecommunications Satellite Corporation (Canada)

LIST OF ACRONYMS—(continued)

TOS	Transfer Orbit Stage
TT&C	Tracking, Telemetry and Control
TVRO	Television Receive Only (Satellite Antenna)
TWT	Travelling Wave Tube
TWTA	Travelling Wave Tube Amplifier
VHF	Very High Frequency
VNL	Via Net Loss
VOA	Voice of America
VSAT	Very Small Aperture Terminal
WARC	World Administrative Radio Conference

GLOSSARY

A

ACTIVE SATELLITE A satellite that carries radio repeaters to amplify and re-transmit a signal. It is distinct from some early satellites, such as ECHO, on which signals were just reflected down to earth.

AF–SAT African regional satellite that has been under study for some years, mostly in Europe.

AMPLIFIER Term used to describe an electronic system for amplification of a signal. In satellites, two forms of amplifiers are currently used, Travelling Wave Tube Amplifiers (TWTA's) and Solid State Power Amplifiers (SSPA's). All radio repeaters are amplifiers. In cable TV systems, fibre optic links, long distance microwave links and conventional terrestrial telephone cables, amplifiers are spaced at intervals along the links to rebuild the strength of signals which weaken as they pass through the cable.

ANALOGUE SIGNAL A continuously variable signal. Commonly, telephone voice circuits and TV channels use this technology. Generally in modern communications technology, digital signals are now being employed, providing a better quality of transmission service.

ANIK Name of the domestic Canadian satellites.

ANTENNA A transponder connected to a transmitter which converts the high-frequency AC signals produced by the transmitter into radio or microwave signals which are propagated to the atmosphere or, conversely, to convert the received signals into corresponding AC signals which the receiver can use. The form of an antenna is dependent on a variety of factors but mainly on the strength and frequency and bandwidth of the radio signal involved. A typical satellite TVRO antenna is very short, it requires a dish (usually parabolic) to reflect the very weak signals coming from the satellite into a concentrated and stronger signal that the antenna is capable of receiving.

APOGEE The point in an elliptical satellite orbit which is farthest from the surface of the earth (*see* PERIGEE). Geosynchronous satellites which maintain circular orbits around the earth are first launched into highly elliptical orbits with apogees of 22,300 miles. When the communication satellite reaches its initial apogee, small on-board thruster rockets are fired to nudge the satellite into its permanent circular orbit of 22,300 miles.

APOGEE KICK MOTOR (AKM) A rocket motor which is part of a geostationary satellite and which is fired to transfer the satellite from its elliptical to its geosynchronous orbit. It is fired when the satellite is at its apogee in the elliptical orbit. It converts the elliptical orbit into a circular orbit.

ARABSAT Abbreviation used for either the Arab Satellite Organisation or the satellites which it operates. The organisation is owned by the 22 members of the Arab League and provides regional and domestic satellite capacity for them.

ARIANE Expendable launch vehicle developed by the European Space Agency to ensure autonomy from the USA. The vehicle is primarily used for launching commercial satellites. The name is the French version of ARIADNE.

ARIANE 5 Proposed European launcher intended to replace the current Ariane vehicle from the mid-1990's. It is a completely new vehicle, although it uses out-dated technology. Unlike the current Ariane vehicle, it will be man-rated as it will carry the Hermes spaceplane. It is a considerably larger vehicle than Ariane.

ARIANESPACE Commercial company which markets and operates the Ariane vehicle and is expected to do so for Ariane 5. Company is predominantly French owned but has shareholders from other European countries. The shareholders are primarily aerospace companies involved in constructing Ariane and banks.

ASTRA Name of the satellite system owned by Société Européenne des Satellites in Luxembourg. This is intended to be used for European broadcasting. The Astra name is also being used as a marketing name for TVRO equipment for reception of signals from the satellite.

ASSIGNMENT A portion of the radio frequency spectrum allocated to a particular use or organisation for a particular use. Assignments are either made at a national level, for example by the Federal Communications Commission in the USA, or at an international level, through the International Telecommunications Union and its International Frequency Registration Board.

ASYNCHRONOUS TRANSMISSION Transmission in which each character in a signal is preceded by a start signal and is followed by one or more stop signals. The start signal triggers off the receiver and, following receipt of the character, the stop signals switch off the receiver in readiness for receipt of the next character.

AT&T Was the principal public switched telecommunications operator in the USA in both the local and long distance fields but has been forced by the US courts to divest itself of its regional activities leaving it in a core business of long distance telecommunications. It owns one of the world's foremost R&D establishments, Bell Laboratories, which, apart from inventing the transistor, played a major role in the early development of satellites. AT&T remains a powerful force in the US telecommunications (and, increasingly, the computer) scenes. It operates the Telstar 3 series of satellites.

ATLAS–CENTAUR US launch vehicle currently used for launch of commercial satellites and manufactured by General Dynamics. It was intended that the vehicle, which originated in the 1950's, be phased out in favour of Shuttle launches. The production life for the vehicle may be reopened.

ATTENUATION The process of lowering the signal level of an incoming signal to prevent a receiver from becoming overloaded or the decrease in magnitude of current, voltage or power of a signal that arises during the course of transmission. Amplifiers are used to correct the position under certain circumstances. In other circumstances, other means are required. The term is often used when discussing the relative merits of Ku- and Ka-band satellite frequencies. These bands suffer from rain "attenuation" where the microwave signal is weakened or "washed out" by heavy rain.

ATTITUDE CONTROL SYSTEM (ACS) A control system on a satellite used to maintain its correct orientation with the Earth whilst in orbit. It usually involves a system of momentum wheels and thrusters.

AUSSAT Either the name of the Australian domestic satellite system or an abbreviation of the company which operates it.

AUTOMATIC FREQUENCY CONTROL (AFC) A circuit which automatically controls the frequency of a signal.

AUTOMATIC GAIN CONTROL (AGC) A circuit which automatically controls the gain of an amplifier so that the output signal level is virtually constant for varying input signal levels.

AZIMUTH ANGLE The angle of rotation (horizontal) that a ground-based parabolic antenna must be rotated through to point to a specific satellite in a geosynchronous orbit. The azimuth angle for any particular satellite can be determined for any point on the surface of the earth given the latitude and longitude of that point.

B

BACKHAUL A terrestrial link connecting a satellite earth station with the local switching centre.

BAND PASS FILTER (BPF) An active or passive circuit which allows signals within the desired frequency band to pass through but impedes signals outside the desired pass band.

BANDWIDTH Either the useful frequency range employed by a system, such as a satellite transponder or the frequency range employed by a signal. The bandwidth is the difference between the highest and lowest frequency used in either case. A telephone voice circuit needs a bandwidth of 3000 cycles per second (3 kHz). A colour TV channel requires about 6 m cycles per second (6 MHz). The bandwidth is also a potential measure of the capacity of a communications although there are other engineering factors involved in measuring the capacity.

BASEBAND A term that usually refers to a transmission signal prior to its modulation onto a carrier. The basic direct 6 MHz output signal from a television camera, satellite television receiver, or video tape recorder is a baseband signal. Baseband signals can be viewed only on studio monitors. To display the baseband signal on a conventional television set, a "remodulator" is required to convert the baseband signal to one of the VHF or UHF television channels which the television set can be tuned to receive. The term is also used more loosely in satellite communications to describe the signal received at an earth station from the terrestrial network prior to transmission via satellite. Under these circumstances, the signal may have passed through one or more stages of modulation prior to modulation onto the radio frequency carrier at the earth station.

BAUD A unit of digital signalling speed measured in bits per second. It is the modulation rate.

BEAMWIDTH The angle or conical shape of the beam the antenna projects. Large antennas have narrower beamwidths and can pin-point satellites more precisely.

BI-PROPELLANT PROPULSION SYSTEM Dual liquid fuelled launch vehicle or satellite motor. On a satellite this usually means that its station keeping motors use the same fuel and tanks as the Apogee Kick Motor. If the launch vehicle and perigee stage place the satellite in a very accurate elliptical orbit, this may mean that fuel is saved on the Apogee motor burn and the operating life of the satellite is increased.

BIRD A colloquial term for a satellite. Mostly used by Americans.

BIT Abbreviation for binary digit. The term is used in Bits per Second (bit/s) as a measure of the speed of transmission of digital information. For example, a voice telephone circuit (digital) needs a bit rate of 64,000 bits per second (64 kbit/s).

BIT ERROR RATE (BER) The proportion of transmitted bits which are received in error.

BLANKING An ordinary television signal consists of 30 separate still pictures or "frames" sent every second. They occur so rapidly the human eye blurs them together to form an illusion of moving pictures. This is the basis for television and motion picture systems. The blanking interval is that portion of the television signal which occurs after one picture frame is sent and before the next one is transmitted. During this period of time data signals, such as teletext, can be sent which can be picked up on an adapted television receiver.

BLOCK In digital communications, a block of characters transmitted as a unit.

BLOCK DOWNCONVERTER (LNB) A device used to convert, for example, a C-band downlinked satellite signal of 3·7 to 4·2 GHz down to UHF or lower satellite frequencies (1 GHz and lower). In a TVRO system this would be followed by tuning and demodulation.

B-MAC *See also* MAC. Transmission standard adopted by Australia for quasi-DBS services on the Aussat system.

BNC CONNECTOR A standard medium-sized twistlock coaxial cable connector used at higher frequencies throughout the television industry and by many satellite TV receivers. Other popular connectors include the larger screw in Type N and the smaller miniature-type connectors.

BROADBAND (WIDEBAND) COMMUNICATIONS Refers to communications systems, such as cable and ISDN, which can carry a lot of frequency spectrum, and consequently a lot of services on one system, such as TV, high data rates for computers and voice traffic.

BROADCASTING The process of transmitting a radio or television signal via an antenna to multiple receivers which can simultaneously pick up the signal. This is the principal application of radio communications and is increasingly the principal role of satellites. It is a form of point-to-multipoint communications. There are several definitions of the term which may or may not include cable TV. The latter has been defined as narrow casting by some in the industry.

BROADCASTING SATELLITE SYSTEM (BSS) Another term for Direct Broadcasting Satellite system usually used by Americans. Some observers use the term to describe satellites using medium power transmitters (50–100 Watts power output) for broadcasting, as distinct from high power (200–260 Watts) which they describe as DBS TV.

BS–2, BS–3 Japanese DBS satellites. The first system, BS–2, is supposed to be operational. The follow on series is due for launch at the end of the 1980's. The systems are products of an ambitious industrial policy involving new broadcasting technologies, such as HDTV, rather than a commercial attempt to establish an operational system.

BUS Almost a colloquialism for a satellite platform which carries and services the communications payload.

BUTTON HOOK FEED A shaped piece of waveguide directing signal from the feed to the Low Noise Amplifier (LNA) behind the antenna.

BYTE A set of eight binary digits (bits) considered as a unit in both digital communications and computer technology.

C

CABLE AUTHORITY UK regulatory agency with responsibility for cable and SMATV systems. Has little to do as UK cable companies cannot afford to bury their cables under the ground as is required in most towns.

C-BAND The 3·7–4·2 GHz (downlink) and 5·975–6·425 GHz (uplink) satellite communication band. Used in North American domestic communications, the Intelsat system and some regional and domestic satellite systems in the rest of the world but not in Europe. The band is frequently shared with terrestrial microwave communications systems causing interference problems.

CARRIER TO NOISE RATIO (C/N) The ratio of received carrier power and noise power in a given bandwidth. It is expressed in decibels. This figure is directly related to G/T and S/N; and in a video signal the higher the C/N the better the received picture will be.

CARRIER The basic radio, microwave or television frequency signal which is used to carry the desired analogue or digital information in a transmission. It is basically a constant frequency which is altered by two basic methods in the process of carrying the information. The deviations from the constant carrier signal represent that information. The carrier is "modulated" by manipulating either its amplitude (making it louder or softer) or its frequency (shifting it up or down) in relation to the signal being carried. Satellite carriers are usually frequently modulated.

CARRIER FREQUENCY The main frequency on which a radio station, television station, or microwave transmitter operates. AM radio stations operate in the frequency band from 535,000 cycles (535 kHz) to 1600 kHz. FM radio stations operate in the frequency band from 88 MHz to 108 MHz. Terrestrial television station transmitters operate in the frequency band whose channels range from 54 MHz (Channel 2) to 890 MHz. Microwave and satellite communications transmitters operate in the band from one to 30 GHz (a gigahertz is one billion cycles per second).

CASSEGRAIN ANTENNA An antenna which uses a subreflector at the focal point to reflect energy to or from a feed located at the apex of the main reflector.

CATV Originally meant Community Antenna Television but now is used to mean Cable Television. The origins of cable television lie in the need in some areas to relay terrestrial off-air TV to households via a cable. This particularly applied in the early days of TV when there were insufficient transmitters and exists in some areas today because local terrain may block the signals from transmitters. In the USA operators of these relay systems began to pick up distant TV stations too far away for household transmitters to pick up. The relay operators saw a market opportunity to develop many channel systems. However, the big growth in cable in the USA came after satellites began to be used to distribute new channels to cable networks. The resulting cable systems are altogether much larger and sophisticated systems than the early relay systems and are now big business in the USA.

CCIR International Radio Consultative Committee of the International Telecommunications Union. Handles technical operating and tariff matters between member countries of the organisation.

CCITT International Telegraph and Telephone Consultative Committee of the ITU. Handles similar matters to the CCITT except that it is for telephony.

CENTAUR Top stage of versions of the Titan and Atlas US launch vehicles. It was intended to use it as a transfer stage with large spacecraft launched with Shuttle but this has been abandoned for safety reasons.

CENTRE NATIONAL D'ETUDES SPATIALES (CNES) French National Space Agency formed in 1962 which has played a major role in both the French national and the European space programmes. It has a major influence on the European Space Agency.

CEPT European Conference of Postal and Telecommunications Administrations. A secretive cartel of European PTT's covering technical and tariff matters of common interest. The Eutelsat organisation grew out of CEPT committees.

CHANNEL An information transfer path. For such traffic as voice, channels are commonly formed by multiplexing together many circuits (up to 2000 on a satellite channel) (*see also* Single Carrier per Channel). For TV, the term channel is used loosely but technically refers to a frequency band in which a specific broadcast signal is transmitted.

Television signals require a 6 MHz frequency band to carry all the necessary picture detail compared with a typical high fidelity audio channel of 15,000 Hz.

CHURN The turnover rate of subscribers to cable and pay TV systems.

CIRCULAR POLARISATION Most satellites launched since the end of the 1970's use a combination of horizontal and vertical polarisation of up and downlinks to use the frequency spectrum twice. However, the Intelsat satellites transmit their signals in a rotating corkscrew-like pattern as they are down-linked to earth. On some satellites, both right-hand rotating and left-hand rotating signals can be transmitted simultaneously on the same frequency; thereby doubling the capacity of the satellite within a given frequency spectrum range to carry communications channels.

CLARKE ORBIT *See* GEOSYNCHRONOUS ORBIT.

CLOSED CYCLE MOTOR or ENGINE Launch vehicle engine in which the turbo-pump exhaust is ducted into the main engine, thereby improving its performance and the ability to throttle the engine. The Space Shuttle main engines and the Japanese H–II vehicle use this technology. All other Western launch vehicles use the older open-cycle technology.

C–MAC *See also* MAC, D–MAC and B–MAC. UK developed transmission standard for satellite television which has yet to be adopted as a European standard.

COAXIAL CABLE Commonly referred to as COAX. Used in nearly all cable TV systems as well as TVRO units. A transmission line in which an inner conductor is surrounded by an outer conductor or shield and separated by a nonconductive material. Coaxial cables have the capacity to carry wide bandwidth and high frequency signals used for TV.

CO-LOCATE To place two or more satellites at the same orbital position.

COMMON CARRIER Any organisation which operates communications circuits used by other organisations or individuals. All PTT administrations are common carriers. Their principal activity is to provide telecommunications capacity, but not the information (such as telephone calls, computer information and TV channels) carried on the system. The dividing line between a common carrier and other carriers is becoming blurred as new services such as Value Added Networks (VANS) are developed. The term mainly applies to the US telecommunications scene where there are a large number of telecommunications companies often offering differentiated services. In Europe, telecommunications is still largely a monopoly of PTT's and therefore it is less necessary to distinguish between different types of carrier. US common carriers include the telephone companies as well as the owners of communications satellites such as RCA Americom, Western Union and AT&T.

COMSAT Either an abbreviation of the term communications satellite or the commonly used name of Communications Satellite Corporation in the USA. The latter effectively set up Intelsat and is the US signatory to the organisation and Inmarsat.

CONUS Continental or Contiguous United States. The USA excluding Hawaii, Alaska and Puerto Rico.

CONVERTER *See* DOWN CONVERTER.

CROSS MODULATION A form of signal distortion in which modulation from one or more radio frequency carrier(s) is imposed on another carrier.

CS–2, CS–3, CS–4 Japanese domestic communications satellite systems operating in the Ka frequency band.

D

dBW Ratio of the power of one watt expressed in decibels. A decibel is the standard unit used to express the ratio of two power levels. It is used in communications to express either a gain or loss in power between the input and output devices. A −3 dB loss is approximately 50% of a decrease in power. A −2 dB loss is approximately 37% of a decrease in power.

DECODER (unscrambler) A set-top device which enables the home subscriber to convert an electronically scrambled picture into a viewable signal.

DELTA US expendable and dated launch vehicle that was expected to be phased out in favour of Shuttle. It is capable of launching lighter weight satellites. The production line may be reopened to make up for the capacity loss caused by the loss of Shuttle as a commercial satellite launcher.

DEMAND ASSIGNMENT MULTIPLE ACCESS (DAMA) An access system to a satellite where the capacity on the satellite is assigned according to demand from a number of users. It is unlike SCPC, TDMA and FDMA techniques which allocate capacity according to preset fixed amounts. DAMA allocates different amounts to different users according to their individual demands. DAMA is used in conjunction with SCPC and TDMA techniques.

DEMODULATOR A receiver circuit which extracts or "demodulates" the information carried in a modulated carrier signal (*see* carrier).

DEUTSCHE BUNDESPOST Western German PTT administration and also the largest employer in Western Europe. One of the more classic examples of engineering led telecommunications monopolies. It is due to be the operator of the DFS–COPERNICUS domestic communications satellite system as well as the TV–SAT system. It is also investing heavily in an ambitious programme to provide cable TV to that country.

DEUTSCHE FORSCHUNG-UND VERSUCHSANATALT FUR LUFT-UND RAUMFAHRT (DFVLR) West German equivalent of NASA funded by the Research and Technology Ministry (BMFT). A reasonably strong force in the European satellite and space R&D field.

DFS–COPERNICUS West German domestic satellite system due to become operational in 1988. It will be operated by Deutsche Bundespost, the German PTT. A product of the bandwagon of the early 1980's which saw many European countries trying to build and operate commercial satellite systems.

DIRECT BROADCAST SATELLITES (*See also* BROADCASTING SATELLITE SERVICES and FIXED SATELLITE SERVICES). Satellites intended to distribute TV channels direct to households using small TVRO terminals with dishes in the 30–90 cm diameter range. Such satellites are technically different from most communications satellites in that they employ high power repeaters in the 200 Watt power output range. Many in the industry believe that they are not a commercial proposition and that their intended role has been taken over by medium power technology (satellites with repeaters in the 50 Watt power range) or by alternative TV technology such as Cable and VCR's.

DIRECTION GENERALE DE TELECOMMUNICATIONS (DGT) The French PTT. It operates the Telecom 1 domestic satellite system but not TDF–1. The organisation has made a substantial improvement to the French telecommunications network over the last decade.

DISH Term used to describe a satellite earth terminal. However, it strictly applies to the main reflector of the terminal, usually a parabolic shape.

DISTRIBUTION CENTRE The central point from which the television signals are distributed from a television programming organisation to its network stations or receivers. The distribution centre for a satellite television programmer will usually consist of a bank of video tape machines, studio facilities, and large uplink (transmit) antenna for transmission of the signal directly to the desired satellite.

DITHERING Normally a television signal requires about 6 MHz of bandwidth. A satellite transponder typically has about 36 MHz of bandwidth, although 72 MHz is common. However, particularly in C-band satellite transmissions, a satellite TV signal is likely to be affected by terrestrial microwave interference. Dithering is the method of overcoming this problem and involves shifting the six megahertz satellite TV signal up and down the 36 MHz satellite transponder spectrum at a rate of 30 times per second (30 Hertz). The satellite signal is "dithered" to spread the transmission energy out over a band of frequencies wider than a terrestrial common carrier microwave circuit operates within, thereby minimising the potential interference that any one single terrestrial microwave transmitter could possibly cause to the satellite transmission.

DOMSAT Abbreviation for domestic (communications) satellite system. The latter is a satellite system intended to provide only capacity for the internal requirements of a country.

DOWN-CONVERTER That portion of the satellite television receiver that converts the signals from the C-band or Ku-band microwave range to (typically) the more readily used 70 MHz range. Down-converters typically were located physically at the front end of the receiver, requiring bulky and expensive coaxial cable feeds from the antenna to the receiver. Nowadays, designs place the down-converter at the antenna, often in combination with the LNA and a simple coaxial cable brings the satellite TV signal into the house.

DOWNLINK The satellite to earth transmission path of radio signals transmitted via satellite (all signals to and from a satellite are radio signals).

DRIFT The tendency of a satellite to drift away from its intended orbital position. It does so because of such things as the gravitational pull of the sun and the moon and solar wind. The drift is corrected by use of thrusters fired perhaps every two weeks.

DROP In cable TV networks, a small branch of a cable which connects the antenna terminals on the back of the subscriber's television receiver to the feeder cable in the street.

DUPLEX TRANSMISSION A transmission which takes place in both directions simultaneously. A normal telephone call is a duplex transmission. TV broadcasts are not as they are one way.

E

EARTH STATION The ground segment part of a satellite system which is either used to transmit signals to a satellite or receive them or both. The size and design of an earth station depends on the volume of traffic passing through it. 10 foot to 20 foot diameter size used for TV reception to as large as 100 feet in diameter for international communications.

EFFECTIVE ISOTROPIC RADIATED POWER (EIRP) Commonly used to describe the strength of a signal from a satellite when it is received on the Earth; the higher the EIRP, the lower the size of dish needed to receive it. The EIRP is determined by both the power output of the satellite transponder and the extent to which the downlinked beam is concentrated over a geographical area. Technically, the term describes the strength of the signal leaving the satellite antenna or the transmitting earth station antenna, and is used to determining the C/N and S/N ratios. For a European DBS satellite, an EIRP figure in excess of 62 dBW is required. For FSS telephony transmissions, where the aim is to keep the cost of the satellite circuit as low as possible at the expense of ground equipment, an EIRP as low as 33 to 38 dBW, is common. Modern Ku-band satellites aimed at the TV market commonly have an EIRP performance in the 40 to 50 dBW range. This is because it is important to keep the cost of the ground equipment low as it is often a point-to-multipoint system using many receive stations. For transmitting video earth stations, a minimum EIRP is required to properly excite the satellite and is of the order of 80 dBW. EIRP may be increased either by increasing the transmitter power, or by using a large and higher gain transmitting antenna.

ELEVATION The upward tilt to a satellite antenna measured in degrees required to aim the antenna at the communications satellite. When aimed at the horizon, the elevation angle is zero. If tilted to point directly overhead, the satellite antenna would have an elevation of 90°. Thus in northerly latitudes, it is often difficult to get line of sight to a satellite, whilst in countries located on the equator, it is only a matter of pointing the satellite dish vertically and swinging it across the arc in that position.

ENCODER (Scrambler) A device used to electronically alter a signal so that it can only be viewed on a receiver equipped with a special decoder. Encoders are used to prevent unauthorised reception of a signal. In satellite communications they are increasingly required to receive satellite TV channels.

ELEVATION ANGLE *See* ELEVATION.

ENCRYPTION (*See also* ENCODER). The process of scrambling a signal so that it cannot be picked up by an unauthorised person or organisation. For satellite TV it has become a major issue. Cable TV companies such as Home Box Office in the USA using satellites to distribute their programming to cable TV networks have begun scrambling their signals to prevent TVRO users picking them up for nothing. The TVRO owners can buy decoding equipment but are concerned at both the cost and the compatibility with other channels encrypting in the future. It has also hit the sales of TVRO's very hard. In Europe some channels scramble to avoid problems of copyright rather than to seek revenue. Sky Channel is an example. The industry in both continents is not applying consistent rules to the technology involved.

EQUIVALENT NOISE TEMPERATURE (ENT) The level of noise temperature interference accompanying a signal is measured in degrees because signal noise, the movement of the electrons, is relative to temperature.

EUROPEAN BROADCASTING UNION (EBU) A cooperative organisation for European broadcasters which deals with technical standards, organises inter-organisation distribution of news programming and special events and has played a role in establishing the Europa satellite TV channel (a public service broadcasting answer to commercial satellite TV), has attempted to have C–MAC adopted as a European DBS TV transmission standard and is a user of the Eutelsat system.

EUROPEAN SPACE AGENCY European space R&D organisation set up in 1975 out of the remnants of the European Space Research Organisation and the European Launcher Development Organisation. It works because control of each of its major projects lies outside its hands and in the hands of the country putting up the largest amount of money for the project. A typically bureaucratic R&D organisation that wastes a lot of its expertise by allowing engineering and scientific matters to control it.

EUTELSAT The European Telecommunications Satellite Organisation. Operates the Eutelsat I satellite system on behalf of 26 European PTT administrations. It is a product of European industrial policy and in terms of both concept and organisation is close to Intelsat. Unlike Intelsat, it entered the satellite communications market after the latter had been well established and is consequently under some pressure to establish itself in a market niche.

F

FEDERAL COMMUNICATIONS COMMISSION (FCC) US communications regulatory body covering telecommunications, cable TV, satellites, mobile communications radio and television broadcasting, but sharing the regulatory role with both individual states and the courts. By a long margin, the FCC is the world's most dominant player in the changing policies towards communications, although it is the instigator rather than the originator of change. The organisation has a reputation for turning out US communications bureaucrats. It shares regulatory power with the courts by virtue of the US constitution and anti-trust regulation. It also shares power with the individual states that make up the union because the USA is a decentralised country. The current chairman of the organisation is Mark Fowler who is strongly in favour of communications liberalisation. Besides Fowler there are six other commissioner's who are political appointees, although because their terms are fixed, the organisation is not a direct reflection of current governmental policy.

F/D Ratio of antenna focal length to antenna diameter. A higher ratio means a shallower dish.

FEED A term used to describe the transmission of video programming from a distribution centre. It is also used to describe the feed system of an antenna. The feed system may consist of a subreflector plus a feedhorn only.

FEEDHORN A satellite TV receive antenna component that collects the signal reflected from the main surface reflector and channels this signal into the low-noise amplifier (LNA).

FIXED SATELLITE SERVICE (FSS) Quasi-bureaucratic term used to describe satellite services for point-to-point communications. The term now appplies to conventional satellites not used for DBS, RDSS or MSS services. It was at one time envisaged that satellites would be primarily used for point-to-point telephony services with occasional use point-to-point distribution of video programming. In practice conventional satellites are being dominated by TV requirements and the same satellite may be used for voice, business or broadcasting requirements. The concept of DBS satellites has not proved to be a commercial proposition and to some extent the role is being taken over by conventional satellites.

FM THRESHOLD That point at which the input signal power is just strong enough to enable the receiver demodulator circuitry successfully to detect and recover a good quality signal from the incoming video carrier. Using threshold extension techniques, a typical satellite TV receiver will successfully provide good pictures with an incoming carrier noise ratio of 7 dB. Below the threshold a random noise, called "sparkles" appears in the video picture.

FOCAL LENGTH Distance from the feed to the centre of the dish.

FOCAL POINT The area toward which the primary reflector directs and concentrates the signal received.

FOOTPRINT A map of the signal strength showing the EIRP contours of equal signal strengths as they cover the earth's surface. Different satellite transponders on the same satellite will often have different footprints of signal strength. Footprints maps are available from the common carriers and from both SST and CATA. The accuracy of EIRP footprint or contour data improves with the operational age of the satellite although the actual EIRP levels decrease slowly as the satellite ages.

FORWARD ERROR CORRECTION A transmission error detection and correction technique used in data communications. It both detects errors and corrects them without recourse to retransmission.

FREQUENCY-AGILE The ability of a satellite TV receiver to select or tune all of the channels (transponders) from a satellite. Receivers not frequency-agile are dedicated to a single channel and are most often used by cable TV operators. Frequency agility can be via continuously variable tuning or discrete step (channel selection) tuning.

FREQUENCY COORDINATION An exercise needed for any activity involving the use of new services using any form of radio communications. There is limited capacity in the radio frequency spectrum and new services may cause interference with existing activities using the radio spectrum or may be affected by existing users of it. Coordination is required both to avoid this and to comply with the complex national and international rules which allocate the spectrum. It is a complex business which has led to the formation of a number of specialist businesses serving organisations using the frequency spectrum as well as a large public sector administration both on national and international bases. The International Frequency Registration Board of the ITU allocates frequencies on an international basis, largely on a first come first served principle with a priori allocation in terms of services (*see also* WORLD ADMINISTRATIVE RADIO CONFERENCES).

FREQUENCY DIVISION MULTIPLE ACCESS (FDMA) In this method of accessing a satellite, the bandwidth of the satellite transponder is divided up into sub-bands, each centred on its own carrier. Transmission signals are multiplexed onto each of these carriers. the system is not an efficient user of the frequency available in that some is wasted in protecting each sub-band from another. However, it requires cheaper ground equipment than TDMA techniques.

FREQUENCY REUSE In satellite communications a method of conserving use of the frequency spectrum by either using such methods as spot beams to reuse the same frequency in different areas or through use of polarisation techniques.

FREQUENCY SPECTRUM The electromagnetic frequency range which includes what are commonly referred to as radio frequencies as well as visible light. Just who owns the electromagnetic frequency band is one of the world's more significant problems. It is a scarce resource yet without a clear form of ownership. It is nominally allocated at either the international level by the ITU or at national level by the political administration. In both cases the allocation is subject to pressure groups rather than being based on a price or economic allocation method.

G

GAIN A measure of amplification expressed in watts or decibels.

GALLIUM ARSENIDE FIELD EFFECT TRANSISTOR (GaAsFET) A semiconductor material used in low noise microwave devices which has had a major impact on the costs of satellite earth terminals.

GATEWAY In satellite communications, this usually refers to an international gateway which is the transmit and receive major earth terminal connecting the domestic telecommunications network with the Intelsat system.

GEOSYNCHRONOUS ORBIT (GSO or Geostationary Orbit) Also known as the Clarke circular orbit. This is an orbit of the Earth where the orbiting body takes one day to complete an orbit of the Earth. As it moves in the same direction as the Earth is spinning, and it takes one day for the Earth to spin through 360°, the satellite appears to be stationary when viewed from the surface of the Earth. As a matter of simple physics, the orbit where this effect is achieved is 22,300 miles above the equator. Any higher and the satellite would have to move so fast to keep up with the spinning of the Earth that it would fly off out into space. Any lower it would go so slow that it would be attracted by the Earth's gravity and its orbit would decline. Strictly speaking, Geosynchronous satellites are not quite geosynchronous. Gravitational anomolies in the Earth and gravitational attractions from the sun and the moon tend to pull them off a perfect orbit as does the pressure of solar wind. Consequently, satellites carry small thrusters to periodically put them back in position. They are also in the course of normal operations allowed to drift slightly.

GEOSYNCHRONOUS TRANSFER ORBIT (GTO) The elliptical orbit that a satellite is placed in before its apogee kick motor is fired to place it in the final geostationary orbit.

GIGAHERTZ (GHz) One billion cycles per second. Signals operating above 1 Gigahertz are known as microwaves, and begin to take on the characteristics of visible light.

GLOBAL BEAM An antenna down-link pattern used by the Intelsat satellites, which effectively covers one third of the globe. Global beams are aimed at the centre of the Atlantic, Pacific and Indian Oceans by the respective Intelsat satellites, enabling all nations on each side of the ocean to receive the signal. Because they transmit to such a wide area, global beam transponders have low EIRP performances on the surface of the Earth. Therefore, earth stations receiving global beam signals need large antennas, up to 30 metres in diameter.

G/T A figure of merit of an antenna and low noise amplifier combination expressed in dB. "G" is the net gain of the system and "T" is the noise temperature of the system. The higher the number, the better the system.

GUARD CHANNEL The wasted part of the frequency spectrum which arises because simultaneous transmissions of several channels need to be protected from interfering with each other by an unused portion of the spectrum known as the guard channel. This is why FDMA transmission techniques are so inefficient in their use of a satellites total transmission capability. FDMA techniques typically divide a satellite transponder up into several channels with a guard channel inserted between each. The guard channel is typically several Megahertz in bandwidth.

H

H–I, H–II Japanese launch vehicles used for launching a variety of satellites, including communications spacecraft. The H–I vehicle which first flew in 1986, is derived from the US Delta launch vehicle and is seen as a test bed for H–II. The latter is an advanced expendable launch vehicle which may be man-rated. It is expected to be available from 1992. Unlike previous Japanese launch vehicles, it has a sufficient payload to launch all commercial satellites and will be offered to non-Japanese satellite operators. H–I is not available because part of it is built under licence from the USA and there are clauses preventing sale outside Japan.

HEADEND Usually termed as the central control and distribution point of a cable TV system which usually includes or is directly connected to satellite and terrestrial TV receive antennas. Under some definitions it will include the cable operators central studios and production facilities.

HERMES Proposed European spaceplane that has originated in France. It is a European answer to Shuttle but is not intended to launch commercial satellites. It is intended to help provide European autonomy in the International Space Station and will be launched with a man-rated version of the Ariane 5 expendable launch vehicle.

HIGH POWER Relative term used in the satellite industry to describe television satellites with transponders in the 200–260 Watt power output range as distinct from satellites using transponders in the 50–120 Watt power output range.

HOTOL Air breathing fully reusable UK concept for a launch vehicle to be available towards the end of the century. Concept is being offered for development within the framework of the European Space Agency. It is seen by some as a European answer to advanced US trans atmospheric vehicles currently under study or as a rival to Hermes and Ariane 5. Hotol uses a Shuttle like main engine together with a heat exchanger where liquid hydrogen cools the incoming air which is used as the oxidiser in the system. Nitrogen from the air is either fed into the combustion system or is dumped overboard, depending on what part of the flight sequence the vehicle is in.

HYDRAZINE Liquid propellant used for station keeping purposes on some designs of satellites.

I

INCLINATION Angle between the Earth's equatorial plane and the orbit plan of a satellite, space station or in-orbit spacecraft, such as Shuttle. Measured in degrees. Geostationry satellites have an inclination of 0°.

INDEPENDENT US term applying to a TV station that has no exclusive affiliation which obliges it to carry programs from one of the national networks (*see* NETWORK).

INDEPENDENT BROADCASTING AUTHORITY UK regulatory agency for commercial radio and television and DBS TV but not cable TV or for the British Broadcasting Corporation. It also operates terrestrial transmitters for radio and television but does not intend to operate DBS satellites. In the latter case it is responsible for issuing franchises and regulating programming and advertising content once the system is operational. It has taken over the major role in DBS from the BBC. The IBA was also responsible for developing the C–MAC transmission standard.

INERTIAL UPPER STAGE (IUS) Boeing built Upper Stage for use on Shuttle or Titan. It is a two stage perigee and apogee liquid fuelled system for use with large satellites such as TDRSS.

INMARSAT The International Maritime Satellite Organisation, an international civil service body with headquarters in London initially set up to provide maritime mobile communications on a commercial basis but now moving into aeronautical satellite communications.

INSAT Indian multipurpose domestic satellite system used for telephony, TV and radio networking, quasi-DBS TV and meteorology. Country is working on a second generation Insat II system for the 1990's. The Insat system is a case study of the application of satellites for third world requirements.

INTELSAT International Telecommunications Satellite Organisation with headquarters in Washington DC. It was formed in 1964 with strong backing from the USA but as an international cooperative of PTT administrations to primarily provide capacity for international public switched telephony systems and for distribution of TV programming. It has had a major impact on the international communications scene but appears to be strongly threatened by new technologies, notably fibre optic communications. It is the largest commercial satellite operator in the world and has sixteen satellites in orbit.

INTERNATIONAL FREQUENCY REGISTRATION BOARD (IFRB) A body of the International Telecommunications Union at which requests for use of the frequency spectrum are registered and coordinated. The five members of the board serve as "custodians of an international public trust". They are responsible for recording assignments of the radio frequency spectrum throughout the world and advising members on technical matters.

INTERNATIONAL INSTITUTE FOR COMMUNICATIONS London based organisation backed by sponsorship from telecommunications equipment manufacturers and operators and concerned with the study of the broader issues of communications. It acts as a non-profit making organisation.

INTERNATIONAL TELECOMMUNICATIONS UNION (ITU) A potpourri of industrialists and bureaucrats representing the telecommunications interests of 158 member countries, both from an industrial and communications point of view. The ITU is part of the United Nations although its origins date back to the second half of the 19th century. It consists of three principal organs, the CCIR, CCITT and the IFRB which between them are said to plan, regulate, standardise and coordinate telecommunications of every kind.

INTERSPUTNIK Soviet backed equivalent of Intelsat with fourteen members, mostly communist. Uses the Stationar satellites for its services which cover voice and TV traffic. Members are the USSR, Bulgaria, Hungary, Czechoslovakia, Poland, East Germany, Romania, Mongolia, Vietnam, South Yemen, Cuba, Nicaragua, Syria and Laos. Several other countries use the system. These are Afghanistan, Algeria, Cuba, India, Iraq, Kampuchea, Libya and North Yemen.

INTERACTIVE CABLE Systems having capabilities to send signals upstream and downstream (*see* TWO-WAY CAPACITY).

INTERACTIVE SWITCHED DIGITAL NETWORK (ISDN) An engineering led concept for the future switched public network capable of carrying broadband communications (TV, videoconferencing, high bit rate computer to computer communications links) as well as narrowband voice traffic. It uses all digital communications technology and pilot systems are now going into operation. It is liked by PTT's as it conceptually requires a monopoly on communications. From an engineering viewpoint it is a major development in communications technology but has caused considerable controversy as to where and when it is needed, if at all. Some proponents of ISDN have argued that telecom administrations should also have a monopoly on computing capacity in the system. It is both argued that satellites may be forerunners to ISDN, allowing services to develop before the terrestrial ISDN system is available or that satellites are an alternative which render ISDN unecessary by allowing by-pass of the terrestrial system.

INTERCONNECT To link CATV headends, usually with microwave, so that subscribers to different cable systems can see the same programming simultaneously.

INTERFERENCE Energy which tends to interfere with the reception of the desired signals, such as fading from airline flights, radio frequency interference from adjacent channels, or "hosting" from reflecting objects such as mountains and buildings.

ISOTROPIC When radiation from a source has equal strength in all directions. An omnidirectional antenna generates isotropic radiation.

K

Ka-BAND That part of the frequency spectrum allocated to satellite communications in the 20–30 GHz range. It is mostly being used for demonstration or experimental satellite communications although has been used in Japan for operational satellites.

KELVIN (K) The temperature measurement scale used in the scientific community. Zero K represents absolute zero, and corresponds to −459°F. Thermal noise characteristics of an LNA are measured in Kelvins. *Note*: Unlike other temperature measurements, the word "degree" (or symbol) is not used when expressing temperatures in Kelvins.

Ku-BAND That part of the radio frequency spectrum allocated to satellite communications in the 10–20 GHz range. The Ku-band was initially exploited for satellite communications by Europe and has the advantage that there is less likelihood of interference from terrestrial microwave communications. However, it suffers from the disadvantage of rain attenuation and that much of the ground segment infrastructure has been built for use with the C-band.

L

LEASAT (ALSO KNOWN AS SYNCOM) US military communications satellite system owned by Hughes Communications and leased to the US military. The satellites use a wide body design and are capable of only being launched on Shuttle.

LOW EARTH ORBIT (LEO) Self explanatory term for orbits typically 180–250 miles above the surface of the earth. Not normally used for communications satellites.

LOW NOISE AMPLIFIER (LNA) This is the preamplifier between the antenna and the earth station receiver. For maximum effectiveness, it must be located as near the antenna as possible, and is usually attached directly to the antenna receive port. The LNA is especially designed to contribute the least amount of thermal noise to the received signal.

LOW NOISE CONVERTER (LNC) A combination Low Noise Amplifier and down converter built into one antenna-mounted package.

LOW POWER SATELLITE Term which refers to the transponder power output of a satellite rather than the total power generated by the satellite platform or used by its communications payload. Typically low power satellites use transponders in the 20 Watt or less power output range and are used for true FSS services.

LOW POWER TV (LPTV) A new television service being established in a number of countries, notably the USA, West Germany and, surprisingly, India. In the USA, it was authorised by the Federal Communications Commission in October 1980. LPTV broadcasting stations typically radiate between 100–1000 Watts of power, covering a geographic radius of 10–15 miles. Some US observers believe that upwards of 5000 LPTV stations will ultimately be licensed for operation in that country. Many of these would obtain their programming from new satellite television networks.

M

MARECS Acronym for Maritime European Communications Satellite. Three satellites of this type have been constructed by European industry for Inmarsat maritime mobile communications services.

MARISAT A satellite system for mobile civil and military maritime mobile communications owned by Comsat and now leased to Inmarsat. The latter uses them as in-orbit spares. They are also used by the US Navy and the Navies of its allies. As a member of Inmarsat, the Soviet Union uses them.

MASTER ANTENNA TELEVISION (MATV) A small cable TV system perhaps serving a single block of flats or a small part of a town. Typically they are used to relay terrestrial TV channels from a single aerial rather than providing specialised cable TV channels to the households connected. Cable TV channel programmers see MATV systems as a potential market outlet but most conventional relay systems require substantial upgrading to provide the capacity to do so.

MICROWAVE Usually refers to a radio frequency in excess of 1 GHz. Although nearly all commercial satellites operate with frequencies above 1 GHz, the term is usually used loosely to describe terrestrial point-to-point line of sight transmission systems. Such systems were developed out of technology first used in the Second World War. They had a major impact of long distance communications which eventually led to the break up of AT&T's monopoly in the USA. Microwave systems are used for a variety of purposes, including long distance voice, local high volume links, TV feeds and data links. They may be used in remote areas without landlines or in urban environments or for trunk intercity routes. They are often a problem in satellite communications as they create interference. However, they are also often used to connect out-of-town satellite stations with urban users, such as TV stations or business users.

MICROWAVE INTERFERENCE Typically in satellite communications, interference which occurs when a TVRO earth station aimed at a distant satellite picks up a second, often stronger signal, from a terrestrial microwave relay transmitter. Microwave interference can also be produced by nearby radar transmitters as well as the sun itself. Relocating the TVRO antenna by only several feet will often completely eliminate the microwave interference. Professional users will often employ specialised radio frequency engineers to establish likely sources of such interference before deciding on the location and position of their earth station.

MOBILE SATELLITE SERVIVES (MSS) Satellite communications services aimed at providing thin route voice, telex and text services to vehicles, aircraft, ships and railway trains. The largest organisation providing these services is Inmarsat. The USA is considering plans for an MSS service for that country in conjunction with the Canadian M–SAT system. MSS overlaps with FSS type communications in that it can be used effectively for communications to permanent or temporary remote locations. It also overlaps with Radio Determination Satellite Services which are a specialised form of MSS providing text and location services only.

MODULATION The process of manipulating the frequency or amplitude of a radio carrier to carry information (*see* CARRIER).

MODULATOR A device which modulates a carrier. Modulators are found as components in radio transmitters and in satellite transponders (*see* CARRIER).

MOLNIYA ORBIT A non-geostationary orbit used by Soviet communications satellites and particularly suitable for communications with Northerly latitudes and land mobile stations.

MULTIPLEXED ANALOGUE COMPONENT (MAC) Family of TV transmission standards developed for satellite TV and intended to allow the introduction of other future developments such as High Definition TV (HDTV). There are several versions of MAC, mostly incompatible with each other, which are being proposed but nearly all satellite TV transmissions still use either SECAM, PAL or NTSC standards used for existing terrestrial TV. The different versions of MAC include B–MAC, C–MAC, D–MAC, D2–MAC and E–MAC. There have been attempts to get one standard adopted on an international basis but it is unlikely that any attempt will succeed. Technically, MAC is a cross between TDMA and FDMA transmission techniques in that the full channel is used sequentially for video and audio information, rather than in the existing method where sound is carried on a sub-carrier in parallel with the audio signal (FDMA type technique). However, with MAC, the two sets of signals appear to be multiplexed together because of the rapid number of times the signal jumps from carrying video to audio information and back.

MULTIPLEXING The process of combining separate signal channels into one composite stream.

MULTIPOINT DISTRIBUTION SYSTEM (MDS) In the USA, a common carrier licensed by the FCC to operate an omnidirectional microwave transmission facility within a given city. MDS carriers often pick up satellite transmissions from an out-of-town earth station for onward distribution within the city and vice versa. MDS systems may also be used for distribution of TV channels direct to private households.

N

NATIONAL AERONAUTICS AND SPACE ADMINISTRATION (NASA) US civilian R&D organisation that has directed the US space programme since the 1950's and operates all of the current US vehicles used to launch commercial satellites. Its principal project since 1973 has been the development of the troubled Shuttle. It is now looking to The International Space Station as its next major project but is suffering from "a major emotional crisis" as a result of the loss of Shuttle.

NATIONAL SPACE DEVELOPMENT AGENCY OF JAPAN (NASDA) Japanese space agency with partial responsibility for setting the country's space programme and handling the R&D work. Also responsible for operating the country's launch vehicles.

NETWORK In the TV industry, this term is used for the distribution arrangements for programming between two or more TV stations. Traditionally the exercise, known as networking, has used terrestrial distribution means to distribute a large amount of programming per day to other stations for retransmission. In the USA, no company is allowed to own more than 12 TV stations, out of a US total of more than 900. In order to have as many outlets for programming as possible a network system is used whereby three dominant New York based companies, ABC, CBS and NBC, distribute their programming to affiliated stations which are nominally independent. Two thirds of TV stations are affiliated in this way. They retransmit most of the programming they receive but typically include some of their own programming, such as local news. The advent of medium power satellites is breaking this arrangement down and the three national networks are facing competition from specialised networks and a new fourth national network being set up by Rupert Murdoch. In the UK, networking is a term applied to commercial TV where 15 companies each have a regional monopoly and share programming for most of their transmission time. There is no one dominant station providing the networked programming although five of the 16 between them dominate the arrangement. In other countries the term networking may simply apply to the distribution of programming between stations all owned by the same broadcaster. Networking has been and is expected to be a major source of revenue for communications satellite operators who are also able to change the structure of networking activity.

NOISE FIGURE (NF) A term which is a figure of merit of a device, such as an LNA or receiver, expressed in dB, which compares the device with a perfect device. The lower the number the better the unit. For example, a 1·5 dB LNA has better noise characteristics than 2·6 dB LNA. The figure of merit of a device may also be expressed as noise temperature in °K or K4. Again the lower the number, the better the unit. GaAsFET's typically have noise temperatures of 120°K (1·5 dB NF) to 238°K (2·6 dB NF), while uncooled parametric amplifiers typically have a noise temperature of 55°K.

NON-PREEMPTIBLE Commercial term covering the conditions of lease or sale of a satellite transponder or capacity where the purchaser cannot be bumped off the satellite to make room for higher priority traffic or to make room for another user whose capacity is no longer available, for example, through failure of another satellite in the system. To be able to offer non-preemptible terms a satellite operator requires at least two satellites in its system.

O

OFF-AIR RECEPTION The reception of a terrestrial TV signal from a conventional TV transmitter.

OLYMPUS European Space Agency designed satellites to test new communications payloads and to develop a large platform needed for DBS TV payloads. The project dates back to 1972 when it was known as H–SAT. Work began in earnest in 1979 and the satellite is expected to be launched in 1988. Olympus is also the generic name given to platforms derived from the first Olympus design.

ORBITAL PERIOD The time it takes a satellite to make one full orbit of the Earth. For a geostationary satellite it is one day. As such satellites orbit the earth in the same direction it is spinning and the earth spins around its own axis once per day, the satellite appears to be stationary in the sky when viewed from the surface of the Earth.

ORBITAL SPACING There is a finite number of orbital locations for geosynchronuos satellites over and above which there will be interference between the radio signals involved. For example, in the Ku-band, satellites cannot be located closer than 2 degrees. This sets an upper limit of 180 satellites world wide in that frequency band. Orbital spacing was considered to be a major national and international problem in the first half of the 1980's but declining interest in satellite communications and the development of new frequency reuse technologies and the opening up of the Ka-band appear to have resolved the lack of enough potential orbits. Satellites can be parked at the same orbital location as long as they are operating with different frequencies. Orbital slots are allocated by the IFRB basically on a first come first served basis which has caused concern amongst developing countries that they would all be used up by the West before they were able to launch their own satellites.

ORION Proposed US FSS private satellite system intended to provide communications across the North Atlantic for business and TV use.

P

PARABOLIC ANTENNA The normal shape of a satellite earth station dish which takes its name from the mathematical term parabola used to determine the shape. The function of the parabolic shape is to focus the weak microwave signal hitting the surface of the dish into a single focal point in front of the dish. It is at this point that the feedhorn is usually located. There are several alternatives to a parabolic dish which offer the advantages of either locating the focal point away from the centre (where it blocks out incoming and outgoing energy) or enabling the dish to see several satellites at different locations at the same time. However, these are more expensive.

PARKING ORBIT Temporary orbit used for a satellite prior to placing it in its working orbit.

PAYLOAD ASSIST MODULE (PAM) McDonnell Douglas manufactured solid fuel transfer stage used to launch satellites by Shuttle, Delta or Titan. There are three versions. PAM–D intended for satellites in the Hughes HS–376 and RCA–3000 class, PAM–DII intended for satellites in the RCA–4000 class and PAM–A intended for launch of Intelsat V satellites via Shuttle. The latter is unlikely to be used.

PAY-PER-VIEW One method of charging a customer to watch a TV channel. Mostly used in the USA. It involves TV signals which are scrambled before distribution and are unscrambled at the homeowner's set with a decoder that responds upon payment of a fee for each programme. Pay TV can also refer to a system when subscribers pay an extra fee for access to a special channel, which might offer sport, first-run films or education.

PAY TV Includes Pay-per-View TV but also periodic pay TV where the consumer pays a monthly charge. Broadly speaking, Pay-TV offers the advantage of TV without advertisements. In the long run many observers believe that Pay-TV will be a major form of TV. In the UK, the government is considering converting the public service broadcaster, the BBC, from finance by an annual tax on receivers to Pay-TV. In France, a successful off-air Pay-TV venture, Canal Plus, has been running for some years. However, most Pay-TV occurs on cable networks. Amongst the most successful of such channels is Home Box Office.

PERIGEE The point in an elliptical satellite orbit which is closest to the surface of the earth.

PERIGEE STAGE Another name for a transfer stage used to place a satellite in its elliptical orbit prior to firing of the apogee motor.

PERUMTEL Indonesian operator of the Palapa satellites. The organisation also handles all of Indonesia's terrestrial domestic telecommunications.

PHASE ALTERNATION SYSTEM (PAL) A European colour television transmission standard. US and Canadian satellites use the NTSC colour television transmission system and the French Telecom 1 satellite uses SECAM. Intelsat satellites usually use the PAL system. Most countries are looking at new transmission standards which will, in the future, be able to offer compatibility with High Definition TV. Such standards include MAC. However, it is unlikely that there will be a standard world system.

PLATFORM The basic structure of a satellite which holds and services its communications payload, maintains the satellite in its correct position and attitude and provides power and thermal control.

POLARISATION A technique used by the satellite transmission channels by using the same frequencies twice. It was introduced into satellite communications technology during the late 1970's. There are two types of polarisation techniques. With linear polarisation, half of the capacity on a satellite transmits signals to earth in a vertically polarised mode; the other half horizontally polarise their down links. Although the two sets of frequencies overlap, they are 90 degrees out of phase, and will not interfere with each other. This has effectively increased the bandwidth on a satellite from 500–1000 MHz and thereby doubling potential capacity. To successfully receive and decode these signals on earth, the TVRO earth station need to be fitted with a polarised feedhorn to select the vertical or horizontally polarised signals as desired. The Intelsat satellites use a technique known as left-hand and right-hand circular polarisation (*see* CIRCULAR POLARISATION).

PREEMPTIBLE Term used to describe the conditions under which a satellite transponder uses leases or purchases a transponder from a satellite operator. Preemptible conditions require that the user hand back the transponder under certain circumstances. If the satellite operator wants the transponder back to use for other services or customers (for example, if another customer's transponder fails), the operator can take back the transponder, leaving the original customer with no satellite capacity. Preemptible terms are not popular with satellite users but occur when a satellite operator wants to sell in-orbit spare capacity needed in the event of failure of its main capacity. The charges for preemptible capacity are usually less than for non-preemptible.

PROCESSOR An electronic device which receives a radio signal from an antenna, cleans or filters noise from the signal, and converts the signal to the desired channel for retransmission over coaxial cable.

PROGRAMMER/PROGRAMME PROVIDER In satellite television, a supplier of television channels or programmes transmitted via satellite. Examples include Home Box Office, Cable News Network, Sky Channel and so forth. There are about 20 such channels in Europe and over a 100 in the USA.

PROTECTED-USE TRANSPONDER Another term for non-preemptible guaranteed transponder. A satellite transponder provided by a satellite operator or owner of transponders to a user where, if the transponder fails, the owner guarantees the user that it will switch over to another transponder, perhaps preempting some other nonprotected user from the other transponder. It is a form of insurance against failure and reflects in the price paid for the use of the transponder.

PROTON Soviet launch vehicle dating from 1965 which is being offered to several non-Soviet satellite operators for launch of their spacecraft, including Inmarsat.

PTT Generic name for postal and telecommunications monopolies. The term is used loosely in the communications industry to describe organisations such as British Telecom which has no postal activities and, before its break up, occasionally to describe AT&T. Despite the examples set by the USA and the UK, most national telecommunications is carried by large state owned monopolies.

PULSE CODE MODULATION Method of transmitting analogue voice signals in a digital form. A voice bandwidth is converted into a 64 kbit/s stream.

R

RADIO DETERMINATION SATELLITES SERVICES (RDSS) A specialised version of Mobile Satellite Services providing one or two-way text communications and position location services. The US Geostar and the European Locstar plans fall into this category. It is possibly that they may be able to replace some parts of the railway signalling system.

RECEIVER An electronic device which enables a particular satellite signal to be separated from all others being received by an earth station and which converts the signal format into a format for onward transmission.

RECEIVER ISOLATION The attenuation between any two receivers connect a satellite system.

REGIONAL ADMINISTRATIVE RADIO CONFERENCE (RARC) A regional version of WARC's (*see* WARCS).

RESIDUAL ERROR RATE The undetected error rate remaining after error detection and correction methods have been used.

RF ADAPTOR In satellite television, an ad-on modulator which interconnects the output of the satellite television receiver to the input (antenna terminals) of the user's television. The RF adaptor converts the baseband video signal coming from the satellite receiver to a radio frequency (RF) signal which can be used by the television set on a VHF channel.

S

SATCOL Name of a proposed Colombian C-band domestic communications satellite system that has now been abandoned. Colombia is now looking at being a partner in the Andean regional satellite system called Condor.

SATCOM Abbreviation for satellite communications and also the trade name of RCA-Americom's US domestic satellites.

SATELLITE Strictly speaking, a satellite is an object that orbits a celestial body. The moon is a satellite (and has been used in the past as a communications satellite by bouncing radio waves off of its surface back to earth). Man made satellites were originally known as artificial satellites. For communications purposes, satellites is a term used to describe a space platform located in geostationary orbit and carrying a collection of radio repeaters and amplifiers. A few Soviet satellites do not operate in Geostationary orbit, rather they operate in a special orbit known as Molniya (*see* MOLNIYA ORBIT).

SATELLITE BUSINESS SYSTEMS (SBS) US domestic satellite operator which began offering specialised services to business in 1980. It was conceived at a time when it was considered that the convergence of computer and communications technology would lead to considerable demand for such services as computer to computer links. This market has remained small and specialised. The company was originally a partnership between IBM, Comsat and Aetna Life. It is now owned by MCI. SBS found that the demand for business services was very limited and attempted to move into simple long distance telephony markets. However, that market is moving to fibre optic technology at a time when SBS still remains in a weak position.

SATELLITE TELEVISION CORPORATION (STC) A subsidiary of Comsat set up to develop and operate a Direct Broadcast Satellite system. Comsat decided that it could not operate the system commercially and announced in 1984 that it was pulling out. The satellites were under construction at the time and the company decided to complete construction and sell them to another DBS operator. It has failed to do this and is now the owner of two unwanted and obsolescent DBS satellites.

SATELLITE TERMINAL A receive-only satellite earth station consisting of an antenna (typically parabolic in shape), a feedhorn, a low-noise amplifier, a down converter, and a satellite receiver.

SCRAMBLER *See* ENCODER.

SECAM A colour television transmission standard developed by France and used in that country, its former territories and the Soviet Union. SECAM operates with 625 lines per picture frame, a PAL television receiver is capable of showing a SECAM transmission but only in black and white (monochrome). Both SECAM and PAL receivers are totally incompatible with US NTSC transmission standards because the latter uses a different number of lines as well. Europe opted for different colour transmission standards on commercial grounds rather than the technical merits involved opting for incompatible standards is a method of protecting markets from foreign competition through use of patent protection. SECAM is no longer protected by patents (*see also* MAC).

SIDEBAND Lower and upper frequency bands which extended beyond the carrier frequency band when it is modulated. Under certain transmission techniques only a single sideband is used.

SIGNAL TO NOISE RATIO (S/N) The ratio of the signal power and noise power. A video S/N of 54–56 dB is considered to be an excellent S/N, that is of best broadcast quality. A video S/N of 48–52 dB is considered to be a good S/N at the headend for Cable TV.

SIMPLEX TRANSMISSION As distinct from duplex transmission, simplex is where the transmission path is in one direction only with no return path. TV transmissions are an example.

SINGLE CHANNEL PER CARRIER (SCPC) SCPC is a technique for transmitting a single voice channel without resort to multiplexing. Each user channel is allocated a specific carrier in the same way that a group of multiplexed channels is allocated a channel in FDMA techniques. The technology allows a very high quality of signal but is an inefficient method of using the radio frequency spectrum. The system essentially provides for fewer channels per transponder but with the benefit of a higher power signal. In the SCPC technology, the ground segment is cheap. There is an engineering trade off between the cost of the space segment and the ground segment. SCPC is particularly useful under conditions where digital rather than analogue transmission techniques are used. Intelsat has a SCPC service marketed under the name SPADE.

SLOT Also known as Orbital Slot. The longitudinal angular position in the geosynchronous orbit at which a communications satellite is located. Above the United States, communications satellites are, typically positioned in slots which are based at three to four degree intervals. This is now being reduced to two degrees. European DBS satellite allocations are at 12 degree intervals. A one degree spacing places satellites at 450 mile intervals.

SMATV (SATELLITE MASTER ANTENNA TV SYSTEM) A small cable TV system, perhaps serving a block of flats, to which a satellite TVRO terminal has been added to provide satellite television programming (*see* MASTER ANTENNA TV SYSTEM).

SNOW A form of noise picked up by a television receiver receiving channels from a TVRO system caused by a weak signal. Snow is characterised by alternate dark and white dots randomly appearing on the picture tube. To eliminate Snow, a more sensitive receive antenna or better amplification must be provided in the receiver (or both).

SOLAR ARRAY Method of providing the electrical power for a satellite using solar cells to collect and turn sunlight into electricity.

SPACE SEGMENT That proportion of a satellite system that is physically located in space as distinct from the ground segment used to either uplink to the satellite or receive from it. In terms of cost, the ground segment, in total, costs about twice that of the space segment although the ground segment is often paid for by different organisations and people than the space segment.

SPACELAB Laboratory carried in the cargo bay of Shuttle. It was developed by the European Space Agency and is the basis of the Columbus elements of the International Space Station. Spacelab relies heavily on the TDRSS satellite system for communicating research data back to earth.

SPACENET US domestic satellite system owned by GTE but originally owned by Southern Pacific Communications Corporation. Satellites operate in both C- and Ku-bands.

SPACING *See* SLOT. In the geostationary orbit one degree is equal to 450 miles. There is therefore very little chance of satellites colliding with each other.

SPARKLIES A form of satellite television "snow" caused by a weak signal.

SPHERICAL ANTENNA The second major form of satellite television earth station antenna, the spherical antenna (unlike the parabolic antenna) has the ability to simultaneously see, and thus receive, several satellites in orbit. Because of this feature, spherical antennas are becoming increasingly popular in the USA for the home satellite TV user who may wish to switch from satellite to satellite easily and conveniently.

SPIN STABILISATION One of two principal methods of maintaining the attitude of a satellite. In the Spin method the satellite rotates and acts like a gyroscope to maintain attitude. Its antenna platform is "despun", maintaining a fixed orientation towards the earth. Satellites built by Hughes Aircraft Company are of this form. Most other manufacturers prefer 3-axis stabilisation. There are advantages and disadvantages to both, but with currently available launch vehicles, the spin-stabilised satellites are limited to a power supply of around 2700 Watts. When Hughes wanted to go above this, they had to design a hybrid spin/3-axis stabilised platform, the HS–394.

SPOT BEAM A focused transponder pattern used by the Intelsat satellites to a limited geographical area. Spot beams are also used by the US domestic satellites to deliver certain transponder signals to Hawaii, Alaska, and Puerto Rico.

SUBSCRIPTION TELEVISION (STV) *See* PAY-TV. A broadcasting television station transmitting pay television (usually first-run movies) in a scrambled mode. Over 50 STV stations are presently in operation or planned for cities throughout the United States.

SUBCARRIER A signal "piggybacked" onto a main signal to carry additional information. In satellite television transmission, the video picture is transmitted over the main carrier. The corresponding audio is sent via an FM subcarrier. Some satellite transponders carry as many as six audio or data subcarriers whose signals may or may not be related to the main programming. Some radio broadcasters use sub-carriers, particularly in the USA, as an alternative to conventional SCPC methods. In Europe, Europa RV uses six sub-carriers on the Eurelsat I F1 satellite to carry the soundtrack to its TV video channel in six different languages.

SUPERBAND The frequency band from 216–600 MHz, used for fixed and mobile radios and additional television channels on a cable system.

SUPERSTATION A US term originally used to describe Red Turners WTBS Channel 17 UHF station in Atlanta, Georgia USA. When Turner took this station over it was an almost bankrupt local TV station but he used the then (1976) emerging satellite opportunity to change it from a local station into a national station carried on cable TV networks. With the addition of several other major independent television stations whose signals are also carried by satellite, the term superstation has come to mean any regional television station whose signal is picked up and retransmitted by satellite to cable companies nationwide in the USA.

SYNCHRONISATION The process of orienting the transmitter and receiver circuits so that information which is sent in relation to a precise instant of time by the transmitter will be perfectly related to that same instant by the receiver. Home television sets are synchronised by an incoming sync signal with the television cameras in the studios 60 times per second. The horizontal and vertical hold controls on the television set are used to set the receiver circuits to the approximate sync frequencies of incoming television picture and the sync pulses in the signal then fine-tune the circuits to the exact frequency and phase.

SYNCOM *See* LEASAT.

T

TDF–1/TDF–2 French DBS satellite due for launch in 1987–88. It was developed along with TV–SAT as a Franco-German attempt to build up expertise in the manufacturing of large platforms for DBS activity. The French have made several unsuccessful attempts to turn TDF–1 into a commercial and operational satellite system. Its future remains in doubt as does that of the already ordered TDF–2 satellite intended as the back-up spare. The position is similar in West Germany where users for TV–SAT have to be found. The future of TV–SAT 2 is also in doubt. It is estimated that in excess of $700 m has been spent on the two satellite systems.

TELECOM 1/2 French domestic C- and Ku-band communications satellite system built by Matra.

TELECONFERENCE An electronic multilocation, multiperson conference using audio, computer or slow-scan video systems. Videoconferencing is one form of teleconference.

TELEPORT In satellite communications usually a collection of large earth stations on the outskirts of a city serving customers in that town by providing access to one or more satellites. The Teleport may be connected to the users in the city through the public switched network or a microwave network or a separate fibre optic system. The Teleport may also be used by the local cable TV company to receive channels or the local off-air broadcaster for receiving networked programming or occasional use traffic. London has a Teleport in its docklands area.

TELESAT CANADA Canadian domestic satellite operator. It operates the Anik series and is the designated operator of the M–SAT MSS system. It is currently procuring its fifth generation system, Anik E. The company considerably overestimated the demand for its satellites in the early and mid-1980's and hence is suffering from overcapacity.

TELETEXT A method of distribution of text information, either down a telephone line or as part of a TV channel. In both cases it is received on a TV set. It can also be transmitted simultaneously over a satellite transponder using a data subcarrier. Several US satellite programmers are now transmitting teletext signals. Most Western national television networks have implemented their own teletext systems. In Europe, teletext has been looked at as a method of providing subtitles in different languages to video programming which is received in several different countries.

TELEVISION RECEIVE ONLY TERMINAL (TVRO) Receive only earth stations that are used exclusively for reception of TV channels and their sub-carriers. The term usually applies to consumer equipment rather than the professional terminals used by TV stations to pick up networked or occasional use transmissions or by cable TV networks to pick up satellite distributed cable TV channels. However, there is little difference except in size and quality and the term may also be applied to the latter two types. A typical consumer TVRO in Europe is currently between 1·2 metres and 1·8 metres in diameter and costs around $1500–2500. In the USA, the preponderence of lower quality C-band signals means that many TVRO units are between 3 and 5 metres in diameter, although costs are still generally lower than in Europe. Most TVRO units are now manufactured or assembled in the Far East. The term TVRO is also loosely applied to DBS receivers between 30 and 90 cm in diameter. These are technically very similar to conventional TVRO's, just smaller as they are intended to pick up transmissions from high power transponders.

TERRESTRIAL TV Ordinary VHF (Very High Frequency) and UHF (Ultra High Frequency) television transmission limited to an effective range of 100 miles or less. Terrestrial TV transmitters operate at frequencies between 54 MHz and 890 MHz.

THREE AXIS STABILISATION Method by which a satellite's attitude toward the earth is maintained. It uses momentum wheels housed within the satellite one or more of which is speeded up to change the attitude of the satellite. It is the act of speeding up which moves the satellites attitude, not the constant speed of the wheel. The method generally allows larger satellites to be built than spin stabilisation because large solar arrays can be attached to the three axis satellites. This allows a larger power supply.

THRESHOLD EXTENSION A technique used by satellite television receivers to improve the signal-to-noise ratio of the receiver by approximately 3 dB (50%). When using small receive-only antennas (10–12 feet), a receiver equipped with the threshold extension feature can make the difference between obtaining a decent picture or no picture at all.

TIER Single or multiple channels offered over and above a basic cable service. It may include additional premium services, specialised services and distant signals.

TIME DIVISION MULTIPLE ACCESS (*See also* FDMA). An access technique where instead of the satellite transponder being divided up to carry a number of multiplexed channels, the whole transponder is given over to a channel. The other channels are also given use of the same transponder. However, each channel is given only a brief period of use of the transponder before another channel is given time. The technique allows very efficient use of the transponder but also requires that all of the ground stations using the transponder are coordinated. This makes the ground segment expensive. Generally TDMA techniques have not proved to be commercially successful.

TITAN The only US expendable launch vehicle that is currently in production although all of the vehicles under production are for the US military. The manufacturer is now attempting to commercialise one version of the vehicle, Titan III. The other two current versions are Titan IV and Titan III.

TRACKING AND DATA RELAY SATELLITE SYSTEM (TDRSS) Complex US satellite system intended to provide communications for Shuttle and other spacecraft in low earth orbit. It has been plagued by problems including the loss of the first two satellites in the series.

TRANSFER ORBIT STAGE (TOS) US privately funded transfer stage for use with Shuttle or Titan. The design has been developed by Orbital Sciences Corporation.

TRANSFER STAGE Upper stage used to transfer a payload from low earth orbit into (usually) an elliptical orbit prior to final placing of the satellite in geostationary orbit. The stage may also be used to take the spacecraft out of the earth's orbit altogether, such as on interplanetary missions. The same transfer stage may be used with several launch vehicles although it has, in the past, largely been used with reference to Shuttle launches. There are a number of different transfer stages around including TOS, PAM, IUS, Centaur and Transtage.

TRANSLATOR An unattached television repeater. A translator picks up a broadcast TV station from a distant city and retransmits the picture locally on another channel.

TRANSMITTER An electronic device consisting of oscillator, modulator, and other circuits which produce a radio or television electromagnetic wave signal for radiation into the atmosphere by an antenna.

TRANSPARENT A loose term used to describe a transmission system that can carry a signal in several different forms. For example, a satellite may be described as transparent if it is capable of carrying TV channels transmitted in PAL, SECAM, MAC and NTSC transmission standards.

TRANSPONDER A combination receiver, transmitter, and antenna package, physically part of a communications satellite. Transponders have a typical power output of five 50 Watts (although for DBS TV the figure may be as high as 230 Watts), operate over a frequency band with a 36 MHz bandwidth in the 4–6 and the 10–14 GHz microwave spectrum. Communications satellites typically have between 12 and 24 on-board transponders.

TRAVELLING WAVE TUBE AMPLIFIER (TWTA) Combination of a travelling wave tube and its power supply. It is a method of amplifying a radio signal by using an electron gun. The technology is related to that of a TV set cathode ray tube. In satellite communications, TWTA's are used in uplink earth stations and on board the satellites where they are part of the transponder. Alternatives to TWTA's are Solid State Power Amplifiers and Klystrons.

TV-SAT *See* TDF-1/2.

TWEEKING The process of adjusting an electronic receiver circuit to optimise its performance.

U

ULTRA HIGH FREQUENCY (UHF) Officially the band of frequencies ranging from 300–3000 MHz. In television use, refers to the set of frequencies starting at 470 MHz.

UPLINK Signal from an earth transmit station to a satellite.

UPPER STAGE *See* TRANSFER STAGE.

V

VERY HIGH FREQUENCIES (VHF) The range of frequencies extending from 30–300 MHz. Used for TV transmissions in many countries as an alternative to UHF frequencies.

VERY SMALL APERTURE TERMINALS (VSAT's) Small satellite earth stations dedicated to specialised business services such as facsimile transmission, private voice networks, videoconferencing and data collection and distribution. Such terminals may be receive only or transmit and receive and typically have a diameter of 1·2–4 metres. The Intersat IBS service and Eutelsat's SMS service are used in conjunction with VSAT's. It was expected that VSAT's would provide a major market opportunity in the satellite industry but the demand has proved to be moderate.

VISTA Intelsat small dish telephony service intended for remote and rural communications. It allows two to four voice circuits using a small dish two-way earth terminal.

W

WORLD ADMINISTRATIVE RADIO CONFERENCE (WARC) Periodic conferences of ITU members convened to discuss and allocate the radio frequency spectrum. The last WARC was held in 1985.

WAVEGUIDE A microwave conductor, typically rectangular in shape, used to carry microwave signals into and out of microwave antennas.